Lecture Notes in Artificial Intelligence   3397

Edited by J. G. Carbonell and J. Siekmann

Subseries of Lecture Notes in Computer Science

Tag Gon Kim (Ed.)

Artificial Intelligence and Simulation

13th International Conference on AI, Simulation,
and Planning in High Autonomy Systems, AIS 2004
Jeju Island, Korea, October 4-6, 2004
Revised Selected Papers

 Springer

Series Editors

Jaime G. Carbonell, Carnegie Mellon University, Pittsburgh, PA, USA
Jörg Siekmann, University of Saarland, Saarbrücken, Germany

Volume Editor

Tag Gon Kim
Korea Advanced Institute of Science and Technology
Department of Electrical Engineering and Computer Science
373-1 Kusong-dong, Yusong-ku, Taejon, Korea 305-701
E-mail: tkim@ee.kaist.ac.kr

Library of Congress Control Number: 2004118149

CR Subject Classification (1998): I.2, I.6, C.2, I.3

ISSN 0302-9743
ISBN 3-540-24476-X Springer Berlin Heidelberg New York

Springer is a part of Springer Science+Business Media

springeronline.com

© Springer-Verlag Berlin Heidelberg 2005
Printed in Germany

Typesetting: Camera-ready by author, data conversion by Olgun Computergrafik
Printed on acid-free paper   SPIN: 11382393   06/3142   5 4 3 2 1 0

Preface

The AI, Simulation and Planning in High Autonomy Systems (AIS) 2004 Conference was held on Jeju Island, Korea, October 4–6, 2004. AIS 2004 was the thirteenth in the series of biennial conferences on AI and simulation. The conference provided the major forum for researchers, scientists and engineers to present the state-of-the-art research results in the theory and applications of AI, simulation and their fusion. We were pleased that the conference attracted a large number of high-quality research papers that were of benefit to the communities of interest.

This volume is the proceedings of AIS 2004. For the conference full-length versions of all submitted papers were refereed by the respective international program committee, each paper receiving at least two independent reviews. Careful reviews from the committee selected 77 papers out of 170 submissions for oral presentation. This volume includes the invited speakers' papers, along with the papers presented in the conference.

In addition to the scientific tracks presented, the conference featured keynote talks by two invited speakers: Bernard Zeigler (University of Arizona, USA) and Norman Foo (University of New South Wales, Australia). We were grateful to them for accepting our invitation and for their talks. We also would like to express our gratitude to all contributors, reviewers, program committee and organizing committee members who made the conference very successful. Special thanks are due to Tae-Ho Cho, the Program Committee Chair of AIS 2004 for his hard work in the various aspects of conference organization.

Finally, we would like to acknowledge partial financial support by KAIST for the conference. We also would like to acknowledge the publication support from Springer.

November 2004   Tag Gon Kim

Conference Officials

Committee Chairs

Honorary Chair   Bernard P. Zeigler
                                     (University of Arizona, USA)
General Chair   Tag Gon Kim (KAIST, Korea)
Program Chair   Tae-Ho Cho
                                     (Sungkyunkwan University, Korea)

Organizing Committee

Sung-Do Chi, Hankuk Aviation University, Korea
Jong-Sik Lee, Inha University, Korea
Jang-Se Lee, Korea Maritime University, Korea
Young-Kwan Cho, ROK Air Force HQ, Korea
Fernando J. Barros, University of Coimbra, Portugal
Hessam Sarjoughian, Arizona State University, USA
Shingo Takahashi, Waseda University, Japan
Adelinde Uhrmacher, University of Rostock, Germany
Ryo Sato, University of Tsukuba, Japan

Program Committee

Jacob Barhen, Oak Ridge National Laboratory, USA
Agostino Bruzzone, Università degli Studi di Genova, Italy
Luis Camarinha-Matos, New University of Lisbon/Univova, Portugal
François E. Cellier, University of Arizona, USA
Etienne Dombre, LIRMM, France
Cuneyd Firat, ITRI of Tubitak-Marmara, Turkey
Paul Fishwick, University of Florida, USA
Norman Foo, University of South Wales, Australia
Claudia Frydman, DIAM-IUSPIM, France
Erol Gelenbe, University of Central Florida, USA
Sumit Ghosh, Stevens Institute of Technology, USA
Norbert Giambiasi, DIAM-IUSPIM, France
Mark Henderson, Arizona State University, USA
David Hill, Blaise Pascal University, France
Mehmet Hocaoglu, ITRI of Tubitak-Marmara, Turkey
Syohei Ishizu, Aoyama Gakuin University, Japan
Mohammad Jamshidi, ACE/University of New Mexico, USA
Andras Javor, Technical University of Budapest, Hungary

Table of Contents

Keynotes

Modeling and Simulation Methodologies I

Intelligent Control

Computer and Network Security I

HLA and Simulator Interoperation

Manufacturing

Agent-Based Modeling

DEVS Modeling and Simulation

Modeling and Simulation Methodologies II

Parallel and Distributed Modeling and Simulation I

Mobile Computer Network

Web-Based Simulation, Natural System

Modeling and Simulation Environments

AI and Simulation

Component-Based Modeling

Watermarking, Semantic

Parallel and Distributed Modeling and Simulation II

Visualization, Graphics and Animation I

Computer and Network Security II

Business Modeling

Visualization, Graphics and Animation II

DEVS Modeling and Simulation

Continuity and Change (Activity) Are Fundamentally Related in DEVS Simulation of Continuous Systems

Bernard P. Zeigler, Rajanikanth Jammalamadaka, and Salil R. Akerkar

Arizona Center for Integrative Modeling and Simulation
Department of Electrical and Computer Engineering
University of Arizona, Tucson, Arizona 85721, USA
zeigler@ece.arizona.edu
www.acims.arizona.edu

Abstract. The success of DEVS methods for simulating large continuous models calls for more in-depth examination of the applicability of discrete events in modeling continuous phenomena. We present a concept of event set and an associated measure of activity that fundamentally characterize discrete representation of continuous behavior. This metric captures the underlying intuition of continuity as well as providing a direct measure of the computational work needed to represent continuity on a digital computer. We discuss several application possibilities beyond high performance simulation such as data compression, digital filtering, and soft computation. Perhaps most fundamentally we suggest the possibility of dispensing with the mysteries of traditional calculus to revolutionize the prevailing educational paradigm.

1 Introduction

Significant success has been achieved with discrete event approaches to continuous system modeling and simulation[1,2,3]. Based on quantization of the state variables, such approaches treat threshold crossings as events and advance time on the basis of predicted crossings rather than at fixed time steps [4,5,6]. The success of these methods calls for more in-depth examination of the applicability of discrete events in modeling continuous phenomena. I have previously proposed that discrete events provide the right abstraction for modeling both physical and decision-making aspects of real-world systems. Recent research has defined the concept of activity which relates to the characterization and heterogeneous distribution of events in space and time. Activity is a measure of change in system behavior – when it is divided by a quantum gives the least number of events required to simulate the behavior with that quantum size. The number of DEVS model transitions, and hence the simulation execution time, are directly related to the threshold crossings. Hence activity is characteristic of continuous behaviors that lower bounds work needed to simulate it on a digital computer. The activity measure was originally formulated in the context of ordinary and partial differential equations as the integral of the magnitudes of the state space derivatives. This paper goes deeper into the activity measure to relate it to the information content of a system behavior and to the very concept of continuity itself.

The activity, re-examined, turns out to be a measure of variation defined on finite sets of events. The value of this measure will tend to increase as we add events. But what is critical is that, once we have enough events to get the qualitative characteristics of the curve, the measure slows down markedly or stops growing at all. Indeed, if we are lucky enough to start with the right set of events then the measure should stay constant from the very start of the refinement process. By qualitative characteristics of the curve we mean the placement of its minima and maxima and we restrict the curves of interest to those for which there are only a finite number of such extreme points in the finite interval of interest. We will show that for a continuous curve, for any initial sample set containing these points, the variation measure must remain constant as we continue to inject new samples. If the sample set does not include these extreme points, then the measure will grow rapidly until points are included that are close enough to these extrema. Since performing simulations with successively smaller quantum sizes generates successive refinements of this kind, we can employ this concept to judge when a quantum size is just small enough to give a true qualitative picture of the underlying continuous behavior. For spatially extended models, employing the measure for successively smaller cell sizes, gives the same characterization for resolution required for continuity in space.

2   Review of Activity Results

Models with large numbers of diverse components are likely to display significant heterogeneity in their components' rates of change. The activity concept that we developed is intended to exploit this heterogeneity by concentrating computing attention on regions of high rates of change – high activity – in contrast to uniformly attending to all component changes with indifference to their activity levels. The concept of activity, informally stated in this manner, applies to all heterogeneous milticomponent models, whether expressed in continuous or discrete formalisms. Our focus here however, is on elucidating the activity concept within the context of continuous systems described by differential equations with the goal of intimately linking the concept to discrete event simulation of such models. We have shown in recent work that activity can be given a very intuitive and straightforward definition in this context, that useful theoretical and practical implications can be derived, and that such implications can be verified with empirical computational results. Indeed, several studies have confirmed that using the quantization method of differential equation solution, DEVS simulation naturally, and automatically, performs the requisite allocation of attention in proportion to activity levels. It does so by assigning time advances inversely to rates of change, so that high rates of change get small time advances, while low rates of change get large time advances. Hence component events are scheduled for execution inversely to their activity levels. Furthermore, this occurs in a dynamic manner, tracking changes in rates in a natural way, "at no extra charge."

2.1 Mapping ODEs to Quantized DEVS Networks

A mapping of ordinary differential equations (ODE) into DEVS integration networks using quantization is detailed in [5], where supporting properties such as completeness and error dependence on quantum size are established. Essentially, an ODE is viewed as a network of instantaneous functions and integrators that are mapped in a one-one manner into an equivalent coupled network of DEVS equivalents. Each integrator operates independently and asynchronously in that its time advance is computed as the quantum divided by the just-received input derivative. Such inputs are computed by instantaneous functions from outputs of integrators that they receive via coupling. Were such inputs to be the same for all integrators and to remain constant over time, then all integrators would be undergoing equal time advances, identical to the time steps of conventional numerical schemes as characterized by the Discrete Time System Specification (DTSS) formalism. However, such equality and constancy is not the norm. Indeed, as just mentioned above, we expect the DEVS simulation to exploit situations where there is considerable heterogeneity.

2.2 Activity Definition

We proceed to review the concept of activity as defined in [3], which should be consulted for more detailed discussion of the material in the next sub-sections. Fig 1 illustrates the concept of activity as a measure of the amount of computational work involved in quantized ODE simulations. Given equally spaced thresholds separated from each other by a quantum, the number of crossings that a continuous non-decreasing curve makes is given by the length of the range interval it has traveled divided by the size of the quantum. This number of threshold crossing is also the number of DEVS internal transitions that an integrator must compute. While the quantum size is an arbitrary choice of the simulationist, the range interval length is a property of the model, thus justifying the designation of this length to underlie our activity measure .Generalizing to a curve that has a finite number of alternating maximum and minima, we have the definition of activity in an interval

$$A = \sum_i |m_{i+1} - m_i|$$
(1)

where $m_0, m_1, m_2 .. m_n$ the finite sequence of extrema of the curve in that interval. The number of threshold crossings, and hence the number of transitions of the quantized integrator, is then given by the activity divided by the selected quantum size

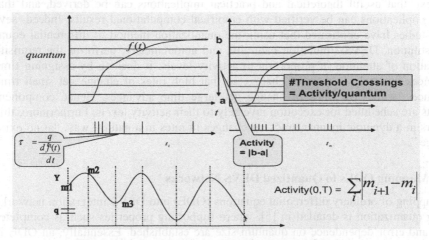

Fig. 1. Activity as a characteristic of continuous functions

2.3 Computing the Activity

The activity measure in Eqn. 1 relates back to quantized integration of an ordinary differential equation. The total number of DEVS transitions is the sum of those of the individual integrators. This number is predicted by the total activity divided by the quantum size, where total activity is the sum of individual activities of the integrators.

We can derive the rates of activity accumulation from the underlying ODE by noting that:

$$\frac{dy_i}{dt} = f_i(y_1...y_n)$$

$$\Rightarrow$$

$$\frac{dA_i}{dt} = |f_i(y_1...y_n)|$$

i.e., the instantaneous rate of activity accumulation at an integrator is the absolute value of its derivative input. When integrated over an interval, this instantaneous differential equation turns out to yield the accumulated activity expressed in (1).

The total activity of all integrators can be expressed as

$$A = \int \bar{f}(y_1...y_n)dt$$

where   (2)

$$\bar{f}(y_1...y_n) = \sum_i |f_i(y_1...y_n)|.$$

We employ (2) to derive activity values where analytically possible in the next section.

2.4 Activity in Partial Differential Equation Models

The activity formula derived so far applies to ordinary differential equations. We extended it to partial differential equations (PDE) by discretizing space into a finite number of cells and approximating the PDE as an ODE whose dimension equals the number of cells. The activity formula (2) is than applied to this family of ODEs with parameter N, the number of cells. For a one-dimensional diffusion example, when the number of cells increases to infinity, the solution converges to that of the PDE and we found the activity likewise converges. Table 1 displays activity formulas for different initial states (diffusant distributions), where each formula gives the accumulated activity over all cells until equilibrium is reached. In each case, the average activity (total divided by number of cells) approaches a constant whose value depends on the initial state.

2.5 Ratio of DTSS to DEVS Transitions

Fig. 2 illustrates how we compare the number of transitions required by a DTSS to that required by a quantized DEVS to solve the same PDE with the same accuracy. We derive the ratio:

$$\frac{\#DTSS}{\#DEVS} = \frac{MaxDeriv*T}{A/N} \geq 1$$

$$\text{where} \quad MaxDeriv = Max_i\left(\left|\frac{dy_i}{dt}\right|\right)$$

(3)

Table 1. Activity calculation for one-dimensional diffusion with different initial conditions

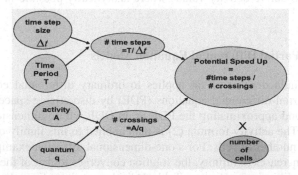

	Initial state	Activity	Activity/N as N→∞
	Rectangular pulse	2HN(W/L)(1 −W/L)	2H(W/L)(1 −W/L)
	Triangular pulse	(N−1)*H/4	H/4
	Gaussian pulse	$\frac{N*H}{L}\left(\frac{2}{\sqrt{2*\pi}*e}\ln\left(\frac{L}{\sqrt{2*c*t_{start}}}\right)\right. - \frac{1}{2}erf\left(\frac{L}{\sqrt{4*c*t_{start}}}\right)$ $+erf(0.707)$	Constant/L

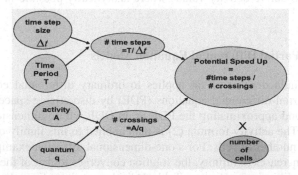

Fig. 2. Illustrating how to compare the number of DTTS transitions with those of DEVS for solutions with the same accuracy

Table 2 shows the result of plugging in the activity calculations in Table 1 into the ratio formula for the initial states discussed earlier. We see that for the rectangular and triangular initial states, the ratio grows with the number of cells; while for the Gaussian pulse, the ratio grows with the length of the space in which the pulse is contained. The increasing advantage for increasing cell numbers was confirmed in work by Alexandre Muzy [7] who compared the execution times of the quantized DEVS with those of standard implicit and explicit methods for the same fire-spread model as shown in Fig 3.

We see that the growth of the quantized DEVS execution time with cell numbers is linear while that of the traditional methods is quadratic. Indeed, the fire-spread model employed in Muzy's results is characterized by a sharp derivative at ignition while otherwise obeying a diffusion law. This suggests that the predictions of the quadratic advantage for the quantized DEVS approach to the rectangular pulse in Table 2 apply.

Table 2. Ratio of DTSS transitions to DEVS transitions for the same accuracy of solution

Initial state	#DTSS/#DEVS
Rectangular Pulse	$\dfrac{TcN^2}{2w(1-w)L^2}$ where w is the width to the length ratio
Triangular Pulse	$\dfrac{4TcN}{L^2}$
Gaussian Pulse	$\dfrac{0.062T}{(c*t_{start}^3)^{1/2}}f(L)$ where f is an increasing function of L

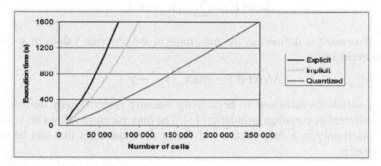

Fig. 3. Comparison of traditional DTSS methods with DEVS quantized integration (taken from [7] with permission)

This concludes our brief review of activity theory. For more details the reader is referred to the references at the end of the paper.

3   Basic Concepts of Event Sets

We now re-examine the activity concept from a more general point of view. We start with a discrete set theoretic formalism that has direct digital computer implementation. We introduce the concept of event set and event set refinement as an implementable approach to continuity. An *event set* is a finite ordered set of such event pairs,

$$E = \{(t_i, v_i) \mid i = 1...n\}$$

where the pairs are ordered in increasing order of the left hand elements of the pairs. Further the event set is a functional relation in that no two event pars have the same left hand value. In other words, we don't allow multiple events to be recorded at the same time or in the same place in the same event set – although we can use more than one event set to capture such simultaneous or co-located events. The following are needed

$$domain(E) = \{t_i \mid i = 1...n\} \text{ and } range(E) = \{v_i \mid i = 1...n\}$$
$$size(E) = n$$

The asymmetry between domain and range, succinctly encapsulated in the mathematical definition of function, will turn out to be an essential fact that we will later exploit to reduce simulation work. We are interested in the intervals that contain the respective domain and range points. Thus we define: $domainInterval(E) = [t_1, t_n]$ and $rangeInterval(E) = [v_{min}, v_{max}]$.

3.1 Measures of Variation

We will work with pairs of successive values $(v_1, v_2), (v_2, v_3), ...(v_i, v_{i+1})...$ On this basis, we define a *measure of variation* that will turn out to be the critical measure in measuring change and continuity. The measure is defined as the sum of the absolute values of successive pairs in the event set:

$$Sum(E) = \sum_i |v_{i+1} - v_i|$$

A second measure is defined as the maximum of the absolute values of successive pairs in the event set:

$$Max(E) = \max_i |v_{i+1} - v_i|$$

The sum of variations turns out to be activity measure as previously defined in the context of differential equation simulation [3]. The max measure allows us to characterize the uncertainty in a data stream and the smallest quantum that can be used in quantization of it.

3.2 Extrema – Form Factor of an Event Set

The *form factor* of an event set consists of the values and locations of its extrema. This information is represented in an event set as follows:

$$extrema(E) = \{(t^*_i, v^*_i)\} \subseteq E$$

where (t^*, v^*) represents a maximum or minimum at location t^* with value v^*.

A subsequence $(t_i, v_i)(t_{i+1}, v_{i+1}), ..(t_{i+m}, v_{i+m})$ is monotonically increasing if $v_i > v_{i+1} > .. > v_{i+m}$. The sequence is non-decreasing if $v_i \geq v_{i+1} \geq .. \geq v_{i+m}$.

A similar definition holds for the terms monotonically decreasing and non-increasing. An algorithm to obtain the form factor proceeds from the following:

Theorem 1. Let E be an event set. Then its minima and maxima alternate, one following the other. The subsequences between successive extrema are either non-increasing or non-decreasing, according to whether they start with a minimum or a maximum.

Corollary 1. An event set can be decomposed into a "disjoint" union of monosets. We use the quotation marks "disjoint" to indicate disjointness except for overlapping end points.

Proposition 1. The sum of variations in a monoset E can be expressed as:

$$Sum(E) = | v_1 - v_n |$$

Theorem 2. For an event set E with $extrema(E) = \{(t*_i, v*_i)\}$, the sum of variations,

$$Sum(E) = \sum_i | v*_{i+1} - v*_i |.$$

In other words, $Sum(E)$ is the sum of its monoset sums of variation.

3.3 Refinement

Refinement is a process by which we add new data to an existing event set without modifying its existing event pairs. The result of refinement is a new event set that is said to refine the original one. We define the relation E *refines* E' iff $E \supseteq E'$, i.e., the set of event pairs in E includes the set of events in E'.

Proposition 2. E *refines* $E' \Rightarrow Sum(E) \geq Sum(E')$

Proof. Since E *refines* E' there is a pair, (t', v') squeezed between some pair $(t_i, v_i), (t_{i+1}, v_{i+1})$, i.e., $t_i < t' < t_{i+1}$. Then $| v_{i+1} - v_i | \leq | v_{i+1} - v' | + | v' - v_i |$.

3.4 Within-the-Box Refinement

We now identify precisely the type of refinement that does not increase the sum of variations. A refinement is *within-the-box* if the added pair (t', v') satisfies the constraints: $t_i < t' < t_{i+1}$ and $v' \in [\min(v_i, v_{i+1}), \max(v_i, v_{i+1})]$. Then we have:

E *refines* wtb E' if E *refines* E' and all refinement pairs are within-the-box.

Within-the-box refinement has a special property – it preserves the sum and tends to decrease the maximum variation. As we shall see, this simple rule characterizes refinement of continuous functions once their form factors have been identified.

Proposition 3. Assume that E *refines* E'. Then $Sum(E) = Sum(E')$ if, and only if, E *refines* wtb E',

Also E *refines* wtb $E' \Rightarrow Max(E) \leq Max(E')$, but the converse is not true.

Theorem 3. The monoset decomposition of an event set is not altered by within-the-box refinement. Consequently, an event set's form factor is invariant with respect to within-the-box refinement.

Proof: A within-the-box single refinement falls into some monoset domain interval. It is easy to see that the within-the-box refinement does not change the non-increasing (or non-decreasing) nature of the monoset sequence. In particular, it does change the values and locations of the extrema.

4 Domain and Range-Based Event Sets

An event set $E = \{(t_i, v_i) \mid i = 1...n\}$ is domain-based if there is a fixed step Δt such that $t_{i+1} = t_i + \Delta t$ for $i = 1...n-1$ We write $E_{\Delta t_s}$ to denote a domain-based event set with equally spaced domain points separated by step Δt .

An event set $E = \{(t_i, v_i) \mid i = 1...n\}$ is *range-based* if there is a fixed quantum q such that $| v_{i+1} - v_i | = q$ for $i = 1...n-1$. We write E_q to denote a range-based event set with equally spaced range values, separated by a quantum q .

For a range-based event set E_q, the measures of variation are simply expressed as $Sum(E_q) = size(E_q)q$ and $Max(E_q) = q$. For a domain-based event set $E_{\Delta t}$, we have $\Delta t = \dfrac{domainIntervalLength}{size(E_{\Delta t})}$

We can map between the two forms of representation. If we are interested in what happens in a behavior at specific times than the domain-based representation is more appropriate. However, usually we are interested more in recording only times of significant changes and inferring that nothing of interest occurs between such events. In this case, the range-based representation is appropriate and has significant economy of representation. We next see the application of this principal to continuous function representation.

5 Event Set Representation of Continuous Functions

Let $f : [t_i, t_f] \to R$ be a continuous function defined everywhere on the closed interval $[t_i, t_f]$ with a finite number of alternating minima and maxima separated by non-increasing, and non-decreasing segments. Surprisingly, perhaps, it can be shown that differentiable continuous functions, except for those with damped infinite oscillations, are of this form. Let $extrema(f)$ denote the extrema of f .

Definition. $E(f)$ is an event set that is a sample of f if $(t, v) \in E(f) \Rightarrow f(t) = v$. . We say $E(f)$ samples f . In addition, if $E(f) \supseteq extrema(f)$ we say $E(f)$ *represents* f .

The smallest event set able to represent f is $E_{extrema}(f) = extrema(f)$. Almost by definition, if any $E(f)$ *represents* f then it refines $E_{extrema}(f)$.

If $E'(f)$ refines $E(f)$ and samples f we say that $E'(f)$ is a *continuation of $E(f)$*. The relation "is a continuation of" is transitive with minimal element $E_{extrema}(f)$, i.e., every event set that represents f is a continuation of $E_{extrema}(f)$.

Our fundamental connection to within-the-box refinement is given in:

Theorem 4. Let $E(f)$ represent f. Then

1. Every extremum of f is an extremum in $E(f)$, and conversely.
2. $Sum(E(f)) = Sum(f) = \sum_i |v^f_{i+1} - v^f_i|$
3. Every continuation of $E(f)$ is a within-the-box refinement.
4. For every $E'(f)$ continuation of $E(f)$, we have:
   a. $Sum(E'(f)) = Sum(E(f))$
   b. $Max(E'(f)) \leq Max(E(f))$.

5.1 Uncertainty Metric

We introduce a metric to facilitate a comparison of the number of samples required to attain a given accuracy of representation. Since samples are assumed to be noise-free, the remaining imprecision is in the uncertainty that a finite set of samples implies about the rest of the points in the function. For a given domain point, we take its uncertainty as the size of the interval in which its mapped value is known to lie.

Definition. The *box* in $E(f)$ containing $(v,t) \in f$ is spanned by the domain sub-interval (t_i, t_{i+1}) containing t and the associated range sub-interval (v_i, v_{i+1}). Define uncertainty$(E(f), t) = |v_i - v_{i+1}|$ where (v_i, v_{i+1}) is vertical side of the box in $E(f)$ containing (v,t). Now

$$\text{uncertainty}(E(f)) = \max\{\text{uncertainty}(E(f), t) \,|\, t \in domainInterval(E(f))\}$$

Proposition 4.
1. uncertainty$(E(f)) = Max(E(f))$.
2. For a range-based continuation with quantum q, uncertainty$(E(f)) = q$
3. If f is differentiable, then for a domain-based continuation with small enough step Δt, we have

$$\text{uncertainty}(E(f)) \approx MaxDer(f) * \Delta t$$

where $MaxDer(f)$ is the magnitude of the largest derivative of f.

Proof. Assertions 1 and 2 follow easily from the definitions. For Assertion 3, we note that for successive values, v_i, v_{i+1} with small enough domain step $t_{i+1} - t_i$ we have $|v_i - v_{i+1}| \approx |f'(t_i)| (t_{i+1} - t_i)$, where $f'(t)$ is the derivative we have assumed to exist. The rest follows easily.

Theorem 5. The sizes of domain-based and range-based continuation representations of a differentiable function f having the same uncertainty are related as follows:

$$R = \frac{size(E_q(f))}{size(E_{\Delta t}(f))} \approx \frac{Avg(f)}{MaxDer(f)} \leq 1 \text{ where } Avg(E(f)) = \frac{Sum(f)}{domainIntervalLength} = Avg(f).$$

We see that the number of points required for a range-based representation is proportional to its sum of variations. This is the general statement of the result proved in previous work for the activity in differential equation systems.

5.2 The Range-Based Representation Can Be Arbitrarily More Efficient Than Its Domain-Based Equivalent

Theorem 6. There are smooth continuous functions whose parameters can be set so that any range-based representation uses an arbitrarily smaller number of samples than its domain-based equivalent (by equivalent, we mean that they achieve the same uncertainty.)

In Table 1 we compare sizes for representations having the same uncertainty. Except for the sine wave, there are parameters in each of the classes of functions that can be set to achieve any reduction ratio for range-based versus domain-based representation. In the case of simulation, this implies that the number of computations is much smaller hence the execution goes much faster for the same uncertainty or precision. Indeed, we can extend the event set concept to multiple dimensions with the appropriate definitions, and obtain:

Theorem 7. The sizes of domain-based and range-based continuation representations of a differentiable n-dimensional function $f = \langle f_1, ..., f_n \rangle$ having the same uncertainty are related as follows: $R = \frac{size(E_q(f))}{size(E_{\Delta t}(f))} \approx \frac{Avg(f)}{n * MaxDer(f)} \leq 1.$

Calculations similar to those in Table 1 show that a performance gain that is proportional to the number of dimensions is possible when there is a marked in- homogeneity of the activity distributions among the components.

For example, for a signal $f_i(t) = a_i \sin(\omega_i t), i = 1..n$, the ratio R$= \frac{2 \sum_{i=1}^{n} \omega_i a_i}{n \pi \max_i \{\omega_i a_i\}}$ varies as $\frac{2}{\pi} \leq R \leq \frac{2}{n\pi}$, For an n-th degree polynomial we have $\frac{1}{n^2} \leq R \leq \frac{1}{n}$. So that potential gains of the order of $O(n^2)$ are possible.

Table 3. Compuation of Range-based to Domain-based Representation Size Ratios

$f(t) =$	domainInterval	$Sum(f)$	$\dfrac{Avg(f)}{= \dfrac{Sum(f)}{domain\ IntervalLength}}$	$MaxDer(f)$	$R = \dfrac{size(E_q(f))}{size(E_{\Delta t}(f))}$ $\approx \dfrac{Avg(f)}{MaxDer(f)}$
$t^n, n \geq 1$	$[0,T]$	T^n	T^{n-1}	nT^{n-1}	$1/n$
$t^{-n}, n \geq 1$	$[1,T]$	$1-T^{-n}$	$\dfrac{1-T^{-n}}{T-1}$	$nt^{-(n+1)} \mid 1$ $= n$	$\dfrac{1-T^{-n}}{n(T-1)}$ $\lim_{T \to \infty} = \dfrac{1}{n}$
$A\sin \omega t$	$[0, \dfrac{2\pi}{\omega}]$	$4A$	$\dfrac{2A\omega}{\pi}$	$A\omega$	$\dfrac{2}{\pi}$
Ae^{-at}	$[0,T]$	$A(1-e^{-aT})$	$A(1-e^{-aT})/T$	Aa	$\dfrac{1-e^{-aT}}{aT}$ $\lim_{T \to \infty} = \dfrac{1}{aT}$

6   Event Set Differential Equations

Although space precludes the development here, the integral and derivative operations can be defined on event sets that parallel those in the traditional calculus. Assuming this development, we can formulate the event set analog to traditional differential equations as follows:

Let $E_i = \{(t_j, v_j) \mid j = 1...n\}$ be required to be an indefinite sequence of within-the-box refinements that satisfies the equation: $D\ (E_i) = f(E_i)$ for all $i = 1, 2,...$ where

- $D\ (E)$ is the derivative of E and
- $f(E) = \{(t_i, f(v_i)) \mid i = 1...n\}$ with $f : rangeInterval(E) \to R$

A solution to this equation is

$$E_q^{solution} = \{(t_{q,j}, jq) \mid j = 1...rangeInterval / q\}$$

where $t_{q,j} = I(G)_{q,j}$ and $I(G)$ is the integral of $g(v) = 1/ f(v)$, i.e., $I(G)_{q,j} = \{(jq, I(G_{q,j}))\}$. We note that the solution is a range-based event-set that can be generated by DEVS quantized simulation. It parallels the solution to $\dfrac{dx}{dt} = f(x)$ written as $\int_{x(t_i)}^{x(t)} \dfrac{dx}{f(x)} = t - t_i$. It turns out that the range-based event set concept is essential to writing explicit solutions that parallel the analytic solutions of classical calculus. Recall that without going to the state description $\dfrac{dx}{f(x)} = dt$,we can only state the recursive solution $F(t) = \int f(F(\tau))d\tau$. Likewise, we can't write an explicit solution for the event set differential equation without going to the range-based, i.e., quantized formulation.

7 Discussion and Applications

Originally developed for application to simulation of ordinary, and later, partial differential equations, the activity concept has herein been given a more general and fundamental formulation. Motivated by a desire to reconcile everyday discrete digital computation with the higher order continuum of traditional calculus, it succeeds in reducing the basics of the latter to computer science without need of analytical mathematics. A major application therefore is to the revamping education in the calculus to dispense with its mysterious tenets that are too difficult to convey to learners. An object-oriented implementation is available for beginning such a journey.

Other applications and directions are:

- **Sensing**– most sensors are currently driven at high sampling rates to obviate missing critical events. Quantization-based approaches require less energy and produce less irrelevant data.
- **Data compression** – even though data might be produced by fixed interval sampling, it can be quantized and communicated with less bandwidth by employing domain-based to range-based mapping.
- **Reduced communication** in multi-stage computations, e.g., in digital filters and fuzzy logic is possible using quantized inter-stage coupling.
- **Spatial continuity**–quantization of state variables saves computation and our theory provides a test for the smallest quantum size needed in the time domain; a similar approach can be taken in space to determine the smallest cell size needed, namely, when further resolution does not materially affect the observed spatial form factor.
- **Coherence detection** in organizations – formations of large numbers of entities such as robotic collectives, ants, etc. can be judged for coherence and maintenance of coherence over time using this paper's variation measures.

References

1. S. R. Akerkar, Analysis and Visualization of Time-varying data using the concept of 'Activity Modeling', M.S. Thesis, , Electrical and Computer Engineering Dept., University of Arizona,2004
2. J. Nutaro, Parallel Discrete Event Simulation with Application to Continuous Systems, Ph. D. Dissertation Fall 2003, Electrical and Computer Engineering Dept, Univerisity of Arizona
3. R. Jammalamadaka,, Act,ivity Characterization of Spatial Models: Application to the Discrete Event Solution of Partial Differential Equations, M.S. Thesis: Fall 2003, Electrical and Computer Engineering Dept., University of Arizona
4. Ernesto Kofman, Discrete Event Based Simulation and Control of Hybrid Systems, Ph.D. Dissertation: Faculty of Exact Sciences, National University of Rosario, Argentina
5. Theory of Modeling and Simulation, 2nd Edition, Academic Press By Bernard P. Zeigler , Herbert Praehofer , Tag Gon Kim ,
6. J. Nutaro, B. P.Zeigler, R. Jammalamadaka, S.Akerkar,,Discrete Event Solution of Gas Dynamics within the DEVS Framework:Exploiting Spatiotemporal Heterogeneity, ICCS, Melbaourne Australia, July 2003
7. A. Muzy, Doctoral Dissertation, (personal communication)

Systems Theory:
Melding the AI and Simulation Perspectives

Norman Foo[1] and Pavlos Peppas[2]

[1] National ICT Australia, and The School of Computer Science and Engineering,
University of New South Wales, Sydney NSW 2052, Australia
norman@cse.unsw.edu.au
[2] Dept of Business Administration, University of Patras, Patras, 26 500, Greece
ppeppas@otenet.gr

Abstract. The discipline of modelling and simulation (MaS) preceded artificial
intelligence (AI) chronologically. Moreover, the workers in one area are typically
unfamiliar with, and sometimes unsympathetic to, those in the other. One reason for this is that in MaS the formal tools tend to center around analysis and
probability theory with statistics, while in AI there is extensive use of discrete
mathematics of one form or another, particularly logic. Over the years however,
MaS and AI developed many frameworks and perspectives that are more similar
than their respective practitioners may care to admit. We will argue in this paper
that these parallel developments have led to some myopia that should be overcome because techniques and insights borrowed from the other discipline can be
very beneficial.

1 Introduction

The mathematical modelling of dynamic systems began with classical mechanics using differential equations, and analog computers were heavily used to compute solutions to these equations. Serious work on the modelling of systems that were not
primarily governed or describable by differential equations did not take off until the
advent of digital computers. Since then many frameworks have been proposed and implemented. Foremost among the the ones that are based on discrete events is DEVS
[Zeigler, et.al. 2000]. The theory underpinning uses classical notions from automata
theory, but overlays it with ideas from simulation processes and object orientation. Its
meta-theory has debts to the philosophy of science, but strikes out in new directions.
Parallel to this work was that of artificial intelligence logics [Reiter 01]. These logics
were designed with the goal of imbuing robots with reasoning facilities about the real
world. It should not suprise anyone that the two disciplines often invented the same
ideas separately, but regretfully they seldom communicated. Because of this insularity
the good ideas from one were not transmitted to the other.

This paper is an attempt to begin the bridging of this intellectual gap.

2 Philosophy of Systems

Zeigler's pioneering work on a philosophy of systems, simulation and fundamental issues about correctness and adequacy of models in the early 70s that eventually led to a

T.G. Kim (Ed.): AIS 2004, LNAI 3397, pp. 14–23, 2005.

formal framework called *DEVS*. This has been very influential in the modelling and simulation community and beyond, and a contemporary summary of *DEVS* and its background is [Zeigler, et.al. 2000]. We use the word *philosophy* without apology. Mathematical systems theory had, until Zeigler's intervention, been an erudite yet arcane field populated by people of subtle formal sophistication. However, meta-modelling questions like when is a particular kind of model suitable, and what is meant by the correctness of a model, were not usually asked. After Zeigler's work, such questions became meaningful and routine. It is only recently that a parallel development arose in AI, in the work of Sandewall [Sandewall 95].

At this point we hope the reader will forgive a bit of digression into personal history. Zeigler imparted his philosophy about modelling to the first author, NF, who completed a doctoral dissertation [Foo 74] in this area under his supervision. In turn NF imparted the central themes of this philosophy to the second author, PP, while he was completing his doctoral dissertation [Peppas 93] in the AI logic of actions with NF as co-supervisor. It is this pedigree, together with the intervening years in which we (NF and PP) were active in the AI community but without forgetting our roots in systems philosophy of the Zeigler variety, that we believe qualifies us to draw attention to the parallels between MaS and AI.

3 Real Systems, Base Systems, etc.

In this section we explicate the terms that are used in the rest of the paper. Some which will be familiar to the *DEVS* community were introduced by Zeigler in his philosophy of systems – *Base System, Lumped System, Experimental Frame* – and we essentially retain his meanings for them[1]. The other terms are part of our attempt to re-work and refine the original Zeigler concepts to accommodate AI perspectives to meld similar notions in MaS and AI.

It is critical at the outset to clarify the terms *model* and *theory*. According to traditional scientific usage, a *model* is some formal "mirror" of a portion of the external world, and often comes with some inferential machinery. Unfortunately, this is also exactly what is called a *theory* in AI, a convention[2] adopted from mathematical logic. In this convention a model is a set-theoretic structure that satisfies the formulas of the theory.

In figure 1 the "real world" is represented as RS. Out of this amorphous entity (whatever it is!) we *select a portion of interest to us*. Then we conceive of a set of input-output experiments of interest on that portion; this can be viewed abstractly as a relation between input and output trajectories over time (or if time is irrelevant just a relation between input and output). As experiments delineate what one can observe or measure, this restriction on what experiments one is prepared to perform is aptly called the *Experimental Frame* by the MaS community, and the conceptualized set of input-output trajectories is just as aptly called the *Base System*. The Base System is represented as BS in the figure, and it determines the *observational language* that describe the observations or measurements associated with the inputs and outputs of the experiments. T

---

[1] We are however responsible for any distortion of his interpretation that is unpalatable.

[2] This is also the usage in physics to describe generic axioms, as in *electromagnetic theory*.

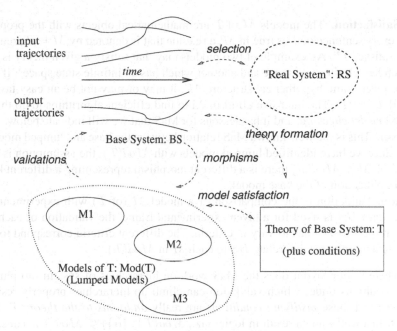

Fig. 1. Theories, Models and Systems.

is an attempt at a formal theory of BS and may contain *hidden variables* (also called *theoretical terms* if they are explicitly named) that do not occur in BS. Thus, this theory T is what less formal practitioners call a model. Here we suggest a distinction between T and the mathematical structures that satisfy T, called the *models* of T in the terminology of logic, and denoted by $Mod(T)$. There could conceivably be more than one such model; we propose to identify any such model M, $M \in Mod(T)$ with the Zeigler concept of a *lumped system*. These models M (in the logical sense) of T will hopefully match many of the input-output experiments of BS, the ideal case being that T has exactly one model M which faithfully reproduces the trajectory pairs in BS. More generally, however, each model M of T is faithful to BS with respect to specialized criteria or conditions. We will say more about those conditions below.

Figure 1 also indicates the relationships among the depicted entities which we now describe. Some are taken directly from the MaS terminology.

Selection. As described above this is driven by the goals of the modelling enterprise. It is informal. A system that is ostensibly "simple" to one modeller can be "complex" to another, e.g., the parable of a stone to a lay observer in contrast to a geologist.

Theory Formation. The theory T is usually crafted by accessing parts of BS, through active experiment, passive observation, inductive learning, inspired guesswork, or a combination of these and iterations thereof. In MaS this is taken for granted as having been done, but this relation is at the heart of much of AI, particularly that branch which has to do with *machine learning*. In traditional engineering the area of *systems identification* is an example of a highly restricted kind of machine learning to fill in parameters of an already assumed form of the theory.

Model Satisfaction. The models M of T are mathematical objects with the property that every sentence of T is true in M, a relation that is denoted by $M \models T$, read as "M satisfies T". An example of two models may that satisfy a given theory is one which has a finite state space and another which has an infinite state space[3]. If one were given some hypothetical structure M', it may or may not be an easy task to see if $T \models M'$. The intensive effort in AI to find efficient algorithms to do this is called *model checking*, and it has lessons for MaS that we will indicate below.

Morphism. This is the well-known MaS relation between the base and lumped models, and since we have identified lumped models with $Mod(T)$, the assumption is that for each $M \in Mod(T)$ there is a different morphism representing a different kind of simplification of the base model.

Validation. Validation is the comparison of a model M (of T) with experiments in BS. Since BS is fixed for a given experimental frame, the validation of each M with respect to BS has to vary in criteria. The different criteria correspond to the different morphisms that relate BS to each M in $Mod(T)$.

So where in our layout does the MaS work on hierarchies of system morphisms and the conditions under which validation can climb up hierarchies properly reside? We suggest that these *justifying conditions* are really *additions to the theory T*. This accords with a well-known result in logic, viz., $Mod(T \cup \{\alpha\}) \subseteq Mod(T)$. Thus, we might begin with an initial theory T that is relatively permssive in having two models M_1 and M_2, e.g., M_1 is just a set of input-output trajectories like BS whereas M_2 has internal state structure. As a justifying condition α that formalizes the existence of states is added, M_1 is eliminated from consideration for validation for the trivial reason that it is not a model of $T \cup \{\alpha\}$ because it cannot interpret α. This is a useful way to think about what is formally entailed by moving up hierarchies, that is really about successive tightening of the criteria to remain a satisfying model.

We believe that this framework is able to accommodate all meta-level questions about modelling of interest to both the MaS and AI communities. The remainder of the paper comprises arguments for this belief.

4   Dynamic Systems in AI

In the last two decades AI logicians have worried about problems of exception handling that appear to be easy for human commonsense but difficult for formal logic. The prototypical example of this in a static domain is when a person is informed that Tweety is a bird, and asked if Tweety can fly. A "normal" answer is Yes. But this is not justified by any simple classical logic for the reason that the informal "rule" that the person has used cannot be simply $Bird(X) \rightarrow Fly(X)$ together with the fact $Bird(Tweety)$, for it may well be the case that Tweety is an emu. This gave rise to *non-monotonic logic*, NML for short. A version of this is the re-formalization of the preceding as $Bird(X) \wedge \neg Abnormal(X) \rightarrow Fly(X)$, and a list of what kinds of birds are abnormal, e.g., $Abnormal(Emu)$, $Abnormal(Kiwi)$. This blocks the inference that Tweety flies if one is also told that Tweety is an emu, but in the absence of that

---

[3] It is easy to write simple theories that do not restrict models to either property.

additional information, the inference goes through, provided that we add a meta-rule saying that unless something can be proved to be abnormal it is not so. The latter is what makes the logic non-classical and it is in fact an instance of Prolog's *negation as finite failure*.

In this paper however, we are interested in dynamic domains. There is a corresponding logic for such domains, and it is there that the interesting parallels with MaS formalisms occur. The AI literature for dynamic systems, like that for static systems, uses the vocabulary of classical logic with inferential rules enhanced with meta-rules that make the logic non-monotonic. A prototypical example is a series battery-driven circuit with two switches that can be either open or closed, and they power a light. The formalization of this simple setting can be done in language of propositional logic, with $sw1$ and $sw2$ to denote the switches and *light* to denote the light. Although they look like propositions and in the technical treatment in one level they behave like them, it is customary in NML to call them *fluents*. A logical formalism that is commonly used in NML is the *situation calculus* [Reiter 01] in which *situation terms* denote sequences of actions, e.g., the sequential actions of closing switches 1 and 2, then opening siwtch 1 is represented as $Do(Open1, (Do(Close2, (Do(Close1, S_0)))$ where S_0 is an initial situation and the names of the actions are as expected. We need a way to say what configurations hold in the initial (or indeed any situation). For this a predicate $Holds(F, S)$ is introduced to encode the assertion that fluent F holds (is true) in situation S. Hence $\neg Holds(sw1, S_0) \land Holds(sw2, S_0)$ essentially says that in the initial situation S_0 switch 1 is open and switch 2 is closed, which fixes our convention here of what configuration (open, closed) we associate with the Boolean *true* for the fluents. A constraint C that captures the naive physics of this circuit can be written as $Holds(sw1, S) \land Holds(sw2, S) \rightarrow Holds(light, S)$. Suppose we begin with initial situation S_0 as above and perform the action of closing switch 1. Presumably we only need to specify this action by its *effect* on the proposition that represents that switch, viz., $Holds(sw1, Do(Close1, S))$. We hope that this will suffice to infer that after the action the light will be on, i.e., we can infer $Holds(light, Do(Close(sw1, S_0))$. Unfortunately the logic specified so far cannot do that unless we also say that closing switch 1 does not affect switch 2! The latter is the essence of *inertia*, which can be formalized at a meta-level, and is non-monotonic. The formal way in which such inertia is actually captured is through *Frame Axioms*. Let us see how this is done in the circuit example here. The frame axiom $Holds(sw2, Do(Close1, S)) \leftrightarrow Holds(sw2, S)$ says that closing switch 1 does not affect switch 2; there is a dual one for closing switch 2. By adding these to the rule, initial state and action specification above, we can now infer $Holds(light, Do(Close1, S_0))$. If such a simple system demands two frame axioms, then what horrors might await us in more complex systems? This is the legendary *Frame Problem*.

At this point MaS readers can be forgiven for thinking that logic is simply not the appropriate methodology for reasoning about such systems. It is trivial to represent such a circuit in DEVS and the computation of effects of actions in it is equally trivial with guaranteed correctness. Moreover, constraints like C above are easily captured in DEVS with the variable influence relation. This is a correct objection, but not because logicians are a perverse lot. Logic has its place, as we shall argue later. It is important to draw the right lesson here. It is this:

DEVS specified systems implicitly assume full inertia, as do all formally specified engineering systems.

This is true of differential equation specified systems, of automata, and all the familiar systems with an engineering flavor. A corollary is this: *Mathematical theories of engineering systems do not have a frame problem.*

If one thinks about it more carefully from the computational perspective, it is precisely full inertia (or equivalently the lack of a frame problem) that permits DEVS state update to be efficient and intuitive. External events are one kind of actions, and internal events are another. Formal specifications say exactly which variables events affect, and it is assumed that no others are affected. In a simulation, which is really successive state updates, the unaffected variables are "carried forward" with no change to the new state. The second lesson is therefore this:

A "balance-carried-forward" state updating is correct if and only if a system is fully inertial.

This is where AI stands to benefit from MaS. If it is known that a dynamic system is inertial with respect to its theoretically specified actions, then the obvious lesson is *Do Not Use Logic – Use DEVS instead.* There is in fact an ancient[4] AI formalism called *STRIPS* [Fikes and Nilsson 71] that is only correct for inertial systems. It is hardly logic even though it uses the vocabulary of logic, but the most telling feature of *STRIPS* is that it uses a "balance-carried-forward" state update procedure called add and delete lists. A persuasive case can be made that in fact *STRIPS* is DEVS in disguise.

5   Query Answering and Planning in AI

In the previous section we identified a class of systems for which DEVS-like theories offer superior computational answers compared with AI logics. On the other hand AI logics has developed techniques for special tasks that should be seriously considered by MaS workers.

One of this is the specialized area of *query answering*. Here is a toy example that illustrates the main points. Suppose we have a *blocks world* of blocks $B_1, B_2, B_3, \ldots, B_N$. The atom $On(A, B)$ means that block A is on block B, and the atom $Table(A)$ means the block A is on the (very large) table. The only actions are to move a block from one position to another, e.g., $Move(C, D)$ means to pick up block C (assuming there is no block on it) and put it on block D (assuming there is no block on it), and $Unstack(B)$ means to pick up block B (assuming there is no block on it) and put it on the table. After a sequence of such actions starting from some initial configuration, some queries that may be of interest are: "where is block B now?", or "how many blocks are on top of block B?". Other forms of queries may be like "Explain how block $B1$ came to be above block $B3$?"

Another area is *planning*. We can use the same blocks world domain to illustrate the key point. Starting with some configuration S we wish to arrive at another configuration S'. A sequence of actions that will take us from S to S' is called a *plan*. Usually

---

[4] In AI this means older than 30 years.

we are interested in not just any plan but one that satisfies some optimality criterion like a shortest plan.

AI action logics of the kind explained in section 4, often using the situation calculus, is good for both query answering and planning. They achieve implementation efficiency by a number of insights, one of which appeals to *localness*. To understand this, consider the query in the blocks world "where is block B now?". It is not hard to see that we do not need to consider the entire action sequence from the beginning, but only the subsequence that can possibly affect the location of block B, i.e., those movements that either move B, or put something on it, etc. In other words, to answer a query we can often *localize* the extent of the system and its history that has to be examined. This insight is widespread in AI planning as well, and the very active work in *model checking* is about using clever data structures and algorithms to exploit localness. Now contrast this with what one has to do if the system is captured in a DEVS-like setting. To answer a query we have to run the entire system until we complete the simulation of the action sequence. Along the way a lot of computation is done to produce state updates on objects that cannot possibly aid in answering the query. Likewise, if we use simulation to devise plans, we either have to re-run the system many times, or else find ways to meta-program the system to discover action sequences that achieve the desired configuration. The lesson from this is:

In query answering or planning, use AI logics when possible.

However this also suggests a direct importation of some facility into MaS frameworks to do "local" simulations to mimic the efficiency of the AI logics for these purposes.

6   Correctness and Galois Correspondence

In this section we re-examine the question of what it means for a theory (with its associated calculus – logic, differential equations, DEVS, etc) to be *correct*.

It does not take much reflection to see that in the framework shown in figure 1 a necessary condition for correctness is that each model M of T does not produce misleading results, i.e., any prediction, inference, etc. L of M is actually in BS, where typically L is an input-output pair of trajectories. More formally we can write this as: *For all M $M \models L \Rightarrow L \in BS$.* This is equivalent to saying that every provable fact L of T is also true in BS. This property is usually called the *soundness of T* (with respect to BS). But we also require a sufficiency condition, that if L is in BS then our models will say so, i.e., *for every M $L \in BS \Rightarrow M \models L$,* or equivalently that any L in BS is provable in T. This is usually called the *adequacy* or *completeness* of T (with respect to BS). Both conditions are desired by MaS and AI workers for their computational realizations of T with respect to their experimental frames.

However, it may be the case that we have a theory T that is unnecessarily powerful in the sense that it also provides correct answers for a much larger class of experiments than BS, i.e., it also handles a wider experimental frame. Although this does no harm it may come at the cost of more elaborate computation. So, what might it mean for T to be "just correct and no more" with respect to some BS? If it is just a single selected system from the "real system" resulting in a particular experimental frame, we already have an answer above – that T has to be sound and complete. To see how we can broaden this question, consider the modelling and simulation of an entire *class* of systems, e.g., the *inertial systems* discussed in section 4. For a more practical example, consider railroad systems. Each particular railroad that we model will have its own rail network, freight cars, schedules, etc. But they all belong to a *class* and share features and procedures. It makes sense to design a theory T that can accommodate any instance of this class rather than just one case. But it also makes sense to not consciously design a T that happens to work for instances of other classes, e.g., an ocean transport network. Thus the fundamental object for conceptualized experiments is not just one base system BS but a *class* $\mathcal{BS}$ of base systems, and it is for this class that we may have to design a theory T to handle. Ideally, what we want for T is the set of truths (trajectory pairs in particular) common to all the base systems in $\mathcal{BS}$. A way of writing this is $Th(\mathcal{BS}) = T$, where the operator $Th(\_)$ "extracts" the common truths. But in order that T should be "just correct and no more" we need $Mod(T) = \mathcal{BS}$, for this guarantees that T will not work outside the members of the class $\mathcal{BS}$. To paraphrase the latter, it says that if some base system BS' is not in $\mathcal{BS}$, then T will make some wrong prediction, inference, etc. for it. The combination of the two equalities $Th(\mathcal{BS}) = T$ and $Mod(T) = \mathcal{BS}$ is called a Galois correspondence (see, e.g., [Cohn 81]). It ensures the tightest possible fit between a theory and a class of base systems for which it is designed.

Are there any known Galois correspondences about systems? The answer is yes. It requires familiarity with a kind of NML called *circumscription* to appreciate its significance. However we can give a paraphrase. The result is about a class of systems that have situations as described in section 4, and the kind of theory that is the tightest possible for them. This class has both the *Markov* and the *inertial* properties – two situations that "look" the same with respect to the fluents that hold in them are indeed the same, and no situation that does not have to make a transition to another (due to an action) does so. It can be shown that the theory (and its logic) is one that uses the circumscription NML in such a way as to enforce situation transitions that cause the least amount of fluent change. A working paper that describes this result is [Foo, et.al. 01].

For posssibly non-inertial systems that need some frame axioms (so that even DEVS-like specifications have to handle them in some procedural way), a natural question that arises, even when there is a theory T that is related to the class $\mathcal{BS}$ via a Galois correspondence, is how succinct this T can be? The intuitive idea is to associate with T a logic or other calculus that is tailored specially for $\mathcal{BS}$ so that it avoids consultation with large sets of frame axioms. Some progress has been made toward an answer in [Peppas, et.al. 01].

7   Causality and Ramification

In this section we translate some AI logics of action terminology into familiar MaS terminology or phrases to further close the gap between AI and MaS.

In AI there was a suspicion that whenever actions were under-specified in effects there ought to be general principles on which states (or situations) should be preferred to others. For instance the principle of inertia is a natural one – that if possible the system should not change state at all. However if a change has to occur, one principle that stood out as highly intuitive is that the new state should differ *minimally* from the previous one. This measure of minimality was itself subject to a variety of interpretation, but ultimately certain properties were agreed to be at least necessary and they were axiomatized as a preference ordering on possible next states (relative to the present one and the action to be taken). As if to confuse matters, a rival approach was developed that used *causal rules* to narrow next-state choices to the point of near-uniqueness. Here is what a causal rule looks like: $sw1 \land sw2 \leadsto light$. As you can see, we have used the series circuit as an example again but this time the fluents are propositions. The interpretation of this rule is not logical implication but truth transmission – if the left hand side is true, it forces the right hand side to be true. This is uni-directional and is not reversible by its contrapositive (unlike material implication). Hence its flavor is not "pure" logic but procedural. The resolution of the comparative experessive power of these two approaches, one using preferences and the other using causality, was achieved recently by Pagnucco and Peppas [Pagnucco and Peppas 01]. The result was elegant. They showed that if the information content conveyed by the chosen fluents was comparable (i.e., no "hidden" fluents), then causal rules are essential in the sense that there are systems whose *a priori* dynamics cannot be completely captured using preference orderings alone.

What does this say about the DEVS state transition mechanism? It is interesting to note that causal rules are in fact built into DEVS. In fact in the co-ordinatized version of its state representation the state transition functions are defined component by component with explicit *influencers* and *influencees* as the fundamental causal graph topologies capture which co-ordinates should feature as the antecedents and consequents of each "mini" transition function. It is an interesting observation that the notion of an influence relation was also (re-?) discovered in AI logics by a number of workers including Thielscher [Thielscher 97].

Ramifications (see also Thielscher, op.cit.) in AI logics are analogous to *internal events* in DEVS, but with possible *chaining*. Recall that such events are those that inevitably follow from some external event without further intervention. A prototypical example is the event of opening a water tap that starts filling a cascade of basins in an Italian style waterfall fountain. The ramifications are the overflow events for each basin in the cascade. One AI logic that handles this with grace is the Event Calculus of Kowalski and Sergot [Kowalski and Sergot] that has uncanny resemblance to DEVS but with none of the latter's highly practical features for interrupts, event aborts, etc. despite much development beyond the pioneering paper cited. The "narrative" facility in some current versions of the Event Calculus permits users to query states at the end of a sequence of events, both external and internal. This should be an easy exercise in DEVS.

8   Conclusion

We have surveyed commonalities between AI and MaS, and highlighted the cross-borrowings that we believe will enrich both disciplines. In particular we suggested that fully inertial systems in AI should just use DEVS instead of logic, and on the other hand query answering and planning in DEVS should call upon localization techniques in AI instead of repeated runs. With increased interaction between the two communities we are hopeful that other cross-borrowings will be identified.

Acknowledgement

The work of the first author was performed in the National ICT Australia, which is funded through the Australian Government's *Backing Australia's Ability* initiative, in part through the Australian Research Council.

References

[Cohn 81] Cohn, P.M., *Universal Algebra (Mathematics and Its Applications)*, Kluwer Academic Publishers, 1981.

[Fikes and Nilsson 71] Fikes, R. E. and Nilsson, N. J., " STRIPS: A New Approach to the Application of Theorem Proving to Problem Solving", *Artificial Intelligence*, 2, 1971, 189-208.

[Foo 74] Foo, N., *Homomorphic Simplification of Systems*, Doctoral Dissertation, Computer and Communication Sciences, University of Michigan, 1974.

[Foo, et.al. 01] Foo, N., Zhang, D., Vo, Q.B. and Peppas, P., "Circumscriptive Models and Automata", working paper, downloadable from http://www.cse.unsw.edu.au/ ksg/Pubs/ksgworking.html.

[Kowalski and Sergot] Kowalski, R.A. and M.J. Sergot. 1986. A Logic-Based Calculus of Events. New Generation Computing 4: 67-95.

[Pagnucco and Peppas 01] Pagnucco, M. and Peppas, P., "Causality and Minimal Change Demystified", Proceedings of the Seventeenth International Conference on Artificial Intelligence (IJCAI'01) , pp 125-130, Seattle, August. Morgan Kaufmann, 2001; downloadable from http://www.cse.unsw.edu.au/ ksg/Abstracts/Conf/ijcai01-DeMystify.html.

[Peppas 93] Peppas, P., *Belief Change and Reasoning about Action – An Axiomatic Approach to Modelling Inert Dynamic Worlds and the Connection to the Logic of Theory Change*, Doctoral Thesis, Computer Science, University of Sydney, 1993; downloadable from http://www.cse.unsw.edu.au/ ksg/Abstracts/Thesis/pavlos.PhD.html.

[Peppas, et.al. 01] Peppas, P., Koutras, C.D. and Williams, M-A., "Prolegomena to Concise Theories of Action", Studia Logica, 67, No 3, April 2002, pp 403-418.

[Reiter 01] Reiter, R., *Knowledge in Action: Logical Foundations for Specifying and Implementing Dynamical Systems*, MIT Press, Cambridge, MA., 2001.

[Sandewall 95] Sandewall, E., *Features and Fluents – The Representation of Knowledge about Dynamical Systems, Volume 1* Clarendon Press, Series: Oxford Logic Guides, Oxford 1995.

[Thielscher 97] Thielscher, M., "Ramification and Causality", Artificial Intelligence Journal, 1997, 89, No 1-2, pp 317-364, 1997.

[Zeigler, et.al. 2000] B.P. Zeigler, H. Praehofer and T.G. Kim, *Theory of Modeling and Simulation :integrating discrete event and continuous complex dynamic systems*, 2nd ed, Academic Press, San Diego, 2000.

Unified Modeling for Singularly Perturbed Systems by Delta Operators: Pole Assignment Case

Kyungtae Lee[1], Kyu-Hong Shim[2], and M. Edwin Sawan[3]

[1] Aerospace Engineering Department, Sejong University,
98 Kunja, Kwangjin, Seoul, Korea 143-747
kntlee@sejong.ac.kr
[2] Sejong-Lockheed Martin Aerospace Research Center,
Sejong University, 98 Kunja, Kwangjin, Seoul, Korea 143-747
kyuhshim@sejong.ac.kr
[3] Electrical Engineering Department, Wichita State University
Wichita, Kansas 67260, USA
edwin.sawan@wichita.edu

Abstract. A unified modeling method by using the δ-operators for the singularly perturbed systems is introduced. The unified model unifies the continuous and the discrete models. When compared with the discrete model, the unified model has an improved finite word-length characteristics and its δ-operator is handled conveniently like that of the continuous system. In additions, the singular perturbation method, a model approximation technique, is introduced. A pole placement example is used to show such advantages of the proposed methods. It is shown that the error of the reduced model in its eigenvalues is less than the order of ε (singular perturbation parameter). It is shown that the error in the unified model is less than that of the discrete model.

1 Introduction

This paper proposes the new methods, i.e., the unified modeling using the δ-operators and the model approximation by the singular perturbation technique, and shows its advantages in illustration of the pole placement design.

1.1 Unified Modeling

One of the main drawbacks of the q-operating discrete system is the truncation and round-off errors caused by the finite word-length pre-assigned. For the continuous-time and the discrete-time models, one uses the differential operator, d/dt, and the forwarding shift operator, q, respectively. The δ-operator is an incremental difference operator that unifies both the continuous and the discrete models together. However, the first disadvantage of using the q-operator is an inconvenience as it is not likes the differential operator, d/dt. The second disadvantage of the normal q-operator models is that there are located the poles near the boundary of the stability circle at small sampling interval. Middleton and Goodwin showed that unified model with the δ-operators has better finite word-length characteristics compared with the discrete model with q-operators when its poles are located closer to $1+j0$ than to the ori-

T.G. Kim (Ed.): AIS 2004, LNAI 3397, pp. 24–32, 2005.
© Springer-Verlag Berlin Heidelberg 2005

gin [1]. This may cause both the instability and the pole/zero cancellation problem due to the low resolution of the stability circle [2]. If the discrete model is converted into the δ-operating unified model, a numbers of the terms of the discrete model are reduced without losing any generality. The finite word-length characteristics are improved by reducing the round-off and truncation errors. Salgado *et al.* illustrated that the unified model with the δ-operators had less relative error than that of the discrete model with the q-operator for rapid sampling for a Kalman filter design [3]. Middleton and Goodwin studied the unified model using the δ-operators in the basic areas of the control systems and the signal processing [4]. Li and Gevers showed some advantages of the δ-operator state-space realization of the transfer function over that of the q-operator on the minimization of the round-off noise gain of the realization [5]. Li and Gevers compared the q-operator and the δ-operator state-space realizations in terms of the effects of finite word-length errors on the actual transfer function [6]. Shim and Sawan studied the linear quadratic regulator (LQR) design and the state feedback design with an aircraft example in the singularly perturbed systems by unified model with the δ-operators [7]-[8].

1.2 Singularly Perturbed Models

The model whose eigenvalues are grouped by the fast and slow sub-models is called the two-time-scale (TTS) model. Where the real parts of eigenvalues of the TTS model are grouped in the distance, it is called the singularly perturbed model, whose non-diagonal terms are weakly coupled. This model is decoupled into the fast and slow sub-models by a matrix diagonalization and becomes an approximate model. This is called the singular perturbation method [9],[10],[11]. Kokotovic *et al.* and Naidu studied the singular perturbation method in the continuous and the discrete models of the time domain, respectively [12],[13]. Naidu and Price studied the singularly perturbed discrete models with illustrations [14],[15],[16]. Mahmoud *et al.* made intensive studies of the singularly perturbed discrete models [17],[18],[19],[20].

2 Unified Modeling

A linear and time-invariant continuous model is considered as

$$\frac{dx}{dt} = Ax(t) + Bu(t) \cdot \tag{1}$$

where x is a $n \times 1$ state vector and u is a $r \times 1$ control vector. A and B are $n \times n$ and $n \times r$ matrices, respectively. The corresponding sampled-data model with the zero-order hold (ZOH) and the sampling interval Δ is given by

$$x(k+1) = A_q x(k) + B_q u(k), \ y(k) = C_q x(k) \cdot$$
$$A_q = e^{A\Delta}, \ B_q = \int_0^\Delta e^{A(\Delta - \tau)} B d\tau. \tag{2}$$

The delta operator is defined as [4],

$$\delta = \frac{(q-1)}{\Delta}. \tag{3}$$

$$qx(k) = A_q x(k) + B_q u(k), \ y(k) = C_q x(k) \cdot \tag{4}$$

$$\delta x(\tau) = A_\delta x(\tau) + B_\delta u(\tau), \ y(k) = C_\delta x(k). \tag{5}$$

The parameter identities of the q- and the δ-operators in Eq. (4) and Eq. (5) as

$$A_\delta = \frac{(A_q - I)}{\Delta}, \ B_\delta = \frac{B_q}{\Delta}, \ C_q = C_\delta. \tag{6}$$

The parameters between the continuous model and the delta model are identified as

$$A_\delta = \Omega A, \ B_\delta = \Omega B \cdot$$

$$\Omega = \frac{1}{\Delta} \int_0^\Delta e^{A\tau} \, d\tau = I + \frac{A\Delta}{2!} + \frac{A^2 \Delta^2}{3!} + \cdots . \tag{7}$$

Therefore, as Δ goes to zero, Ω becomes the identity matrix, thus, A_δ is identified as A. The continuous and the discrete models are written as a comprehensive form using the δ-operators as

$$\rho x(\tau) = A_\rho x(\tau) + B_\rho u(\tau), \ y(\tau) = C_\rho x(\tau) \cdot \tag{8}$$

$$A_\rho = \left\{ \begin{matrix} A \\ A_q \\ A_\delta \end{matrix} \right\}, \ B_\rho = \left\{ \begin{matrix} B \\ B_q \\ B_\delta \end{matrix} \right\}, \ \rho = \left\{ \begin{matrix} d/dt \\ q \\ \delta \end{matrix} \right\}, \ \tau = \left\{ \begin{matrix} t \\ k \\ \tau_\delta \end{matrix} \right\} \cdot$$

It is noted that row one, row two, and row three of A_ρ, B_ρ, ρ and τ denote the continuous model, the discrete model, and the delta model, respectively. When $\Delta \to 0$, then $A_\delta \to A$, $B_\delta \to B$. This means that, when the sampling time goes to zero, the discrete-like δ-model becomes identical with the continuous model. The stability regions for the various operators are introduced as below. For the continuous models, the operator is d/dt and the transform variable is s. For the discrete models, the operator is q and the transform variable is z. For the unified models, the operator is δ and the transform variable is γ. The stability regions are as follow. For the continuous model: $\text{Re}\{\gamma\} < 0$, for the discrete model: $|z| < 1$, for the unified model: $\frac{\Delta}{2}|\gamma|^2 + \text{Re}\{\gamma\} < 0$. As Δ approaches zero, the stability inequality of the unified model becomes that of the continuous model.

3   Singularly Perturbed Unified Model

3.1   Block Diagonalization

Consider the model (8), and assume that the model satisfies the stability conditions above, and then one can write the *two-time-scale*(TTS) model as

$$\begin{bmatrix} \rho x(\tau) \\ \varepsilon \rho z(\tau) \end{bmatrix} = \begin{bmatrix} A_{\delta 11} & A_{\delta 12} \\ A_{\delta 21} & A_{\delta 22} \end{bmatrix} \begin{bmatrix} x(\tau) \\ z(\tau) \end{bmatrix} + \begin{bmatrix} B_{\delta 1} \\ B_{\delta 2} \end{bmatrix} u(\tau) \cdot \tag{9}$$

where x and z are n and m dimensional state vectors, u is an r dimensional control vector, and $A_{\delta ij}$ are matrices of the appropriate dimensionality. Also, it is required that $A_{\delta 22}$ be non-singular. Model (9) has a TTS property, if

$$0 < |E_{s1}| < |E_{s2}| \cdots < |E_{sn}| < |E_{f1}| < |E_{f2}| \cdots < |E_{fm}| . \tag{10}$$

$$\varepsilon = |E_{sn}|/|E_{f1}| \ll 1. \tag{11}$$

where E denotes eigenvalues of the model. From the model (9), the slow and fast sub-models are obtained as

$$\begin{bmatrix} \rho x(\tau) \\ \rho z(\tau) \end{bmatrix} = \begin{bmatrix} A_{\delta s} & 0 \\ 0 & A_{\delta f} \end{bmatrix} \begin{bmatrix} x(\tau) \\ z(\tau) \end{bmatrix} + \begin{bmatrix} B_{\delta s} \\ B_{\delta f} \end{bmatrix} u(\tau) . \tag{12}$$

$$A_{\delta s} = A_{\delta 11} - A_{\delta 12} L, \quad A_{\delta f} = A_{\delta 22} + L A_{\delta 22} . \tag{13}$$

$$B_{\delta s} = B_{\delta 1} - M B_{\delta 2} - M L B_{\delta 1}, \quad B_{\delta f} = B_{\delta 2} + L B_{\delta 1} .$$

Here, L and M are the solutions of the nonlinear algebraic Riccati-type equations as

$$L A_{\delta 11} + A_{\delta 21} - L A_{\delta 12} L - A_{\delta 22} L = 0 . \tag{14}$$

$$A_{\delta 11} M - A_{\delta 12} L M - M A_{\delta 22} - M L A_{\delta 12} + A_{\delta 12} = 0 .$$

Its initial conditions are

$$A_{\delta 0} = A_{\delta 11} - A_{\delta 12} L_0, \quad B_{\delta 0} = B_{\delta 1} - M_0 B_{\delta 2} . \tag{15}$$

$$L_0 = A_{\delta 22}^{-1} A_{\delta 21}, \quad M_0 = A_{\delta 12} A_{\delta 22}^{-1} .$$

The sequences to obtain the solution are defined by

$$L_{k+1} = A_{\delta 22}^{-1}(A_{\delta 21} + L_k A_{\delta 11} - L_k A_{\delta 12} L_k) . \tag{16}$$

$$M_{k+1} = (A_{\delta 11} + A_{\delta 12} - A_{\delta 12} L_k M_k - M_k L_k A_{\delta 12}) A_{\delta 22}^{-1} .$$

We can use L_0 as an order of ε (for $\varepsilon \ll 1$) approximation of L [11].

$$L = L_0 + O(\varepsilon) = A_{\delta 22}^{-1} A_{\delta 21} + O(\varepsilon) . \tag{17}$$

3.2 Pole Placement

The feedback control inputs of the fast and the slow sub-models, where G_0 and G_2 are state feedback gains, are given as

$$u_s = G_{\delta 0} x_s, \quad u_f = G_{\delta 2} z_f .$$

$$u_c = G_{\delta 0} x_s + G_{\delta 2} z_f = G_{\delta 0} x + G_{\delta 2}[z + A_{\delta 22}^{-1}(A_{\delta 21} x + B_{\delta 2} G_{\delta 0} x)] . \tag{18}$$

$$u_c = G_{\delta 1} x + G_{\delta 2} z, \quad G_{\delta 1} = (I_r + G_{\delta 2} A_{\delta 22}^{-1} B_{\delta 2}) G_{\delta 0} + G_{\delta 2} A_{\delta 22}^{-1} A_{\delta 21} .$$

If $A_{\delta 22}^{-1}$ exists and if the slow model pair $(A_{\delta 0}, B_{\delta 0})$ and the fast model pair $(A_{\delta 22}, B_{\delta 2})$ are each controllable, and $G_{\delta 0}$ and $G_{\delta 2}$ are designed to assigned to assign distinct eigenvalues $\xi_i, i = 1, \cdots, n$ and $\xi_j, j = 1, \cdots, m$, to the matrices $A_{\delta 0} + B_{\delta 0} G_{\delta 0}$ and $A_{\delta 22} + B_{\delta 2} G_{\delta 2}$ respectively, then there exists an $\varepsilon^* > 0$ such that for all $\varepsilon \in (0, \varepsilon^*]$

the application of the composite feedback control (18) to the model (9) results in a closed-loop model containing n small eigenvalues $\{\xi_1^c, \xi_2^c, \cdots, \xi_n^c\}$ and m large eigenvalues $\{\xi_{n+1}^c, \xi_{n+2}^c, \cdots, \xi_{n+m}^c\}$, which are approximated by

$$\xi_i^c = \xi_i(A_{\delta 0} + B_{\delta 0}G_{\delta 0}) + O(\varepsilon), \; i = 1,2,\cdots, n \cdot$$

$$\xi_j^c = [\xi_j(A_{\delta 22} + B_{\delta 2}G_{\delta 2}) + O(\varepsilon)]/\varepsilon, \; i = n+j, \; j = 1,2,\cdots, m \cdot$$ (19)

ξ_i is formulated as (where λ_i is an eigenvalue of the continuous model.)

$$\xi_i = \frac{e^{\lambda_i\Delta} - 1}{\Delta} \cdot$$ (20)

The poles of the fast sub-model remain and only the poles of the slow sub-model are shifted to design a state feedback controller. Therefore, letting $G_{\delta 2} = 0$ in Eq. (18) results in $G_{\delta 1} = G_{\delta 0}$. The feedback gain of the slow sub-model, $G_{\delta 0} = [g_{\delta 01}, g_{\delta 02}, \cdots, g_{\delta 0n}]$, is obtained by comparing the coefficients of Eq. (21).

$$Desired\ characteristic\ polynomial = (\gamma - \xi_1)(\gamma - \xi_2)\cdots(\gamma - \xi_n) \cdot$$

$$Characteri\ stic\ polynomial = \det[\gamma I_r - (A_{\delta 0} + B_{\delta 0}G_{\delta 0})] \cdot$$ (21)

where ξ_n is eigenvalue of the unified model by the δ-operators. The feedback gain, $G_{\delta r}$, is used to compute the eigenvalues of the actual model (9) as

$$\xi_r = eig(A_\delta - B_\delta G_{\delta r}) = \xi_{r1}, \xi_{r2}, \cdots, \xi_{rn} \cdot$$

$$G_{\delta r} = [g_{\delta 01}, g_{\delta 02}, \cdots, g_{\delta 0n}, 0_{n+1}, 0_{n+2}, \cdots, 0_{n+m}] \cdot$$ (22)

The error between the exact solution and the reduced solution should be as $|\xi_i| = |\xi_r| + O(\varepsilon)$.

4   Singularly Perturbed Discrete Model

4.1   Block Diagonalization

The general form for a linear, shift-invariant, and singularly perturbed discrete model with $(n+m)$ by $(n+m)$ order is given as

$$x(k+1) = A_{q11}x(k) + A_{q12}z(k) + B_{q1}u(k), x(0) = x_0 \cdot$$

$$z(k+1) = A_{q21}x(k) + A_{q22}z(k) + B_{q2}u(k), z(0) = z_0 \cdot$$ (23)

Eigenvalues of the discrete model are arranged as

$$|1| > |p_{s1}| > |p_{s2}| > \cdots > |p_{sn}| > |p_{f1}| > |p_{f2}| \cdots > |p_{fm}| \cdot$$ (24)

The singular perturbation parameter ε is defined as in Eq. (25) and it is included in A_{q21}, A_{q22}, B_{q2} of model (23).

$$\varepsilon = |p_{f1}|/|p_{sn}| << 1 \cdot$$ (25)

The model (23) is decoupled as

$$\begin{bmatrix} x_s(k+1) \\ z_f(k+1) \end{bmatrix} = \begin{bmatrix} A_{qs} & 0 \\ 0 & A_{qf} \end{bmatrix} \begin{bmatrix} x_s(k) \\ z_f(k) \end{bmatrix} + \begin{bmatrix} B_{qs} \\ B_{qf} \end{bmatrix} u(k) \cdot$$ (26)

The parameters are identified as

$$A_{qs} = A_{q11} - A_{q12}D = A_{q11} + A_{q12}A_{q21}A_{q11}^{-1} = A_{q11} + O(\varepsilon)$$

$$A_{qf} = A_{q22} + DA_{q12} = A_{q22} - A_{q21}A_{q11}^{-1}A_{q12}$$

$$B_{qs} = B_{q1} + EDB_{q1} + EB_2 = B_{q1} - A_{q11}^{-1}A_{q12}A_{q21}A_{q11}^{-1}B_{q1} + A_{q11}^{-1}A_{q12}B_{q2} = B_{q1} + O(\varepsilon)$$ (27)

$$B_{qf} = LB_{q1} + B_{q2} = B_{q2} - A_{q21}A_{q11}^{-1}B_{q1}$$

D and E are computed from Eq. (28).

$$D_{i+1} = (A_{q22}D_i + D_i A_{q12}D_i - A_{q21})A_{q11}^{-1}.$$

$$E_{i+1} = A_{q11}^{-1}(E_i A_{q22} + E_i DA_{q12} + A_{q12}DE_i + A_{q12}).$$ (28)

$$D_0 = -(I - A_{q22})^{-1}A_{q21} = -A_{q21}A_{q11}^{-1}, \quad E_0 = A_{q0}^{-1}A_{12}.$$

$$D = D_0 + O(\varepsilon), E = E_0 + O(\varepsilon).$$

4.2 Pole Placement

The feedback control inputs of the fast and the slow sub-models are given as [21]-[22].

$$u_s(k) + u_f(k) = G_{q0}x_s(k) + G_{q2}z_f(k).$$

$$u_s(k) + u_f(k) = \{(I - G_{q2}(I - A_{q22})^{-1}B_{q2})G_{q0} - G_{q2}(I - A_{q22})^{-1}A_{q21}\}x_s(k)$$

$$+ G_{q2}(I - A_{q22})^{-1}(A_{q21} + B_{q2}(k)G_{q0})x_s(k) + x_f(k)\}.$$ (29)

$$u = G_{q1}x(k) + G_{q2}z(k), G_{q1} = (I - G_{q2}(I - A_{q22})^{-1}B_{q2})G_{q0} - G_{q2}(I - A_{q22})^{-1}A_{q21}$$

The poles of the fast sub-model remain and only the poles of the slow sub-model are shifted to design a state feedback controller. The poles of the fast sub-model remain and only the poles of the slow sub-model are shifted to design a state feedback controller for the reduced solution. Therefore, letting $G_{q2} = 0$ in Eq. (29) results in $G_{q1} = G_{q0}$. The feedback gain of the slow subsystem, $G_{q0} = [g_{q01}, g_{q02}, \cdots, g_{q0n}]$, is obtained by comparing the coefficients of Eq. (30).

$$\textit{Desired characteristic polynomial} = (z - p_1)(z - p_2)\cdots(z - p_n).$$

$$\textit{Characteristic polynomial} = \det[zI_r - (A_{q0} + B_{q0}G_{q0})].$$ (30)

The feedback gain, $G_{qr} = [g_{q01}, g_{q02}, \cdots, g_{q0n}, 0_{n+1}, 0_{n+2}, \cdots, 0_{n+m}]$, is used to compute the eigenvalues of the actual system (23) as

$$p_r = eig(A_q - B_q G_{qr}) = p_{r1}, p_{r2}, \cdots, p_{rn}.$$ (31)

The error between the exact solution and the reduced solution should be as $|p_i| = |p_r| + O(\varepsilon)$.

5 Numerical Example

The model (23) is given as a linear time-invariant continuous model as following for $\varepsilon = 0.25$. $A = [-0.2\ 0.2\ 0\ 0\ ;\ 0\ -0.5\ 0.5\ 0\ ;\ 1\ 0\ 0\ 1;\ -1\ -4\ -1\ -2]$, $B = [0\ ;\ 0\ ;\ 0\ ;\ 1]$. Eigenvalues of A are -0.2, -0.5, -4.0, -4.0. One places the slow eigenvalues -0.2, -0.5 to $-0.707\pm0.707j$. Let the gain of the fast sub-model in the closed-loop model G_2 be zero. The subscripts d, cu, and du denote *discrete model, continuous-like unified model* and *discrete-like unified model*, respectively. Δ_d is a sampling interval when converting the continuous model into the discrete model as shown in Eq. (2). Δ_{cu} is a sampling interval to obtain the continuous-like unified model as shown in Eq. (20). Δ_{du} is a sampling interval to obtain the continuous-like unified model as shown in Eq. (6).

5.1 Continuous Model

Ao= [-0.2 0.2 ; 0.5 -2.5], Bo=[0; 0.5]. With the poles $-0.707\pm0.707j$. One obtains Go=8.5690, -2.5720, Aco=[-0.2 0.2 ; -3.7845 -1.2140]. Gcr=8.5690, -2.5720, 0, 0, λcr=-5.3373, -2.2972, $-0.5328\pm1.0103j$. The error of the absolute values between the desired poles and the resulting poles is 0.1423 that is smaller than ε. That is, $|\lambda_{desired}|=|\lambda_{actual}|+O(\varepsilon)$ holds.

5.2 Discrete Model

For discretization, let the sampling time Δ_d be *0.01* by the ZOH method. *Ad=[0.9980 0.0020 0 0 ; 0.0001 0.9950 0.0050 0.0001 ; 0.0392 -0.0031 0.9992 0.0384 ; -0.0392 -0.1534 -0.0388 0.9224], Bd=[0 ; 0 ; 0.0008 ; 0.0384], Ado=[0.9980 0.0020 ; 0.0049 0.9753], Bdo=[0 ; 0.0049].* The poles in the continuous model, $-0.707\pm0.707j$ are converted in the discrete model as p=$0.9929\pm0.0070j$. Using these poles, one obtains *Gdo=8.6087, -2.5479, Adco=[0.9980 0.0020 ;-0.0376 0.9879], Gdr=8.6087, -2.5479, 0, 0, Pr=0.9483, 0.9770, 0.9946\pm0.0100j*. Converting p_r into the continuous model gives λpr=-5.3057, -2.3224, $-0.5358\pm1.0096j$. The error of the absolute values between the desired poles and the resulting poles is *0.1431* that is smaller than ε. That is, $|p_{desired}|=|p_{actual}|+O(\varepsilon)$ holds.

5.3 Continuous-Like Unified Model

Let the sampling time Δ_{cu} be *0.01*. *Acu=[-0.1998 0.1993 0.0005 0 ; 0.0098 -0.4993 0.4986 0.0097 ; 3.9171 -0.3071 -0.0784 3.8432 ; -3.9176 -15.3377 -3.8820 -7.7647], Bd=[0 ; 0.0001 ; 0.0779 ; 3.8432], Acuo=[-0.1993 0.1973 ; 0.4933 -2.4686], Bcuo=[0.0005 ; 0.4938].* The poles in the continuous model, $-0.707\pm0.707j$ are converted in the continuous-like unified model as ξcu=-0.7070\pm0.7020j$. Using these poles, one obtains *Gcuo=8.6087, -2.5479, Acuo = [-0.2036 0.1986 ; -3.7578 - 1.2104], Gcur=8.6087, -2.5479, 0, 0, ξcur=-5.1674, -2.2956, -0.5394\pm1.0042j*. The error of the absolute values between the desired poles and the resulting poles is *0.1401* that is smaller than ε. That is, $|\xi_{cu-desired}|=|\lambda_{cu-actual}|+O(\varepsilon)$ holds.

5.4 Discrete-Like Unified Model

Let the sampling time Δ_{du} be 0.01. $Adu=[-0.2\ 0.2\ 0\ 0\ ;\ 0\ -0.5\ 0.5\ 0\ ;\ 1\ 0\ 0\ 1\ ;\ -1\ -4\ -1\ -2]$, $Bdu=[0\ ;\ 0\ ;\ 0\ ;\ 1]$, $Aduo=[-0.2\ 0.2\ ;\ 0.5\ -2.5]$, $Bduo=[0\ ;\ 0.5]$. The poles of the slow sub-model in the continuous model, $-0.707\pm0.707j$, are converted in the discrete-like unified model as $-0.7070\pm0.7020j$. Using these poles, we obtain $Gduo=8.5185$, -2.5920, $Aduo=[-0.2\ 0.2\ -3.7593\ -1.2040]$, $Gdur=8.5185$, -2.5920, 0, 0, $\xi dur=-5.3271$, -2.3175, $-0.5277\pm1.0040j$. The error of the absolute values between the desired poles and the resulting poles is 0.1344 that is smaller than ε. That is, $|\xi_{du-desired}|=|\xi_{du-actual}|+O(\varepsilon)$ holds.

For $\Delta_{du}=0.1$, $\xi_{du}=-0.0664$, for $\Delta_{du}=0.01$, $\xi_{du}=-0.1344$, for $\Delta_{du}=0.001$, $\xi_{du}=-0.1415$, and for $\Delta_{du}=0.0001$, $\xi_{du}=-0.1423$.

6 Conclusion

The results obtained as in the section 5.1-5.4 shows that the error bound between the exact and approximate solutions is satisfied for all the four kinds of models, and that the unified solutions using the δ-operators have less error than the discrete solution. As shown at the end of the section 5.4, the solution of the discrete-like unified model becomes the same as that of the continuous model as sampling interval approaches zero. It is shown that the unified model unifies the both continuous model and the discrete model, and that has an improved finite word-length characteristics.

Acknowledgement

This research was supported by the Korea Research Foundation under grant No. 2002-005-D20002

References

1. Middleton, R.H., Goodwin, G.C.: Improved finite word length characteristics in digital control using delta operators. IEEE Trans. on Automatic Control, **31** (1986) 1015-1021
2. Janecki, D.: Model reference adaptive control using delta operator. IEEE Trans. on Automatic Control, **33** (1988) 771-775
3. Salgado, M. Middleton, R.H., Goodwin, G.C.: Connection between continuous and discrete Riccati equations with application to Kalman filtering. IEE Proceedings Pt. D, **135** (1988) 28-34
4. Middleton, R.H., Goodwin, G.C.: Digital Control and Estimation: A Unified Approach. Prentice-Hall, Englewood Cliffs, NJ, (1990)
5. Li, G. and Gevers, M.: Round-off noise minimization using delta-operator realizations, IEEE Trans. on Signal Processing, **41** (1993) 629-637
6. Li, G., Gevers, M.: Comparative study of finite word-length effects in shift and delta operator parameterizations. IEEE Trans. on Automatic Control, **38** (1993) 803-807
7. Shim, K.H., Sawan, M.E. Linear quadratic regulator design for singularly perturbed systems by unified approach using delta operators, International Journals of Systems Science, **32** (2001) 1119-1125

8. Shim, K.H., Sawan, M.E.: Near-optimal state feedback design for singularly perturbed systems by unified approach. Int'l J. of Systems Science, **33** (2002) 197-212
9. Chang, K.W.: Diagonalization method for a vector boundary problem of singular perturbation type. J. of Mathematical Analysis and Application, **48** (1974) 652-665
10. Chow, J., Kokotovic, P.V.: Eigenvalue placement in two-time-scale systems. IFAC Symposium on Large Scale Systems, (1976) 321-326
11. Kokotovic, P.V.: A Riccati equation for block diagonalization of ill-conditioned systems. IEEE Trans. on Automatic Control, **20** (1975) 812-814
12. Kokotovic, P.V., Khalil, H., O'Reilly, J.: Singular perturbation methods in control analysis and design. Academic Press, Orlando, FL, (1986)
13. Naidu, D.S.: Singular Perturbation Methodology in Control Systems. Peter Peregrinus, London, United Kingdom, (1988)
14. Naidu, D.S., Rao, A.K.: Singular perturbation analysis of the closed-loop discrete optimal control system. Optimal Control Application & Methods, **5** (1984) 19-37
15. Naidu, D.S., Price, D.B.: Time-scale synthesis of a closed-loop discrete optimal Control system. J. of Guidance, **10** (1987) 417-421
16. Naidu, D.S., Price, D.B.: Singular perturbations and time scales in the design of digital flight control systems. NASA Technical Paper 2844 (1988)
17. Mahmoud, M.S.: Order reduction and control of discrete systems. IEE PROC. **129** Pt.D, (1982) 129-135
18. Mahmoud, M.S., Chen, Y.: Design of feedback controllers by two-time-stage methods. Appl. Math. Modelling, **7** (1983) 163-168
19. Mahmoud, M.S. and Singh, M.G.: On the use of reduced-order models in output feedback design of discrete systems. Automatica, **21** (1985) 485-489
20. Mahmoud, M.S. Chen, Y., Singh, M.G.: Discrete two-time-scale systems. Int'l J. of Systems Science, **17** (1986) 1187-1207
21. Tran, M.T., Sawan M.E.: Reduced order discrete-time models, Int'l J. Systems Science, **14** (1983) 745-752
22. Tran, M.T., Sawan, M.E.: Decentralized control for two time-scale systems. Int. J. Systems Science, **15** (1984) 1295-1300

A Disaster Relief Simulation Model of a Building Fire

Manabu Ichikawa[1], Hideki Tanuma[2], Yusuke Koyama[1], and Hiroshi Deguchi[3]

[1] Department of Computational Intelligence and Systems Science,
Tokyo Institute of Technology, 4259 Nagatsuta-cho,
Midori-ku Yokohama, 226-8502 Japan
{ichikawa,koyama}@degulab.cs.dis.titech.ac.jp
[2] The Institute of Medical Science, University of Tokyo
htanuma@ims.u-tokyo.ac.jp
[3] Department of Computational Intelligence and Systems Science,
Tokyo Institute of Technology
deguchi@dis.titech.ac.jp

Abstract. We will introduce a basic simulation model, using SOARS. We designed the simple simulation model of a building fire to describe the agents' escaping action from a fire. In this simulation, a fire brakes out in a building suddenly. A fire rapidly spreads across the floor, and even to the next floor. The people have to escape from the building when they find or notice the fire. We show the algorithm and the result, how the fire spreads the floor and the people escape from the fire, of this model and the possibility of future model.

Keywords: Agent Based Modeling, SOARS, Simulation of a building fire

1 Introduction

After the Hanshin-Awaji (Kobe) Earthquake in 1995, the research of disaster relief has become active in Japan. The RoboCup Rescue Project shows the disaster relief simulator and the rescue strategy. Especially, the model of the fire simulation and the life rescue simulation in case of the earthquake are developed.

In this research, we target the disaster relief from the building especially on the case when a fire in the building breaks out. As a first step for developing the simulation model of the life rescue, the extinction activity, and the escaping inducement, we made a very simple model using a new language SOARS. SOARS: Spot Oriented Agent Role Simulator is a new type agent based simulation framework. In this language, agents move on the spots rather than the square cells, and play plural roles. In this paper, we make a model of a fire in the shopping mall of two floors and prepare an easy escaping route for agents. In this description, we analyze the escaping activities of agents.

2 Abstract of SOARS

SOARS is designed to describe agent activities under the roles on social and organizational structure. Role taking process can be described in our language. SOARS is also designed depending on the theory of agent based dynamical systems. Decomposition of multi-agent interaction is most important characteristics in our framework. The

T.G. Kim (Ed.): AIS 2004, LNAI 3397, pp. 33–41, 2005.
© Springer-Verlag Berlin Heidelberg 2005

notion of spot and stage gives special and temporal decomposition of interaction among agents. New information of SOARS can be seen on web site <http://degulab.cs.dis.titech.ac.jp/soars/index.html>.

3   A Simulation Model of a Building Fire

In this section, we will show the framework of this simulation model.

3.1   A Map of a Shopping Mall

We assume a simple shopping mall which has 40 shops, 4 main entrances, 4 stairs, and many spaces where agents can move. Fig 1.shows the place of the main entrance, stairs, shops and spaces in the shopping mall.

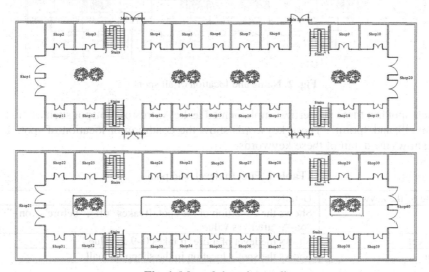

Fig. 1. Map of shopping mall

This shopping mall has two floors and each floor has 20 shops. In the first floor, there are 4 main entrances, 4 stairs and spaces where agents can move. In the second floor, there are also 4 stairs and spaces but no main entrances. Each shop has at least one entrance and agents have to go through the entrance whenever they go into or out the shop. Main entrances are used when agents go into or out the shopping mall. When agents go up or down the floor, stairs are used. Agents go through some spaces when they move from a shop to shop, stairs, main entrance, and so on.

3.2   The Definition of the Spot

In this simulation model we make about 200 spots. Each shop except bigger shops such as Shop1, Shop20, Shop21 and Shop40, is assigned to one spot whose name is shop's name. The bigger shop is divided into 5 spots, for example, Shop1 is divided into Shop1a, Shop1b, Shop1c, Shop1d, and Shop1e. There is one entrance spot for each small shop spot and are two entrance spots for bigger one. We also make 4 main entrance spots and 8 stairs spots. For example, spot EntraceA means one of main entrance and StairsA1 means one of four stairs which is located on the first floor. StairsA2 and StairsA1 are same stairs but not located on the same floor. Spaces where people can move are divided into several spots, and named such as Space1-2-4, Space2-3-4, and so on. The number which followed by "Space" is used to indicate the location of space spot. For example all spots' name of the first floor is shown in Fig.2. First line show the name of the spots and second line show the located place of the spot.

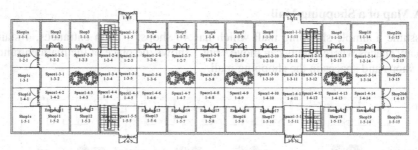

Fig. 2. Name and location of all spots

Each spot has three properties implemented as keywords of the spot such as "fire", "firelevel", and "point". These keywords show the condition or location of spot. Table1. shows the detail of these keywords.

Table 1. Fire, firelevel, and point

Name of the keyword	Details
fire	Shows the condition of the spot. It takes "off", "before", "on+", "on-", "after" as value.
firelevel	Shows the level of a fire. It takes 0~9 as value.
point	Shows the spot's location in the shopping mall.

3.3 The Definition of the Agent

There are some agents in this simulation model. These agents are in Station spot at first and move into Main Entrance spot, then move some spot in the shopping mall. Each agent has four properties, implemented as keywords such as "condition", "escape", "shopping", and "go". The details of these keywords are shown as follows.

The keyword "condition" means the condition of the agent. The keyword "condition" takes "alive" or "dead" as value. An initial value is "alive". When an agent is in the spot whose keyword "fire" is "on+" or "on-", agent's keyword "condition" will change from "alive" to "dead" at a certain probability.

The keyword "escape" means whether the agent has to escape from a fire or not. The keyword "escape" can take "yes", "no" or "end" as a value. An initial value is "no". When an agent finds a fire or knows information that a fire break out in the shopping mall, keyword "escape" will change from "no" to "yes" and begins to escape from a fire. If an agent finishes escaping from a fire, the keyword "escape" will change to "end".

The keyword "shopping" takes "before" or "after" as value and it shows whether an agent has to go into a shop or not. If an agent has been to a certain shop, the keyword "shopping" will change to "after". An initial value of this keyword is "before" and an agent can go out the shopping mall when the keyword "shopping" becomes "after", this keyword "shopping" prohibits an agent to go out the shopping mall without going into some shops.

The keyword "go" is referred when an agent is in the Stairs spot. This keyword can take "up" or "down" as a value. If an agent is on the first floor, the keyword "go" is set to "up" and if on the second floor, the keyword "go" is set to "down".

3.4 The Normal Movement Rule of the Agent

In this subsection, we explain about the movement rule of an agent under the situation of no fire. The rule is decided according to the spot where an agent is. If an agent is in the Station spot, the agent moves to one Main Entrance spot of four at a probability of 25%. When the agent is in one Main Entrance spot, then he moves to the space spot which is in neighborhood. For example, when an agent is in EntranceA spot, he will move to Space1-1-5 spot in the next step. This rule is adjusted to all agents when they enter the shopping mall.

The movements rules of the agent are in the shopping mall are shown as follows. When an agent is in a certain Space spot, the agent will move to a certain spot such as Shop's entrance spot, another Space spot, Stairs spot or Main Entrance spot which are in neighborhood of the spot where an agent is now in. If there are three spots in neighborhood, the agent will move to one spot of three at a probability of 33%. If there are four spots, the probability will become 25%.

We will explain the case which an agent is in the Shop's Entrance spot. In this case, the movement rules change according to the value of the agent's keyword "shopping". If keyword "shopping" is "before", the agent moves to the Shop spot and changes the keyword "shopping" from "before" to "after". If the keyword "shopping" is "after", the agent moves to the Space spot which is in neighborhood of the Entrance spot (Fig.3). When an agent moves to the Shop spot, he can move to the Shop's Entrance spot at a probability of 33%.

We describe the movement rules when an agent is in the Stairs spot. In this case, the movement rule changes according to the agent's keyword "go". There are two spots for each stairs. If an agent is in the Stairs spot which is located on the first floor and the keyword "go" is "up", he moves to the Stairs spot which is located on the second floor. After he reaches the second floor, he moves to Space spot which is in neighborhood and changes the keyword "go" from "up" to "down". If he is in the Stairs spot of the first floor and the keyword "go" is "down", he moves to Space spot and changes the keyword "go" from "down" to "up". The case on the second floor is quite opposite to the case on the first floor (Fig. 4).

At last, we describe the case when an agent is in the Main Entrance spot. The keyword "shopping" is also affected by the rule. In this case, if the keyword "shopping" is "before", which means that the agent has not been to any shops, he can not go out the shopping mall and has to move to Space spot. If the keyword "shopping" is "after", which means the agent has been to some shops, he can go out the shopping mall and moves to the Station spot (Fig.5).

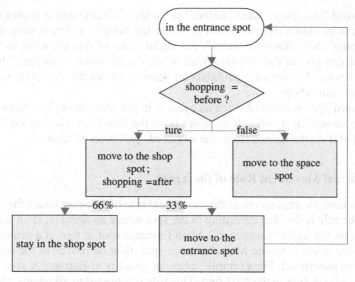

Fig. 3. The flow chart of the agent's movement rule of shopping

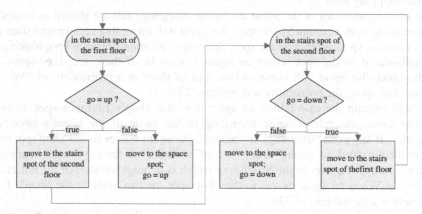

Fig. 4. The flow chart of the agent's movement rule of go up and down

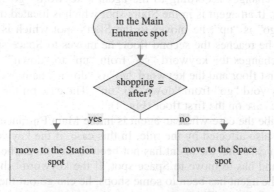

Fig. 5. The flow chart of the agent's movement rule of go out the shopping mall or not

3.5 The Escaping Rule from a Fire

In this subsection, we describe the movement rules of an agent escaping from a fire. The escaping rule is very simple. If an agent's keyword "escape" is "yes", the agent follows this rule. The agent moves by the shortest route to the nearest Main Entrance spot (Fig.6). In escaping from a fire, there is a possibility of facing the situation where a fire on a spot is large and an agent has a risk to go through. In this simulation model, when an agent faces the spot of a high value of "firelevel", the agent takes panic role and moves to neighborhood at random.

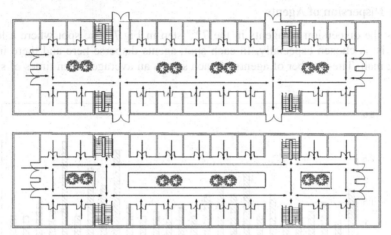

Fig. 6. The shortest route to the nearest main entrance

3.6 The Model of a Fire

When a fire breaks out somewhere in the shopping mall, the keyword "fire" of the spot where a fire happened change to "on+" and change the keyword "fire" of the spot next to burning spot to "before". This means that the spot gets information of a fire. If the spot gets information of a fire, information of a fire is expanded to the spot one after another in each step.

A fire also spreads on the spot. If the spot's firelevel is 4, a fire spreads and the keyword "fire" and "firelevel" of the neighborhood changes to "on+" and 1. A firelevel increases from 1 to 9 during the keyword "fire" is "on+" and it decreases to 0 during the keyword "fire" is "on-". When the keyword "firelevel" changes from 8 to 9, the keyword "fire" changes to "on-" and when the keyword "firelevel" reaches to 0, the keyword "fire" changes to "after" that means a fire is extinguished.

3.7 The Death Algorithm

The definition of the agent's death is very simple. The probability of the death depends on the spot's keyword "firelevel". The probability is calculated from the following formula. When an agent dies, he never moves again.

$$\text{The probability of dying (\%)} = \text{firelevel} \times 100 / 10 \qquad (1)$$

4   The Result of Simulation

We will show the result of breaking out a fire in a specific spot by using this simulation model. We prepare 200 agents and they move in the shopping mall according to the normal movement rule of agents for 72 steps. In the 73rd step from the simulation beginning, a fire is happened in Shop2 spot and agents start escaping from a fire according to the escaping rule of an agent. This simulation is ended by 144th steps. We simulate this model ten times and introduce some results.

4.1   The Dispersion of Agents

We show the dispersion of agents in the 72nd step in Fig.7. The spot where a lot of agents exist is chosen especially. In each spot, agents disperse here and there in the shopping mall. The number of agents of each spot is an average of ten times of simulation.

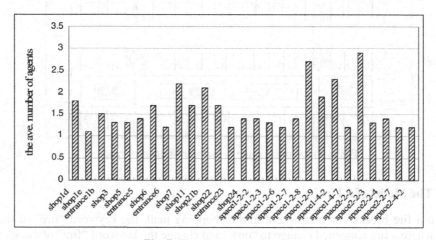

Fig. 7. The dispersion of agents

4.2   Agents' Mortality Rate

The mortality rate of agents is estimated from the number of agents whose keyword "condition" is "dead". The mortality rate and the number of dead agents are shown in Fig.8. The number and percentage are the result of each simulation of ten times.

4.3   The Dangerous Spots

The dangerous spot is specified from the rate of dead agent who stayed there when a fire breaks out. We will show the dangerous spot in Fig.9. This show each spot's percentage of agents who are going to die.

5   Conclusion

In this simulation, we assumed a fire which breaks out in a specific spot and a very simple shortest escaping route from a fire. This escaping route is often seen on the

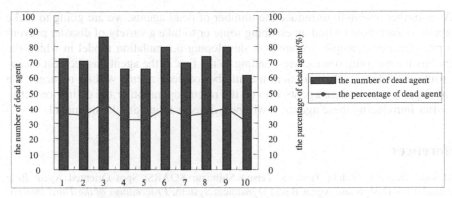

Fig. 8. The number and percentage of death agent

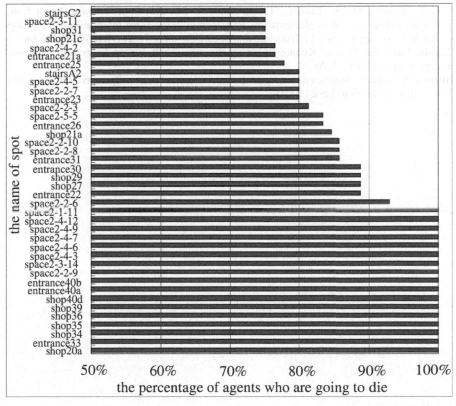

Fig. 9. The dangerous spots' name and percentage of the agents who are going to die

room information of hotel etc. In this simulation, the given escaping route is not necessarily safe. This route has a possibility of danger according to the place where a fire breaks out. It is necessary to prepare the best escaping route according to the situation.

As a further research, to reduce the number of dead agents, we are going to make an agent to learn how to find an escaping route or to take a variety of disaster prevention policies. For example, we are now developing a simulation model in which the agent has the escaping route corresponding to the fire. If the agent has the best escaping route according to the situation, the number of dead agents will be reduced. We are also going to introduce agents who do the rescue operation or the extinction activity. After introducing these agents, it will be a disaster relief simulation model.

References

1. Hiroshi Deguchi, Hideki Tanuma, Tetsuo Shimizu: SOARS: Spot Oriented Agent Role Simulator – Design and Agent Based Dynamical System, *Proceedings of the Third International Workshop on Agent-based Approaches in Economic and Social Complex Systems (AESCS'04)*49–56
2. Hiroshi Deguchi, Hideki Tanuma, Tetsuo Shimizu: SOARS: Spot Oriented Agent Role Simulator – Program and Implementation, *Proceedings of the Third International Workshop on Agent-based Approaches in Economic and Social Complex Systems (AESCS'04)*57–64
3. Satoshi Tadokoro, Hiroaki Kitano: RoboCup-Rescue: The RoboCup Federation. Kyoritsu Shuppan, Japan (2000) 3–90
4. Hiroshi Deguchi: Economics as an Agent-Based Complex System: Springer-Verlag, Berlin Tokyo Heidelberg New York (2004)

Evaluation of Transaction Risks of Mean Variance Model Under Identical Variance of the Rate of Return – Simulation in Artificial Market

Ko Ishiyama[1], Shusuke Komuro[2], Hideki Tanuma[3],
Yusuke Koyama[1], and Hiroshi Deguchi[1]

[1] Interdisciplinary Graduate School of Science and Engineering, Tokyo Institute of Technology
ishiyama@degulab.cs.dis.titech.ac.jp
{koyama,deguchi}@dis.titech.ac.jp
[2] Department of Industrial and Systems Engineering, Chuo University
s566jp@hotmail.com
[3] The Institute of Medical Science, The University of Tokyo
htanuma@ims.u-tokyo.ac.jp

Abstract. Mean Variance (MV) model has spread through institutional investors as one of the most typical diversified investment model. MV model defines
the investment risks with the variance of the rate of return. Therefore, if any
variances of two portfolios are equal, MV model will judge that the investment
risks are identical. However, even if variances are equal, two different risk
cases will occur. One is just depended on market volume. The other is fully depended on speculators who raise stock prices when institutional investors are
purchasing stocks. Consequently, the latter makes institutional investors pay
excessive transaction costs. Development of ABM (Agent Based Modeling) in
recent years makes it possible to analyze this kind of problem by simulation. In
this paper, we formulate a financial market model where institutional investors
and speculators trade twenty stocks simultaneously. Results of simulation show
that even if variances are equal, investment risks are not identical.

1 Introduction

Nowadays, when an institutional investor construct the portfolio according to Mean
Variance (MV) model, they have to take care of not only market volume but also
speculators' behavior, if they prefer to avoid paying excessive transaction costs. MV
model employ the variance of the rate of return as a measure of investment risk.
However, sometimes the variance is fully depended on speculators' behavior.

In this paper, we apply agent based modeling framework SOARS (Spot Oriented
Agent Role Simulator) and formulate the financial market model where institutional
investors and speculators trade twenty stocks simultaneously and simulate the relation
between speculators' behavior and the variance of the rate of return.

Results of simulation show that the variance of the rate of return is not enough as
the measure of investment risk in order to avoid the risk of paying excessive transaction costs.

2 Mean Variance Model with Transaction Costs

In Mean Variance (MV) model, the rate of return is assumed as the random variables.
Let define R as the rate of return for the stocks. Let x be the vector of investment

T.G. Kim (Ed.): AIS 2004, LNAI 3397, pp. 42–49, 2005.
© Springer-Verlag Berlin Heidelberg 2005

proportion. Then MV model minimizes the variance of portfolio $V[R(x)]$ in condition that fixes the expected return $E[R(x)]$ as a parameter ρ. The standard MV model is formulated as follows:

$$\min V[R(x)]$$

$$subject \quad to \quad E[R(x)] = \rho$$

$$\sum_{j=1}^{n} x_j = 1, \quad x \geq 0, \qquad j = 1,...,n$$

(1)

When we evaluate the performance of MV model by simulation, we calculate the return of portfolio using historical data in a particular period. In concrete, when we determine such period as 19yy/mm – 19y'y'/m'm', we solve MV model using a few years of historical date before 19yy/mm and decide the proportion of portfolio. Then we assume that we can purchase the portfolio at the price of 19yy/mm and calculate the return of the portfolio using the date from 19yy/mm to 19y'y'/m'm'.

In MV model with transaction cost, we have to subtract the transaction cost from the return of portfolio. Therefore MV model with transaction costs is formulated as follows:

$$\min V[R(x)]$$

$$subject \quad to \quad E[R(x) - c(x)] = \rho$$

$$\sum_{j=1}^{n} x_j = 1, \quad x \geq 0, \qquad j = 1,...,n$$

(2)

Where $c(x)$ represents the transaction costs of the portfolio. However, the unit transaction cost increases beyond some point, due to the "illiquidity" Effect, which means that large demand arises the price of asset [Konno, 2001]. It is simply lead by the relation between demand and supply in the market. Therefore the expected return may be less than ρ due to increasing of transaction costs. This is very important problem for institutional investors who have to manage large amount of money.

3   Hidden Risks Under Identical Variance of the Rate of Returns

Recently information technology and deregulation enable many speculators to participate in capital markets. Then speculator's behavior affects transaction costs for institutional investors.

Typical strategy of speculators is shown as follows. They, in advance, purchase the stocks, which institutional investors are purchasing for portfolio construction. Then they sell the stocks after the institutional investor has finished buying.

It is difficult for MV model to avoid this kind of risks. Because the investment risk is defined as the variance of the rate of return, same variance is regarded as same investment risk.

Following simple simulation show the impact of market volume and speculator's behavior. We assume two types of traders. Both of them order "buy" or "sell" in stock market. One is "random trader" who sells or buys randomly. The other sells or buys following the market trend captured by difference between the latest price and the moving average of the price chart. Latter trader's behavior is a model of typical speculator's behavior who raises market price.

We simulate above model in four conditions as follows:

1. There are 50 random traders in the stock market (Fig. 1).
2. There are 500 random traders in the stock market (Fig. 2).
3. There are 47 random traders and 3 speculators in the stock market (Fig. 3).
4. There are 470 random traders and 30 speculators in the stock market (Fig. 4)[1].

Results of simulation are shown in the following 4 figures. If there are only random traders in the stock market, increase of market participants make the volatility and the variance of the rate of return smaller because increase of market participants increase market volume. However, if there are a few speculators with random traders in the stock market, the impact of the increase of market volume are not remarkable. Reduction of the volatility and the variance of the rate of return $V[R(x)]$ are smaller than the former case. Fig.4's variance of the rate of return is almost equal to Fig.1's. This shows that risk evaluation based on the variance is unnatural if we regard the transaction costs. In the next chapter, we will investigate the relation between speculator's behavior and institutional investor's risks of paying excessive fee by more large-scale simulation.

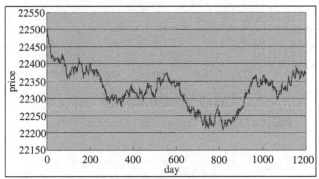

$V[R(x)]= 0.0000000507905$

Fig. 1. 50 Random Traders Case

There are 50 random traders in the stock market. The variance of the rate of return is bigger than Fig.2 because market volume is smaller than Fig.2.

There are 500 random traders in the stock market. The variance of the rate of return is smaller than Fig.1 because market volume is bigger than Fig.1.

There are 47 random traders and 3 speculators in the stock market. In spite of the numbers of the market participants are same, the variance of the rate of return is bigger than Fig.1 because of the impact of 3 speculator's behavior.

There are 470 random traders and 30 speculators in the stock market. In the simulation of Fig.1 and Fig.2, reduction of the variance of the rate of return are remarkable

---

[1] Because the past market prices before the simulations start are given randomly as initial value, each case has different market data.

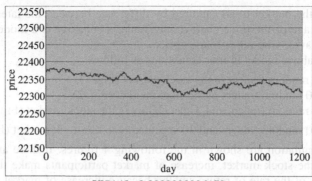

$V[R(x)]= 0.0000000036179$

Fig. 2. 500 Random Traders Case

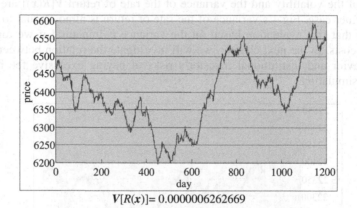

$V[R(x)]= 0.0000006262669$

Fig. 3. 47 Random Traders and 3 Speculators Case

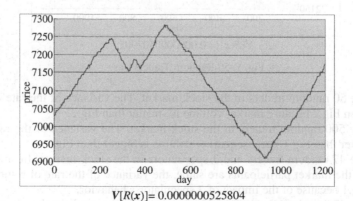

$V[R(x)]= 0.0000000525804$

Fig. 4. 470 Random Traders and 30 Speculators Case

because market volume increase. However, in Fig.3 and Fig.4 only noises of prices disappear. Moreover, if we compare the two variances of Fig.1 and Fig.4, they are quite similar. That shows even if the two variances of the rate of return is equal, the characteristics of market participants and volatilities are not indifferent.

4   Modeling Framework of Market Simulation

We formulate a financial market model where an institutional investor and speculators trade twenty stocks simultaneously and compare the results of simulation in two conditions to investigate whether the investment risks are truly identical or not.

We apply Itayose' algorithm [2], a kind of sealed bid double auction protocol, to calculate the contract prices of each stock [Ishinishi 2002].

We also apply SOARS (Spot Oriented Agent Role Simulator) that provides the conceptual framework for agent-based modeling, seamless spot-oriented extension to application object equipped by Java programming language and agent based dynamical system (ADBS) [Deguchi, 2004]. We formulate a market model based on SOARS framework as follows:

1. Spot Structure
The "Spot" is a place on which agents interact each other. Spot is not only a concrete physical place but also an abstract place for interaction such as market. There are two types of spots in this model.
1) Market Spot: There are 20 Market Spots in this model, and Itayose algorithm is equipped to each Market Spot.
2) Strategy Spot: There is a Strategy Spot in this model, and algorithm of Mean Variance model is equipped to the Strategy Spot.

2. Role Structure
In the stock market, agents buy or sell under their roles on social and organizational structure like the institutional investor or speculators.
1) Random Trader: Random Trader is in the Market Spot and sells or buys a stock randomly. The order price is limit price and set randomly around the latest stock price.
2) Speculator: Speculator is in the Market Spot and sells or buys a stock following the market trend captured by difference between the latest price and the moving average of the price chart. The order price is limit price and set randomly around the moving average. The reference term of the moving average is a week.
3) Institutional Investor: Institutional Investor is in the Strategy Spot and they "buy" stocks according to Mean Variance model to construct the portfolio. Institutional Investor dose not sell stocks because the purpose of Institutional Investor role is to construct the portfolio. An institutional investor's amount of orders is 1000 times Random Trader's or Speculator's amount of orders.

---

[2] The organizing committee of the U-Mart has developed a virtual market simulator of a futures market that include Itayose algorithm. We refer it to apply Itayose algorithm. The U-Mart simulator is a system that traders access to the U-Mart servers using TCP/IP protocol the Internet, and its salient feature is that the U-Mart system enables hybrid simulation involving both human traders and the software trading agents.

3. Stage Structure

We formulate the process of this market model as follows:

Simulation starts when random traders and speculators are selling or buying every market. They sell or buy once a day, and the period of simulation is 80 days[3]. In the last month, institutional investor starts to buy stocks according to MV model during a month. Institutional Investor orders once a week so that they will order 4 times in this simulation. Random traders and speculators also sell or buy during the month when institutional investor buys stocks. Then the traders will make markets prices go up or down according to traders' characteristics.

The process of this market model is divided to three stages.

1) Traders' Order Stage: Random Trader and Speculator buy or sell a stock in Traders' Order Stage.
2) Institutional Investor's Order Stage: After 60 days passed, Institutional Investor starts to buy stocks according to Mean Variance model during a month.
3) Contract Stage: Itayose algorithm calculates the contract price according to the orders of Random Traders, Speculators and Institutional Investor.

4. Model Assumption

When the institutional investor buys stocks according to MV model to construct the portfolio, they will give much demand for the stocks. We assume that prices charts will move as follows:

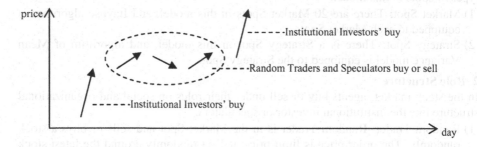

To investigate the above situation, we formulate the following two cases of same variance, which are extended models of above simple simulation of Fig.1 and Fig.4.

1) Random Traders Model

We assume that there are 20 stocks. 50 Random Traders are buying or selling every market spot, and the sum of the traders is $20 \times 50 = 1000$ in this model.

2) Speculators Model

We assume that there are 20 stocks. 470 Random Traders and 30 Speculators are buying or selling every market spot, the sum of the traders is $20 \times (470 + 30) = 10000$ in this model.

---

[3] The markets are closing on Saturday and Sunday, and a month is 20 days in this simulation. 80 days represents quarter.

5   Results of Simulation

The following figures show the results of the simulation (Fig.5 & 6). We investigate the impact of speculators' behavior for MV model.

In the case of Random Traders Model, the institutional investor buys 5 stocks according to MV model[4]. Simultaneously, Random Traders also buy or sell the stocks, and it causes the market prices go up or down randomly.

However, in the Speculators Model, speculator "buys" stocks and they make the market prices only go up and never go down. Therefore, the institutional investor pays excessive transaction costs when they buy the 3 stocks, which they choose.

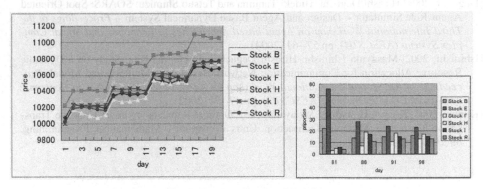

Fig. 5. Random Traders Model

Above two figures are the price chart for a month when the institutional investor constitute portfolio and institutional Investor's proportion of the stocks they purchase in a case of Random Traders Model.

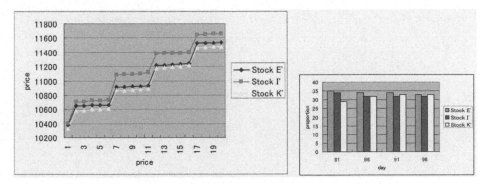

Fig. 6. Speculators Model

Above two figures are in a case of Speculators Model.

---

[4] The institutional investor buys only 5 stocks because the expected return of some of the 20 stocks are negative in this simulation.

6 Conclusion

We start to analyze hidden risks caused by speculators under identical variance of the rate of returns by simple simulation, and then we investigate the impact of speculators' behavior by more large-scale simulation. The results of simulation show that even if variances are equal, investment risks are not identical, and the institutional investor pays excessive transaction costs due to speculators' behavior.

References

[Deguchi, 2004] Hiroshi Deguchi, Hideki Tanuma and Tetsuo Shimizu: SOARS: Spot Oriented Agent Role Simulator – Design and Agent Based Dynamical System -, *Proceedings of the Third International Workshop on Agent-based Approaches in Economic and Social Complex Systems (AESCS'04)*, pp.57--64 (2004)

[Ishinishi, 2002] Masayuki Ishinishi, Hiroshi Deguchi, and Hajime Kita: Study on a Dynamic Resource Allocation for Communication Network Based on a Market-based Model, *Proceedings of the Second International Workshop on Agent-based Approaches in Economic and Social Complex Systems (AESCS'02)*, pp.47--54 (2002)

[Konno, 2001] Konno, H. and A., Wijayanayake, "Optimal Rebalancing under Concave Transaction Costs and Minimal Transaction Units Constraint", Mathematical Programming, 89(2001) 233-250

Association Rule Discovery in Data Mining by Implementing Principal Component Analysis

Bobby D. Gerardo[1], Jaewan Lee[1], Inho Ra[1], and Sangyong Byun[2]

[1] School of Electronic and Information Engineering, Kunsan National University
68 Miryong-dong, Kunsan, Chonbuk 573-701, South Korea
{bgerardo,jwlee,ihra}@kunsan.ac.kr
[2] Faculty of Telecommunication & Computer Engineering, Cheju National University
66 Jejudaehakno, Jeju-si, Jeju-do, South Korea 690-756
byunsy@cheju.ac.kr

Abstract. This paper presents the Principal Component Analysis (PCA) which is integrated in the proposed architectural model and the utilization of apriori algorithm for association rule discovery. The scope of this study includes techniques such as the use of devised data reduction technique and the deployment of association rule algorithm in data mining to efficiently process and generate association patterns. The evaluation shows that interesting association rules were generated based on the approximated data which was the result of dimensionality reduction, thus, implied rigorous and faster computation than the usual approach. This is attributed to the PCA method which reduces the dimensionality of the original data prior to the processing. Furthermore, the proposed model had verified the premise that it could handle sparse information and suitable for data of high dimensionality as compared to other technique such as the wavelet transform.

1 Introduction

Given that data mining can be performed on heterogeneous databases, Knowledge Discovery on Databases or KDD is one of the best tools for mining interesting information in enormous and distributed databases. The discovery of such information often yields important insights into business and its client may lead to unlocking hidden potentials by devising innovative strategies. The discoveries go beyond the online analytical processing (OLAP) that mostly serves reporting purposes only.

One of the most important and successful methods for finding new patterns in data mining is association rule generation. The present trends show that vendors of data management software are becoming aware of the need for integration of data mining capabilities into database engines, and some companies are already allowing for integration of database and data mining software.

Frequently, there are numbers of variables contained in the database, and it is possible that subsets of variables are highly associated with each other. The dimensionality of a model is determined according to the number of input variables used. One of the key steps in data mining is finding ways to reduce dimensionality without sacrificing the correctness of data. One popular method in dimensionality reduction is integration and transformation to generate data cubes. As explained by Margaritis et. al. [8], data cubes may be used in theory to answer query quickly, however, in practice

T.G. Kim (Ed.): AIS 2004, LNAI 3397, pp. 50–60, 2005.
© Springer-Verlag Berlin Heidelberg 2005

they have proven exceedingly difficult to compute and store because of their inherently exponential nature. To solve this problem, several approaches have been proposed by other studies. Some suggest materializing only a subset of views and propose a principled way of selecting which ones to prefer. Many studies claimed that data reduction will hasten processing tasks because mining on reduced data is more efficient and faster while producing the same results. However, some predicament that various researchers say about popular compression techniques is that a quantity of information has to be thrown forever which imply considerably loss of data.

Some other constraints that most researchers observed in the data mining tasks were computing speed, reliability of the approach for computation, heterogeneity of database, and vast amount of data to compute. Often these are restraints that defeat typical and popular mining approach. In this paper, we wish to investigate some techniques and propose a model in data mining to process, analyze and generate association patterns. This study will examine the data reduction technique as integrated component of the proposed model to reconstruct the approximation of original data prior to determination of the efficiency and interestingness of association rules which will be generated through the use of a popular algorithm.

2   Related Works

Among the essential components of data mining is association rule discovery rendered on from simple database repositories to complex database in a distributed system. Association rule mining tasks are finding frequent patterns, associations, or causal structures among sets of items or objects in transactional databases, relational databases, and other information repositories. Data mining uses various data analysis tools such as from simple to complex and advanced mathematical algorithms in order to discover patterns and relationships in dataset that can be used to establish association rules and make effective predictions.

2.1   The Data Reduction Method

The data reduction technique in data mining is used to reduce the data into smaller volume and preserves the integrity of such data. This implies that mining on reduced data is more efficient and faster while producing the same mining results. The lossy technique is a compression method that reconstructs the approximation of the original data. The known popular and efficient method among the lossy data compression techniques are wavelet transform and principal component analysis. This study will explore the use of the latter technique in data mining.

2.2   Data Mining

Numerous data mining algorithms have been introduced that can perform summarization, classification, deviation detection, and other forms of data characterization and interpretation. There are varieties of data mining algorithms that have been recently developed to facilitate the processing and interpretation of large databases. One example is the association rule algorithm, which discovers correlations between items in transactional databases. In Apriori algorithm, candidate patterns that receive sufficient support from the database are considered for transformation into a rule. This type of algorithm works well for complete data with discrete values. One limitation of many association rule algorithms, such as the Apriori is that only database entries that exactly match the candidate patterns may contribute to the support of the candidate pattern. Some research goals are to develop association rule algorithms that accept partial support from data. In the past years, there were lots of studies on faster, scalable, efficient and cost-effective way of mining a huge database in a heterogeneous environment. Most studies have shown modified approaches in data mining tasks which eventually made significant contributions in this field.

3 Architecture of the Proposed Data Mining System

Based on the earlier premise, the researchers developed the proposed architecture of the data mining system which will be presented in the subsequent models.

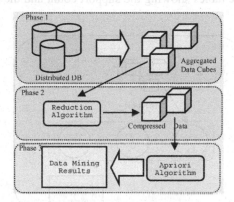

Fig. 1. Proposed Data Mining Model

Fig. 1 shows the proposed three phase implementation architecture for the data mining process. The first phase is the data cleaning process that performs data extraction, transformation, loading and refreshing. This will result to an aggregated data cubes as illustrated in the same figure. Phase two of the architecture shows the implementation of the reduction algorithm using the Principal Component Analysis.

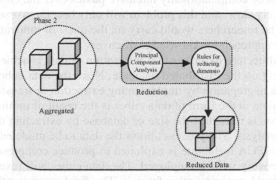

Fig. 2. Data Reduction Model Using Principal Component Analysis (PCA)

The refined process is illustrated by Fig. 2. This explains that the reduction algorithm will be rendered on the aggregated data and thus, the PCA algorithm and the rule for dimensionality reduction which was set by the researcher during the experiment shall be employed. This further means that phase 2 shall generate a reduced data as a result of the reduction algorithm. This will be in accordance with the main premise of this study to generate meaningful association rules on reduced data in order to perform faster computation.

Phase 3 is the final stage in which the Apriori algorithm will be employed to generate the association rules. Further refinements is illustrated by the model in Fig. 3, which should calculate for the frequent itemsets then compute for the association rules as given by the algorithms in section 4.4, these are in equations 2 and 3, respectively. Finally, the discovered rules will be generated based on the assumptions on support and confidence threshold set by the researcher in this study. The output is given by the last rectangle showing the discovered rules. In this study, the discovered rules are provided in the tables showing the support count and the strength of its confidence.

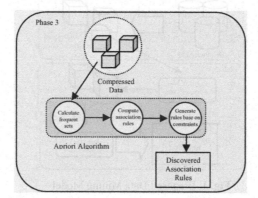

Fig. 3. The Association Rule Discovery Model

As mentioned by Bronnimann et. al. [7], "the volume of electronically accessible data in warehouses and on the Internet is growing faster than the speedup in processing times which was predicted by Moore's Law and classical data mining algorithms that require one or more computationally intensive passes over the entire database are becoming prohibitively slow, and this problem will only become worse in the future". With this premise, the researchers would carry-on the quest of innovating methods to ascertain interesting approach for knowledge discovery.

The proposed architecture describes that integration of data is achieve by first performing data cleaning on the distributed database. Next, the data cubes generation is the result of the data aggregation by implementing extraction or transformation. Note that the aggregated data in the form of data cubes is the result of mining process. The purpose of such cubes is to reduce the size of database by extracting dimensions that are relevant to the analysis. The process allows the data to be modeled and viewed in multiple dimensions. PCA technique is exploited to produce compressed data. Then, association rule discovery will be employed. The data cubes will reveal the frequent dimensions, thus, could generate rules from it. The final stage is utilization of the result for decision-making or strategic planning. The proposed architecture will implement the association rule algorithms on a compressed database and would expect faster but efficient and interesting data mining results.

4 The PCA Technique and Apriori Algorithm

The **Principal Component Analysis** (PCA) will be utilized to execute the data reduction as part of data mining process. PCA is similar to Karhunen-Loeve transform which is a method for dimensionality reduction by mapping the rows of data matrix into 2 or 3 dimensional points and that can be plotted to reveal the structure of the dataset such as in cluster analysis and linear correlations [9]. PCA searches for c k-dimensional orthogonal vectors that can be used to represent the data, where $c \leq k$. The original data are thus projected into smaller space, thus, results to data compression [4]. Consequently, this technique can be utilized for dimensionality reduction. Now, let us define Principal Components Analysis as follows.

Consider a data matrix: $X = | x_{ij} |$

in which the columns represent the p variables and rows represent measurements of n objects or individuals on those variables. The data can be represented by a cloud of n points in a p-dimensional space, each axis corresponding to a measured variable. We can then look for a line OY1 in this space such that the dispersion of n points when projected onto this line is a maximum. This operation defines a derived variable:

$$Y_1 = a_1 x_1 + a_2 x_2 + ... + a_p x_p$$

with coefficients a_i satisfying the condition: $\sum_{i=1}^{p} a_i^2 = 1$

After obtaining OY1, consider the (p-1)-dimensional subspace orthogonal to OY1 and look for the line OY2 in this subspace such that the dispersion of points when projected onto this line is a maximum. This is equivalent to seeking a line OY2 perpendicular to OY1 such that the dispersion of points when they are projected onto this line is the maximum. Having obtained OY2, consider a line in the (p-2)-dimensional subspace, which is orthogonal to both OY1 and OY2, such that the dispersion of points when projected onto this line is as large as possible. The process can be continued, until p mutually orthogonal lines are determined. Each of these lines defines a derived variable:

$$Y_i = a_{1i} X_1 + a_{2i} X_2 + a_{3i} X_3 + ... + a_{ni} X_p \tag{1}$$

where the constants a_{ij} are determined by the requirement that the variance of Yi is a

maximum, subject to the constraint of orthogonality as well as: $\sum_{k=1}^{p} a_{ik}^2 = 1$ for each i.

The Yi thus obtained are called Principal Components of the system and the process of obtaining them is called Principal Components Analysis. The principal components are reduced form of complex multivariate data which choose the first q principal component (q <p) that explains most of the variation in the original variables.

Steps for the Computation of Principal Components

1. Enter attributes to be computed.
2. Combine two or more correlated attributes into one factor. Extract principal components based on the amount of variance maximizing (known as varimax) rotation of the original variable (attribute) space.
3. Extract factors based on the criterion of Eigenvalues ≥ 1.
4. Include attributes with high and positive factor loadings (strong correlation). This means to remove the variables that do not meet the threshold.
5. Generate and reconstruct the approximation of the original data

4.1 Rules to Determine the Strength of Principal Components

The determination of how many factors to extract are based on the assumption that states that we can retain only factors with Eigenvalues ≥ 1. Attributes in the given database shall only be retained if it will show a stronger principal component value. In our assumption, attributes that show a positive value of 0.5 or better are strong components, so other dimensions with less than the assumed threshold could be classified as weaker components. Therefore those items which show weaker component value will be eliminated and this will result to a reconstructed approximation of the original data which include only components and its corresponding attributes with stronger significance.

4.2 Implementation of the Desired Models

An example of an association rule algorithm is the Apriori algorithm designed by Agrawal and Srikant [1]. The use of such algorithm is for discovering association rules that can be divided into two steps: (1) find all itemsets (sets of items appearing together in a transaction) whose support is greater than the specified threshold. Itemsets with minimum support are called frequent itemsets, and (2) generate association rules from the frequent itemsets. All rules that meet the confidence threshold are reported as discoveries of the algorithm. Let T be the set of transactions where each transaction is a subset of the itemset I. Let C be a subset of I, then the support count of C is given by:

$$Suppport\_count(C) = \left|\{t \mid t \in T, C \subseteq t\}\right| \qquad (2)$$

C is a set containing element t such that t belongs to T and C is the subset of t [16]. For example we have the transaction on electronic products given by Table 1. The association rule is in the form of $(X \Rightarrow Y)$, where $X \subseteq I$ and $Y \subseteq I$. The support of the rule $(X \Rightarrow Y)$ is defined by:

$$Support(X \Rightarrow Y) = \frac{(X \cup Y)}{T} \qquad (3)$$

In Table 1, the support count of each candidate is given by set L= [DigitalCamera:7, FlashMemory:5, VideoPlayer:6, VideoCamera:7, Flatmonitors:7, ElectronicBook:6, AudioComponents: 6]. Support of the rule (DigitalCamera^ FlashMemory)$\Rightarrow$ FlatMonitors is (DigitalCamera, FlashMemory, FlatMonitors)/12 is equal to 4/12 or 0.33. While its confidence is (DigitalCamera, FlashMemory, Flat-

Monitors)/ (DigitalCamera, FlashMemory) is equal to 4/5 or 0.8 (80%). The algorithms for the confidence and output rules are given in section 4.4 by equations 4 and 5, respectively.

Table 1. Transactions on electronics products*

TID	List of Items
1	DigitalCamera, FlashMemory, VideoPlayer, Flatmonitors
2	DigitalCamera, FlashMemory, VideoCamera, Flatmonitors, ElectronicBook, AudioComponents
3	DigitalCamera, VideoPlayer, VideoCamera, AudioComponent
4	VideoPlayer, VideoCamera, ElectronicBook, AudioComponent
5	DigitalCamera, FlashMemory, VideoPlayer, VideoCamera
6	DigitalCamera, VideoCamera
7	VideoPlayer, VideoCamera, Flatmonitors
8	ElectronicBook
9	VideoPlayer, VideoCamera, Flatmonitors, AudioComponent
10	DigitalCamera, FlashMemory, Flatmonitors, ElectronicBook
11	DigitalCamera, FlashMemory, VideoCamera, Flatmonitors, ElectronicBook, AudioComponents
12	VideoCamera, Flatmonitors, ElectronicBook, AudioComponents

* Example of transactions on electronic products [11]

Now, let us converse on the approach on the implementation of the desired models on the given database. Apriori algorithm is a level-wise search strategy used in Boolean association rule for mining frequent itemsets. This algorithm has an important property called Apriori property which is used to improve the efficiency of the level-wise generation of frequent itemsets. There are two steps in the implementation of Apriori property, namely the **join step** which will find L_k, a set of candidate k-itemsets by joining L_{k-1} with itself. The next step is the **prune step** in which C_k is generated as a superset of L_k, that is, its members may or may not be frequent, but all of the frequent k-itemsets are included in C_k. The Apriori property implies that any (k-1)-itemset that is not frequent cannot be a subset of a frequent k-itemset; hence, the candidate can be removed.

4.3 Generating Rules from Frequent Itemsets

From the frequent itemsets, the association rules could be generated based on the two important criteria: (a) the rules satisfy the assumed minimum support threshold and (b) the rule has greater confidence limit compared to the assumed minimum confidence threshold. The conditional probability illustrated by the equation (4) was used to calculate for the confidence based on itemset support count.

$$confidence(X \Rightarrow Y) = \frac{Support\_count(X \cup Y)}{Support\_count(X)} . \tag{4}$$

Where $Support\_count(X \cup Y)$ is the number of transactions containing the itemset $(X \cup Y)$. And the $Support\_count(X)$ is the number of transactions containing the itemset X. The association rules were generated by means of the following procedures: (a) generating all nonempty subsets of l, for every frequent itemset; and (b) for every nonempty subsets of l, the output rule is given by:

$$"s \Rightarrow (l\text{-}s)" \text{ if } \frac{Support\_count(l)}{Support\_count(s)} \geq Min\_confidence\_threshold. \tag{5}$$

The above equation implies that it satisfies the minimum support threshold because the rules are generated from frequent itemsets.

5   Experimental Evaluations

The experiment was performed on the database containing 30 attributes comprising six (6) major dimensions and a total of 1,220 tuples of e-commerce and transactional types of data. The evaluation platforms utilized in the study were IBM compatible PC, Window OS, C++, Java, and applications like DBminer, Python, and SPSS.

For the purposes of illustrating the database used in the experiment, we present the Database D showing partially the data as revealed in Table 2. The abbreviated notations for the attributes stand as follows: A_n= books and its corresponding subcategories, B_n = Electronics, C_n = Entertainment, D_n= Gifts, E_n = Foods, and F_n = Health. Furthermore, A_n Book attribute is consist of subcategories like A_1= Science, A_2=social, A_3=math, A_4=computer, A_5=technology, A_6=religion, and A_7=children books. Other dimensions are written with notations similar to that of A_n. The discrete values indicated by each record are corresponding to the presence or absence of the attribute in the given tuples.

Table 2. Database D on Consumer Products

Tuples	Attributes														
	A1	A2	A3	A4	A5	A6	A7	A8	B1	B2	B3	B4	B5	...	F4
1	0	0	0	1	0	0	0	0	0	0	1	0	0	...	0
2	0	0	0	0	0	0	0	0	1	0	0	0	0	...	0
3	0	0	0	0	0	0	0	0	1	0	1	0	0	...	0
4	0	0	0	0	0	0	1	0	0	0	0	0	0	...	0
5	0	0	0	0	0	0	0	0	0	0	0	0	0	...	0
6	0	0	0	0	0	0	0	0	0	0	0	0	0	...	0
7	1	0	0	0	0	0	0	1	0	0	0	0	0	...	1
8	0	0	0	0	0	0	0	0	1	0	0	0	0	...	0
9	0	0	0	1	0	0	0	0	0	0	1	0	0	...	0
:	:	:	:	:	:	:	:	:	:	:	:	:	:	...	:
N	0	0	0	0	0	0	0	0	0	0	1	0	0	0	0

5.1   The PCA Technique and the Discovered Rules

It is remarkable to observe that the use of PCA generated a simpler and understandable association patterns as presented in Table 4. The result implies that there is a considerable trouncing of rules due to the use of PCA technique. Unlike in the other results where PCA was not used, this shows that more association rules were produced. However, the former shows only significant and interesting results.

Table 3 shows the component matrix of database D using principal component analysis. The negative values indicate inverse correlation among the dimensions. The dimensions included are those with positive correlation that met the strong component constraints set by the researcher (factor load $\geq$ 0.5). The retention of factors (principal components) is based on the criterion proposed by Kaiser in 1960 which is the most widely used. It states that we can retain only factors with Eigenvalues greater than 1. In this study, a total of 12 principal components (factors) were extracted.

Table 3. The Data Attributes and PCA Values

Item	Attribute	Parent Attribute	Factor Load	Item	Attribute	Parent Attribute	Factor Load
A1	Science		0.106	D1	Balloon		0.863*
A2	Social		0.684*	D2	Jewelry		0.710*
A3	Math		0.593*	D3	Toys	Gifts	0.775*
A4	Computer	Books	0.735*	D4	Flowers		-0.109
A5	Technology		0.897*	D5	Others		0.159
A6	Religion		0.150	E1	Cakes	Foods	0.842*
A7	Children		0.824*	E2	Groceries		0.903*
B1	Cameras		0.660*	E3	Snacks		0.605*
B2	Cellular		0.119	E4	Others		0.819*
B3	PC	Electro-nics	0.799*	F1	Toiletries		0.824*
B4	HomeApp		0.192	F2	Cosmetics	Health	0.792*
B5	OfficeApp		0.792*	F3	MedicalDent		0.815*
B6	Videogames		0.058	F4	Supplements		0.054
C1	Educational		0.603*				
C2	Movies	Entertain-	0.394				
C3	Music	ment	0.273				
C4	Tickets		0.746*				

* attributes showing strong significance, ≥ 0.5

An approximation of the original data showing only the components with stronger value was produced. For instance, only attributes with significance higher that assumed threshold were retained and became part of the approximated dataset. The compressed data has a total of 20 attributes retained.

5.2 Discovered Association Rules Computed on Reduced and Original Data

Table 4 shows that there were a total of 104 rules that were generated when data reduction was used. This entails a more compact, efficient and faster computing results than the other approach. The table only shows the first ten rules discovered.

Table 4. The Discovered Association Rules

With dimensionality reduction 104 rules with support higher than or equal to 0.850 found. Showing the first 10 rules.			Original data 226 rules with support higher than or equal to 0.850 found. Showing the first 10 rules.		
supp	conf	rule	supp	conf	rule
0.944	0.984	A5=Buy -> C4=Buy	0.975	0.995	A2=buy -> A6=buy
0.944	0.959	C4=Buy -> A5=Buy	0.975	0.980	A6=buy -> A2=buy
0.939	0.985	A3=Buy -> C4=Buy	0.954	0.994	A5=buy -> A6=buy
0.939	0.954	C4=Buy -> A3=Buy	0.954	0.960	A6=buy -> A5=buy
0.935	0.984	A7=Buy -> C4=Buy	0.948	0.995	A3=buy -> A6=buy
0.939	0.985	A3=Buy -> C4=Buy	0.954	0.994	A5=buy -> A6=buy
0.939	0.954	C4=Buy -> A3=Buy	0.954	0.960	A6=buy -> A5=buy
0.935	0.984	A7=Buy -> C4=Buy	0.948	0.995	A3=buy -> A6=buy

The same table shows a total of 226 rules that were generated based on the original data. This implies that there were many rules that were discovered than using the other approach, however, some of these rules are uninteresting. The result in Table 4 used the same range of dataset for comparison purposes. Higher ranges of selected data imply more rules that were generated.

5.3 Computing Time

Comparison of computing time shows that the dimensionality reductions approach in conjunction with the association rule discovery was faster than the typical approach. Below shows two figures comparing time, range and the rules discovered.

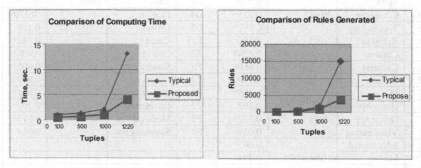

Fig. 4. Comparison between proposed and typical approach

6 Conclusions

The results used the data mining algorithm and the proposed model for data dimensionality reduction which generated the association patterns observed considering the synthetic data illustrated here. We have provided examples and generated interesting rules but more rigorous treatment maybe needed if dealing with more complex and real world databases. The model using the PCA showed that it generated fewer but improved association rules than the other method, and only shows significant and interesting results. By principle, the computation to discover association rules is faster because it exploits only the reduced and approximated dataset. However, there is a matter of concern on the accuracy and reliability of predicting association rules. Because of the attempt to reduce the dimensionality of the dataset, thus, it leads sacrificing and losing some of its components.

Moreover, the proposed model that transform the given data into approximate of the original data prior to association mining had verified the premise that it could handle sparse data as shown by the result in section 5 and is suitable for data of high dimensionality as compared to other technique such as the wavelet transform. For future studies, the researchers recommend a modified architecture and thorough treatment of real world databases in distributed networks.

References

1. Agrawal, R. and Srikant, R. Fast Algorithms for Mining Association Rules. Proc. of International Conference on Very Large Databases VLDB, 1994, 487-499.
2. Han J. & Kamber M. Data mining concepts & techniques. USA: Morgan Kaufmann (2001).
3. Hellerstein, J.L., Ma, S. and Perng, C. S. Discovering actionable patterns in event data. IBM Systems Journal, Vol. 41, No. 3, 2002.
4. Multi-Dimensional Constrained Gradient Mining.
   ftp:// fas.sfu.ca/pub/cs/ theses/ 2001/ JoyceMan WingLamMSc.pdf.

5. Agrawal, R., Imielinski, T, and Swami, A. Mining association rules between sets of items in large databases. Proc. of ACM SIGMOD International Conference on Mngt. of Data, 1993.
6. Chen, B., Haas, P., and Scheuermann, P. A new two-phase sampling based algorithm for discovering association rules. Proceedings of ACM SIGKDD International Conference on Knowledge Discovery & Data Mining, 2002.
7. Bronnimann, H., Chen, B., Dash M, Hass, P., Qiao, Y., & Scheuermann, P. Efficient Data-Reduction Methods for On-Line Association Rule Discovery. In Data Mining: Next Generation Challenges & Future Directions. In Press, 2004.
8. Margaritis D., Faloutsos, C., S. Thrun, and. NetCube: A Scalable Tool for Fast Data Mining and Compression. 27th Conference on Very Large Databases (VLDB), Roma, Italy. Sept. 2001.
9. Korn, F., Labrinidis, A., Kotidis, Y., Faloutsos, C., Kaplunovich, A., and Perkovic, D. Quantifiable Data Mining Using Principal Component Analysis Technical Report, University of Maryland, College Park, Number CS-TR-3754, February 1997.
10. Han, E. H., Karypis G., Kumar, V., and Mobasher, B. Clustering in a high-dimensional space using hypergraph models. Available at: http://www.informatik.uni-siegen.de/~galeas /papers/general/Clustering_in_a_High-Dimensional_Space_Using_Hypergraphs_Models_ %28 Han 1997b%29.pdf, 1998.
11. Gerardo B., Lee J.W., Lee, J.S., Park, M.G., and Lee, M.R. The association rule algorithm with missing data in data mining. Proceedings of International Conference on Computer Science and its Applications (ICCSA 2004), Vol.1, 97-105, May 2004.
12. Principal Component Analysis.
http://www.unesco.org/ webworld/ idams/ advguide/ Chapt6_2.htm.

Reorder Decision System Based on the Concept of the Order Risk Using Neural Networks

Sungwon Jung[1], Yongwon Seo[2], Chankwon Park[3], and Jinwoo Park[1]

[1] Department of Industrial Engineering, Seoul National University,
Seoul, 151-744, South Korea
jsw25@ultra.snu.ac.kr, autofact@snu.ac.kr
[2] Department of Management, Dankook University,
Cheonan, 330-714, South Korea
seoyw@dankook.ac.kr
[3] Department of e-business, Hanyang Cyber University,
Seoul, 133-791, South Korea
chankwon@hycu.ac.kr

Abstract. Due to the development of the modern information technology, many companies share the real-time inventory information. Thus the reorder decision using the shared information becomes a major issue in the supply chain operation. However, traditional reorder decision policies do not utilize the shared information effectively, resulting in the poor performance in distribution supply chains. Moreover, typical assumption in the traditional reorder decision systems that the demand pattern follows a specific probabilistic distribution function limits practical application to real situations where such probabilistic distribution function is not easily defined. Thus, we develop a reorder decision system based on the concept of the order risk using neural networks. We train the neural networks to learn the optimal reorder pattern that can be found by analyzing the historical data based on the concept of the order risk. Simulation results show that the proposed system gives superior performance to the traditional reorder policies. Additionally, managerial implication is provided regarding the environmental characteristics where the performance of the proposed system is maximized.

1 Introduction

The improvement of modern information technologies allows many companies to implement the information management system. The information management system (e.g. POS) makes it possible for cooperating companies to track the sales information and share inventory status information in real time. In this environment, the use of a reorder policy based on the shared stock information updated in real time becomes a major issue.

Traditional reorder policies can be classified into installation stock policies and echelon stock policies. In echelon stock policies, the reorder time is determined based on the sum of the inventory at the subsystem consisting of the considered facility itself as well as of all the downstream facilities whereas in

T.G. Kim (Ed.): AIS 2004, LNAI 3397, pp. 61–70, 2005.

the installation stock policies, the reorder time is determined based on the inventory at the considered facility only. In serial and assembly systems, echelon stock policies show better performance than installation stock policies [4]. On the other hand, in distribution systems, the echelon stock policies do not always outperform installation stock policies [3], and both policies may be far from optimal [1]. Nevertheless, both policies have been commonly used for distribution systems [2]. Furthermore, the echelon stock policies have often been used in the situations where the shared stock information is available [5]. The reason echelon stock policies tend to fail in certain situations is related to the way in which they utilize centralized stock information. In one-warehouse multi-retailer systems, the echelon stock of the warehouse is defined as the sum of the stocks at the warehouse and all the retailers. Thus the method of the evaluating the stock cannot capture inventory unbalance among retailers, since the details of the retailers' stock information are lost when calculating the sum of the individual stocks. To use the shared information more effectively, a new reorder decision policy called as the 'order risk policy' was introduced [7]. The order risk represents the relative cost increase due to immediate orders, as compared to that due to delayed orders. The order risk policy determines the reorder time based on the value of the order risk. The detailed descriptions of the concept of the order risk will be presented in section 2.2.

Although the order risk policy has proven itself to be as an effective inventory control policy, there are some limits to its application in real practice. First, the time required to compute the order risk increases exponentially with the size of the problem. Since the problem size in real practice tends to be large, the practicability of the order risk policy is limited. Second, the order risk policy requires the assumption to be made that the demand follows a specific distribution function. However, it is undesirable to apply this assumption to real situations, because the demand information in the supply chain tends to be too distorted for it to be matched to a specific distribution. To overcome this weakness, we propose a reorder decision system based on the concept of the order risk using neural network.

The rest of the paper is organized as follows; In Section 2, we describe the backgrounds to explain the proposed system. Section 3 introduces the proposed reorder decision system using neural networks. The experimental result is presented in Section 4, and in Section 5 are our conclusions and some directions for future research.

2   Backgrounds

2.1   Two-Echelon Distribution System Model

In this study, we focus on the development of the reorder decision method in the two-echelon distribution system. The general two-echelon distribution system model consists of one warehouse and n retailers facing time variant demand as shown in Figure 1. The retailers order in batches, and the lead time(transportation times) is constant. Similarly, the warehouse replenishes its

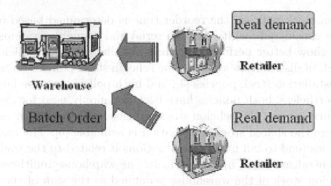

Fig. 1. Two-echelon distribution system model.

stock by ordering batches from an outside supplier with the constant lead time. Unfilled demand at the retailers is backordered and a shortage cost is incurred in proportion to the time remaining until delivery. Unfilled demand at the warehouse is ultimately delivered to a retailer on a first come-first-serve basis and a shortage cost is incurred in the same manner as for the retailers. There are linear holding costs at all locations.

2.2 Optimal Reorder Decision Based on the Order Risk

In this section, we briefly explain the concept of the order risk. At each moment, one should determine whether to order immediately or to delay ordering. In order to make this decision, one needs to quantify the risk associated with an immediate order. The order risk represents the relative cost increase due to an immediate order, compared to that associated with a delayed order. If the value of the order risk is positive, it is beneficial to delay ordering, whereas if the order risk is negative, an order should be issued immediately.

The order risk is derived from marginal analysis. Since the cost increase due to an immediate order is equivalent to the cost reduction obtained by delaying the order, it is necessary to consider the marginal cost savings resulting from delaying the order, denoted by $\Pi(i_0, \Omega_0)$, which is a function of the current warehouse inventory level, i_0, and the sum of the future orders from the retailers within the warehouse lead time, denoted by Ω_0. Let us assume that h_0, p_0 and Q_0 are the holding cost, penalty cost and ordering quantity at the warehouse respectively. The marginal cost savings can be calculated using equation 1.

$$\Pi(i_0, \Omega_0) = \begin{cases} h_0 Q_0 & if \quad \Omega_0 < i_0 \\ h_0(i_0 + Q_0 - \Omega_0) + p_0(i_0 - \Omega_0) & if \quad i_0 \leq \Omega_0 < i_0 + Q_0 \\ -p_0 Q_0 & if \quad \Omega_0 \geq i_0 + Q_0 \end{cases} \quad (1)$$

To explain the concept of the order risk policy, let us define the reorder decision support function (RDSF) as the function whose value is the basis for the reorder decision. Any continuous-review batch ordering policy can be said to be one in which an order is issued when the value of the RDSF is below

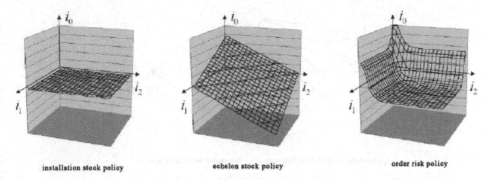

installation stock policy   echelon stock policy   order risk policy

Fig. 2. The reorder decision surface function of three policies.

a certain threshold value, i.e. the reorder point. Installation stock policies use
the installation stock level as their RDSF, while echelon stock policies use the
echelon stock level as the RDSF and the order risk policy use the value of order
risk as RDSF. In the case of a distribution system with one-warehouse and two
retailers in which the inventory levels of these three facilities are denoted as i_0, i_1
and i_2, repectively, each RDSF of the three policies are shown in Figure 2

To apply the order risk policy to real practice, the estimation of the value
of Ω_0 in equation 1 is required, since Ω_0 is not known at the decision time.
Hence, then, the expected marginal savings $E(\Pi(i_0, \Omega_0))$ is to be calculated,
based on the demand distribution assumption at each retailer. However, since
the computation time to calculate the exact value of the expected marginal
savings $E(\Pi(i_0, \Omega_0))$ increases exponentially with the problem size, the real-
time calculation of the expected marginal savings is not practical for the large-
sized problems. On the other hand, the calculation of the value $\Pi(i_0, \Omega_0)$ for
the historical data can be easily calculated, since the sum of the orders from
retailers is already a known value. By analyzing the relationship between the
inventory status and the corresponding $\Pi(i_0, \Omega_0)$ values, we can find the pattern
of the marginal savings for a given inventory status without assuming a specific
probabilistic distribution for the customer demands occurring at retailers.

Based on this idea, we use a neural network in order to apply the concept
of the order risk to real practice. The proposed reorder decision method using
the neural network consists of three steps. First, we find the patterns of the
$\Pi(i_0, \Omega_0)$ values for a given inventory status by analyzing the historical data.
Second, we train the neural network to learn the patterns and finally we use the
trained neural network for the reorder decision in real time.

3   Reorder Decision System Using Neural Networks

A straightforward architecture for the reorder decision system using the neural
network can be one having the system inventory status as the inputs and the
corresponding marginal savings $\Pi(i_0, \Omega_0)$ as the output, as depicted in Figure 3.
However, through experiments with various numbers of the hidden layers and
neurons, we found that the learning speed is extremely slow, and furthermore,

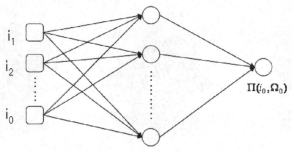

i_0 : inventory level at warehouse , i_k : inventory level at retailer k

Fig. 3. Straightforward architecture for the reorder decision system.

the resulting performance of the system is very poor. This phenomenon can be explained by the fact that the size of the training set should be large enough to learn the patterns of the marginal savings for all the possible system inventory status, especially when the system size is large. For example, if there is one warehouse and 4 retailers whose batch ordering quantity is 50, then the resulting number of the possible system inventory status to be learned will be $50^4\eta$, where η represents the possible inventory status of the warehouse. Thus, the number of the system inventory status increases exponentially with the number of retailers in the considering system. However, since the amount of the historical data set is finite, it seems not possible to train the network to learn the patterns for all the possible system status, especially when the system size is large.

Thus, we adopt the modular concept. By analyzing the calculation procedure of the marginal savings, we can discover that the decision structure can be divided into 2 parts. One is to estimate the sum of the retailer orders, i.e. Ω_0. The other is to estimate the marginal savings of the warehouse, $\Pi(i_0, \Omega_0)$. Therefore, based on the decision structure, we can consider the modular neural network architecture, as described in the following section.

3.1 System Architecture

Based on the rationale mentioned above, we propose the reorder decision system architecture which is composed of two submodule groups, called as the ROE (Retailer Order likeliness Estimator) module and the MSE (Marginal Savings Estimator) module as shown in Figure 4.

ROE Module. The ROE module investigates the inventory status of the retailer, and estimates the retailer's order likeliness. Thus each retailer has its own ROE module. The input node of the ROE module is the inventory status of the retailer, and the output node is the retailer's order likeliness. Through experiments with various network configurations, we found that the configuration of one hidden layer with three hidden neurons shows the best performance.

MSE Module. The role of the MSE module is to estimate the marginal savings for a given system inventory status. Thus the input nodes of the MSE module are the warehouse's inventory status and the all the retailers' order likeliness

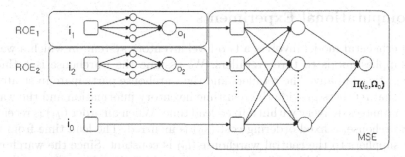

Fig. 4. Modular architecture for the reorder decision system.

values transferred from the ROE modules, and the output node is the marginal savings. The complexity of the neural network of the MSE module depends on the number of the retailers in the system. Thus, in each case, the actual configuration of the MSE module should be determined. To select the suitable network configuration of the MSE module, the cross-validation technique is used. We divide the data set into the training data set and the test data set, and then a number of experiments with various network configurations are executed. In each experiment, we train the neural network using training data and evaluate it by test data. The network configuration which shows the best performance is selected for the MSE module.

3.2   Training ROE and MSE Modules

ROE Module. To train the ROE_k module whose role is to estimate the order likeliness of the retailer k (o_k), the training data is obtained in following ways. Let $i_{k,t}$ and $i_{k,t+l_0}$ mean the inventory status of the retailer k at time t and at time $t + l_0$ respectively, in which l_0 represents the lead time from the outside supplier to the warehouse. The training data consists of one input value (i_k) and output value (z_k) as expressed in equation 2 where R_k means the reorder point at retailer k.

$$[X_{ROE_k}, Y_{ROE_k}] = [(i_k), (z_k)]$$

$$where \quad z_k = \begin{cases} 1 & , if \quad i_{k,t+1} < R_k \\ 0 & , otherwise \end{cases} \tag{2}$$

MSE Module. The input values are the inventory status at warehouse (i_0) and the retailers' order likeliness values ($o_1, o_2, .., o_n$) transferred from the ROE modules. The output value is the marginal savings ($\Pi(i_0, \Omega_0)$) for the given system inventory status. The testing data set can be expressed as in equation 3 where q_k means the base order quantity of the retailer k

$$[X_{MSE}, Y_{MSE}] = [(i_0, o_1, o_2, ..., o_n), (\Pi(i_0, \Omega_0))]$$

$$where \quad \Omega_0 = \sum q_k \times z_k \tag{3}$$

4   Computational Experiments

Our experimental model involves a two-level inventory system, in which a warehouse supplies goods to three retailers. We assume the warehouse is a third-party logistic company. The retailers and the warehouse participate in strategic alliance that retailers provide the real-time inventory information and the warehouse guarantees delivery within a fixed lead time. When an order (q_0) is received at the warehouse, a fixed ordering cost (s_0) is incurred. The lead time from the outside suppliers to the central warehouse (l_0) is constant. Since the warehouse guarantees delivery within a fixed lead time, the lead time from the warehouse to retailer k (l_k) is also constant. There is a linear holding cost at the warehouse (h_0) and at retailer k (h_k). For experimental purpose, we generate the customer demands based on a compound Poisson process, in which the number of the customers of the retailer k per time unit follows poisson distribution with λ_k and the order quantity from each customer follows normal distribution with average m_k and standard deviation σ_k. Excessive customer demand at retailer k is fully backordered and incurs a linear penalty cost (p_k). Excessive retailer orders at the central warehouse are satisfied by an emergency operation that incurs a linear penalty cost (p_0) at the warehouse, and the additional goods required are assumed to be subtracted from the future replenishment to the warehouse. Retailer k uses an ordinary (r_k, q_k) policy based on the local inventory position. The base order quantity is calculated through the EOQ process shown in equation 4, where $\tilde{d}$ means the average demand.

$$q_k = \sqrt{2\tilde{d}s_k/h_k} \tag{4}$$

In the experiment, we select the four experimental factors – 1) the lead time (l_0) at the warehouse, 2) the penalty cost (p_0) at the warehouse, 3) the customer demand variance (σ_k) at retailer k, and 4) the order quantity (q_k)) at the retailer k. We choose three levels of each factor to configure the various experimental environments. The detailed experimental settings are described in Table 1.

The experimental model was developed using C++. We use Matlab neural network tool box for the ROE and MSE module. In the simulation experiments, each of the factor combinations are considered. At each case, the simulation is run for 1000 time units. Following an initial warm-up period of 200 time units, we calculate the average cost difference between the NN-based reorder system and the systems based on the echelon stock policy and between the NN-based order policy and based on the installation stock policy.

Table 1. Parameter setting for experiment.

h_0	p_0	l_0	q_0	h_k	p_k	l_k	q_k	m_k	σ_k
1	10	1	$5q_k$	2	50	2	50	4	1
	30	2					75		2
	50	3					100		3

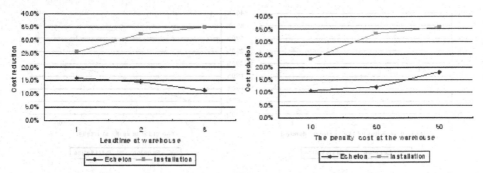

Fig. 5. Cost reduction with varying l_0.   **Fig. 6.** Cost reduction with varying p_0.

The simulation result shows that the average cost reduction obtained by adopting the proposed NN based reorder system rather than the system based on the echelon stock policy and the installation stock policy is 12.89 and 29.87%, respectively.

Figure 5 shows the variation in the cost reduction when the lead time from the outside supplier to the warehouse changes. As can be seen in this figure, the NN-based inventory reorder system is superior to the systems based on the traditional policies in all cases. The cost reduction gap between the reorder systems based on the installation stock and the echelon stock grows as the the warehouse lead time increases, as indicated in [3]. The cost difference between the NN-based reorder system and the echelon stock based reorder system decreases as the warehouse lead time increases. This result is reasonable since a longer warehouse lead time causes a higher variance in the sum of the retailer orders, Ω_0, during warehouse lead time and, consequently, the value of the shared stock information decreases. Thus, the superiority of the NN-based reorder system is more prominent for short warehouse lead times. Figure 6 shows the variation in the cost reduction as a function of the penalty cost at the warehouse. It shows that the cost difference between the NN-based reorder system and the other systems grows as the penalty cost at warehouse increases. This result is also reasonable, since the higher penalty cost at the warehouse reflects the importance of the order decision at warehouse and the NN-based reorder system uses the centralized information more effectively than the other systems. Thus, the NN-based reorder system is better suited to the supply chain environment, in which the penalty cost at the warehouse is high.

Figure 7 shows the variation in the cost reduction when the standard deviation of the customer demand changes. The echelon stock based system is relatively superior to the installation stock based system regardless of the demand rate. The cost difference between the NN-based reorder system and the other systems decreases when the standard deviation of the customer demand increases. Again, this result is intuitively reasonable, since higher variance of the customer demand brings about a higher variance of the orders from the retailers and, therefore, the value of the individual stock information decreases as the demand rate increases. Thus, the superiority of the NN-based reorder system is

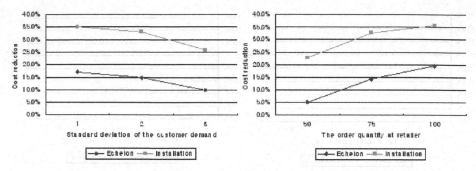

Fig. 7. Cost reduction with varying σ_i.   **Fig. 8.** Cost reduction with varying q_i.

more prominent when the standard deviation of the customer demand at the retailer is low. Figure 8 shows the variation in the cost reduction as a function of the base order quantity of the retailer. This result shows that the cost difference between the NN-based reorder system and the other systems grows as the base order quantity increases. It can be explained in the following ways. As we mentioned above, the echelon stock based system does not use the shared information effectively, because it cannot capture inventory unbalance among retailers. When the base order quantity increases, the range of the inventory position at the retailers also broadens and the inventory unbalance becomes more complex. Since our proposed NN-based reorder system is already trained to capture the inventory unbalance and make a decision based on this information, the cost difference increases as the base order quantity increases.

5   Conclusion

In this study, we proposed the reorder decision system based on the concept of the order risk using the neural network, which can be applied to real practice. We developed the modular architecture for the reorder decision system, consisting of the ROE module and the MSE module. The training procedure for each module is provided. The experimental results show that the proposed reorder decision system shows superior performance to the systems which make the reorder decision based on the traditional policies such as the installation stock policies or the echelon stock policies. Through sensitivity analysis, we found that the benefits from adopting the proposed system can be maximized when the warehouse lead time is long, the warehouse penalty cost is high, the variance of the demand incurred at retailers is low, and the base order quantities of the retailers are large. The reorder decision system using neural networks developed in this research can be extended to the VMI environment that is recently one of the popular strategic alliances in the supply chains. In this case, the system architecture should be modified to enable the simultaneous reorder decision for all of the retailers as well as the warehouse. Furthermore, it can be also extended to more complex system models. For example, the extension to the systems consisting of more than two-echelons or including the floor level scheduling is worth to be studied.

References

1. Axäster S.: Comparison of echelon stock and installation stock policies for two-level inventory systems. International Journal of Production Economics. **19** 109-110 (1997)
2. Axäster S. and Zhang W.: A joint replenishment policy for multi-echelon inventory control. International Journal of Production Economics. **59** 243-250 (1999)
3. Axäster S. and Juntti L.: Comparison of echelon stock and installation stock policies for two-level inventory systems. International Journal of Production Economics. **45** 303-310 (1996)
4. Axäster, S. and Rosling L.: Installation vs. echelon stock policies for multi-level inventory control. Management Science. **39** 1274-1280 (1993)
5. Chen F, Zheng Y, S.: One-warehouse multiretailer systems with centralized stock information. Operations Research. **45** 275-287 (1997)
6. Haykin S.: *Neural Networks* (Maxwell Macmillan Publishing Company. 1998)
7. Seo Y., Jung S. and Hahm J.: Optimal reorder decision utilizing centralized stock information in a two-echelon distribution system. Computers and Operations Research. **29** 171-193 (2002)

Simulation Modeling with Hierarchical Planning: Application to a Metal Manufacturing System

Mi Ra Yi and Tae Ho Cho

School of Information & Communication Eng.,
Sungkyunkwan University, Suwon, 440-746, Korea
{yimira,taecho}@ece.skku.ac.kr

Abstract. There has been an increasing volume of research that combines artificial intelligence (AI) and simulation in the last decade to solve the problems of various kinds, some of which are related to manufacturing systems. In modern manufacturing industries, automatic systems composed of computers are common, and these systems are continuing to enhance the efficiency of the manufacturing process by analyzing the overall production process – from design to manufacturing. This paper deals with the problem regarding how to improve the productivity of a metal grating manufacturing system. To solve this problem, we proposed and applied Hierarchical RG-DEVS formalism, which is a modeling methodology for incorporating the hierarchical regression planning of AI and simulation, for constructing an environment for sound modeling. This research presents not only an improvement of the metal production design environment that can predict efficiency in the manufacturing process, but also a cooperation technique for AI planning and simulation.

1 Introduction

There has been an increasing volume of research that combines artificial intelligence (AI) and simulation in the last decade [1,2]. The simulation formalisms are concerned with the time related changes in entities' states, while those of AI are concerned with relationships that do not take time into account. The distinguishing feature of AI schemes is their ability to represent knowledge in declarative, as opposed to procedural form, in which states of an entity are presented in propositional form as facts, amenable to general purpose logical manipulation. In other words, such knowledge is explicitly isolated as data rather than intermingled with the code that uses it [3,4]. Simulation models are representations that also package knowledge about a particular system domain to meet specific objectives. It is not surprising, therefore, that the research that exploits both dynamic knowledge of simulation and the declarative knowledge of AI are of great concern [1,3,5].

In modern manufacturing industries, automatic systems composed of computers are common. These systems enhance the efficiency of the manufacturing process by analyzing the overall production process – from design to manufacturing [6,7,8,9]. Most of the early CAD systems performed drafting-first functions to take the place of a drawing board. However, as time passes, people are starting to recognize that CAD systems could also be used to include specification data and thus further help to automate the design process [7]. Moreover, many CAD systems focus on the program

T.G. Kim (Ed.): AIS 2004, LNAI 3397, pp. 71–80, 2005.

environment, which computes the quantity of materials by analyzing CAD data or includes a FMS (Facility Management System) for other processes in addition to design. Next-generation systems will enable the capture of a broader variety of product information, and will support a wider range of product development activities than do existing tools [10].

This paper deals with the problem of how to improve the productivity of a metal grating manufacturing system. To solve this problem, we proposed and applied Hierarchical RG-DEVS formalism, which is a modeling methodology for incorporating the hierarchical regression planning of AI and simulation, for constructing an environment for sound modeling. This research presents not only an improvement of the designing system that can predict efficiency in the manufacturing process, but also a cooperation technique for AI planning and simulation.

The paper is organized as follows. Section 2 reviews RG-DEVS formalism and introduces Hierarchical RG-DEVS, and Section 3 describes problems in our target system, a metal grating manufacturing system. To solve the problem, the planning and simulation parts of Hierarchical RG-DEVS application are explained in Section 4 and Section 5 respectively. Finally, Section 6 gives the conclusions.

2   Hierarchical Planning and Simulation

2.1   AI Planning and DEVS

AI planning systems have wide application areas, such as motion planning of robot, production planning in manufacturing, collaborative planning on multi-agent systems, network control, object-oriented module development, etc. Simulation can play an important role in designing, analyzing and testing such systems. Especially, a model with intelligent planning capability is a valuable aid to simulation in the above listed areas.

The DEVS (Discrete EVent system Specification) formalism is a theoretical, well-grounded means of expressing hierarchical, modular discrete-event models [11]. In DEVS, a system has a time base, inputs, states, outputs, and functions. The system functions determine the next states and outputs based on the current states and input. The formalism has two types of models: basic models and coupled models. The structure of the basic model is described in Figure 1.

The RG-DEVS (ReGression – Discrete EVent system Specification) gives planning capability to DEVS models [12]. RG-DEVS, an extension of classic DEVS, expands the classes of system models that can be represented in DEVS. The regression mechanism of AI production systems is applied in selecting the rules or actions. These selected rules are the desired sequence of actions that form action plans for achieving goals. The sequential states of a model are defined during the regression process by incorporating regressed subgoals of predicate formulas with other attributes predefined for describing the model's states. So, the model's sequential states are dynamically generated according to the goal, which is just one of many possible goals that the model can achieve. The mechanism for detecting miss execution of plans and building amended, or new, plans for performing corrective actions are also given to provide easy modeling of the miss execution of planned actions in the real system counterpart.

DEVS

$$M = \, <X, S, Y, \delta_{int}, \delta_{ext}, \lambda, ta>$$

Where,

X : set of external input event types
S : sequential state set
Y : set of external event types
    generated as output
δ_{int} : internal transition function ;
    $\delta_{int} : S \rightarrow S$
δ_{ext} : external transition function ;
    $\delta_{ext} : Q \times X \rightarrow S$
    $Q = \{(s,e) \mid s \in S, 0 \leq e \leq ta(s)\}$
λ : output function; $\lambda : S \rightarrow Y$
ta : time advance function;
    $ta : S \rightarrow R_0^+$

RG-DEVS

$$M_{RG} = \, <X, S_R, Y, \delta_{ext}, \delta_{int}, \lambda, ta, F>$$

Where,

X, Y : same as in classic DEVS
S_R : $S_R = S^C \times S^D_R$
    S^C : S in classic DEVS
    S^D_R : set of regressed state
    descriptions of a goal
δ_{int} : $S_R \times (Fu \cup \emptyset) \rightarrow S_R$
δ_{ext} : $Q \times X \rightarrow S_R$
    $Q = \{(s,e) \mid s \in S_R, 0 \leq e \leq ta(s)\}$
λ : $S_R \rightarrow Y$
ta : $S_R \rightarrow R_{0,\infty}$
F : set of production rules

Hierarchical RG-DEVS

$$M_{HRG} = \, <X, S_R, Y, \delta_{ext}, \delta_{int}, \lambda, ta, F>$$

$F = F_1 \cup F_2 \cup \ldots \cup F_N$, F_k : set of operation rules for operations in F_{k-1}

$S_R = S^C \times S^D_R$, $S^D_R = S^D_{R1} \times S^D_{R2} \times \ldots \times S^D_{RN}$,

                    S^D_{Rk} : set of states regressed by operations in F_k

Fig. 1. Basic Model Structures of DEVS, RG-DEVS, and Hierarchical RG-DEVS

2.2 Hierarchical RG-DEVS

Ever since the conception of AI, hierarchical problem solving has been used as a method to reduce the computational cost of planning. The idea of hierarchical problem-solving is to distinguish between goals and actions of different degrees of importance, and solve the most important problem first. Its main advantage derives from the fact that by emphasizing certain activities while temporarily ignoring others, it is possible to obtain a much smaller search space in which to find a plan. Today, a large number of problem-solving systems have been implemented and studied based on the concept of hierarchical planning [13,14].

In this paper, we expand RG-DEVS formalism in order to embed the hierarchical planning concept in the DEVS formalism, and name it as Hierarchical RG-DEVS. In Figure 1, the structure of M_{HRG} describes the model in Hierarchical RG-DEVS. In M_{HRG}, the set of operations for planning is the union of operation sets for each level in the hierarchical planning process, and the set of states is the Cartesian product of state sets that belong to each level. The hierarchy concept embedded in the simulation model helps make it easy to analyze the target system according to each of its levels or different points of view.

3 Problem in Metal Grating Manufacturing

In this section, we describe the problem in a metal grating manufacturing system, and discuss the background of the use of Hierarchical RG-DEVS to solve the problem.

3.1 Domain and Problem: Grating Manufacturing System

A grating is a grid or net shaped product used to cover water drains along the sides of roads. It is also used as a flooring material for various types of production plants and ships. Referring to Figure 2 (b), the grid of a grating is composed of *Bearing-Bars (BB)* in the metal length direction) and *Cross-Bars (CB)* in the width direction. *BB-Pitch* and *CB-Pitch* are regular gaps between BBs and CBs, respectively. Figure 2 (a) and (b) show the usage and the construction of a grating.

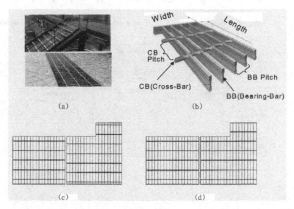

Fig. 2. Grating Usage (a), Construction (b), Before (c) and After (d) CB-Line Matching

The grating production system that we face has been using computerized systems (GDS and CPS) to improve efficiency of production process. *GDS (Graphic automatic Drawing System)* is an automatic design and drawing system that was developed based on problems provided by the design workers of a factory located near the city of Seoul, South Korea. Within GDS, a user (designer) initially draws frame lines manually in a cyber drawing sheet, and then the grating items are automatically allocated inside the designated area, bordered by the frames, with the input of additional parameters. *CPS (Cutting Planner System)* is an automatic cutting planner system that arranges grating items in a grating panel to be cut. CPS was developed for the purpose of reduction of the material loss, but the cutting plan affects other parts of the manufacturing process.

The standard requirement of *a grating item* (the product unit) has the constraint that the width of a grating item must be an integral multiple of the BB-Pitch, while the length of a grating item is not constrained by the CB-Pitch. So, *CB-Line Matching* among grating items, as shown in (c) and (d) of Figure 2, is one of the important issues in the design process of a grating production system. However, CB-Line Matching is not just a problem of the design stage.

CB-Line Matching is closely related to the efficiency of the manufacturing process: perfect matching of CB-Lines for all items may reduce the efficiency. For CB-Line Matching, designers should estimate how to match the CB lines in a way that it does not seriously reduce the efficiency. So, the CB-Line Matching function in GDS has options for designers to try executing it in various ways. However, it is difficult to find the appropriate rate of CB-Line Matching, because it is not easy to predict the efficiency of the manufacturing process (especially if the manufacturing is being processed dynamically). Until the application of our modeling methodology, the workers at the factory had just followed their intuition based on their experience of matching the CB lines.

3.2 Approach: Hierarchical Planning and Simulation

To solve this problem, designers should be able to predict the efficiency of the CB-Line Matching, and make higher quality drawings (considering the efficiency from the point of view of suppliers) for grating design using the proper prediction.

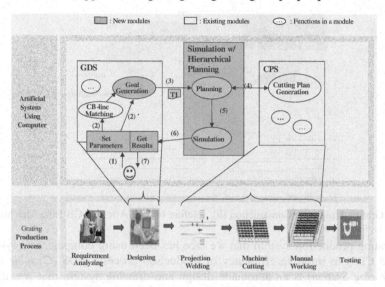

Fig. 3. Grating Manufacturing System and CB-Line Matching

In order to predict the efficiency of the manufacturing process under the dynamically changing environment, we use the AI planning technology to make a manufacturing plan from the design drawing, and then use the simulation technology to evaluate the efficiency of the plan. Especially, hierarchical planning is used to reduce the cost of planning, and to reflect the properties of the target system according to the different points of views. So, we use our Hierarchical RG-DEVS formalism that embeds the hierarchical planning capability to the simulation model.

Figure 3 presents the overall structure of the grating manufacturing system and the flow of CB-Line Matching with the simulation module. The parenthesized numbers are the sequence of activities for the CB-Line Matching process.

4 Hierarchical Planning in the Application

Now, we present how a plan is generated in our domain. Figure 4 describes planning factors and Figure 5 shows operations for planning.

We defined three levels for the hierarchical planning in this domain. The planning units of Level 1, Level 2 and Level 3 are for the task, group and item, respectively.

A *task* is a set of one or more orders, which have the same use of grating. A *group* is a set of one or more grating *items*, which have the same manufacturing-specification that describes the types of materials (BB and CB), BB-Pitch and CB-Pitch. So, a task consists of several groups. Grating items in the same group are manufactured in the same way; this depends on the method of welding types for the generation of a grating panel: *1-weld, 2-weld,* and *2-weld-comp.* The type of welding is determined by the composition of the manufacturing-specification and the manufacturing resources.

Factor	Description		
	Level 1	Level 2	Level 3
Unit of a plan	Task	1-weld, 2-weld, 2-weld-comp; for a group	1wm1, 1wm2, 2wm1, 2wm2; for a group
Objects	Group1, Group2, ... , GroupN	A Group	Panel1($I_1,I_2,...I_k$), Panel2($I_{k+1}, I_{k+2}...$), ..., PanelJ(.....,I_N)
Operator	Machine	Machine	Machine
Operations	1-weld, 2-weld, 2-weld-comp	1w-m1, 1w-m2, 2w-m1, 2w-m2	InsertQ,Alloc,Cut,Convey, Dealloc
State Variables	GsToBePlan : group-list to be planned GsPlanned : group-list planned TimeSpent : time reasonable CostSpent : cost reasonable Q1W, Q2W: queues for welding-types	IsSpecified : is the group specified? Q1wm1, Q1wm2, Q2wm1, Q2wm2 : queues for each machine	PansToBeCut : panels to be cut PansCut : panelss cut done ItemsToBeCut : items to be cut ItemsCut : items cut done MachStates: states of machines currPOSs : current positions cutPOSs : cut positions
Initial State	GsToBePlan : all groups for a task GsPlanned : empty TimeSpent : 0 CostSpent : 0 Q1W : Q2W = 1: 1	IsSpecified: false Q1wm1 : empty Q1wm2 : empty Q2wm1 : empty Q2wm2 : empty	PansToBeCut : all panels for a group ItemsToBeCut : all items for a group ItemsCut, PansCut : empty MachStates: idle currPos, cutPos: 0
Goal State	GsToBePlan: empty GsPlanned: all groups for a task TimeSpent : maximum allowed time CostSpent : maximum allowed cost Q1W : Q2W = 1 : 2 (default rate)	IsSpecified: true There is the group in one of the four queues. Q1WM1: Q1WM2 = 1:1 Q2WM1: Q2WM2 =1:1	PansToBeCut , ItemsToBeCut : empty PansCut : all panels for a group ItemsCut : all items for a group MachStates: idle currPos, cutPos: 0

Fig. 4. Planning Factors: State Variables and States

The planning in Level 1 generates the sequence of actions for determining the type of welding for each group within a task; the planning in Level 2 generates actions for deciding which type of machines should be applied for the groups processed in Level 1; and finally the planning in Level 3 generates the sequential actions indicating which item of a group should be processed by which commands of a machine.

The planning engine approximately calculates the efficiency based on the planned operations. The values when calculating the efficiency are updated whenever an operation is selected, and those are used to select another operation in the next step of reasoning. In the planning process, conflict resolution follows the order of operation rules and the order of groups of which numbers are assigned in a task. Figure 5 describes the operations for each level. The rules for operations are described by several predicates.

As mentioned before, we used the hierarchical and regression planning technique via Hierarchical RG-DEVS, and we used depth-first planning to perform the hierarchical planning because the lower level's plan influences the higher level's plan in the next step. Figure 6 shows the planning procedure by the application using these planning techniques. The series of ellipses that are connected by arrows with solid lines is the sequence of actions from the goal state (S_G) to the start state (S_S). The circled numbers represent the procedure of the planning: depth-first regression planning. The actions of upper levels become constraints for the actions of lower levels, and the states of lower levels, which are updated by the actions, gives feedback to the states of upper levels.

Class	Operation	Pre-condition	Action	
			Delete	Add
Level 1	2-weld(g)	H_ERSpec(g) ShortQL(Q2w, Q1w)	WaitForPlan(g) NotGInQ(g,Q2w)	Planned(g) IsGInQ(g, Q2w)
	2-weld-comp(g)	M_ERSpec(g) ShortQL(Q2w, Q1w)	WaitForPlan(g) NotGInQ(g, Q2w)	Planned(g) IsGInQ(g, Q2w)
	2-weld-comp(g)	NoTimeToSpare NoCostToSapre	WaitForPlan(g) NotGInQ(g, Q2w)	Planned(g) IsGInQ(g, Q2w)
	1-weld(g)	L_ERSpec(g) ShortQL(Q1w, Q2w) TimeToSpare CostToSapre	WaitForPlan(g) NotGInQ(g, Q1w)	Planned(g) IsGInQ(g, Q1w)
Level 2	2w-m3(g)	ShortQL(Q2wm3, Q2wm4) NotSpec(g)	NotSpec(g)	IsGInQ(g,QL2wm3) IsSpec(g)
	2w-m4(g)	ShortQL(Q2wm4, Q2wm3) NotSpec(g)	NotSpec(g)	IsGInQ(g,QL2wm4) IsSpec(g)
	1w-m1(g)	ShortQL(Q1wm1, Q1wm2) NotSpec(g)	NotSpec(g)	IsGInQ(g,QL1wm1) IsSpec(g)
	1w-m2(g)	ShortQL(Q1wm2, Q1wm1) NotSpec(g)	NotSpec(g)	IsGInQ(g,QL1wm2) IsSpec(g)
Level 3	Alloc(m, p)	IsMIdle(m) WaitForCut(m, p)	IsMIdle(m)	IsMBusyByP(m, p) IsNotRCut(m, p)
	Dealloc(m, p)	IsMBusyByP(m, p) IsRCut(m, p) WaitForCut(m, p) IsPanAtEnd(m, p)	IsMBusyByP(m, p) IsPanAtEnd(m, p) WaitForCut(m, p)	IsMIdle(m) IsNotRCut(m, p) CutDone(m, p)
	Cut(m, p, x)	IsMBusyByP(m, p) IsRCut(m, p) WaitForCut(m, p, x)	IsRCut(m, p) WaitForCut(m, p, x)	IsNotRCut(m, p) CutDone(m, p, x)
	Convey(m, p)	IsMBusyByP(m, p) IsNotRCut(m, p)	IsNotRCut(m, p)	IsRCut(m, p)

Fig. 5. Planning Factors: Rules

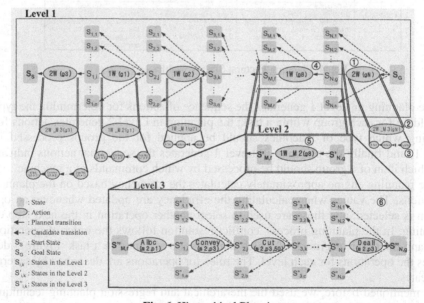

Fig. 6. Hierarchical Planning

5   Simulation in the Application

The sequential actions of the hierarchical planning are sequential functions to translate the state of models in simulation, and the sequential states corresponding to the actions are sequential states to be translated in the simulation model. So, a simulation model is dynamically generated by planning. Now, we examine the simulation procedure of the model defined by planning.

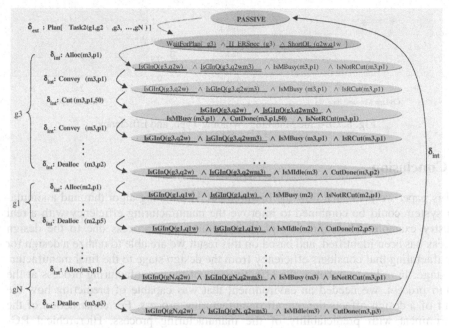

Fig. 7. State Transition Diagram Based on Hierarchical RG-DEVS

Figure 7 represents a part of the state transition diagram of the simulation model –
sequential actions and states – formed by planning. Each state is composed of several
predicates, and the thick-underlined predicates, thin-underlined predicates and non-
underlined predicates are generated by planning Levels 1, 2 and 3, respectively. In
Section 2.2, we already described these properties: a state is the Cartesian product of
states of each level in planning.

Now, we explain the results of simulations executed as per the above procedures.
Figure 8 shows the analysis of the three tasks for which manufacturing-specifications
are different.

The left graph of Figure 8 shows the analysis of the relation between CB-Line
Matching rate and efficiency. The efficiency value of 1 means that a task uses the
given resources without surplus or shortage, and a value lower than 1 means that
resources are wasted or insufficient. The graph shows that the three tasks use the
given resources most effectively when their CB-Line Matching rates are 60%, 80%
and 70%, respectively. The regions rising efficiency mean that resources are sufficient, while the regions of falling efficiency mean that resources are deficient. The
right graph of Figure 8 shows the analysis of the relation between welding type and
efficiency; this graph illustrates the case when CB-Line matching rates of each task
are close to the efficiency value of 1. According to this graph, we can identify the
contribution of each welding type with respect to the efficiency.

The efficiency related to the CB-Line Matching rate is determined based on the
state values generated by planning Level 3, while the efficiency related to welding
type is based on planning Level 1. Therefore, this fact means that the Hierarchical
RG-DEVS environment makes it easy to analyze the target system from different
points of view.

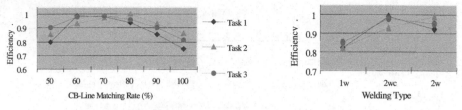

Fig. 8. CB-Line Matching, Welding Type, and Efficiency

6   Conclusion

In this paper we presented how the hierarchical AI planning algorithm and a simulation system could be combined to improve the manufacturing efficiency with a real industry example. The efficiency of the manufacturing process due to the design process has been identified, and based on this result we are able to realize a design for manufacturing that considers efficiency from the design stage to the final manufacturing stage. In order to enhance the effectiveness of the manufacturing process at the design process, we needed an environment that was capable of predicting how the result of a design affects the overall manufacturing process. For construction of the environment with predictability of the manufacturing process, Hierarchical RG-DEVS modeling methodology was presented and applied. Hierarchical RG-DEVS contributes to simulation modeling in the following three aspects. First, it reduces the computational cost of planning, second, it shows how to apply the hierarchical operations of a target system within the simulation model, and finally it helps to analyze the target system from different points of view as shown in Figure 8. Future work includes the generalization of embedding various AI algorithms into simulation models and intensive applications to real industries.

References

1. Zeigler, B.P., Cho, T.H., and Rozenblit, J.W.: A Knowledge-Based Simulation Environment for Hierarchical Flexible Manufacturing. IEEE Transactions on Systems, Man and Cybernetics, Vol. 26, No. 1 (1996)
2. Cho, T.H., and Zeigler, B.P.: Simulation of Intelligent Hierarchical Flexible Manufacturing: Batch Job Routing in Operation Overlapping. IEEE Transactions in Systems, Man, and Cybernetics, Vol. 27., No. 1, (1997)
3. Zeigler, B.P: Object-Oriented Simulation with Hierarchical, Modular Models, Academic Press, San Diego, California. (1990)
4. Allen, J.F.: Towards General Theory of Action and Time. Artificial Intelligence, Vol. 23 (1984)
5. Kwon, Y., Park, H., Jung. S. and Kim, T.: Fuzzy-DEVS Formalism: Concepts, Realization and Applications. AI, Simulation and Planning in High Autonomy Systems. EPD, University of Arizona, San Diego. (1996)
6. Kunwoo Lee: Principles of CAD/CAM/CAE Systems, Addison Wesley Longman. (1999)
7. Cho, Y.S.: Construction of efficient CAD environment. (2000)
8. V. Hetem: Communication: Computer Aided Engineering in the Next Millennium. Computer-Aided Design, Vol 32, No 5-6 (2000)

9. U. S. Kalyan-Seshu and B. Bras: Towards Computer Aided Design for the Life Cycle. IEEE International Symposium on Electronics and the Environment, Oak Brook, Illinois, USA (1998)
10. S. Szykman, Steven J. Fenves, Walid Keirouz, Steven B. Shooter: A Foundation for Interoperability in Next-generation Product Development Systems. Computer-Aided Design, Vol 33, No. 7 (2001)
11. Zeigler, B.P., Praehofer, H., and Kim, T.G.: Theory of Modeling and Simulation: Integrating Discrete Event and Continuous Complex Dynamic Systems. San Diego, California (2000)
12. Cho, T.H.: Embedding Intelligent Planning Capability to DEVS Models by Goal Regression Method. Simulation Transactions of The Society for Modeling and Simulation International, Vol. 78, No. 12, Dec. (2002)
13. Nilsson, N.J.: Artificial Intelligence: A New Synthesis. Morgan Kaufmann, San Francisco, California. (1998)
14. Qiang Yang: Intelligent Planning. Springer-Verlag, Berlin (1997)

Vulnerability Modeling and Simulation
for DNS Intrusion Tolerance System Construction

Hyung-Jong Kim

Korea Information Security Agency(KISA), Garak-Dong, Songpa-Gu, Seoul, Korea
hjkim@kisa.or.kr

Abstract. To construct the ITS(Intrusion Tolerance System), we should concern not only the FTS(Fault Tolerant System) requirements but also intrusion and vulnerability factors. But, in the ITS, we can not take into account the intrusion and vulnerability as they are, because the characteristics and pattern of them is unknown. So, we suggest vulnerability analysis method that enable ITS to know the pattern of vulnerability exploitation more specifically. We make use of the atomic vulnerability concept to analyze the vulnerability in DNS system, and show how to make use of the analysis result as monitoring factors in our DNS ITS system. Also, this analysis result is used in modeling and simulation to see the dynamics of computer network for vulnerability and external malicious attack. This paper shows simulation execution examples making use of the vulnerability analysis result.

Keywords: Vulnerability Analysis, Intrusion Tolerance, DEVS formalism, Atomic Vulnerability, DNS

1 Introduction

In the security research area, the survivability related work is urgently required to preserve the continuity of essential services such as DNS, DHCP and so on. One of those works is the ITS (Intrusion Tolerant System) which enables client users to access the service though severe attack is appeared. Since ITS system's functions are for defending the attacks that cannot be handled by the IDS and Firewall's defense mechanism, it is almost impossible to apply the known vulnerability information to the ITS system's monitoring rules. In our work, we make use of the atomic-vulnerability based vulnerability analysis and we utilize it to monitor the system and network's attack symptoms. Atomic-vulnerability is undividable unit that used to compose the vulnerability usually named as CVE or CAN ID.

In this paper, we show vulnerability analysis of DNS system, and how to apply the analysis result to our ITS construction. Especially, we use the analysis result to construct the monitoring knowledge of ITS middleware. At the end of this work, we show some modeling and simulation examples, which make use of our analysis result.

At the second section, the background works of this research are introduced. Third section shows our main research contents consist of vulnerability analysis method, DNS vulnerability analysis and deployment in our ITS system. In the fourth section, we will show modeling and simulation using our analysis result. At fifth section, we will make our conclusion and show some future works.

T.G. Kim (Ed.): AIS 2004, LNAI 3397, pp. 81–89, 2005.

2 Backgrounds

Matt Bishop[4] suggests primitive condition that is used to represent the system's vulnerability. When a system is represented by primitive condition, there are some duplicated primitive conditions in different vulnerabilities. So, it is economic to represent and analyze vulnerability using this concept. In his work, when for each two vulnerability V_a and V_b, the primitive condition sets of two vulnerabilities are defined as $C(V_a)$ and $C(V_b)$, if $C(V_a) \cap C(V_b) \neq \emptyset$, the two vulnerabilities have common primitive conditions.

Among the ITS related work, MAFTIA[11] is representative research project which is conducted by EU's IST. MAFTIA's principle about the intrusion is that all intrusion cannot be protected and some of them are inserted into the system, and system must prepare about those uncontrolled attacks. Based on this principle, middleware type ITS framework which contains five main modules such as intrusion detection, group communication protocol, encryption module, data fragmentation/scattering, and user access control is suggested.

The OASIS[12] project that is carried out by DARPA proceeds 12 research projects which are categorized four research topics such as server architecture, application program, middleware, network fundamental technology. Some main concepts of the 12 project in OASIS are as follows. First, the diversity of application and OS platform enhances the availability of the service and system because usually intrusions exploit vulnerabilities that exist in a specific system platform simultaneously. Second, the system and service should be redundant and if there is intrusion or fault that cause problem in system or service, the redundant system or service do the work of main system and service during the recovery time. Third, there are some ITS specific mechanisms to guarantee the integrity and availability of services, and those mechanisms cooperate with the conventional security mechanisms of intrusion prevention and detection system. To enhance the availability, load-balancing facility is applied and to enhance the integrity of service, voting mechanism, service member isolation and restoration are used. Fourth, there are monitoring facility used to see the abnormal status of services and systems. In ITS, as monitoring is done after the prevention and detection mechanism is applied, the monitoring scheme should be specialized to detect the intrusions that are unknown to conventional security system.

In this work, we defined monitoring factors that are appropriate the DNS intrusion tolerant system. For selection of them, we should define the vulnerability analysis method using atomic vulnerability analysis method. Although the atomic vulnerability concept is for the modeling and simulation, it is well applied to intrusion tolerance.

In the next section, we will show the atomic vulnerability concept and we applied it to the DNS vulnerability. Also, we will show how to deploy the analysis result.

3 DNS Vulnerability Analysis for Intrusion Tolerance

In this section, vulnerability analysis method based on the atomic vulnerability concept is introduced. The atomic vulnerability concept is for the vulnerability modeling and simulation of network [7]. In this paper, we make use of it to define monitoring factors in DNS intrusion tolerance system.

Vulnerability Analysis Method for Intrusion Tolerance

In this research, to extract the monitoring factor in ITS system we defined two concepts. The one is AV(Atomic Vulnerability) and the other is CV(Compound Vulnerability). AV is the vulnerability that cannot be divided any more and CV consists of AV set. We can define our Node using the AV and CV concept. The definition is as follows.

Node = $\{T, X, Y, \Omega, D, \Lambda\}$,
where
T: Time.
X: $\{x_1, x_2, x_3....x_n\}$
  x_i : attacker's action input.
Y: $\{y_1, y_2, y_3....y_m\}$
  y_i : reaction output of nodes.
Λ: $D \times X \rightarrow Y$(or Λ: $D \rightarrow Y$) is the output function
D is a set, set of CV's reference.

The Nodes has component D that contains a set of CV's reference. The component D represents all vulnerabilities in the nodes and it determines dynamics of the Nodes. In the Node definition, it has output function Λ, and the Node generates its output at the Consequence state. Especially, in our case, the component D of Nodes refers the CVs that are represented by the AVs and operators used to define the relations among AVs.

The definitions of CVs and AVs are as follows:

Compound Vulnerability: CV = $\{I_{cv}, Q_{cv}, \delta_{cv}, WSX, VX\}$
where,
I_{cv} = $\{I_{cv1}, I_{cv2}, ..., I_{cvn}\}$
Q_{cv}= $\{$Normal, Intermediate, Warning, Consequence$\}$
δ_{cv} : $I_{cv} \times Q_{cv} \rightarrow Q_{cv}$
WSX : warning state vulnerability expression
VX : vulnerability expression

In the definition of CV, I_{cv} is a set of attack input sequences and it means the external inputs (X) in the Node. Q_{cv} has four essential states that are meaningful in the its dynamics. Normal state is a state in which a target system is waiting for input packets. When the target system is under attack, system's state is Intermediate. The Warning state means that probability of exploitation occurrence is beyond a specific level, and the system can transit to an abnormal state by a simple attack input. Consequence state is a goal state, which means the target system is exploited by attacker. δ_{cv} is state transition function and each state transition is defined as shown in Fig.1.

A CV is represented by logical composition of AVs. VX holds the expression. An expression is composed of AVs and four binary logical operators. If this expression is evaluated as TRUE, it means that the vulnerability is exploited by attack action sequence and state transition to compromised state occurs in the model. WSX is warning state vulnerability expression. Syntax of WSX is the same as VX's. If this expression is TRUE, state transition to Warning state occurs.

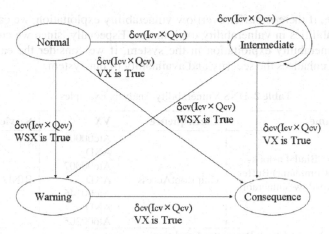

Fig. 1. State and state transition of CV

Table 1. Logical operators for VX

Operator	Description
AND	To represent vulnerability exploited if both AVs are true
OR	To represent vulnerability exploited if either or both AVs are true
POR	To represent vulnerability exploited if either or both AVs are true. But each AV has weight value that accounts for vulnerability of target system for each AV(from 0 to 1).
SAND	To represent vulnerability exploited if one AV at front is true and then the other AV is true sequentially.

Atomic Vulnerability : $AV = \{I_{av}, Q_{av}, \delta_{av}, Type, Category\}$
where,
$I_{av} = \{I_{av1}, I_{av2}, \ldots, I_{avn}\}$
$Q_{av} = Q(initial\ state) \cup Q(final\ state)$
$\delta_{av} : I_{av} \times Q(initial\ state) \rightarrow Q(final\ state)$
Type : {Fact, NonProb, Prob}
Category : {Generic, Application-Specific, System-Specific}

In the definition of AV, Q_{av} is a set of states. I_{av} is a set of attack input sequences to AV and it is also same as the I_{cv}. δ_{av} is a state transition function. An AV is one of three type; Fact, NonProb or Prob. A Fact AV has no input(NONE) and no state(NONE). Therefore, a Fact AV's δ_{av} is δ_{av}(NONE, NONE) = NONE. This type explains the corresponding AV's origin. NonProb and Prob type AVs explain whether these AVs are exploited probably or deterministically. Category is Generic, Application-Specific for application specific vulnerability, System-Specific for OS or H/W specific vulnerability.

DNS system vulnerability analysis examples: Based on analysis method in this section, we analyze the two popular DNS vulnerabilities. In our analysis, we defined atomic vulnerabilities for each vulnerability unit CVE ID that is used usually in vulnerability definition. Since the atomic vulnerability means undividable one to express the compound vulnerability, we can define monitoring factor more specifically.

For example, if there is buffer overflow vulnerability exploitation, we can monitor atomic vulnerabilities in vulnerability exploitation. Especially, since we consider the undetected vulnerability exploitation in the system, if we consider the each step of exploitation, it enhances the security and availability of our system.

Table 2. DNS Vulnerability Analysis Examples

CVE ID	Name	Consequence	VX	Service	Ver.
CVE-2001-0011	ISC Bind 4 nslook-upComplain() Buffer Overflow Vulnerability	GainRootAccess	Ato00006 AND Ato00007 AND Ato00146 AND Ato00205	BIND	4.9 4.9.3 4.9.5 4.9.6 4.9.7
Ato00006	Doesn't Check Parameter Condition				
Ato00007	Stack's return address can be altered				
Ato00146	Stack can be overflowed by malicious input				
Ato00205	named's nslookupComplain function call using long input				
CVE ID	**Name**	**Consequence**	**VX**	**Service**	**Ver.**
CVE-2002-0400	ISC Bind 9 Remote DoS Vulnerability	Denial of Service	Ato00206 AND Ato00207	BIND	9.0 9.1 9.1.1 9.1.2 9.1.3 9.2
Ato00206	There is no exception handle for unexpected malicious input				
Ato00207	Send rdataset parameter as not NULL value to named				

DNS Vulnerability and Monitoring Factor Extraction

In this section, we analyze the DNS vulnerability for our aim to extract the monitoring factor for the intrusion tolerant system. The analysis method that is shown in previous section is applied in the analysis process. As DNS system is used to translate the domain name to ip address and vice versa, if it is crashed by an intrusion, there are big problem in network usage of internet users. So, DNS system is categorized as essential ones of which availability and integrity should be guaranteed.

To guarantee the availability and integrity of DNS service, we analyze the vulnerability and extract the monitoring factor for the ITS system. Of course, there are some other mechanism to enhance the availability and integrity such as load balancing, group management and voting mechanism. But, in this work, we focus on extraction of monitoring factors that are used by monitoring modules located behind the intrusion prevention and detection system. In this situation, we should see the phenomena (or consequence) that is presented, when the vulnerabilities are exploited, as shown in Fig. 2. Also, based on the monitored phenomena, the monitor should infer the causes of phenomena such as vulnerability and attack. Since we define the countermeasure for each phenomena and vulnerability, if the abnormal phenomena are monitored, the defined countermeasure is conducted.

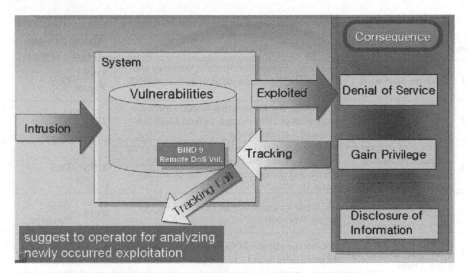

Fig. 2. Relation among vulnerability, exploit, and phenomena

In our analysis, we categorized the vulnerabilities based on the several standards such as attack target, aim, and vulnerability type for DNS and DHCP system. Our analysis target vulnerabilities are from the ICAT meta-base which provides meta-data based search engine. In the case of DNS system, we analyze the 81 vulnerabilities. As shown in Table 3, we categorized the 10 types of vulnerabilities based on their monitoring target.

Deployment of Vulnerability Analysis Result

Fig. 3 shows our system deployment. SITIS is a kind of middleware to support the ITS facility such as redundant member management using voting mechanism, reliable multicast protocol and recovery scheme adequate to situation. The analysis method suggested in this paper is for Member Monitor's knowledge in SITIS. Our monitoring module contains rules and facts based on the analysis in previous section. Analysis result of vulnerabilities contains the monitoring factors based on its own atomic vulnerabilities. Since we make use of the atomic vulnerability in our monitoring module, we can monitor more microscopic level than other detection schemes.

4   Simulation Modeling and Execution

Our analysis result is inserted into the VDBFS (Vulnerability DataBase For Simulator) that is used to simulate the vulnerabilities assessment in the network. The simulation system is constructed using MODSIM III and Oracle Database. The simulator has host model that has dynamics based on the vulnerability information analyzed by our approach. So, when we execute the simulation model we can see the status of the systems, attack path, and exploited vulnerabilities in each path.

Table 3. 10 DNS Vulnerability Categories and Monitoring Factors

No.	Vulnerability	Monitoring Factor	Attack Target	Conse-quence
1	Resource Starvation	Number of Connection Request, Memory and CPU Usage, Resource status in device of DNS access paths	DNS, Router	DoS
2	Bypass the Access Control Path	Inspect the log information whether there are evidences of DNS control without Identification and Authentication	DNS	Gain Privilege
3	Absence of inappropriate packet handling capability in DNS System	See the packet payload if it is appropriate to the DNS service request regularly Request the service and see the current service providing status	DNS	DoS
4	Vulnerability in DNSSEC-Key Generation Protocol	Check the Vulnerability of DNSSEC Key Generation Protocol	DNS	Gain Privilege
5	Access Control Bypassing by Disguising DNS Traffic	Monitor the client's DNS service requests which are indifferent with DNS service	Client	DoS, Gain Info.
6	DoS of Client's application caused by inappropriate DNS response	If there are abnormal responses from DNS Server to Client, find the relation between the response and status of client. Based on the above process, construct the rules to prevent such situation.	Client	DoS
7	Buffer overflow of Client application by abnormal DNS response	If there are abnormal responses from DNS Server to Client, find the relation between the response and status of client. Based on the above process, construct the rules to prevent such situation.	Client	Gain Privilege
8	Traffic amplifying using DNS traffic	See the DNS traffic whether there are repeated service request from same client host	Client	DoS
9	Malicious Updates of DNS Records	See the existence of vulnerability to bypass the access control of DNS record update Inspect the log information whether there are evidences of DNS control without Identification and Authentication	NS	Gain Privilege
10	Configuration Error of Security System for DNS Protection	See the service status using remote service request test	DNS	DoS

Our simulation environment consists of experimental frame and network model. Experimental frame is used to set testing environment and feed testing input to the network model and get back response from the model. Fig. 4 shows execution example of our simulation system. Our analysis results are in the VDBFS (Vulnerability DataBase For Simulator) and it is extracted during simulation execution. If there is the exploitation of vulnerability, attack path is created as shown in Fig. 4. The simulator can show the vulnerability in a microscopic level named atomic vulnerability. Fig. 5. shows atomic vulnerabilities that are displayed when we select a composed vulnerability.

Actually our vulnerability analysis approach based on the atomic vulnerability concept is for this simulation system. Since atomic vulnerability can be used to represent states of each system, it is well suited in our simulation system. This atomic vulnerability characteristic is well explained in [6][7].

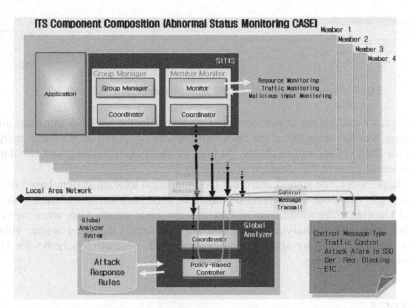

Fig. 3. ITS Component Composition in Abnormal Status Monitoring

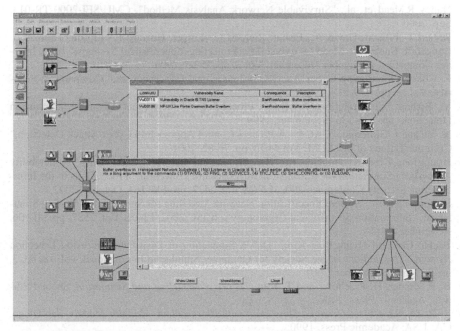

Fig. 4. Modeling and Simulation using MODSIM III – An execution result

5 Conclusion

In this work, we make use of atomic vulnerability concept to extract the monitoring factor of DNS intrusion tolerant system. When we make use of the atomic vulnerabil-

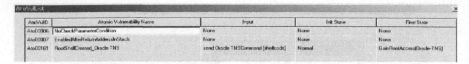

Fig. 5. Atomic Vulnerabilities of a Selected Compound Vulnerability

ity concept, we can get more microscopic monitoring factor for DNS intrusion tolerant system. We also categorized the DNS vulnerability types and shows some analysis sample for DNS related vulnerability. At last, we show our modeling and simulation environment to present the meaning of our analysis.

As a future work, we should construct DNS intrusion tolerant system based on our analysis result. Also, our analysis result should be used for intrusion tolerant system's essential knowledge for monitoring of system's abnormal status. Also, when we compile more vulnerability analysis results that are represented based on atomic vulnerabilities, we can get more specific and unduplicated monitoring scheme for DNS intrusion tolerance system.

References

1. Nancy R.Mead et. al., "Survivable Network Analysis Method", CMU/SEI-2000-TR-013, Sep. 2000
2. Robert J. Ellison, David A. Fisher, Richard C. Linger, Howard F. Lipson, Thomas A. Longstaff, Nancy R. Mead "Survivability: Protecting Your Critical Systems," IEEE Internet Computing, November December, Vol 3, pp. 55-63, 1999
3. F. Cohen, "Simulating Cyber Attacks, Defenses, and Consequences," Computer & Security, Vol.18, pp. 479-518, 1999
4. M. Bishop, "Vulnerabilities Analysis," Proceedings of the Recent Advances in Intrusion Detection," pp. 125-136, September, 1999
5. N. Ye and J. Giordano, "CACA - A Process Control Approach to Cyber Attack Detection," Communications of the ACM, Vol.44(8), pp. 76-82, 2001.
6. HyungJong Kim, KyoungHee Ko, DongHoon Shin and HongGeun Kim, "Vulnerability Assessment Simulation for Information Infrastructure Protection," Proceedings of the Infrastructure Security Conference 2002, LNCS Vol. 2437, pp. 145-161, October, 2002.
7. HyungJong Kim, "System Specification Network Modeling for Survivability Testing Simulation," Information Security and Cryptology – ICISC 2002, LNCS Vol. 2587, pp. 90-106, November, 2002.
8. TaeHo Cho and HyungJong Kim, "DEVS Simulation of Distributed Intrusion Detection System," Transactions of the Society for Computer Simulation International, vol. 18, no. 3, pp. 133-146, September, 2001.
9. B. P. Zeigler, H. Praehofer and T. Kim, *Theory of Modeling and Simulation*, Second Edition, Academic Press, 2000.
10. B. P. Zeigler, *Object-Oriented Simulation with Hierarchical, Modular Models*, San Diego, CA, USA: Academic Press, 1990.
11. Adelsbach, A., et. Al, "Conceptual Model and Architecture of MAFTIA," Project MAFTIA IST-1999-11583 deliverable D21. 2002.
12. M. Cukier, J. Lyons, et. Al, "Intrusion Tolerance Approaches in ITUA" FastAbstract in Supplement of the 2001 International Conference on Dependable Systems and Networks, Göteborg, Sweden, July 1-4, 2001, pp. B-64 to B-65.

NS-2 Based IP Traceback Simulation Against Reflector Based DDoS Attack

Hyung-Woo Lee[1], Taekyoung Kwon[2], and Hyung-Jong Kim[3]

[1] Dept. of Software, Hanshin University, Osan, Gyunggi, 447-791, Korea
hwlee@hs.ac.kr
[2] School of Computer Engineering, Sejong University, Seoul, 143-747, Korea
tkwon@sejong.ac.kr
[3] Korea Information and Security Agency, Garak, Songpa, Seoul, 138-803, Korea
hjkim@kisa.or.kr

Abstract. Reflector attack belongs to one of the most serious types of Distributed Denial-of-Service (*DDoS*) attacks, which can hardly be traced by traceback techniques, since the marked information written by any routers between the attacker and the reflectors will be lost in the replied packets from the reflectors. In response to such attacks, advanced IP traceback technology must be suggested. This study proposed a *NS-2 based traceback system* for simulating *iTrace* technique that identifies DDoS traffics with multi-hop iTrace mechanism based on TTL information at reflector for malicious reflector source trace. According to the result of simulation, the proposed technique reduced network load and improved filter/traceback performance on distributed reflector attacks[1].

Keywords: NS-2, Reflector Attack, DDoS, IP Traceback, Simulation.

1 Introduction

In a *distributed denial-of-service* (DDOS) attack, the attacker compromises a number of *slaves* and installs flooding *servers* on them, later contacting the set of servers to combine their transmission power in an orchestrated flooding attack [1,2]. The dilution of locality in the flooding stream makes it more difficult for the victim to isolate the attack traffic in order to block it, and also undermines the potential effectiveness of common *traceback* techniques for locating the source of streams of packets with spoofed source addresses [3,4].

In *reflector attack*, one host (*master*) sends control messages to the previously compromised slaves, instructing them to target a given *victim*. The slaves then generate high volume streams of traffic toward the victim, but with fake or randomized source addresses, so that the victim cannot locate the slaves [5,6]. *The problem of tracing back such streams of spoofed packets has recently received considerable attention.*

With considerably higher probability the router marks the packets with highly compressed information that the victim can decode in order to detect

---

[1] This work is supported by University IT Research Center (ITRC) Project from Korea.

T.G. Kim (Ed.): AIS 2004, LNAI 3397, pp. 90–99, 2005.

the edges (pairs of packet-marking routers) traversed by the packets, again enabling recovery of the path back to the slave. This scheme can trace back potentially lower-volume flows than required for traceback using iTrace (ICMP Traceback) [7].

The use of hundreds or thousands of slaves can both greatly complicate traceback (due to the difficulty of disentangling partial traceback information relating to different sources, and/or having to contact thousands of routers) and greatly hinder taking action once traceback succeeds (because it requires installing hundreds of filters and/or contacting hundreds of administrators) [4].

Attackers can do considerably better still by structuring their attack traffic to use *reflectors*. A reflector is any IP host that will return a packet if sent a packet. So, for example, all Web servers, DNS servers, and routers are reflectors, since they will return SYN ACKs or RSTs in response to SYN or other TCP packets. *Thus currently available technologies do not provide active functions to cope with reflector attack such as tracing and confirming the source of DoS hacking attacks.* Thus it is necessary to develop a technology to cope actively with such DDoS reflector attacks. Even if the trace-route technique is applied to identify the source address, the technique cannot identify and trace the actual address because the address included in reflector based DDoS (Distributed Denial of Service) is spoofed [5].

When a DDoS attack has happened, methods like ingress filtering filter and drop malicious packets at routers on the network, so they are passive to DDoS attacks. In traceback methods such as [9,10], routers generate information on the traceback path while transmitting packets are sent by reflector attack on slaves, and insert traceback information into the packets or deliver it to the IP address of the target of the packets.

On existing Reverse iTrace [8], common routers send ICMP messages to the source of the just-processed packet rather than its destination (unlike iTrace). Routers on the path between slave and the reflector will send ICMP messages to Victim to enable trace back to the slaves. But, in this study we propose *a new reflector traceback scheme* which combine Pushback module on reflector traceback. This study proposes a technique to simulate traceback the source IP of spoofed DDoS packets using NS-2 [11] by combining the existing method, which provide a control function against DDoS reflector attacks [12], with a traceback function. Therefore, a router performs the functions of *identifying/controlling traffic*, and when a DDoS attack happens it sends packet to its previous hop router by marking router's information on the header with advanced ICMP traceback mechanism.

2   NS-2 for DDoS Simulation

2.1   Introduction to Network Simulation: NS-2

The *NS-2* simulator [11] is a discrete event simulator widely used in the networking research community. It was developed at the University of California at

Berkeley and extended at Carnegie Mellon University to simulate wireless networks. These extensions provide a detailed model of the physical and link layer behavior of a wire/wireless network and allow arbitrary movement of nodes within the network. NS-2 Simulator is usually a software package that simulates a real system scenario. Through the simulation we can test how a device or a system will perform in terms of timing and result. In addition to that it can be used to explore new policies, operating procedure without interrupting the system in real time. Network simulation allows us to check the system compressing the time, or expanding it.

Network simulation is very important because the network designer can test a complex network and make the right decisions about the designing in order the network will not face any problems in the future. New network devices can be added and testing without disturbing the existing network. Also during the simulation the designer can test how he can improve the network bandwidth or the current data speed. Finally network simulation can be used for tutorial so a network engineer can solve problems with real devices. NS-2 has high performance and it is very easy to use because of the combination of the above languages. NS architecture follows the OSI model.

Node in a network is a point that connects other points, either a distribution point or an end point for data transmissions. A node can sent or receive data. All kind of nodes in ns-2 are separated in two types of nodes. A unicast node that sends packets to only one node and a multicast node that sends packets to more than one node. *Attack traffic source* node is a node that sends malicious data (spoofed data) to other nodes. Traffic agents such as TCP or UDP are assigned on those nodes.

The receiving node is called sink node. A sink node can be an end node of the network. It can receive from different type of traffic source node. In case of a sink receives data form a TCP traffic node is defined as Agent/TCP Sink and as Agent/Null if it is received from a UDP traffic node. Two types of agent can be assigned on the same node.

2.2   DDoS Simulation Modeling Architecture with NS-2

We motivate our discussion with an example of a DDoS network illustrated. The main goal of our work is to recreate such network in a simulation environment where the behavior of the network can be analyzed. In our simulation environment, a typical network scenario will consist of three types of nodes as Fig. 1.: 1) *traceback nodes* that monitor their immediate environment, 2) *target nodes* that generate the various traceback stimuli that are received by multiple nodes over different channels. 3) *user nodes* that represent clients and administrators of the network.

Shown in Fig. 1, three type of node models make up the key building blocks of our simulation environment. The *traceback nodes* are the key active elements, and form our focus in this section. In our model, each traceback node is equipped with one network protocol stack and one or more traceback stacks. The role of the traceback protocol stacks is to detect and process traceback stimuli on the

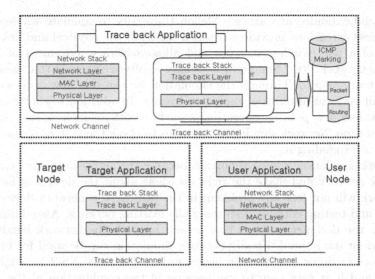

Fig. 1. Simulation Architecture on NS-2.

network channel and forward them to the application layer which will process them and eventually transmit them to a user node in the form of DDoS reports. In addition to the protocol and traceback stacks that constitute the algorithmic components, each node is also equipped with a ICMP information corresponding to the packet transmission path components.

We used *gTraceBack* module for simulate DDoS packet in NS-2. DDoS can be implemented with an UDP flow through bottle-neck traffic. The default is to use an UDP connecton woth a CBR traffic. By default the functions uses an UDP Agent with a CBR traffic.

We use a class named *offender* which give some base for people who want to develop their own offender classes. We derivate this class first in a class named *offender_proto* which contain the protocol used by the offender (DDoS attacker). And then an *offender_app* class which derivated from the *offender_proto*. It contains the application source and destination of the offender.

There is one function remaining to create attackers against victims. We can use one or more targer (victim) and one more attackers. The default is to use an UDP connection with a CBR Traffic. In this case we have made a special function to set the rate of the offender with some arguments.

$ns multi-create-agent-offender. *targets* < *nodes* − 1 >< *pktClass_0* >< *array_srcsrc* >< *array_dstdst* >: This function return an *offender_app* list with all the needed parameters to work.

- *targets:* is a list of target. A target is a node that we must created before (a simple node is sufficient with the command set *n(0) [ns node]* for example).
- *node:* if we give a lost of (or one) node, they will be used to attach the attackers on it/them. If we give as nodes as targets then this will make a one

against one topology. If we don't give any nodes, then they are created for convenience. But be careful, in this case we must link them with its topology.
- *pktClass:* if we give a list (or one value), then the flow ID will be what we give.
- *array_src and array_dst:* these are the name of an array which contains a list of argument as follow.
  - name of the protocol (UDP – default for the sender), Null (default for the receiver), TCP/Sack1, TCPSink/Sack1,...).
  - name of the application (under the *app* hierarchy) (Traffic/CBR – default for the sender), FTP,...).
  - a list of arguments to pass to the application constructor.
  - the name of an already initialized procedure to pass to the agent (Fig. 2).
  - a list with the arguments to pass when the function calls the initizlized procedure.

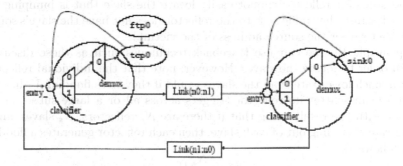

Fig. 2. Traceback Agent with TCP Connections.

Based on these simulation architecture, we can generate DDoS simulation network and evaluate the overall traffics by randomly selected network node as Fig. 1. This architecture can also be applied into reflector based DDoS attack model for simulating proposed mechanism as Fig. 3.

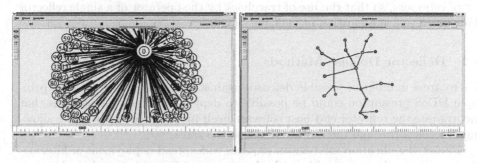

Fig. 3. DDoS Attack Simulation on NS-2.

3   Reflector Based DDoS Attacks

3.1   Reflector Attack Mechanism

Using these library in NS-2 simulator, we can construct a simulated reflector based DDoS attack system. At first, we consider reflector based attack mechanism.

Reflector Attack: *The attacker first locates a very large number of reflectors. They then orchestrate their slaves to send to the reflectors spoofed traffic purportedly coming from the victim, V. The reflectors will in turn generate traffic from themselves to V. The net result is that the flood at V arrives not from a few hundred or thousand sources, but from a million sources, an exceedingly diffuse flood likely clogging every single path to V from the rest of the Internet.*

The operator of a reflector cannot easily locate the slave that is pumping the reflector, because the traffic sent to the reflector does not have the slave's source address, but rather the source address of the victim.

In principle the we can use traceback techniques such as those discussed above in order to locate the slaves. However, note that the individual reflectors send at a much lower rate than the slaves would if they were flooding V directly. Each slave can scatter its reflector triggers across all or a large subset of the reflectors, with the result being that if there are N_r reflectors, N_s slaves, and a flooding rate F coming out of each slave, then each reflector generates a flooding rate as follows.

$$F' = \frac{N_s}{N_r} F \tag{1}$$

So a local mechanism that attempts to automatically detect that a site has a flooding source within it could fail if the mechanism is based on traffic volume.

In addition, common traceback techniques such as iTrace [7] and PPM (probabilistic packet marking) [9] will fail to locate any particular slave sending to a given reflector. If there are N_r reflectors, then it will take N_r times longer to observe the same amount of traffic at the reflector from a particular slave as it would if the slave sent to the victim directly. Against a low-volume traceback mechanism like SPIE, the attacker should instead confine each slave to a small set of reflectors, so that the use of traceback by the operator of a single reflector does not reveal the location of multiple slaves.

3.2   Reflector Defense Methods

There are a number of possible defenses against reflector attacks. But, in principle DDoS prevention could be possible to deploy traceback mechanisms that incorporate the reflector end-host software itself in the traceback scheme, allowing traceback through the reflector back to the slave.

Packet classification mechanism requires widespread deployment of filtering, on a scale nearly comparable with that required for widespread deployment of anti-spoof filtering, and of a more complicated nature. Common traceback mechanism has enormous deployment difficulties, requiring incorporation into a large number of different applications developed and maintained by a large number of different software vendors, and requiring upgrading of a very large number of end systems, many of which lack any direct incentive to do so.

In addition, traceback may not help with traceback in practice if the traceback scheme cannot cope with a million separate Internet paths to trace back to a smaller number of sources. *So we need an advanced new mechanism against reflector-based DDoS attack by using combined technique both packet classification and advanced traceback mechanism.*

4   Advanced Traceback Against Reflector Attacks

4.1   IP Traceback Against Reflector Attack

In this study, we propose a new iTrace mechanism against reflector attacks by using modified pushback [13] module as follow Fig. 4, which shows overall structure of proposed scheme.

Let's say A_x is the IP address of R_x, P_x is IP packet arrived at R_x, and M_x is 24 bits on the header of P_x in which marking information can be stored. In packet P_x, M_x is composed of 8-bit *TOS (type of service)* field, and 16-bit *ID field*. The use of TOS field does not affect the entire network. This study defines the unused 2 bits out of TOS field as *TM (traceback marking flag)* and *CF (congestion flag)*. In TOS field, the first 3 bits are priority bits, and next three bits are minimum delay, maximum performance and reliability fields but not used currently.

TTL (time to live) in all packets is an 8-bit field, which is set at 255 in ordinary packets. The value of TTL field is decreased by 1 at each router until the packet reaches the target. Specifically because the maximum network hop

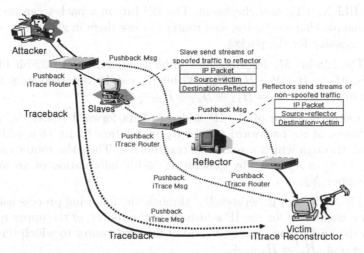

Fig. 4. Pushback based iTraceback Against Reflector Attack.

count is 32 in general, the distance of packet transmission can be calculated only with the lower 6 bits out of the 8 bits of TTL field in packet P_x arrived at router R_x.

Step 1: The router extracts information of the lower 6 bits from the TTL field of packet P_x, names it $T_x = TTLofP_x \wedge 00111111$ and stores it in TOS 6-bit field P_x^{TF} of the packet.

T_x value indicates the distance of the packet from the attack system. When informed of the occurrence of abnormal traffic, router R_x performs marking for packet P_x corresponding to congestion signature classified by decision module.

Step 2: After the router received a packet, it resets TM field in TOS field as 1. Then it calculates T_x for 8-bit TTL field of packet P_x and stores it in the 6 bits of TOS field. Then the router calculates 8-bit hash value for A_x the address of router R_x and T_x calculated earlier using hash function $H(\cdot)$, and marks the value on P_x^{MF1}, the first 8 bits of ID field. The marked packet is delivered to R_y, the next router on the routing path to the target address.

Step 3: Now when router R_y checks P_x^{TM} the value of TM field in the packet and finds it is 1, the router applies the hash function to the value obtained by subtracting 1 from P_x^{TM}, which is corresponding to the 6 bits of TOS field in the packet, and router IP address A_x and marks the resulting value on $P_x^{MF1} = H(T_x|A_x)$, $P_x^{MF2} = H(P_x^{TF} - 1|A_y)$.

After marking, the router set CF at 1 and sends the packet to the next router. The next router, finding TM and CF are set at 1, does not perform marking because the packet has been marked by the previous router.

4.2   Generate ICMP Traceback Message Against Reflector Attack

We generate the suspicious packet into ICMP packet and send it by iTrace module to the victim host. In an IP header excluding the option and the padding, the length of the unchanging part to the bit just prior to the option is 128 bits excluding HLEN, TTL and checksum. The 128 bits in a packet can be used to represent unique characteristics, so a router can use them in generating a ICMP traceback message for the packet.

Step 4: The 128 bit M_x information can be divided into four 32-bit blocks as follows. $M_x = H_{x1}|H_{x2}|H_{x3}|H_{x4}$. 32-bit H_x can be obtained from the four 32-bit sub-blocks. $H_x = H_{x1} \oplus H_{x2} \oplus H_{x3} \oplus H_{x4}$

Now the router is aware of A_y IP address of its forward router R_y and A_x the IP address of its backward router R_x in relation to its own address A_x on the path through which a packet is transmitted. Then the router calculates $A_x' = A_x \oplus A_y \oplus A_z \oplus N_x$ by generating 32-bit information of an arbitrary random number N_x.

Step 5: The the router generates H_x' through the following process using A_x', which is calculated for the IP addresses of the router, of the upper router to which the packet has been sent, and of the next router to which the packet is to be sent. $H_x' = H_x \oplus A_x'$

H_x' is generated by XOR operation on information unique to the IP packet in addition to the 32-bit IP address of the router and information related to the path. Specifically, H_x' is bit-interleaved with N_x and produces 64-bit information. It is included in 64-bit information in an ICMP traceback packet and sent to the target IP address. Of course, transmitted ICMP message I_x is not delivered to the source IP address but to the target IP address.

Step 6: From ICMP message I_x and packet P_x arrived at the target IP address, the victim system identifies path information. First it obtains H_x' and N_x for 64-bit information included in the ICMP message.

Here, we can calculate H_x' is $H_x \oplus A_x \oplus A_y \oplus A_z \oplus N_x$, $H_x' \oplus N_x$, therefore, we can get $H_x \oplus A_x \oplus A_y \oplus A_z$. Now it is possible to obtain H_x by generating M_x', which is information corresponding to 128 bits in packet P_x. Finally the victim system can obtain A_x the 32-bit IP address of the router as well as the addresses of the routers before and after that through $H_x' \oplus N_x \oplus M_x'$ operation. $A_x \oplus A_y \oplus A_z = H_x' \oplus N_x \oplus M_x'$.

4.3   Reflector DDoS Traceback Simulation on NS-2

In order to evaluate the functionality of the proposed method, we simulated its traceback procedure using NS-2 in Linux as shown Fig. 5. In the method proposed, a classification technique is adopted in classifying and control DDoS traffic and as a result the number of marked packets has decreased. We can control the DDoS traffic by issuing traceback message to upper router and marking router's own address in IP packet. The method proposed in this study runs in a way similar to existing iTrace/PPM mechanism, so its management load is low. Furthermore, because it applies identification/control functions to packets at routers it reduces load on the entire network when hacking such as DDoS attacks occurs. The method proposed in this study uses an packet marking with iTrace for providing reflector traceback and control/filter function and marks path information using the value of TTL field, which reduces the number of packets necessary for restructuring a traceback path to the victim system.

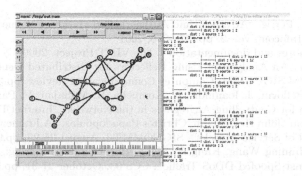

Fig. 5. Reflector based DDoS Traceback Simulation on NS-2.

5   Conclusions

The dilution of locality in the DDoS flooding reflector stream makes it more difficult for the victim to isolate the attack traffic in order to block it. When a DDoS attack has happened, methods like ingress filtering filter and drop malicious packets at routers on the network, so they are passive to DDoS attacks. In traceback methods, routers generate information on the traceback path while transmitting packets are sent by reflector attack on slaves, and insert traceback information into the packets or deliver it to the IP address of the target of the packets. Thus this study proposes a technique to trace back the source IP of spoofed DDoS packets by combining the existing both pushback and iTrace method, which provide a control function against DDoS reflector attacks, with a traceback function.

References

1. John Elliott, "Distributed Denial of Service Attack and the Zombie and Effect", IP professional, March/April 2000.
2. L.Garber, "Denial-of-Service attacks trip the Internet", Computer, pages 12, Apr. 2000.
3. Andrey Belenky, Nirwan Ansari, "On IP Traceback", IEEE Communication Magazine, pp.142-153, July, 2003.
4. Tatsuya Baba, Shigeyuki Matsuda, "Tracing Network Attacks to Their Sources", IEEE Internet Computing, pp. 20-26, March, 2002.
5. Vern Paxson, "An Analysis of Using Reflectors for Distributed Denial-of-Service Attacks", ACM Comp. Commun. Rev., vol.31, no.3, July 2001, pp. 3-14.
6. Chang, R.K.C., "Defending against flooding-based distributed denial-of-service attacks: a tutorial", IEEE Communications Magazine, Volume: 40 Issue: 10 , Oct 2002, pp. 42 -51.
7. Steve Bellovin, Tom Taylor, "ICMP Traceback Messages", RFC 2026, Internet Engineering Task Force, February 2003.
8. C. Barros, "[LONG] A Proposal for ICMP Traceback Messages," http://www.research.att.com/lists/ietf-itrace/2000/09/msg00044.html, Sept. 18, 2000.
9. K. Park and H. Lee, "On the effectiveness of probabilistic packet marking for IP traceback under denial of service attack", In Proc. IEEE INFOCOM '01, pages 338-347, 2001.
10. D. X. Song, A. Perrig, "Advanced and Authenticated Marking Scheme for IP Traceback", Proc, Infocom, vol. 2, pp. 878-886, 2001.
11. K. Fall, "ns notes and documentation", The VINT Project, 2000.
12. Vern Paxson, "An Analysis of Using Reflectors for Distributed Denial-of-Service Attacks", ACM SIGCOMM, Computer Communication Review, pp.38-47, 2001.
13. S. Floyd, S. Bellovin, J. Ioannidis, K. Kompella, R. Mahajan, V. Paxson, "Pushback Message for Controlling Aggregates in the Network", Internet Draft, 2001.
14. Alefiya Hussain, John Heidemann, Christos Papadopoulos, "A Framework for Classifying Denial of Service Attacks", SIGCOMM'03, August 25-29, pp.99-110, 2003.
15. Cheng Jin, Haining Wang, Kang G. Shin, "Hop-Counter Filtering: An Effective Defense Against Spoofed DDoS Traffic", CCS'03, October 27-31, pp.30-41, 2003.

Recognition of Human Action for Game System

Hye Sun Park[1], Eun Yi Kim[2], Sang Su Jang[1], and Hang Joon Kim[1]

[1] Department of Computer Engineering, Kyungpook National Univ., Korea
{hspark,ssjang,hjkim}@ailab.knu.ac.kr
[2] Department of Internet and Multimedia Engineering, Konkuk Univ., Korea
eykim@konkuk.ac.kr

Abstract. Using human action, playing a computer game can be more intuitive and interesting. In this paper, we present a game system that can be operated using a human action. For recognizing the human actions, the proposed system uses a Hidden Markov Model (HMM). To assess the validity of the proposed system we applied to a real game, Quake II. The experimental results verify the feasibility and validity of this game system.This system is currently capable of recognizing 13 gestures, corresponding to 20 keyboard and mouse commands for Quake II game.

Keywords: Gesture Recognition, Game System, HMM (hidden Markov model).

1 Introduction

Human actions can express emotion or information, either instead of speaking or while human is speaking. The use of human actions for computer-human interaction can help people to communicate with computer in more intuitive way. Recently, almost of a popular computer game involves the human player directly controlling a character. If the game players operate game through human actions, they should be more intuitive and interesting during playing such computer game [1–3].

This paper presents a new game system using human actions. Human actions have many variations, such variations occur even if the same person performs the same gesture twice. Thus we use a hidden Markov model (HMM) for recognizing user gestures. A HMM has a rich mathematical structure and serves as the theoretical basis for a wide range of applications. HMM based recognizer has been proposed by many researchers and become quite common for game systems (e.g. [4–6]).

The rest of the paper is organized as follows. We explain a proposed game system in Section 2 and describe an each module of the system more detail in subsection of Section 2: *a feature extraction*, *a pose classification* and *recognition of human action*. The experimental results and performance evaluation are presented in Section 3. Finally, Section 4 summarizes the paper.

T.G. Kim (Ed.): AIS 2004, LNAI 3397, pp. 100–108, 2005.

2   The Proposed Game System Using Human Actions

The proposed game system is *Quake II* game controlled by human actions. *Quake II* game is one of action game, which involve the human player controlling a character in a virtual environment.

Our game system can be controlled *Quake II* game with 13 human actions. The 13 actions are frequently used command in *Quake II*: *walk forward, back pedal, attack, turn left, turn right, look up, look down, step left, step right, center view, up/jump, down/crouch and run.* So we represent those commands using gestures, which are described in Fig. 1.

Quake Command	Gesture	Quake Command	Gesture	Quake Command	Gesture
Walk Forward (WF) Shake a right hand in a forward direction.		*Back Pedal (BP)* Shake a left hand in a backward direction		*Attack (A)* Stretch out a right hand in front.	
Turn Left (TL) Stretch out a left hand to the left.		*Turn Right (TR)* Stretch out a right hand to the right.		*Look Up (LU)* Move a head to the left.	
Step Left (SL) Move a right hand down and a left hand to the left.		*Step Right (SR)* Move a left hand down and a right hand to the right		*Look Down (LD)* Move a head to the right.	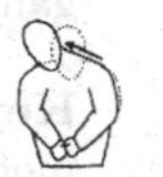
Center View (CV) Stretch out both hands horizontally, then return to the same position.		*Up/Jump (U/J)* Move both hands down and tilt head back.		*Down/Crouch (D/C)* Lift both hands simultaneously.	
Run (R) Swing arms alternately.					

Fig. 1. Thirteen types of gestures used in *Quake II*.

The proposed system consists of an input device, a processing unit, and an output device, and the overall configuration is shown in Fig. 2. A video camera, which is located above the user at an angle of sixty-two degrees, captures gestures of a user in real-time. The input image sequences are sent to a processing unit that is core of the proposed game system. The processing unit recognizes human actions from the sequences. This processing unit is performed by four steps: *a feature extractor, a pose classifier, a gesture recognizer,* and *a game controller.* Firstly, a feature extractor extracts a feature that is represented by positions of user head and hands. Subsequently, a pose classifier classifies a pose using symbol table and a gesture recognizer recognizes human actions from the classified pose symbol sequence. Finally, a game controller translates that sequence into the game commands. The result of experiments shows that the proposed system is suitable for application of real game. The game situations that are operated by user gestures are appeared in a big screen again through a projector.

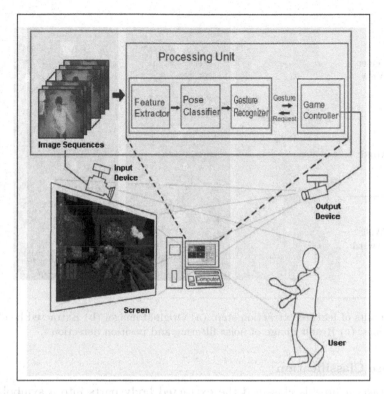

Fig. 2. Computer game setup using the gesture recognition method.

2.1   Feature Extraction

To extract the positions of head and hands as features from image sequence, feature extraction in the proposed method is performed by three steps: skin color detection, noise filtering, and position detection. Skin color detection identifies body parts using skin color model, where the color distribution of human skin is clustered within a small area of chromatic color space and then it can be approximated using a 2D-Gaussian distribution [7]. Therefore, the skin color model can be approximated by a 2D-Gaussian model, $N(m, \sum^2)$, where the mean and variance are as follows:

$$m = (\bar{r}, \bar{g}), \; where \; \bar{r} = \frac{1}{N} \sum_{i=1}^{N} r_i \; and \; \bar{g} = \frac{1}{N} \sum_{i=1}^{N} g_i, \tag{1}$$

$$\sum = \begin{bmatrix} \sigma_{rr} & \sigma_{rg} \\ \sigma_{gr} & \sigma_{gg} \end{bmatrix}. \tag{2}$$

Then, noise filtering eliminates noise and fills out any holes in the image. Finally, position detection identifies the position of the head and hands by labeling a region, making a box based on the labeled region, and discriminating body parts using heuristic rules. Fig. 3. shows an example of feature extraction.

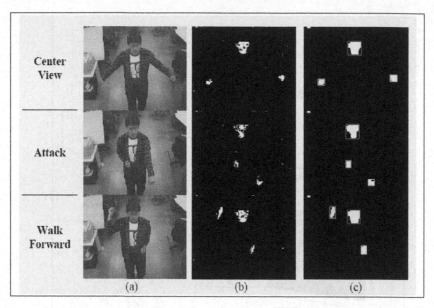

Fig. 3. Results of feature extraction step: (a) Original image. (b) Extracted head and hands regions. (c) Result image of noise filtering and position detection.

2.2   Pose Classification

In this stage, a pose is classified the extracted body parts into a symbol in a symbol table according to pose classification. Here, we assumed that each gesture consists of start pose, intermediate poses between a start pose and a distinctive pose, a distinctive pose and end pose.

Fig. 4 shows the symbol table used in our system that includes 23 poses such as *a start (end pose), distinctive poses, and intermediate poses.* A symbol is approximated for an input feature using the predefined symbols in the symbol table. An input feature is classified to a symbol that has a smallest norm between input feature and the predefined symbol table.

Fig. 5 shows an example of pose classification.

2.3   Gesture Recognition

Since gestures are presented in 3D spatio-temporal space in the real world, many variations occur although the same person performs the same gesture twice. Thus, the recognition model needs to be robust to such variations in time and shape [8]. An HMM has a rich mathematical structure and serves as the theoretical basis for a wide range of applications. It can model spatio-temporal information in a natural way, and includes elegant and efficient algorithms for learning and recognition, such as the *Baum-Welch algorithm* and *Viterbi search algorithm* [9].

Therefore, we use a HMM for recognize gestures. Every time a symbol is given, a gesture recognizer determines whether the user is performing one of the

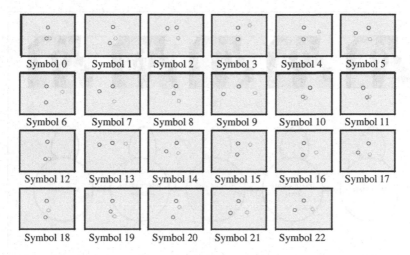

Fig. 4. A symbol table: it includes 23 poses such as *a start (end pose), distinctive and intermediate poses.*

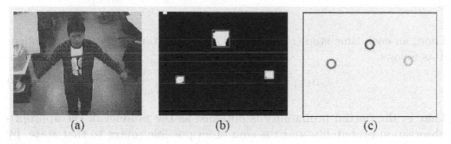

Fig. 5. Results of pose classification step: (a) Original image. (b) Result of Feature Extraction. (c) An approximated Symbol.

thirteen gestures predefined above, or not. If he or she is, the gesture recognizer returns the gesture that the user is performing.

For many applications, especially in speech recognition, left-right model has been widely used. So we create a discrete HMM for each gesture. Fig. 6 shows an example of the structure used for the HMMs. For each of the 13 gestures, a 5-state HMM was trained separately using the *Baum-Welch algorithm.*

To determine the probability of an observed sequence given an HMM when the parameters (A, B, π) are known, a forward algorithm is used to calculate the probability of a T long observation sequence,

$$Y^{(K)} = \{Y_{K_1}, \cdots, Y_{K_r}\},\tag{3}$$

where each belongs to the observable set. Intermediate probabilities (α') are calculated recursively by initial α. Then the initial α for each state is calculated using the following equation.

$$\alpha_1(j) = \pi(j) \cdot b_{jk},\tag{4}$$

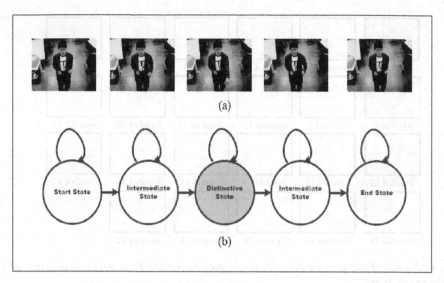

Fig. 6. An HMM model: (a) the gesture for the command *ATTACK*, (b) a left-right HMM.

Then, for each time step $(t = 2, 3, \cdots, T)$, the partial probability α is calculated as follows:

$$\alpha_{t+1}(j) = \sum_{t=1}^{n} \alpha_t(i) a_{ij} \cdot b_{jk}, \qquad (5)$$

That is, the partial probability is obtained as the product of the appropriate observation probability and the sum of all possible routes to that state, by exploiting recursion based on knowledge of these values for the previous time step.

Finally, the sum of all partial probabilities gives the probability of the observation, given the HMM, λ.

$$\Pr(Y)^{(K)} = \sum_{j=1}^{n} \alpha_T(j). \qquad (6)$$

Eventually the model with the highest probability is selected as the objective one.

3   Experimental Results

To show the impressive effect of our gesture based game system using HMM in game entertainment, we combine it with an interactive game, *Quake II*. This game system is developed in PC platform, the operating system is Windows XP, CPU is *Pentium IV*-2.0 *GHz*, and memory size is 512*M*. The gesture based game system is demonstrated Fig. 7.

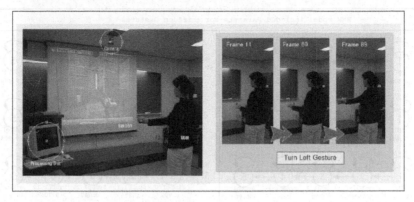

Fig. 7. Environment of the proposed game system.

Table 1 shows the mean and covariance matrix of the skin color model obtained from 200 sample images. Since the model only has six parameters, it is easy to estimate and adapt to different people and lighting conditions.

Table 2 shows the performance of the proposed HMM in recognizing 13 commands. The result shows recognition rate of about 90.87% for the thirteen gestures.

Table 1. Actual 2D-Gaussian parameters.

Parameters	Values	Parameters	Values
μ_r	117.588	$\rho_{x,y}\sigma_g\sigma_r$	-10.085
μ_g	79.064	$\rho_{x,y}\sigma_r\sigma_g$	-10.085
σ_r^2	24.132	σ_g^2	8.748

Table 2. A gesture recognition results.

Gesture Types	Recognition Rates
Attack (A)	94.25%
Walk Forward (WF)	88.75%
Back Pedal (BP)	90.00%
Turn Left (TL)	91.85%
Turn Right (TR)	91.88%
Run (R)	80.62%
Step Left (SL)	96.12%
Step Right (SR)	96.25%
Look Up (LU)	86.75%
Look Down (LD)	76.25%
Center View (CV)	96.25%
Up·Jump (UJ)	96.15%
Down·Crounch (DC)	96.25%

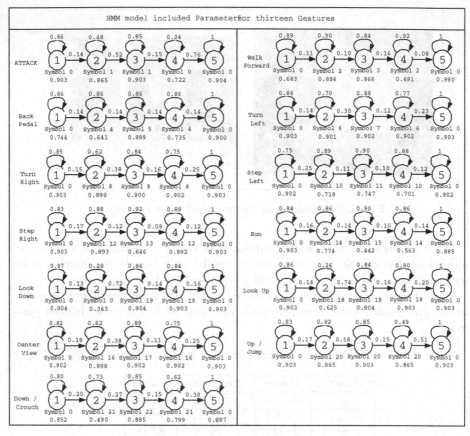

Fig. 8. HMM of each 13 gestures: the arrows represent a_{ij}, circles are state, $b_{ij}'s$ results are represented symbol x, which is located under each circles and also is showed that probability values under that.

Each gesture was recognized by the HMM. Fig. 8 shows the parameter results for the HMM and HMM models. π is the initial state distribution, a_{ij} is the state transition probability matrix and b_{ij} is the output probability matrix.

Consequently, our experimentation shows that the proposed HMM have a great potential to a variety of multimedia application as well as computer games.

4   Conclusions

We developed a game system using human action that can provide a more convenient and intuitive user interface. For recognition of human action, we use a HMM. Experimental results show reliability of about 90.87% with false recognition of 9.13% and the proposed system is applicable for real game system as generalized user interface.

Acknowledgements

This research was supported by grant No.R05-2004-000-11494-0 from Korea Science & Engineering Foundation.

References

1. I. Cohen, N. Sebe, A. Garg, L.S. Chen, and T.S. Huang: Facial expression recognition from video sequences: temporal and static modeling. Computer Vision and Image Understanding, Vol. 91 (2003) 160–187
2. H. S. Park et al.: A vision-based interface for interactive computer games. CIRAS, (2003) 80–84
3. C. S. Lin, C. C. Huan and C. N. Chan et al.: Design of a computer game using an eye-tracking device for eye's activity rehabilitation. Optics and Lasers in Engineering, (2004) 90–108
4. S. Eickeler, A. Kosmala and G. Rigoll.: Hidden Markov Model Based Continuous Online Gesture Recognition. International Conference on Pattern Recognition, Brisbane, (1998) 1206-1208
5. A. Wilson: Adaptive Models for Gesture Recognition. PhD Thesis, MIT, (2000)
6. A. Wilson and A. Bobick: Realtime Online Adaptive Gesture Recognition. International Conference on Pattern Recognition, Barcelona, Spain, (2000)
7. J. Yang, A. Waibel: A real-time face tracker, Applications of Computer Vision. WACV, vol. 15, no. 1 (1996) 142–147
8. B. W. Min, H. S. Yoon, S. Jung, T. Ohashi and T. Ejima: Gesture based edition system for graphic primitives and alphanumeric characters. Engineering applications of Artificial Intelligence, (1999) 429–441
9. X. D. Huang, Y. Ariki and M. A. Jack: Hidden Markov Models for Speech Recognition. Edinburgh. Edinburgh Univ. Press, (1990)

The Implementation of IPsec-Based Internet Security System in IPv4/IPv6 Network

So-Hee Park, Jae-Hoon Nah, and Kyo-Il Chung

Information Security Infrastructure Research Group,
Electronics and Telecommunications Research Institute
161, Gajeong-Dong, Yuseong-Gu, Daejeon, Korea
{parksh,jhnah,kyoil}@etri.re.kr

Abstract. IPsec has now become a standard information security technology throughout the Internet society. It provides a well-defined architecture that takes into account confidentiality, authentication, integrity, secure key exchange and protection mechanism against replay attack also. For the connectionless security services on packet basis, IETF IPsec Working Group has standardized two extension headers (AH&ESP), key exchange and authentication protocols. It is also working on lightweight key exchange protocol and MIB's for security management. IPsec technology has been implemented on various platforms in IPv4 and IPv6, gradually replacing old application-specific security mechanisms. In this paper, we propose the design and implementation of controlled Internet security system, which is IPsec-based Internet information security system in IPv4/IPv6 network and also we show the data of performance measurement. The controlled Internet security system provides consistent security policy and integrated security management on IPsec-based Internet security system.

1 Introduction

The popularity of Internet has given rise to the development of security solution to protect IP datagrams. The traditional approaching methodology of offering information security service in the network is finding the independent solution that do not influence the application program on the upper layer of the protocol stack. To add to this, from the view of the network protocol designers, it is also most effective to offer the security service from the network layer.

So, in November 1992, the members of IETF started to design the IP layer security protocol, which is suitable for the large scaled Internet environment. As a result of this, swIPe was born and its design concept was proven that the security service from the IP layer is possible [1]. After this, IETF IPsec WG has started to write the specification of the IP layer security. During the 34th IETF Meeting (December 1995), the interoperability test of IPsec system was performed for the first time. The system, which is based on only the IPsec documents, implemented by the Internet device manufactures and researchers independently. After this, IPsec security architecture, transform algorithms, AH (Authentication Header), ESP (Encapsulation Security Payload), IKE (Internet Key Exchange) were confirmed as a RFC [2,3,4,5,6,7,8]. Up to date, new RFCs and drafts related to IPsec are made. The IPv6 WG adopted IPsec, and it becomes mandatory requirement of the next generation Internet and optional requirement of IPv4.

T.G. Kim (Ed.): AIS 2004, LNAI 3397, pp. 109–116, 2005.
© Springer-Verlag Berlin Heidelberg 2005

IPsec provides a transparent information security services to the Internet users because it is offered by the IP layer and is needless to modify the application programs of the upper layer. And the consistent security service is possible in the system because IPsec provides the same information security service to the application layer and the transport layer. Also, IPsec has an open architecture, therefore it does not depend on the specific algorithm or authentication mechanism and it easily can adopt the existing technology or a new technology.

IPsec is implemented on various OS platform (i.e., Linux, FreeBSD and Windows2000). And it is researching in open projects FreeS/Wan on based Linux and KAME on based FreeBSD, but not completed [9,10]. IPsec is used broadly all network system, especially in VPN (Virtual Private Network) equipment. IPsec is understood to be the only Internet security protocol to solve scalability and compatibility in VPN when adopt to the large scaled network. So in this paper, we propose the design and implementation of controlled Internet security system in IPv4/IPv6, which is IPsec based Internet information security service system and provides consistent security policy and integrated security management. Also we will show the data of performance measurement.

2   The Controlled Internet Security System

2.1   The Architecture of Our System

Controlled Internet security system is composed of secure host/gateway system containing IPsec engine, Internet key exchange system and key management system, security management system, security policy system and security evaluation system. Figure 1 shows the architecture of our system.

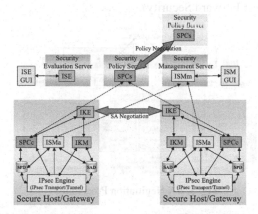

Fig. 1. The Architecture of Controlled Internet Security System

2.1.1   IPsec Engine

IPsec provides a standard, robust and extensible mechanism in which to provide security to IP and upper layer protocols (i.e., UDP or TCP). The method of protecting IP datagrams or upper layer protocols is by using on of the IPsec protocols, AH and ESP. AH offers integrity of IP packet and authentication of packet origin. ESP is used to encrypt the upper layer information of packet payload.

Two protocols are implemented in the IPsec engine. IPsec engine is divided into two different functions as secure host and gateway. The two functions are similar in some ways but have much difference in the system role and position. Secure host function is ported generally on the user terminal with single network interface, however secure gateway function is ported on the system with multiple network interfaces such as router, firewall and VPN server. Figure 2 shows the IPsec processing procedure of IP packet.

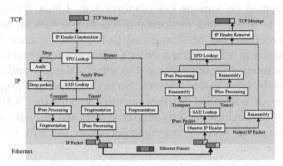

Fig. 2. IPsec Processing Procedure

2.1.2   Internet Key Exchange (IKE)

IKE offers automated key negotiation, and it is a mixed type of protocols ISAKMP, Oakley and SKEME. ISAKMP protocol provides two phases of processing and the functions are the authentication and key exchange. Oakley provides definition of key exchange mode and SKEME provides key sharing and re-keying mechanism. IKE is designed to defense DoS (Denial of Service) and man-in-the-middle attack, and also satisfy the PFS (Perfect Forward Security).

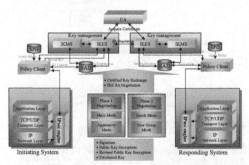

Fig. 3. SA Negitiation Procedure

To negotiate and create key, IKE processing is divided into two phases. In phase 1, ISAKMP SA negotiation and key material creation are performed for protection of ISAKMP messages. In phase 2, IPsec SA negotiation and key material creation are performed for security services of IP packet. IKE has 4 exchange modes, which are main, aggressive, quick and new group. The ISAKMP SA created from phase 1 and IPsec SA created from phase 2 are stored in ISAKMP SADB and IPsec SADB, respectively. Figure 3 shows the interaction of CA and IKE and SA negotiation between IKEs.

2.1.3 Internet Key Management (IKM)

IKM is the key management system, which manages the negotiated SAs and keys from IKE. IKM stores, deletes and updates SA using pf_key protocol [11].

Also, when the lifetime of SA is expired and key information is compromised, IKM requests IKE to renegotiate SA. Another function is to store and manage the certificate from CA and to provide API of crypto-library. Figure 4 shows the functions of IKM.

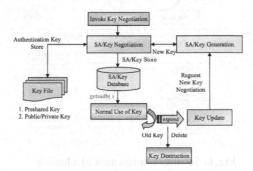

Fig. 4. The Functions of IKM

2.1.4 Security Policy System (SPS)

Security policy system decides the appropriate security policy between two systems. The security policy is the one of three mechanisms – apply IPsec, bypass IPsec and discard.

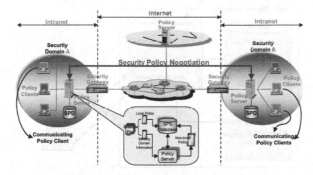

Fig. 5. Policy Negotiation Procedure

If security policy is the apply IPsec, security policy system decides hash and encrypt algorithm, the size of the key, a term of validity and connection type of the domains or the systems and also manages them [12]. The security policy is manually set up by the security manager or set up automatically by the policy negotiation between SPSs. Figure 5 shows the security policy negotiation procedure between policy servers in two different domains.

2.1.5 Security Management System (SMS)

SMS offers control mechanism to security manager. The functions of the SMS are monitoring of security service status, collection of audit information and manual configuration of SA and SP. To monitor the system status, the definition of MIB (Management Information Base) is necessary. MIB is not standardized yet and IETF is working on it. MIB at present are IPsec monitoring MIB, IKE monitoring MIB, ISAKMP DOI-Independent MIB and IPsec DOI textural conventions MIB [13,14,15,16]. Figure 6 shows security management mechanism of SMS.

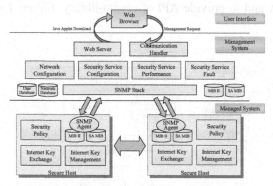

Fig. 6. Security Management Mechanism

2.1.6 Security Evaluation System(SES)

SES (Security Evaluation System) estimates the system safety and finds the threat factor before the threat occurs.

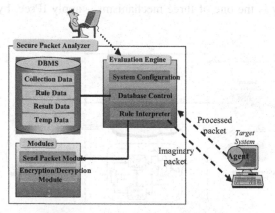

Fig. 7. Security Evaluation Mechanism

The functions of SES are collecting network information using sniffer, searching evaluation rule database (ERD) to evaluate specific system, analyzing the result and reporting the result to security manager. ERD has evaluation method and attack technique of how to evaluate and attack the security of the system. Figure 7 show security evaluation mechanism of SES.

2.2 The Procedure of Our System

Secure host/gateway is containing IPsec Engine, SAD, SPD and IKEB. IPsec Engine, SAD and SPD operate in the kernel layer and IKEB operates in the application layer.

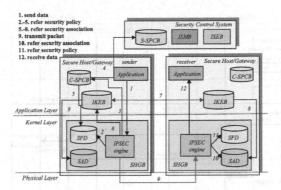

Fig. 8. The Connectivity of Controlled Internet Security System

And security control system is composed of SPCB, ISMB and ISEB. Figure 8 shows the connectivity of blocks in controlled Internet security system.

When secure host/gateway establishes secure connection between systems - hosts, gateways and host and gateways, IPsec engine requests the appropriate policy to SPS. When the appropriate security policy and security association are already exist, IPsec engine must reference SPD and SAD on the procedure of outbound packet process and inbound. Otherwise, SPS and peer SPS negotiate security policy by using SPP to make a new policy between two end-to-end systems [17]. Then SPS invokes IKE and IKE and peer IKE negotiate SA. SA has the information of key, hash function, encryption algorithm and lifetime and is stored in SADB. The negotiated SA and key are managed by the IKM. After procedure of negotiating and storing SPD and SAD, IPsec engine can encrypt/decrypt the IP datagrams.

In outbound process, IPsec engine decides whether to apply IPsec or how to apply it from SPD. Then IPsec engine must observe SAD including SA and encrypts the sending IP packets using SAD. In inbound process, IPsec engine decrypts the received IP packets using SAD, which is obtained in IKM. Also IPsec engine verifies if the security service is correctly adopted from SPD.

The SMS monitors the system status in each step and reports the collected security information to the security manager. Figure 9 shows the procedure of our system.

3   Performance

IPsec performance parameters of our interest include latency and throughput. We measured latency using ping test. The measurement configuration consists of two machines running over C-ISCAP software. Two machines were 1GHz Pentium equipped with 100Mbps Ethernet card. We did the test for different packet size (512, 1024, 2048 and 4096 bytes of payload) and different IPsec transform(HMAC-MD5, 3DES-MD5) between each other. The results can be seen in Figure 10. The graph shows that the cost of authenticating packets does not downgrade response time, but that encryption (especially 3-DES) is major bottleneck.

In the second test, we transferred 20MB of Image data from Pentium PC to SUN Enterprise 450 with 100Mbps Ethernet card. We used ttcp to measure throughput, with TCP as the transport protocol. Figure 11 shows the results.

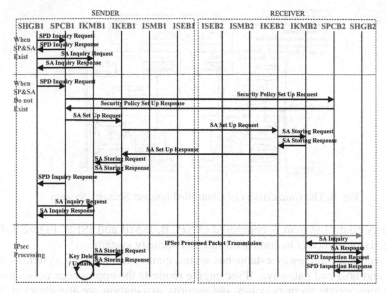

Fig. 9. The Procedure of Controlled Internet Security System

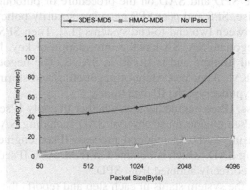

Fig. 10. Ping Performance of IPsec

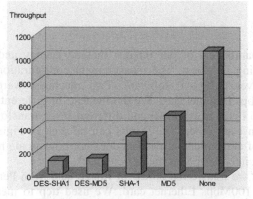

Fig. 11. Throughput of TCP Transfer

4 Conclusions and Future Works

In this paper, we have mentioned about the architecture and implementation of controlled Internet security system, Internet information security system based on IPsec. IPsec is considered as a successful Internet standard protocol with IKE. We can see it from the fact that in spite of VPN equipment manufactures have their own security protocol, such as PPTP and L2TP, they adopt IPsec as a VPN security standard. However, to deploy IPsec and IKE, the supply of PKI(Public Key Infrastructure) must be advanced. Also for the performance enhancement of IPsec engine of massive packet processing in large-scaled network, hardware-based encryption algorithm is necessary.

The future works must be focused on IPsec and IKE adaptation in remote and Mobile IP environment, which are already discussed in IETF. So our study would be in the direction of defining SAD and SPD extension fields related mobility and simplifying heavy IKE.

References

1. J. Ioannidis and M. Blaze, The Architecture and Implementation of Network-Layer Security Under Unix, *Fourth USENIX Security Symposium Proceedings*, October 1993.
2. S. Kent, R. Atkinson, Security Architecture for the Internet Protocol, RFC2401, November 1998.
3. S.Kent, R.Atkinson, IP Authentication Header, RFC2402, November 1998.
4. C. Madson, R. Glenn, The use of HMAC-MD5-96 within ESP and AH, RFC2403, November 1998.
5. C. Madson, R .Glenn, The use of HMAC-SHA-1-96 within ESP and AH, RFC2404, November 1998.
6. C. Madson, N. Doraswamy, The ESP DES-CDC Cipher Algorithm With Explicit IV, RFC2405, November 1998.
7. S. Kent, R. Atkinson, IP Encapsulating Security Payload(ESP), RFC2406, November 1998.
8. D. Harkins, D. Carrel, Internet Key Exchange (IKE), RFC2409, November 1998.
9. FreeS/Wan from http://www.freeswan.org
10. KAME from http://www.kame.net
11. D. McDonald, C. Metz, B. Phan, PF_KEY Key Management API, Version 2, RFC2367, July 1998.
12. M. Blaze, A. Keromytis, M. Richardson, L. Sanchez, IPsec Policy Architecture, Internet draft, July 2000.
13. T. Jenkins, J. Shriver, IPsec Monitoring MIB, Internet draft, July 2000.
14. T. Jenkins, J. Shriver, IKE Monitoring MIB, Internet draft, July 2000.
15. T. Jenkins, J. Shriver, ISAKMP DOI-Independent Monitoring MIB, Internet draft, July 2000.
16. J. Shriver, IPsec DOI Textual Conventions MIB, Internet draft, June 2000.
17. L. Sanchez, M. Condell, Security Policy Protocol, Internet draft, July 2000.

Describing the HLA Using the DFSS Formalism

Fernando Barros

Dept. Eng. Informática, Universidade de Coimbra, 3030 Coimbra, Portugal
barros@dei.uc.pt

Abstract. The High Level Architecture (HLA) is a standard for enabling the interoperability of simulation systems in a distributed computer environment. The
HLA has been established by the US/DMSO and it is aimed to promote the
communication among complex simulators. In this paper, we present a formal
representation of the modeling aspects of the HLA using the *Discrete Flow System Specification* (DFSS), a modeling formalism aimed to represent dynamic
structure systems. The HLA has introduced concepts not typically used in the
modeling and simulation field. A formal representation of the HLA permits its
description using standard constructs, making the HLA more easily intelligible
by a larger community of users. A formal description also permits to highlight
the current limitations of the HLA when compared with a more complete formal framework like the DFSS formalism that supports a wider range of dynamic structure systems.

1 Introduction

The *High Level Architecture* (HLA) is a standard created by the US Defense Modeling and Simulation Office (DMSO) to permit the interoperability of simulators communicating through a computer network [5]. HLA main advantaged is to allow the
interoperability of simulators to create complex scenarios. A typical situation can
involve the use of several aircraft simulators to study platoon behavior. For achieving
interoperability, the HLA requires that simulators collaborate in a common protocol
for message exchanging. Although the HLA is a complex architecture using some
features not currently found in simulation environments, its comparison with other
tools has not been reported. This paper intends to bridge the gap between HLA and
current modeling formalisms. For achieving this goal we present a formal representation of the HLA using the *Discrete Flow System Specification* (DFSS) [2]. This formalism is used due to its abilities to describe the dynamic structure characteristics of
the HLA. The formal description of the HLA permits to highlight its key characteristics and limitations. In this study, we give particular attention to the communication
protocols existing among HLA components, to the kind of modeling constructs provided and to the support offered to structural changes. The description of the HLA is
made at its two levels, namely at the federate level and at federation level, corresponding to the DFSS atomic and dynamic structure network levels. This comparison
is not always straightforward. One of the reasons is that separated concepts in the
DFSS formalism can be merged into one single notion in the HLA. In particular,
DFSS precludes a separation between models and simulators for both atomic and
network models. On the contrary, the HLA uses the *Run Time Infrastructure* (RTI) as
a blender for these different aspects. Another limitation of the HLA is the lack of
support to *peer-to-peer* communication (p2p). Although DFSS was designed to repre-

T.G. Kim (Ed.): AIS 2004, LNAI 3397, pp. 117–127, 2005.
© Springer-Verlag Berlin Heidelberg 2005

sent p2p, we show that the broadcast protocol supported by the HLA can be described by the formalism. Many features of the HLA like changing the broadcast topology are described in the realm of dynamic structure models. By providing a broad perspective of adaptive models, we highlight the HLA limitations. Not all HLA characteristics will be considered and we left aside some aspects not considered fundamental. These aspects include HLA class/subclass relationship, a concept that helps defining filtering and broadcast topologies. Other features not directly related to simulation, like the support for real time are not discussed being only considered time regulating and time constrained policies. A major problem encountered in this work was the representation of the RTI. Although the major role of the RTI is to control the network model (federation), the communication involving the RTI is non-modular. This kind of protocol is a source of non-determinism and limits modeling by imposing a fixed set of messages that can exchanged between the RTI and federates. To overcome these problems we have choose to represent the non-modular RTI/HLA protocol by the DFSS modular protocol.

2   Discrete Flow Components

The *Discrete Flow System Specification* (DFSS) provides the extension of the DEVS formalism [6] to support dynamic structure models. The DFSS formalism has introduced a novel representation of model networks, while using the DEVS formalism to represent basic models.

2.1   Basic Model

A basic model in the DFSS formalism is defined by

$$DFSS = (X, Y, S, \tau, q_0, \delta, \lambda)$$

where

X is the set of input values
Y is the set of output values
S is the set of partial states (p-states)
$\tau: S \rightarrow \mathbf{R}_0^+$ is the time-to-output function
$Q = \{(s,e)| s \in S, 0 \leq e \leq \tau(s)\}$ is the state set
$q_0 = (s_0, e_0) \in Q$, is the initial state
$\delta: Q \times X^\phi \rightarrow S$ is the transition function, with

$$X^\phi = X \cup \{\phi\}$$

$\lambda: S \rightarrow Y^\phi$ is the partial discrete output function

The output function, $\Lambda: Q \rightarrow Y^\phi$, is defined by

$$\Lambda(s,e) = \begin{cases} \lambda(s) & \text{if } e = \tau(s) \\ \phi & \text{if } e < \tau(s) \end{cases}$$

Basic models have modular characteristics and offer a good representation of HLA federates as described in Section 3.1. When a basic component is in p-state s it changes to a new p-state whenever a value arrives or the time elapsed in the p-state s reaches the value $\tau(s)$. Component new p-state s' is given by $s' = \delta(s,e,x)$. Component output can be non-null when the elapsed time in the current p-state $e = \tau(s)$. The output value is given by $\lambda(s)$. DFSS components require non instantaneous propagation [3], i.e., a component receiving an input at time t will only change its state at time t^+, where $t^+ = t + \varepsilon$, $\varepsilon > 0$ and $\lim \varepsilon \to 0$. Non-instantaneous propagation makes possible to support self-loops and, as a consequence, to ensure that the p-state of a component is always well defined. This feature is also crucial to support structural changes [3].

2.2 Network Model

The DFSS formalism can represent models with a time-varying structure. Dynamic structure network models offer a more intuitive representation of real systems for they are able to mimic the dynamic creation and destruction of entities, and the dynamic nature of the relationship among entities. Formally, a Discrete Flow System Specification Network is a 4-tuple

$$DFN_N = (X_N, Y_N, \eta, M_\eta)$$

where

N is the network name
X is the set of input values
Y is the set of output values
η is the name of the dynamic structure network executive
M_η is the model of the executive η

The model of the executive is a modified DFSS model, defined by

$$M_\eta = (X_\eta, Y_\eta, S_\eta, \tau_\eta, q_{0,\eta}, \delta_\eta, \lambda_\eta, \Sigma^*, \gamma)$$

where

Σ^* is the set of network structures
$\gamma: S_\eta \to \Sigma^*$ is the structure function

The network structure $\Sigma_j \in \Sigma^*$, corresponding to the p-state $s_{j,\eta} \in S_\eta$, is given by the 4-tuple

$$\Sigma_j = \gamma(s_{j,\eta}) = (D_j, \{M_{i,j}\}, \{I_{i,j}\}, \{Z_{i,j}\})$$

where

D_j is the set of component names associated with the executive p-state $s_{j,\eta}$
for all $i \in D_j$
$\quad M_{i,j}$ is the model of component i
$\quad I_{i,j}$ is the ordered set of components influencers of i
$\quad Z_{i,j}$ is the input function of component i

These variables are subject to the following constraints for every $s_{j,\eta} \in S_\eta$:

$\eta \notin D_j, N \notin D_j, N \notin I_{N,j}$
$M_{i,j} = (X_{i,j}, Y_{i,j}, S_i, \tau_i, q_{0,i}, \delta_{i,j}, \lambda_{i,j})$ is a basic DFSS model, for all $i \in D_j$, with
$\qquad \delta_{i,j}: Q_i \times X_{i,j} \to S_i$
$Z_{i,j}: \underset{k \in I_{i,j}}{\times} V_{k,j} \to X_{i,j}$, for all $i \in D_j \cup \{\eta\}$

where

$$V_{k,j} = \begin{cases} Y_{k,j} & \text{if } k \neq N \\ X_N & \text{if } j = N \end{cases}$$

The network output function is given by

$$Z_{N,j}: \underset{k \in I_{N,j}}{\times} Y_{k,j} \to Y_N$$

Adaptations in the network structure include the ability to modify network composition and coupling and they are achieved by changing the executive state. The mapping from executive state into network structure is made by function γ. Network semantics enable incremental operators to adapt dynamically the structure [4]. The key reason to choose the DFSS formalism is its ability to represent adaptive structures making possible to describe the HLA dynamic topologies.

2.3 Determinism in Dynamic Structural Networks

As seen in the last section DFSS networks define their structure based on the state of the executive. To achieve determinism it is crucial to define the executive state non ambiguously. A possible source of ambiguous behavior can be the occurrence simultaneous events. The DFSS uses parallel transitions to impose determinism, i.e., all the transitions scheduled for the same time are taken simultaneously. This procedure avoids the use of a complex selection function involving not only the scheduled events but also involving the order that messages are sent to the executive. We emphasize that the traditional selection function including only events is not enough to guarantee determinism. Given that the order messages are delivered to the executive plays a key role in structure, message selection is also crucial. The parallel behavior of the DFSS formalism enables determinism without the complexity of specifying two selection functions. The major problem of parallelism is to guarantee that the order transitions are taken is not important. In static structure networks of modular components, like in DEVS networks, this can be achieved easily. However, in dynamic structure networks, components are not independent. In fact, a change in the network executive can influence other components. Namely, components can be removed or the connections can be changed when the executive changes its p-state. A practical solution taken in the DFSS formalism is to guarantee that the executive is the last component to be trigged [4].

3   The HLA

We describe the HLA at two levels. At the federate level, the HLA will be considered equivalent to an atomic DEVS model. At the federation level, the HLA will be studied as a DFSS dynamic structure network. We consider here a modeling perspective, and issues like real time or the choice of conservative/optimistic simulation strategies will not be discussed.

3.1   HLA Federate

Although HLA federates are not strictly modular models, the choice of a modular formalism like DEVS provides a good approximation. A federates exchange messages with the other federates through a well defined set values and except for the communication with the RTI it can be regarded as a DFSS basic model. A federate can be described by the model

$$M = (X, Y, S, \tau, q_0, \delta, \lambda)$$

Sets X and Y can be computed based on the classes subscribed and published by a federate. These classes play a key role on structure definition and they will be described in the next Section. The other functions can easily be mapped to RTI operations, as we show next.

HLA federates can operate under two different policies: a parallel behavior policy where values produced at the same time instant are sent to the federate, or in alternative, a federate can choose to obtain just one of the values at each transition. The latter policy can lead to non-deterministic behavior and it will not be described here. The parallel behavior involve that sets X and Y are usually bags and not simple objects.

Actually, the HLA Federate Ambassador can be considered to combine a model with its simulator, making harder the modeling task. Federates need to be defined in respect to the simulator, i.e., in a federate definition the modeler needs to make explicit calls to the RTI, that can be regarded as implementing the simulator in the DEVS perspective. On the contrary, DEVS models are not aware of their simulators making models easier to build.

To illustrate how DEVS models can be defined in the HLA/RTI framework we consider how the interaction-based communication could be defined.

A HLA-DEVS simulator needs to keep a buffer to store all the input values and handle simulation time. The simulator is responsible to issue a `nextEventRequest()` call and to handle the callbacks `timeAdvanceGrant()` and `receiveInteraction()`. This last method can be defined in pseudo-code by

```
void receiveInteraction(InteractionClassHandleAndValues msg) {
    buffer.add(msg);
}
```

When an interaction is received, it is added to the buffer. We describe now how the separation between models and simulators can be achieved. The RTI `timeAdvanceGrant()` is sent to the federates when federation time advances. We consider that Federate Ambassador has the role of simulator and a model can be built using the DEVS protocol, supporting namely, transition, time-to-output, and output functions. The `timeAdvanceGrant()` callback method can be defined in pseudo-code by

```
void timeAdvanceGrant(const FedTime& time) {
    e = time - timeLast;
    if (e == model.timeToOutput())
        out = model.output();
    else
        out = null;
    model.transition(e,buffer);
    buffer.empty();
    if (out != null) RTI.sendInteraction(out);
    RTI.nextEventRequest(time + model.timeToOutput());
    timeLast = time.
}
```

In this definition, it is shown that models can be built without any reference to the RTI, making them independent of the simulation environment. In this simplified version using only interaction communication, the simulator, implemented here as a Federate Ambassador, is the only responsible to interact with the RTI, requiring from the modeler little knowledge about the HLA framework.

Given that the time advance is handled by the nextEventRequest() we guarantee that all events with the same time stamp will be delivered simultaneously to the federate. The timeAdvanceGrant() callback determines if the model can send an output, storing the component output, computed by the call model.output(), in variable *out*. It then changes model state by invoking model transition() function. Finally, outputs are sent using the call RTI.sendInteraction(out), and a nextEventRequest() is issued. Thus, in principle we can separate models and simulators in the HLA environment obtaining a DEVS like framework.

The HLA also uses objects for communication. To support this kind of communication the HLA requires non-modular communication between federates and the RTI, making the separation between models and simulators virtually impossible. This kind of communication will be treated in Section 3.3 since it requires a more detailed description of the RTI.

Another problem in representing a federate is motivated by federate ability to request for structural changes. The changes require access to the RTI that is achieved through non-modular communication described in the next Section.

3.2 HLA Federation

A federation is a composition of HLA federates; and for description purposes we represent a HLA federate as a DFSS network. This choice is mandatory since a federation has a dynamic structure nature that cannot be captured by a static DEVS network.

A federation can be viewed as particular case of a dynamic structure network with several limitations related to network composition and coupling. A HLA federation have a fixed set of federates, i.e., the destruction and creation of federate are currently supported as real time operations and do not increase HLA modeling capabilities since no federate can add or remove other federates. Federations support mainly a broadcast type of communication and *peer-to-peer* (p2p) communication is not supported in the HLA. Given that no access is granted to the RTI, except for a limited configuration through the FED file, the RTI provide a limited set of (fixed) services.

The choice of a broadcast communication protocol (bdc) is well suited to scenarios where all simulation entities can possibly interact, like in many military simulations. Actually, the HLA permits to define multicast topologies using *Data Distribution Management* (DDM) services [5].

We define a HLA federation by the autonomous DFSS network given by

$$DFN_N = (\eta, M_\eta)$$

The choice of an input/output free network reflects the non hierarchical nature of the HLA that permits only one level of hierarchy. We consider that the DFSS network executive can be used has a model of the HLA/RTI. The model of the executive is defined by

$$M_\eta = (X, Y, S, \tau, q_0, \delta, \lambda, \Sigma^*, \gamma)$$

Several simplifications can be made when modeling the RTI. The structure function is restricted to an invariant set of components D. The RTI can be regarded as a passive model that only reacts to external stimuli by going into a transitory state of zero time duration in order to send immediately an output. Then it goes again into the passive state waiting for the next message.

Since the modeler has no access to the inside of the RTI, and because the RTI is not a federate, the HLA provides a fixed set of messages that can be used by federates to communicate with the RTI. Thus, the transition function δ is actuality structured by the set of messages $\{\delta_m\}$, where m represents a message name. Some of these messages are related to structure management. For example, the message `publishinter-actionClass(RTI::InteractionClassHandle aClass)` will link the publishing federate to all federates subscribing to aClass.

RTI output values correspond to the messages sent to federates. Given that the RTI is not a federate, this set of messages is fixed and corresponds to callback messages.

To allow the description of the network structure we consider that the RTI defines a set of interaction values V. Each value can be assigned to a single class of the set of interaction classes Γ. These classes are actually defined in the HLA/FED that can be regarded as the only access to the RTI definition.

To define network structure we recur to the publishing function π and the subscribing function σ. These functions map federates into a set of classes and they are given by

$$\pi\colon S \times D \to 2^{\Gamma}$$
$$\sigma\colon S \times D \to 2^{\Gamma}$$

where S is the partial state set of the network executive representing the RTI and D is the set of federates.

With these functions, one can describe the behavior of the RTI messages. The subscribing message, for example, issued by a federate d to subscribe to a class c can be defined by

$$\delta_{subscribe}((s,e),(d,c)) = s' \therefore \sigma(s',d) = \sigma(s,d) + \{c\}$$

where the operator $\therefore$ *cond* is a short notation for stating that all properties remain the same except for condition *cond* that now holds. For example, if $\delta_{subscribe}((s,e),(d,c)) = s'$, then $\pi(s',d) = \pi(s,d)$, i.e., the subscribe operation does not change the published classes.

Similarly, the unsubscribe function can be defined by

$$\delta_{unsubscribe}((s,e),(d,c)) = s' \therefore \sigma(s',d) = \sigma(s,d) - \{c\}$$

The influencers class c of a component i when the executive is in p-state s_j can be defined by

$$I_{i,j,c} = \{d \mid d \in D \wedge d \neq i \wedge \pi(s_j,d) = c\}$$

The condition "$d \neq i$" is imposed by the HLA that does not consider self-loops.

From this set we can define the set of influencers of component i by

$$I_{i,j} = \cup\{I_{i,j,c} \mid c \in \Gamma\}$$

The input function of a component i returns a bag with the output values produced by the influencers of i. To illustrate federation description we consider a network defined by

$\Gamma = \{A,B,C\}$
$D = \{F1,F2,F3\}$

where at the executive current p-state s the following conditions hold:

$\sigma(s,F1) = \{A,B\},\ \pi(s,F1) = \{A,B\}$
$\sigma(s,F2) = \{A,B\},\ \pi(s,F2) = \{A,C\}$
$\sigma(s,F3) = \{C,B\},\ \pi(s,F3) = \{A,C\}$

This network can be described by Figure 1, where the links highlight the broadcast topology joining federates F1, F2 and F3. If F2 produces the values [A:a1,A:a2] and simultaneously F3 produces the values [A:a1, A:a3] then F1 receives the bag [A:a1,A:a1,A:a2,A:a3].

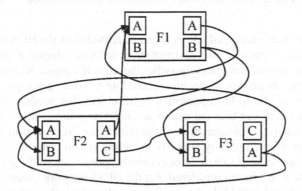

Fig. 1. Broadcast network linking federates

Support for structural changes is provided by the HLA by means of (un)publishing and (un)subscribing calls, namely, `publishInteractionClass()`, `unpublishInter-actionClass()`, `subscribeInteractionClass()` and `unsubscribeInteraction-Class()`. Figure 2 depicts the new communication network after F1 stops publishing class B using. All the links between F1 and federates F2 and F3 that involve class B are removed. We note that given the broadcast protocol used it is not possible to remove the link form F1:B to F2:B without removing also the corresponding link from F1 to F3. Would this be necessary and a different class should be used to allow this discrimination.

Actually, the HLA supports the class/subclass relationship and the rules for interaction broadcast are enlarged to handle this relationship. Thus, if a federate subscribes class X it will receive also interactions of class Y derived from class X, being the values of Y coerced to class X, loosing possibly some of its attributes. We note that the term class has little connection with the concept of class used in the object paradigm, where besides data, classes also define methods. On the contrary, classes in the HLA serve as a construct to carry information and to define constraints in the broadcast topology.

3.3 HLA Objects

Objects are another mechanism to provide communication among federates. The main difference between objects and interactions is that objects persist, while interactions

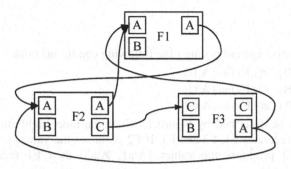

Fig. 2. Changed topology in interaction communication

are destroyed after being received. Objects can be regarded as the HLA solution to the inexistent support for creation and destruction of federates during a simulation. For example, when an aircraft lunches a missile, the missile cannot be represented as a federate responsible to update its position according to some guidance strategy. The HLA requires the missile to be represented by an object that needs to be updated by some federate. We consider that the use of an object as a surrogate to a federate is a poor solution. A federate representation of the missile, i.e., representing the missile as an autonomous entity, will greatly simplify modeling [1]. In order to account for object communication we need to extend our current representation of the HLA.

In the last section, we have considered that the DFSS network executive provides a representation of the HLA/RTI. In this section, we show how the executive can represent HLA/RTI object management.

One of the features of the RTI is the ability to create and destroy objects. These objects persist in memory and a set of messages is provided by the HLA to manage them. Operations include the right to update objects and the choice of what federates will be informed of the updates. To illustrate the behavior of object communication we use the federation of Figure 3 with federates F1..3 and objects A:a and B:b that have been previously created by the call registerObjectInstance(). Object ownership is represented by a solid line and a dashed line represents object reflection. The owner federate has the privilege to change object attributes using the call updateAttributeValues(). When attributes are updated, the subscribing federates will be informed of the updates by so-called *reflections* in the HLA terminology. In our example, federate F1 owns object A:a and F3 owns B:b. When the attributes of B:b are changed by federate F3, changes will be "reflected" using the callback reflectAttributeValues() sent to federates F1 and F2 that have previously discovered object B:b through the callback discoverObjectInstance() and have previously subscribe to object class using the call subscribeObjectClassAttributes(). Similarly, when object A:a is updated Federate F3 will be notified of the changes. In reality the picture is more complex because federates can acquire the ownership of some object attributes and they can be interested in receiving updates of only a few attributes.

Ownership and *reflectionship* relations can be changed by a negotiation process, making the network structure dynamic. If, for example, i) F1 divests the ownership of A:a by invoking the call negotiatedAttributeOwnershipDivestiture(), ii) F3 divests the ownership of B:b using the call unconditionallyAttributeOwnership-

Divesture() and unsubscribes to class A using the call unsubscribeObjectClass(),
iii) F2 accepts the ownership of objects A:a and B:b by invoking the attributeOwn-
ershipAcquisition() call and after obtaining confirmation through the callback
attributeOwnershipAcquisitionNotification(), the new network topology can be
depicted by Figure 4.

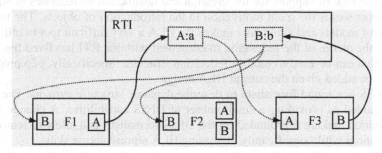

Fig. 3. Object communication

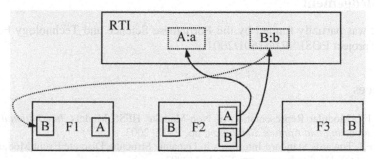

Fig. 4. Changed topology in object communication

Object negotiation can cause non-deterministic behavior in the HLA. When a federate gives away the ownership of one object the other federates interested in the
object class will receive the callback message requestAttributeOwnership-
Assumption(). When two or more federates decide to acquire object ownership the
result of the operation cannot be anticipated because it will depend on the order messages are delivered to the HLA. Although a parallel behavior can be imposed to federates, as discussed in Section 3.1, parallelism cannot be forced into the RTI that is
not as a federate, and non-determinism arises. We note that adding HFSS capabilities
to the HLA, namely, the ability to create and destroy federates in run time would
virtually remove the need for object support, considerably simplifying the modeling
task.

4 Conclusions

We have described the modeling features of the HLA using the DFSS formalism. This
description has highlighted many of the key aspects of the HLA and has shown HLA
advantages and limitations. The HLA was mainly designed to represent scenarios
composed by spatially distributed entities where the use of broadcast or multicast

communication protocols can be successfully applied, offering a rich set of primitives that facilitates the management of structural changes in these types of simulations. The DFSS has proven to be a sound framework to describe the HLA. All key aspects of the HLA could be described and the comparison made possible to evidence some of HLA weaknesses, namely, lack of determinism and limited support for structural changes. The lack of support for the creation and destruction of federates in simulation run-timer seems the main motivation to the introduction of objects. The unclear separation of models and simulators makes the HLA a very difficult tool to utilize. In particular, the merge of the federation management with the RTI has fixed the set of operators that can be used to modify federation structure. Specifically, p2p protocols can hardly be added given the current HLA design.

The DFSS is a sound formalism to describe dynamic structure systems. The HLA can be regarded as providing a small subset of DFSS capabilities. A change in the HLA definition making it grounded on the DFSS formalism will allow to remove its main limitations while significantly increasing HLA representation skills.

Acknowledgement

This work was partially funded by the Portuguese Science and Technology Foundation under project POSI/SRI/41601/2001.

References

1. Barros, F.J. Modular Representation of Non-Modular HFSS Models. *International Workshop on Modeling and Applied Simulation*, pp. 10-15, 2003.
2. Barros, F.J. Towards Standard Interfaces in Dynamic Structure Discrete Event Models. *Proceedings of the AIS Conference*, pp. 170-174, 2002.
3. Barros, F.J. Handling Simultaneous Events in Dynamic Structure Models. *Proceedings of SPIE 12th Annual International Symposium on Aerospace/Defense Sensing, Simulation and Controls: Enabling Technology for Simulation Science*, Vol. 3369, pp. 355-363, 1998.
4. Barros, F.J. Abstracts Simulators for the DSDE Formalism. *Proceedings of the Winter Simulation Conference*, pp. 407-412, 1998.
5. US Defense Modeling and Simulation Office. *High Level Architecture Run Time Infrastructure*. 2001.
6. Zeigler, B.P. *Theory of Modeling and Simulation*, Kluwer, 1976.

Proposal of High Level Architecture Extension

Jae-Hyun Kim and Tag Gon Kim

Dept. of EECS,
Korea Advanced Institute of Science and Technology (KAIST),
373-1 Guseong-dong, Yuseong-gu, Daejeon, Republic of Korea
jhkim@smslab.kaist.ac.kr, tkim@ee.kaist.ac.kr

Abstract. The paper proposes three dimensional extension to High
Level ARchitecture (HLA) and Runtime Infrastructure (RTI) to solve
several issues such as security, information hiding problem and interoperability and performance of RTI software. The hierarchical and modular
design of RTI software provides natural way to form complex distributed
simulation systems and methods to tune performance of federates with
selective and replaceable modules. The extension of specification level
from application programming interface (API) to message-based protocols makes RTI software communicate each other and even with the
protocol-talking hardware. The extension includes new APIs to the Federate Interface Specification, improving reusability of federates.

1 Introduction

The High Level Architecture (HLA) [1–3] is the specification for interoperation
among heterogenous simulations. The HLA also focus on reusability of partici-
pating simulations.

Under the HLA, a combined simulation system is called a federation, and the
individual simulation components are called federates. The Runtime Infrastructure (RTI) is software that implements IEEE 1516.1 Federate Interface Specification [2]. It provides a set of services available to the federates for coordinating
their operations and data interchange during an execution.

The HLA has been applied successfully to a military application, especially
for interoperation of distributed training simulators. The HLA is applied not
only to the field of distributed simulation but also to various applications including virtual reality, voice over IP, and other generic network applications. A
lot of interoperability and performance related issues have been raised from the
experience of large-scale interoperation between different organizations.

The current HLA does not support multi- or hierarchical federations. The
HLA assumes that there is a single federation. All federates in a single federations
are able to access Federation Object Model (FOM) data. This single federation
does not suffice applications with multiple security levels. This is called Information Hiding Problem [5–7].

Besides this problem, a single flat federation is not adequate to model complex systems with hierarchical components [8, 9]. Hierarchical structure of models
or simulators are essential to simulate complex and large systems. To form a hierarchical federation, many methods such as a federation gateway, proxy, bridge

T.G. Kim (Ed.): AIS 2004, LNAI 3397, pp. 128–137, 2005.

or brokers have been introduced [6–11]. However, these approaches requires additional interfacing entities that are not part of RTI software. To improve the whole performance of RTI, hierarchical federation scheme should be supported by RTI itself. Discrete Event System Specification (DEVS) formalism [4] demonstrates how to model and simulate complex systems with hierarchial structures. From the concept of DEVS formalism, we define hierarchical architecture of RTI and functionality of processors in the hierarchy.

Another big limitation is that HLA only specifies a standard services (application programming interfaces). The implementation methods or architectural design of RTI are not part of HLA standard. Although RTI developers are able to apply their own technology to implement RTI software, lack of standard prohibits interoperation between various RTI software from different vendors. This is one big drawback because one major goal of HLA is to achieve interoperation between heterogeneous simulations.

Open RTI protocol will make it possible for different RTIs to communicate with each other. In addition, a hardware-in-the-loop simulation becomes more efficient because this protocol enables direct communication between hardware and RTI. Open RTI Protocol Study Group of SISO is now working on the proposal of an open, message-based protocol. However, the target architecture of RTI is flat and fully distributed. We propose different RTI protocol designed to fit in hierarchical architecture.

The performance – speed or size – of RTI software is always a hot issue. RTI is a kind of middleware so that the performance of RTI greatly affects that of total system. To meet the requirement of target system, developers should have methods tune the performance of RTI software.

Normally RTI software is too heavy because it is designed to accommodate all kinds of services in one library. However, not all applications require all kinds of management services. Some may not need data distribution management services, and some only uses receive-order messages. Some applications require light-weight software to fit into an embedded systems.

Modular architecture of a local RTI component (LRC) makes the federate lighter and faster. If a federate does not need data distribution management services, the federate will not load the data distribution management module at run-time. In addition, the module is replaceable as long as the interface of the module is the same. A third party is able to develop its own modules with the open interface of the modules. Users will choose modules that meet the performance requirements of target federates.

2   Three Dimensional Extension of HLA/RTI

The paper proposes a three dimensional extension of HLA/RTI. Figure 1 shows the proposed extension. The first dimension is Runtime Infrastructure Software Design. The original HLA standard does not include any specific descriptions about implementation of RTI software. However, a detailed standard about architecture or protocol is essential to overcome previously discussed issues.

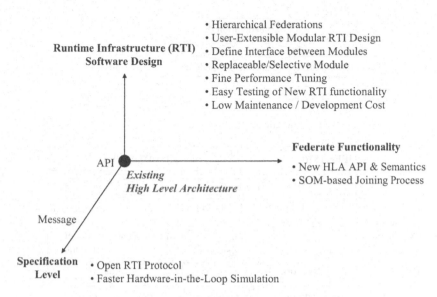

Fig. 1. Three Dimensional Extension to HLA/RTI.

The paper proposes the hierarchical and modular design of RTI software. The hierarchical design of federations provides natural ways to simulate complex systems. Each federation controls flows of information using an extended FOM that defines object or interaction classes open to higher level federations.

The modular design of RTI software suggests that RTI software consists of a set of interconnected modules and the modules are easily replaceable by any developers. Building a RTI software as a combination of modules from different venders of various performance makes it possible to meet the performance requirements of the target application.

The second dimension is specification level. The existing HLA specifies only APIs. The proposal extends this standard to message-based protocol level. Combined with the fixed hierarchical architecture of RTI, this protocol defines messages between entities in the federation hierarchy. Each modules in LRC is responsible for handling delivered messages destined to the module.

The last dimension is federate functionality. The proposed extension includes new APIs to give more functionality to federates. The SOM-based joining process with the extended join federation execution service [5] is adopted to increase reusability and modularity of a federate. This method is essential to the hierarchical structure because a federation is able to change its data filter without changing FOM of upper-level federation.

3   Hierarchical and Modular Runtime Infrastructure

3.1   Overall Architecture

Figure 2 shows the overall architecture of the proposed hierarchical and modular RTI. The whole simulation system forms a tree structure and is composed of

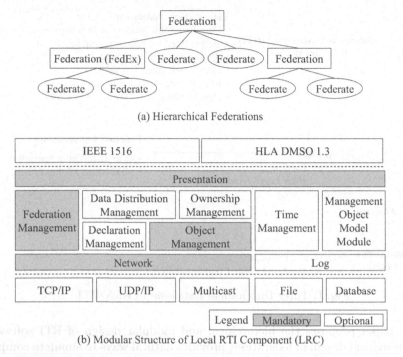

(a) Hierarchical Federations

(b) Modular Structure of Local RTI Component (LRC)

Fig. 2. Proposed Architecture of Hierarchical and Modular Runtime Infrastructure.

two types of simulation processes – Federation Execution (FedEx) processes and federates. All leaf nodes are federates and others FedEx processes. Federates only talk to its associated FedEx process, while the FedEx process exchange data with its parent, child FedEx processes as well as federates. A FedEx process coordinates and represents a federation. In addition to the traditional role of FedEx processes in DMSO RTI, the FedEx process acts as a federate to the FedEx of higher level.

Figure 2 (b) shows the modular structure of a local RTI component (LRC). The key idea of this structure is that not all modules are mandatory. Although HLA provides various kinds of services, most federates only need a partial set of services. With unnecessary modules eliminated, a federate will have a lighter code and better performance.

3.2   The Federation Execution Process

The main role of a federation execution process is scheduling and routing of events.

A federation becomes a federate to the higher level federation. This means that a FedEx process does not distinguish its child processes. A FedEx process acts like a federate with minimum lower bound time stamp (LBTS) of the associated federation. The FedEx process exchanges timing information with its parent FedEx process as well as its child federates.

The hierarchical federation requires extended FOM. The FOM not only contains the internal data inside the federation, but also includes the filtering information that which data to send to or receive from higher level federation. According to extended FOM information, the FedEx process automatically control the flow of events. The FedEx process forwareds allowed object updates and interactions to its parent FedEx process.

3.3   Modules of the Local RTI Components

Presentation Module. The presentation module maps HLA APIs to inside modules. There are two types of APIs – IEEE1516 and DMSO 1.3. DMSO 1.3 version is preliminary to IEEE1516, however, currently more in common due to free-distribution of RTI software. IEEE1516 and DMSO 1.3 are similar in functionality but function names, data types and some semantics differ. Therefore, the presentation module for each specification is necessary in order to accommodate two HLA specifications,

Network Module. HLA specification requires two types of network transportation – reliable and best effort service. TCP/IP is currently available for reliable services and UDP/IP for best effort service. Third type of network transportation, i.e., multicast, is very useful to deliver data to specified sets of receivers. The network module should provide APIs for reliable, best effort, multicast transportation to its upper modules. The module is easily extensible to accommodate new functionality such as quality of service (QoS) support.

Log Module. Logging of internal data or activity is the most valuable tool for developers. The log module provides an API to produce text outputs to a file and/or screen. Sometimes, a file output is not enough for a large-scale system. The database is a good choice to manage large amount of log data. Detailed implementation issues are up to module developers.

Federation Management Module. There are 13 RTI ambassador services and 9 federate ambassador callback functions that support federation management. Federation management module deals with federation-wide synchronization services. Therefore, federation management module controls time and object management modules for synchronization purpose. These two modules play a main role in simulation and are responsible for time advancement and data exchange. Other modules, however, are rather passive and contain information for reference. When a synchronizing service begins, the federation management module notifies it to the two modules. These modules then stop processing and wait until the completion notification arrives. The federation management module keeps the current status of the federation and this federate. It also keeps statuses of other joined federates.

Declaration Management Module. A declaration management module should keep the publication and subscription data of the current federate and subscription data from other federates. Declaration management module calculates the mapping between published data and its subscribers. We assume the sender-side message filtering. Sender-side message filtering means that a sender selects receivers before it sends. Therefore, sender-side filtering actually reduces network usage while it requires more computational power. The declaration management module provides subscriber information of published objects and interactions to the object management module. The object management module requires the information when it sends published data. If there is no declaration module at run-time, the object management module will work as if all joined federates subscribes all published object in the federate.

Object Management Module. An object management module includes SOM-based joining process [5]. The module keeps object and interaction class hierarchies and updates them when the federate joins to the federation. The constructed class structure is referenced by the declaration management module to compare relative positions of classes in the hierarchy.

The object management module retrieves subscribers information from the declaration management and stores it. It also stores registered and discovered object instances. Management Object Model (MOM) Management services are initiated by calling object management services with MOM handles. MOM objects and interactions are treated as same as other objects at the API level. When the object management module receives service call from presentation manager, it checks if the specified object instance or interactions belong to MOM. If it is a MOM request, the object management module forwards the requested service to the MOM management module. The object management module should manage a receive-order queue and a time-stamped order queue. The object instance updates and interactions with time stamp are delivered to time-stamped order queue. The messages without time stamp are delivered to the receive-order queue. According to the setting of time management module, it determines when the messages in the queue are delivered to the federate via federate ambassador callback services.

Time Management Module. Time management includes time regulation / constrained option settings and various time update services. Time management provides 4 combinations of regulation / constrained options and 3 different time advancement services to the federate. Time management module updates its logical time whenever the federate requests to advance its time. The time management module sends time update message to its FedEx process to notify its logical time. Each federate calculates lower bound time stamp (LBTS) whenever it receives time update from the FedEx process. The time management module is responsible to deliver TSO messages. The time module makes the object management deliver proper TSO messages in TSO queue. Receive order (RO) messages are delivered when time management module is in time-advancing mode with asynchronous delivery option disabled. If the asynchronous delivery option is enabled or the time module is not in time-constrained mode, the RO messages will be delivered whenever the federate invoke tick service. Also, if there is no time management module, the object management module will work as if the federate is in non-regulating and non-constrained mode.

Ownership Management Module. The ownership management module handles ownership information of all registered and discovered object instances. The module also processes the algorithm for ownership transfer between joined federates. The ownership management module notifies the changes in ownership to the object management so that the object management decides whether to allow object modification or not. Without the ownership management module loaded, the object management module allows all attempts to modify any objects known to the federate.

Data Distribution Management Module. The data distribution management module handles the routing space attached to published and subscribed data classes. Every time the value in the routing space changes, the data distribution management module calculates the connection between published and subscribed federates. The connection information is referenced by the object management module. The object management module uses the connection data as well as data from declaration management module to decide whether it delivers data or not. The messages and protocol about data distribution management is not yet fully specified. Without the data distribution management module loaded, the object management module decides data delivery only based on the data from declaration management module.

Management Object Model (MOM) Management Module. The Management Object Model (MOM) consists of a number of object and interaction classes through which federates monitor and tune the operation of active federation. MOM module collects status and activities of other management modules. The MOM module periodically updates or sends MOM events if it is requested to do so. All MOM events are delivered via object management module. Therefore, the direct access to network manager is not required. If there is no active MOM module, MOM services are not available to all federates in the federation and MOM events from other federates are treated as ordinary objects or interactions.

4   Open RTI Protocol

This section introduces the proposed Open RTI Protocol that includes message formats, sequences and their handling algorithms between federates and its associated FedEx, and between a parent and a child FedEx processes.

A fixed size and format message header, shown in Table 1, precedes each message. A message content is followed by the header, and is depends on the message type. The version field indicates the version of the protocol. Sending federate handle and Receiving federate handle represent federate handle of sender and receiver, respectively. A federation handle and a federate handle forms an unique address for a specific federate.

Module field indicates which module should handle the message (see Table 2). Message Type becomes unique only with Module field. Detailed message types per modules are presented in the following sections. Message Length field means total length of message contents not including the message header.

Table 1. Message Header.

Bits	Field	Value
16	Version	1
16	Sending federation handle	1
16	Sending federate handle	1
16	Receiving federation handle	1
16	Receiving federate handle	0xFF (Broadcast)
8	Module	2 (Federation Management)
8	Message Type	5 (JOIN_FEDERATION_EXECUTION)
16	Message Length	20 (Bytes)

Table 2. Modules.

ID	Module
1	Network
2	Federation Management
3	Declaration Management
4	Object Management
5	Time Management
6	Ownership Management
7	Data Declaration Management

Figure 3 depicts a sample message sequence for object management services.

For simplicity, a federate talks to only its associated FedEx process. The FedEx process communicate with its parent FedEx process or its child federates (or FedEx processes).

To register an object, a federate sends REGISTER_OBJECT_INSTANCE message to its FedEx process. If the object class is published by the FedEx process, the message is forwarded to its parent FedEx process. Topmost FedEx process assigns appropriate object handle and replies the message. Also, Topmost and middle FedEx processes generate DISCOVER_OBJECT_INSTANCE messages to subscribed federates. By this way, all subscribed federates will receive DISCOVER_OBJECT_INSTANCE messages.

A message sequence for update attribute value service is simple. FedEx processes will forward the message according to publication and subscription status.

5   Extension of API

One way to extend HLA specification is to introduce new APIs to the Federate Interface Specification. There has been many efforts to add new functionality to HLA. We have already introduced new join federation execution service to increase reusability of a federate and to eliminate information hiding problems [5]. Real-time extension to HLA includes new semantics on APIs as well as extension of Framework and Rules and Object Model Template (OMT) [12]. However, care must be taken because introducing new functionality to the federate may cause complete re-design of internal structure of RTI.

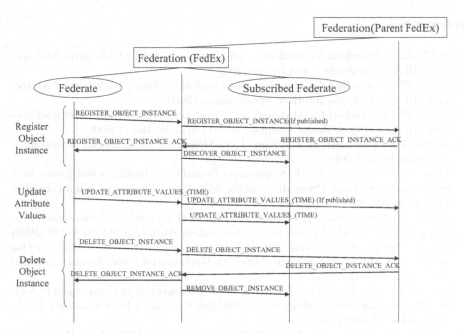

Fig. 3. Message Sequence for Object Management Services between federates and FedEx processes.

6   Conclusion

The paper proposes an extension of High Level Architecture (HLA). The three dimensional extension is proposed to solve several issues. The hierarchical architecture of federations controls flow of information (or events) so that each federation in the hierarchy has different level of security. The modular structure of RTI gives federate developers more flexible designs. Developers are free to replace modules and even unload unnecessary ones for fine performance tuning. The open RTI protocol, together with the fixed architecture, help RTI software from different vendors collaborate each other. A RTI protocol-talking hardware is able to participate a federation, and the hardware-in-the-loop simulation becomes more efficient. The modified join federation execution service is applied to this proposal. We are planning to add other API extensions to give more functionality to federates.

Not all RTI protocol and module interfaces are specified so far. We continue to specify and implement a complete specification of RTI protocol with message types, formats and their handling algorithms, and interface of modules. The full implementation of RTI software with hierarchy and modularity leads us to the base line of research about RTI performance enhancement issues.

References

1. IEEE: IEEE Standard for Modeling and Simulation (M&S) High Level Architecture (HLA) – Framework and Rules. (2000)
2. IEEE: IEEE Standard for Modeling and Simulation (M&S) High Level Architecture (HLA) – Federate Interface Specification. (2000)
3. IEEE: IEEE Standard for Modeling and Simulation (M&S) High Level Architecture (HLA) – Object Model Template (OMT) Specification. (2000)
4. Zeigler, B.P., Praehofer, H. and Kim, T.G.: "Theory of Modeling and Simulation," Academic Press (2000)
5. Kim, J.H. and Kim, T.G.: Federate-Level Reusability: Joining a Federation with SOM Document Data. Proceedings of the 2004 Spring Simulation Interoperability Workshop. 04S-SIW-141 (2004)
6. Myjak, M.D. and Sharp, S.T.: Implementations of Hierarchical Federations. Proceedings of the 1999 Fall Simulation Interoperation Workshop. 99F-SIW-180 (1999)
7. Cai, W., Turner, S.J. and Gan, B.P.: Hierarchical Federations: An Architecture for Information Hiding. Proceedings of the 15th Workshop on Parallel and Distributed Simulation. (2001)
8. Cramp, A., Best, J. and Oudshoorn, M.: Time Management in Hierarchical Federation Communities. Proceedings of the 2002 Fall Simulation Interoperability Workshop. 02F-SIW-031 (2002)
9. Aoyama, K., Ninomiya, S., Takeuchi, Y., Miyajima, S. and Tsutai, A.: Hierarchical Multi-Federation Structure of the Sensor Data Fusion Simulation in JUSA. 98F-SIW-045 (1998)
10. Dingel, J., Garlan, D. and Damon, C.: Bridging the HLA: Problems and Solutions. Proceedings of the 6th IEEE International Workshop on Distributed Simulation and Real-Time Application. (2002)
11. Granowetter, L.: RTI Interoperability Issues – API Standards, Wire Standards, and RTI Bridges. Proceedings of the 2003 European Simulation Interoperability Workshop. 03E-SIW-077 (2003)
12. Zhao, H. and Georganas, N.D.: HLA Real-Time Extension. Proceedings of the Fifth IEEE International Workshop on Distributed Simulation and Real-Time Applications. (2001) 12–21

High Performance Modeling for Distributed Simulation*

Jong Sik Lee

School of Computer Science and Engineering
Inha University, Incheon 402-751, South Korea
jslee@inha.ac.kr

Abstract. This paper presents a modeling of a distributed simulation system with a data management scheme. The scheme focuses on a distributed simulation concept, which is load balancing, suggests distribution of a different functionality to each distributed component, and assigns various degrees of communication and computation loads in each component. In addition, this paper introduces a design with an inter-federation communication on HLA-compliant distributed simulation. The design focuses on integration among multiple federations and constructs the larger distributed simulation system by suggesting a HLA bridge connection. The integration supports system simulation flexibility and scalability. This paper discusses design issues of a practical system with a HLA bridge for inter-federation communication. This paper analyzes and evaluates performance and scalability of the data management scheme with load balancing on distributed simulation, especially with inter-federation and inside-federation communication configurations. The analytical and empirical results on a heterogeneous OS distributed simulation show improvement of system performance and scalability by using data management and inter-federation communication.

1 Introduction

There is a rapidly growing demand of distributed simulation which includes a variety of system simulations such as process control and manufacturing, military command and control, transportation management, and so on. Most of distributed simulations are complex and large in their size. In order to execute those complex and large-scale distributed simulations within reasonable communication and computing resources, a development of a large-scale distributed modeling and simulation environment has drawn attention of many distributed simulation researchers. A large-scale distributed simulation requires an achievement of real-time linkage among multiple and geographically distant simulation components, and thus has to execute a complex large-scale execution and to share geographically dispersed data assets and simulation resources collaboratively. This paper proposes a data management scheme with load balancing that supports reduction of interactive messages among distributed simulation components and communication flexibility for modeling and simulation of a complex system. The proposed scheme focuses on a load balancing which indicates communication load distribution to each distributed simulation component. The

---

* This work was supported (in part) by the Ministry of Information & Communications, Korea, under the Information Technology Research Center (ITRC) Support Program.

T.G. Kim (Ed.): AIS 2004, LNAI 3397, pp. 138–146, 2005.

scheme is extended from communication data management schemes [1], [2], [3], [4], [5], [6] to model and simulate complex and large-scale distributed systems with reasonable communication resources. This paper applies the data management scheme to a satellite cluster management [7], [8]. The scheme improves simulation performance and scalability of a satellite cluster management through communication data reduction and computation synchronization. In addition, this paper suggests a HLA [9], [10]-compliant distributed simulation environment that allows HLA bridge-based inter-federation communications [11], [12], [13], [14], [15] among multiple federations and improves flexibility in modeling and simulation. This paper provides a design of HLA bridge federate to connect multiple federations. The design allows to execute a complex and large-scale distributed simulation, gives a promise of simulation flexibility and scalability, and creates useful simulation-based empirical data. This paper is organized as follows: Section 2 introduces the data management scheme with load balancing and presents how to apply the scheme to a satellite cluster management. Section 3 discusses a modeling of satellite cluster management with a HLA bridge. Section 4 discusses performance analysis of the data management scheme with load balancing. Section 5 illustrates a testbed for experiment and discusses performance evaluation of the data management scheme with load balancing and the inter-federation communication with a HLA bridge. The conclusion is in Section 6.

2 Data Management with Load Balancing

Data management improves simulation performance of a complex and large-scale distributed simulation since data management reduces data transmission among distributed simulation components. This paper proposes a data management scheme with load balancing which reduces transmission-required data by assigning different communication and computation load to each transmission-related component. This scheme reduces simulation cost by reducing communication resources among simulation components with separated communication loads. In addition, it reduces local computation load of each distributed simulation component. The scheme improves simulation flexibility and performance through communication data reduction, computation synchronization, and scalability for a large-scale distributed simulation. This paper introduces a satellite cluster management [7, 8] as a case study and applies to the data management scheme with load balancing to it.

Satellite Cluster Management
Construction and execution of autonomous constellations system follow distributed system construction concepts: 1) functionality balancing in multiple distributed satellites;2) increasing system robustness a nd maintainability;3) reduction of communication and computation resources. Distributed satellite functionality includes command and control, communications, and payload functions. For effective execution of constellations system, a cluster paradigm with a central cluster manager is modeled and simulated in this paper. A central manager controls functionality of each satellite inside cluster and communication among satellites. Separated satellites in a cluster occupy their distributed space assets in a constellations system. A cluster management is essential to progress a satellite cluster mission with cluster functionalities such as resource management, navigation, guidance, fault protection, and so on. Sat-

ellite is also called spacecraft in constellations system. While a centralized management approach is defined, a cluster manager provides the cluster functionalities. The functionalities consist of four categories: spacecraft command and control, cluster data management, flying formation and fault management. For a cluster management, a cluster manager should keep information of each spacecraft including position, velocity, attitude quaternion, system time, spacecraft mode, fuel level, sensor states, and so on.

This paper introduces a ground system operation as a case study to discuss non-data management and data management with load balancing and evaluate system performance. A ground system commands and controls a cluster of spacecrafts. Basically, a ground system requires operations and manpower to monitor a cluster, makes a decision, and sends proper command strings. For a small cluster, a centralized approach is cost effective and accepted to command and control spacecrafts individually. As Fig. 1 (a) illustrates, a ground system sends its command strings to each spacecraft. The command strings include commands to "observe a specified region, take a picture, and send image data of a picture." The command should contain a region location. Each spacecraft receives different region location from a ground station.

To improve total system performance by reducing transmission- required data, a data management scheme with a load balancing of ground operations is proposed in this paper. The scheme indicates that it separates ground functions and distributes a set of functions to spacecrafts. Fig. 1 (b) illustrates a data management with load balancing. A ground station separates four regions to be observed, makes four different command strings, and sends them to a cluster manager. The cluster manager parses the command strings and forwards them to each proper spacecraft. The parsing and forwarding assigns light loads in the cluster manager. The cluster manager gets the lighter loads, while the heavier communication data are required between the cluster manager and the ground station. Here, this paper classifies a degree of load balancing: low and high. There is the higher load balancing with the cluster manager over parsing and forwarding. The ground station does not separate four regions to be observed and sends a total region to the cluster manager. The cluster manager should include the load for division of region. The cluster manager with the division load should understand technologies including region division, image capturing, image visualization, image data transmission, and so on. The cluster manager includes the heavier loads, while the lighter communication data are required between the cluster manager and the ground station.

3   Modeling of Satellite Cluster Management with HLA Bridge

In this section, this paper discusses an inter-federation communication architecture on a multi-federation platform and applies an inter-federation communication to a data management of a satellite cluster system. With a viewpoint of an extension to more practical satellite cluster system, this paper proposes an inter-federation communication architecture. Eventually, a satellite cluster management system can be separated as two divisions: space cluster and ground station. The separation indicates a division for geography, functionality, and different communication groups. Actually, a ground station has various connections to organizations on the earth, thus it is a part of a

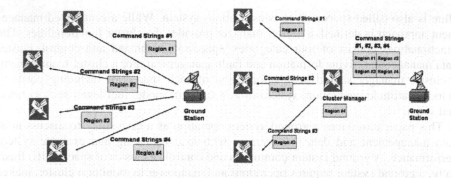

(a) Non-Data Management   **(b) Data Management with Load Balancing**

Fig. 1. Comparison between Non-Data Management and Data Management

different communication group. This paper applies two divisions (e.g. space cluster and ground station) to an advanced HLA architecture, which is an inter-federation communication architecture. To execute an inter-federation communication, this paper uses a bridge federate which is physically located in each federation and plays a role of RTI message passing between federations. As Fig. 2 illustrates, this paper develops two federations: cluster and ground. A cluster federation includes multiple federates. Each federate is assign in each spacecraft. A ground federation includes two federates: cluster manager and ground station. Both federations include a cluster manager federate which is assigned in a bridge federate for an inter-federation communication. Notice that cluster manager federates in both federations have different functionalities, respectively. A cluster manager federate in a cluster federation works cluster management and RTI message passing inside/outside federation. A cluster manager federate in a ground federation only concentrates communication to a cluster federation.

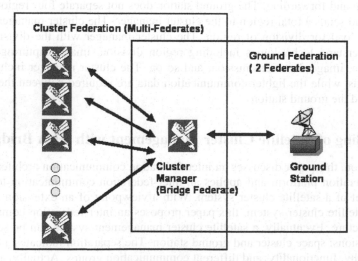

Fig. 2. Inter-Federation Communication

4 Performance Analysis

To analyze performance of the data management with load balancing, this paper takes an amount of transmission-required data, which are communicated between ground station and spacecrafts. Notice that transmission-required data among spacecrafts inside a cluster are ignored. In this analysis, this paper assumes five conditions: 1) there exist multiple clusters;2) a cluster includes a finite number of spacecrafts (N); 3) a region is square-shaped and has 4 points $((x_1, y_1), (x_2, y_2), (x_3, y_3), (x_4, y_4))$;4) two of 32 double precision bits (e.g. 64 bits) are needed to represent a point (x_1, y_1); 5) this analysis is based on one cycle transmission.

As Table 1 shows, the data management with load balancing significantly reduces the number of messages passed and the number of bits passed. Basically, there occurs overhead bits (H) needed for satellite communication when a ground station sends a command. The centralized approach without data management causes an amount of overhead messages and bits since it makes a ground station send each spacecraft messages individually. Compared with a degree of load balancing, a high load balancing significantly reduces transmission-required data bits since it transmits a total region information irrelevant to the number of spacecrafts (N) in a cluster. Specially, as the number of spacecrafts (N) goes infinity, transmission-required data bits in a low load balancing increases linearly. The increasing slope is (4*64)*M. However, a high load balancing still requires the same lower transmission-required data bits. The analysis in Table 1 reveals that, especially the large numbers of spacecrafts working in a cluster, the greatest transmission-required data reduction is expected with a high load balancing of data management.

Table 1. Analysis of Transmission Data Reduction

Approach		Transmission Required Messages	Transmission Required Bits	Coefficient of N as N-> ∞
Non-Data Management		R*N*M	(H4*64)*R*N*M	(H4*64)*R*M
Data Management (Degree of Load Balancing)	Low	M	(H4*64*R*N)*M	(4*64*R)*M
	High	M	(H +4*64) *M	None

(Note: N: Number of spacecrafts in a Cluster;M: Number of Clusters;H: Number of overhead bits in satellite communication (160 bits assumed), R: Number of regions at one spacecraft on one transmission (40 assumed))

5 Experiment and Performance Evaluation

5.1 Testbed

This paper defines a scenario of satellite cluster management and evaluates system performance of the data management with load balancing. A cluster of 4 spacecrafts flies on pre-scheduled orbits. One of spacecrafts acts as a cluster manager that communicates with a ground station. A cluster manager gathers states of each spacecraft and sends telemetry information back to a ground station. At any given time, a ground station can send an observation request to a cluster manager. In turns, a cluster manager will coordinate with other spacecrafts in a cluster to perform the requested observation in synchronization. A cluster manager then aggregates data collected from the other spacecrafts in a cluster and send them back to a ground station. There are three assumptions: 1) A cluster manager always communicates with a ground station without interruption;2) A position representation of each spacecraft is relative to a reference circular orbit;3) A spacecraft flies with height of 600km on a reference circular orbit. This yields a period of 5810 sec.

To execute the scenario, this paper develops a testbed with an inter-federation communication. For an inter-federation communication with a HLA bridge, this paper develops two federations: cluster and ground. As Fig. 3 illustrates, a cluster federation includes four spacecraft federates, including a cluster manager, and a ground federation includes two federates: cluster manager and ground station. Both federations have a cluster manager federate which is called bridge federate. A HLA bridge implementation supports a bridge federate functionality for an inter-federation RTI message passing, thus it makes an inter-federation communication executable. In a platform setting of the testbed, this paper develops a heterogeneous distributed simulation system which includes various operating systems including SGI Unix, Linux, Sun Unix, and Windows. Five federates are allocated to five machines, respectively, and they are connected via a 10 Base T Ethernet network.

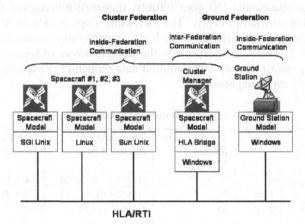

Fig. 3. Simulation Testbed of Inter-Federation Communication

5.2 Performance Evaluation

In order to evaluate system execution performance of the data management with load balancing, this paper compares system execution time between non-data management and data management with load balancing. A comparison is achieved with variation of number of satellites. A system execution time considers communication and computation reduction. A non-data management requires a large amount of communication data, however it does not need local computation. A system execution time for a non-data management is mostly caused from an amount of communication data. The data management with load balancing reduces an amount of communication data and uses local operations for load balancing. A system execution time for load balancing is caused by both of data communication time and load operation time. Especially, a high load balancing requires the more load operation time than that for low load balancing. Fig. 4 compares system execution time in two cases: non-data management and data management with load balancing. A degree of load balancing is separated with low and high. The system execution time of Fig. 4 is provided from an execution on only one federation with inside-federation communication. The data management with load balancing apparently reduces system execution time. The reduction indicates that execution time reduction from transmission-required data reduction is greater than time expense from load operation. In comparison between high and low load balancing, there exists a tradeoff between transmission-required data reduction and degree of load balancing. In inside-federation communication system of Fig. 4, the low load balancing shows the lower execution time in the lower task load. The smaller number of satellites presents the lower task load. As the task load increases, the high load balancing shows the lower execution time.

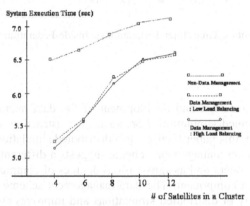

Fig. 4. System Execution Time on Inside-Federation Communication (Non-Data Management vs. Data Management with Load Balancing (Low and High))

Inter-federation vs. Inside-Federation Communication

To evaluate system execution performance of an inter-federation communication system, this paper compares system execution time of an inter-federation communication with that of an inside-federation communication with only one federation. An inter-federation communication system is operated with a bridge federate between two federations: cluster and ground. Basically, we understand that an inter-federation communication system reduces its local computation time. An inter-federation communication system includes multiple federations, thus it separates its tasks and assigns sub-tasks in each federation. However, an inter-federation communication system increases its communication time since an inter-federation message passing time would be greater than an inside-federation message passing time in a federation. Meanwhile, we can expect that an inside-federation communication system performs the higher local computation time and the lower communication time. Fig. 5 compares system execution time of two communications: inside-federation and inter-federation. Fig. 5 measures system execution time in case of the data management of high load balancing. The execution time of inter-federation communication system is lower in all the tasks, but not much. As higher is task, the execution times of the two communication systems are closed.

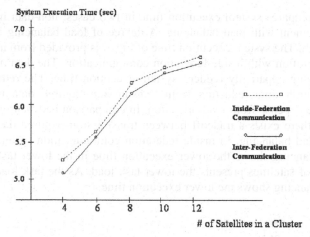

Fig. 5. System Execution Time (Inter-Federation vs. Inside-Federation Communication)

6 Conclusion

This paper presents a design and development of the data management with load balancing in a distributed simulation. For practical construction and execution of a distributed simulation, this paper focuses on a distributed simulation concept which is load balancing. The data management scheme suggests a different load balancing to each distributed component and assigns various degrees of communication and computation loads in each component. The data management scheme allows a complex execution for a variety of distributed simulations and improves system performance through reduction of communication data and local computation load. This paper analyzes system performance and scalability of the data management with load balancing. The empirical results show favorable reduction of communication data and overall execution time and prove usefulness in a distributed simulation.

This paper considers distributed simulation concepts, including data communication management, HLA-compliant inter-federation communication, functionality balancing, simulation flexibility, and scalability and provides a high performance modeling for a satellite cluster management paradigm with a cluster manager. Especially, a HLA bridge-based inter-federation communication for a satellite cluster management in a HLA-compliant distributed simulation improves system modeling flexibility and scalability by allowing multiple connections not only among satellites inside a cluster federation but also among multiple cluster federations. The simulation flexibility through a HLA bridge-based inter-federation communication allows to represent and simulate topologies of a variety of autonomous constellations systems, to analyze a complex large-scale space mission system, and to provide analytical and empirical results. The results show inter-federation communication on a HLA-compliant distributed simulation would be useful while flexibility and scalability of system modeling and simulation are focused.

References

1. Katherine L. Morse: Interest management in large scale distributed simulations. Tech. Rep., Department of Information and Computer Science, University of California Irvine (1996) 96-127
2. Katherine L. Morse, Lubomir Bic, Michael Dillencourt, and Kevin Tsai: Multicast grouping for dynamic data distribution management in Proceeding of the 31st Society and Computer Simulation Conference SCSC"99 (1999)
3. J. Saville: Interest Management: Dynamic group multicasting using mobile java policies in Proceedings of the 1997 Fall Simulation Interoperability Workshop, number 97F-SIW-020. (1997)
4. Boukerche and A. Roy: A Dynamic Grid-Based Multicast Algorithm for Data Distribution Management. 4th IEEE Distributed Simulation and Real Time Application (2000)
5. Gary Tan et. al.: A Hybrid Approach to Data Distribution Management. 4th IEEE Distributed Simulation and Real Time Application (2000)
6. Bernard P. Zeigler, Hyup J. Cho, Jeong G. Kim, Hessam Sarjoughian, and Jong S. Lee: Qøntization based filtering in distributed simu lation: experiments and analysis in Journal of Parallel and Distributed Computing, Vol. 62, number 11 (2002) 1629-1647
7. D.M. Surka, M.C. Brito, and C.G. Harvey: Development of the Real-Time Object-Agent Flight Software Architecture for Distributed Satellite Systems. IEEE Aerospace Conf., IEEE Press, Piscataway N.J. (2001)
8. P. Zetocha: Intelligent Agent Architecture for Onboard Executive Satellite Control. Intelligent Automation and Control, vol. 9. TSI Press Series on Intelligent Automation and Soft Computing, Albuquerque N.M. (2000) 27-32
9. Draft Standard for Modeling and Simulation (M&S) High Level Architecture (HLA) - Federate Interface Specification, Draft 1. DMSO (1998)
10. High Level Architecture Run-Time Infrastructure Programmer's Guide 1.3 Version 3, DMSO (1998)
11. M. Myjak, R. Carter, D. Wood, and M. Petty.: A taxonomy of multiple federation executions in 20th InterserviceIndustry Training Systems and Education Conference (1998) 179-189
12. T. Lake.: Time Management over Inter-federation Bridges in Simulation Interoperability Workshop (1998)
13. J. Dingel, D. Garlan, and C. Damon: A feasibility study of the HLA bridge. Technical Report CMU-CS-01-103, Department of Computer Science, Carnegie Mellon University (2000)
14. J. Dingel, D. Garlan, and C. Damon: Bridging the HLA: Problems and Solutions in Sixth IEEE International Workshop on Distributed Simulation and Real Time Applications DS-RT '02 (2002)
15. J. Filsinger: HLA Secure Combined Federation Architecture (Part One). Technical Report Trusted Information Systems (1996)

The Hierarchical Federation Architecture
for the Interoperability of ROK and US Simulations

Seung-Lyeol Cha[1], Thomas W. Green[1], Chong-Ho Lee[1], and Cheong Youn[2]

[1] The Combined Battle Simulation Center, ROK-US Combined Forces Command,
Po-Box 181, YongSan-Dong 3-Ga, YongSan-Gu, Seoul, 140-701, Rep. of Korea
chasl@passmail.to, GreenT@usfk.korea.army.mil,
lchongho@lycos.co.kr
[2] Department of Information and Communication, Chungnam National University
220, Gung-Dong YungSung-Gu, DaeJeon, 305-764, Rep. of Korea
cyoun@cs.cnu.ac.kr

Abstract. This paper presents the hierarchical federation architecture as it applies to the ROK-US combined exercises, such as Ulchi Focus Lens (UFL). We have analyzed and extracted the necessary improvements through the review of the simulation architecture currently used for ROK-US combined exercises and from the current ROK Armed Forces modeling and simulation (M&S) utilization. We have designed an advanced federation architecture based on a multi-federation architecture. Moreover, we have validated the usability and technical risks of our proposed architecture through development of a pilot system and its testing. Finally, we expect that this architecture will provide an enhancement in the ROK-US combined exercises while reducing costs. Furthermore, we believe that this architecture is an example to interoperate simulations with other allies.

1 Introduction

The defense of Korea is the central mission of the ROK-US Combined Forces Command (CFC). This overriding mission requires the conduct of combined exercises that train and maintain the combat readiness of ROK and US forces. This training increasingly relies on the use of modeling and simulation (M&S) applications that provide great realism to the highly digitized battlefields of today. Many of those exercises are conducted in a joint and combined context at the ROK-US Combined Battle Simulation Center (CBSC) in Seoul, Korea. Stemming from its premier position in developing military applications for M&S, the simulation support for ROK-US combined exercises are currently led by the US and the costs are shared by the ROK [1].

US models such as the Corps Battle Simulation (CBS)(ground warfare), Research Evaluation and Systems Analysis (RESA)(naval warfare), and Air Warfare Simulation (AWSIM)(air warfare) were introduced to the ROK military services in the early 1990s. This introduction provided simulation support to each service. Further, in the late 1990s, that introduction formed the basis for the ROK Army's indigenous development of its ChangJo-21 (CJ21), a ground forces model. The development of CJ21 as an original ROK Army model stimulated an earnest development of other models germane to the other ROK military services. To achieve its missions with better fidel-

T.G. Kim (Ed.): AIS 2004, LNAI 3397, pp. 147–156, 2005.
© Springer-Verlag Berlin Heidelberg 2005

ity, the CBSC is taking steps to ensure interoperability of ROK and US models in an advanced simulation architecture for combined exercises.

Until recently, the interface protocol between each model of the Joint Training Confederation (JTC), which has been applied to combined exercises, had been Aggregate Level Simulation Protocol (ALSP). It has now been replaced with a new open system called High Level Architecture /R un-Time Infrastructure (HLARTI). This step was implemented first during a joint/combined exercise; Reception, Staging, Onward movement and IntegrationFoal Eagle 2004 (RSOIFE 04). The indigenous development of ROK military models as HLA compliant means the ROK models will be able to confederate with the US models and execute a seamless interface for combined exercises.

To execute this improvement, the ROK military established a master plan [2] that incorporates the development of its models in accordance with HLA concepts. The CBSC is also developing a model confederation system to expand the participation of ROK models during combined exercises. Ultimately, these steps will lead to the development of the Korean Simulation System (KSIMS), which will include each service model applicable to the ROK-US combined exercises.

To develop and apply a new federation that interoperates ROK and US models, there are many factors to be considered. Two such factors would be; a combined development team based on a ROK and US mutual agreement, and the confederations between similar models that would simulate the same battlefield medium. The design of a confederation architecture for efficient ROK and US models interoperability is a major issue.

This paper will present an advanced simulation architecture for ROK-US combined exercises. Chapter 2 will review current simulation architecture for ROK-US combined exercises. Chapter 3 will describe the existing research results of a multi-federation. Chapter 4 will present the design of hierarchical federation that will be applied for future combined exercise. Finally, chapter 5 will review the results of prototyping.

2 Simulation Architecture for ROK-US Combined Exercises

ROK Armed Forces have been extending its application of M&S continuously, since it started simulation support in the early 1990s by bringing US models to each service's training. But in the late 1990s, to apply the future battle field environment, the new weapon systems and the tactics, techniques and procedures of ROK Armed Forces, it was realized that the ROK required the development and application of its own models. An improvement of the simulation architecture is also required to provide the necessary ROK and US models interoperability. Through interoperability, we will be able to provide a suite of combined models which will further enhance training for the defense of the peninsula.

Currently, the simulation architecture for ROK-US combined exercises, for which the US is the leading provider, has a complex structure applied by various confederation protocols. As shown in Figure 1, the simulation components are sorted out as the JTC main models, special models, interfaces, and C4ISR systems. The models and C4ISR systems are linked to each other by interfaces such as Gamers (human interface), Master Interface (MI), Tactical Simulation (TACSIM) ALSP Translator (TAT),

Run Time Manager (RTM), and Virtual Downlink (VD) Network. These are defined and developed according to the individual characteristic of each system. The confederation between the JTC main models is supported by HLA based JTC Infrastructure Software (JIS) with a single federation architecture [3].

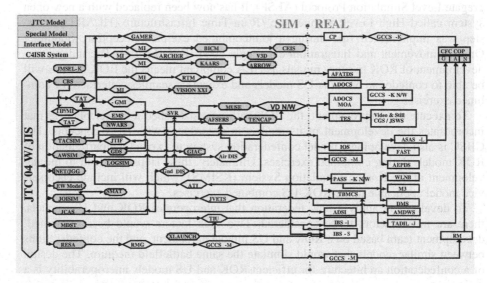

Fig. 1. Simulation Architecture for 04 UFL Exercise

For the ROK Armed Forces war-game models to be participants in the combined exercise simulation architecture, the single federation needs to be expanded to allow for other federates to join. One method of expansion would be to create a single Federation Object Model (FOM) sharing both ROK and US federates. Although the simplest way to approach, considering the currently used RTI functionality, this could cause many issues such as degradation in system performance, security classification, and workload to develop and operate such a large single federation.

The system performance impact could be created through an excessive amount of system tuning as too many federates join only one federation. Aggravating this impact is in the RTI NG version, where the stability is only guaranteed when all federates is organized in the same Local Area Network (LAN) [5]. Unfortunately, there is no solution suggested and it's difficult to find an alternative for solving this issue when the combined exercises require extensive distribution.

Security could be considered compromised with all the data confederated in a single FOM. Therefore, additional considerations on individual security mechanisms have to be applied in order to protect the information that one country doesn't want exposed to other countries [12] [13].

The workload necessary in a combined ROK-US team would be herculean. In order to define confederation data, both countries would have to understand a certain amount of the other side's design concepts as it relates to differences in each military organization, weapon system characteristics, and operational doctrines. The personnel who actually execute the combined development are separated by long distances and by discrepancies in languages, systems, policies, regulations and even cultural environments. Greater effort will be required to develop one federation rather than each country developing a separate federation.

A multi-federation architecture that can be interoperated between ROK and US federations must be eventually considered in the future.

3 Multi-federation Architecture

The HLA concept is expressed as, a federation made up of a set of federates, RTI services and a FOM. Original developers of HLA considered that a single federation would be adequate for all models and therefore only expected to have a single FOM and RTI.

There are many limitations, mentioned above, in using a single federation architecture for integration of existing federations. It is insufficient to accomplish the interoperability and reusability, which are HLA's main objectives, in the reorganization of existing models and simulations to meet new requirements.

To overcome these limitations, we should consider an advanced federation architecture that is interoperable with multiple federations. By doing so, will enhance the interoperability and reusability of existing federations.

3.1 HLA Based Multi-federation

According to general HLA definition, it seems that there are no interoperations among federations during federation executions. But in the interpretation of basic HLA rules, there is a factor that makes the interoperation among federations possible. That is, the interoperation would be possible if a special federate which has a data sharing mechanism, simultaneously joins in several federations. The level of mechanism that supports the data sharing function among joined federations will determine the capability of interoperability.

Two or more interoperating federations, with a special federate simultaneously joined in them, would be defined as a multi-federation. The intra-federation communication uses a designated unique FOM for each federation. While the inter-federation communication applies another mechanism such as a super FOM extracted from each federation's FOM, and it interoperates as one federation's events effect the other federations.

3.2 Multi-federation Types

Combinations of FOM and RTI types, based on integration schemes of existing federations, can be sorted into 4 kinds: homogeneous FOM and RTI, homogeneous FOM and heterogeneous RTI, heterogeneous FOM and homogeneous RTI, heterogeneous FOM and RTI [10]. Primary consideration will be given to the heterogeneous FOM and RTI scheme that can apply in general application environments.

The gateway, proxy federate, RTI broker, and RTI protocol are the four kinds of connections between federations that will also make it possible to construct a multi-federation [8]. The details of each connection type follows.

Gateway is an internal connection device that transmits information between federations through another communication channel, instead of a HLA connection. It supports transmitting the information offered, within the range of the joined federate transformation ability, to the other joined federates. This process provides an interoperation between two individual HLA based federates or a HLA based federate and an external system which is not a HLA based federate.

A proxy federate is a translation device that interconnects two (or more) federations using the Federate Interface specification, unlike the gateway. The HLA rules state a specific federation can only have a single FOM and permits information transmission through a RTI. It neither allows nor forbids connecting 2 federations. A proxy federate uses only the defined service within API articles supported by the RTI, but it is more complex than a gateway because it has to apply numerous numbers of federate ambassador's and FOM's.

The RTI broker is a translation device that uses the RTI-to-RTI API to pass not only federate level data but to also communicate RTI internal state information between two (or more) RTIs. Potentially, it could provide a communication medium from different vendors or conform to different communication architectures.

Lastly, RTI protocol is capable of transporting federate information and RTI internal state between RTI implementations. This could also be done between local RTI components in a manner independent of a RTI vendor.

Gateway and proxy federate can be applied to an advanced simulation architecture design for ROK-US combined exercises, so called KSIMS-JTC confederation which would include a special purpose interface with C4ISR systems, and a Distributed Observation Tool (DOT) that executes the remote monitoring of the whole system.

4   An Advanced Simulation Architecture Design

The architecture design requirements necessary to enhance ROK-US simulation interoperability, to be supported in future ROK-US combined exercises, are as follows [4]. First, the interoperability between ROK and US models, especially from an operational perspective, should accomplish the combined exercise objectives. Secondly, the system development and maintenance should be easy. Thirdly, the resource reusability should be maximized through minimizing the modification of existing systems. Fourthly, each country should be able to apply their specific security regulations. Fifthly, the impact of specific model's constraints to the operation of the whole system should be minimized.

Even though there are many issues to be considered in simulation architecture design to fulfill these requirements, this research is focused on the simulation interoperability and reusability which are the HLA objectives, and obeyed current HLA concept and RTI functionality.

4.1   Integration of KSIMS and JTC

Considering the limitations of current RTI functionality, there are 4 possible methods (Figure 2) that should be considered to interoperate the ROK federation, KSIMS and US federation, JTC. To link multiple federations, a new component, Confederation

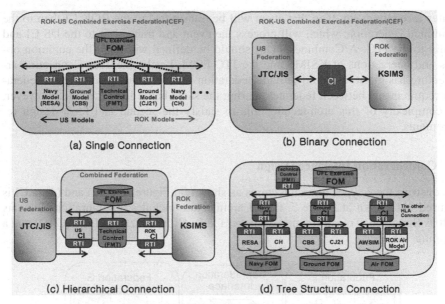

Fig. 2. KSIMS-JTC Integration Schemes

Interface (CI) is introduced. The CI will function like a proxy federate, facilitating the joining of multiple federations [11] [14].

Figure 2(a) is a Single Connection architecture, where all federates are joined to a single federation, called a Combined Exercise Federation (CEF). Figure 2(b) is a Binary Connection architecture directly connecting KSIMS and JTC through a CI. Figure 2(c) is a Hierarchical Connection architecture indirectly connecting KSIMS and JTC through a new Combined Federation and two CIs. Figure 2(d) is a Tree Connection architecture that integrates the federations, after constructing each federation that consists of federates sharing the same battle field medium.

The peculiarity of each architecture is as follows. First, the Single Connection architecture is based on the JTC currently used in simulation support for ROK-US combined exercises. In this architecture, the JTC would be expanded with the inclusion of the ROK models. However, this architecture will not allow autonomous development and would also limit what specific KSIMS applications could be used. ROK models would effectively have to be added in stages. Although this architecture is simple in design, it does not satisfy the second, forth and fifth design requirements. Secondly, the Binary Connection architecture applies an interface, like a gateway or proxy federate that directly connects the two federations. There is merit in a simple architecture to interoperate KSIMS and the JTC. However, a concentration phenomenon will exist with a single CI enduring the extreme load from each federation. Thirdly, the Hierarchical Connection architecture is an architecture that applies an additional upper federation, a Combined Federation with a Combined FOM. Using this upper federation, the architecture will be able to support the confederation requirements through the maximum utilization of current HLA standard and the functionality of the RTI. However, this architecture could induce internal delay with three steps being required in each federation's event transmission. For example, if KSIMS creates a certain event, then that event will be transmitted to the ROK CI then to the Combined Federation, which will process the event and transmit it to the US CI and finally to the JTC. A Combined FOM should be defined with only the common objects and interactions of KSIMS and the JTC FOM. Lastly, the Tree Connection architecture has merit, it would provide federation performance by settling confederation requirements between federates sharing the same battle field medium. However, the complexity of this architecture fails to satisfy the second and forth design requirements.

4.2 Confederation Interface Design

To construct the 3 types of architectures as shown on Figure 2(b), (c) and (d), a CI is required. As shown at Figure 3, the CI architecture includes configuration elements such as Surrogate, Transformation Manager (TM), CI Initialization Data (CIID) and a Mapping Table.

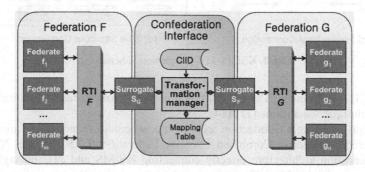

Fig. 3. Confederation Interface Structure

The CI conducts the role of connecting Federation F and G as a bridge federate. Surrogate S_G has surrogating functions to join federation F, collecting federation F's information, interfacing between federation F and the TM, and transferring federation G's information to federation F. The TM functions to transfer data between Surrogate S_G and S_F, and conducts data conversion. Surrogate S_F's functions are same with that of Surrogate S_G in reverse.

The CI also synchronizes the simulation time between two federations and may also play a role as security guard to prevent unnecessary or classified information transmission. CIID contains the initialized information of the CI operation. The Mapping Table saves and refers the object handle value, interaction handle value and data conversion information necessary to inter-map between the two federations.

4.3 Hierarchical Federation Scheme

In order to develop a ROK Armed Forces federation that reflects the various future operational requirements, and to ensure the best use of ROK simulation resources for ROK-US combined exercises, consideration is given to the Hierarchical Connection architecture as shown at Figure 2(c);development is currently in progress [4].

A new federation using Hierarchical Connection architecture is defined as the Hierarchical Federation. This scheme can support each countries use of its own federation with its FOM in combined exercises, which would minimize additional modification of existing federations. CI would support each countries application of its own security regulations, by not fully exposing its FOM to the other participants. The CI would also be capable of relaying information even though each federation uses a different kind of vendor and version of RTI. Compared with a single federation, the whole federation's execution efficiency would be maximized since the number of federates to be controlled by one RTI decreases.

One consideration that must be taken into account is the concentration phenomena at each of the CI's. Each event transmission would have to pass through each CI and the upper federation. There may also be some technical risk in the development of the CI, especially for a heterogeneous environment.

5 Prototyping and Testing

The CBSC developed a pilot system of the hierarchical federation architecture using CI as shown in Figure 4.

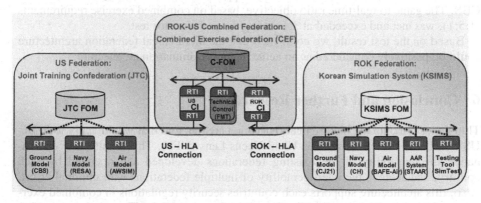

Fig. 4. System configuration of a pilot system

The pilot system consisted of three federations;the Joint Training Confederation (JTC) for US Federation, the Korean Simulation System (KSIMS) for ROK Federation and the Combined Exercise Federation (CEF) for ROK-US Combined Federation.

The development environment was embodied with C+and JAVA language on Windows 2000 and Linux RH 7.3. RTI-1.3NG v6.4.3, currently used in ROK-US combined exercises, was applied. CJ21 (Ground model), Chung-Hae (Naval model), SAFE-Air (Aerial model), SimTest (events generator) and STAAR (System for Theater level After Action Review) participated in KSIMS as federates. CBS, RESA and AWSIM participated in the JTC, and FMT (Federation Management Tool) and KFMT (Korean Federation Management Tool) participated in the upper CEF, which was confederated through CIs.

The CI simultaneously joins two different federations and then executes the Transformation Manager (TM) function, which undertakes the RTI service relays and necessary data conversion. The Federation, Declaration, Object and Time Management services [7] are relayed, while the relay functions of Ownership and Data Distribution Management services are not.

The hierarchical federation successfully executed in several stages such as in initialization, federation joining, time control policy establishment, Mapping Table register, and declaration management. It then transmitted events between KSIMS and JTC through CEF until destroyed by the ending order.

The actual test was divided into an integration test, functional test, and performance test which was conducted at the ROK-US Confederation Test [5]. Several different size log files in various testing environments were used in the execution of these tests. The maximum size of a log file was approximately 680,000 messages with 12,000 object instances for 30 hours of simulation time.

In the integration test, KSIMS, JTC and CEF were successfully integrated into a hierarchical federation using a ROK CI and US CI. In the functional test, the combined operational functions such as interoperation among air, ground and sea, were properly confederated in a whole federation. However, ground to ground confederation such as CJ21 to CBS was not fully supported due to the current limitations of CBS. The game to real-time ratio objective, based on combined exercise requirements (1.5:1), was met and exceeded at 3:1 during the performance test.

Based on the test result, we confirmed that the hierarchical federation architecture could be positively considered as an actual exercise simulation architecture.

6   Conclusion and Further Research

This paper presents the hierarchical federation architecture that would apply to ROK-US combined exercises, such as Ulchi Focus Lens (UFL). This architecture can enhance the reusability of the existing federations developed and operated by each country, and achieve the interoperability of multiple federations with minimal effort. Also, this architecture supports each countries security regulations in combined exercises without modifying their federation. The impact of specific model's constraints to the whole system operation is also minimized.

This study's goal is to provide an improved simulation architecture for ROK-US combined exercises that would enhance the training for the defense of Korea. The architecture would also provide a great level of enhancement in the exercises while reducing costs. This same architecture would be applied to simulation interoperability with other allies as a useful precedent.

For this architecture to apply to a real domain there are several issues to be reviewed, including enhancement of quality factors such as completeness, robustness, availability, and maintainability. Further research must be conducted in the following areas. First, pertaining to completeness, all RTI services based on HLA Interface Specification should be covered in the CI relay functions. Secondly, pertaining to robustness, solutions concerning the bottleneck phenomena of the CI should be considered in the design of the CI structure and upper federation's FOM. Thirdly, focusing on availability, disparate RTI execution should be guaranteed to support operating environments with a Wide Area Network (WAN) around the world. Lastly, focusing on maintainability, the Federation Development and Execution Process (FEDEP) should be advanced to respond to new requirements on multi federation environments.

References

1. Republic Of Korea (ROK) Joint Chief of Staff (JCS): Memorandum of Understanding (MOU) between United States Forces Korea (USFK) and ROK JCS concerning ROK-US Combined Forces Command (CFC) Major Combined Exercises (1998)
2. ROK Ministry of National Defense: Master Plan for War-game Training System (2003)
3. ROK-US Combined Forces Command: UFL 04 Simulation Control Plan (2004)
4. ROK-US Combined Forces Command (CFC) Combined Battle Simulation Center (CBSC): Korean Simulation System (KSIMS) System/Subsystem Specification (SSS) & Operational Concept Description (OCD) (2002)
5. ROK-US Combined Forces Command (CFC) Combined Battle Simulation Center (CBSC): 3rd and 4th ROK-US Confederation Test After Action Review Reports, (2003)
6. IEEE Standard for Modeling and Simulation (M&S), "High Level Architecture (HLA) Federate Interface Specification", IEEE Std 1516.1 (2000)
7. DoD Defense Modeling and Simulation Office: High Level Architecture Run-Time Infrastructure: RTI-1.3 Next Generation Programmer's Guide, Version 3.2 (2000)
8. Michael D. Myjak: Implementations of Hierarchical Federations, Fall Simulation Interoperability Workshop, SISO paper 99F-SIW-180 (1999)
9. Frederick Kuhl, Richard Weatherly, Judith Dahmann: "Creating Computer Simulation Systems", Prentice Hall (1999)
10. Michael D. Myjak, Duncan Clark, Tom Lake: RTI Interoperability Study Group Final Report, Fall Simulation Interoperability Workshop, SISO paper 99F-SIW-001 (1999)
11. Gerry Magee, Graham Shanks: Hierarchical Federations, Spring Simulation Interoperability Workshop, SISO paper 99S-SIW-085 (1999)
12. LouAnna Notargiacomo, Scott Johnston, etc.: The High-Level Architecture Multilevel Secure Guard Project, Fall Simulation Interoperability Workshop, SISO paper 01F-SIW-016 (2001)
13. Benoit Breholee, Pierre Siron: Design and Implementation of a HLA Inter-federation Bridge, Euro Simulation Interoperability Workshop, SISO paper 03E-SIW-054 (2003)
14. Wesley Braudaway, Reed Little: The High Level Architecture's Bridge Federate, Fall Simulation Interoperability Workshop, SISO paper 97F-SIW-078 (1997)

PPSS: CBR System for ERP Project Pre-planning*

Suhn Beom Kwon[1] and Kyung-shik Shin[2]

[1] School of e-Business, Kookmin University
861-1 Chongnung-dong, Sungbuk-ku, Seoul 136-702, South Korea
sbkwon@ieee.org
[2] College of Business Administration, Ewha University
11-1 Daehyun-dong, Seodaemun-ku, Seoul 120-750, South Korea
ksshin@ewha.ac.kr

Abstract. At the initiation stage of project, a project pre-planning is an essential job for project success, especially for large-scale information system projects like ERP (Enterprise Resource Planning). Systematic estimation of resources like time, cost and manpower is very important but difficult because of the following reasons: 1) it involves lots of factors and their relationships, 2) it is not easy to apply mathematical model to the estimation, and 3) every ERP project is different from one another. In this article, we propose a system named PPSS (Project Pre-planning Support System) that helps the project manager to make a pre-plan of ERP project with case-based reasoning (CBR). He can make a project pre-plan by adjusting the most similar case retrieved from the case base. We adopt rule-based reasoning for the case adjustment which is one of the most difficult jobs in CBR. We have collected ERP implementation cases by interviewing with project managers, and organized them with XML (Extensible Markup Language)

1 Introduction

In the late 1990's, several articles have reported failures and budget run-over of ERP projects, and some of them analyzed what have made the ERP implementations more risky [8], [10], [13]. Ross asserts that characteristics of ERP projects are far different from those of previous information system(IS) projects in a sense that an ERP project is a kind of BPR(Business Process Reengineering) and it involves changes of individual line of work, business process, and even company organization [13]. Krasner stressed management problem of ERP project such as integrated project team planning, a formal decision-making, managed communication and top-management involvement [10]. Especially, he suggested applying lessons learned from earlier implementations to later implementations. Hawking proposed that an ERP project is a large scale complex information system and requires careful planning of time and budget to avoid project disaster [8]. Due to those potential risks, companies hesitate to invest large money into ERP implementation project. In short, ERP project is complex, project management is more critical than software development efforts, and its impact into organization is usually huge, so more careful project planning is highly recommended.

Relative to the other phases in information system (IS) project life cycle, the importance of the initiation phase has been emphasized by many field practitioners and

---

* This work was supported by Research Program 2004 of Kookmin University.

T.G. Kim (Ed.): AIS 2004, LNAI 3397, pp. 157–166, 2005.
© Springer-Verlag Berlin Heidelberg 2005

academic researchers. It has been demonstrated that a poor planning takes more time and workforce afterwards, and this phenomenon is usually getting worse as time goes by. Poor project planning is revealed as one of the most common factors of IS project failure [5], [9], [15], [17]. Dvir showed positive correlation between project planning efforts and project success [5]. ERP projects, a kind of IS project, are no exception.

By the way, project manager cannot take much enough time for scoping and planning in the initiation phase, because management often presses him to start project work instead of spending time to generate a project detail plan [16]. At the initiation phase, management wants to know resource requirements for ERP implementation approximately not with exact figures. One of the most effective ways for project manager to persuade management is to show real figures of previous projects of similar size companies in the same industry. He could somewhat justify his project plan (resource plan, time plan, man-month plan, project team plan, implementation methodology) with those information. But he has got no systematic support for the planning, and has no choice but to depend upon his own experiences and knowledge. His individual experiences and knowledge is, however, very confined and unorganized, so it is hard to apply in systematic manner.

There have been two supporting tools for project manager's planning job. One is a project management tool like PERT and COCOMO model that help making activity plan and estimation of project effort, respectively. However, both are not suitable for the early stage of IS project, because they require quantified data to generate detailed activity plans and exact estimations. The other is a knowledge management system that stores previous project implementation experiences into knowledge base and provides information from the knowledge base to project managers with search based manner. Though knowledge management systems can serve useful information, applying searched information to the planning job is another matter. In other words, making a pre-plan is still done by project manager's artwork.

Case-based reasoning (CBR) is a research paradigm in machine learning, which has been well applied to problems with success and failure cases and hard to find analytic solving heuristics. CBR is based upon the idea that previous solutions, whether successful or not, can lead to useful solutions for new problem. CBR system can provide a solution by adaptation of previous solutions.

We suggest that CBR can be well applied to ERP project pre-planning problems: ERP project sizing and resource estimation. Project pre-planning is usually done at the end of 'Scope the project' and before 'Developing project plan' which means full-out undertaking of project (Fig. 1). At this time point, management wants to know pre-planning results: rough estimation of resource requirement for project. Since 1990's, a plethora of companies have implemented ERP, we could get some cases to use in CBR for ERP project pre-planning.

We developed a prototype system PPSS (Project Pre-planning Support System) using CBR approach for ERP project manager's pre-planning job. The case base of PPSS is organized by XML(eXtensible Markup Language), and 4 R's of CBR reasoning cycle: retrieve, reuse, revise, retain of cases are performed in PPSS with ASP and Visual Basic program. The PPSS would be helpful for both project managers of companies implementing ERP and contractors like EDS, IBM, etc. Theses consulting firms can garner their ERP project experiences and knowledge.

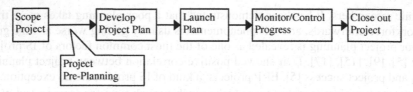

Fig. 1. Pre-planning in Project Development Life Cycle

The remainder of paper is organized as follows. The next section describes the related researches: CBR applications to IS or ERP planning. In section 3 and 4, we explain how to organize cases of PPSS by using XML and reasoning process of PPSS to generate a plan. Section 5 considers contributions and further researches.

2 Related Researches

CBR is a generic methodology and has been applied to many problems in diverse areas such as medical diagnosis, engineering product sales, electromechanical device design, robotic navigation, to mention but a few [2], [11], [14].

The use of analogies for IS project have also been suggested by many researchers and applied successfully. Grupe explored CBR as a mechanism to improve software development productivity and quality at each stage of software development process: requirement definition, effort estimation, software design, troubleshooting, and maintenance process [7]. Among software development stages, effort estimation is one of the most frequently mentioned issues since Boehm suggested at first [4], [9], [12].

When applying CBR, we should consider several decision parameters such as feature subset selection, analogy adaptation, and similarity measure selection. In this paper, we describe how to configure these decision parameters for ERP pre-planning problem in PPSS. Other contributions of PPSS at Table 1, XML-based case organization and rule driven adaptation will be explained in section 3, and 4.

Table 1. CBR Applications to Information System Project

Author	Domain Problem	Contribution
Grupe (1998)	Software development processes	- Application exploration
Kadoda (2001)	Software project effort prediction	- CBR configuration for application
Mendes (2002)	Web project cost estimation	- CBR configuration for application
Kwon (2004)	ERP project	- CBR configuration for application - XML-based case organization - Rule driven adaptation

3 Case Base of PPSS

In this section, we explain the structure and building process of the PPSS case base. A case is usually a collection of attribute value pairs. We derived case attributes from literatures and we elaborated them with ERP project manager's interviews. The selected case attributes are categorized into two groups: company characteristics and project determinants. Project managers want to examine previous ERP project experiences of other similar companies, so company characteristics play a key role for

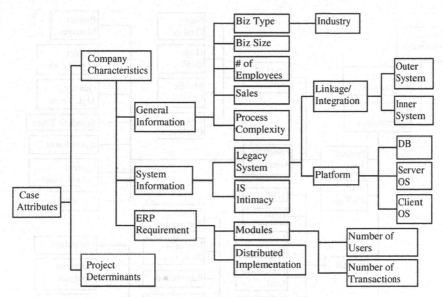

Fig. 2. Company Characteristics Attributes

searching similar cases in reasoning process. Resource requirements for ERP project are also dependent upon company characteristics like business type, information system, ERP requirements, etc.

Company characteristics have three sub-categories of attributes: company general facts, information system and ERP requirements. Fig. 2 shows the structure of company characteristic factors. Company general facts include business type, size, revenue, number of employees, process complexity. Information system sub-category includes legacy system information and company's intimacy level of information system. If a company has legacy systems and want to use some of them, project manager should recognize them as systems to be linked or integrated with ERP. As legacy system integration or linkage often become a technology-barrier and time-consuming job, how many this kind of legacy systems exist and integration level are significant for ERP implementation. ERP requirements include modules of ERP to be implemented with number of users and transactions and distributed ERP implementation according to geographic locations or business divisions.

Second group of attributes is ERP project determinants, which are of resource invested to complete project such as project team, budget, time period, and project management methodology. Project manager can guess resource requirements for his own project from this information. Fig. 3 shows the structure of ERP project determinant attributes.

Project manager's one of the most pressing jobs in the early stage is building project team [3]. Project team is usually organized with functional areas with various skill level members. Labor cost usually depends on skill level of ERP consultants, programmers, and other members. Project budget is a management's top concern, and project manager needs previous project's budget items and their allotments. Project time varies according to resources put in and project scope. Should a company expand the scope of ERP implementation, then it will increase the time or other re-

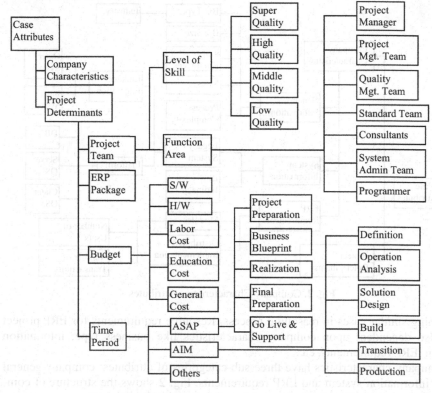

Fig. 3. ERP Project Determinant Attributes

sources. And project time period decreases, as more man-month input. The information of how much time spent at each stages of project life cycle is another concern and helpful for project manager. Most ERP package vendors have their own ERP implementation methodology, so determination of ERP package often means selection of ERP implementation methodology. Each methodology has 4 to 6 stages, and the time period of each stage will be a good reference for project manager.

We use XML (eXtensible Markup Language) framework to represent and organize a case base. Benefits that XML framework provides are flexibility to organize and re-organize case structure, independence between content and representation, easiness to search data from case base, reusability by modularizing case base. As data organization of a case usually has a hierarchical tree structure, XML is one of the most suitable framework to both organize and represent a case. Abidi has shown how electronic medical records can be readily transformed into XML format and used as cases in medical diagnostic system [1].

In order to organize a case, we use DTD (Document Type Definition) of XML. DTD is a declaration of data structure, a kind of metadata. When a new case is created, a XML parser checks whether the data structure is accord with DTD or not. For representation of a case, we use XSL (eXtensible Style-sheet Language), a kind of transformer XML contents into a well-formed format for Internet browser. Fig. 4 and Fig. 5 show a part of DTD and XML content of a case respectively.

```
<?xml version="1.0" encoding="EUC-KR"?>
<!ELEMENT Cases (Case+)>
<!ELEMENT Case (Characteristics,Determinants)>
<!ATTLIST Case id CDATA #REQUIRED>
<!-- Company Characteristics   -->
<!ELEMENT Characteristics (GeneralInfo,SysInfo,Requirement)>
<!ELEMENT GeneralInfo (Biz-
Type,Size,Employee,Sales,ComplexityOfBizProcess)>
<!ELEMENT BizType (#PCDATA)>
<!ATTLIST BizType PubORPrivate (Public|Private) "Private">
<!ELEMENT Size (#PCDATA)>
<!ELEMENT Employee (#PCDATA)>
<!ELEMENT Sales (#PCDATA)>
<!ELEMENT ComplexityOfBizProcess (#PCDATA)>
<!-- -->
<!ELEMENT SysInfo (LegacySystem,Intimacy)>
<!ELEMENT LegacySystem (Link?, Platform?)>
<!ATTLIST LegacySystem Usage (Y|N) "Y">
<!ELEMENT Link (Internal|External)>
<!ELEMENT Internal (#PCDATA)>
<!ELEMENT External (#PCDATA)>
<!ELEMENT Platform (DB,ClientOS,ServerOS)>
<!ELEMENT DB (#PCDATA)>
<!ELEMENT ClientOS (#PCDATA)>
<!ELEMENT ServerOS (#PCDATA)>
<!ELEMENT Intimacy (#PCDATA)>
```

Fig. 4. Part of DTD of Cases.xml

```
<?xml version="1.0" encoding="EUC-KR"?>
<!DOCTYPE Cases SYSTEM "cases.dtd">
<Cases>
<Case id="1">
  <Chracteristics>
    <GeneralInfo>
       <BizType PubORPrivate="Public">Leisure</BizType>
       <Size>Middle</Size>
       <Employee>700</Employee>
       <Sales>1000000000</Sales>
       <ComplexityOfBizProcess>Low</ComplexityOfBizProcess>
    </GeneralInfo>
    <SysInfo>
       <LegacySystem Usage="Y">
         <Link>
           <Internal>45</Internal> </Link>
         <Platform>
           <DB>Oracle</DB>
           <ClientOS>Win98</ClientOS>
           <ServerOS>Win2000</ServerOS> </Platform>
       </LegacySystem>
       <Intimacy>High</Intimacy>
    </SysInfo>
    <Requirement>
        <Module No="2">
          <Name>FI</Name>
          <User>200</User>
          <Transaction>100</Transaction>
          <Name>CO</Name>
          <User>200</User>
          <Transaction>100</Transaction> </Module>
        <Decentralization>4</Decentralization>
    </Requirement>
  </Chracteristics>
```

Fig. 5. Part of Case Representation with XML

4   Reasoning Process of PPSS

Reasoning process of PPSS follows the general process of CBR: retrieve, reuse, revise and retain. Fig. 6 shows the reasoning process of PPSS. In order to retrieve the most similar case, a project manager should input facts about his problem, ERP project pre-planning. Company characteristics in Fig. 2 are major information to be input into PPSS.

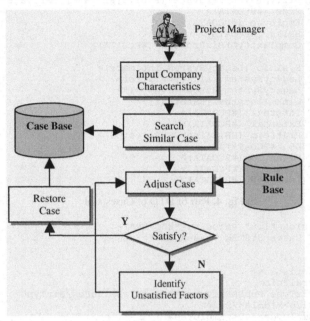

Fig. 6. Reasoning Process of PPSS

Based upon input information, PPSS retrieves the similar case by using the nearest neighbor algorithm. Among cases in case base, the case with the biggest similarity index value is selected as the most similar case. Equation (1) is a formula for computing similarity index value in PPSS with the nearest neighbor algorithm.

$$\text{Similarity (T, S)} = \sum_{i=1}^{N} f_i\big(T_i, S_i\big) * w_i$$

$$f_i = \begin{cases} Min(T_i, S_i) \big/ Max(T_i, S_i), & \text{if attribute i has numeric or scale value} \\ 1, & \text{if attribute i has descriptive value and } T_i = S_i \\ 0, & \text{if attribute i has descriptive value and } T_i \neq S_i \end{cases} \tag{1}$$

T: Target Case,   S: Source case
i: ith Attribute   N: Number of Attributes
f_i: Similarity Function for Attribute i, w_i: Importance Weight of Attribute i

For numeric and scale value attributes, we adopt comparative magnitude of two values as similarity function. The function can give a normalization effect to numeric value attributes with different scales. For descriptive value attributes, we adopt zero or one dichotomy function: 1 if two values are equal and 0 otherwise. Weight is applied to reflect comparative importance of attributes. Most interviewee replied that 'business type', 'decentralization', 'number of ERP modules introduced', 'transaction volume', and 'ERP package' are more important factors than the others. Therefore, we put double weight to 'decentralization' and triple weight to 'number of ERP modules introduced', 'transaction volume', and quadruple weight to 'business type' and 'ERP package'.

Then, we need to adjust the most similar case to the current problem. PPSS system has a rule-base which contains knowledge for ERP pre-planning. For example, 'If ERP is implemented at the multiple places and to be integrated, cost for consulting manpower usually rises by 10 ~ 20%.' There exist some causal relationships among attributes and we can get knowledge of this kind during interviews with project managers who have experiences on ERP projects. PPSS displays the relevant rules to project manager when he adjusts and determines attribute values by himself. We can also make PPSS to adjust the similar case automatically, that means rule-based reasoning is started not by user but by PPSS. But for the most pre-planning job, human judgment is crucial and rule-based knowledge just supports human judgment like most rule-based systems. Fig. 7 shows the screen shot of case adjustment. At the bottom, you can see the relevant rule associated with the attribute 'Labor Cost'.

Marling surveyed CBR integrations with other methodologies such as rule-based system, constraint satisfaction problem solving, genetic algorithm, information retrieval [11]. They gave several example hybrid systems of CBR and rule-based system, but two systems were in equivalent position for problem solving. PPSS uses rule-based system as a case adjustment tool, a part of CBR in order to help project manager adjust the attribute value, one of the most difficult manual jobs.

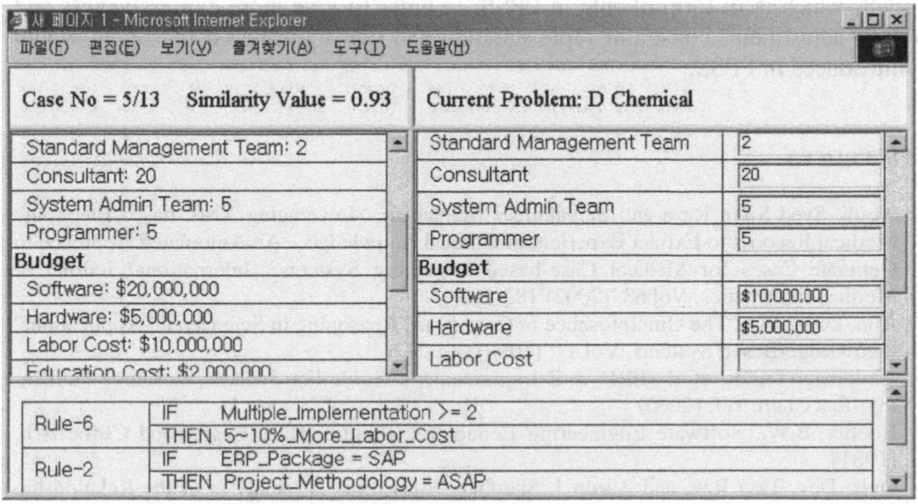

Case No = 5/13 Similarity Value = 0.93	Current Problem: D Chemical	
Standard Management Team: 2	Standard Management Team	2
Consultant: 20	Consultant	20
System Admin Team: 5	System Admin Team	5
Programmer: 5	Programmer	5
Budget	**Budget**	
Software: $20,000,000	Software	$10,000,000
Hardware: $5,000,000	Hardware	$5,000,000
Labor Cost: $10,000,000	Labor Cost	
Education Cost: $2,000,000		

Rule-6	IF Multiple_Implementation >= 2 THEN 5~10%_More_Labor_Cost
Rule-2	IF ERP_Package = SAP THEN Project_Methodology = ASAP

Fig. 7. Sample Screen of Case Adjustment

According to the general CBR process, the adjusted case is restored into case base as a new case. However, PPSS stores the adjusted case as an 'in-progress case' which means the case should be readjusted with the real figures at the end of ERP project. So, cases in progress status are excluded in default, when PPSS retrieves similar case. When a new case is stored in case base, the case representation follow XML format defined by DTD.

5 Conclusion

ERP is one of the most important information systems for corporate, so whether ERP project succeed or not is crucial for corporate. Project pre-planning is far more important for ERP project, because ERP project is not a matter of software development but a matter of project management such as business process reengineering, change management, project team making.

We proposed a framework and system that supports project manager to pre-plan ERP project by using CBR method. We surveyed and organized attributes which are factors project manager should consider. Two things are methodological improvements from normal case-based reasoning. First, we adopt XML scheme as representation and organization tool for case content. Case structure can be easily re-organized by DTD, and represented in the web environment without change of case contents by XSL. Second thing is hybrid framework of case-based reasoning and rule-based reasoning. We adopt rule-based reasoning as an adjustment tool of the selected case. By using rule-based reasoning, we can systematically help PPSS users to adjust case, which is the most difficult job in case-based reasoning.

The more cases does PPSS store in case base, the more feasible solution PPSS provide. Current PPSS is a prototype system with 8 cases in a way that shows the proposed framework. So PPSS needs more cases stored in order to help ERP project in the field. Another thing to be stored up to meaningful level is knowledge of ERP project, which is in form of rule in PPSS. In order to give more expressiveness and smooth adjustment, constraint representation and constraint-based reasoning would be introduced in PPSS.

References

1. Abidi, Syed Sibte Raza and Selvakumer Manickam, "Leveraging XML-based Electronic Medical Records to Extract Experiential Clinical Knowledge – An Automated Approach to Generate Cases for Medical Case-based Reasoning Systems," International Journal of Medical Informatics, Vol.68, (2002) 187-203
2. Aha, David W., "The Omnipresence of Case-based Reasoning in Science and Application," Knowledge-Based Systems, Vol.11, (1998) 261-273
3. Anderegg, Travis, et al., ERP: A-Z Implementer's Guide For Success, Resource Publishing: Eau Claire WI, (2000)
4. Boehm, B.W., Software Engineering Economics, Prentice-Hall: Englewood Cliffs, N.J., (1981)
5. Dvir, Dov, Tzvi Raz, and Aaron J. Shenhar, "An Empirical Analysis of the Relationship between Project Planning and Project Success," International Journal of Project Management, Vol. 21, (2003) 89-95

6. Flowers S., Software Failure: Management Failure, Chichester, UK: Johh Wiley, (1996)
7. Grupe, Fritz H., Robert Urwiler, Narender K. Ramarapu, Mehdi Owrang, "The Application of Case-based Reasoning to the Software Development Process," Information and Software Technology, Vol. 40, (1998) 493-499
8. Hawking, Paul and Andrew Stein, "E-Skills: The Next Hurdle for ERP Implementations," Proc. of 36th HICSS, (2003)
9. Kadoda, Gada, Michelle Cartwright, and Martin Shepperd, "Issues on the Effective Use of CBR Technology for Software Project Prediction," Proceeding of ICCBR 2001, (2001) 276-290
10. Krasner, H., "Ensuring e-business Success by Learning from ERP Failures" IT Professional, Vol.2 No.1, (2000) 22–27
11. Marling, Cynthia, et al., "Case-Based Reasoning Integrations," AI Magazine, spring (2002) 69-86
12. Mendes, Emilia, Nile Mosley and Steve Counsell, "The Application of Case-Based Reasoning to Early Web Project Cost Estimation," Proceeding of 26th International Computer Software and Application Conf., (2002)
13. Ross, Jeanne W., "Surprising Facts about Implementing ERP," IT Professional, Vol.1 No.4, (1999) 65–68
14. Watson, I., "Case-based Reasoning Is a Methodology Not a Technology," Knowledge-Based Systems, Vol.12, (1999) 303-308
15. Whittaker, Brenda, "What Went Wrong? Unsuccessful Information Technology Projects," Information Management & Computer Security, Vol.7 No.1, (1999) 23-29
16. Wysocki, Robert K., Robert Beck Jr., David B. Crane (Eds.), Effective Project Management, John Wiley & Sons, (2001)
17. Yeo, K.T., "Critical Failure Factors in Information System Project," Project Management, Vol. 20, (2002) 241-246

A Scheduling Analysis in FMS
Using the Transitive Matrix

Jong-Kun Lee

LIS/Computer Engineering Dept. Changwon national University
Salim-dong 9, Changwon ,Kyungnam, 641-773 Korea
jklee@sarim.changwon.ac.kr

Abstract. The analysis of the scheduling problem in FMS using the transitive matrix has been studied. Since the control flows in the Petri nets are based on the token flows, the basic unit of concurrency (short BUC) could be defined to be a set of the executed control flows in the net. In addition, original system could be divided into some subnets such as BUC of the machine's operations and analyzed the feasibility time in each schedule. The usefulness of transitive matrix to slice off some subnets from the original net, and the explanation in an example will be discussed.

1 Introduction

In FMS, one of the most important subjects is to formulate the general cyclic state-scheduling problem. Especially, scheduling problems have been arisen when a set of tasks has to be assigned to a set of resources in order to optimize a performance criterion [15]. Since the state space explosion occurs during the analysis of the system including concurrent modules, it is difficult to analyze and understand the large system composing of several modules. Various scheduling methods to solve these problems have been proposed [1-11,15-17], and classic cyclic schedules represented by a particular class of Petri nets (short PN)[5].

The transitive matrix could be explained on all relationships between places and transitions in PN. In this paper, we focuses on developing a more simple and efficient method for the analysis of cyclic scheduling problem in FMS. After slicing some subnets (BUCs) with transitive matrix, a new simple method is developed.

Petri Nets and place transitive matrix in section **2** and basic unit of concurrency in section **3** will be defined. In section **4**, an extraction of optimal sequence after slicing some BUC based on the transitive matrix in an example will be discussed, and in section **5**, the benchmarking resultants after the analysis with the performance evaluation factors will be shown. Finally, a conclusion will be given in section **6**.

2 Time Petri Nets and Transitive Matrix

In this section, certain terms, which are often used in the latter part of this paper, are defined [9, 12-14].

Let N=<P,T,I,O,M$_o$,τ>be a Time Petri Nets, where P is the set of places, T is the set of transitions, and $P \cap T = \phi$, I:T->P$^\infty$ is the input function, O:T->P$^\infty$ is the output

T.G. Kim (Ed.): AIS 2004, LNAI 3397, pp. 167–178, 2005.
© Springer-Verlag Berlin Heidelberg 2005

function. $M_0 \in M = \{M \mid M : P \rightarrow N\}$, M_0 is an initial marking, N: N is the set of positive integers. $\tau(t) : T \rightarrow N$. $\tau(t)$ denotes the execution time (or the firing time) taken by transition t.

The number of occurrences of an input place P_i in a transition t_j is $\#(P_i,I(t_j))$, also the number of occurrences of an output place P_i in a transition t_j is $\#(P_i,O(t_j))$.

The matrix of PN structure, C is C=<P, T, C$^-$, C$^+$>, where P,T are a set of places and transitions, respectively. C$^-$ and C$^+$ are matrices of m rows by n columns defined by

$$C^- = [I,j] = \#(P_i,I(t_j)), \text{ matrix of input function,}$$
$$C^+ = [O,j] = \#(P_i,O(t_j)), \text{ matrix of output function.}$$

And an incidence matrix B, $B = C^+ - C^-$.

(Def. 2.1): Invariant
A row vector $X = (x_1, x_2, ..., x_m)$ is P-invariant if $X \bullet B = 0$, where $X \neq 0$. A column-vector $Y = (y_1, y_2, ..., y_n)^T \geq 0$ is called a T-invariant, if $B \bullet Y = 0$, where Y is an integer solution x of the homogeneous matrix equation and $Y \neq 0$.

(Def. 2.2): Place transitive and Transition matrix
Place transitive and transition matrix are as follows;

$$C_P = C^-(C^+)^T$$
$$C_T = (C^+)^T C^-$$

This time, we may extend a definition (Def. 2.2) to show the relation between transitions and places, and call this new definition as "labeled place transitive matrix":

(Def. 2.3): labeled place transitive matrix,

Let L_{CP} be the labeled place transitive matrix:

$$L_{CP} = C^- diag(t_1, t_2, ..., t_n)(C^+)^T$$

The elements of L_{CP} describe the direct transferring relation that is from one place to another through one or more transitions.

(Def. 2.4): Let L_{CP} be the m×m place transitive matrix. If a transition t_k appears s times in the same column of L_{CP}, then we replace t_k in L_{CP} by t_k / s in L_{CP}.

3   BUC (Basic Unit of Concurrency)

3.1   BUC Choose Algorithm

Since the control flows is based on the token flows, the independent tokens status can be explained by a control flow. If a token is divided into several tokens after firing a transition, a control flow can be allocated to several flows by the same manner. Accordingly, we define that BUC is a set of the executed flow control based on the behavioral properties in the net. Herein we would like to propose an algorithm to slice a Petri net model in resource shared, after defining P-invariants, which can be used in constructing concurrent units [9,12] as follows:

(Def. 3.1): BUC
A P-invariant, which is not contained in other P-invariants, is called BUC.

In the FMS model, there are several resource-shared situations in the machine so we slice BUC based on the resource shared of machine. This means that the sliced BUC is a P-invariant based on the machine.

We chose the BUC in the resource shared model after foaming the place transitive matrix, which was defined in previews section. In this paragraph we show a choice algorithm.

Algorithm: Choose BUC algorithm
Input: N = <P,T,I,O,M>
Output: BUC of N, N_S=<P_S,T_S,I_S,O_S,M_S>

(1) Define L_{CP} in the Place transitive matrix.

(2) Find BUC based on the resource shared places(machines).

(3) Find the all-relational places in each column L_{CP} and make an element of own BUC with this initial marking place (machine). Also, link the same place after finding the place in row.

Example: BUC
In this concurrency model, to partition the independently concurrent sub-net is important in the slice of the nets. The transitive matrix of place L_{CP} of this model is as follows (Table 1):

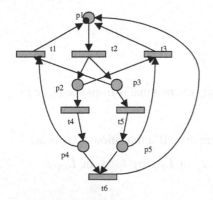

Fig. 1. A model of Petri net

Table 1. Transitive matrix of (Fig. 1)

$$L_{CP} = \begin{bmatrix} 0 & t2 & t2 & 0 & 0 \\ t3/2 & 0 & 0 & t4 & 0 \\ t1/2 & 0 & 0 & 0 & t5 \\ t1/2 & & & & \\ t6/2 & 0 & 0 & 0 & 0 \\ t3/2 & & & & \\ t6/2 & 0 & 0 & 0 & 0 \end{bmatrix} \begin{matrix} P1 \\ P2 \\ P3 \\ P4 \\ \\ p5 \end{matrix}$$

with column headers: p1   p2   p3   p4   p5

In this table, we consider one place p1 as the initial token. In the first column, if we choose p1, we can find p2 and p3 using transition t2. The two places p2 and p3 are shown the concurrences. Now, if we select p2 with transition t2 and choose p2 in the row, we can find p1 with t3 and p4 with t4. In the case of p4 in row, p1 can be selected using t1 and t6. If these places and transitions are connected, it becomes a concurrence as shown in Fig. 2.a. On the other way, if we choose p3 with t2 then we will obtain another BUC$_2$ (b) in (Fig. 2).

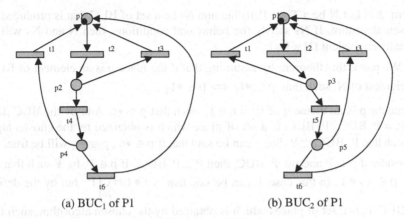

(a) BUC₁ of P1   (b) BUC₂ of P1

Fig. 2. BUC of model Petri net

3.2 BUC Properties

The system is divided into BUC by the method of the partitioning based on the behavioral properties in the net. The Time Petri nets slice produced using the chosen algorithm is defined as follows:

(Def.3.2): BUC slice

Let $N=(P,T,F,M,\tau)$ be a Time Petri net, where the places set P is divided by BUC choose algorithm, where $F \subseteq (P \times T) \cup (T \times P)$ is the flow relations. BUC are defined as $(BUC_i \mid i=1, \ldots, n)$ and each $BUC_i = (P_i,T_i,F_i,M_i,\tau_i,)$ satisfies the following conditions:

$P_i = P\_BUC_i$: a set of place sets which are obtained by the choose algorithm
$T_i = \{ t \subset T \mid s \subset P_i, (s, t) \in \Gamma \text{ or } (t, s) \in F\}$,
$F_i = \{(p, t) \in F, (t, p) \in F \mid p \in P_i, t \in T_i \}$,
$\forall \tau_i \in \tau, \tau_i(t) = \tau(t) \text{ and } \forall p \in M_i, M_i(p) = M(p)$.

In the Petri net model, the behavioral condition of transition t is all predecessors' places (⁎t), which are connected with transition t, has a token. Petri net slices are divided into subnets based on BUC by transition level. A behavioral condition is defined as follows:

(Def. 3.3) Behavioral condition

Let $BUC_i = (P_i,T_i,F_i,M_i,\tau_i,)$ be a set of BUC which is obtained by the chosen algorithm.

$\forall\ t_i \in T_i$, transitions t_i should be satisfied one of the following behavioral conditions:

(1) if transition t_i is not shared: satisfy the precondition of transition t_i only.
(2) if transition t_i is shared by several Slices: satisfy the preconditions in all BUC,
   which have transition t_i.

(Def.3.4) Behavioral Equivalent:

Let P and Q are two Time Petri Nets, if two reachability graphs of P and Q are one-to-one mapping, then P and Q can be defined as: $P \equiv Q$ since they are functionally equivalent.

(Theorem. 3.1) Let N be a Time Petri net and Ns be a set of BUC that is produced by the chosen algorithm. If Ns satisfy the behavioral conditions, then N and Ns will be functionally equivalent (N ≡ Ns).

(Proof) We prove this theorem by assuming that if the place p is an element of BUC, p is an element of N, such that $p \in \bullet t_i \Leftrightarrow p \in \bullet t_T$.

($\Rightarrow$) Since the p is an element of P_i, $\exists\ t_i \in T_i$, such that $p \in \bullet t_i$. And by the BUC definition, $P_i = P\_BUC_i$, P_BUC_i is a set of place which is obtained by the chosen algorithm, such that $P\_BUC_i \subseteq P$. So, it can be said that if $p \in \bullet t_i$, $p \in \bullet t_T$ will be true.

($\Leftarrow$) Consider if $p \in P$ and $p \notin P\_BUC_i$ then $P \not\subset P\_BUC_i$. If $p \in P$, t∈ T such that p∈ •t. And $p \notin \underset{t \in T_i}{\cup} \bullet\, t$, in this case, it can be said that $\underset{t \in T_i}{\cup} \bullet\, t = \overset{n}{\underset{i=1}{\cup}} P_i$ but by the definition of BUC, P_i is a set of places which is obtained by the chosen algorithm, such that $P_i \subseteq P$. So if $p \in P$ and $p \notin P\_BUC_i$ then $P \not\subset P\_BUC_i$ is not true.

4   Extraction of Sub-net of Petri Nets Using Transitive Matrix

We consider a system with two machines such as M1, M2 and two jobs such as OP$_A$ and OP$_B$ as shown in Fig. 3 [3]. Let us assume that the production ratio is 1:1, which means that the goal is to manufacture 50% of OP$_A$ and 50% of OP$_B$. OP$_A$: t1(5) t2(5) t3(1), OP$_B$: t4(3) t5(4) t6(2), and M1: OP$_{A1}$(t1), OP$_{A3}$(t3), OP$_{B2}$(t5) M2: OP$_{A2}$(t2), OP$_{B1}$(t4),OP$_{B3}$(t6), where ( ) is an operation time.

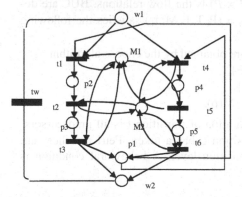

Fig. 3. A model, which has two resources, shared

Table 2. Transitive matrix of (Fig. 3)

$L_{CP} =$

	p1	p2	p3	p5	p6	M1	M2	
	0	t1/2	0	t4/2	0	t1/2	t4/2	P1
	0	0	t2/2	0	0	0	t2/2	P2
	t3/2	0	0	0	0	t3/2	0	P3
	0	0	0	0	t5/2	t5/2	0	P5
	t6/2	0	0	0	0	0	t6/2	P6
						t1/2		
	t3/2	t1/2	0	0	t5/2	t3/2	0	M1
						t5/2		
							t6/2	
	t6/2	0	t2/2	t4/2	0	0	t2/2	M2
							t4/2	

We are ignored the places w1 and w2 and transition tw in this example for easy work. Now, the transitive matrix of example net is described in Tab. 2. In this table, the initial tokens exist in M1 and M2. The cycle time of OP$_A$ and OP$_B$ is 11 and 9, respectively.

Now, we select row and column of p1, p3, p5, and M1 in (Tab. 2), and make one BUC of M1 based on the selected places and transitions. The selected BUCs of machines M1 and M2 are shown in Fig. 4 and Fig. 5.

Based on these BUC, the optimal solution of this example can be obtained as shown in Fig. 6. Specially, the optimal schedule of two cycles is also obtained.

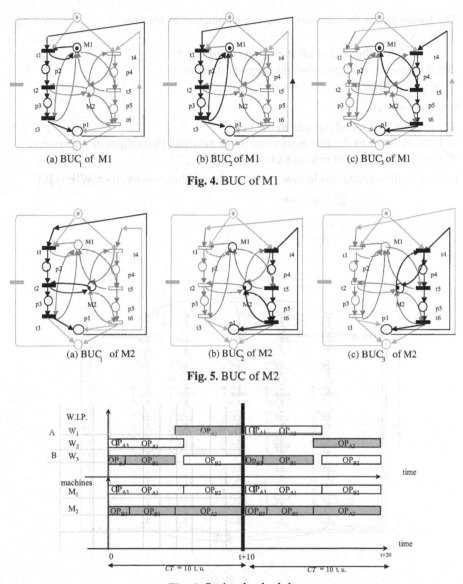

(a) BUC₁ of M1   (b) BUC₂ of M1   (c) BUC₃ of M1

Fig. 4. BUC of M1

(a) BUC₁ of M2   (b) BUC₂ of M2   (c) BUC₃ of M2

Fig. 5. BUC of M2

Fig. 6. Optimal schedule

5   Bench Mark

5.1   Notations

In this section, one example taken from the literature is analyzed in order to apply three cyclic scheduling analysis methods such as Hillion[4], Korbaa[6], and the previously presented approach. The definitions and the assumptions for this work have been summarized [6].

The formulations for our works, we can summarize as follows:

$\mu(\gamma) = \sum_{\forall t \in \gamma} D(t)$, the sum of all transition timings of γ

$M(\gamma)$ ($=Mo(\gamma)$), the (constant) number of tokens in γ,

$C(\gamma) = \mu(\gamma)/M(\gamma)$, the cycle time of γ,

Where γ is a circuit.

$C^* = Max(C(\gamma))$ for all circuits of the net,

CT the minimal cycle time associated to the maximal throughput of the system:

CT $= Max(C(\gamma))$ for all resource circuits $= C^*$

Let CT be the optimal cycle time based on the machines work, then WIP is [6]:

$$WIP = \sum_{\text{pallets type } i} \left[\frac{\sum \text{Operating time } OS \text{ to be carried by } i}{CT} \right]_i$$

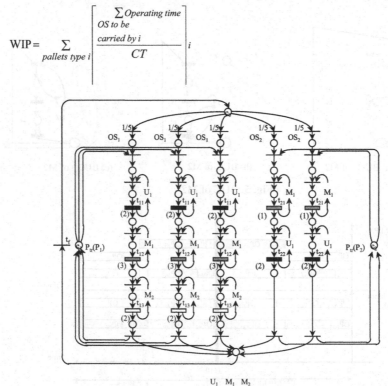

Fig. 7. Illustrative example

We introduce an illustrative example in Camus[1], two part types (P_1 and P_2) have to be produced on three machines U_1, M_1 and M_2. P_1 contains three operations: u_1(2 t.u.) then M_1(3 t.u.) and M_2(3 t.u.) P_2 contains two operations: M_1(1 t.u.) and U_1(2 t.u.). The production horizon is fixed and equalized to E={3P_1, 2P_2}. Hence five parts with the production ratio 3/5 and 2/5 should be produced in each cycle. We suppose that there are two kinds of pallets: each pallet will be dedicated to the part type P1 and the part type P2. Each transport resource can carry only one part type. The operating sequences of each part type are indicated as OS1 and OS2. In this case, the cycle time of OP_{11}, OS12 and OS_{13} are all 7 and Op_{21} and OS_{22} all 3, also the machines working time of U1 is 10, M1 is 11 and M2 is 6. So the cycle time CT is 10. The minimization WIP is:

$$WIP = \left\lceil \frac{\sum \text{Operating Times of OS}_{p1}}{CT} \right\rceil$$

$$+ \left\lceil \frac{\sum \text{Operating Times of OS}_{p2}}{CT} \right\rceil$$

$$= \left\lceil \frac{7+7+7}{11} \right\rceil + \left\lceil \frac{3+3}{11} \right\rceil = 3$$

5.2 Benchmark

By the example, we can obtain some results like as the following figures (Fig. 8-10).

1) Optimization

The Hillion's schedule[4] has 6 pallets and the Korbaa's schedule[3] 3 ones and the proposed schedule 4 ones. This solution showed that the good optimization of Korbaa's schedule could be obtained and the result of the proposed schedule could be better than that of the Hillion's.

Also, the solutions of the proposed approach are quite similar to (a) and (c) in Fig. 10 without the different position.

2) Effect

It's very difficult problem to solve a complexity value in the scheduling algorithm for evaluation. In this works, an effect values was to be considered as the total sum of the numbers of permutation and of calculation in the scheduling algorithm to obtain a good solution. An effected value of the proposed method is 744, i.e. including all permutation available in each BUC, and selecting optimal solution for approach to next BUC. An effect value to obtain a good solution is 95 in the Korbaa's method; 9 times for partitions, 34 times for regrouping, and 52 times for calculation cycle time. In the Hillion's method, an effected value is 260; 20 times for machine's operation schedule and 240 times for the job's operation schedule.

3) Time

Based on the three algorithms, we can get time results for obtaining the good solution. Since this example model is simple, they need very small calculation times; 1 sec for the Korbaa's approach and 1.30sec for both of the Hillion's and the proposed approaches. The Korbaa's approach has minimum 1 minute and maximum 23 hours in the 9 machines and 7 operations case in Camus[1], while the prposed approach 3 minutes. Meanwhile the Hillion's and the Korbaa's approaches belong to the number of the operation and the machines, the proposed method to the number of resource shares machines. This means that the Hillion's and the Korbaa's approaches analyzing times are longer than the proposed one in the large model. As the characteristic resultants of these approaches are shown in Fig. 11, the Korbaa approach is found out to be good and the Hillion approach is to be effectiveness in the time. And on the effort point, the proposed approach is proved to be good.

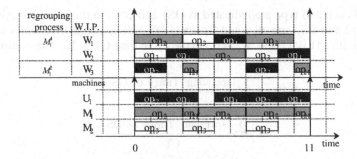

Fig. 8. Hillion's schedule

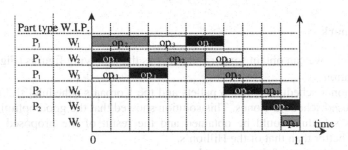

Fig. 9. Korbaa's schedule

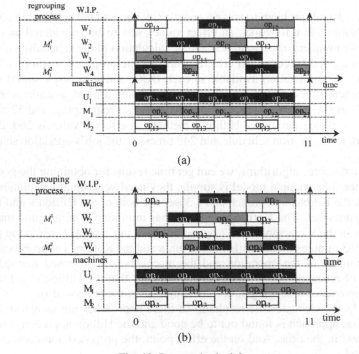

(a)

(b)

Fig. 10. Proposed schedule

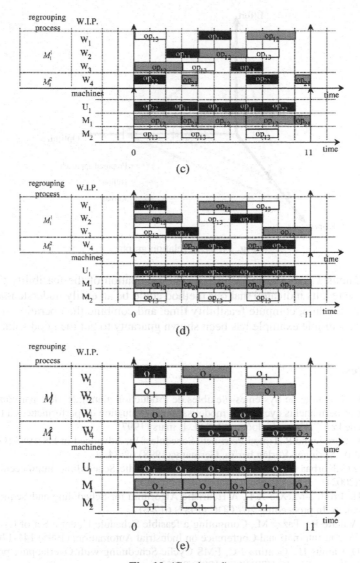

Fig. 10. (Continued)

6   Conclusion

The analysis of the schedule for the determination of the optimal cycle time after slicing BUC using the transitive matrix has been studied. The scheduling problem in the resource shared system was strong NP-hard. That meant that if the original net were complex and large, then it could be difficult to analyze. To solve this problem, an algorithm to analyze the scheduling problem after slicing off some BUCs used transitive matrix based on resource share place has been proposed. After applying to a system with two machines and two jobs some of the best solutions including another

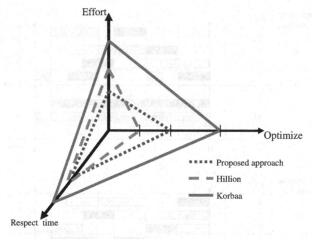

Fig. 11. Total relation graph

type with Camus[1] have been identified. The calculation of the feasibility time was simple. These results indicated that our method could be an easily understandable tool to find good schedule, compute feasibility time, and combine the operation schedules also. Finally, a simple example has been shown guaranty to get the good solution.

References

1. Camus H., Conduite de Systèmes Flexibles de Production manufacturière par composition de Régimes permanents cycliques : modélisation et évaluation de performances à l'aide des Réseaux de Petri",These de doctorat USTL 1, mars (1997)
2. Carlier J., Chretienne P.,Timed Petri nets Schedules, In, Advanced in PN'1988,(G. Rozenberg, (Ed.)) , Springer-Verlag,berlin, Germany,(1998) 62-84.
3. Gentina J.C.,Korbaa O.,Formal approach of FMS Cyclic Scheduling, In:proceeding IEEE SMC'02(2002)
4. Hillion H., Proth J-M, Xie X-L, A Heuristic Algorithm for Scheduling and Sequence Job-Shop problem, In proceeding 26th CDC (1987).612-617.
5. Julia S., Valette R., Tazza M., Computing a feasible Schedule Under a Set of Cyclic Constraints, In: 2nd International Conference on Industrial Automation, (1995) 141-146.
6. Korbaa O, Camus H., Gentina J-C, FMS Cyclic Scheduling with Overlapping production cycles, In :Proceeding ICATPN'97, (1997) 35-52.
7. Korbaa O,Benasser A.,Yim P.,Two FMS Scheduling Methods based on Petri Nets: a global and local approach, International Journal of Production Research,41(7),(2003), 1349-1371
8. Lee DY, DiCesare F., Petri Net-based heuristic Scheduling for Flexible Manufacturing, Petri Nets in Flexible and Agile Automation, (Zhou MC(Ed.)), , Kluwer Aca. Pub., USA. (1995).149-187
9. Lee J. Korbaa O, Modeling and Scheduling of ratio-driven FMS using Unfolding time Petri nets, Jr. of Computer & Industrial Engineering, 46(4),(2004),639-653
10. Lee J., Korbaa O., Gentina J-C , Slice Analysis Method of Petri nets in FMS Using the Transitive Matrix", In: Proceeding INCOM01, (2001).
11. Lee J., Benchmarking Study of the Cyclic Scheduling Analysis Methods in FMS, In:Proceeding IEEE SMC02,(2002).

12. Lee WJ, S.D.Cha, Kwon YR, A slicing-Based Approach to Enhance Petri net reachability Analysis, Jr. of Research and Practice in Information technology, Vol. 32, No2 (2000) 131-143.
13. Liu J., Itoh T., Miyazawa I., Seikiguchi T., A Research on Petri nets Properties using Transitive matrix, In: Proceeding IEEE SMC99, (1999).888-893.
14. Murata T., Petri Nets: Properties, Analysis an Applications, Proceedings of the IEEE, 77(4), IEEE,USA (1989) 541-580.
15. Ohl H., Camus H., Castelain E. and Gentina JC, Petri nets Modelling of Ratio-driven FMS and Implication on the WIP for Cyclic Schedules, In:Proceeding SMC'95 (1995).3081-3086.
16. Richard P., Scheduling timed marked graphs with resources: a serial method, accepted in proceeding INCOM'98.(1998)
17. Valentin C., Modeling and Analysis methods for a class of Hybrid Dynamic Systems, In: Proceeding Symposium ADPM'94, (1994) 221-226.

Simulation of Artificial Life Model in Game Space*

Jai Hyun Seu[1], Byung-Keun Song[1], and Heung Shik Kim[2]

[1] Dept. of Computer Engineering, Inje University,
Kimhae, Kyungnam 621-749, Korea
{jaiseu,amidala}@cs.inje.ac.kr
[2] Dept. of Computer Engineering,
Institute of Basic Sciences, Inje University,
Kimhae, Kyungnam 621-749, Korea
hskim@cs.inje.ac.kr

Abstract. Game designer generally fixes the distribution and characteristics of game characters. These could not possibly be changed during playing a game. The point of view at online game player, usually the game playing time is getting longer. So online game is bored because of fixed distribution and characteristics of game characters. In this study, we propose and simulate the system about distribution and characteristics of NPCs. NPCs' special qualities can be evolved according to their environments by applying gene algorithm. It also produces various special quality NPCs from a few kinds of NPCs through evolution. Game character group's movement can be expressed more realistically by applied Flocking algorithm.

1 Introduction

Types of Computer games have been changing currently. Moreover people prefer to enjoy games with others not alone. As results, people who interested in same types of game build community. Commonly game planner design and place specific characteristic NPC (Non Player Character) group on specific location in the on-line game worlds. By using this kind of method, designer can compose the map and regulate NPC's level according to player's level. Also, they can create easily NPCs behavior pattern using Finite State Machine [1]. But, this kind of game design tends to be monotonous and inform the players what the NPCs range in specific region. Also, NPC attributes are stationary if they are experienced once.

This paper proposes a system that NPCs' special qualities can be evolved according to their environments by using gene algorithm. This system can produce various special quality NPCs from a few kinds of NPCs through evolution [2]. NPC group movement can be expressed more realistically by using Flocking algorithm. If we create NPC groups that move and react against surrounding groups' actions. The players may have impressions that seem to receive new game whenever they play the game.

---

* This work was supported by the 2000 Inje University Research Grant.

T.G. Kim (Ed.): AIS 2004, LNAI 3397, pp. 179–187, 2005.

2   Related Researches

In this chapter, we introduce about flocking algorithm that defines the movement of groups. Gene algorithm is used on artificial creature's special qualities. Flocking is action rule algorithm that analyze and defines actions for real lives. We can simulate movement for NPC groups by using this algorithm. Craig Reynolds introduced flocking algorithm in 1987 and defined by four basis rules of flocking [3]. These four rules are as following.

- Separation: Turn direction not to collide with surrounding *Boid.*
- Alignment: Indicate direction same as surrounding *Boid.*
- Cohesion: Turn direction to average position with surrounding *Boid* .
- Avoidance: Avoid collision with surrounding enemy or stumbling blocks.

Genetic Algorithm is one of optimization techniques based on biologic evolution principle by Darwin's 'Survival of the fittest' [4]. Reproduction process can create a next generation based upon selection, hybridization and mutation processes.

1. Selection operator: This arithmetic process selects genes in hybridization. Some genes that adapt well in their environments are alive. Others are weeded out.
2. Hybridization operator: This process exchanges their chromosomes in selected pair of gene codes. If there are genes A and B, certain part of chromosome will be taken from A and rest part will be taken from B. The specific parts are decided by random value. This process creates new object. At this time, characteristics of parents are inherited properly to their children. Hybridization operator can be divided into more detailed types of hybridization. Such as simplicity, one point, plural point and uniformity hybridization operator.
3. Mutation operator: This process changes a bit in gene code. In binary bit string, mutation means that toggles bit value. A mutation is just a necessary process in order not to lose a useful latent ability. A mutation process is accomplished based on mutation possibility. Generally mutation possibility is below 0.05.

3   Evolution Compensation Model of an Artificial Creature

In this chapter, evolution compensation models for NPC will be discussed. The NPCs move around in Game World. It competes with other NPCs near by, and tries to be an upper grade one.

In compensation model 1, the predator and food NPC groups will be applied same compensation model for evolution. Therefore, if certain NPC defeats opponent NPC whether in the same group or not, it will receive same evolution weight value such as +1. In the other way, defeated NPC's weight value will be decreased by −1. So it uses same compensation model and calculates evolution weight in NPC's own point of view.

Compensation model 2 is divided into following conditions by attributes of surrounding environments.

1. Friendly Relation
2. Opposite Relation
3. None

These are applied equally to all NPCs that include competing NPCs as well as their own. Therefore, if NPC has same attributes value as surrounding environment attribute, it will have bonus weight value by +1. The other hand, if it has opposite relation with location of map, it receives bonus weight value of −1. Finally, in condition that do not influence, it will have as same evolution weight as in compensation model 1. Differences between compensation model 1 and model 2 is whether it takes effect from surrounding environment or not. In compensation model 2, if predator NPC has opposite and opponent NPC has friendly relation, there is possibility that predator NPC loses in competition with food NPC.

4   System Design and Implementation

In this chapter, we are going to discuss how the system sets characteristic value for NPCs. How it distributes NPCs on the map automatically. The same rules are applied to distribution and characteristic grant for NPCs. The rules are as follows: food chain hierarchy and using map properties. These two rules let NPC have distinctive distribution and property. Also, distribution and properties will be continuously changed.

4.1   Distribution Environment

In this paper, four rules of flocking algorithm are applied to our simulation system. These rules are separation, align, cohesion, and avoidance. NPC group movement is based on above four rules. Also it makes NPC groups keep independent scope even though they are mixed together. However, there is no hierarchical relation between NPCs. Another words, NPC does not affect other NPCs in same class. This kind of relation is unsuitable for applying various NPC's characteristics to game. Therefore, competition rule should be applied for same class NPCs. Also we need chasing and evasion rule for relation between predator and prey. It is our goal to express real creatures' movement in our system.

We set elements in base level very simple in our simulation environment. However, movements will be expressed in various ways according to their interactions among group members and position in map.

Size of a NPC is 1x1 and size of map is 100x100 in simulation environment. This grating environment has been used for simulation of a capture hover action [5–8]. In our simulation environment, NPC groups are divided into three hierarchical levels. There is only one NPC group exist at each class. A number of groups at each class can be altered according to the experimental environment.

Table 1. Set Value of Each Class Object.

	number of object	striking power	energy	speed
1st Consumer	100	15	50	10
2nd Consumer	25	20	100	20
3rd Consumer	6	30	200	30

But in our environment system, the number of NPC groups should not go over three groups in total. The number of objects, striking power, energy, and moving speed values for each class are shown in Table 1.

4.2 Distribution Rules

Each NPC group may move around in the given world (100x100 grating map) freely. Each NPC do collision check in its 3x3 scope. Each NPC movement is generated basically by four rules of flocking algorithm. Chase, evasion, and competition rules are added. As a result, the movement will be expressed realistically as in the natural world. Initially each NPC is distributed at random on the map. It competes to secure group scope and to evolute. There is limitation and a rule for competition. They are speed of NPC and its visual limit. When each NPC moves around in a map, a speed and a visual limit are not fixed in our environment system. Because they are set up differently at each experiment. However, if high class NPC recognizes low class NPC as food, it cannot run away from the high class NPC. Because the low class NPC is always slower than a high class NPC.

There are ways to run away from a high class NPC. First case is when high class NPC blocked by barriers. So high class NPC is slowed down. Second case, if the low class NPC defeats high class NPC, it can run away from a high class NPC. Then how it defeats high class NPC. The energy of a high class NPC becomes zero, results of attacked by low class objects. All NPCs have hunger value. When the hunger value is getting higher than certain point, NPC starts hunting to bring down hunger value. There is limit for hunting. The number of NPC that may hunt is three at one hunting time. This prevents extermination of low class NPC, because the total number of low class NPC is four times of the total number of high class NPC. All NPCs will reproduce and transfer genetic code to their children at the extinction. Also, the newly born NPC will have the protection period that does not get attacked.

4.3 Characteristic Environment

Initially, designer decides characteristics for NPCs based on game character. The basic characteristic will be developed by competition and influenced by surrounding environment. Through this process, NPCs mutually evolve in game space. This is another subject of this paper.

Simulation environment for characteristic of NPCs is not much different from distribution environment. The main difference is setting on each NPC, not on

Table 2. Required Weight and Awarded Points.

	rise weight	increase energy	increase power
1st Consumer	1	2.5	0.15
2nd Consumer	3	5	0.2
3rd Consumer	3	10	0.3

NPC groups. Each NPC have value of {energy, striking power, level, attribute, inclination} for its own character. These characteristic values are used to construct genetic code by genetic algorithm. NPCs are awarded certain points whenever take victory in competition as shown in Table 2. NPCs will be leveled up according to those weight value.

The energy and striking power value are influenced by NPC's level. Hence physical strength and striking power value goes up along with NPC's level rise as shown in Table 2.

Inclination value displays whether NPC's inclination is offensive or defensive. NPC's inclination value could be positive or negative. Positive value expresses degree of offensive inclination. Negative value expresses degree of defensive inclination. If it beats surrounding NPC in competition, then 1 inclination weight is given. On the other hand, if it is defeated, -1 inclination weight is given. NPC's inclination value is inherited via generation. This NPC's inclination influences its group's inclination.

4.4 Characteristic Rules

In this paper, genetic algorithm transfers attribute of NPC to next generation. At this point, characteristic of NPC is distinguished by index such as Table 3. The degree of inclination is presented as numerical value.

The NPC group, which wins in competition continuously, is going to have offensive inclination. The defeated group is going to have defensive inclination. To inherit NPC inclination to next generation, NPC's inclination classified into attack-first or not-attack-first. They are distinguished by index. The degree of inclination is expressed as numerical value such as Table 3.

4.5 Genetic Composition

Our simulation system uses genetic algorithm to deliver changes of NPC to next generation. Genetic code is composed with value of factors. The factors are energy, striking power, attribute, and inclination. The value of each factor is composed in 5 digits. Detailed code example is shown in Table 4.

Table 3. Index Numbers for Attribute and Inclination.

ATTRIBUTE	INDEX		INCLINATION	INDEX
AQUA	1		ATTACK FIRST	1
FIRE	2		NO ATTACK FIRST	2
EARTH	3			

Table 4. Genetic Code Of Characteristic.

FACTOR	ENERGY	ATTACK POWER	ATTRIBUTE	INCLINATION
Factor Value	200	30	Fire 30	Attack First 150
Genetic Code	00200	00030	20030	10150

The genetic code is used to deliver parents' characteristic to its child. It also uses average value from its group characteristic value too. NPC will reproduce when it disappears. In hybridization, it uses 3 points hybridization operator, because hybridization point is 3.

5   Experiment and the Analysis Results

5.1   Simulation

In this simulation, the number of NPCs in each group is 100, 25, and 6. The mutation rate is fixed on 5% for experiment. The experimental model A is comparative model that is not applied any compensation model. Each compensation model contains weight value for character evolution. Compensation model 1 is not concerning any surrounding environment. This model only cares about special qualities of character. Compensation model 2 concerns surrounding environment as well as special qualities of character. This difference will show how NPCs evolve by themselves at what kind of environment.

The distribution rate between groups and characteristic data of NPCs are being recorded at every frame. Distribution rate is used to calculate how far each other. It also is used to find out common distributed regions for groups. The movement of largest group in each class is recorded. Each NPC's characteristic is recorded also. The character of each NPC group is presumed upon these data.

As a result of comparing and analyzing special quality data and the distribution rate which recorded in each experiment model. We can evaluate how the experiment model has evolved spontaneously.

5.2   Graph Analysis

In the graph, you can see the sharp differences in each experiment model. These differences are detected more clearly between experiment model A and experiment model C. In experiment model A, it ignores all special qualities of NPCs, but simply uses food chain relation. So it shows that all NPC groups are mixed and exist all over the map.

On the contrary, in experiment model C, there are many NPCs are gathering at the particular region which is friendly environment in the map. Also the primary consumer class NPCs are mainly distributed at the place with friendly relation. Also we can see that the secondary consumers hesitate to move into the location where relation is opposite. Even though there exist primary consumers as food.

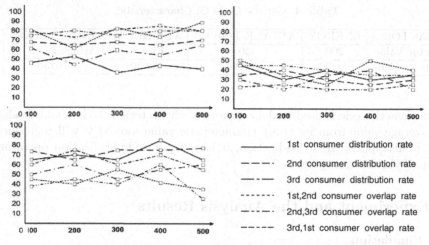

Fig. 1. Experimental Results of Model A, B, C.

A difference between experiment model A and experiment model B shows not much. In experiment model B, each NPC group has been distributed just by flocking algorithm, because they are designed to use only characteristic of NPC to compete with. This phenomenon breaks out in experiment model A as well. But, in experiment model B, overlapped territories does not appear much as in experiment model A. In experiment model B and experiment model C, the difference of distribution is similar to the difference in experiment model A and experiment model C. However, because of the map property, overlapped territories for groups has been showing as same as in comparison between experiment model A and experiment model C. Fig. 1 is graphs that show each NPC group's distribution and overlap according to their experiment model.

The graph of upper left corner in Fig. 2 is showing how actively each NPC group moves in the map. The size of 100x100 map is divided into the size of 10x10 gratings. Every grating has its own numbers. The area code in this graph is the grating number that has the largest quantity of NPC in the same class. Each NPC group actively moves around according to the movement change of surroundings. Meanwhile it maintains the movement of every NPC in the group. Data for the graph of upper left corner in Fig. 2 was taken after twenty generation. It shows how each group chase opponent group. It follows well enough as designed. This proves that each generation's special quality has been transmitted to next generation as special quality by gene algorithm. But, we could analogize that the attacking power will be changed. Because last graph in Fig. 2 shows that the physical strength change for individuals by generation.

6   Conclusion and Future Research

Game types have been changing to on-line game environment which massive number of players do enjoying together at same time. Also it is difficult to

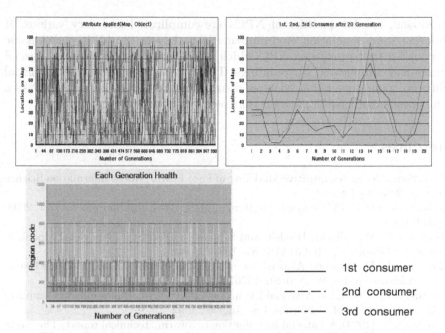

1st consumer

2nd consumer

3rd consumer

Fig. 2. Group Movement, Attack Value, and Energy Value Change.

apply ordinary non-network game element to on-line games to keep up with development of spreading fast Internet. These problems become an issue more and more Increasing maintenance expenses could not be ignored. Therefore, it is not right to decide how to distribute artificial creature and special quality of game at game designing stop. The system better include that distribution and special quality must be kept on changing by surrounding environment during the game.

Key points for this system are flocking and gene algorithm that describes action of living things and how living things evolve in ecosystem according to their environment. The proposed experiment model in this paper is as following.

- Model A: Experiment model that NPC does not evolve through competition
- Model B: Experiment model that NPC evolves without effect of surrounding environment
- Model C: Experiment model that is NPC evolves with effect of surrounding environment

In experiment model A, NPC is distributed by flocking algorithm. There is no fact about evolution by competition with other living things. But, the model B and C are competing to evolve. In this model, NPCs have new form of special quality other than initial special quality. It also shows different group distribution than the model A. This means that distribution and special quality of NPC has been altered according to surrounding environment. In this paper, NPC simulation system has limitation. The number of special qualities and individuals were minimized to simulate distribution and special quality. But, in actual game environment, special qualities of NPCs are complicated and very various. It is different from simulation environment. In real game world, it makes game server loaded that processing NPC which is influenced by such various special qualities and complicated surroundings. Therefore, It needs separated artificial intelligence game server that controls distribution and evolution for NPC. As a result, players can have impression that seems to play new game always.

References

1. B. K. Song, Design for Intuitive Modeling of FSM Editor, Korea Information Science Society, 2003: pp458 460
2. B. K. Song, NPC NPC developing system at MMORPG, Inje Journal vol 18.1, 2003: pp411 418
3. Reynolds, C. W. "Flocks, Herdsm and Schools: A Distribute Behavioral Model, in Computer Graphics", SIGGRAPH 87, 1987: 25 34
4. Glodberg, D. E., "Genetic Algorithms in Search, Optimization and Machine Learning", Addison-Wesley, ISBN 0-201-15767-5, 1989.
5. Collins, R.J., "Studies in Artificial Evolution", Phd Thesis, Philosophy in Computer Science, University of California, Los Angeles, 1992
6. Gutowitz, H. (1993) A Tutorial introduction to Swarm. Technical report, The Santa Fe Institute. Santa Fr Institute Preprint Series.
7. Kaza, J.R, "Genetic Programming: On the programming of computers by means of natural selection", MIT Press, ISBN 0-262-11170-5, 1992
8. Menczer93, F. & Belew, R.K. (1993) Latent energy environments: A tool for artificial life simulations. Technical Report CS93 298.

An Extensible Framework for Advanced Distributed Virtual Environment on Grid*

Seung-Hun Yoo, Tae-Dong Lee, and Chang-Sung Jeong**

School of Electrical Engineering in Korea University,
1-5ka, Anam-Dong,
Sungbuk-Ku, Seoul 136-701, Korea
{friendyu,lyadlove}@snoopy.korea.ac.kr, csjeong@charlie.korea.ac.kr

Abstract. This paper describes a new framework for Grid-enabled advanced DVE (Distributed Virtual Environment) which provides a dynamic execution environment by supporting discovery and configuration of resources, mechanism of security, efficient data management and distribution. While the previous DVEs have provided *static execution environment* only considering communication functions and efficient application performance, the proposed framework adds resource, security management, extended data management to static execution environment using Grid services, and then brings *dynamic execution environment* which result in QoS (Quality of Service) enhanced environment better performance.The framework consists of two components: Grid-dependent component and Communication-dependent component. Grid-dependent component includes RM (Resource Manager), SDM (Static Data Manager), DDM (Dynamic Data Manager), SYM (Security Manager). Communication-dependent component is composed of SNM (Session Manager) and OM (Object Manager). The components enhance performance and scalability through the DVEs reconfiguration considering resources, and provides mutual authentication mechanism of both servers and clients for protection of resources, application and user data. Moreover, effective data management reduces overhead and network latency by data transmission and replication.

1 Introduction

Research and development in DVEs have mainly progressed in two complementary directions, one addressing the application performance, the other addressing communication. While application and communication aspect have been well recognized and studied by traditional network traffic analysis and performance evaluation methods, DVE society didn't consider that the utilization and availability of resources may change computers and network topology when old components are retired, new systems are added, and software and hardware on existing systems are updated and modified. In large-scale DVE, a large number of

---

* This work has been supported by KOSEF and KIPA-Information Technology Research Center, University research program by Ministry of Information & Communication, and Brain Korea 21 projects in 2004.
** Corresponding author.

T.G. Kim (Ed.): AIS 2004, LNAI 3397, pp. 188–197, 2005.

participants interact with each other, and many entities are simulated simultaneously. Therefore, the continuing decentralization and distribution of software, hardware, and human resources make it essential to achieve the desired qualities of service (QoS). The effects make DVE society require new abstractions and concepts that allow applications to access and share resources and services across distributed networks. It is rarely feasible for programmers to rely on standard or default configurations when building applications. Rather, applications need to discover characteristics of their execution environment dynamically, and then either configure aspects of system and application behavior for efficient and robust execution or adapt behavior during program execution. Therefore, an application requirement for discovery, configuration, and adaptation is fundamental to the rapidly changing dynamic environment. We define an execution environment only considering DVE aspects as *static execution environment*, and the one considering both DVE and Grid [1] aspects as *dynamic execution environment*. The dynamic execution environment must support the following attributes:

 – Heterogeneity: Optimizing the architecture for performance requires that the most appropriate implementation techniques be used.
 – Reconfiguration: The execution environment must allow hardware and software resources to be reallocated dynamically. During reconfiguration, the application data must remain consistent and real-time constraints must be satisfied.
 – Extended data management: Reducing the volume of the traffic of data exchanged during communication of the object. For the management of a dynamic data, the movement and replication of data will be included probably.
 – Security: Both resources and data are often distributed in a wide-area network with elements administered locally and independently, which needs the intra or inter execution environment security.

This paper is organized as follows. Section 2 outlines CAVERNsoft G2 [4] and Globus (Grid middleware) [5]. Section 3 presents the architecture of proposed framework including four managers. Section 4 measures the performance on DVE for tank simulation as simple application, and compare it with previous DVE. Section 5 gives a conclusion.

2   Related Works

CAVERNsoft G2 (now called Quanta) was developed with consideration of Grid which supports the sharing and coordinated use of diverse resources and security in dynamic, distributed virtual organizations (VOs) [2]. CAVERNsoft G2 is an Open Source C++ toolkit for building Distributed networked applications, whose main strength is in providing networking capabilities for supporting high throughput Distributed applications.

Regardless of DVE society, Globus toolkit [5] was designed and implemented to support the development of applications for high-performance distributed computing environments. The Globus toolkit is an implementation of a bag of

Grid services architecture, which provides application and tool developers not with a monolithic system but rather with a set of standalone services. Each Globus component provides a basic service, such as authentication, resource allocation, information, communication, fault detection, and remote data access. Information services are a vital part of Grid infrastructure, providing fundamental mechanisms for discovery and monitoring using Globus Resource Allocation Manager (GRAM), and hence for planning and adapting application behavior [2]. Grid also provides Grid Security Infrastructure (GSI) which offers secure single sign-on and preserves site control over access policies and local security [3]. GSI provides its own versions of common applications, such as FTP and remote login, and a programming interface for creating secure applications.

3 Architecture of the Proposed Framework

The architecture of our proposed framework is shown in figure 1. It has been developed to provide a common framework for high-performance distributed applications. It supports managements for inter-application interactions, and provides services to applications in a way that is analogous to how a middleware provides services to applications. These management are arranged into six basic managers. The six managers describe the interface between the applications and CAVERNsoft G2 or Globus or both, which provides a framework for dynamic execution environment. Managers are divided into two categories as shown in figure 1: Communication-dependent component (CavernSoft G2) and Grid-dependent component. The former includes Session Manager (SNM), Object Manager (OM) and the latter Resource Manager (RM), Static Data Manager (SDM), Dynamic Data Manager (DDM), Security Manager (SYM). The managers are implemented based on the architecture of stub and skeleton in proxy design pattern. Clients create objects in server through object stub/skeleton and communicate with the objects in server, which is useful wherever there is a need for a more sophisticated reference to an object than a simple pointer or simple reference.

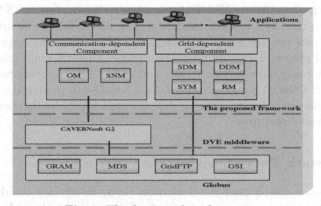

Fig. 1. The framework architecture.

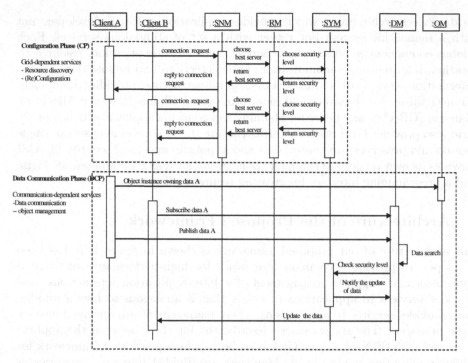

Fig. 2. Scenario.

Figure 2 shows scenario using the proposed framework. We divide the phases into two phases: Configuration Phase (CP), Data Communication Phase (DCP). The scenario is started as follows: First, ClientA requests the connection to SNM, and then SNM forwards the request to RM. RM requests resource information to MDS, and receives the response from MDS. RM can lock, unlock, allocate, co-allocate, and synchronize resources. Moreover, RM may specify Quality of Service (QoS) requirements with clients (applications) may declare the desired bandwidth using Globus Network Library in CAVERNsoft G2. After RM receives the security level from SYM, RM responses to the request to SNM. Finally, ClientA receives the best server information from SNM. When ClientB requests the connection, the process is repeated identical. After the configuration phase is ended, the next step is data communication phase. ClientA makes the instance of object owning any data by using OM, and then ClientB subscribes the data by using DM. If ClientA publishes the data, ClientB will receive the data. The relationship between CP and DCP is 1:N because DCP can happen N times when CP happened one time.

3.1   SDM (Static Data Manager)

For the large-scale distributed simulation on Grid, we divides the data model into static and dynamic data models. Static data model deals with data for a

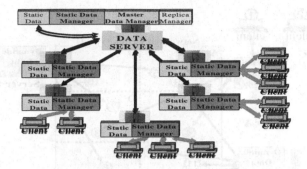

Fig. 3. client-server hierarchical model.

simulated environment such like terrain data, entity data and generally statically handled data. Dynamic data model uses data for the presentation of object state and behavior which are exchanged by the clients. For the management of the static data, static data model uses shared distributed model using client-server database. The overall structure composes of client-server hierarchies based on geographical location, and provides a scalable infrastructure for the management of storage resources across Grid environments. Figure 3 shows the hierarchical structure. SDM is connected to one or more clients, which in turn is connected to Master SDM called Data server which runs on grid environment. SDM provides the service about static data with clients by transmitting and storing them among clients. Master SDM sends distributed static data and shows information about them through a data catalog to clients. The data catalog does not store real distributed and replicated data but stores metadata for the data. Master SDM updates new information stored in data catalog, and manages it in transparent way. Data catalog contains the information of the replicated data. In large-scaled Grid environment, many clients are scattered geographically, and there arises the need for the replication of data due to the network bandwidth or long response time. Replica manager in SDM creates or deletes replicas at a storage site only to harness certain performance benefits of policies like communication efficiency.

3.2   DDM (Dynamic Data Manager)

DVE provides a software system through which geographically dispersed people all over the world may interact with each other by sharing in terms of space, presence, time [8]. The key aspect is scalability for interactive performance, because a large numbers of objects likely impose a heavy burden especially on the network and computational resources. Dynamic data occurs continuously for maintenance and consistency of the virtual world, and largely occupy the bandwidth of the network. Therefore, the dynamic data greatly determines the scalability of DVE and the number of participants. So, we proposes a multiple-server structure for a dynamic data model based on the virtual world. Each server is responsible for handling a segmented region of the virtual world for a characteristic of activity or event that happen in the virtual world. Figure 4

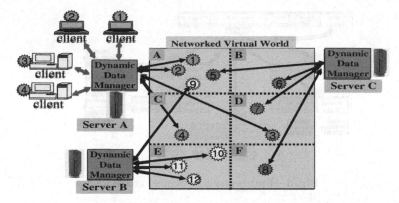

Fig. 4. Region segmentation and allocated multicast address.

shows the features how to divide a network virtual environment. Statically a virtual world is divided based on the visibility of each objects, each region is allocated one multicast address. Each server has a DDM which manages dynamic objects generated by clients connected to it, and share the information of with the clients connected to other DDMs. DDM operated in a server A, is managing four objects (1,2,3,4). Object 1 and 2 are active in region A, and object 3 in region D. Each of those object is managed by the same DDM regardless of its location until becomes extinct. Four objects (1,2,3,4) send their own dynamic data to their corresponding DDM which in turn distributes the data into other regions where the objects resides by sending the multicast address of each region. In case of DDM in Server A, data on 1 and 2 are sent to the multicast address of region A, while object of 3 and 4 are sent to the multicast address of region D and C respectively. DDM must join all the multicast groups which exist on the network virtual environment, and receives dynamic data from other DDMs. DDM filters the information, and only receives those from the regions where its managing objects resides, delivery them to the corresponding object. Therefore, each object receives dynamic information from other objects in the same region as it resides. Such region-based connection reduces network latency and bottleneck phenomenon which might be concentrated to the server.

3.3 SYM (Security Manager)

We are interested in DVE applications based on the proposed framework that integrates geographically distributed computing, network, information, and other systems to form a virtually networked computational resources. Computations span heterogeneous collections of resources, often located in multiple administrative domains. They may involve hundreds or even thousands of processes having security mechanism provided by GSI modules in Globus. Communication costs are frequently critical to performance, and programs often use complex computation structures to reduce these costs. The development of a comprehensive solution to the problem of ensuring security in such applications is clearly a complex problem.

4   Experiments

The experiment evaluates DVE based on the proposed framework, and compare it with the existing DVE using only CavernSoft G2. Figure 5 shows the specification of six Linux servers which are used in the experiment and other test conditions. First experiment is that a tank moves the same path on each server, and measure the execution time. Second experiment is that dynamic interactions of objects using proposed framework happen, and we measure data number.

	Server-1	Server-2	Server-3	Server-4	Servier-5	Server-6
OS	Redhat 9.0	Redhat 8.0	Redhat 9.0	Redhat 8.0	Redhat 9.0	Redhat 8.0
CPU	Pentium IV	Pentium III	Pentium IV	Pentium III	Pentium IV	Pentium III
Clock (MHz)	2400	2200	1700	1000	850	500
Memory (Mbytes)	1024	512	1024	512	256	128

Test Conditions :
 * One object transmits 100byte packet per update → 100 byte/object
 * 10 update/sec per one object are happened → 1kbyte/sec
 * 800 Object are experimented → 800Kbyte/sec
 * Network : 100 Mbps(LAN)

Fig. 5. Specification of Servers and Test Condition.

4.1   Evaluation of Configuration Phase (CP) and Data Communication Phase (DCP) Between a Server and a Client

Our scenario is divided into two parts: CP and DCP. The purpose of CP is to find a server which has the best performance. Performance difference according to the proper server assignment server has been examined by measuring the response time before the measurement of the time required in CP. The scenario is that the clients connect to each server on Grid environments and execute a tank simulation game. In DVEs, tank moves the fixed path on same map and we measure the time at fixed location (L1, L2, L3 and L4) on each path. Screen shot is shown in Figure 6. Figure 7 shows the result of execution that measures the time in fixed four locations on moving path. For security, authentication

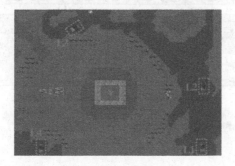

Fig. 6. Experiment 1.

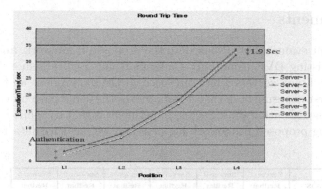

Fig. 7. The result of Experiment 1.

time is spent in L1. Figure 7 shows the difference of 1.9 second according to
server performance. As a result, authentication and information service spend a
little time. It is much smaller than the total time consumed in DVEs, and the
authentication and server assignment at that time interval plays an important
role to enhance the overall performance.

4.2   Evaluation of SDM and DDM on Many Servers

SDM and DDM are related to the data communication of objects. We measure
the number of packets from group A to group D shown in Figure 8. The shorter
object distance is, the more data communication happens. The more it takes
time, the more dynamic data happen. The experiment scenario is as follows:
800 tanks are located at fixed places. Each tank does not interact with another
tank because the distance between two tanks is longer than interaction distance
like shown figure 8(a). As time goes on, tanks move. First, the group with two

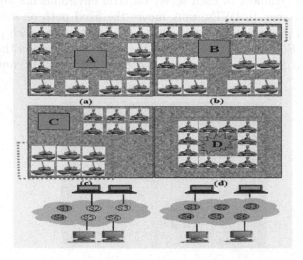

Fig. 8. Scenario of Tank Simulation.

interacting tanks is shown in figure 8(b). As interactions increase, processing time is increased, which decreases the data packet transfer rate. After group B, six tanks are grouped into one interaction group. Group C makes the data traffic more, but the processing time is needed more. Finally, 12 tanks are gathered into one group, which makes the data traffic busy with maximum processing time. As a result, more processing time reduces the data transfer rate. The testbed is composed of six servers and four clients. In testbed with only CavernSoft G2, four clients connects to fixed four servers (Server-2, 3, 5, 6). In testbed with proposed framework, four clients connect to session manager. After connection, each client connects to best servers.

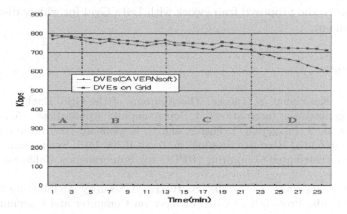

Fig. 9. The result of DVEs and DVEs on Grid.

Figure 9 shows the number of packets measured in DVEs using CAVERNsoft for both cases without and with our proposed framework. The number of packets with proposed framework is identical as time goes on, but that without proposed framework goes down as time goes on. The main reason of performance difference is that the proposed framework selects and allocates the best server among many servers, and it provides management of separately dynamic data.

5   Conclusion

This paper has described a new framework for Grid-enabled Distributed Virtual Environment which supports both the discovery, configuration of resources and the mechanism of security. The proposed framework has provided dynamic execution environment which consists of two components: Grid-dependent component and Communication-dependent component. Grid-dependent component includes RM (Resource Manager), SYM (Security Manager), SDM (Static Data Manager) and DDM (Dynamic Data Manager). Communication-dependent component composed of SNM (Session Manager) and OM (Object Manager). Enabling dynamic execution environment for large-scaled DVE with components: RM has knowledge of all hosts in the same VO (Virtual Organization) which will be allocated to the clients with rather better policies. SYM provides for enabling secure authentication and communication over the DVE, including mutual authentication and single sign-on. According to the classification of data model, SDM and DDM provide efficient data management about the static data and dynamic data. To support that the proposed framework increases performance, we did two experiments. First, we measured the execution time between a server and a client. The result showed that the server with better specification provides the better performance. In second experiment, we investigated the number of packets by increasing the interactions among objects. It showed that the method with proposed framework in this paper has an advanced performance than method which uses only communication (CavernSoft G2) component. In future, we are going to improve the proposed framework with Data Grid for widely distributed data management.

References

1. I. Foster, C. Kesselman, S. Tuecke. The Anatomy of the Grid: Enabling Scalable Virtual Organizations. International J. Supercomputer Applications, 15(3), 2001.
2. K. Czajkowski, S. Fitzgerald, I. Foster, and C. Kesselman. Grid Information Services for Distributed Resource Sharing. in Tenth IEEE Int'l. Symposium on High-Performance Distributed Computing (HPDC-10), San Francisco, California, August 2001.
3. I. Foster, C. Kesselman, G. Tsudik, S. Tuecke. A Security Architecture for Computational Grids. Proc. 5th ACM Conference on Computer and Communications Security Conference, pp. 83-92, 1998.
4. K. Park, Y.J. Cho, N. Krishnaprasad, C. Scharver, M. Lewis, J. Leigh, A. Johnson. CAVERNsoft G2: a Toolkit for High Performance Tele-Immersive Collaboration. ACM 7th Annual Symposium on Virtual Reality Software & Technology (VRST), Seoul, Korea, 2000.
5. I. Foster and C. Kesselman. The Globus Project: A Status Report. In Proceedings of the Heterogeneous Computing Workshop, pages 4-18, 1998.
6. Allcock, W., et al. Data Management and Transfer in High-Performance Computational Grid Environments. Parallel Computing, 2001.
7. A. Boukerche and A. Roy, In search of data distribution management in large scale distributed simulations, in "Proc. of the 2000 Summer Computer Simulation Conference, Canada," pp. 21-28.
8. S. Han and M. Lim, ATLAS-II: A Scalable and Self-tunable Network Framework for Networked Virtual Environments, The Second Young Investigator's Forum on Virtual Reality (YVR2003), Phoenix Park, Kangwon Province, Korea, 12-13 February, 2003.

Diffusion of Word-of-Mouth in Segmented Society: Agent-Based Simulation Approach

Kyoichi Kijima[1] and Hisao Hirata[2]

[1] Tokyo Institute of Technology, Graduate School of Decision Science and Technology,
2-12-1 Ookayama, Meguro-ku, Tokyo 152-8552, Japan
kijima@valdes.titech.ac.jp
http://www.valdes.titech.ac.jp/~kk-lab/
[2] Booz Allen and Hamilton Inc, Japan
http://www.boozallen.jp/contactus/contactus.html

Abstract. The present research examines, by using agent-based simulation, how word-of-mouth about a new product spreads over an informal network among consumers. In particular, we focus on clarifying relationship between diffusion of word-of-mouth and network structure of the society. Whether or not there is any essential difference of diffusion process between in a mosaic and in an oligopolistic society is one of our main questions. The findings obtained not only are insightful and interesting in academic sense, but also provide useful suggestions to marketing practice.

1 Introduction

The present research examines, by using agent-based simulation, how word-of-mouth about a new product spreads over an informal network among consumers, with particular emphasis on the relationship between social structure and the diffusion power of the information. Agent-based simulation is a remarkable approach to observe an entire event that emerges out of local interactions between agents and comprehend the mechanism of emergence [1]. Some authors have applied it to investigation of diffusion process: Axelrod [1] proposes a disseminating culture model, in which agents in a population are culturally integrated or segmented through interactions with neighbors. The model successfully shows the diffusion of cultural attributes. Midgley et al [7] investigate the relation between the diffusion of innovations and the network structure to find that the communication links have a significant impact on the diffusion process.

We are interested particularly in diffusion process of word-of-mouth about such a product that has not been advertised over the major mass media yet, like a new movie. Since "authorized" or "official" information is not available, potential consumers have to decide according to evaluations, or word-of-mouth, from their friends, neighbors and so on. If they purchase the product, the consumers may transmit their evaluation about the product to others. We will argue relationship between diffusion of word-of-mouth and the network structure of society by investigating whether or not there is any essential difference of diffusion process between in a mosaic and oligopolistic society.

We believe that the findings obtained not only are insightful in academic sense, but also provide useful suggestions to marketing practice.

T.G. Kim (Ed.): AIS 2004, LNAI 3397, pp. 198–206, 2005.
© Springer-Verlag Berlin Heidelberg 2005

2 WOM Model and Its Assumptions

2.1 Structural Assumptions

Social Structure

Our word-of-mouth (WOM) model assumes the society is segmented into several groups. We assign group ID, say, 1, 2,..., to each group. We specify structure of society by using the idea of bipartite network model, instead of the traditional model (See Fig. 1).

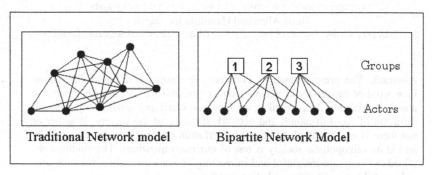

Fig. 1. Traditional and Bipartite Network Model

Each agent belongs to at least one group, which we call his/her primal group. The characteristics of the primal group determine his/her attributes to some extent. Each agent can contact other members beyond his/her primal group with the maximum of five. Hence each agent is associated with a link code composed of five-digit numbers, such as '11213', where each digit represents group ID. The order of five numbers is not important, but how many times the group ID appears in the code is crucial. For example, if the five-digit link code of an agent is "11213", then it implies that the agent contacts the members whose primal group is 1 with probability of 3/5 since "1" appears three times in it. Similarly, the agent interacts the members of group 2 with probability of 1/5.

The link code of each agent is determined by the interaction rate between groups, which is a characteristic of the society (See Fig. 2). If the society is composed of cohesive and mutually exclusive groups, then the rate is small. On the other hand, if the society shows animated interactions between groups, then the rate is high. For example, if the group interaction rate of the society is 0.40, in the link code of an agent the digit of his/her primal group ID appears with probability of 0.60, while other digits come with probability of 0.40. It implies that an agent is likely to transmit information to the members of the same primal group with probability of 0.60.

Diffusion of Information

An agent is assumed to transmit word-of-mouth to another agent based on the SIR model [6], where agents are classified into three types: I-, S- and R-type (Refer to Fig. 3). I-agent (infected agent) is an agent who knows the information and is willing to let others know it because the agent is excited by the information. S-agent (susceptible agent) has not been infected yet but can be so through interaction with I-agent. I-agent shifts to R- agent (removed agent) after a certain period passes. R-agent stops diffusing information and can never be infected again because the agent has lost interest in the information. On the contrary, I-agent can be infected again before the agent changes to be removed. Each I-agent tries to transmit information to other susceptible or infected agents who share at least one group ID in their link codes, so that possible targets for infection are limited by the link codes.

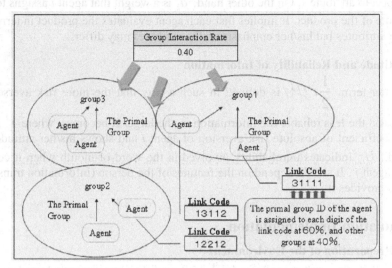

Fig. 2. Scheme for assigning link codes based on the group interaction rate

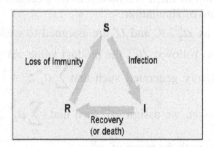

Fig. 3. Three states of SIR model

2.2 Assumptions on Decision Activity of Agents

Agent i calculates his or her *enthusiasm about the product caused by word-of-mouth from j* by

$$E^{ij} = \sum_k a_k^i z_k - \frac{1}{2} r^i U^j \qquad (1)$$

and makes a decision on purchase of the product based on it. In the simulation algorithm, the value is adjusted between 0.0 and 1.0, so as to be used as probability of the purchase. The value is determined by the feature of the product (the former term) as well as by the agent's attitude towards risk and reliability of the information transmitter (the latter term).

Utility Due to the Product

The former term, $\sum_{k} a_k^i z_k$, represents utility due to the product itself for agent i,

where the product is characterized by k attributes and z_k is the value of the product with respect to attribute k. On the other hand, a_k^i is a weight that agent i assigns to the kth attribute of the product. It implies that each agent evaluates the product in terms of the same attributes but his/her emphasis on the attributes may differ.

Risk Attitude and Reliability of Information

In the latter term, $\dfrac{1}{2} r^i U^j$, is defined in such a way that the more risk averse the

agent is, and the less reliable the information is, the larger it becomes, where r^i indicates a coefficient of absolute risk aversion of agent i and shows his/her attitude toward risk. U^j indicates unreliability involved in the word-of-mouth when it comes from an agent j. It should depend on the features of the person (information transmitter) who provides it.

3   Simulation Implementation

Step 1: Preparation of the Environment

1) First of all, the attribute values $\{z_k\}$ of the product are generated randomly in such a way that the sum of them is adjusted to be a certain fixed value. They are unchanged until the end of simulation.

2) The parameters such as a_k^m, r^i, and U^j are assigned to each agent depending on his/her primal group as follows: a_k^m, the product value with respect to attribute k for group m, is randomly generated such that $\sum_{k} a_k^m z_k = 1.0$. Then, to agent i whose primal group is m, we assign a_k^{im} such that $\sum_{k} a_k^{im} z_k = 1.0$ according to the normal distribution with the mean of a_k^m.

3) U^m for group m is derived by using the uniform distribution. Then, to agent i whose primal group is m, we assign U^{im} according to the normal distribution such that the mean is equal to U^{im}.

4) The coefficient of risk attitude r^m for group m is derived from the normal distribution with the mean of -1.0 and the standard deviation is 0.5. It means negative coefficients are assigned to groups with probability of 0.25. (Such groups consist of risk-prone agents, who prefer uncertainty of information.) The value 0.25 corresponds to what Rogers [9] mentions as the probability of innovators. To agent i whose primal group is m, r^{im} is assigned by the normal distribution in such a way that the mean is equal to r^{im}.

5) The link code is also generated for each agent in accordance with the group interaction rate.

Step 2: Running of Simulation (See Fig. 4)

The simulation begins with choosing a seeding way. Seeding is an activity to start to incubate word-of-mouth among particular agents. Roughly, seeding can be divided into two types; dense seeding and sparse seeding. The former is seeding a specific group densely, while the latter is seeding over multiple groups in a scatter way.

1) Initial infected agents, or seeded agents, are selected. Only infected agents are motivated to contact with another and to transmit information.
2) Each of them randomly picks up one agent as a target and chooses a digit of the target's link code. The activated agent also chooses one digit from its own link code. If both the digits are the same, the agent contacts with the target, while, if not, the agent looks for another agent.
3) The agent repeats picking up a target over and over again until the agent finds a valid one. According to this process, the agent contacts only with those who belong to the same group with positive probability. If the group interaction rate is very low, they interact almost only within their primal group.
4) After the agent selects a valid target, the SIR status of the target is checked. If the target is R, the activated agent cannot make the target infected. Otherwise, the agent makes the target infected in accordance with the value calculated by (1).
5) If the target gets infected, the agent transmits information about the product to the target. At the same time, the infection period of the target is reset even if another agent has already infected the target.
6) At the end of each period, for all infected agents it is checked whether their infection periods are expired or not. If so, their status change to R and they stop transmitting information to others. The information never infects them again.

Step 3: The End of Simulation

Simulation finishes when all the agents become either S or R. Since all infected agents must be changed to R sometime, the transmission stops within finite repetitions.

4   Evaluations and Analysis of Simulation Results

4.1   Evaluation Criteria

We will evaluate word-of-mouth effect by two criteria, *i.e.*, Maximum Instantaneous I Population Rate (MIIPR) and IR Experienced Population Rate (IREPR). They are measured by MIIPR = $Max_t N_{it} / N$, and IREPR = N_{Rt*} / N, respectively, where N denotes the number of population in the society. N_{it} shows the number of infected agents at period t while N_{Rt*} denotes that of removed agents at period $t*$, where $t*$ is the end of the simulation where all infected agents have changed to R.

Since the MIIPR indicates the maximum momentary rate of infected population over all periods, it shows how big enthusiasm occurs through word-of-mouth communication. If a product is associated with high MIIPR, it is worth considering possibility of broadcasting it, say, by mass media as publicity in the society. On the other hand, the IREPR indicates effect of word-of-mouth itself, since it is a rate of all R agents against the whole population at the end. If it is high, we may say that information prevails in the society through word-of-mouth communication.

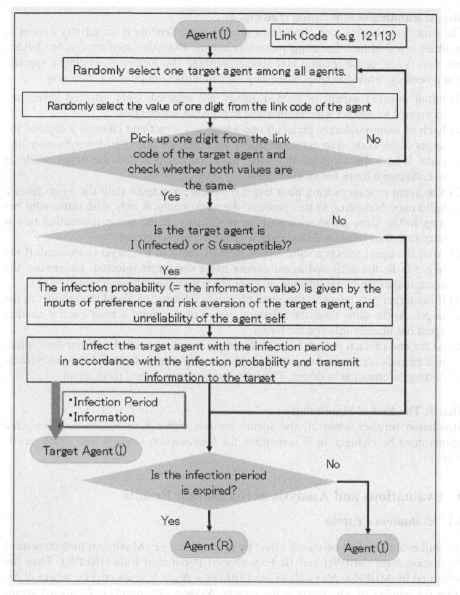

Fig. 4. Flow chart of the WOM model

4.2 Discussions on Simulation Results

1) Our first observation from the simulation is that "The effect of the group interaction rate is limited; too much interaction between groups does not necessarily contribute to the further spread of word-of-mouth" (See Figs. 5 (a) and (b)).

Each figure compares the MIIPR of both oligopolistic and mosaic societies, where the MIIPR is measured at every 0.05 group interaction rate. The parameters of the

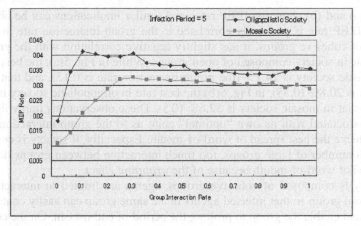

Fig. 5a. MIIPR of oligopolistic and mosaic society with short infection period

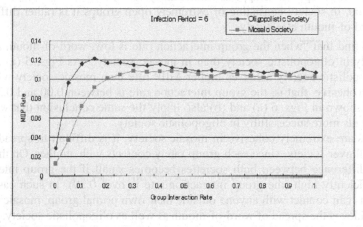

Fig. 5b. MIIPR of oligopolistic and mosaic society with long infection period

seeded group are randomly assigned, and dense seeding is adopted in both societies. Fig. 5(a) shows the case where the infection period is short, namely 5, while Fig. 5(b) does that where it is long, *i.e.,* 6.

When groups in society are collective (the group interaction rate is between 0.00 and about 0.30), the MIIPR is positively correlated to the group interaction rate. Namely, the more groups communicate with one another, the easier it is for the society to enthuse over word-of-mouth. As far as oligopolistic society is concerned, the best MIIPR is 4.1% in the case of Fig. 5(a) and it is 12.1% in the case of Fig. 5(b). Similarly, for a mosaic society the best MIIPR is 3.3% in the case of Fig. 5(a), and it is 10.8% in the case of Fig. 5(b).

On the other hand, if groups in the society are sufficiently open (the group interaction rate is over 0.30), effect of the group interaction rate is almost indifferent. If anything, the MIIPR slightly has negative correlation with the group interaction rate especially when society is oligopolistic.

2) Figs. 6 (a) and (b) show the IREP rates and similar implications can be observed. While the IREP rate is positively correlated to the group interaction rate in society composed of cohesive groups, it has slightly negative correlation with the group interaction rate in society composed of open groups. While in Fig. 5(a), the best IREPR of oligopolistic society is 24.1% (the group interaction rate is 0.15), and that of mosaic society is 20.8% (0.35), in Fig. 5(b) the best rate in oligopolistic society is 52.5% (0.25), and that in mosaic society is 52.6% (0.5). These observations imply that each society is associated with its own "optimal" point as to the group interaction rate in order to achieve the best spread of word-of-mouth. Especially, if society is composed of the small number of large groups, too much interaction between groups decreases the diffusion of word-of-mouth because of the *wrapping effect*.

If society is composed of cohesive groups, agents are limited to interact almost within its own group so that infected agents in the same group can easily contact with each other and enable the group to prolong the period of enthusiasm. On the contrary, an extremely open group cannot mature enthusiasm within it, and infection quickly spreads over all groups but disappears immediately.

To sum up, in society consisting of extremely open groups it is rather difficult to diffuse word-of-mouth effectively.

3) We also find that "when the group interaction rate is low, word-of-mouth spreads more widely in oligopolistic society than in mosaic society." As Figs. 5 (a) and (b) show, oligopolistic society achieves higher MIIP rate than mosaic society when the groups are cohesive, that is, the group interaction rate is between 0.00 and 0.30. The IREP rates shown in Figs. 6 (a) and (b) also imply the same conclusion that word-of-mouth prevails more successfully in oligopolistic society.

If groups are extremely cohesive in mosaic society, it is difficult to spread word-of-mouth all over society, since each group rarely contacts with another. On the other hand, the difference between both societies becomes small if the group interaction rate is sufficiently high (the group interaction rate is over 0.30). In such cases, because agents can contact with anyone outside their own primal group, mosaic society can also achieves the spread of word-of-mouth as well as oligopolistic society.

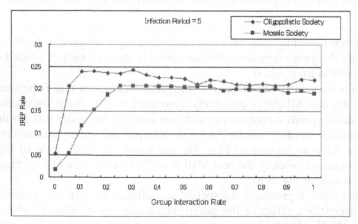

Fig. 6a. IREPR of oligopolistic and mosaic society with short infection period

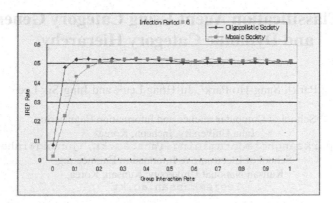

Fig. 6b. IREPR of oligopolistic and mosaic society with long infection period

5 Conclusions and Future Studies

This study examined diffusion process of word-of-mouth in a group-segmented society by agent-based simulation. Some of unique and interesting insights obtained are as follows: First, cohesiveness of each group gives big influence on the process. When the interaction rate is relatively low, *i.e.*, when society is segmented into cohesive groups, interaction rate between groups is positively correlated to spread of word-of-mouth. However, the correlation changes to be slightly negative if the group interaction rate is over the optimal point. These suggest that there is the optimal cohesiveness for spread of word-of-mouth in a market. Secondly, in oligopolistic society it is easier to prevail word-of-mouth than in mosaic society, especially when the groups are cohesive. It implies that if a market is segmented into large groups, the marketers have rooms for playing an active role.

These implications can be useful for marketers who try to spread word-of-mouth about a product. What is the most important for them is to know what kinds of groups are in the market and to know the best seeding strategy.

References

1. Axelrod, Robert. *The Complexity of Cooperation,* Princeton University Press, 1997
2. Epstein, Joshua M. and Axtel, Robert. *Growing Artificial Societies: Social Science from the Bottom Up.* MIT Press, 1996
3. Gilbert, Nigel and Troitzsch, Klaus G. *Simulation for the Social Scientist,* Open University Press, 1999
4. Hirata, Hisao. *Word-of-mouth Effect: Simulation Analysis based on the Network Structure,* Master Thesis, Dept. of Value and Decision Science, Tokyo Institute of Technology, 2004
5. Huhns, M. and Singh, M.P. *Readings in Agents,* Morgan Kaufman, 1998
6. Kermack, W.0. and McKendrick, A.G. A contribution to the mathematical theory of epidemics, *Proceedings of the Royal Society of London, Series A,* 115:700-721, 1927
7. Midgley, David F., Morrison, Pamela D., and Roberts, John H. The effect of network structure in industrial diffusion processes, *Research Policy,* 21:533-552, 1992
8. Milgrom, Paul and Roberts, John. *Economics, Organization and Management,* Prentice Hall, 1992
9. Rogers, Everett M. *Diffusion of Innovations,* Fifth Edition, Flee Press, 2003
10. Schellillg, T.C. Models of segregation, *American Economic Review,* 59(2):488-493, 1969.

E-mail Classification Agent Using Category Generation and Dynamic Category Hierarchy

Sun Park[1], Sang-Ho Park[1], Ju-Hong Lee[1], and Jung-Sik Lee[2]

[1] School of Computer science and Information Engineering,
Inha University, Incheon, Korea
{sunpark,parksangho}@datamining.inha.ac.kr, juhong@inha.ac.kr
[2] School of Electronic & Information Engineering,
Kunsan National University, Kunsan, Korea
leejs@kunsan.ac.kr

Abstract. With e-mail use continuing to explode, the e-mail users are demanding a method that can classify e-mails more and more efficiently. The previous works on the e-mail classification problem have been focused on mainly a binary classification that filters out spam-mails. Other approaches used clustering techniques for the purpose of solving multi-category classification problem. But these approaches are only methods of grouping e-mail messages by similarities using distance measure. In this paper, we propose of e-mail classification agent combining category generation method based on the vector model and dynamic category hierarchy reconstruction method. The proposed agent classifies e-mail automatically whenever it is needed, so that a large volume of e-mails can be managed efficiently

1 Introduction

The e-mail user spends most of their time on organizing e-mails. Many tools have been developed, with the aim of rendering help in the task of users to classify their e-mail, for example defining filters, i.e., rules that allow the classification of a message. Such tools, however, are mainly human-centered, in the sense that users are required to manually describe rules and keyword list that can be used to recognize the relevant features of messages. Such approaches have the following disadvantages: 1) the efficiency of classifying e-mails declines when a large number of e-mails contain too many overlapping concepts.; 2) it is necessary for users to reorganize their e-mails or filters[5].

The previous work on classifying e-mails focused on detecting spam messages. Cohen[2] described rule-based systems using text mining techniques that classify e-mails. Androutsopoulos[1] and Sakkis[8] described Bayesian classifier that filters out anti-spams. Drucker[4] used support vector machines(SVM's) for classifying e-mail as spams or non-spams. But, these approaches make users directly construct message folders that can contain received messages. The learning and testing of classifications is also called for before classifying e-mails practically.

Manco and Masciari[5] used clustering for managing and maintaining the received e-mails efficiently. This method has the complexity of computation, requiring preprocessing over several steps and using the similarity between various extracted information.

T.G. Kim (Ed.): AIS 2004, LNAI 3397, pp. 207–214, 2005.
© Springer-Verlag Berlin Heidelberg 2005

In this paper, we propose the novel e-mail classification agent for the purpose of solving these problems. The proposed method combines the generation method of category label based on the vector model and the dynamic category hierarchy reconstruction method. The proposed method in this paper has the following advantages: First, they don't require user intervention because the subject-based category labels is automatically generated; second, Categories are reclassified whenever users want to reconstructed e-mails using a dynamic category hierarchy reconstruction method; third, a large number of e-mails can be managed efficiently; and fourth, our proposed method doesn't require learning, thus it is adequate for an environment that it has to classify e-mails quickly.

The rest of this paper is organized as follows. In Section 2, we review the vector model. In Section 3, the dynamic reconstruction of category hierarchy is described. Section 4, we propose the e-mail classification agent using a category label generation method and dynamic category hierarchy reconstruction method. Section 5, some experimental results are presented to show efficiency of the proposed method. Finally, conclusions are made in Section 6.

2 Vector Model

In this section, we give a brief introduction to the vector model[7] that is used in this paper. The vector model is defined as follows.

Definition 1. For the vector model, the weight $w_{i,j}$ associated with a pair(k_i, d_j) is positive and non-binary. Further, the index terms in the query are also weighted. Let $w_{i,q}$ be the weight associated with the pair $[k_i, q]$, where $w_{i,q} \geq 0$. Then, the query vector $\vec{q}$ is defined as $\vec{q} = (w_{1,q}, w_{2,q}, \dots , w_{t,q})$ where t is the total number of index terms in the system. As before, the vector for a document d_j is represented by $\vec{d} = (w_{1,j}, w_{2,j}, \dots, w_{t,j})$.

The vector model proposes to evaluate the degree of similarity of the document d_j with regard to the query q as the correlation between the vectors $\vec{d}_j$ and $\vec{q}$. This correlation can be quantified, for instance, by the cosine of the angle between these two vectors. That is,

$$sim(d_j, q) = \frac{\vec{d}_j \bullet \vec{q}}{|\vec{d}_j| \times |\vec{q}|} = \frac{\Sigma_{i=1}^{t} w_{i,j} \times w_{i,q}}{\sqrt{\Sigma_{i=1}^{t} w_{i,j}^2} \times \sqrt{\Sigma_{j=1}^{t} w_{i,q}^2}} \tag{1}$$

Definition 2. Let N be the total number of documents in the system and n_i be the number of document in which the index term k_i appears. Let $freq_i$ be the raw frequency of term k_i in a document d_j (ie., the number of times the term k_i is mentioned in the text of a document d_j). Then, the normalized frequency $f_{i,j}$ of term k_i in a document is give in by

$$f_{i,j} = \frac{freq_{i,j}}{max_l \times freq_{l,j}} \tag{2}$$

where the maximum is computed over all terms which are mentioned in the text of a document d_j. If the term k_i does not appear in a document d_j then $f_{i,j}=0$. Further, let idf_i, inverse document frequency for k_i, be given by

$$idf_i = \log \frac{N}{n_i} \qquad (3)$$

The best known term-weighting scheme uses weights which are given by

$$w_{i,j} = f_{i,j} \times \log \frac{N}{n_i} \qquad (4)$$

Or by a variation of this formula. Such a term-weighting strategy is called the *tf-idf* scheme.

3 Dynamic Restructuring of Category Hierarchy

We define dynamic restructuring of category hierarchy[3]. In this section, we give a brief introduction to the Fuzzy Relational Products that is used in Dynamic restructuring of Category Hierarchy. The fuzzy set is defined as follows.

Definition 3. α-cut of a fuzzy set A, denoted by A_α, is a set that contains all elements whose membership degrees are equal to or greater than α. $A_\alpha = \{ x \in X \mid \mu_A(x) \geq \alpha \}$.

A fuzzy implication operator is an extended crisp implication operator to be applied in the fuzzy theory. A crisp implication operator is defined as $\{0,1\} \times \{0,1\} \rightarrow \{0,1\}$, while a fuzzy implication operator is defined as $[0,1] \times [0,1] \rightarrow [0,1]$ to be extended in multi-valued logic. We use the implication operator defined as follows [6].

$$a \rightarrow b = (1-a) \vee b = \max(1-a, b), \ a = 0 \sim 1, b = 0 \sim 1 \qquad (5)$$

In set theory, "$A \subseteq B$" is equal to "$\forall x$, $x \in A \rightarrow x \in B$", and it is also equal to "$A \in \wp(B)$". Here $\wp(B)$ is the power set of B. Thus, in the fuzzy set, the degree of $A \subseteq B$ is the degree of $A \in \wp(B)$, so it is denoted by $\mu_{\wp(B)}A$ and defined as follows:

Definition 4. Given the fuzzy implication operator $\rightarrow$, a fuzzy set B of a crisp universe set U, the membership function of the power set of B, $\mu_{\wp(B)}$ is $\mu_{\wp(B)}A = \bigwedge_{x \in U} (\mu_A x \rightarrow \mu_B x)$.

Definition 5. Let U_1, U_2, U_3 be finite sets, R be a fuzzy relation from U_1 to U_2, S be a fuzzy relation from U_2 to U_3. That is, R is a fuzzy subset of $U_1 \times U_2$ and S is a fuzzy subset of $U_2 \times U_3$. Fuzzy relational products are fuzzy operators that represent the degree of fuzzy relation from a to c for a $\in U_1$, c $\in U_3$. The fuzzy triangle product as a fuzzy relation from U_1 to U_2, $\triangleleft$ is defined as follows

$$(R \triangleleft S)_{ik} = \frac{1}{N_j} \sum_j (R_{ij} \rightarrow S_{jk})$$

This is called *Fuzzy Relational Products*.

Definition 6. The fuzzy implication operators vary in the environments of given problems. The afterset aR for a $\in U_1$ is a fuzzy subset of U_2 such that y is related to a, for $y \in U_2$. Its membership function is denoted by $\mu_{aR}(y) = \mu_R(a,y)$. The foreset Sc for $c \in U_3$ is a fuzzy subset of U_2 such that y is related to c, for $y \in U_2$. Its membership function is denoted by $\mu_{Sc}(y) = \mu_S(y,c)$ for $y \in U_2$. The mean degree that aR is a subset of Sc is meant by the mean degree such that the membership degree of $y \in aR$ implies the membership degree of $y \in Sc$, so it is defined as follows:

$$\pi_m(aR \subseteq Sc) = \frac{1}{N_{U_2}} \sum_{y \in U_2} (\mu_{aR}(y) \rightarrow \mu_{Sc}(y)) \tag{6}$$

Here, π_m is a function to calculate the mean degree ♦

The above mean degree denoted by $R \lhd S$ can be regarded as the mean degree of relation from a to c [9].

By applying the fuzzy implication operator of the formula (5) to the fuzzy relational products of the formula (6), we can get the average degree of fuzzy sets inclusion for categories, $\pi_m(C_i \subseteq C_j)$. We interpret this degree as the *similarity relationship degree* of C_i to C_j , it is the degree to which C_i is similar to C_j , or we interpret it as the *fuzzy hierarchical degree* of C_i to C_j , it is the degree to which C_j can be a subcategory of C_i. An attention is required from the fact that the hierarchical ordering is reverse to the concept of set inclusion in that C_j is the superset of C_i, but C_j is the hierarchical subcategory of C_i. Intuitively, it is conjectured that the category comprising many indices likely inherits the properties from super-class categories. However, $\pi_m(C_i \subseteq C_j)$ have some problems representing the fuzzy hierarchical degree of C_i to C_j. That is, if C_i had many element x's of which membership degrees, $\mu_{Ci}(x)$, are small, we could have a problem in which the fuzzy relational products tend to be converged to 1 regardless of the real degree of fuzzy sets inclusion of $C_i \subseteq C_j$. Thus, we define *Restricted Fuzzy Relation Products* as follows to calculate the real degree of fuzzy sets inclusion of two categories, $\pi_{m,\beta}(C_i \subseteq C_j)$.

$$\pi_{m,\beta}(C_i \subseteq C_j) = (R^T \lhd_\beta R)_{ij} = \frac{1}{|C_{i\beta}|} \sum_{K_k \in C_{i\beta}} (R^T_{ik} \rightarrow R_{kj}) \tag{7}$$

Here, K_k is the k'th category, and C_i, C_j are the i'th and the j'th category respectively, $C_{i,\beta}$ is C_i's β-restriction, that is, $\{x \mid \mu_{C_i}(x) \geq \beta\}$, $|C_{i,\beta}|$ is the number of elements in $C_{i,\beta}$. R is $m \times n$ matrix such that R_{ij} is $\mu_{C_j}(K_i)$,that is, the membership degree of $K_i \in C_j$. R^T is the transposed matrix of R such that $R_{ij} = R^T_{ji}$.

4 E-mail Classification Agent

The e-mail classification agent consists of a preprocessing, automatic category generation and dynamic category hierarchy reconstruction steps. The Preprocessing step performs the task of extracting the index terms, such as representative subjects and contents from e-mails. The automatic category label generation generates category label according to a similarities between index terms using category generation method. Then an e-mail is classified to a category. The dynamic category hierarchy reconstruction reconstructs the category hierarchy of e-mails using a dynamic cate-

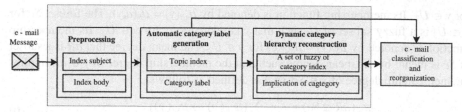

Fig. 1. E-mail classification agent system

gory reconstruction method whenever it is needed. Figure.1 represents the diagram of e-mail classification agent system.

4.1 Category Generation Method

In this section, we propose a category generation method. Using formula (1), this method generates the topic index terms that means the index terms having the highest degree of representativeness. The next step generates a category label automatically. The message is then classified according to the categories.

But, in the case of topic index terms not containing the meaningful information, the method cannot represent the actual meaning, therefore formula (1) is unnecessary or classify e-mails incorrectly. In order of to resolve this problem, we use the dynamic category hierarchy reconstruction method to reorganize the category hierarchy made by classification through category generation method.

4.2 Reorganization of E-mails Using the Dynamic Restructuring of Category Hierarchy

In this paper, the relationship between indices and categories can be decided by normalized term frequency values between 0 and 1 (meaning a fuzzy degree of membership). That is, a category can be regarded as a fuzzy set comprising the indices appearing in the e-mails pertaining to the corresponding category. To check the similarity of two categories, we may have to use a more sophisticated method rather than the similarity check method based on the vector models. Actually, in the domain of categories, we must consider the hierarchical inclusion between categories as well as the similarity between categories. That is why we rely on the fuzzy relational products, which is able to describe the inclusive relation of two objects. The relationship of two categories can be decided by calculating the average degree of inclusion of a fuzzy set to another one using the fuzzy implication operator. An average degree of fuzzy set inclusion can be used for making a similarity relationship between two categories. By using this, similarity relations of categories can be obtained dynamically. The following example is derived from formula (7).

Example 1. When $\beta = 0.9$, $\pi_{m,\beta}(C_2 \subseteq C_4)$ denoted by $(R^T \lhd_\beta R)_{24} = 0.05$, and $\pi_{m,\beta}(C_5 \subseteq C_3)$ denoted by $(R^T \lhd_\beta R)_{53} = 0.60$. *Fuzzy hierarchical* relations of any two categories are implemented by *Restricted Fuzzy Relational Products* (7) as shown in the tables (a) and (b).

We transform values of $(R^T \triangleleft_\beta R)$ to crisp values using α-cut. (a) of table 2 is the final output of $(R^T \triangleleft_\beta R)$ using $\alpha = 0.94$. That is, the ones lower than 0.94 are discarded as 0, the ones greater than or equal to 0.94 are regarded as 1. (b) of table 2 is the final output using $\alpha = 0.76$.

Table 1. Restricted Fuzzy Relational Products of categories and indices

	I_1	I_2	I_3	I_4	I_5
C_1	0.9	1.0	1.0	1.0	1.0
C_2	0.0	1.0	0.1	0.0	1.0
C_3	1.0	0.8	0.0	1.0	1.0
C_4	0.0	0.0	1.0	0.0	0.1
C_5	0.0	1.0	1.0	0.8	1.0

(a) R^T

	C_1	C_2	C_3	C_4	C_5
C_1	0.98	0.44	0.76	0.24	0.78
C_2	1.00	1.00	0.90	0.05	1.00
C_3	0.97	0.33	1.00	0.03	0.60
C_4	1.00	0.10	0.00	1.00	1.00
C_5	1.00	0.70	0.60	0.37	1.00

(b) $(R^T \triangleleft_\beta R)$

Table 2. The final output of $(R^T \triangleleft_\beta R)$ using α-cut

	C_1	C_2	C_3	C_4	C_5
C_1	1	0	0	0	0
C_2	1	1	1	0	1
C_3	1	0	1	0	0
C_4	1	0	0	1	1
C_5	1	0	0	0	1

(a) $\alpha = 0.94$

	C_1	C_2	C_3	C_4	C_5
C_1	1	0	1	0	1
C_2	1	1	1	0	1
C_3	1	0	1	0	1
C_4	1	0	0	1	1
C_5	1	1	1	0	1

(b) $\alpha = 0.76$

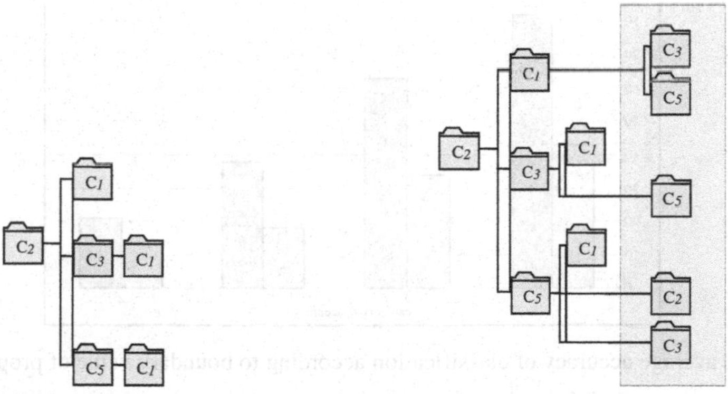

(a) $\alpha = 0.94$　　　　　　　　(b) $\alpha = 0.76$

Fig. 2. Fuzzy hierarchy as final results

Figure 2 shows the relation diagram of categories as the final results obtained from table 2. Figure 2 (a) shows the diagram regarding the fuzzy hierarchical relations of categories in the case of $\alpha = 0.94$, the category C_1 is a subcategory of all categories presented except C_4, C_3 and C_5 are subcategories of C_2. Figure 2 (b) shows the diagram in the case of $\alpha = 0.76$, C_2 is located at the highest of the hierarchy, while C_2, C_3 and C_5 are at the lowest. It can be indicated that the case in (b) has a broader diagram than the case of (a), including all relations in the case of (a).

5   Experimental Results

In this paper, we implemented the prototype using Visual C++. We performed two experiments on the prototype. Experiment 1 evaluated the accuracy of e-mail classification using category generation method. Experiment 2 evaluated the accuracy of reorganizing category of e-mails using the dynamic reconstruction of category hierarchy.

Experiment 1. Our test data was 267 messages files received over a 2 month period. The Experiment evaluated the accuracy of e-mail classification with category label constructed automatically according to classification proposed method. The accuracy of the classification was evaluated by classification expert whether messages or not belong to the correct category. Figure 3 shows the results of this experiment. Here, we can see that by increasing the number of category labels, the accuracy improves and the classification boundary value becomes lower.

Experiment 2. We evaluated the accuracy of reorganizing according to the α value. The test data was the results from Experiment 1. Table 3 shows the results of this experiment. Here, we can see that increasing a number of category labels improves accuracy as the α value becomes higher.

We can therefore conclude that accuracies are improved to 82% ~ 93% using reorganizing e-mail categories.

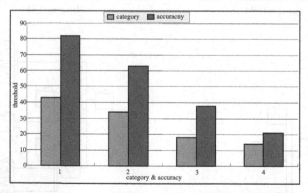

Fig. 3. The average accuracy of classification according to boundary value of proposed method

Table 3. The average accuracy of the reorganize e-mail using dynamic reconstruction of category hierarchy

α-cut value	The number of Category Label	Accuracy
90 %	43	82 %
70 %	52	86 %
50 %	67	93 %

6 Conclusions

In this paper we proposed an e-mail classification agent that will generate category labels for classifying e-mails automatically and reconstruct the category hierarchy when necessary. The proposed method uses the similarity on vector model to generate category labels of e-mails. It reconstructs the category hierarchy when it is needed.

The proposed method can always be adjusted to reclassify e-mails according to the user demands. Thus, the proposed method has the following advantages: The user can manage e-mails efficiently; it improves the accuracy of classification of e-mails; and finally, it does not require the learning for classification.

Acknowledgements

This research was supported by University IT Research Center Project.

References

1. Androutsopoulos, I. et al.: An Evaluation of Naïve Bayesian Anti-Spam Filtering. In Proc. Workshop on Machine Learning in the New Information Age, (2000)
2. Cohen, W.W.: Learning Rules that classify E-mail. In Proc. AAAI Spring Symposium in Information Access, (1999)
3. Choi, B. G., Lee, J. H., Park, Sun.: Dynamic Reconstruction of Category Hierarchy Using Fuzzy Relational Products. In proceddings of the 4th International Conference On Intelligent Data Engineering and Automated Learing. Hong Kong, China, (2003) 296-302
4. Drucker, H. Wu, D. and Vapnik, V. N.: Support Vector Machines for Spam Categorization. IEEE Trunsuctions on Neural network, 10(5), (1999)
5. Manco, G. Masciari, E.: A Framework for Adaptive Mail Classification. In Proceedings of the 14th IEEE International Conference on Tools with Artificial Intelligence, (2002)
6. Ogawa, Y. Morita, T. and Kobayashi,K.: A fuzzy document retrieval system using the keyword connection matrix and a learning method. Fuzzy Sets and System, (1991) 163-179
7. Ricardo, B. Y. Berthier, R. N.: Modern Information Retrieval. Addison Wesley, (1999)
8. Sakkis, G. et al.: Stacking classifiers for anti-spam filtering of e-mail. In Proc. 6th Conf. On Empirical Methods in Natural Language Processing, (2001)
9. Bandler, W. and Kohout, L.: Semantics of Implication Operators and Fuzzy Relational Products. International Journal of Man-Machine Studies, vol. 12. (1980) 89-116

The Investigation of the Agent in the Artificial Market

Takahiro Kitakubo, Yusuke Koyama, and Hiroshi Deguchi

Department of Computational Systems and Science,
Interdisciplinary Graduate Scholl of Science and Engineering, Tokyo Institute of Technology
kitakubo@degulab.cs.dis.titech.ac.jp,
{koyama,deguchi}@dis.titech.ac.jp

Abstract. In this paper, we investigate the investment strategy in the artificial market called U-Mart, which is designed to provide a common test bed for researchers in the fields of economics and information sciences. UMIE is the international experiment of U-Mart as a contests of trading agents. We attended UMIE 2003 and 2004, and our agent won the championship in both experiments. We examin why this agent is strong in UMIE environment. The strategy of this agent is called "on-line learning" or "real-time learning". Concretely, the agent exploits and forecasts futures price fluctuations by means of identifying the environment in reinforcement learning.
We examined an efficiency of price forecasting in the classified environment. To examine the efficacy of it, we executed experiments 1000 times with UMIE open-type simulation standard toolkits, and we verified that forecasting futures price fluctuation in our strategy is useful for better trading.

Keywords: U-Mart, Artificial Market, Reinforcement Learning

1 Introduction

The U-Mart project is a research program which aims at establishing a methodology for artificial market studies. The U-Mart project has provided an artificial market simulation system as a common test bed, which are used by researchers in the fields of economics and information science, which is called U-Mart System [1].

International experiments of U-Mart as contests of trading agents, called UMIE have been held since 2002 [2]. The contests called for participation of trading software agents. Participants trade for a futures to gain from the futures market. Agents are evaluated and ranked by some criterions. These are average of final profit, maximum profit, etc. We challenged to this contest, we modeled AI-type agent which exploits futures price fluctuations by means of identifying the environment in reinforcement learning [3]. This agent won the championship in the UMIE2003 and UMIE2004 experiments; however, we have not been understood the efficacy of predicting futures prices, and why the agent won. And so we decide to analyze them.

2 U-Mart Details

2.1 Rules of U-Mart Standard Toolkit Experiment

We use U-Mart Standard Toolkit to examine the agent behavior. U-Mart Standard Toolkit is a simulation toolkit provided to UMIE participants. This toolkit has 20

T.G. Kim (Ed.): AIS 2004, LNAI 3397, pp. 215–223, 2005.
© Springer-Verlag Berlin Heidelberg 2005

sample agents which are called standard agent set. These agents are also participate in the experiments. All participants trade 4 times a day. Trades are continued for 60 days. Prices list is selected from the sequence of J30, an index of Japanese stock Market, published by Mainichi Newspapers. All agents are given one billion virtual Japanese yen for the initial cash. Deposit for one unit of futures is uniformly 300,000 yen. No charge has to pay for contract and settlement.

2.2 U-Mart System from an Agent's View

We show an overview form an agent. (Fig.1)

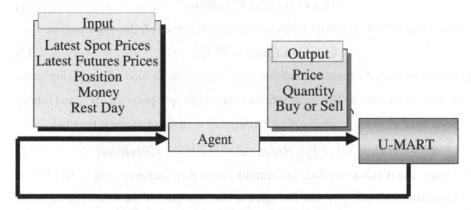

Fig. 1. An agent's overview of U-Mart System

An agent receives input information shown at Fig.1. Latest Spot Prices $SP_t = (s_1, s_2, ..., s_{120})$ contains latest spot prices from time $t-119$ to time t. Latest Futures Prices $FP_t = (s_1, s_2, ..., s_{60})$ contains latest futures prices at time t. Position is latest position of the agent. Money is latest money of the agent. Rest day is remaining time to final settlement.

3 Agent Trade Strategy

Our agent exploits input information, learns it, and forecasts futures price. Then, the agent makes an order decision according to the forecasted price.

3.1 Agent Learning Model

In this section, we show how our agent exploits information from the market and predict futures price.

We explain the learning model of the agent that participated in U-Mart. The agent applies four rules to classify current environments into 128 conditions. An environment means latest input information (Fig.1). Each condition is expressed by bits of sequence. Each rule encodes information from environment, and creates bits of sequence. Each condition is a bit sequence connected from these encoded bit sequences.

A condition has price fluctuation rate. The agent incrementally changes expected next futures price fluctuation at a certain condition. The learning algorithm is as follows. Fig. 2 shows the overview.

Learning Algorithm

This algorithm prepares futures price fluctuations map whose key is a condition and value is a futures price flucuation. And then, it updates the map according to the input information.

0. Initialize: Initialize futures price fluctuation map $f : Condition \rightarrow \Re$ for all condition.

$$f(C) \leftarrow 0, \forall C \in Condition \tag{1}$$

Each condition C is 7 binary bits of sequence defined as following equation:

$$C \in Condition = \{0,1\}^7 \tag{2}$$

1. Get condition: Get recent condition $C_t \in Condition$ at time t by applying rules to input information. The input information is both spot price list SP_t, and futures price list FP_t. This condition $C_t \in Condition$ is saved to use next time t+1.

$$C_t \leftarrow Rules(SP_t, FP_t), Rules : Z^{120} \times Z^{60} \rightarrow Condition \tag{3}$$

2. Learn fluctuation: Update fuluctuation map $f : Condition \rightarrow \Re$. The key is previous condition C_{t-1} and the value is the right part of the following equation. Let f' be the updated map. Learn futures price fluctuation into previous condition C_{t-1} from latest futures price fluctuation $(p_t - p_{t-1})/ p_{t-1}$. α is a learning parameter.

$$f'(C_{t-1}) \leftarrow (1-\alpha) \times f(C_{t-1}) + \alpha \times \frac{p_t - p_{t-1}}{p_{t-1}} \tag{4}$$

3. Return 1.

Forecast Price

Forecast next futures price $\hat{p}_{t+1|t}$ from futures price fluctuations map . The key is the latest condition C_t at time t. C_t is obtained from equation (3).

$$\hat{p}_{t+1} = p_t + f(C_t) \times p_t \tag{5}$$

And also we show 4 rules.

Rule 1 Spot- Futures Spread
This rule compares last spot price and last futures price, and calculates spread between last spot price and last futures price by following equation.

$$SpotFuturesSpread = (s_t - p_t) / p_t \tag{6}$$

After that, it creates 3 bits of sequence $condition_{Rule1}$ according to an area which contains the value of $SpotFuturesSpread$.

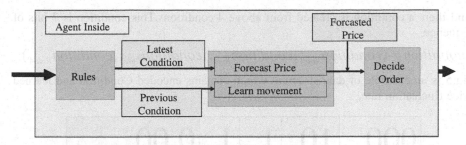

Fig. 2. Agent inside overview

Areas are divided by thresholds {-0.05,-0.02,-0.01, 0, 0.01, 0.02, 0.05}. (Fig.3)

For Example, if SpotFuturesSpread = 0.012 and then, Rule1 creates a bits of sequence "101".

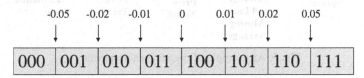

Fig. 3. Rule1

Rule 2 Short-Long Moving Average

Rule2 compares short moving average and long moving average that are calculated following equation.

Short moving average S = 3 step moving average of futures price
Long moving average L = 7 step moving average of futures price
Comparison Value SL = S / L

And Rules2 creates 2 bits of sequence $condition_{Rule2}$ according to an area which contains the value of SL. Areas are divided by thresholds {0.95, 1.00, 1.05} (Fig.4).

Rule 3 Futures Price Trend

This rule compares last futures price p_t and previous futures price p_{t-1} , and creates 1 bit sequence by following equation.

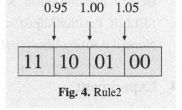

Fig. 4. Rule2

$$condition_{Rule3} = 0(if \quad p_t > p_{t-1}), 1(if \quad p_t \leq p_{t-1}) \tag{7}$$

Rule 4 Spot Price Trend

This rule compares last spot price s_t and previous spot price s_{t-1}, and creates 1 bit sequence by following equation.

$$condition_{Rule4} = 0(if \quad s_t > s_{t-1}), 1(if \quad s_t \leq s_{t-1}) \tag{8}$$

And then, a condition is created from above 4 conditions.This condition is 7 bits of sequence.

$$condition = (condition_{Rule1}, condition_{Rule2}, condition_{Rule3}, condition_{Rule4})$$

Fig.5 is an example of data structure which contains encoded Condition and learned price fluctuation rate.

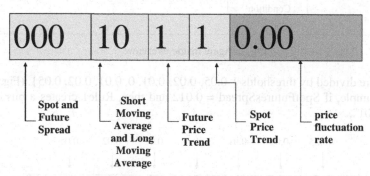

Fig. 5. An example of encoding environment

3.2 How to Decide Order

The agent decides Order (**Price, Quantity, BuySell**) by following equations.

$$PositionAdjust = orderGain \times (\hat{p}_{t+1} - p_t) - positionReductionRatio \times Position \quad (9)$$

$$Price = \hat{p}_{t+1} \quad (10)$$

$$Quantyty = min(|PositionAdjust|, maxQuant) \quad (11)$$

$$BuySell = \begin{cases} BUY(if \ PositionAdjust > 0), \\ SELL(if \ PositionAdjust < 0), \\ None(if \ PositionAdjust = 0) \end{cases} \quad (12)$$

4 Experiment

4.1 Supposition

To evaluate the efficiency of learning, we put forward following hypothesis.

Forecasted Price Fluctuations by Learning Algorithm Are Better Predictions Than Previous Price Fluctuations

We defined distance between actual price fluctuations and forecasted price fluctuations to investigate this hypothesis by following equation.

$$d(\hat{F}, A) = \frac{1}{T} \sum_{t=1}^{T} |\hat{f}_t - a_t| \quad (13)$$

$A = (a_1, a_2, ..., a_T)$ is an actual futures prices list. T=240 is trading period of itayose. $\hat{F} = (\hat{f}_1, \hat{f}_2, ..., \hat{f}_T)$ is a forecasted prices fluctuations list.

Let $\hat{F}_{Exploited}$ be a forecasted prices fluctuations list obtained form Learning and Forecast Algorithm. $\hat{F}_{EasyForcasted}$ is a forecasted prices fluctuations list calculated following equation:

$$\hat{F}_{EasyForcasted} = (\hat{f}_1, \hat{f}_2, ..., \hat{f}_T,) = \left(0, \frac{p_1 - p_0}{p_0}, ..., \frac{p_{T-1} - p_{T-2}}{p_{T-2}} \right) \quad (14)$$

And, we suppose

$$d(\hat{F}_{Exploited}, A) < d(\hat{F}_{EasyForcasted}, A) \quad (15)$$

That is, $\hat{F}_{Exploited}$ is better performance than $\hat{F}_{EasyForcasted}$.

4.2 Simulation Settings

1000 times of simulations are executed with UMIE2003 Standard Toolkit. We remain simulation settings default values. Prices lists are selected from J30. Trading start day is randomly selected.

Then we set the agent parameters as follows.

Table 1. Agent Parameters

maxQuant	200
orderGain	4
positionReductionRatio	0.1
α	0.5

5 Simulation Results

The simulation results are shown as Table 2. Averages of two distance and histogram of two distance show that our hypothesis is correct. Thus, we can find forecasted price fluctuations by learning algorithm are better predictions than previous price fluctuations.

Table 2. Simulation Results

average profit	50450187
average of distance $d(\hat{F}_{Exploited}, A)$	0.003456
average of distance $d(\hat{F}_{EasyForcasted}, A)$	0.005681

The histogram of the profit is shown in Fig.7. The agent can gain profit from various kind of environment with Standard Toolkit.

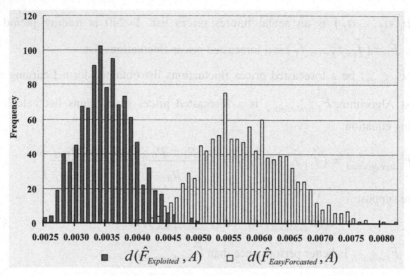

Fig. 6. Histogram of two distances

And also, we show a simulation result of relation between forecasted price fluctuations and actual price fluctuations in Fig.8. This is a simulation result. Other results are similar result. This result shows some good performance of forecast algorithm.

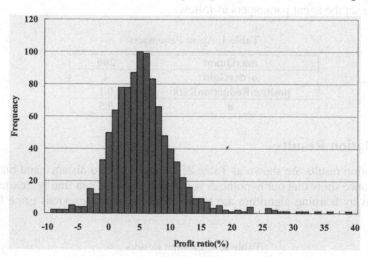

Fig. 7. Histogram of profit ratio form simulation results

And then, we show relation between the distance and profit from simulation results in additional experiments. (Fig.9.) In additional experiments, We modify equation (4) in our leaning algorithm to reveal when accuracy of predicting future prices are bad.

$$f'(C_{t-1}) \leftarrow (1-\alpha) \times f(C_{t-1}) + \alpha \times \frac{p_t - p_{t-1}}{p_{t-1}} + Noize, Noize \in U(-w, w) \qquad (16)$$

$U(-w, \quad w)$ means uniform distribution whose range is $[-w, \; w]$. $w \in N(0,0.02^2)$ is randomly selected in each simulation. $N(0,\sigma^2)$ means regular distribution whose average is 0 and variance is σ^2. We can see their correlation on the whole. The accuracy of predicting future price movement is important in our investment strategy.

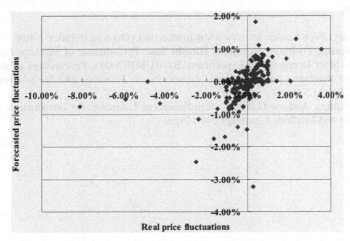

Fig. 8. A simulation result: Comparison of forecasted and real price fluctuations

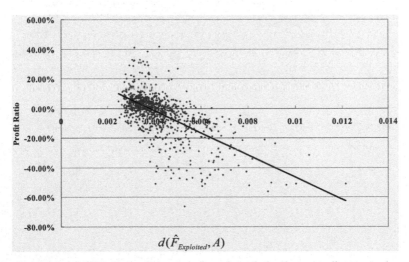

Fig. 9. Additional 1000 times of simulation results: relation between distance and profit

6   Conclusion

The goodness of the learning algorithm is verified by the above experiment. And also, we showed the relation between the accuracy of predicting future price movement and profit by the additional simulation. However, it is revealed that our agent may suffer a loss when future price movement is so hard. To improve agent more robustly, we think that the agent's invest strategy should count on risk avoidance according to the volatility of learned price movement list. As a next research program, we will think robust investment strategy.

References

1. U-Mart Project Web Server: http://www.u-mart.econ.kyoto-u.ac.jp/index-e.html
2. Hiroyuki Matsui, Yoshihiro Nakajima, Hiroshi Sato: Performance of Machine Agents Submitted to U-Mart International Experiment 2004(UMIE2004), *Proceedings of the Second International Workshop on Agent-based Approaches in Economic and Social Complex Systems (AESCS'04)* (2004)
3. Richard S.Sutton, Andrew G.Barto: "Reinforcement Learning: An Introduction (Adaptive Computation and Machine Learning)", Mit Press

Plan-Based Coordination of a Multi-agent System for Protein Structure Prediction

Hoon Jin and In-Cheol Kim

Department Of Information Science, Kyonggi University
Suwon-si, Kyonggi-do, South Korea
{jinun,kic}@kyonggi.ac.kr

Abstract. In this paper, we describe the design and implementation of a multi-agent system that supports prediction of three dimensional structure of an unknown protein from its sequence of amino acids. Some of the problems involved are such as how many agents should be coordinated to support prediction of protein's structures as well as how we devise agents from multiple resources. To address these problems, we propose a plan-based coordination mechanism for our multi-agent system. In our proposed system, MAPS, The control agent coordinates other agents based on a specific multi-agent plan, which specifies possible sequences of interactions among agents. This plan-based coordination mechanism has greatly increased both coherence and flexibility of our system.

1 Introduction

Management Information Systems are often intended to achieve high performance by modeling the business processes within an enterprise. Even though there are so many processes different from each other depending on human experiences, in most cases, those models that are implemented can be defined as the best computerized processes. And while the operators can be human or automated programs at the operating level, most of these processes are stereotypes, or similar to each other. It is not the case for biology, however, in which research processes are more dependent upon specialists' experience and knowledge and for the more, the processes taken can not be regarded as the best methods, as each steps in the course of research requires judgement based on expertise. And in case of predicting protein structure, the operations at each step are not uniform in their nature even though they appear to be all similar. This is caused by the fact that each databases and software tools are developed in different places with different goals and within different time frame as different nations and organizations are involved in the development. Taken all these characteristics into consideration, the solution requires systemic approaches. First of all, it needs to analyze whether the processes can be computerized, and then can be executed within the constraining time and other conditions. Following that, we should decide what methods to be used to achieve the goal of predicting structures. In this case, processes are composed of message querying and answering operations in most of processes. Of course, these message executions practically require considerations of so many factors. From the perspective of actions, the goal is a sequence of actions, in other words, composed of succeeding results of message actions. Therefore, there is a need for special types of multi-agent system that are capable of supporting the strong

T.G. Kim (Ed.): AIS 2004, LNAI 3397, pp. 224–232, 2005.

modulation, reliability, and extensibility on the basis of messaging processes. In this paper, we describe the design architecture and implementation of MAPS(Multi-Agent system for Protein Structure prediction). MAPS is a multi-agent system which is suitable for operating in the prediction of protein's three dimensional structures. Using intelligent techniques, it can change its course of actions in reaction to stimulus imposed from the environment while performing various actions. MAPS also meets a variety of needs by connecting to Agentcities network, accommodating a great deal of changes in biological databases both on quantitative and qualitative aspects, while sharing the data at the same time being operated as an integrated unit[9].

2 Backgrounds

Currently there is a great emergence of resources (which are databases and software tools) that deals with biological information at numbers and volumes on the Internet. Particularly, starting from the latest period of the genome project, protein-related resources have been growing up at surprising rates [1]. Among them there exist the databases that are directly related to protein structure prediction, utilized by biologists for computational prediction processes. These include PDB, 3D-pssm, SwissProt, Psipred, Scop, PredictProtein, HMMSTR, and others [2]. In addition, there are many different types of software tools which facilitates such tasks as creating structure, searching similar sequences, aligning multiple sequences, or analyzing sequences, etc., which are related with protein sequence analysis. These tools include NCBI-Blast which searches similar sequences, Clustalw that is used for multiple sequences alignment, SWISS-MODEL that creates three dimensional structures with querying sequences and templates.

The research on agent and multi agent system has reached the phase that some of the research results are conducive to the development of agent-based system, and agent-based system is being regarded as the method of development for the next generation. BioMAS uses DECAF for genome annotation, which is a toolkit that provides multi-agent environment based on RETSINA, and is a prototype system containing repository databases to store automated annotation and sequences [4,6]. The ECAF is a system suitable for biological domain in that it includes a planner, which carries out the roles of plan writer, plan coordinator, plan executer, and scheduler and executor. MeLiSA is a system that uses ontology-based agent techniques in medical science [5]. It is an agent system, which uses a simple agent that searches medline database based on the MESH system. BioAgent is a multi and mobile agent system developed for genome analysis and annotation process by biologists [3]. It was developed to support researchers by completely decentralizing the tasks needed for research, and reducing network overloads, providing conveniences to researchers. GeneWeaver is a system designed for genome analysis and protein structure prediction [1]. Its special features indicate that it was developed for analysis of biological data resulting from combined use of heterogeneous databases and software tools. It is noted that GeneWeaver employed the concept of multi-agents to solve the problems occurring in distributed and dynamically changing environment.

The most important problem related to a multi-agent system comprised of multiple agents, is the coordination which deals with how to guarantee the autonomy of each

Fig. 1. A taxonomy of coordination mechanisms

component agent and how to maintain the global coherence of system [8]. Until now, various methods to handle the coordination of multi agents have been developed in many application areas. These are summarized in figure 1. Firstly, coordination mechanism is divided into cooperation and competition. Employing cooperation mechanism, all component agents try to cooperate for one global goal, and in competition mechanism, conflict or competition among the agents take place to secure benefits. For cooperation, it can be applied the method of coordinating the component agents according to the plan meticulously devised ahead. On the other hand, to conflict or competition, coordination through the self-controlled negotiation between the components can be applied. Especially in plan-based coordination, either distributed planning method in which all the component agents together participate in making plans, or the centralized planning method in which a system administrator or an agent is fully responsible for can be chosen.

3 System Design

We have developed a system to facilitate diverse researches about proteins through transforming the protein resources into agents and administrating these agents regardless of platforms or locations. This paper describes the design of the system that performs protein structure prediction. MAPS utilizes an agent, named CODY, as mediator and coordinator, then, which is capable of operating the steps of protein structure prediction comparatively easily and changing the prediction process autonomously. To materialize it, we used the techniques of plan-based intelligent agent. The data produced in the middle of processing stage need to be processed or changed so that the data can be approached easily, or in some cases the data should be processed through some specified software. The agents in MAPS decide whether or not to perform these jobs through interactions among agents.

3.1 The Process of Protein Structure Prediction

There are three types of methods in predicting protein structures which are defined by CASP(Critical Assessment of techniques for protein Structure Prediction). Homology modeling method predicts three dimensional structures by comparing the three dimensional structure of similar sequence with that of query sequence, when sequence similarity is higher than roughly 25%. On the other hand, fold recognition is used when the similarity is lower than 25%. It is the method, which finds fold structure on the basis of secondary structures predicted through comparison and alignment with the other secondary ones, that predicts three dimensional structures. If similar sequences can not be found because similarities are too low, ab-initio method is used. This method predicts three dimensional structures by measuring and analyzing physical and chemical properties computationally, based on simple sequence information [7].

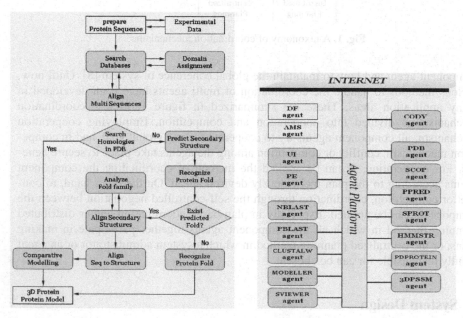

Fig. 2 A typical process of protein structure prediction

Fig. 3. The architecture of MAPS

3.2 System Architecture

In MAPS, there are four types of agent that can be categorized according to their roles and functions.

− Interface agents: UI, PE
− Brokering agents: AMS, DF, CODY
− Task agents: NBLAST, PBLAST, CLUSTALW, MODELLER, SVIEWER
− Resource agents: PDB, 3DPSSM, SPROT, PPRED, SCOP, PDPROTEIN, HMMSTR

Interface agents play roles as the interface in the system to assist users in using an entire system. Brokering agents control and coordinate the operations between agents, and save those information as needed. Task agents compute the information produced about by resource agents, analyze the processed results. Resource agents practically connect to the protein structure databases which should be used for protein structure prediction, query, and find the information, replacing human browsing and searching.

Actually, the operating agents, which take the role of connecting to databases and responding to user's queries, find and provide the information requested by users. In the case of PDB agent, it fetches the files of protein structures and the related additional information at the request of users, and 3DPSSM and SPROT agents bring various types of information that are secondary, tertiary, annotation information or structure files. Task agents filter the essential information needed for analyzing information based on what users query and predicting of protein structure. NBLAST and PBLAST agents search similar sequences and analyze results of the search. CLUSTALW agent analyzes information produced through multiple sequence alignment with a queried sequence and found sequences, acquire from that process the information of evolutionary relationship needed for family analysis, origination analysis, and domain analysis. MODELLER agent connects to SWISS-MODEL server with user's query data and derivates a structure. Queried data can be only a sequence or sequence and template file, and the result is a predicted structure file along with related information. SVIEWER agent utilizes a Swiss-PdbViewer program in producing the datum for MODELLER agent. PE (Plan Editor) is an agent that will be developed so as to support user by himself in designing and describing the process of protein structure prediction. The users of system are mostly the researchers in biology or in related fields even though they are not specialists on agent systems. It is not possible for them to write plans themselves, and therefore we intend to develop a plan editor which is devoted to writing plans, and through which users can write plans easily without having knowledge on detailed processes and related syntaxes. CODY agent carries out the most important role in all of the process which is planning, coordinating, communicating, mediating. AMS and DF agents perform administration and directory services that are essential in operating a multi-agent system.

3.3 Plan-Based Coordination

In MAPS system using multiple agents, the role of CODY agent is very important. CODY agent processes users' requests from UI agent, or mediates the results and data. The most important task is which coordinates agents based on plans, and it must be processed separately in aspect of plan and coordination. First of all, in the aspect of plan, in case that there is no single comprehensive path that is standardized and precisely defined to achieve a global goal, it can be a solution that the whole process is divided into primitive plan units, and achieve the goal by carrying out these primitive plans consecutively. But because these primitive plan units are not performed repeatedly in the same sequence, it is necessary to specify the preconditions for plan execution and selectively choose the plans to execute in accordance with these conditions. That is, it requires the description of conditions upon which the plans are judged whether to be executed or not.

Table 1 describes some part of the process executed by the CODY agent for protein structure prediction. The process reveals that most of the forms of interactions are a request- and a response-based message communications. The problems arise from the multiplicity of agents. Generally, agents in multi-agent systems have plans for the individual goal and its operation, pursuing the 'selfish' objectives. But the agents comprising MAPS should be cooperative considering domain characteristics and most of the processes of agents operation are performed in a serial manner rather than the parallelized one. Therefore, it is desirable to pursue a goal through cooperation in the form of request and response between the leading and the other following agents in accordance with the plans, rather than to compete among selfish agents. Consequently, CODY agent comes to have more complex architecture than other agents.

Table 1. Some part of the process executed by the CODY agent

```
if(protein ID exists){
  PDBID = ask for SPROT(protein ID);
  if(PDBID exists){
    3D_structure_ file = ask for PDB(PDBID);
    respond to UI(3D_structure_file);
  } else{   // if PDBID not exists
    sequences = SPROT(protein ID);
    ask for NBLAST to search the similar sequences(sequence);
    if(sequences exists lower than E value 1) getting m number of sequences;
    else if(sequences exists higher than E value 1 and lower than 10){
      getting n number of sequences or
      ask for PBLAST to search the similar sequences (sequence);
      ...
} else{ ... }              //if not exists protein ID
//m, n, p, q are user defined variables
```

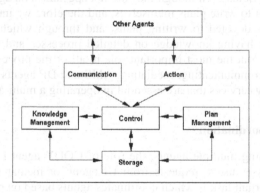

Fig. 4. The structure of the CODY agent

The structure of CODY agent is given in figure 4. CODY agent has five modules on the whole. Communication module mainly handles the control process among agents based on messaging. Action module communicates with other agents based on messaging as well, and it is to transfer data which are produced overwhelmingly large in quantity due to the characteristics of biological domain. The data that are processed and collected at the directions of Communication are transferred to CODY agent and saved through Storage module. Knowledge Management is a module for mainly analyzing the meaning of data utilizing ontology, and is planned to be implemented in the next phase of our research. Control module manages and mediates the processes among all the internal modules. Aside from these, in a multi-agent system, there is a need for supporting agents that assist and manage the control of calling and communication. MAPS has AMS(Agent Management System) to manage agents and DF(Directory Facilitator) agent which provides the directory service for the names of agents and the service lists.

4 Implementation and Execution

MAPS is a system that is operated to predict protein structure based on multiple agents. Accordingly it needs to support various types of agents that are distributed in many different platforms. And the system can not operate based on a complete, single plan that is predefined. In this paper, we make a connection to Agentcities network to support openness and decentralization of agent system utilizing JADE (Java Agent Development Framework) for the purpose of openness, and JAM architecture for the purpose of decentralization.

Table 2. An example plan written in JAM script language

```
//--------------------------------------------
//PLAN    find_3dSTR_byPID
//--------------------------------------------
PLAN : {
NAME :
          "find 3D structure process by PID"
GOAL :
          ACHIEVE got_3dSTR "true";
PRECONDITION :
          FACT is_PID $is_pid;
          (== is_pid "true");
CONTEXT :
          FACT is_3dSTR $is_3dSTR;
          (== $is_3dSTR "false");
          FACT is_PDBID "false";
BODY :
          RETRIEVE 3dSTR $3dSTR;
          RETRIEVE PID $pid;
          RETRIEVE PDBID $pdbid;
          RETRIEVE is_PDBID $is_pdbid;
          EXECUTE Search_SPROT $pid;          //function 1
          WHEN : TEST (== $is_pdbid "true")
          {
                  ACHIEVE find_3dMODEL_by_PDBID;
          }
          WHEN : TEST (== $is_pdbid "false")
          {
                  ACHIEVE find_3dMODEL_byNo_PDBID;
          }
          SUCCEED;
}
```

JADE is a platform developed strictly following specifications proposed by FIPA, and is a middle ware system supporting the development of multi-agent application system based on peer-to-peer network architecture. Using JADE, one can devise the agents capable of negotiating, coordinating among agents, while being distributed, which are equipped with such traits as pro-activity, multi-party, interaction, openness, versatility, usability, and mobility. Agentcities network provides connection services among various and heterogeneous agents following the FIPA standard, and is an open-based network system supporting the worldwide online cooperation of agents. The reason for adopting Agent cities network environment is that it can increase the flexibility in integrating services of heterogeneous agent platforms by providing the agent mechanism that is highly adaptable and intelligent owing to its capability of interchanging and sharing through ACL messages. Until now, 97 platforms have been registered and we can use their various services through a platform directory, an

Table 3. Protein's id and sequence retrieved from SwissProt

PID : P08588
Fasta format Sequence :
>sp|P08588|B1AR_HUMAN Beta-1 adrenergic receptor (Beta-1 adrenoceptor) (Beta-1 adrenoreceptor) - Homo sapiens (Human).
  MGAGVLVLGASEPGNLSSAAPLPDGAATAARLLVPASPPASLLPPASESPEPLSQQWTA
GMGLLMALIVLLIVAGNVLVIVAIAKTPRLQTLTNLFIMSLASADLVMGLLVVPFGATIVV
WGRWEYGSFFCELWTSVDVLCVTASIETLCVIALDRYLAITSPFRYQSLLTRARARGLVCT
VWAISALVSFLPILMHWWRAESDEARRCYNDPKCCDFVTNRAYAIASSVVSFYVPLCIMA
FVYLRVFREAQKQVKKIDSCERRFLGGPARPPSPSPSPVPAPAPPPGPPRPAAAAATAPLAN
GRAGKRRPSRLVALREQKALKTLGIIMGVFTLCWLPFFLANVVKAFHRELVPDRLFVFFN
WLGYANSAFNPIIYCRSPDFRKAFQGLLCCARRAARRRHATHGDRPRASGCLARPGPPPSP
GAASDDDDDDVVGATPPARLLEPWAGCNGGAAADSDSSLDEPCRPGFASESKV

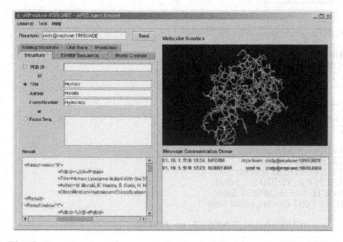

Fig. 5. A screenshot of an examplary execution processed by MAPS

agent directory, and a service directory. The representative agent platforms that are accessible to Agentcities, include April, Comtec, FIPA-OS, JADE, ZEUS, and others.

JAM is an intelligent agent architecture, equipped with a function of supporting plan-based reasoning. Especially, it has a merit in that it achieves a goal by performing a hierarchy of small plan units, leading to the fulfillment of one global plan unlike rule-based systems. JAM is composed of the plan library which is a collection of plans that an agent can use to achieve its goals, the interpreter which is the agent's "brains" that reason about what the agent should do and when it should do it, the intention structure which is an internal model of the goals and activities the agent currently has and keeps track of progress the agent has made toward accomplishing those goals, and the observer which is a lightweight plan that the agent executes between plan steps in order to perform functionality outside of the scope of its normal goal or plan-based reasoning. Table 2 gives an example plan written in JAM script language.

An example of the protein structure prediction process using a prototype is described below. We try to find three dimensional structure of a protein about an amino acids sequence with a locus id named 'ADRB1'. This sequence is a material that resides in human chromosome 10. To begin with the process, a query is sent to SPROT agent to find its protein id and sequence with 'ADRB1'. The result is shown in table 3. But the information returned from a PDB agent when queried with a PID is not what has been searched for. When queried with the sequence again, we obtain 14 structures in total as the result of the search. And all the structures are downloaded, templates are constructed from each of them using SVEIWER agent, and then transferred to MODELLER agent to retrieve predicted structure file. An example of predicted structures is depicted in figure 5.

5 Conclusion

In this paper, we have described our research on the prediction of protein's three dimensional structures using a multi-agent system. Being an implemented system, MAPS has the capability of operating the steps of protein structure prediction by coordinating CODY agents on the basis of plans, unlike the conventional agent-based systems. Our proposal has two aspects: One is cooperation, which is suitable as coordinating type in biological domain. The other is the method which is the plan-based approach for archiving the global goal composed of a succession of small plans, not as a monolithic and complete single plan. Also as MAPS is capable of joining and operating on heterogeneous platforms whenever the system requires, or wherever it is located, by connecting to agent cities network, it can immensely increase its openness, decentralization, and extensibility.

References

1. Bryson, K., Luck, M., Joy, M. and Jones, D.T.: Applying Agents to Bioinformatics in GeneWeaver, Proceedings of the Fourth International Workshop on Cooperative Information Agents, Lecture Notes in AI, Springer-Verlag. (2000).
2. Cynthia Gibas, Per Jambeck: Developing Bioinformatics Computer Skills, O'Reilly. (2001).
3. Emanuela Merelli and Rosario Culmone.: "BioAgent: A Mobile Agent System for Bioscientists", Proceedings of NETTAB02. Agents in Bioinformatics. (2002).
4. John R. Graham, Keith S. Decker, Michael Mersic.: "DECAF: A Flexible Multi Agent System Architecture". Autonomous Agents and Multi-Agent Systems. Vol. 7, No. 1-2. (2003) 7-27.
5. J. M. Abasolo, M. Gomez: MELISA: An ontology-based agent for information retrieval in medicine. Proceedings of the First International Workshop on the Semantic Web (SemWeb2000). Lisbon, Portugal, (2000) 73-82.
6. Keith Decker, Salim Khan, Carl Schmidt, Gang Situ, Ravi Makkena, Dennis Michaud.: "Biomas: A multi-agent system for genomic annotation", International Journal of Cooperative Information Systems, Vol. 11, No.3-4, (2002) 265-292.
7. Rost, B. et al.: "Protein structure prediction in 1D, 2D, and 3D", The Encyclopaedia of Computational Chemistry, No. 3, (1998) 2242-2255.
8. Sycara, Katia.: "Multiagent Systems", AI Magazine. Vol.19, No.2, (1998) 79-92.
9. Willmott, S.N. and Dale, J., Burg, B. and Charlton, C. and O'brien, P.: "Agentcities: A Worldwide Open Agent Network", Agentlink News, Vol.8, (2001) 13-15.

Using Cell-DEVS for Modeling Complex Cell Spaces

Javier Ameghino[1] and Gabriel Wainer[2]

[1] Computer Science Department, Universidad de Buenos Aires
Ciudad Universitaria (1428), Buenos Aires, Argentina
[2] Department of Systems and Computer Engineering, Carleton University
1125 Colonel By Dr. Ottawa, ON. K1S 5B6, Canada

Abstract. Cell-DEVS is an extension to the DEVS formalism that allows the definition of cellular models. CD++ is a modeling and simulation tool that implements DEVS and Cell-DEVS formalisms. Here, we show the use of these techniques through different application examples. Complex applications can be implemented in a simple fashion, and they can be executed effectively. We present example models of wave propagation, a predator following prey while avoiding natural obstacles, an evacuation process, and a flock of birds.

Introduction

In recent years, many simulation models of real systems have been represented as cell spaces [1, 2]. Cellular Automata [3] is a well-known formalism to describe these systems, defined as infinite n-dimensional lattices of cells whose values are updated according to a local rule. Cell-DEVS [4] was defined as a combination of cellular automata and DEVS (Discrete Events Systems specifications) [5]. The goal is to improve execution speed building discrete-event cell spaces, and to improve their definition by making the timing specification more expressive.

DEVS is a formalism proposed to model discrete events systems, in which a model is built as a composite of basic (behavioral) models called **atomic** that are combined to form **coupled** models. A DEVS atomic model is defined as:

$$M = < X, S, Y, \delta_{INT}, \delta_{EXT}, \lambda, D > \tag{1}$$

where X represents a set of input events, S a set of states, and Y is the output events set. Four functions manage the model behavior: δ_{INT} the internal transitions, δ_{EXT} the external transitions, λ the outputs, and D ▯the duration of a state. When external events are received, the external transition function is activated. The internal events produce state changes when the lifetime of a state is consumed. At that point, the model can generate outputs, and results are communicated to its influencees using the output ports.

Once an atomic model is defined, it can be incorporated into a coupled model:

$$CM = < X, Y, D, \{Mi\}, \{Ii\}, \{Zij\} > \tag{2}$$

Each coupled model consists of a set of **D** basic models **Mi**. The list of influences **Ii** of a given model is used to determine the models to which outputs (**Y**) must be sent, and to build the translation function **Zij**, in charge of converting outputs of a

T.G. Kim (Ed.): AIS 2004, LNAI 3397, pp. 233–242, 2005.
© Springer-Verlag Berlin Heidelberg 2005

model into inputs (**X**) for the others. An index of influences is created for each model (**Ii**). For every j in **Ii**, outputs of model **Mi** are connected to inputs in model **Mj**.

In Cell-DEVS, each cell of a cellular model is defined as an atomic DEVS. Cell-DEVS atomic models are specified as:

$$TDC = < X, Y, S, \theta, N, delay, d, \delta_{INT}, \delta_{EXT}, \tau, \lambda, D > \qquad (3)$$

Each cell will use the **N** inputs to compute the future state **S** using the function τ. The new value of the cell is transmitted to the neighbors after the consumption of the delay function. **Delay** defines the kind of delay for the cell, and **d** its duration. The outputs of a cell are transmitted after the consumption of the delay.

Once the cell atomic model is defined, they can be put together to form a coupled model. A Cell-DEVS coupled model is defined by:

$$GCC = < X_{list}, Y_{list}, X, Y, n, \{t_1,...,t_n\}, N, C, B, Z > \qquad (4)$$

The cell space **C** defined by this specification is a coupled model composed by an array of atomic cells with size $\{t_1 \ x...x \ t_n\}$. Each cell in the space is connected to the cells defined by the neighborhood **N**, and the border (**B**) can have different behavior. The **Z** function allows one to define the internal and external coupling of cells in the model. This function translates the outputs of output port m in cell C_{ij} into values for the m input port of cell C_{kl}. The input/output coupling lists (**Xlist**, **Ylist**) can be used to interchange data with other models.

The CD++ tool [6] was developed following the definitions of the Cell-DEVS formalism. CD++ is a tool to simulate both DEVS and Cell-DEVS models. Cell-DEVS are described using a built-in specification language, which provides a set of primitives to define the size of the cell-space, the type of borders, a cell's interface with other DEVS models and a cell's behavior. The behavior of a cell (the τ function of the formal specification) is defined using a set of rules of the form: *VALUE DELAY CONDITION*. When an external event is received, the rule evaluation process is triggered to calculate the new cell value. Starting with the first rule, the CONDITION is evaluated. If it is satisfied, the new cell state is obtained by evaluating the VALUE expression. The cell will transmit these changes after a DELAY. If the condition is not valid, the next rule is evaluated repeating this process until a rule is satisfied.The specification language has a large collection of functions and operators. The most common operators are included: boolean, comparison, and arithmetic. In addition, different types of functions are available: trigonometric, roots, power, rounding and truncation, module, logarithm, absolute value, minimum, maximum, G.C.D. and L.C.M. Other available functions allow checking if a number is integer, even, odd or prime. In addition, some common constants are defined.

We will show how to apply Cell-DEVS to simulate different systems. We describe different models that are implemented and executed using the CD++ tool, focusing on particular characteristics of each system. We will show how complex applications can be implemented in a simple fashion.

A Model of Wave Propagation

The state of a wave on water is characterized by its phase, intensity, direction and frequency. We built a model of wave propagation, which is based on sine waves [7].

If two waves with the same frequency are combined, there is a constant interference pattern caused by their superposition. Interference can be either constructive (meaning that the strength increases), or destructive (strength is reduced). In destructive interference, after superposition, the resultant wave will correspond to the sum of both waves. As phases are different, the wave will be canceled in those zones where there is overlap. The amount of interference depends of the phase difference in a point. When a wave encounters a change in the medium, some or all the changes can propagate into the new medium (or it can be reflected from it). The part that enters the new medium is called the transmitted portion, and the other the reflected portion. The reflected portion depends on the characteristic of the incident medium: if this has a lower index of refraction, the reflected wave has an 180° phase shift upon reflection. Conversely, if the incident medium has a larger index of refraction, the reflected wave has no phase shift. In order to simulate the interference between waves and the propagation in CD++, we defined a multidimensional model, in which each plane was defined by every direction of wave spread (four directions for this example). We defined an integration plane, which is a composition of the direction planes, and contains the value or intensity of the wave corresponding to this position. Every cell in the cell space represents the minimal possible portion of the medium in which the wave propagates. Each cell has two values, phase and intensity. An integer value between zero and eight represent phase and a fractional value between zero and one (intensity).

```
rule: {0} 100 {trunc(0,0,0)=#maxFase or (fractional(0,0,0)<.0001 and
            fractional((0,0,0))>0)}
rule: {trunc(1,0,0)+fractional(1,0,0)*#attenuation} 100 {(1,0,0)!=0}
rule: {trunc(0,0,0)+1+fractional(0,0,0)} 100 { (0,0,0)!=0}
```

Fig. 1. Rules for wave propagation

Figure 1 shows the rules used in CD ++ to specify the spread behavior. The first rule governs the attenuation of the wave. If the wave intensity is below of 0.0001 the wave propagation stops. The second rule contemplates the spread of the wave towards its neighbors, which preserve the phase of the wave but attenuate the intensity. The third rule contemplates the spread of the wave in the current cell. In this case, the cell intensity value does not change (only the phase).

```
rule: {((sin(PI/4*trunc((0,0,1))) * fractional((0,0,1)) + sin(PI/4 *
trunc((0,0,2))) * fractional((0,0,2)) + sin(PI/4 * trunc((0,0,3))) *
fractional((0,0,3)) + sin(PI/4*trunc((0,0,4))) * fractional((0,0,4))
)) * 10 } 100 {(0,0,1)!=0 or (0,0,2)!=0 or (0,0,3)!=0 or (0,0,4)!=0}
```

Fig. 2. Integration Rule

Figure 2 shows the rule describing the values of the direction planes used in order to obtain the wave value in the medium in which is traveling (the value corresponds to the discretization in eight phases of sine wave). Figure 3 shows the simulation results of the wave model. Only the integration plane is showed (direction and intensity). It is possible to appreciate that the wave propagating produce attenuation.

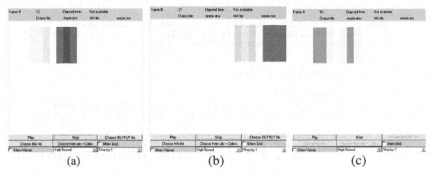

Fig. 3. Result of wave propagation. (a) A wave traveling from left to right. (b) The wave reflects with the right border. (c) The wave before reflecting with the left border

Predator-Prey Model

In this model, a predator seeks a prey, which tries to escape [8]. With predators always searching, prey must constantly avoid being eaten. A predator detects prey by smelling, as prey leave their odor when moving. The predator moves faster than the prey, and when there is a prey nearby, it senses the smell, and moves silently towards the prey. Each cell in the space represents an area of land (with vegetation, trees, etc.). Dense vegetation does not allow animals to advance, and thick vegetation form a labyrinth. When a prey is in a cell, it leaves a track of smell, which can be detected for a specific period (depending on weather conditions).

The cell's states are summarized in the following table. For example, the cell value 214 represents a prey following vegetation to find the exit in E direction.

Table 1. The following table describes the cell state codification using a natural number

Cell Value	Description
1..4	An animal moving towards N (1), W (2), S (3) or E (4).
200	Prey looking for the labyrinth's exit.
210	Prey following the forest to find the exit.
300	Predator looking for the labyrinth's exit.
310	Predator following the forest to find the exit.
400	Thick forest (does not allow animal movement).
0	Thin forest (allows animal movement).
101..104	Smell.

Figure 4 shows the specification of the model using CD++.

In Table 1, cell's value equals to 400 represents forest where the prey and predator can't move (labyrinth). To specify this cell behavior we use a special rule (first rule in Figure 4), which is evaluated if only if the cell's value is equal to 400. The second and third rules govern the predator movement towards N. In this model the cell's value from 200, 210, 300 and 310 finished in 1 represents *toward N*, finished in 2 represents *toward East*, finished in 3 represents *toward S* and finally cell's value finished in 4 represents *toward W*. So, as you can see in the second rule, a cell's

```
[pred-prey]
type : cell            dim : (20,20)           delay : inertial
neighbors: (-1,-1)(-1,0) (-1,1) (0,-1) (0,0) (0,1) (1,-1) (1,0) (1,1)
border : nowrapped         localtransition : movement

[movement]
rule : { (0,0) } 100000 { (0,0)=400}

%Predator moving towards N
rule: 0 800 {(-1,0)>100 and (-1,0)<250 and ((0,0)=311 or (0,0)=301)}
rule: 311 800 {(0,0)>100 and (0,0)<250 and ((1,0)=311 or (1,0)=301)}

%Predator moving towards East
rule: 0 800 {(0,1)>100 and (0,1)<250 and ((0,0)=313 or (0,0)=303)}
rule: 313 800 {(0,0)>100 and (0,0)<250 and((0,-1)=311 or (0,-1)=301)}
...
%Rules from smell path
rule: {(0,0)-1} 1000 {(0,0)>101 and (0,0)<105}
rule: 0 1000 {(0,0)<=101}
rule: {(0,0)} 100 {t}
```

Fig. 4. Specification of prey-predator's model

Fig. 5. Execution result of prey-predator's model [8]

value equals to 301 o 311 represents a predator following the labyrinth toward N. In this model, the movements rules are in pairs, because one rule is used to move the prey o predator to the new position and the other is used to actualize the cell where the animals was. In the movement rules we use a 800 msec delay, which represents the speed of the predator moving in the forest. The last 2 rules represent the smell path for a prey. As weather conditions disperse the smells, so we use four different values to represent different dispersion phases. The first line in the *smell* rules govern the smell attenuation by subtracting 1 to the actual cell's value every second. The last rule change the actual cell's value to zero (no smell in the cell).

Figure 5 shows the execution results of the model. A prey is trying to escape from the predator. Finally, the predator eats the prey.

Evacuation Processes Model

The simulation of evacuation processes has originally been applied to buildings or the aviation industry. Recently, however, the models have been transferred to ship evacuation taking into account special circumstances [9,10,11]. Our model represents people moving through a room or group of rooms, trying to gather their belongings or related persons and to get out through an exit door. The goal is to understand where the bottlenecks can occur, and which solutions are effective to prevent congestion [12].

The basic idea was to simulate the behavior and movement of every single person involved in the evacuation process. A Cell-DEVS model was chosen with a minimum set of rules to characterize a person's behavior:

- A normal person goes to the closest exit.
- A person in panic goes in opposite direction to the exit.
- People move at different speeds.
- If the way is blocked, people can decide to move away and look for another way.

```
[deck]
dim : (18,18,2)           delay : inertial        border : wrapped
localtransition : EvaRule
neighbors : (-1,-1,0) (-1,0,0) (-1,1,0) (0,-1,0) (0,0,0) (0,1,0)
(1,-1,0) (1,0,0) (1,1,0) (-1,-1,1) (-1,0,1) (-1,1,1) (0,-1,1) ...

[EvaRule]
% Rules to govern people movement
rule : {trunc((0,0,0)/10)*10+1} {1000 / remainder(trunc((0,0,0)
/10),10) } {((0,0,0)>0 AND  remainder(trunc((0,0,0)/1),10) =0 AND
remainder(trunc((0,0,0)/100000),10) =0 AND ((0,-1,0)=0 OR (0,-1,0)=-
2) AND cellPos(2)=0 )AND (((0,-1,1) <= (1,-1,1) OR (1,-1,0)>0 OR (1,-
1,0)=-1 OR (randint(5)=0) ) AND ((0,-1,1) <= (1,0,1) OR (1,0,0)>0 OR
(1,0,0)=-1 OR (randint(5)=0) ) AND ((0,-1,1) <= (1,1,1) OR (1,1,0)>0
OR (1,1,0)=-1 OR (randint(5)=0) ) AND ((0,-1,1) <= (0,1,1) OR
(0,1,0)>0 OR (0,1,0)=-1 OR (randint(5)=0) ) AND ((0,-1,1) <= (-1,1,1)
OR (-1,1,0)>0 OR (-1,1,0)=-1 OR (randint(5)=0) ) AND ((0,-1,1) <= (-
1,0,1) OR (-1,0,0)>0 OR (-1,0,0)=-1 OR (randint(5)=0) ) AND ((0,-1,1)
<= (-1,-1,1) OR (-1,-1,0)>0 OR (-1,-1,0)=-1 OR (randint(5)=0) ))}
...

% Rules to control panic behavior
rule : {trunc((0,0,0)/10)*10+1} {1000/remainder(trunc((0,0,0)/10),10)
} {((0,0,0)>0 AND remainder(trunc((0,0,0)/1),10)=0 AND remain-
der(trunc((0,0,0)/100000),10)>0 AND ((0,-1,0)=0 OR (0,-1,0)=-2) AND
cellPos(2)=0)AND (((0,-1,1)>= 1,-1,1) OR (1,-1,0)>0 OR (1,-1,0)=-1)
AND ((0,-1,1)>=(1,0,1) OR (1,0,0)>0 OR (1,0,0)=-1) AND ((0,-1,1)>=
(1,1,1) OR (1,1,0)>0 OR (1,1,0)=-1) AND ((0,-1,1)>=(0,1,1) OR
(0,1,0)>0 OR (0,1,0)=-1) AND ((0,-1,1)>=( 1,1,1) OR (-1,1,0)>0 OR (-
1,1,0)=-1) AND ((0,-1,1)>=(-1,0,1) OR (-1,0,0)>0 OR (-1,0,0)=-1) AND
((0,-1,1)>=(-1,-1,1) OR (-1,-1,0)>0 OR (-1,-1,0)=-1))}
```

Fig. 6. Specification of evacuation model

Figure 6 shows the main rules of evacuation model. We have to different planes to separate the rules that govern the people moving among walls or isles from the orientation guide to an exit. The following table describes the encoding of the cell state, in which each position of the state is represented by natural number, in which each digit represent:

Digit	Meaning
6	Next movement direction. 1:W; 2:SW; 3:S; 4:SE; 5:E; 6:NE; 7:N; 8:NW
5	Speed (cells per second: 1 to 5)
4	Last movement direction, it can vary from 1 to 8 (as digit #6)
3	Emotional state: the higher this value is the lower the probability that a person gets in panic and find a wrong path.
2	Number of Movement that increase the potential of a cell
1	Panic Level, represent the number of cells that a person will move increasing the cell potential.

We used two planes: one for the floor plan of the structure and the people moving, and the other for orientation to an exit. Each cell in the grid represents 0.4 m² (one person per cell). The orientation layer contains information that serves to guide persons towards emergency exits. We assigned a potential distance to an exit to every cell of this layer. The persons will move for the room trying to minimize the potential of the cell in which they are (see Figure 6).

The first set of rules in Figure 6 serves to define what path a person should follow using the orientation plane. The basic idea is to take the direction that decrease the potential of a cell, building a path following the lower value of the neighbors. We used the *remainder* and *trunc* functions to split the different parts of a cell's value. Also, we use the CD++ function *randint* to generate integer random values. We have 8 rules to control the people's movement, one for each direction. In all cases the rule analyze the 8 near neighbors to understand what direction the person should take. We use *randint* for the case that all the 8 near neighbors has the same value.

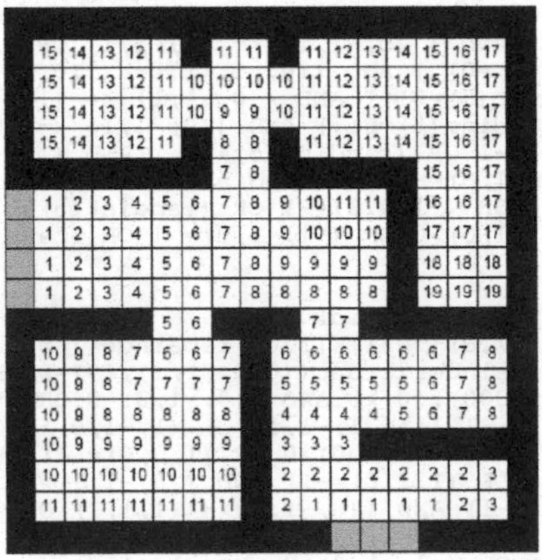

Fig. 7. Orientation layer: potential value (people try to use a path decreasing the potential)

The second set of rules governs the panic behavior: a person will take a wrong path or will not follow the orientation path. In that case, the direction will be calculated taking the path where the cell's potential is increased. In this case also we analyze the 8 near neghibors. This model doesn't allow people collitions, so every time that a person move, the destination cell must be empty.

Figure 8 shows the simulation results. The gray cells represent people who want to escape using the exit doors. The black cells represent walls. Note that the leftmost part in the figure shows people waiting in the exit door.

Flock of Birds

The motion of a flock of birds seems fluid, because of the interaction between the behaviors of individual birds. To simulate a flock we simulate the behavior of an individual bird (or at lE that portion of the bird's behavior that allows it to participate in a flock), based on the following behavior rules [13]:

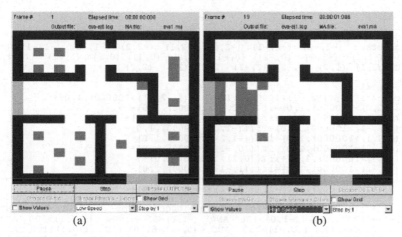

Fig. 8. (a) People seeking an exit. (b) after 15 seconds, people found the exit

- Collision Avoidance: avoid collisions with nearby flock mates.
- Velocity Matching: attempt to match velocity with nearby flock mates.
- Flock Centering: attempt to stay close to nearby flock mates.

  Characteristics of the birds:

- The birds fly in certain direction at a certain speed.
- The field of vision of the birds is 300 grades, but only they have good sight forward (in a zone from 10 to 15 grades).

Based on those rules we built a Cell-DEVS model to simulate the birds' fly [14]. Each cell of the model represents a space of 4 m², which can fit a medium size bird (~18-23 cm.). A second of simulation's time represents a second of real time. Therefore, a bird that moves in the simulation with a speed of 7 cells per second, represents to a bird that flies to 14 m/s. The cell state codification represents with a natural number the direction of the bird (1:NW; 2:N; 3:NE; 4:W; 6:E; 7:SW; 8:S; 9:SE) and the speed. For example, the cell value 10004 represents a bird flying to W (unit value equal to 4) at 1 second per cell.

To avoid the collision of birds, when two or more birds want to move to the same place, they change their direction using a random speed variable. Figure 9 describes the model specification in CD++.

Figure 9 shows the execution of the model using CD++. Basically the rules represents the fly behavior. Birds fly in a freedom way, instinctively when a bird detect others birds, try to follow them. The bird change the direction and the velocity to avoid collitions or lost the flock. In this model, we using different time conditions to simulate the change of bird's velocity.

We show the birds flying, and when one bird sees the other, they start flying together.

```
[boids]
type : cell          dim : (20,20)        delay :  transport
border : wrapped
neighbors :  (-2,-2) (-2,-1) (-2,0) (-2,1) (-2,2) (-1,-2) (-1,-1) (-
1,0)  (-1,1)  (-1,2)  (0,-2)  (0,-1)  (0,0)  (0,1)  (0,2)  (1,-2) (1,-1)
(1,0) (1,1) (1,2) (2,-2) (2,-1) (2,0) (2,1) (2,2)
...
[fly-rule]
rule : { 1+if(((-2,-2)>100000),1,0)+if(((-2,-1)>100000),1,0)+
if(((-2,0)>100000),1,0)+if(((-2,1)>100000),1,0)+
if(((-2,2)>100000),1,0)+if(((-1,-2)>100000),1,0)+
if(((-1,-1)>100000),1,0)+if(((-1,0)>100000),1,0)+
if(((-1,1)>100000),1,0)+if(((-1,2)>100000),1,0)+
if(((0,-2)>100000),1,0)+if(((0,-1)>100000),1,0)+
if(((0,1)>100000),1,0)+if(((0,2)>100000),1,0)+
if(((1,-2)>100000),1,0)+if(((1,-1)>100000),1,0)+
if(((1,0)>100000),1,0)+if(((1,1)>100000),1,0)+if(((1,2)>100000),1,0)+
if(((2,-2)>100000),1,0)+if(((2,-1)>100000),1,0)+
if(((2,0)>100000),1,0)+if(((2,1)>100000),1,0)+if(((2,2)>100000),1,0) }
{90+trunc((0,0)/10-10000)*10}   {(0,0)>100000    AND   (((-1,-1)>100  AND
(-1,-1)=9100)   OR   ((-1,0)>100   AND   (-1,0)=8100)   OR   ((-1,1)>100  AND
(-1,1)=7100)   OR   ((0,-1)>100   AND   (0,-1)=6100)   OR   ((0,1)>100  AND
(0,1)=4100)    OR    ((1,-1)>100    AND    (1,-1)=3100)    OR    ((1,0)>100
AND (1,0)=2100) OR ((1,1)>100 AND (1,1)=1100)) }
...
```

Fig. 9. Specification of the Flock of birds model

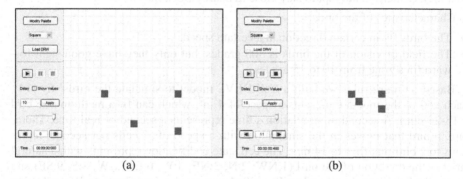

(a)   (b)

Fig. 10. Joining behavior (a) four birds flying isolated; (b) birds flying together

Conclusion

Cell–DEVS allows describing physical and natural systems using an n-dimensional cell-based formalism. Input/output port definitions allow defining multiple interconnections between Cell-DEVS and DEVS models. Complex timing behavior for the cells in the space can be defined using very simple constructions. The CD++ tool, based on the formalism entitles the definition of complex cell-shaped models. We also can develop multidimensional models, making the tool a general framework to define and simulate complex generic models.

The tool and the examples are the public domain and they can be obtained in:
http://www.sce.carleton.ca/faculty/wainer/

References

1. M. Sipper. "The emergence of cellular computing". IEEE Computer. July 1999. Pp. 18-26.
2. D. Talia. "Cellular processing tools for high-performance simulation". IEEE Computer. September 2000. Pp. 44 –52.
3. S. Wolfram. "Theory and applications of cellular automata". Vol. 1, Advances Series on Complex Systems. World Scientific, Singapore, 1986.
4. G. Wainer; N. Giambiasi,. "Timed Cell-DEVS: modelling and simulation of cell spaces". In Discrete Event Modeling & Simulation: Enabling Future Technologies. 2000. Springer-Verlag.
5. B. Zeigler, H. Praehofer, T. Kim. Theory of Modeling and Simulation: Integrating Discrete Event and Continuous Complex Dynamic Systems. 2000. Academic Press.
6. Wainer, G. "CD++: a toolkit to define discrete-event models". Software, Practice and Experience. Wiley. Vol. 32, No.3. pp. 1261-1306. November 2002.
7. D'Abreu, M.; Dau, A.; Ameghino, J. "A wave collision model using Cell-DEVS". Internal report, Computer Science Department. Universidad de Buenos Aires. 2002.
8. Baranek, A.; Riccillo, M.; Ameghino, J. "Modelling prey-predator using Cell-DEVS". Internal report, Computer Science Department. Universidad de Buenos Aires. 2002.
9. Klüpfel, H.; Meyer-König, T.; Wahle J.; Schreckenberg, M. "Microscopic Simulation of Evacuation Process on Passenger Ships". In "Theoretical and Practical Issues on Cellular Automata", Springer-verlag 2001.
10. Meyer-König, T.; Klüpfel, H.; Schreckenberg, M. "A microscopic model for simulating mustering and evacuation processes onboard passenger ships". In Proceedings of TIEMS 2001, Oslo, Norway. 2001.
11. Hongtae Kim, Dongkon Lee, Jin-Hyoung Park, Jong-Gap Lee, Beom-Jim Park and Seung-Hyun Lee. "Establishing the Methodologies for Human Evacuation Simulation in Marine Accidents". Proceedings of 29th Conference on Computers and Industrial Engineering. Montréal, QC. Canada. 2001.
12. Brunstein, M; Ameghino, J. "Modeling evacuation processes using Cell-DEVS". Internal report. Computer Science Department. Universidad de Buenos Aires. 2003.
13. Reynolds Craig, W. "Flocks, Herds, and Schools: A Distributed Behavioral Model". Computer Graphics 21(4), pp. 25-34, July 1987.
14. Dellasopa J; Ameghino, J. "Modelling a Flock of birds using Cell-DEVS". Internal report. Computer Science Department. Universidad de Buenos Aires. 2003.

State Minimization of SP-DEVS

Moon Ho Hwang and Feng Lin

Dept. of Electrical & Computer Engineering, Wayne State University,
Detroit, MI, 48202, USA
mhhwang@kalman.eng.wayne.edu, flin@ece.eng.wayne.edu

Abstract. If there exists a minimization method of DEVS in terms of behavioral equivalence, it will be very useful for analysis of huge and complex DEVS models. This paper shows a polynomial-time state minimization method for a class of DEVS, called schedule-preserved DEVS (SP-DEVS) whose states are finite. We define the behavioral equivalence of SP-DEVS and propose two algorithms of compression and clustering operation which are used in the minimization method.

1 Introduction

From the discrete event system specification (DEVS) [1] schedule-preserved DEVS (SP-DEVS) has been modified so that the behavior of coupled SP-DEVS can be described as atomic SP-DEVS whose states space is finite [2]. Even though there exists an atomic SP-DEVS whose behavior is equivalent to that of SP-DEVS networks, if we can minimize the state space of atomic SP-DEVS, it will be practically useful to identifying its qualitative and quantitative properties [2] of huge and complex systems.

When trying to show decidability of behavioral equivalence between regular languages, testifying their structural equivalence of two finite state automata (FSA) generating the regular languages is more practical than comparing two languages themselves, because the languages may have the infinite number of words [3], [4]. This approach seems to be applicable to state-minimization of DEVS (SP-DEVS) whose states are finite.

But there is one big difference between FSA and SP-DEVS. Every single state transition of FSA is invoked by an external event so a state transition is observable. In SP-DEVS [2] however, there is an internal state transition which occurs according to the *time schedule*. Moreover, when an internal state transition happens there may be no output generated so the transition is *unobservable* outside. Since the behavior of SP-DEVS is defined as a set of observed event sequences with its happening time, such an unobservable internal event is obstructive when applying direct comparison of states as in FAS.

Thus we propose a two-step procedure for state minimization as shown in Fig. 1[1]. The first step in the proposed procedure is compression in which SP-DEVS is modified so that unobservable internal state transitions are eliminated

---

[1] The proposed procedure assumes that its input is an atomic SP-DEVS. If the target we want to minimize is a coupled SP-DEVS model, then we can apply time-translating equivalent (TTE) minimization introduced in [2].

T.G. Kim (Ed.): AIS 2004, LNAI 3397, pp. 243–252, 2005.
© Springer-Verlag Berlin Heidelberg 2005

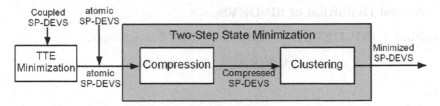

Fig. 1. Two Steps Procedure of State Minimization.

as much as the behavior is preserved (Section 3). The compressed SP-DEVS is used as the input to a clustering step in which behaviorally equivalent states are clustered as one, then the minimized SP-DEVS whose states are clusters can be constructed (Seciont 4). Finally, this article shows that we will always achieve the state-minimized SP-DEVS in polynomial time (Section 5).

2   Behavioral Equivalence of SP-DEVS

2.1   Timed Language

Given an arbitrary event set A, we can then consider a situation that an event string $\bar{a} \in A^*$ occurs when time is $t \in T = \mathbb{R}_0^{+\infty}$ (non negative real numbers with infinity) where A^* is the *Kleene closure* of A [3]. A *timed trajectory* is $\omega : T \to A^*$ and a *timed word* is a timed trajectory restricted to an observation interval. We write it as $\omega_{[t_i, t_f]}$ or $\omega : [t_i, t_f] \to A^*$, in which $t_i, t_f \in T$ s.t. $t_i \leq t_f$. For representing the boundary condition, we use '[' or ']' for a closed boundary while '(' or ')' for the open. In this paper, the *timed empty word* within $[t_i, t_f]$, denoted by $\epsilon_{[t_i, t_f]}$, is that $\omega(t) = \epsilon$ for $t \in [t_i, t_f]$ where ϵ is the *nonevent* or the *empty string*.

A pair of segments $\omega_1 \in \Omega_{A[t_1, t_2]}$ and $\omega_2 \in \Omega_{A[t_3, t_4]}$ are said to be *contiguous* if $t_2 = t_3$. For contiguous segments ω_1 and ω_2 we define the *concatenation operation* $\omega_1 \cdot \omega_2 : [t_1, t_4] \to A^*$ such that $\omega_1 \cdot \omega_2(t) = \omega_1(t)$ for $t \in [t_1, t_2)$, $\omega_1(t) \cdot \omega_2(t)$ for $t = t_2$, $\omega_2(t)$ for $t \in (t_3, t_4]$ where $\omega_1(t) \cdot \omega_2(t)$ is the concatenation of the event string. If there is no confusion, we will omit '$\cdot$' so $\omega_1 \omega_2$ is the same as $\omega_1 \cdot \omega_2$.

For each $t_d \in T$, we define a unary operator on the segment, *translation operator* $TRANS_{t_d}$ such that if there is $\omega_{[t_i, t_f]} = (t_0, \bar{a}_0)(t_1, \bar{a}_1) \in \Omega_A$ then $TRANS_{t_d}(\omega_{[t_i, t_f]}) = \omega_{[t_i + t_d, t_f + t_d]} = (t_0 + t_d, \bar{a}_0)(t_1 + t_d, \bar{a}_1)$. For concatenation of two *discontiguous* segments such as $\omega_{1[t_1, t_2]}$, $\omega_{2[t_3, t_4]}$ where $t_2 \neq t_3$, we can apply the translation operator to ω_2 for making them contiguous and then apply the contiguous concatenation.

A *timed language* over A in $[t_i, t_f]$ is a set of timed words over A in $[t_i, t_f]$. The *universal language* over A in $[t_i, t_f]$ is the set of all possible timed words over A in $[t_i, t_f]$, denoted by $\Omega_{A[t_i, t_f]}$. We will omit the time range of a timed word such as Ω_A when the time range is the same as $[0, \infty)$.

2.2 Formal Definition of SP-DEVS

Definition 1 (SP-DEVS). *A* model of SP-DEVS *is a 9-tuple,*

$$M =< X, Y, S, ta, \delta_x, \delta_\tau, \lambda, S_0, S_F >$$

where,

- $X(Y)$ is a *finite set of input (output) events.*
- S is a *non-empty and finite states set.* S can be partitioned into two sets: *rescheduling states set* S_τ and *continued scheduling states set* S_c such that $S_\tau \cap S_c = \emptyset$ and $S_\tau \cup S_c = S$.
- $ta : S \rightarrow \mathbb{Q}_0^{+,\infty}$ is the *maximum sojourning time* to the next scheduled state where $\mathbb{Q}_0^{+,\infty}$ denotes a set of non-negative rational numbers with infinity.
- $\delta_x : S \times X \rightarrow S_c$ is the *partial external transition function.*
- $\delta_\tau : S \rightarrow S_\tau$ is the *partial internal transition function.*
- $\lambda : S \rightarrow Y^\epsilon$ is the *partial internal output function* where Y^ϵ means $Y \cup \{\epsilon\}$.
- $S_0 \subseteq S_\tau$ is a *set of initial states.*
- $S_F \subseteq S$ is a *set of acceptable states.*

2.3 Timed Languages of SP-DEVS

Given $M =< X, Y, S, ta, \delta_x, \delta_\tau, \lambda, S_0, S_F >$, the *total states set* $Q = \{(s, r) | s \in S, 0 \leq r \leq ta(s)\}$ considers the remaining time r at $s \in S$. Based on the total state set, the *state transition function* $\delta : Q \times X^\epsilon \rightarrow Q$ is represented by two state transition functions: for $(s, r) \in Q$, $x \in X^\epsilon$, $\delta((s, r), x) = (\delta_x(s, x), r)$ if $x \in X \wedge \delta_x(s, x) \perp;$[2] $(\delta_\tau(s), ta(\delta_\tau(s)))$ if $x = \epsilon \wedge r = 0 \wedge \delta_\tau(s) \perp;$ (s, r) otherwise.

Then the *active state trajectory function* $\hat{\delta} : Q \times \Omega_X \rightarrow Q$ is a partial function such that for $q \in Q$, $\omega = (t, x)$ where $x \in X^\epsilon$ and $t \in T$, $\hat{\delta}(q, \omega) = q$ if $x = \epsilon;$ $\delta(q, x)$ if $x \in X \wedge \delta(q, x) \neq q;$ is undefined otherwise. Now suppose that $\omega_1 = \omega_2(t, x)$, $x \in X^\epsilon$, $\omega_1, \omega_2 \in \Omega_X$ and $t \in T$. Then for $q \in Q$, $\hat{\delta}(q, \omega_1) \overset{def}{=} \hat{\delta}(\hat{\delta}(q, \omega_2)(t, x)))$.

Finally, the event trajectory over $X \cup Y$ is described by a *IO trajectory function* $\hat{\gamma} : Q \times \Omega_X \rightarrow \Omega_{X \cup Y}$ such that for $x \in X, (s, r) \in Q$ and $\omega, \omega_2 \in \Omega_X$, $\hat{\gamma}(q, \omega) = (t, x)$ if $\omega = (t, x);$ $(t, \lambda(s))$ if $\omega = (t, \epsilon) \wedge r = 0;$ $\hat{\gamma}(q, \omega_2)(t, x)$ if $\omega = \omega_2(t, x) \wedge \hat{\delta}(q, \omega_2) \perp;$ $\hat{\gamma}(q, \omega_2)(t, \lambda(s'))$ if $\omega = \omega_2(t, \epsilon) \wedge \hat{\delta}(q, \omega_2) \perp \wedge r' = 0;$ is undefined otherwise, where $\hat{\delta}(q, \omega_2) \perp$ then $\hat{\delta}(q, \omega_2) := q' = (s', r')$.

Now we are ready to define the two languages associated with SP-DEVS. The *language generated from s of M*, denoted by $L(M(s)) \subseteq \Omega_{X \cup Y}$, is that

$$L(M(s)) = \{\hat{\gamma}((s, ta(s)), \omega_x) | \omega_x \in \Omega_X, \hat{\delta}((s, ta(s)), \omega_x) \perp\}. \tag{1}$$

For example, for M_1 shown in Fig. 2, $L(M_1(s_1)) = \{(t_0, ?x_1), (t_0, ?x_1)(t_1, !y_1)$ s.t. $0 \leq t_1 - t_0 \leq 2\}$. Then the *generated language of M*, denoted by $L(M) \subseteq \Omega_{X \cup Y}$, is that $L(M) = \bigcup_{s \in S_0} L(M(s))$.

---

[2] $\perp$ indicates that the associated function is defined. For example, $\delta_x(s, x) \perp$ means that $\delta_x(s, x)$ is defined.

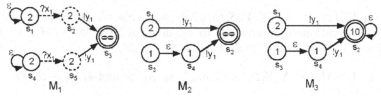

In this paper, circles denote states (solid: $s \in S_\tau$ and dashed: $s \in S_c$), double circles indicate acceptable states, a number inside a state s is $ta(s)$, arcs are state transitions (solid: internal, dashed: external) with ?(!) which indicates an input (output) event.

Fig. 2. SP-DEVS Models.

The *language marked from s of M*, denoted by $L_m(M(s)) \subseteq L(M)$, is that

$$L_m(M(s)) = \{\hat{\gamma}((s, ta(s)), \omega_x)|\omega_x \in \Omega_X, \hat{\delta}((s, ta(s)), \omega_x) \in Q_F\}. \quad (2)$$

where $Q_F = \{(s, r)|s \in S_F, 0 \le r \le ta(s)\}$. For example, for M_1 shown in Fig. 2, $L_m(M_1(s_1)) = \{(t_0, ?x_1)(t_1, !y_1) \text{ s.t. } 0 \le t_1 - t_0 \le 2\}$. Then the *marked language of M*, denoted by $L_m(M) \subseteq L(M)$, is that $L_m(M) = \bigcup_{s \in S_0} L_m(M(s))$.

Before discussing minimization based-on observational equivalence, we would like to focus our interest on a class of SP-DEVS as follows. Our interesting SP-DEVS here is *proper* (1) if $ta(s) = 0$, for $s \in S_\tau$ then $\lambda(s) \ne \epsilon$ and (2) using $ta(s) = \infty$ instead of an internal self-looped state $s \in S_\tau$ such that $\delta_\tau(s) = s \wedge \lambda(s) = \epsilon \wedge \forall x, \delta_x(s, x)$ is undefined. For example, M_1 and M_2 shown in Fig. 2 are proper, but M_3 is not because it violates (2) condition. We are recommended to use M_2 rather than M_3. From now on, all SP-DEVS is supposed to be proper in this paper.

2.4 Behavioral and Observational Equivalence

From the definitions of languages associated with SP-DEVS, we define the behavioral equivalence another equivalence, called observational equivalence.

Definition 2. *Let $M = \langle X, Y, S, ta, \delta_x, \delta_\tau, \lambda, S_0, S_F \rangle$ and $s_1, s_2 \in S$. Then s_1 is said to be* behavioral equivalent *to s_2, denoted by $s_1 \cong^b s_2$ if $L(M(s_1)) = L(M(s_2))$ and $L_m(M(s_1)) = L_m(M(s_2))$.*

For example, for M_1 of Fig. 2, $L(M_1(s_1)) = L(M_1(s_4))$ and $L_m(M_1(s_1)) = L_m(M_1(s_4))$ so $s_1 \cong^b s_4$.

Another way to show the behavioral equivalence is showing observational equivalence. For symbolic representation of an active state trajectory with an associated word, we use $s_1 \xrightarrow{\omega} s_2$ denoting that $\hat{\delta}((s_1, ta(s_1)), \omega_x) = (s_2, r')$ and $\hat{\gamma}((s_1, r), \omega_x) = \omega$ where $s_1, s_2 \in S$, $r \in T$, $\omega_x \in \Omega_X$ and $\omega \in \Omega_{X \cup Y}$ while $s_1 \not\xrightarrow{\omega}$ denotes that neither $\hat{\delta}((s_1, ta(s_1)), \omega_x)$ nor $\hat{\gamma}((s_1, ta(s_1)), \omega_x)$ is defined. For example, M_2 of Fig. 2, $s_1 \xrightarrow{(2, !y_1)} s2$ but $s_4 \not\xrightarrow{(2, !x_1)}$. By symbolic representation of active state trajectories, we define the observational equivalence as follows.

Definition 3. *Let* $M =< X, Y, S, ta, \delta_x, \delta_\tau, \lambda, S_0, S_F >$. *For* $s_1, s_2 \in S$, s_1 *is said to be* observationally equivalent *to* s_2, *denoted by* $s_1 \cong^o s_2$ *if* $\omega \in \Omega_{X \cup Y}$, $s_1 \xrightarrow{\omega} s_1' \Leftrightarrow s_2 \xrightarrow{\omega} s_2'$ *and* $s_1' \in S_F \Leftrightarrow s_2' \in S_F$.

Lemma 1. *Let* $M =< X, Y, S, ta, \delta_x, \delta_\tau, \lambda, S_0, S_F >$ *and* $s_1, s_2 \in S$. *The* $s_1 \cong^b s_2$ *if and only if* $s_1 \cong^o s_2$.

Proof of Lemma 1. (If Case:) Suppose $s_1 \cong^o s_2$ then $\omega \in \Omega_{X \cup Y}$, $s_1 \xrightarrow{\omega} s_1' \Leftrightarrow s_2 \xrightarrow{\omega} s_2'$. That is, $L(M(s_1)) = L(M(s_2))$. Moreover, if $s_1 \cong^o s_2$ then $s_1' \in S_F \Leftrightarrow s_2' \in S_F \Rightarrow L_m(M(s_1)) = L_m(M(s_2))$. Thus $s_1 \cong^o s_2 \Rightarrow s_1 \cong^b s_2$

(Only If Case:) Let $L(M(s_1)) = L(M(s_2))$. Suppose $\exists w \in \Omega_{X \cup Y} : s_1 \xrightarrow{\omega} s_1'$ but $s_2 \not\xrightarrow{\omega}$ then $L(M(s_1)) \supset L(M(s_2))$. Similarly, if $\exists w \in \Omega_{X \cup Y} : s_2 \xrightarrow{\omega} s_2'$ but $s_1 \not\xrightarrow{\omega}$ then $L(M(s_1)) \subset L(M(s_2))$. Thus for $\omega \in \Omega_{X \cup Y}$, $s_1 \xrightarrow{\omega} s_1' \Leftrightarrow s_2 \xrightarrow{\omega} s_2'$.

Suppose that $\omega \in \Omega_{X \cup Y}$, $s_1 \xrightarrow{\omega} s_1' \Leftrightarrow s_2 \xrightarrow{\omega} s_2'$ but $s_1' \in S_F \not\Leftrightarrow s_2' \in S_F$. This means $L_m(M(s_1)) \neq L_m(M(s_2))$ and it contradicts. Thus $s_1 \cong^b s_2 \Rightarrow s_1 \cong^o s_2$. ∎

So sometimes we will use the observational equivalence instead of the behavioral equivalence.

3   Compression

An internal state transition without generating any output can not be observed outside so it should be eliminated as far as the behavior of SP-DEVS can be preserved. This section addresses compression operation, the condition of compression for preserving behavior, completeness, and complexity of a compression algorithm.

3.1   Identifying Compressible States

Compression is merging two states connected with an internal transition. Given $M =< X, Y, S, ta, \delta_x, \delta_\tau, \lambda, S_0, S_F >$, a function $\delta_\tau^{-1} : S \to 2^S$ is used for the *internal source states* to $q \in S$ such that $\delta_\tau^{-1}(q) = \{p \in S | \delta_\tau(p) = q\}$. For example, in Fig. 3(a), $\delta_\tau^{-1}(s_3) = \{s_1\}$ and $\delta_\tau^{-1}(s_4) = \{s_2\}$.

Definition 4 (Compression). *Given* $M =< X, Y, S, ta, \delta_x, \delta_\tau, \lambda, S_0, S_F >$ *and* $s \in S$, *Compression* (M, s) *is that* $\forall s^{-1} \in \delta_\tau^{-1}(s)$ *s.t.* $s^{-1} \neq s$
(1) $\delta_\tau(s^{-1}) := \delta_\tau(s)$; *(2)* $ta(s^{-1}) := ta(s^{-1}) + ta(s)$; *(3)* $\lambda(s^{-1}) := \lambda(s)$;
(4) remove s *from* S *(and* S_F *if* $s \in S_F$);

From this operation, we introduce a condition in which even after applying the compression operation, the behavior is preserved.

Definition 5 (Compressibility). *Let* $M = <X, Y, S, ta, \delta_x, \delta_\tau, \lambda, S_0, S_F>$. *Then* $s \in S, s \notin S_0$ *is said to be* compressible, *if* $\forall s^{-1} \in \delta_\tau^{-1}(s)$ *s.t.* $s^{-1} \neq s$, *for* $\omega_1 \in \Omega_{X[0, ta(s^{-1}))}, \omega_2 \in \Omega_{X[0, ta(s))}, P^U(\omega_1) = P^U(\omega_2)$

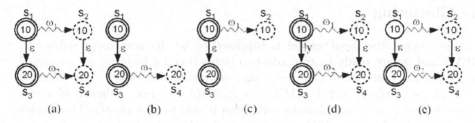

Fig. 3. Compressible or Not.

1. $s^{-1} \xrightarrow{\omega_1} s_x^{-1} \Leftrightarrow s \xrightarrow{\omega_2} s_x$
2. $\delta_\tau(s_x^{-1}) = s_x$ and $\lambda(s_x^{-1}) = \epsilon$
3. $s_x^{-1} \in S_F \Leftrightarrow s_x \in S_F$

where $P^U : \Omega_A \to A^*$ is a *untimed projection* such that for ω, ω' and $\omega'' \in \Omega_A$, $\bar{a} \in A^*$, $P^U(\omega) = \bar{a}$ if $\omega = (t, \bar{a}); P^U(\omega) = P^U(\omega')P^U(\omega'')$ if $\omega = \omega'\omega''$. For example, given $A = \{a, b\}$ and $\omega_{[0,20]} = (7.3, a)(9.6, b)$ then $P^U(\omega_{[0,20]}) = ab$.

The state s_3 and s_4 shown in Fig. 3 (a) are compressible while s_3 and s_4 in Fig. 3 (b), (c) are not because they don't satisfy conditions 1 and Fig. 3 (d), (e) either because of violation of condition 2 and 3 of Definition 5, respectively.

Now we define a function $R : S_\tau \to 2^S$ is *the transited states from* $s \in S_\tau$ such that for $R(s) = \{s\}$ initially and it grows recursively $R(s) = \bigcup_{s_x \in R(s)} \delta_x(s_x, x)$ if $\exists x \in X$ s.t. $\delta_x(s_x, x) \perp$. For example, $R(s_3) = \{s_3, s_4\}$ in Fig. 3 (a).

Theorem 1. *If s is compressible and M_c is achieved by Compression(M, s_x) $\forall s_x \in R(s)$. Then $L(M) = L(M_c)$ and $L_m(M) = L_m(M_c)$.*

Proof of Theorem 1. See Appendix B.

3.2 Completeness and Complexity of a Compression Algorithm

Now we consider algorithms for implementing compression operation. Prior to proposing an algorithm of testing compressible states, we first define an *active input events of a state* defined by $\alpha_x : S \to 2^X$ such that for $s \in S$, $\alpha_x(s) = \{x | \exists x \in X, \delta_x(s, x) \perp\}$. Using this function, an algorithm for identifying compressible states is as follows.

SP-DEVS Compression($M :< X, Y, S, ta, \delta_x, \delta_\tau, \lambda, S_0, S_F >$)
1    $\forall s \in S_\tau$, $\forall s_x \in R(s)$, $\forall s_x^{-1} \in \delta_\tau^{-1}(s_x)$
2   if $(\alpha_x(s_x^{-1}) = \alpha_x(s_x) \wedge \lambda(s_x^{-1}) = \epsilon \wedge s_x^{-1} \in S_F \Leftrightarrow s_x \in S_F)$
3   Compression(M, s_x);
4   return M;

Compression(M) is terminated because the compressible states are eliminated one by one until no more compressible states exist. So we call the result of Compression of SP-DEVS *compressed SP-DEVS* or *non-compressible SP-DEVS*. And for each state $s \in S_\tau, s_x \in R(s)$ and $s_x^{-1} \in \delta_\tau(s_x)$, testing of $\alpha(s_x) = \alpha(s_x^{-1}) \wedge \lambda(s_x^{-1}) = \epsilon \wedge S_F \Leftrightarrow s_x \in S_F$ satisfies all conditions of compressibility. And the pessimistic complexity of Compression(M) is $O(|S_\tau||S_c||S|)$.

4   Clustering

In this section, the input model is supposed to be the non-compressible SP-DEVS and we are ready to compare two states p and q in terms of associated active events, time delay, and acceptance state categorization to test whether $L(M(p)) = L(M(q))$ and $L_m(M(p)) = L_m(M(q))$ or not. A bunch of states whose behaviors are equivalent to each other is said to be a *cluster*. Thus states in a cluster will be merged into one state in the clustering step.

4.1   Identifying Equivalent States

Before introducing an identifying algorithm of equivalent states, we first define two functions. One is an *active events set of a state* defined by $\alpha : S \rightarrow 2^{(X \cup Y)}$ such that for $s \in S$, $\alpha(s) = \{x | \exists x \in X, \delta_x(s, x) \perp\} \cup \{\lambda(s) | \lambda(s) \perp\}$. The other is a *set of input states pairs* $\delta^{-2} : S \times S \rightarrow 2^{S \times S}$ such that for $s_1, s_2 \in S$, $\delta^{-2}(s_1, s_2) = \{(p_1, p_2) | \delta_\tau(p_1) = s_1, \delta_\tau(p_2) = s_2\} \cup \{(p_1, p_2) | \exists x \in X : \delta_x(p_1, x) = s_1, \delta_x(p_2, x) = s_2\}$.

To find those states that are equivalent, we make our best effort to find pairs of states that are distinguishable and then cluster indistinguished state pairs as one state. Following is the algorithm.

```
ClusterSet Finding_Cluster(M :< X, Y, S, ta, δₓ, δτ, λ, S₀, S_F >)
1      ∀p, q ∈ S, add (p, q) to IP;
2      ∀(p, q) ∈ IP, if¬(ta(p) = ta(q) ∧ α(p) = α(q) ∧ p ∈ S_F ⇔ q ∈ S_F)
3          move (p, q) from IP to DP;
4      ∀(s₁, s₂) ∈ DP, ∀(p, q) ∈ δ⁻²(s₁, s₂), if ((p, q) ∈ IP)
5          move (p, q) from IP to DP;
6      NCL := S;
7      ∀p ∈ NCL, CL := ∅; move p from NCL to CL;
8          ∀q ∈ S, if ∃q s.t. (p, q) ∈ IP
9              move q from NCL to CL;
10          append CL to CLS;
11     return CLS;
```

Completeness and Complexity of Finding_Cluster Algorithm. In order to show the completeness of Finding_Cluster(M) algorithm, we use the following theorem.

Theorem 2. *Let $M =< X, Y, S, ta, \delta_x, \delta_\tau, \lambda, S_0, S_F >$ and be a compressed SP-DEVS model. Then $L(M(p)) = L(M(q))$ and $L_m(M(p)) = L_m(M(q))$ if $p, q \in CLS$ of Finding_Cluster(M).*

Proof of Theorem 2. Initially all pairs of all states are treated as indistinguished (line 1). As we can see in lines 2 and 3, for each pair (p, q) in IP (indistinguished pairs), if (1) they have identical maximal sojourning times, (2) have identical active events set and (3) if p is an acceptance state then so q is and vice versa then IP can remain in IP. That is, if one of conditions (1),(2), (3) is violated, (p, q)

moves to DP (distinguished pairs). So after testing lines 2-3, for $(p,q) \in IP$, $p \xrightarrow{\omega} p' \Leftrightarrow q \xrightarrow{\omega} q'$ and $p' \in S_F \Leftrightarrow q' \in S_F$ where $\omega \in \Omega_{X \cup Y[0,ta(p))}$.

As we can see lines 4-5, for $(p,q) \in IP$, $\exists \omega \in \Omega_{X \cup Y} : p \xrightarrow{\omega} p'$ and $q \xrightarrow{\omega} q'$ such that $(p',q') \in DP$, then (p,q) is moved from IP to DP. Thus, after testing 4-5 lines, $(p,q) \in IP$ satisfies $p \xrightarrow{\omega} p' \Leftrightarrow q \xrightarrow{\omega} q'$ and $p' \in S_F \Leftrightarrow q' \in S_F$ where $\omega \in \Omega_{X \cup Y}$ thus $p \cong^o q$. By Lemma 1, $p \cong^o q \Rightarrow p \cong^b q \Rightarrow L(M(p)) = L(M(q))$ and $L_m(M(p)) = L_m(M(q))$.

For finding the cluster, initially NCL (non-clustered states) is assigned as all states S (line 6). For each state p in NCL, a cluster CL is initialized as $\{p\}$ (line 7). For each state $q \in S$, if $p \cong^o q$, then move q from NCL to the cluster (lines 8 and 9). After that, the new cluster CL is added to the clusters set CLS (line 10). According to lines 6-10, each state can be a member of only one cluster in which all states are behavioral equivalent to each other. ∎

The complexity of Finding_Cluster(M) is $O(|S|^2)$ because the testing is based on state pairs.

4.2   Merging States in Cluster

Since we can find the clusters from a SP-DEVS M by Finding_Cluster, the procedure for clustering based on Definition 6 is so straight-forward, that is omitted here and it will be easily designed in $O(|S|)$.

Definition 6 (Clustered SP-DEVS).

Suppose SP-DEVS $M = < X, Y, S, ta, \delta_x, \delta_\tau, \lambda, S_0, S_F >$ is a compressed SP-DEVS. Then $M^m = < X, Y, S^m, ta^m, \lambda^m, \delta^m, S_0^m, S_F^m >$ is *clustered from* M if $S^m = \{s \subset S | \forall s' \in S, s \cong^o s'\}$, $S_\tau^m = S_\tau \cap S^m$; $ta = ta|_{S_\tau^m \to \mathbb{Q}_0^{+,\infty}}$; $\lambda^m = \lambda|_{S^m \to Y^\epsilon}$; $\delta_x^m = \delta_x|_{S^m \times X \to S^m}$; $\delta_\tau^m = \delta_\tau|_{S^m \to S^m}$; $S_0^m = S^m \cap S_0$; $S_F^m = S^m \cap S_F$.

5   Two-Step Minimization Procedure

Figure 4 illustrates what happens in each step of the proposed whole procedure. In this example, 6-state compression is performed after the first step and two clusters consisting of 4 states remain after the second step so the number of states in the minimized model is two. The following theorem addresses the completeness and complexity of the proposed procedure.

Theorem 3. *The two-step minimization method minimizes the states of SP-DEVS in polynomial time.*

Proof of Theorem 3. The compression eliminates unobservable internal state transitions which can be compressed, without changing its behavior. In other words, only unobservable internal state transitions which should be preserved for identical behavior can remain in the process (See Appendix Lemma 2). Therefore, if two states are determined as indistinguished in Finding_Cluster of Section 4.1, they can be merged into one state by Theorem 2. Since each computational

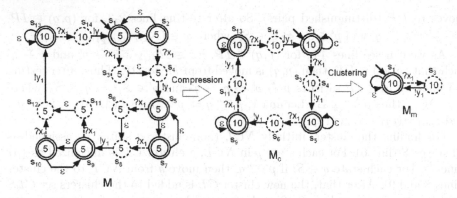

Fig. 4. State Minimization Example.

complexity of compression (Section 3.2) and clustering (which consists of finding clusters (Section 4.1) and merging states in a cluster (Section 4.2)) are polynomial, the whole procedure has also polynomial-time complexity. ■

6 Conclusion and Further Directions

This paper proposed a state-minimization method of proper SP-DEVS and showed the polynomial-time decidability of behavioral equivalence in SP-DEVS models are guaranteed. We first defined the behavioral equivalence of SP-DEVS, and proposed two algorithms for compression and clustering, which are used in the proposed minimization method.

Tough minimization experiments for real systems such as manufacturing systems, embedded software systems etc, should be tested in the future. And one possible improvement will be easily achieved in the clustering step by employing the $O(|S|log|S|)$ algorithm used in finite state automata [4].

References

1. B.P. Zeigler, H.Praehofer, and T.G. Kim. *Theory of Modelling and Simulation: Integrating Discrete Event and Continuous Complex Dynamic Systems*. Academic Press, London, second edition, 2000.
2. M.H. Hwang and Su Kyoung Cho. Timed analysis of schedule preserved devs. In A.G. Bruzzone and E. Williams, editors, *2004 Summer Computer Simulation Conference*, pages 173–178, San Jose, CA, 2004. SCS.
3. J.E. Hopcroft, R. Motwani, and J.D. Ullman. *Introduction to Automata Theory, Languages, and Computation*. Addison Wesley, second edition, 2000.
4. J.E. Hopcroft. An n log n algorithm for minimizing states in a finite automaton. In Z. Kohavi, editor, *The Theory of Machine and Computations*, pages 189–196, New York, 1971. Academic Press.

Appendix: Proof of Theorem 1

Lemma 2. *Suppose that* $M = < X, Y, S, ta, \delta_x, \delta_\tau, \lambda, S_0, S_F >$. *Then* $s \in S_\tau$ *is compressible if and only if* $\forall s^{-1} \in \delta_\tau^{-1}(s)$, $s^{-1} \xrightarrow{\omega} s' \Leftrightarrow s_c^{-1} \xrightarrow{\omega} s_c^{-1\prime}$ *and* $s' \in S_F \Leftrightarrow s_c^{-1\prime} \in S_F$ *where* $\omega \in \Omega_{X[0, ta(s^{-1}) + ta(s))}$, s_c^{-1} *denotes* $s^{-1} \in S$ *of* M_c *and* M_c *is the SP-DEVS in which* $s_x \in R(s)$ *is compressed by* $Compression(M, s_x)$.

Proof of Lemma 2. (If Case) Assume that for $s \in S_\tau$ s is compressible. Let $\omega = \omega_1 \omega_2$ such that $\omega_1 \in \Omega_{X[0, ta(s^{-1}))}, \omega_2 \in \Omega_{X[0, ta(s))}$. Then $\exists (s_x^{-1}, s_x$ and $s')$ s.t. $s^{-1} \xrightarrow{\omega_1} s_x^{-1}$, $\delta_\tau(s_x^{-1}) = s_x$ and $s_x \xrightarrow{\omega_2} s'$. By condition 2 of Definition 5, $s^{-1} \xrightarrow{\omega_1} s_x^{-1} \xrightarrow{\omega_\epsilon} s_x \xrightarrow{\omega_2} s'$ where $\omega_\epsilon = (0, \epsilon)$. Since ω_ϵ is a timed nonevent so $s^{-1} \xrightarrow{\omega_1 \omega_2} s'$.

Let's check the preservation of ω in the compressed model M_c. Since $s \xrightarrow{\omega_1} s_x$ so $s_c^{-1} \xrightarrow{\omega_1} s_{cx}^{-1}$. By condition 1 of Definition 5, $\exists s_c^{-1\prime}$ s.t. $s_{cx}^{-1} \xrightarrow{\omega_2} s_c^{-1\prime}$ so we can say that $s_c^{-1} \xrightarrow{\omega_1 \omega_2} s_c^{-1\prime}$. By the third condition of Definition 5, it is true that $s' \in S_F \Leftrightarrow s^{-1\prime} \in S_F$. Moreover, in the $Compression(M, s)$ there is no change of $s^{-1\prime}$ from the viewpoint of S_F so we can say that $s' \in S_F \Leftrightarrow s_c^{-1\prime} \in S_F$. Therefore, $\forall s^{-1} \in \delta_\tau^{-1}(s)$, $s^{-1} \xrightarrow{\omega} s' \Leftrightarrow s_c^{-1} \xrightarrow{\omega} s_c^{-1\prime}$ and $s' \in S_F \Leftrightarrow s_c^{-1\prime} \in S_F$.

(Only If Case) Assume that for $\forall s^{-1} \in \delta_\tau^{-1}(s)$, $s^{-1} \xrightarrow{\omega} s' \Leftrightarrow s_c^{-1} \xrightarrow{\omega} s_c^{-1\prime}$ and $s' \in S_F \Leftrightarrow s_c^{-1\prime} \in S_F$, but s is not compressible. Let's consider $\omega_1 \in \Omega_{X[0, ta(s^{-1}))}, \omega_2 \in \Omega_{X[0, ta(s))}$ s.t. $P^U(\omega_1) = P^U(\omega_2)$.

First, suppose $s^{-1} \xrightarrow{\omega_1} s_x^{-1}$ but $s \xrightarrow{\omega_2}$ and $\omega = \epsilon_{[0, ta(s^{-1}))} \omega_2$. Since $s \xrightarrow{\omega_2}$ so $s^{-1} \xrightarrow{\omega}$ however $s_c^{-1} \xrightarrow{\omega} s_{cx}^{-1\prime}$ because $s^{-1} \xrightarrow{\omega_1} s_x^{-1\prime}$. This contradicts to the assumption of $s^{-1} \xrightarrow{\omega} s' \Leftrightarrow s_c^{-1} \xrightarrow{\omega} s_c^{-1\prime}$. By contrast, assume that $s^{-1} \xrightarrow{\omega_1}$ but $s \xrightarrow{\omega_2} s_x$ and $\omega = \epsilon_{[0, ta(s^{-1}))} \omega_2$. In this case, $s^{-1} \xrightarrow{\omega} s_x$ but $s_c^{-1} \xrightarrow{\omega}$ so it also contradicts to $s^{-1} \xrightarrow{\omega} s' \Leftrightarrow s_c^{-1} \xrightarrow{\omega} s_c^{-1\prime}$.

Now check the second condition. Suppose that $s^{-1} \xrightarrow{\omega_1} s_x^{-1} \Leftrightarrow s \xrightarrow{\omega_2} s_x$ but $\delta_\tau(s_x^{-1}) \neq s_x$. Let $\omega = \omega_1 \epsilon_{[0, ta(s))}$ then $s^{-1} \xrightarrow{\omega}$ but $s_c^{-1} \xrightarrow{\omega} s_{cx}^{-1}$ so it contradicts to $s^{-1} \xrightarrow{\omega} s' \Leftrightarrow s_c^{-1} \xrightarrow{\omega} s_c^{-1\prime}$. In addition, let $\lambda(s_x^{-1}) \neq \epsilon$ then if $\omega = \omega_1 (0, \lambda(s_x^{-1})) \epsilon_{[0, ta(s))}$ then $s^{-1} \xrightarrow{\omega} s_x^{-1}$ but $s_c^{-1} \xrightarrow{\omega}$, so it contradicts to the assumption.

Finally, assume that $s^{-1} \xrightarrow{\omega_1} s_x^{-1} \Leftrightarrow s \xrightarrow{\omega_2} s_x$ but $\delta_\tau(s_x^{-1}) = s_x$ and $\lambda(s_x^{-1}) = \epsilon$ but $s_x^{-1} \in S_F \nLeftrightarrow s_x \in S_F$. In this case, for $\omega = \omega_1 \epsilon_{[0, ta(s))}$, $s^{-1} \xrightarrow{\omega} s_x^{-1} \Leftrightarrow s_c^{-1} \xrightarrow{\omega} s_{cx}^{-1}$ but $s_x^{-1} \in S_F \nLeftrightarrow s_{cx}^{-1} \in S_F$ so it contradicts to the assumption. ∎

Theorem 1. *If* s *is compressible and* M_c *is achieved by* $Compression(M, s_x)$ $\forall s_x \in R(s)$. *Then* $L(M) = L(M_c)$ *and* $L_m(M) = L_m(M_c)$.

Proof of Theorem 1. By Lemma 1 and Lemma 2. ∎

DEVS Formalism: A Hierarchical Generation Scheme

Sangjoon Park[1] and Kwanjoong Kim[2]

[1] Information and Media Technology Institute, Soongsil University
lub@archi.ssu.ac.kr
[2] Dept. of Computer and Information, Hanseo University
kimkj@hanseo.ac.kr

Abstract. System reproduction model to the growing system structure can be provided to design modeling formalisms for variable system architectures having historical characteristics. We introduce a DEVS (Discrete Event System Specifications)-based extended formalism that system structure gradually grows through self-reproduction of system components. As extended-atomic model of a system component makes virtual-child atomic DEVS models, a coupled model can be derived from coupling the parent atomic model and virtual-child atomic models. When a system component model reproduces its system component, a child component model can receive its parent model characteristics including determined role or behavior, and include different structure model characteristics. A virtual-child model that has its parent characteristics can also reproduce its virtual-child model, which may show similar attributes of the grand-parent model. By self-reproducible DEVS (SR-DEVS) modeling, we provide modeling specifications for variable network architecture systems.

1 Introduction

Modeling formalism methodologies describing system specifications are being developed for more complex system structures. System modeling formalisms are used to analyze dynamic complex systems which have their system structure and proper behavior. Klir [1] introduced a system framework, which hierarchically constructs levels of system knowledge to a real source system. The level of system knowledge is divided into four levels: *source level*, *data level*, *behavior level*, and *structure level*. The lowest level to the system knowledge is the *source level* that identifies variables and determines the observation means to a source system. At the above level, the *data level* collects data from a source system. The next level is the *behavior level* that reproduces data by using formulas or means. As the highest level, the *structure level* represents the knowledge that is about subsystems coupled to a source system. For the modeling formalism and the simulation, Zeigler [2] represents a methodology which specifies discrete event models for simulation implementations. To establish a framework for modeling and simulation, entities such as *source system, model,* and *simulator* are defined. Mapping from a source system to a model can be constructed by observing system and behavior database gathered by system experiments. After the modeling formalism, the model may need a simulator to generate behavior of the source system. Note that the simulator established by the model must represent valid behavior to the model and the source system. That is, actually the validity to relationships among entities is important to analyze a source system correctly.

T.G. Kim (Ed.): AIS 2004, LNAI 3397, pp. 253–261, 2005.
© Springer-Verlag Berlin Heidelberg 2005

Zeigler [2] presents three basic systems modeling formalisms: *differential equation system*, *discrete time system*, and *discrete event system*. The *differential equation system* and the *discrete time system* have been traditionally used to analyze source systems in numerous research areas. The *discrete event system*, named the *DEVS*, is a new modeling paradigm that formalizes the discrete event modeling, and models specified by other formalism methods newly. By coupling the *atomic DEVS*, *coupled DEVS models* can be expressed to specify components that are linked in a system. After the *classic DEVS*, various extended DEVS methodologies have been developed to formalize and simulate more complex dynamic source systems [3]-[6]. As an example, *parallel DEVS* that component DEVS models are interconnected for the specification of a parallel system is able to implement multi-component executions concurrently. A component model in a *parallel DEVS* model may receive and output events from more than one other component. In particular, a parallel DEVS model can be constructed in hierarchical or distributed DEVS pattern to represent dispersed coupling systems [8]. The *dynamic structure DEVS* model [6] [7] represents another extended DEVS model that can dynamically change its system structure through the input segment. In the *dynamic structure DEVS*, the system change includes the deletion of a system component as well as the addition of a system component.

In this paper, we propose a new extended DEVS model, called *Self-Reproducible DEVS (SR DEVS) model*, that a component model can reproduce its child DEVS model to grow a system structure. A child DEVS model may receive the structural characteristic or the structural inheritance from its parent DEVS model so that the behavior of child DEVS model can be similar to the behavior of parent models. We expect that the *SR DEVS model* is applied to develop simulators for social, technical and biological growing systems.

This paper is organized as follows. In Section 2, we summarize the *classic DEVS model*. We describe the *SR DEVS model* in Section 3. In Section 4, we conclude this paper with some remarks.

2 DEVS Formalism

A *classic DEVS* model is represented in the basic formalism specifying the behavior of an atomic system. The *classic DEVS* model for discrete event systems is given by

$$ADEVS = \left(IP, OP, S, \delta_{int}, \delta_{ext}, \lambda, ta\right) \tag{1}$$

where IP is the set of input values, OP is the set of output values, S is the set of states, $\delta_{int} : S \to S$ is the internal transition function, $\delta_{ext} : \Gamma \times IP \to S$ is the external transition function, where $\Gamma = \{(s, e) \mid s \in S, 0 \le e \le ta(s)\}$ is the total state set and e is the time elapsed since last transition, $\lambda : S \to OP$ is the output function, and $ta : S -> R^+_{0,\infty}$ is the time advance function.

If a basic system stay in a state $s (s \in S)$, and no external events occurs, it will stay in the same state s for time $ta(s)$. If the elapsed time $e = ta(s)$, the system operates the output function λ to obtain the output value, and changes to a state $\delta_{int}(s)$. If an external event $x \in IP$ occurs when the system is in a state (s, e) with $e \le ta(s)$, it changes to a state $\delta_{ext}(s, e, x)$. Note that the time base is the nonnegative real numbers including subsets such as integers, and the output function Λ of a dynamic system is given by

$$\Lambda(s,e) = \begin{cases} \lambda(s) & if \quad e = ta(s) \quad and \quad \omega(t) = \varnothing \\ \lambda(s) \quad or \quad \varnothing & if \quad e = ta(s) \quad and \quad \omega(t) \neq \varnothing \\ \varnothing & otherwise \end{cases}$$

where $\omega(t)$ is the time segment.

3 Self-reproducible DEVS Model

A component in growing system structure can produce a child component which may show same or similar behavior fashion and component structure since it receives system characteristics from a parent component. As examples, the social, industrial and population growth, and the cell multiplication in the biology are all about the growing or evolving systems. On the computer simulation, same or similar component models by the model reproduction can be used to increase the simulation efficiency and describe the relations with other component models elaborately. Hence, in this paper, we propose an extended DEVS model, *SR DEVS model* that makes its child DEVS model by using reproduction mechanism. We define the reproduction of a DEVS model that includes a partial reproduction as well as an entire reproduction. The *SR DEVS model* can reproduce a child DEVS component model from a parent DEVS component model virtually. In particular, a child DEVS model receives the structural/functional inheritance (e.g., values/functions) from one parent DEVS model or from multiple parent DEVS model, and reproduce a grand child DEVS model. Hence, if a DEVS family model is constructed by reproducing child DEVS models, it is called a *DEVS cluster* that may have one more parent DEVS models. Note that a *self-reproducible DEVS model* concept can be applied to a coupled DEVS model as well as an atomic DEVS model. Therefore, in this paper, we introduce three *SR DEVS types*: *Atomic SR DEVS, multiple parent SR DEVS*, and *coupled SR DEVS model*.

3.1 Atomic SR DEVS Model

A *SR DEVS model* has a structure as follows

$$SRDEVS_{\Psi} = (IP_{\Psi}, OP_{\Psi}, S_{\Psi}, C_{\Psi}, \delta_{\Psi}, \lambda_{\Psi}, ta)$$

where Ψ is the cluster name of the *SR DEVS model*, IP_{Ψ} is the input set of the *DEVS cluster* including SR trigger, OP_{Ψ} is the output set of the *DEVS cluster*, S_{Ψ} is the set of states of the *DEVS cluster*, $C_{\Psi} = \left(P_{\Psi}, \left\{R_{\Psi'} \mid R_{\Psi'} \in {}^{*} P_{\Psi}\right\} RES_{\Psi'}, SELECT, CON_{R_{\Psi'}}\right)$ where C_{Ψ} is the *connector* of Ψ, P_{Ψ} is the parent DEVS model, $R_{\Psi'}$ is the set of child DEVS models, the symbol $\in {}^{*}$ means the relation that the child component inherits characteristics from the parent component, $RES_{\Psi'}$ is the inheritance function set from the parent DEVS model to the child DEVS model, *SELECT* is the select function of the *DEVS cluster*, and $CON_{R_{\Psi'}} = \left\{\left(\left(P_{\Psi}, out_{\Psi}\right)\left(R_{\Psi'}, in\right)\right), \left(\left(R_{\Psi'}, out\right), \left(P_{\Psi}, in_{\Psi}\right)\right) \mid out_{\Psi} \text{ and } out \in OutCo, in_{\Psi} \text{ and }$

$in \in InCon$ $\}$ means the connection set of P_Ψ and $R_{\Psi'}$, δ_Ψ is the transition function of the *DEVS cluster*, and λ_Ψ is the output function of the *DEVS cluster*.

If the parent DEVS model receives an input event which triggers a child DEVS model reproduction, it reproduces its child DEVS model within the time segment t. While the parent DEVS model reproduces a child DEVS model, the child model receives the structural inheritance from its parent model by the inheritance function. The set $RES_{\Psi'}$ to inheritance functions for the child DEVS model is given by

$$RES_{\Psi'} = \left(r_{\Psi'}, IS_{\Psi'}, OS_{\Psi'}, SS_{\Psi'}, \delta S_{\Psi'}, \lambda S_{\Psi'}\right)$$

where $r_{\Psi'}$ is a child DEVS model reproduced from the parent DEVS model, $IS_{\Psi'}$ is the element inheritance function to IP_Ψ $\left[f\left(\in {}^*IP_\Psi\right)\right]$, $OS_{\Psi'}$ is the element inheritance function to OP_Ψ $\left[f\left(\in {}^*OP_\Psi\right)\right]$, $SS_{\Psi'}$ is the element inheritance function to S_Ψ $\left[f\left(\in {}^*S_\Psi\right)\right]$, $\delta S_{\Psi'}$ is the element inheritance function to $\delta S_{\Psi'}$ $\left[f\left(\in {}^*\delta_\Psi\right)\right]$, and $\lambda S_{\Psi'}$ is the element inheritance function to λ_Ψ $\left[f\left(\in {}^*\lambda_\Psi\right)\right]$. Note that inter-relationships among element inheritance functions should be considered to implement proper behaviors of the parent DEVS model.

Therefore, we obtain the child DEVS model as follows

$$R_{\Psi'} = \left(IP_{\Psi'}, OP_{\Psi'}, S_{\Psi'}, C_{\Psi'}, \delta_{\Psi'}, \lambda_{\Psi'}, ta\right)$$

where $IP_{\Psi'} = \{\alpha_1, \alpha_2, \cdots, \alpha_l \mid \forall \alpha_i \in IP_\Psi, 1 \le i \le l\}$ is the input set of the child DEVS model $R_{\Psi'}$, $OP_{\Psi'} = \{\beta_1, \beta_2, \cdots, \beta_m \mid \forall \beta_j \in OP_\Psi, 1 \le j \le m\}$ is the output set of $R_{\Psi'}$, $S_{\Psi'} = \{\gamma_1, \gamma_2, \cdots, \gamma_n \mid \forall \gamma_k \in S_\Psi, 1 \le k \le n\}$ is the state set of $R_{\Psi'}$, $C_{\Psi'}$ is the *connector* of $R_{\Psi'}$, $\delta_{\Psi'}(\in {}^*\delta_\Psi)$ is the transition function of $R_{\Psi'}$ and $\lambda_{\Psi'}(\in {}^*\lambda_\Psi)$ is the output function of the $R_{\Psi'}$.

If a *SR DEVS model* does not reproduce a child DEVS model $(R_{\Psi'} = \varnothing)$, the *SR DEVS model* can be a classic atomic DEVS model. Otherwise, if a *SR DEVS model* has a child DEVS model, the child DEVS model receives inheritance values/functions from the parent SR DEVS model. Hence, when a child DEVS model is reproduced, and it has relations with its parent DEVS model, we call such a relationship as the *inheritance type* or the *cluster morphism*. We define the following *cluster morphism* types from a parent DEVS model to a child DEVS model. Fig. 1 shows two *cluster morphism* properties.

1) *Integrity*: If a *SR DEVS model* $(SRDEVS_A)$ produce a child DEVS model $(R_{A'})$ in integrity type, its child DEVS model has same DEVS model $(R_{A'} = SRDEVS_A)$ as shown in Fig. 1-(a). All of structural/functional elements in a child DEVS select all of values/functions to related elements in a parent DEVS model so that the child DEVS model shows same behavior to the parent DEVS model.

2) *Fragment*: If a *SR DEVS model* $(SRDEVS_A)$ produces a child DEVS model by the *fragment type*, the child DEVS model is a sub model of its parent model $(R_{A'} \subset SRDEVS_A)$ as shown in Fig. 1-(b). In the *fragment type*, a child DEVS model shows sub operations to its parent DEVS model because it selects

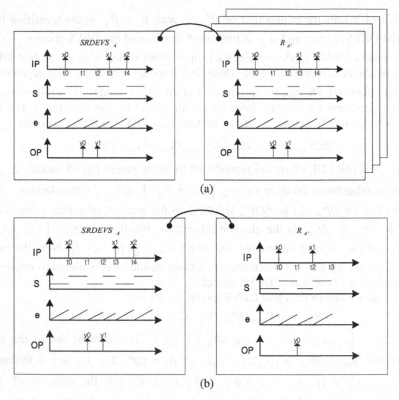

Fig. 1. Reproduction property. (a) *Integrity*. (b) *Fragment*

subsets of values/functions as a sub-model of parent DEVS model. The *fragment* type is divided into two sub-types: *regular* and *irregular cluster morphisms*. In the *regular cluster morphism*, the child DEVS model selects values/functions of its parent DEVS model regularly. In the *irregular cluster morphism*, values/functions of a parent DEVS model are irregularly selected by selection functions of a child DEVS model.

Since each element in a child DEVS model has same or similar values/functions to those of elements of a parent DEVS model $\left(R_\omega \subseteq SRDEVS_\psi \right)$, its behavior shows the similarity to the parent DEVS model. Also, if a child DEVS model is reproduced and connected to a parent DEVS model, a coupled DEVS model can be constructed from the cluster structure. Furthermore, a coupled DEVS model by the reproduction of DEVS model can be constructed for distributed or parallel simulation. However, note that all of DEVS clusters does not have the coupled structure because child DEVS models can independently operate without connecting their parent DEVS models. Fig. 2 shows a *DEVS cluster* that a parent DEVS model has three generation models showing various *cluster morphism types*. In the Fig. 3, after three reproductions, the root parent DEVS model has nine child DEVS models which connect their parent DVES models or other child DEVS models of same level. We assume that the interconnection between two DEVS models is bidirectional connection. In the first

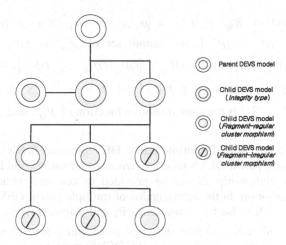

Fig. 2. Structured *DEVS cluster*

generated reproduction level, the child DEVS model showing *integrity type* has same behaviors such like the parent DEVS model and connects neighbor child DEVS model without connecting its parent DEVS model.

In the second reproduction level, the child DEVS model of *integrity type* can show all of behaviors of its parent DEVS model (the first reproduction level). Hence, reproducing child DEVS model by *integrity type* can pass all of inheritance properties to descendant models. A child DEVS model of *fragment type* may show entirely different behaviors compared with other child DEVS models of *fragment type* in same reproduction level. Also, child DEVS models of *fragment type* in lower level may have entirely different behaviors to child DEVS models of *fragment type* in higher level with the exception of parent DEVS models.

3.2 Multiple Parent SR DEVS Model

Multiple Parent DEVS models can produce a child DEVS model which receives composite structural inheritance from its parent DEVS models. The *connector* to child DEVS models is given by

$$C_M = (P_M = \{P_{\alpha_1}, P_{\alpha_2}, \cdots, P_{\alpha_n}\}, \{R_{M'} \mid R_{M'} \in {}^*P_M\}, RES_{M'}, SELECT, CON_{R_{M'}})$$

where RES_M is the inheritance function set from parent DEVS models, the inheritance function set is given by $RES_{M'} = (r_{m'}, IS_{M'}, OS_{M'}, SS_{M'}, \delta S_{M'}, \lambda S_{M'})$, and the connection set is given by $CON = \{((P_{\alpha_1}, out_{\alpha_1}), (R_{M'}, in)), ((P_{\alpha_2}, out_{\alpha_2}), (R_{M'}, in)), \cdots, ((P_{\alpha_n}, out_{\alpha_n}), (R_{M'}, in)), ((R_{M'}, out), (P_M, in_M)) \mid \forall out \in OutCon, \forall in \in InCon\}$

Hence, the child DEVS model is given by

$$R_{M'} = (IP_{M'}, OP_{M'}, S_{M'}, C_{M'}, \delta_{M'}, \lambda_{M'}, ta)$$

where the input set of $R_{M'}$ is $IP_{M'} = \{\alpha_{l_1}, \alpha_{l_2}, \cdots, \alpha_{l_i} \mid \forall \alpha_{l_d} \in IP_M, 1 \le d \le i\}$ where $IP_M = \{IP_{\alpha_1}, IP_{\alpha_2}, \cdots, IP_{\alpha_a}\}$, the output set of $R_{M'}$ is $OP_{M'} = \{\beta_{m_1}, \beta_{m_2}, \cdots, \beta_{m_j} \mid \forall \beta_{m_e} \in OP_M, 1 \le e \le j\}$ where $OP_M = \{OP_{\beta_1}, OP_{\beta_2}, \cdots, OP_{\beta_b}\}$, the state set is $S_{M'} = \{\gamma_{n_1}, \gamma_{n_2}, \cdots, \gamma_{n_k} \mid \forall \gamma_{n_f} \in S_M, 1 \le f \le k\}$ where $S_M = \{S_{\gamma_1}, S_{\gamma_2}, \cdots, S_{\gamma_c}\}$, $C_{M'}$ is the *connector* of $R_{M'}$, $\delta_{M'}$ is the state transition function of $R_{M'}$, and $\lambda_{M'}$ is the output function of $R_{M'}$.

Note that the reproduction of multiple parent DEVS models needs the inheritance co-relationship among parent DEVS models. Hence, for creating child DEVS models, the inheritance co-relationship should be provided to construct structural/function association or composition. In the reproduction of multiple parent DEVS models, the *cluster morphism* type has also two types: *integrity* and *fragment types*.

1) *Integrity type* is divided into two subtypes: *local* and *global cluster morphism*. *Local cluster morphism* is that a child DEVS model receives all of values/functions to specific parent DEVS models in the set of parent DEVS models. Hence, a child DEVS model shows behaviors only to specific parent DEVS models. *Global cluster morphism* is that a child DEVS model selects all of values/functions to all of parent DEVS models.

2) *Fragment type* is divided into *regular* and *irregular types* as above mentioned. In *regular cluster morphism*, a child DEVS model regularly selects values/functions at each parent DEVS model. In the *irregular cluster morphism*, each parent DEVS model irregularly passes values/functions to a child DEVS model.

Hence, from the inheritance of multiple parent DEVS models, a child DEVS model can show *integrity* or *fragment type* as its inheritance property. Furthermore, a child DEVS model may be constructed based on the multi-type composition which *local cluster morphism* is associated with *fragment type*. That is, a child DEVS model receives all of values/functions of a parent DEVS model (*integrity type*), and selectively chooses values/functions to another parent DEVS model (*fragment type*).

If there are parent DEVS models such as $SRDEVS_A$, $SRDEVS_B$ and $SRDEVS_C$, and reproduce a child DEVS model R_m, R_m receives structure inheritance from them $(R_m \in {}^*PSET = \{SRDEVS_A, SRDEVS_B, SRDEVS_C\})$. The connector is represented as $C_m = (P = \{SRDEVS_A, SRDEVS_B, SRDEVS_C\} \mid \{R_m \mid R_m \in {}^*P, RES_m, SELECT, CON_{R_m})$ and the inheritance function set is represented as $RES_m = (r_m, f(\in {}^*IS_{pset}), f(\in {}^*OS_{pset}), f(\in {}^*SS_{pset}), \delta S_m, \lambda S_m)$. Here, we assume that all of parent DEVS models have connected to R_m so that the connection is given by $CON_m = \{((PSET, out_{pset}), (R_m, in)), ((R_m, out), (SRDEVS_A, in)), ((R_m, out), (SRDEVS_B, in))((R_m, out), (SRDEVS_C, in)) \mid \forall out \in OutCon, \forall in \in InCon\}$. Hence, the structure of R_m is given by $R_m = (IP_m, OP_m, S_m, C_m, \delta_m, \lambda_m, ta)$, where $IP_m = \{\alpha_1, \alpha_2, \cdots, \alpha_n \mid \forall \alpha \in$

IP_{pset} }, $OP_m = \{\beta_1, \beta_2 \cdots, \beta_{m'} \mid \forall \beta \in OP_{pset}\}$, $S_m = \{\gamma_1, \gamma_2, \cdots, \gamma_{k'} \mid \forall \gamma \in$
$S_{pset}\}$, $\delta_m \in {}^*\delta_{pset}$ and $\lambda_m \in {}^*\lambda_{pset}$.

Fig. 3 shows the *DVES cluster* that three parent DEVS models hierarchically reproduce child DEVS models which have proper inheritance types. After three reproductions, the parent DEVS model has nine child DEVS models. In the first reproduction level, the child DEVS model representing *global cluster morphism* receives all of structural/functional inheritances from three parent DEVS modes. Hence, the child DEVS model shows all of behaviors of three parent DEVS models. Also, in the first level, a child DEVS model of *local cluster morphism* that receives the inheritance from one or two parent DEVS models will show specific behaviors only to the subset of parent DEVS models. In the second reproduction level, a child DEVS model shows the multi-type composition of *local-irregular cluster morphism*. If the child DEVS model showing the multi-type composition selects the upper child DEVS model of *global cluster mophism* in the first level as *local cluster morphism*, it can show all of behaviors of root parent DEVS models. Note that the *fragment type* is *irregular cluster morphism* if a child DVES model receives values/functions from one parent DEVS model regularly and another parent DEVS model irregularly. In the third reproduction level, the child DEVS model of irregular cluster morphism connects to another child DEVS model in same level as well as upper child DEVS models in the second level. Note that the connection between two child DEVS models in same level is not related to the structural inheritance, but it is only provided for communications of inter DEVS models. In this paper, we don't consider the structural inheritance of inter DVES models in same level.

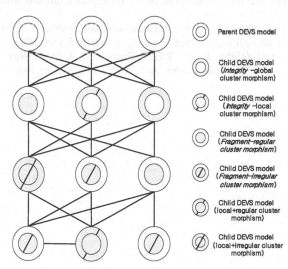

Fig. 3. Multiple parent DEVS models

4   Conclusions

For the modeling formalism and the simulation, the DEVS formalism provides a methodology which specifies discrete event models for simulation implementations.

In this paper, we propose *SR DEVS formalism* that can be used to reproduce a component or a coupled structure describing elaborate inheritances to a parent component or a parent coupled component. *SR DEVS modeling* provide two representative inheritance types: *Integrity* and *Fragment*. When a child component is reproduced, its inheritance type will be determined. Hence, a child component to its parent component can represent same or similar behaviors. Therefore, based on *SR DEVS modeling*, a network structure having inter-relationship between components can be built for the system modeling and simulation. We expect that the *SR DEVS model* can be applied to develop computer simulators about social, technical and biological growing systems.

References

1. Klir, G. J., Architecture of Systems Problem Solving, Plenum Press, NY, 1985.
2. Zeigler, B. P., Praehofer , H., and Kim, T. G., Theory of Modeling and Simulation, Academic Press, 2000.
3. Zeigler, B. P., and Chi, S. D., "Symbolic Discrete Event System Specification," IEEE Transactions on System, Man, and Cybernetics, vol. 22, no. 6, pp. 1428-1443, Nov, 1992.
4. Kwon, Y. W., Park, H. C., Jung, S. H., and Kim, T. G., "Fuzzy-DEVS Formalism: Concepts, Realization and Applications," In Proc. of AIS'96, pp. 227-234, Aug, 1996.
5. Cho, S. M., and Kim, T. G., "Real-time DEVS Simulation: Concurrent, Time-Selective Execution of Combined RT-DEVS Model and Interactive Environment," In Proc. of SCSC'98, pp. 410-415, July, 1998.
6. Barros, F. J., "Dynamic Structure Discrete Event System Specification: Formalism, Abstract Simulators and Applications," Transactions of the Society for Computer Simulation, vol. 13, no.1, pp.35-46, 1996.
7. Barros, F. J., "Modeling Formalisms for Dynamic Structure Systems," ACM transactions on Modeling and Computer Simulation, vol. 7, no. 4, pp. 501-515, Oct, 1997.
8. Concepcion, A. I., and Zeigler, B. P., "DEVS Formalism: A Framework for Hierarchical Model Development," IEEE Transactions on Software Engineering, vol. 14, no.2, pp.228-241, Feb, 1988.

Does Rational Decision Making Always Lead to High Social Welfare?

Dynamic Modeling of Rough Reasoning

Naoki Konno and Kyoichi Kijima

Graduate School of Decision Science and Technology, Tokyo Institute of Technology,
2-12-1 Ookayama, Meguro-ku, Tokyo, 152-8552, Japan
{nkonno,kijima}@valdes.titech.ac.jp

Abstract. The purpose of this paper is two-fold: The first is to propose a dynamic model for describing rough reasoning decision making. The second is to show that involvement of some irrational decision makers in society may lead to high social welfare by analyzing the centipede game in the framework of the model. In perfect information games, though it is theoretically able to calculate reasonable equilibria precisely by backward induction, it is practically difficult to realize them. In order to capture such features, we first develop a dynamic model assuming explicitly that the players may make mistakes due to rough reasoning. Next, we will apply it to the centipede game. Our findings include there is a case that neither random nor completely rational, moderate rational society maximize the frequency of cooperative behaviors. This result suggests that society involving some rough reasoning decision-makers may lead to socially more desirable welfare, compared to completely rational society.

1 Introduction

This paper investigates influences of rough reasoning by developing a dynamic decision making models. We then apply it to the centipede game. For illustrating discrepancy between equilibrium obtained by backward induction and actual experimental results. Finally, we examine possibility that society involving some rough reasoning decision-makers may lead to socially more desirable welfare, compared to completely rational society where all players reason completely.

In the traditional game theory, it is usually assumed that players completely recognize the game situation so as to compare all the results without error and to choose rational strategies. However, as Selten pointed out by using chain store paradox, even subgame perfect Nash equilibrium may lead to strange outcomes. In perfect information games, though it is theoretically able to calculate reasonable equilibria by backward induction, it is practically difficult to realize them due to various complexity and the limitation of abilities.

For the static decision situations, Myerson proposed a concept of trembling hand equilibrium. He assumed that players try to take best response, while errors are inevitable. He also argued that according to the difference of payoff,

T.G. Kim (Ed.): AIS 2004, LNAI 3397, pp. 262–269, 2005.
© Springer-Verlag Berlin Heidelberg 2005

the player takes worse strategies with positive probability. Mckelvey and Palfrey proposed substitute quantal response for best response in the sense that the players are more likely to choose better strategies than worse strategies but do not play a best response with probability 1. He developed the quantal response functions by using logit functions.

On the other hand, as far as dynamic decision situations are concerned, Mckelvey and Palfrey transformed extensive form games into normal form games, and examined quantal response equilibria.

We characterize player's rough reasoning by following two elements. One is the payoff, while the other is the depth of situations. First, along with Mckelvey and Palfrey, we assume reasoning accuracy depends on the difference of the payoffs in such a way that error rate is a decreasing function of the difference of payoffs. Second, as the depth of decision tree is increased, reasoning accuracy tends to decrease. This property describes that it is difficult to compare actions in the far future.

Centipede game is known for the discrepancy between equilibrium obtained by backward induction and that by actual experimental results. Due to Mckelvey and Palfrey, the subgame perfect Nash equilibrium strategies are observed only less than 30%. They tried to rationalize the result by mathematical model in which some of the players have altruistic preferences. Aumann insisted that incompleteness of common knowledge causes cooperative behaviors. Although these factors may work for the centipede game, we claim rough reasoning is also an essential factor leading to cooperative behaviors.

In order to reveal properties of rough reasoning, we propose two specific models; rough reasoning model based on logit function and rough reasoning model based on exponential error. Then we derive common properties from the two by numerical simulations.

This article is organized as follows. Section 2 presents a general rough reasoning model. In Section 3, we propose two specific reasoning models and apply them to the centipede game. We examine influences of irrationality on the centipede game by numerical simulations in Section 4. Finally conclusions and remarks are given in Section 5.

2   General Rough Reasoning Model

In the traditional game theory, it is usually assumed that all players perceive situation precisely, and essentially compare all the strategies without error. However, perfect reasoning is quite difficult in most actual decision situations due to the players' reasoning abilities. We first define true game.

Definition 1. *True game is a finite perfect information extensive form game given by*

$$G = (I, N, A, \alpha, (N_i)_I, (r_i))$$

where I is the set of players, while N is the set of nodes. N_T and N_D are partitions of N, where N_T is the set of terminal nodes and N_D is the set of

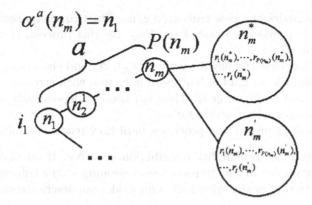

Fig. 1. General reasoning rule.

decision nodes. A is the set of actions α: $N - \{n_1\} \rightarrow N_D$ is the function from nodes except initial node n_1 to the prior nodes. P: $N_D \rightarrow I$ is the player function that determines the player who chooses an action at the node. (r_i): $N_T \rightarrow \mathbf{R}^I$ is the payoff function that determines the payoffs of each agent.

Since G is a perfect information game, subgame perfect equilibria are obtained by backward induction. However, since the players can not compare all the result without error in the actual situations, we assume that players choose actions by the following heuristics. To implement it, we need some notations:

N_2: The set of attainable nodes from n_1 i.e.
  $N_2 = \{n | n \in N, \alpha(n) = n_1\}$.
N_m^1: The set of the last decision nodes of G. i.e.
  $N_m^1 = \{n | n \in N_D, \exists n_t \in N_T$, s.t. $\alpha(n_t) = n$, and $\neg(\exists n_d \in N_D$, s.t. $\alpha(n_d) = n)\}$.
$n*$: A Best node at $n \in N_D$ for $P(n)$. i.e. $n* \in argmax_{n'}\{r_{P(n)}(n') | \alpha(n') = n\}$

Denote by $r(n_d) = (r_1(n_d), ..., r_j(n_d), ..., r_I(n_d))$ a payoff vector that the player reasons to achieve if the optimal choices are taken at every stage after $n_d \in N_d - \{n_1\}$. Then the heuristics are as follows. (Refer to Fig.1).

1. Let i_1 be the player that chooses an action at the initial node n_1. i_1 tries to reason the estimated payoff vector at $n_2 \in N_2$ by backward induction.
2. Indeed, i_1 tries to reason estimated payoff vector at node $n_m \in N_m^1$. Let a be the depth form the initial node to n_m. Let $b(n'_m)$ be the difference between $r_{P(n_m)}(n_m*)$ and $r_{P(n_m)}(n'_m)$. i.e. $b(n'_m) = r_{P(n_m)}(n_m*) - r_{P(n_m)}(n'_m)$, where $n'_m \in \{n | \alpha(n) = n_m\}$.
3. i_1 assigns $r(n_m*)$ to estimated payoff vector $r(n_m)$, while it may occurs an error with a certain probability. We assume that the error probability is an increasing function of a and a decreasing function of b. If there are some best responses, each best action is taken with same probabilitiy.
4. When the above operations have been finished for every $n_m \in N_m^1$, i_1 identifies every $n_m \in N_m^1$ with terminal nodes. Then i_1 generates N_m^2 as a set of

last decision nodes of a new truncated game. Start to reason next reasoning process. This process is iterated until n_2. By this process, i_1 generates a payoff vector at n_2.

5. Finally, i_1 compares the payoff vector of $n_2 \in N_2$ and chooses a best action. (This heuristics is an kind of backward induction with errors.)
6. Let i_2 be a next player after i_1. Then i_2 reasons independently of reasoning of i_1 and chooses a best action for i_2.
7. The players implement these processes until they reach a terminal node.

This process produces probability distribution over N_T. If an player chooses actions more than once in the decision tree, reasoning at the subsequent nodes may contradict to that at the prior node. Our model can describe such situations.

3 Two Specific Models and Their Applications to the Centipede Game

To examine systematic deviation from Nash equilibrium, we focus on the Rosenthal's centipede game by using more specific models. Centipede game is well known as an example illustrating differences between results by backward induction and those by actual experiments.

The centipede game is two person finite perfect information game. We call player 1 is "she", and player 2 is "he". Each player alternately chooses Pass(P) or Take(T) in each decision node. If she chooses action P, her payoff decreases while his payoff increases by more than his decrease. If she chooses action T, the game is over and they receive payoffs at that node. Symmetrically if he chooses action P, his payoff decreases while her payoff increases by more than his decreases. If the game has n decision nodes, we call the $n - move$ centipede game.

The pair of strategies that both the players choose T at every decision node is only subgame perfect equilibrium because the centipede game is finite. This equilibrium leads to the result that the game is over at the first period.

The centipede game has many variants about payoff structures. However we adopt the original Rosenthal's structure, where if she chooses P, her payoff is reduced by 1 and his payoff is increased by 3.

Now, we propose two specific models, rough reasoning model based on logit function and rough reasoning model based on exponential error model.

3.1 Rough Reasoning Model Based on Logit Function

Suppose that player i at node n_k reasons about the decision node n_l. First, we need the following notations:

j: The decision player at node n_l.
N_s: A set of attainable nodes from n_l. i.e. $N_s = \{n | n \in N, \alpha(n) = n_l\}$.
σ: A reasoning ability parameter.

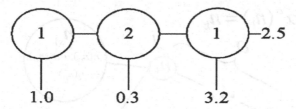

Fig. 2. *three − move* centipede game.

We should notice that σ works as a fitting parameter with respect to the unit. For example, if description about payoffs change from dollar to cent, σ will be $\frac{1}{100}$. Furthermore, if unit is fixed, as the rationality of agent is increased, σ will be increased.

Suppose that $n_{sx} \in N_s$, the rough reasoning model based on logit function with parameter σ, as follows:

Definition 2. *Rough reasoning model based on logit function is a reasoning model that assigns $r(n_{s1})$ to $r(n_l)$ with probability*

$$\frac{e^{\frac{r_j(n_{s1})}{a}\sigma}}{\Sigma_{n_s \in N_s} e^{\frac{r_j(n_s)}{a}\sigma}}$$

The probability essentially depends on the ratio of payoff against a in such a way that if a is sufficiently large, then the choice can be identical with random choice. If a is sufficiently small and b is sufficiently large, the choice can be seem as by the best response.

3.2　Rough Reasoning Model Based on Exponential Error

Suppose that player i at node n_k reasons about the decision node n_l. We need the following notations.

　N_s*: The set of n_s*. i.e. $N_s* = argmax_{n'}\{r_{P(n_s)}(n')|\alpha(n') = n_s\}$
　$b(n_s)$: $b(n_s) = r_j(n_s*) - r_j(n_s)$ for $n_s \in N_s$ and $n_s \notin N_s*$.
　　k: $k = |N_s|$.
　　c: $c = |N_s*|$.
　　ϵ: a reasoning ability parameter.

We should notice that ϵ works as a fitting parameter with respect to the unit. Furthermore, if the unit is fixed, as the rationality of agent is increased, ϵ will be decreased.

We propose rough reasoning model based on exponential error with parameter ϵ, as follows:

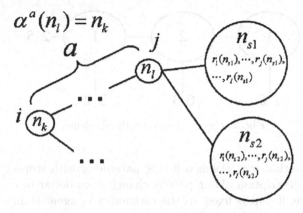

Fig. 3. Two specific reasoning rules.

Definition 3. *Rough reasoning model based on exponential error is a reasoning model that assigns $r(n_{s1})$ to $r(n_l)$ with probability*

$$\begin{cases} min\{\dfrac{1}{k}, e^{\frac{-b(n_{s1})}{a}}\epsilon\}, & if\ n_{s1} \notin N_s* \\ \dfrac{1 - \Sigma_{n_{s1} \notin N_s*} min\{\frac{1}{k}, e^{\frac{-b(n_{s1})}{a}}\epsilon\}}{c}, & if\ n_{s1} \in N_s* \end{cases}$$

The definition explicitly represents probability with which non-optimal node is mis-estimate as optimal by backward induction.

It should be notice that in the both models the probability that non-optimal node is mis-estimated as optimal is an increases function of a and decreasing function of b.

4 Simulation Results and Their Implications

In order to examine frequency of noncooperative behavior T with relation with FCTF, we calculated several simulations, where FCTF denotes frequencies of choice T at first period. We focus on the choice at the first period, because if P is chosen at the first period, the remained subgame can be considered as the $(n-1) - move$ centipede game. Figures 4 and 5 show the simulation results of FCTF on the both models respectively.

Note that in the Figure 4, larger σ means more rationality, while in Figure 5, smaller ϵ implies more rationality.

First, we investigate relation between FCTF and the reasoning ability. For every n in the both models, it is observed that there is a turning point. Until the turning point, as the rationality is increased, noncooperative behavior T tends to decrease. However if reasoning ability exceeds the turning point, as the rationality is increased, noncooperative behavior T tends to increase.

Figures 4 and 5 give the following two implications about the relation between FCTF and the reasoning ability: Moderate rationality tends to bring more

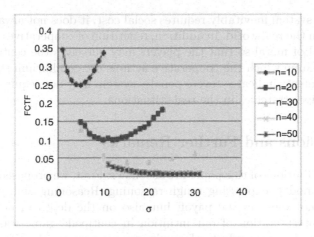

Fig. 4. FCTF logit function model.

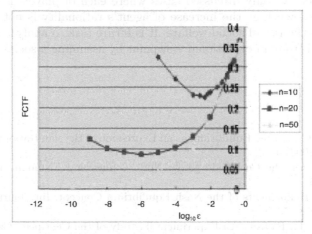

Fig. 5. FCTF exponential error function model.

irrational behaviors than random choice since P is interpreted as an irrational action. On the other hand, even if players are not completely rational, their rough reasoning may leads to socially more desirable outcomes than those of completely rational reasoning, since low frequency of FCTF implies high social welfare.

We next examine the relation between FCTF and the value of n. As n increases, FCTF tends to decrease. Furthermore, the turning point shifts to the direction of higher rationality.

Centipede game can be considered as a kind of situation that cooperation is desired. Since cooperative behavior is not always increase their payoffs, Pareto efficiency is not guaranteed. To implement Pareto optimal results with certainly, we need to introduce a certain penalty system. However, since introduction of such a penalty system inevitably requires social cost, it does not always increase social welfare in the real world. In addition, repetition of cooperative actions may generate a kind of moral so that the players may perceives the centipede game as if it were a game which the cooperative actions are equilibrium strategies.

These arguments indicates severe penalty system may not required to implement cooperative strategies in the real stituations.

5   Conclusions and Further Remarks

The main contributions of this paper are as follows: First, we proposed a dynamic mathematical models expressing rough reasoning. Reasoning ability is defined as dependent not only on the payoff but also on the depth of decision tree. Second, new interpretation of our intuition in centipede game was proposed. Third, we pointed out the effects of rough reasoning on social welfare from two sides, reasoning ability and scale of the problem.

In this paper, we only discussed cases where each of players is equally rational. It was shown that the increase of agent's rationality is not necessarily connected with the rise of social welfare. It is future task to analyze what strategy is stabilized from an evolutionary viewpoint by assuming a social situation is repeated.

References

1. R., Aumann: Correlated Equilbrium as an Expression of Beysian Rationality. Econometrica **55** (1992) 1-18.
2. R., Aumann: On the Centipede game: Note. Games and Economic Behavior **23** (1998) 97-105.
3. R., Myerson: Refinements of the Nash Equilibrium Concept. Int. Journal of Game Theory **7** (1978) 73-80.
4. R., Mckelvey, T., Palfrey: An Experimental Study of the Centipede Game. Econometrica **60** (1992) 803-836.
5. R., Mckelvey, T., Palfrey: Quantal Response Equilibria in Normal Form Games. Games and Economic Behavior **7** (1995) 6-38.
6. R., Mckelvey, T., Palfrey: Quantal Response Equilibria in Extensive Form Games. Experimental Economics **1** (1998) 9-41.
7. R., Rosenthal: Games of Perfect Information, Predatory Pricing and the Chain-Store Paradox. Journal of Economic Theory **25** (1981) 92-100.
8. R,. Selten: The Chain-Store Paradox. Theory and Decision **9** (1978) 127-159.

Large-Scale Systems Design: A Revolutionary New Approach in Software Hardware Co-design

Sumit Ghosh

Department of Electrical & Computer Engineering Stevens Institute of Technology,
Hoboken, NJ 07030
Tel: 201-216-5658, Fax: 201-216-8246
sumit.ghosh@ieee.org, sghosh2@stevens.edu

Abstract. The need for a revolutionary new approach to software hardware co-design stems from the unique demands that will be imposed by the complex systems in the coming age of networked computational systems (NCS). In a radical departure from tradition, tomorrow's systems will include analog hardware, synchronous and asynchronous discrete hardware, software, and inherently asynchronous networks, all governed by asynchronous control and coordination algorithms. There are three key issues that will guide the development of this approach. First, conceptually, it is difficult to distinguish hardware fro m software. Although intuitively, semiconductor ICs refer to hardware while software is synonymous to programs, clearly, any piece of hardware may be replaced by a program while any software code may be realized in hardware. The truth is that hardware and software are symbiotic, i.e., one without the other is useless, and the difference between them is that hardware is faster but inflexible while software is flexible and slow. Second, a primary cause underlying system unreliability lies at the boundary of hardware and software. Traditionally, software engineers focus on programming while hardware engineers design and develop the hardware. Both types of engineers work off a set of assumptions that presumably define the interface between hardware and software. In reality, these assumptions are generally ad hoc and rarely understood in depth by either types of engineers. As a result, during the life of a system, when the original hardware units are upgraded or replaced for any reason or additional software functions are incorporated to provide new functions, systems often exhibit serious behavior problems that are difficult to understand and repair. For example, in the telecommunications community, there is serious concern over the occurrence of inconsistencies and failures in the context of "feature interactions" and the current inability to understand and reason about these events. While private telephone numbers are successfully blocked from appearing on destination caller Id screens under normal operation, as they should be, these private numbers are often unwittingly revealed during toll-free calls. It is hypothesized that many of these problems stem from the continuing use of legacy code from previous decades where timer values were determined corresponding to older technologies and have never been updated for today's much faster electronics. In TCP/IP networking technology, the values of many of the timer settings and buffer sizes are handed down from the past and the lack of a scientific methodology makes it difficult to determine their precise values corresponding to the current technology. The mismatch at the hardware software interface represent vulnerabilities that tempt perpetrators to launch system attacks. Third, while most traditional systems employ synchronous hardware and centralized soft-

T.G. Kim (Ed.): AIS 2004, LNAI 3397, pp. 270–274, 2005.
© Springer-Verlag Berlin Heidelberg 2005

ware, complex systems in the NCS age must exploit asynchronous hardware and distributed software executing asynchronous on geographically dispersed hardware to meet performance, security, safety, reliability, and other requirements. In addition, while many complex systems in the future including those in automobiles and space satellites will incorporate both analog and discrete hardware subsystems, others will deploy networks in which interconnections may be dynamic and a select set of entities mobile.

The NCSDL Approach

This research aims to develop a new approach, networked computational systems design language and execution environment (NCSDL), that will consist of a language in which complex systems may be described accurately and an execution environment that will permit the realistic execution of the executable description on a testbed to assess the system correctness, reliability, safety, security, and other performance parameters. To obtain results quickly for large systems and use them in iterating system designs, the testbed will consist of a network of workstations configured as a loosely-coupled parallel processor. The research is on-going and is organized into two major phases. Under the first phase, the most important features of VHDL (Waxman and Saunders, 1987), including the ability to describe asynchronous behavior, will be integrated with a recent research finding to develop nVHDL which will permit the description and simulation of analog and discrete subsystems of a hybrid system within a single framework. Key language constructs of VHDL will also be modified to enable fast, distributed execution of the executable models on the testbed. Under the second phase, the intermediate nVHDL will be modified to incorporate dynamic interconnection links between entities which stem from the need for some entities to migrate from one geographical region to another and an evolving interconnection topology.

Characteristics of nVHDL

Current mixed signal tools are restricted to maintaining two distinct and incompatible simulation environments, each with its unique mechanism for debugging, analyzing results, and controlling the progress of the simulation execution. As a result, the designer frequently finds it difficult to accurately assess in subsystem A the impact of a design change in subsystem B, and vice versa. While this impacts on the quality of the resulting overall design, it does not lend support to the idea of a system on a chip (Clark, 2000). For the spacecraft design example, stated earlier, integration of the antenna and analog amplifier subsystem with the microprocessor subsystem, is likely to be complex effort, given that they are designed separately. Clark (Clark, 2000) notes the incompatibility of the digital and analog development process and stresses the need to develop a new design methodology and tools for mixed signal designs. A key reason underlying the current difficulty lies in the absence of sound mathematical theory to represent the digital and analog components in a unified framework and to provide a scientific technique for accurate hybrid simulation. Logically, a breakthrough in mixed signal simulation would require the discovery of a new scientific principle that would permit the resolution of times of the analog and discrete subsystems, of any given system, to be unified into a common notion of time. This would then constitute the foundation of a new simulator in which both analog and discrete subsystems of any given system would be uniformly represented and simultaneously executed.

Key characteristics of nVHDL include a fundamentally new approach to the modeling and simulation of analog and discrete subsystems and an error-controlled, provably correct, concurrent simulation of digital and analog subsystems. This will also enable extending the scope of nVHDL into complex, future NCS system designs consisting of digital and analog components such as Mixed-Signal chips, miniaturized control systems, intelligent transportation, banking, and biomedical systems, and other network-enabled devices. Conceptually, this breakthrough approach may be organized into three steps. In step I, each of the analog and digital subsystem, at any given level of abstraction, is analyzed individually to yield the corresponding resolution of time (Ghosh, 1999). This unit of time or timestep, T, guarantees the complete absence of any activity occurring faster than T. Assume that T_A and T_D represent the resolution of times for the analog and discrete subsystems, respectively. Although T_A and T_D should strictly be specified in the form of a range, the upper extreme is of greater interest. For today's discrete electronics technology, T_D has already approached 0.1 ns, and new technologies promise to push T_D down even lower. Clock frequencies in today's analog electronics domain range from 44.1 Khz for audio systems and 2.4 Mhz for DSL systems to 4 ~10 Ghz for satellite systems.

The use of visible light, UV, gamma rays, and hard X rays in the near future may push analog subsystem design beyond 10^{12} Hz and up to 10^{24} Hz. The goal under step II is to determine the "universal time" (Ghosh, 1999) as the common denominator between T_A and T_D, which is essential to achieving uniform simulation of both analog and discrete subsystems. The universal time will constitute the basic timestep (Ghosh and Lee, 2000) in the unified nVHDL simulator. A total of three scenarios are conceivable. Either (i) $T_A \ll T_D$, (ii) $T_A \gg T_D$, or (iii) T_A and T_D are comparable but not identical. For case (i), where $T_A \gg T_D$, T_A may adequately serve as the universal time for the total system design and, thus, the timestep of the corresponding nVHDL simulation. Every timing in the discrete subsystem may be expressed, subject to a small and measurable error margin, as an integral multiple of T_A, resulting in a simulation with acceptable accuracy. Of course, the simulation speed, relative to wall clock time, will be governed by the choice of the timestep, available computational resources, and the nature of the simulation algorithm, i.e. whether centralized or distributed. The nVHDL simulation speed for the complete system is likely to be much slower than that of the individual discrete simulator executing the discrete subsystem. The scenario is similar for case (ii) where $T_A \gg T_D$, except that T_D is utilized as the universal time for the complete system design. Also, the execution speed of the nVHDL simulator is likely to be much slower than that of the individual analog simulator executing the analog subsystem. Cases (i) and (ii) warrant research into new, distributed algorithms to enhance the execution speed of nVHDL, along the lines of the recent development (Ghosh, 2001). The scenario where T_A and T_D are comparable but not identical, is the most interesting case for two reasons. First, the speed of the unified simulator, where successfully developed, is likely to be comparable to those of the individual analog and discrete simulators executing the analog and discrete subsystems, respectively. This would imply a practical benefit to the mixed signal designer.

Second, in the past, the field of electronics had witnessed concurrent improvements to the underlying digital and analog technologies and this trend is likely to continue into the future.

Recently, GDEVS, a Generalized Discrete Event Specification (Escude et al, 2000) has been introduced in the literature to enable the synthesis of accurate discrete event models of highly dynamic continuous processes. As validation of the principles underlying GDEVS, a laboratory prototype simulator, DiamSim, has been developed, executed for two representative systems, and the results compared against those from utilizing the industrial grade MATLAB/Simulink software package. GDEVS builds on DEVS (Zeigler, 1976) (Zeigler et al, 2000) in that it utilizes arbitrarily higher order polynomial functions for segments instead of the classic piecewise constant segments. A logical consequence of GDEVS is that, for a given analog subsystem and a specified desired accuracy, an arbitrary value for the time step may be determined, subject to specific limits, by utilizing piecewise polynomial segments and controlling the order of the polynomials. Thus, for case (iii) described earlier, unification may be achieved by controlling the order of the polynomial, thereby modifying the value of the timestep, T_A, until it matches T_D. Then, T_D will constitute the resolution of time in the nVHDL simulation, permitting both the analog and digital models to be uniformly executed by the underlying nVHDL scheduler. The conversion of the continuous and discrete signal values between the analog and digital models, will be dictated by the order of the polynomial. Although the use of polynomial coefficients in themselves is not new, the combination of the core GDEVS principle and its use in Mixed-Signal system design represents a fundamentally new thinking.

In step III, the analog and digital models of the analog and digital subsystems, respectively, plus the timing and signal exchange between the two subsystems, are represented in nVHDL. The representation is simulated uniformly using a single framework with the goal of validating the correctness of the overall Mixed-Signal system design.

Characteristics of NCSDL

The mobility of a subset of entities in NCSDL, the dynamic interconnection topology, and the need to accommodate distributed asynchronous software, will require the complete reorganization of the internal VHDL database from a centralized, static representation to distributed localized representations, with some local databases continuing to represent static information, while in others the continuously changing interconnections between the entities require them to dynamically update the database through techniques such as flooding. Furthermore, VHDL's centralized scheduler must be replaced by a distributed event driven mechanism to permit fast and accurate execution of a geographically dispersed complex system.

Since complex systems will comprise of computational engines, networking infrastructure, and control and coordination algorithms, NCSDL's role transcends that of VHDL in that it must not only represent the constituent hardware and software subsystems individually but capture the intelligence inherent in networking and, most important, the underlying control algorithm. The presently available network modeling tools including Opnet are imprecise, erroneous, and execute slowly on a uniprocessor, implying significant challenge for the design of NSCDL. A significant characteristic of NCSDL will consist in enabling the design and evaluation of new measures of performance to reflect the behavior of complex systems. NCSDL's greatest advantages may be described as follows. Given that it employs an asynchronous approach and utilizes an asynchronous testbed, complex systems are exposed, during simulation in NCSDL, to timing races and other errors similar to those in the real world. Consequently, simulation results are realistic. Also, hypothetical failures may be injected into NCSDL descriptions and their impact on system behavior assessed, yielding a new approach to testing system vulnerabilities. Most important, however, is that significant portions of the NCSDL description may be directly transferred into operational systems.

References

1. Waxman, R. and Saunders, L., 1987, VHDL Language Reference Manual: IEEE Standard 1076. Technical Report, The Institute of Electrical and Electronic Engineers, New York, U.S.A.
2. Clark, D., 2000, "Handhelds Drive Mixed-Signal Chip Development," IEEE Computer, Vol. 33, No. 11, November, pp. 12-14.
3. Ghosh, S., 1999, "Hardware Description Languages: Concepts and Principles," IEEE Press, New Jersey, U.S.A.
4. Ghosh, S. and Lee, T.S., 2000, "Modeling and Asynchronous Distributed Simulation: Analyzing Complex Systems," IEEE Press, New Jersey , U.S.A.
5. Ghosh, S., 2001, "P2EDAS: Asynchronous, Distributed Event Driven Simulation Algorithm with Inconsistent Event Preemption for Accurate Execution of VHDL Descriptions on Parallel Processors," IEEE Transactions on Computers, Vol. 50, No. 1, January, pp. 1-23.
6. Escude, B., Giambiasi, N., and Ghosh, S., 2000, "GDEVS: A Generalized Discrete Event Speci_cation for Accurate Modeling of Dynamic Systems," Transactions of the Society for Computer Simulation (SCS), Vol. 17, No. 3, September, pp. 120-134.
7. Zeigler, B., 1976, "Theory of Modeling and Simulation," John Wiley & Sons, New York, U.S.A.
8. Zeigler, B., Praehofer, H., and Kim, T.G., 2000, "Theory of Modeling and Simulation," Second Edition, Academic Press, New York, U.S.A.

Timed I/O Test Sequences
for Discrete Event Model Verification

Ki Jung Hong and Tag Gon Kim

Dept. of EECS,
Korea Advanced Institute of Science and Technology (KAIST),
373-1 Guseong-dong, Yuseong-gu, Daejeon, Republic of Korea
kjhong@smslab.kaist.ac.kr, tkim@ee.kaist.ac.kr

Abstract. Model verification examines the correctness of a model implementation with respect to a model specification. While being described from model specification, implementation prepares to execute or evaluate a simulation model by a computer program. Viewing model verification as a program test this paper proposes a method for generation of test sequences that completely covers all possible behavior in specification at an I/O level. Timed State Reachability Graph (TSRG) is proposed as a means of model specification. Graph theoretical analysis of TSRG has generated a test set of timed I/O event sequences, which guarantees 100% test coverage of an implementation under test.

1 Introduction

Model verification examines the correctness of a model implementation with respect to a model specification. As discrete event simulation models are getting more and more complicated verification of such models is extremely complex. Thus, automatic verification of such a simulation model is highly desirable [1].

Since a model specification is implemented in a simulation program model verification can be viewed as a program test. Thus, model verification starts from generation of input/output sequences for an implementation, which covers all possible behaviors of a specified model. Untimed discrete event model can be specified by finite state machine (FSM). FSM can be verified by conformance test [5]. Test sequences of conformance test can be built by the UIO method [3, 4], and others. Test sequences generation of timed discrete event models can be obtained by timed Wp-method [6], which is based on timed automata and region automata.

This paper proposes a new approach to select test cases for a module-based testing of a discrete event simulation program at an I/O level. We assume that specification of each module is known and an implementation is unknown as a black box. Time State Rechability Graph (TSRG) is proposed to specify modules of a discrete event model. TSRG represents a discrete event model in terms of nodes and edges. Each node represents a state of discrete event model associated with which is a time interval. On the other hand, each edge represents transition between nodes with input, output or null event in the specified time

T.G. Kim (Ed.): AIS 2004, LNAI 3397, pp. 275–284, 2005.

interval. Graph theoretical analysis of TSRG generates all possible timed I/O sequences from which a test set of timed I/O sequences with 100 % coverage can be constructed.

An obstacle of the test method lies in different numbers of states between specification and implementation. This is because we assume that an exact number of states in implementation is unknown. However, an assumption on a maximum number of states used in an implementation can overcome the obstacle. Note that the number does not need to be exact. Instead, it may be any number that is greater than or equal to one used in implementation, which only determines complexity of testing. This paper is organized as follows. Section 2 proposes TSRG. Section 3 introduces TSRG related definitions and theorems, and proposes the generation method of test input/output sequences. Finally, conclusions are made in section 5.

2   Timed State Reachability Graph

Timed State Reachability Graph (TSRG) is a modeling means which specifies a discrete event system in timed input/output behavior. Nodes represent time constraints states; edges represent conditions for transitions between nodes. The graph starts from an initial state of a discrete event model and generates edges/nodes which are reachable from the state. The formal semantics of TSRG is given below:

$$TSRG = \langle N, E, \theta_N, \theta_E \rangle$$

$$N : \text{States set : Node}$$

$$E : N \times N : \text{Edge}$$

$$\theta_N : N \rightarrow \Re^+_{(0,\infty)} \times \Re^+_{(0,\infty)}$$

$$\qquad : \text{Node attribute function}$$

$$\theta_E : E \rightarrow (X \cup Y \cup \{\tau\}) \times Boolean$$

$$\qquad : \text{Edge attribute function}$$

$$X : \text{Input events set}$$

$$Y : \text{Output events set}$$

$$\tau : \text{null output event}$$

$$Boolean \in \{true, false\} : \text{CONTINUE or NOT}$$

TSRG has two types of an edge: input transition edge and output transition edge. While one or none output transition edge can be attached at a node, one or more input transition edges can be attached at the same node. An attribute associated with a node is interpreted as a waiting time for the state to be transit. An edge attribute include a Boolean of **CONTINUE**. Meaning of **CONTINUE** is that (1) an input transition occurs at a state before a deadline of a maximum elapsed time defined at the state, and (2) a new state continues keeping the deadline defined at the previous state.

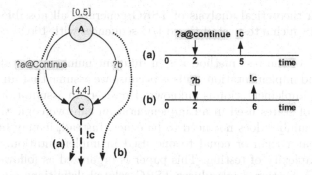

Fig. 1. Event time diagram of continue example.

Let us explain **CONTINUE** in more detail using an example shown in Fig 1. In the Fig, there are two paths from state A to state C. The first path is A $\xrightarrow{?a@[0,5],\textbf{continue}}$ C, and the second one is A $\xrightarrow{?b@[0,5]}$ C. A waiting time of state A is the same value 5 for both paths, but a waiting time of state C is different for each path, because of **continue** in the first path. If ε is an elapsed time for an arrival event ?a of state A, a waiting time of state C for the first path is $5\text{-}\varepsilon$, $0 < \varepsilon \le 5$, and a waiting time of state C for the second path is 4. Briefly, **CONTINUE** means that the waiting timer is continued from the previous state, without resetting for the next state. Due to such timer semantics of **CONTINUE**, a pair of nodes connected by an edge with **CONTINUE** should have a finite waiting time and an output edge.

3   Theorem and Definition: TSRG

The following equivalent node's definition in TSRG is prerequisite to define the minimization of TSRG.

Definition 1 (Equivalent node). *Let $s_1, s_2 \in N$ be nodes. Node s_1 and s_2 are equivalent, i.e., $s_1 \equiv s_2$, when the following condition is satisfied: $\theta_N(s_1) = \theta_N(s_2) \wedge (\forall e_1 = (s_1, \acute{s}_1) \in E, \exists e_2 = (s_2, \acute{s}_2) \in E, \theta_E(e_1) = \theta_E(e_2) \wedge \acute{s}_1 \equiv \acute{s}_2) \wedge (\forall e_2 = (s_2, \acute{s}_2) \in E, \exists e_1 = (s_1, \acute{s}_1) \in E, \theta_E(e_1) = \theta_E(e_2) \wedge \acute{s}_1 \equiv \acute{s}_2).$*

Definition 2 (Minimization of TSRG). *TSRG is minimized if and only if there is no equivalent relation for any two nodes in the node set.*

Figure 2 shows an example of equivalent nodes. Timed input/output (TIO) event trace of model (a) is repetitive sequences of $?a@[0,t_A]\cdot!b@[t_B,t_B]\cdot?c@[0,t_C]$, which is extracted from a state trace with a TIO event A$\xrightarrow{?a@[0,t_A]}$B $\xrightarrow{!b@[t_B,t_B]}$C $\xrightarrow{?c@[0,t_C]}$A. The TIO event trace of model (b) is repetitive sequences of $?a@[0,t_A]\cdot$ $!b@[t_B,t_B]\cdot ?c@[0,t_C]$, extracted from A$\xrightarrow{?a@[0,t_A]}$B $\xrightarrow{!b@[t_B,t_B]}$C $\xrightarrow{?c@[0,t_C]}$D $\xrightarrow{?a@[0,t_A]}$B. Thus, model (a) and (b) in figure 2 have the same TIO event trace.

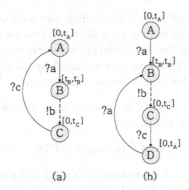

Fig. 2. Equivalent states.

While model (a) is minimized, model (b) has equivalent nodes, $A \equiv D$. In addition, the number of nodes of each model is different.

Test sequences for TSRG visit all edges in a TSRG model through a path from a start node to an end node. Such a path is defined as:

Definition 3 (Path). *Let $i, n \in Z$ be integers with $i < n$, $s_i \in N$ be a node, $e_i \in E$ be an edge with $e_i = (s_i, , s_{i+1})$, and $t_i = \theta_N(s_i)$ be a node attribute function. A path P from s_0 to s_n expressed as $P(s_0, s_n) = (e_0, t_0)(e_1, t_1) \cdots (e_{n-1}, t_{n-1})$, which is the sequence of all pairs of a visited edge e_i and its waiting time t_i from the start node to the end node.*

Each small paths can concatenate to the big one. Path concatenation operator is defined formally to describe such concatenation behavior.

Definition 4 (Path concatenation operator). *Let $s_i, s_j, s_k \in N$ be nodes and $P(s_i, s_j), P(s_j, s_k)$ be paths. Path concatenation operator $\bullet$ is defined as $P(s_i, s_j) \bullet P(s_j, s_k) = P(s_i, s_k)$ with the following properties.*

$$P(s_j, s_k) \bullet P(s_i, s_j) = \phi$$
$$P(s_i, s_j) \bullet \phi = P(s_i, s_j)$$
$$\phi \bullet P(s_i, s_j) = P(s_i, s_j)$$

For full coverage of states and transitions in model verification all states and edges in TSRG should be reachable, or strongly connected, defined in the following.

Definition 5 (Reachable). *TSRG is reachable if and only if there exists one or more path between any two nodes in TSRG.*

If TSRG is reachable, it is possible to visit all edges and nodes to verify TSRG. Otherwise, TSRG is not verifiable. The visit of all nodes can be covered through the visit of all edges in reachable TSRG. Thus, a loop is introduced for the visit of all edge in reachable TSRG. If TSRG is a strongly connected graph, there exist one or more loop paths which can visit any edge and node in the graph. There exists a set of loop paths which can visit all edges and nodes in the graph. If there exists an intersection between two loops an adjacency set to be defined contains the intersection relation between loops. A traversal from an initial state to any other state can be contained by some interconnected loops. Interconnected loops can be bound by an adjacent loop chain.

Theorem 1 (Loop). *Let TSRG be reachable. Any node in TSRG can be traversed by one or more loop paths.*

Proof. $\forall s_i, s_j \in N, \exists P(s_i, s_j)$ *and* $P(s_j, s_i)$ *s.t.* $P(s_i, s_i) = P(s_i, s_j) \bullet P(s_j, s_i)$

Definition 6 (Loop set). *In a reachable TSRG, a loop set is a collections of loops which cover all edges in TSRG.*

All edges of TSRG is covered from the definition of a loop set. If a test target has an equivalent node, visiting routes of all possible edges are made from the loop set of TSRG by the following loop joining concatenation.

Definition 7 (Adjacency set). *Any two loops in a loop set L has adjacent relation if and only if these loops visit the same node. An adjacent relation set, called an adjacency set, has the condition:* $M_{adj} \subseteq L \times L$.

Definition 8 (Loop joining concatenation operator). *Let L be a loop set, M_{adj} be an adjacency set, p_1, p_2 be paths in TSRG, s_1, s_2, s_i be in N, s_i be the first met adjacent node, and $l_1, l_2 \in L$ be loop paths. Then, the following operators are hold.*

$$l_1 \bullet \phi = l_1, \phi \text{ is identity}$$
$$\phi \bullet l_1 = l_1$$
$$l_1 \bullet l_2 = \phi, \text{ if } (l_1, l_2) \notin M_{adj}$$
$$l_1 \bullet l_2 = l_3, \text{ if } (l_1, l_2) \in M_{adj}$$
$$l_1 = p_1(s_1, s_i) \bullet p_1(s_i, s_1)$$
$$l_2 = p_2(s_2, s_i) \bullet p_2(s_i, s_2)$$
$$l_3 = p_1(s_1, s_i) \bullet p_2(s_i, s_2) \bullet p_2(s_2, s_i) \bullet p_1(s_i, s_1)$$

Definition 9 (Loop set joining concatenation operator). *Let L_0, L_1 be loop sets.*

$$L_1 \bigotimes L_2 = \{l_0 \bullet l_1 | \forall l_0 \in L_0, \forall l_1 \in L_1\}$$

Definition 10 (All possible I/O sequences). *Let TSRG be reachable and L be all possible loop sets of TSRG. All possible input/output sequences, $\Omega(h)$, is* $\Omega(h) = \bigotimes_{i=0}^{h} L$.

The function $\Omega(h)$ of all possible I/O sequences is to explore all possible state trajectories from an initial state through a loop set under adjacency relation. h in $\Omega(h)$ represents a depth of adjacency tree exploration. Figure 3 shows an example of adjacency tree exploration. In figure 3, since loops 'A' and 'B' have

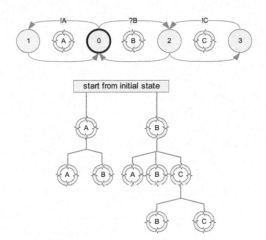

Fig. 3. Adjacency tree exploration.

an initial state at their paths, they only can be selected at the first time when the tree is just explored. After selection of all such loops, next loops may be selected only by the adjacency relation until finding synchronized event sequences.

In the view of test sequences, all possible sequences $\Omega(h)$ can cover all nodes and edges with time intervals. However, $\Omega(h)$ has very high complexity both in time and in space. This is because the size of all possible loop set of TSRG is increased at an exponential rate by the number of edges. To reduce the complexity of $\Omega(h)$, a basic loop set is introduced as the following definition.

Definition 11 (Path inclusion). *Let A, B be paths in TSRG. The path A includes the path B, i.e. $A \supseteq B$, if and only if the path A visits all edges in the path B.*

Definition 12 (Loop independency). *Let L be a loop set and $l_i \in L$. The loop set L is independent if only if there is no $l_i \in \{l_1, l_2, \cdots, l_n\}$, such that $l_i \subseteq \bullet_{k=1, k \neq i}^{n} l_k$.*

Definition 13 (Basic loop set). *Let L be a loop set. L is a basic loop set if and only if a loop set $l_1 \in L$ has an independent path for other loop paths in the loop set, i.e., the remaining loop paths without any selected loop path l_1 can not cover all nodes and edges in TSRG.*

A basic loop set passes through all edges in TSRG. However, since a basic loop set is a subset of a loop set its size is less than that of a loop set. It implies that redundancy of visiting edges is reduced by using loops in a basic loop set. Since a previous edge with **CONTINUE** affects time scheduling of a current edge, a minimal independent loop set to visit all edge can not verify time scheduling of all nodes. To solve this, a TSRG model with **CONTINUE** is modified by the following rules. If TSRG has one or more edges with **CONTINUE**, all edges with **CONTINUE** are merged into a newly inserted equivalent node, and they

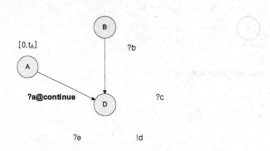

Definition 15 (Mergeable node). *Assume that a node s_0 with waiting time $T(s_0)$ connected to edges with transition rules. Then, the node s_0 is mergeable if and only if there exists a node s_i in TSRG with the following conditions :* $\forall e_1 = (s_0, s_1), e_2 = (s_i, s_j) \in E, s_i \neq s_0 \wedge \theta_N(s_0) = \theta_N(s_i) \wedge \theta_N(s_1) = \theta_N(s_j) \wedge ((\theta_E(e_1), \theta_E(e_2) \in ((Y \cup \{\tau\}) \times boolean) \wedge \theta_E(e_1) = \theta_E(e_2)) \vee ((x_1, \bot) = \theta_E(e_1), (x_2, \bot) = \theta_E(e_2) \in (X \times boolean) \wedge (x_1 \neq x_2 \vee s_1 \equiv s_j)))$.

To cover mergeable nodes, the following defines the concept of a unique timed I/O (UTIO) which is similar to UIO and UIOv [3].

Definition 16 (UTIO). *An UTIO sequence which discriminates a node s_0 from the other one is an input or time-out sequence, x_0. The output sequence produced in response to x_0 from any node other than s_0 is different from that responded from s_0. That is, for $y_i, y_0 \in Y, e_{ij} = (s_i, s_j), e_{01} = (s_0, s_1) \in E,$ $(y_i, false) = \theta_E(e_{ij}), (y_0, false) = \theta_E(e_{01}), \forall s_i \in N \wedge s_i \neq s_0, \theta_N(s_0) \neq \theta_N(s_0) \vee y_i \neq y_0 \vee \theta_N(s_j) \neq \theta_N(s_1)$. And, an ending point of the UTIO sequence at node s_0 is s_0, which means UTIO sequence at node s_0 is a loop path.*

4 Simple Example: Verification of Buffer

The UTIO sequence of a mergeable node should be a loop path, which can be easily attached to basic loop paths. Consider an example of a buffer model of length 2 in figure 5. As shwon in the figure state space of the buffer is two dimensional: processing status and queue length. Processing status, either in Busy or Free, is a status of a processor cascaded to the buffer. Queue length is a maximum size of the buffer. Let us find mergeable nodes and UTIO sequences. Mergeable nodes S_4, S_6, UTIO sequences of nodes S_4 and S_6 are shown in table 1. Two UTIO sequences are attached into the basic loop set of figure 5. The attached basic loop set is shown in table 1. The four basic loop paths can visit all nodes and edges of the buffer model shown in figure 5. Let us further consider another example of a buffer model of length 2 with equivalent nodes. Figure 6 shows two such models: (a) one with three equivalent states and (b) the other

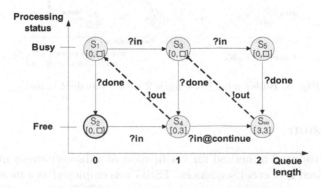

Fig. 5. TSRG of buffer model of length 2.

Table 1. Basic loop paths for buffer of length 2.

UTIO Node	UTIO Seq.		
S_4	$!out@[3,3]\cdot?done@[0,\infty]\cdot?in@[0,3]$		
S_6	$\overline{!out@[3,3]\cdot?done@[0,\infty]\cdot?in@[0,3]}$		

No.	Loop paths	Nodes	UTIO Nodes
l_1	$S_2 \rightarrow?in@[0,3]\cdot!out@[3,3]\cdot?done@[0,\infty]$	S_2,S_4,S_1	NA
l_2	$S_4 \rightarrow !out@[0,3]\cdot\ ?in@[0,\infty]\cdot\ ?done@[0,\infty]\cdot$ $!out@[3,3]\cdot?done@[0,\infty]\cdot?in@[0,3]$	S_4,S_1,S_3	S_4
l_3	$S_4 \rightarrow?in@[0,3], \mathbf{C}\cdot$ $!out@[3,3]\cdot?done@[0,\infty]\cdot?in@[0,3]\cdot$ $!out@[3,3]\cdot?done@[0,\infty]\cdot$ $!out@[3,3]\cdot?done@[0,\infty]\cdot?in@[0,3]$	S_4,S_6,S_3	S_6,S_4
l_4	$S_6 \rightarrow!out@[3,3]\cdot\ ?in@[0,\infty]\cdot\ ?done@[0,\infty]\cdot$ $!out@[3,3]\cdot?done@[0,\infty]\cdot?in@[0,3]$	S_6,S_3,S_5	S_6

with three equivalent nodes and one transition fault in $S_1' \rightarrow S_5$. The variable h of $\Gamma(h)$ is assigned to 1. Test sequences set, generated from $\Gamma(1)$ of figure 5, is $\{l_1 \bullet l_1, l_1 \bullet l_2, l_1 \bullet l_3, l_2 \bullet l_1, l_2 \bullet l_2, l_2 \bullet l_3, l_3 \bullet l_1, l_3 \bullet l_2, l_3 \bullet l_3, l_3 \bullet l_4, l_4 \bullet l_2, l_4 \bullet l_3, l_4 \bullet l_4\}$. The test sequence $l_3 \bullet l_2$ detects the difference between (a) and (b) of figure 6. The test sequence $l_3 \bullet l_2$ visits the nodes in figure 6(a) through the following order: $S_4 \rightarrow S_1' \rightarrow S_3 \rightarrow S_4 \rightarrow S_6 \rightarrow S_3 \rightarrow S_4$. However, for figure 6(b), the order of node visits of $l_3 \bullet l_2$ is $S_4 \rightarrow S_1' \rightarrow S_5 \rightarrow S_6 \rightarrow S_3 \rightarrow S_4$. Consequently, the UTIO sequence of S_4 detects transition fault of figure 6(b).

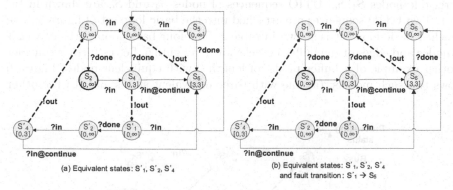

(a) Equivalent states: S'_1, S'_2, S'_4

(b) Equivalent states: S'_1, S'_2, S'_4
and fault transition: $S'_1 \rightarrow S_5$

Fig. 6. Buffer model of length 2 with equivalent states.

5 Conclusion

This paper introduced a method for verification of a discrete event model using a test set of timed I/O event sequences. TSRG was employed as a means of modeling of discrete event systems; an implementation of TSRG was assumed to be a black box with a maximum number of states known. A graph-theoretical analysis of TSRG showed all possible timed I/O sequences $\Omega(h)$ which covers all edges and nodes of TSRG. However, due to some redundancy in visiting nodes/edges of TSRG complexity of $\Omega(h)$ is too high to apply practical verification problems. To solve the problem a basic loop set of TSRG was introduced based on which a minimal test set $\Gamma(h)$ of timed I/O event sequences was extracted. Introduction of predefined UTIOs attached to basic loop paths for mergeable nodes guaranteed that a test coverage of discrete event models using $\Gamma(h)$ was 100 %.

References

1. J. Banks, D. Gerstein, and S. P. Searles, "Modeling process, validation, and verification of complex simulations: A survey", *S.C.S Methodology and Validation, simulation series*, Vol. 19, No. 1, pp 13-18, 1988.
2. Concepcion, and B. P. Zeigler, "DEVS formalism: a framework for hierarchical model development," *IEEE Trans. Software Eng.*, vol. 14, no. 2, pp. 228-241, Feb. 1988.
3. W. Y. L. Chan, C. T. Vuong, M. R. Otp, "An improved protocol test generation procedure based on UIOS," *ACM SIGCOMM Comp. Commun. Review,Symposium proc. communi. arch. & protocols* , vol. 19, pp. 283-294, Aug. 1989.
4. A. T. Dahbura, K. K. Sabnani, and M. Ümit Uyar, "Formal nethods for generating protocol conformance test sequences," *Proc. of the IEEE*, vol. 78, pp. 1317-1325, Aug. 1990.
5. D. Lee, and M. Yannakakis, "Principles and methods of testing finite state machines-A survey," *Proc of the IEEE*, vol. 84, pp. 1090-1123, Aug. 1996.
6. A. En-Nouaary, R. Dssouli, and F. Khendek, "Timed Wp-method: Testing real-time systems," *IEEE Trans. Software Eng.*, vol. 28, pp. 1023-1038, Nov. 2002.
7. K. J. Hong, and T. G. Kim, "A Verification Of Time Constrained Simulation Model Using Test Selection:A New Approach", *In preparation*, 2004.

A Formal Description Specification for Multi-resolution Modeling (MRM) Based on DEVS Formalism

Liu Baohong and Huang Kedi

College of Mechatronics Engineering and Automation,
National University of Defense Technology, Changsha, Hunan, 410073, China
lbh_nudt@sina.com

Abstract. Multi-Resolution Modeling (MRM) is a relatively new research area. With the development of distributed interactive simulation, especially as the emergence of HLA, multi-resolution modeling becomes one of the key technologies for advanced modeling and simulation. There is little research in the area of the theory of multi-resolution modeling, especially the formal description of MRM. In this paper, we present a new concept for the description of multi-resolution modeling, named multi-resolution model family (MF). A multi-resolution model family is defined as the set of different resolution models of the same entity. The description of MF includes two parts: models of different resolution and their relations. Based on this new concept and DEVS formalism, we present a new multi-resolution model system specification, named MRMS (Multi-Resolution Model system Specification). And we present and prove some important properties of MRMS, especially the closure of MRMS under coupling operation. MRMS provides a foundation and a powerful description tool for the research of MRM. Using this description, we can further study the theory and implementation of MRM.

1 Introduction

Multi-Resolution Modeling (MRM) is a relatively new research area. With the development of distributed interactive simulation, especially as the emergence of HLA, multi-resolution modeling becomes one of the key technologies for advanced modeling and simulation [1, 2]. MRM has deep influence on the development of modeling and simulation. However the research on MRM is now on its very initial stage.

By far, there is no formal method to describe multi-resolution model and multi-resolution modeling. Without this, it is difficult to establish a common langue among different researchers and model developers, and it is impossible to develop multi-resolution modeling framework and tools. In this paper, we proposed a new multi-resolution model specification based on the concept of multi-resolution family which we proposed in this paper first and the DEVS which is developed by Zeigler. We hope our work can be helpful to the development of multi-resolution modeling.

This paper is organized into five sections. In section 2, we summarize the general modeling formalism DEVS developed by B. P. Zeigler and a specific model specification for dynamic structure discrete event system developed by F. J. Barros. In section 3, we give the definition and the specification of multi-resolution model family - and prove the related theorems. In section 4, we proposed our specification for multi-resolution model system and summarize some of its key properties, especially its

T.G. Kim (Ed.): AIS 2004, LNAI 3397, pp. 285–294, 2005.
© Springer-Verlag Berlin Heidelberg 2005

closure under coupling. We also give an example of describing multi-resolution model system using our proposed specification. In the last section, we sum up the whole paper and introduce our future work.

2   Foundations

In this section, we will introduce DEVS and Dynamic Structure DEVS briefly. Our MRM specification is based on these two specifications.

2.1 DEVS Specification

DEVS is a system theory based model description specification. Here we only give basic concept of DEVS for the convenience of our specification on multi-resolution model. A detailed description can be found in [3]. A basic discrete event system specification is a structure

$$M = < X, s_0, S, Y, \delta_{int}, \delta_{ext}, \lambda, ta >$$

Where:

X : is the set of inputs; Y : is the set of outputs;

s_0: is the initial state; S : the set of sequential states;

$\delta_{int} : S \rightarrow S$, is the internal state transition function;

$\delta_{ext} : Q \times X \rightarrow S$, is the external state transition function,

where $Q = \{(s,e) \mid s \in S, 0 \leq e \leq ta(s)\}$

$\lambda : S \rightarrow Y$, is the output function;

$ta : S \rightarrow R_{0,\infty}^+$, is the time advance function;

The coupled DEVS model can be described as:

$$N = < X, Y, D, \{M_d\}, \{I_d\}, \{Z_{i,d}\}, Select >$$

Where:

X: the set of input events; Y: the set of output events;
D: a set of component references;

For $\forall d \in D, M_d$ is a DEVS model;

For $\forall d \in D \cup \{N\}, I_d$ is the influencer set of d, i.e. $I_d \subseteq D \cup \{N\}, d \notin I_d$;

and for $\forall i \in I_d \square Z_{i,d}$ is a function, the i-to-d output translation with:

$$Z_{i,d} : X \rightarrow X_d, \ if \ i = N;$$
$$Z_{i,d} : Y_i \rightarrow Y, \ if \ d = N;$$
$$Z_{i,d} : Y_i \rightarrow X_d, \ if \ d \neq N \ and \ i \neq N;$$

2.2 Dynamic DEVS Specification

Dynamic Structure DEVS (DSDEVS) specification strengthens DEVS with the ability to describe the dynamic structure change of a model. In DSDEVS, introduced by Barros [4, 5], basic models are the same as classic DEVS basic models, but the structure of coupled models can change over time.

A DSDEVS coupled model is defined as

$$DSDEN_N =< X_N, Y_N, \chi, M_\chi >$$

Where: N: the DSDEVS network name;

X_N: the input events set of the DSDEVS network;

Y_N: the output events set of the DSDEVS network;

χ: the network executive name;

M_χ: the model of χ.

The M_χ can be defined with the following 9-tuple:

$$M_\chi =< X_\chi, s_{0,\chi}, S_\chi, Y_\chi, \gamma, \Sigma^*, \delta_\chi, \lambda_\chi, \tau_\chi >$$

Where:

S_χ: is the set of the network executive states;

Σ^*: is the set of network structure;

$\gamma: S_\chi \to \Sigma^*$, is call structure function;

Assume $s_{a,\chi} \in S_\chi$ and $\Sigma_a \in \Sigma^*$, we get:

$$\Sigma_a = \gamma(s_{a,\chi}) =< D_a, \{M_{i,a}\}, \{I_{i,a}\}, \{Z_{i,a}\} >$$

The meaning of it element is similar as coupled DEVS model. A complete description of DSDEVS semantics can be found in [4, 5].

3 Multi-resolution Model Families

3.1 The Concept of Multi-resolution Model Families

In multi-resolution modeling, different resolution models of the same entity are not isolated. They are related each other. During the running of multi-resolution models, different resolution models should be coordinated to maintain consistent description of different resolution models. So we call the set of different resolution models of the same entity Multi-resolution model Family (MF).

The description of MF includes two parts: models of different resolutions and their relations. The interface specification of MF is shown below:

$$MF =< \gamma, \{M_r\}, \{R_{i,j}\} >$$

Where:

γ: is the set of model resolutions, which can be regard as the index of models. For example, for $r \in \gamma$, M_r means the models with the resolution of r.

M_r : represents the model with resolution r, $\{M_r\}$ means the set of all models of some entity. M_r can be specified by DEVS:

$$M_r =< X^r, s_0, S^r, Y^r, \delta_{int}^r, \delta_{ext}^r, \lambda^r, ta^r >$$

$R_{i,j}$ is used to describe the relationship between different resolution models.

$R_{i,j}: Y_R^i \to X_R^j$, where $i, j \in \gamma, X_R^j \subset X^j, Y_R^i \subset Y^i$ are multi-resolution related inputs and outputs.

For modularization simulation design, the modules in MF should not access the inner information of each other and models of different resolution can only coordinate through input and output interface.

3.2 Key Properties of MF

Now, let's introduce some important properties of MF.

Theorem 3.1. When $|\gamma| = 1$, MF degenerates into normal DEVS specification.

Proof: when $|\gamma| = 1$, there is only one model M, so $R_{i,j} = \varnothing$. Obviously, MF can be described by normal DEVS specification.   □

Theorem 3.2. MF can be described as DEVS coupled model.

Proof: Though theorem 3.1, we know when $|\gamma| = 1$, the conclusion is obvious. Now, let $|\gamma| \neq 1$, we prove this theorem from two respects:

(1) When models in MF are described by basic DEVS coupled model:

$$MF =< \gamma, \{N_r\}, \{R_{i,j}\} >,$$

$$N_r =< X_r, Y_r, D_r, \{M_{d,r}\}, \{I_{d,r}\}, \{Z_{i,d}^r\}, Select_r >.$$

Accordingly, let the coupled model be

$$N =< X, Y, D, \{M_d\}, \{I_d\} \{Z_{i,d}\}, Select >.$$

We can divide the input and output of each model into two parts, i.e. MRM-related part and MRM-unrelated part. For input, we have $X_r = X_r^M \cup X_r^R$, which the former means the MRM-unrelated input of N_r, the later means the MRM-related input of N_r. Similarly for output, we have $Y_r = Y_r^M \cup Y_r^R$.

Obviously, we have $X = \cup_{r \in \gamma} X_r^M, Y = \cup_{r \in \gamma} Y_r^M, D = \times_{r \in \gamma} D_r$.

In order to rewrite MF using coupled DEVS specification, the key is to construct the relations between different resolution models. Models of different resolution models are inter-effective in MF.

$$\text{For each } d \in D, I_d = \{I_{d,r} \mid d \in D_r\} \cup \{N_i \mid R_{i,d} \neq \varnothing\}$$

$$Z_{i,d} = \bigcup_{r \in \gamma} \{Z_{i,d}^r\} \qquad if \ i = N \vee d = N,$$

$$= (\bigcup_{r \in \gamma} \{Z_{i,d}^r\}) \cup R_{i,d} \quad if \ i \neq N \wedge d \neq N.$$

The design of select function can be divided into two steps: first, select the resolution of the model to be running; second, use the select function of the model to be running to decide which module to be running, i.e.

$$Select : f \circ g$$

$$f : \gamma \to \bigcup_{r \in \gamma} 2^{D_r}$$

$$g = Select_r : 2^{D_r} \to D$$

(2) When models in MF are described by parallel DEVS coupled model:

$$MF = < \gamma, \{M_r\}, \{R_{i,j}\} >,$$

$$N_r = < X_r, Y_r, D_r, \{M_{d,r}\}, \{I_{d,r}\}, \{Z_{i,d}^r\} >.$$

Accordingly, let the parallel DEVS coupled model be

$$N = < X, Y, D, \{M_d\}, \{I_d\}\{Z_{i,d}\} >.$$

All components are same with above except for the select function.
From the above all, we can see that MF can be described as DEVS coupled model.

□

Theorem 3.3. Each MF specification can be described as DEVS atomic specification.

Proof: From theorem3.2, each MF model can be described as DEVS coupled model. And we already know that the classic DEVS specification is closed under coupling. So we get that each MF specification can be described as DEVS atomic specification.

□

Corollary 3.1. the subset of Multi-resolution family is a multi-resolution family.

The above theorems show that: (1) when what we concerned is not the resolution of the models, we can regard MF as a normal DEVS model; (2) the MF also has a structure of hierarchy. In principle, multi-resolution model family can be described by normal DEVS, but this description is very complex and can't show the relations between different resolution models clearly.

4 Multi-resolution Model System Specification

4.1 Introduction to Multi-resolution Model System Specification

In this section, we will give a new multi-resolution model system specification, named MRMS (Multi-Resolution Model system Specification). This specification is based on MF which we introduced in section 3 and DSDEVS. We will also prove its closure under coupling.

The atomic model of MRMS is the same as normal DEVS specification. The coupled specification is shown as following:

$$MRMS =< X_S, Y_S, \kappa, \{\mathcal{M}_k\}, \chi, M_\chi >$$

Where:

X_S: is system inputs; Y_S : is system outputs; κ: is the set of entities;

$\mathcal{M}_k$: $\mathcal{M}_k \subset MF_k$, the subset of multi-resolution model family of entity $k, k \in \kappa$, MF_k means the multi-resolution model family of entity k;

$\mathcal{M}_k =< \gamma_k, \{M_r^k\}, \{R_{i,j}^k\} >$.

χ: is model resolution controller;

M_χ: is the model of χ.

$$M_\chi =< X_\chi, s_{0,\chi}, S_\chi, Y_\chi, \pi, \psi, \{M_\varphi\}, \delta_\chi, \lambda_\chi, \tau_\chi >,$$

Where:

X_χ: is the input of χ; $s_{0,\chi}$: is the initial state of χ;

S_χ: is the set of χ' states; Y_χ: is the output of χ;

$\psi = \underset{i \in \kappa}{\times} \{2^{\gamma_i} - \varnothing\}$: is called the collection of resolution mode of the model.

$\pi: X_\chi \times S_\chi \to \psi$; $\pi(x_\chi, s_\chi) = \varphi \in \psi$;

$M_\varphi =< D_\varphi, \{I_{\varphi,d}\}, \{C_{\varphi,d}\}, \{Z_{\varphi,d}\}, \{R_{\varphi,d}\} >$, represents running model of the system when the resolution mode of the model is φ, where:

D_φ: the set of modules of the system;

$I_{\varphi,d}$ the set of fluencers of module d;

$C_{\varphi,d}$: the set of modules which should maintain consistency with module d;

$Z_{\varphi,d}$: the internal relations in module d;

$R_{\varphi,d}$: the relations between different resolution models including d;

δ_χ: the state transfer functions of χ;

λ_χ: the output functions of χ;

τ_χ: the time advance function of χ;

The resolution mode of a system at time t is marked as $\varphi(t)$.

There are two categories of MRM problems: model abstraction problem and aggregation/dissaggregation problem. The first one can be described by atomic DEVS specification, i.e. $M_r^k =< X_r^k, Y_r^k, S_r, \delta_r, \lambda_r, ta_r >$; the second one can be described by coupled DEVS specification, i.e. $M_r^k =< X_r^k, Y_r^k,$

$D_r^k, \{M_{k,d}^r\}, \{I_{k,d}^r\}, \{Z_{k,d}^r\} >$. Unless explanation, we usually do not distinguish this two kinds of specifications.

This specification shows the idea of separating the model from model control. In our specification, we use a special module named resolution controller χ to support multi-resolution modeling. The resolution related information and resolution control information are viewed as the state of χ. All resolution switches are transferred to state transition of states in χ.

4.2   The Key Properties of MRMS

To use a specification for representing large simulation models, one must guarantee that models can be constructed in a hierarchical and modular manner. In order to describe complex systems such as multiple federations HLA simulation systems using MRMS, MRMS specification should be closed under coupling. If a system specification is closed under coupling, we can describe a system with a hierarchical manner. For a multi-resolution model system, its coupled model specification should be equivalent to basic DEVS formalism, and then a MRMS model can be viewed as an atom basic model to construct more complex MRMS models. Since MRMS basic formalism is DEVS, closure under coupling can be accomplished by demonstrating that the resultant of a MRMS coupled model is equivalent to a basic DEVS specification.

Theorem 4.1. When each entity has only one model in MRMS, MRMS degenerates into normal DEVS specification.

Proof: we need prove that when the model of each entity only has one resolution, MRMS can be simply described by normal DEVS. We suppose the corresponding DEVS is $N = < X, Y, J, \{M_j\}, \{I_j\} \{Z_j\} >$.

Obviously, when each entity only has one resolution, i.e. $\forall i \in \kappa, |\gamma_i| = 1$, we have

$$MRMS = < X_S, Y_S, \kappa, \{\mathcal{M}_k\} >.$$

We can rewrite, $M_r^k = < X_r^k, Y_r^k, D_r^k, \{M_{k,d}^r\}, \{I_{k,d}^r\}, \{Z_{k,d}^r\} >$ to

$$\mathcal{M}_k = < X^k, Y^k, D^k, \{M_{k,d}\}, \{I_{k,d}\}, \{Z_{k,d}\} >,$$

Because of the closure of DEVS models under coupling, the above specification can be written as:

$$M_k = < X, s_0, S, Y, \delta_{int}, \delta_{ext}, \lambda, ta >.$$

Let the resolution of each model is r_0, then $X_\chi, s_{0,\chi}, S_\chi, Y_\chi$ are constant. φ is also a constant. So $M_\varphi = \{D, \{I_d\}, \{Z_d\}\}$. Because $|r| = 1, D = \kappa$, replacing $\mathcal{M}_k$ with these constant, we have $MRMS = < X_S, Y_S, \kappa, \{M_k\}, \{I_k\}, \{Z_k\} >$.

So we have

$$X = X_S, Y = Y_S, J = D = \kappa, M_j = \mathcal{M}_k = M_k, I_j = I_k, Z_j = Z_k. \quad \square$$

From this Theorem, we can see that MRMS is the extension of normal DEVS. The normal DEVS can be regarded as a special case when all entities are simulated in only one resolution.

Theorem 4.2. MRMS is closed under the operation of coupling.

Theorem 4.3. A MRMS coupled model is equivalent to a DSDEVS coupled model.

The proof of theorem 4.2 and Theorem 4.3 are omitted because of the space limitation.

Actually, a DSDEVS coupled model is equivalent to a DEVS basic model, and MRMS coupled model is equivalent to a DSDEVS coupled model, so a MRMS coupled model is equivalent to a DEVS basic model. This is what theorem 4.2 mean. So from theorem 4.3, theorem 4.2 can be derived.

Though, MRMS model can be described as a DSDEVS model, even a DEVS model, DSDEVS and DEVS can't exhibit the characters of multi-resolution models.

4.3 An Example

In this section, we will give an example to illustrate how to describe a multi-resolution model system using MRMS. Considering the following example: we want to model a system composed of an aircraft formation on the blue side, and an antiaircraft gun company and an air surveillance radar station on the red side. The aircraft formation is modeled with two different granularities: formation and single aircrafts. When the distance between radar and aircraft is less than some certain value, the aircraft should be modeled at the resolution of single aircraft, otherwise, the aircraft should be modeled at the resolution of formation. When the distance between the aircraft and the antiaircraft gun company is less than some distance, the antiaircraft gun can attack the aircraft and the aircraft can retaliate upon it. Because the antiaircraft company is model at low resolution, the aircraft should also be modeled at low resolution when interacting with the antiaircraft gun company.

According to MRMS formalism, we get:

$$MRMS =< X_N, Y_N, \kappa, \{\mathcal{M}_k\}, \chi, M_\chi >$$

where: $X_N = Y_N = \varnothing, \kappa = \{R, G, A\}$, here, A represent aircraft.

$\mathcal{M}_R$ and $\mathcal{M}_G$ represent the single resolution DEVS model of radar and antiaircraft gun company respectively, whose resolution is r_R and r_G respectively.

$M_A = \{\gamma_A, \{M_r^A\}, R_{i,j}^A\}$, where $\gamma_A = \{r_P, r_F\}$, M_P^A and M_F^A are classic DEVS model.

χ is the resolution control module, M_χ is the model of χ.

$$M_\chi =< X_\chi, s_{0,\chi}, S_\chi, Y_\chi, \pi, \Psi, \{M_\phi\}, \delta_\chi, \lambda_\chi, \tau_\chi >$$

For this simple example, we can enumerate all of its resolution modes:

$$\{\{r_G, r_R, r_F\}, \{r_G, r_R, r_P\}, \{r_G, r_R, \{r_F, r_P\}\}\}$$

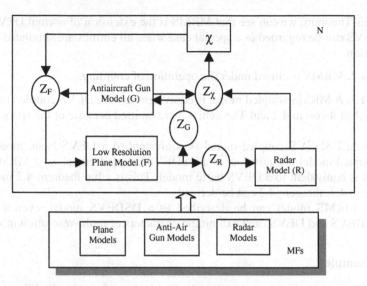

Fig. 1. MRMS-based multi-resolution model description (before resolution change)

At the initialize state $s_{0,\chi}$, the aircraft is model with the resolution of formation, as showed in Fig.1. According to MRMS, we have:

$$\varphi = \pi(x_\chi, s_{0,\chi}) = \{r_G, r_F, r_F\},$$

$$M_\varphi = < D_\varphi, \{I_{\varphi,d}\}, \{C_{\varphi,d}\}, \{Z_{\varphi,d}\}, \{R_{\varphi,d}\} >,$$

$$D_\varphi = < F, R, G >,$$

$$I_{\varphi,F} = \{G, \chi\}, I_{\varphi,R} = \{F\}, I_{\varphi,G} = \{F\},$$

$$I_{\varphi,\chi} = \{F, G, R\},$$

$$C_{\varphi,i} = \varnothing, i \in \{F, R, G\}, \ R_{\varphi,i} = \varnothing, i \in \{F, R, G\},$$

$$Z_{\varphi,F} : Y_\chi \times Y_G \to X_F, \ Z_{\varphi,R} : Y_F \to X_R, \ Z_{\varphi,\chi} : Y_R \times Y_G \to X_\chi.$$

At some point, the distance between the aircraft and radar is less than some value, the radar need the high resolution model of the aircraft and the antiaircraft gun need the low resolution model of aircraft. As showed in Fig.2, So:

$$x_\chi = Z_{\varphi,\chi}, \ s'_\chi = \delta_{ext}(s_{0,\chi}, x_\chi), \ \varphi' = \pi(s'_\chi) = \{r_R, r_G, \{r_F, r_P\}\},$$

Accordingly,

$$M_\varphi = < D_\varphi, \{I_{\varphi,d}\}, \{C_{\varphi,d}\}, \{Z_{\varphi,d}\}, \{R_{\varphi,d}\} >,$$

$$D_\varphi = < F, P, R, G >,$$

$$I_{\varphi,F} = \{G, \chi\}, \ I_{\varphi,R} = \{P\}, \ I_{\varphi,G} = \{F\}, and \ I_{\varphi,\chi} = \{F, P, G, R\}.$$

$$C_{\varphi,F} = \{P\}, \ C_{\varphi,P} = \{F\}, \ R_{F,P} : Y_F \to X_P, \ and \ R_{P,F} : Y_P \to X_F.$$

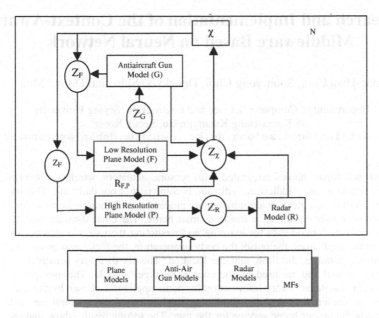

Fig. 2. MRMS-based multi-resolution model description (after resolution change)

5   Conclusion and Future Work

In this paper we represent a new concept called multi-resolution model family and established a new multi-resolution model description formalism called multi-resolution model system specification. Our MRMR specification has the following characters: it clearly describes the fact that there are different resolution models of an entity in system; it describes the relations between different resolution models; it has the ability to describe the dynamic change of model resolutions; it can be used to describe different modeling method; it has the property of closure under coupling.

Our future work include: designing multi-resolution modeling support system based on MRMS, exploring new methods for multi-resolution modeling and designing simulation model base support multi-resolution model storing and querying using MRMS.

References

1. Liu Baohong and Huang Kedi. Multi-Resolution Modeling: Present Status and Trends. Journal of System Simulation. Vol.16 No.6 Jun. 2004. pp.1150-1154
2. Liu Baohong, Huang Kedi. The Concept and Some Design Issues about Multi-Resolution Modeling in HLA. ICSC'2002. Shanghai, China.
3. B. P. Zeigler, H. Praehofe and Tag Gon Kim. Theory of Modeling and Simulation. Academic Press. 2000.pp.138-150
4. F.J. Barros, Modeling formalisms for Dynamic Structure Systems. ACM Transactions on Modeling and Computer Simulation. Vol.7, No.4, 1997 pp.501-515.
5. F.J. Barros. Dynamic Structure Discrete Event System Specification: Formalism, Abstract Simulators and Applications. Transactions of the Society for Computer Simulation Vol.13 No.1, 1996, pp.35-46.

Research and Implementation of the Context-Aware Middleware Based on Neural Network

Jong-Hwa Choi, Soon-yong Choi, Dongkyoo Shin, and Dongil Shin*

Department of Computer Science and Engineering, Sejong University,
98 Kunja-Dong Kwangjin-Gu, Seoul, Korea
{com97,artjian}@gce.sejong.ac.kr, {shindk,dshin}@sejong.ac.kr

Abstract. Smart homes integrated with sensors, actuators, wireless networks and context-aware middleware will soon become part of our daily life. This paper describes a context-aware middleware providing an automatic home service based on a user's preference inside a smart home. The context-aware middleware utilizes 6 basic data for learning and predicting the user's preference on the home appliances: the pulse, the body temperature, the facial expression, the room temperature, the time, and the location. The six data sets construct the context model and are used by the context manager module. The user profile manager maintains history information for home appliances chosen by the user. The user-pattern learning and predicting module based on a neural network predicts the proper home service for the user. The testing results show that the pattern of an individual's preferences can be effectively evaluated and predicted by adopting the proposed context model.

1 Introduction

Since the computing devices are getting cheaper and smaller, we are dealing with ubiquitous computing as Mark Weiser stated in [1]. The original concept of home intelligence was mostly focused on network connections. Researchers claim that smart homes will bring intelligence to a wide range of functions from energy management, access monitoring, alarms, medical emergency response systems, appliance controlling, and even interactive games [2].

Appliances installed in a smart home should be able to deliver enhanced or intelligent services within the home. A fundamental role for "Artificial Intelligence" in smart homes is to perform the underlying monitoring, management, and allocation of services and resources that bring together users and information [3].

Moreover a context-aware middleware is needed to offer an unobtrusive and appealing environment embedded with pervasive devices that help its users to achieve their tasks at hand; technology that interacts closely with its occupants in the most natural ways to the point where such interaction becomes implicit [4].

We propose a context-aware middleware that utilizes 6 basic data for learning and predicting the user's preference of the content: the pulse, the body temperature, the facial expression, the room temperature, the time, the location. This middleware offers a automated and personalized services to the users.

Section 2 gives related research works on context-awareness. Section 3 addresses how the context is constructed in the middleware. In Section 4 we introduce the de-

---

* Corresponding Author.

T.G. Kim (Ed.): AIS 2004, LNAI 3397, pp. 295–303, 2005.
© Springer-Verlag Berlin Heidelberg 2005

tailed architecture of the middleware. Section 5 we introduce the context visualization appliance. Section 6 presents implementation and experimental results. We conclude in section 7.

2 Related Works

In order to enable natural and meaningful interactions between the context-aware smart home and its occupants, the home has to be aware of its occupants' context, their desires, whereabouts, activities, needs, emotions and situations. Such context will help the home to adopt or customize the interaction with its occupants. By context, we refer to the circumstances or situations in which a computing task takes place. Context of a user is any measurable and relevant information that can affect the behavior of the user.

Meaningful context information has to be derived from raw data acquired by sensors. This context processing aims at building concepts from environmental and human data sensed by sensors. This intelligence processing is also know as context interpretation and should contain two sub-steps: modeling and evaluation [5,6]. Raw data is modeled to reflect physical entities which could be manipulated and interpreted. Propositions from the modeling module need to be evaluated against a particular context. Evaluation mechanisms often use artificial intelligence techniques.

A context-aware system can be constructed with several basic components. Most of all the middleware gathers context information, processes it and derives meaningful (re)actions from it [7]. Ranganathan and Campbell argued that ubiquitous computing environments must provide middleware support for context-awareness. They also proposed a middleware that facilitates the development of context-aware agents. The middleware allows agents to acquire contextual information easily, reason about it using different logics and then adapt themselves to changing contexts.

Licia Capra proposed the marriage of reflection and metadata as a means for middleware to give applications dynamic access to information about their execution context [8].

Stephen S. Yau developed a reconfigurable context-sensitive middleware for pervasive computing. Reconfigurable context-sensitive middleware facilitates the development and runtime operations of context-sensitive pervasive computing software [9].

3 Context Definitions

Context's definition is important in context aware middleware. Researcher of context aware proposed a model in which a user's context is described by a set of roles and relations [10]. To attain a user's goal the system must process the user related data along with the environmental data. We proposed a user context of 6 basic data: the pulse, the body temperature, the facial expression, the room temperature, the time, and the location. The pulse and the body temperature are detected by sensors attached on a PDA(Personal Digital Assistant). The user's facial image is also attained by a small camera from the PDA and transmitted wirelessly. Room temperature is measured by a wall-mounted sensor. We installed 4 cameras to detect the user's location inside a room.

Pulse			Body Temperature			Facial Expression		
0.1 : 41-60	0.2 : 61-70	0.3 : 71-80	0.1 : 34	0.2 : 35.0-35.5	0.3 : 35.6-36.0	0.1 : Blank	0.2 : Surprise	0.3 Fear
0.4 : 81-90	0.5 : 91-100	0.6 : 101-110	0.4 : 36.1-36.5	0.5 : 36.6-37.0	0.6 : 37.1-37.5	0.4 : Sad	0.5 : Angry	0.6 : Disgust
0.7 : 111-120	0.8 : 121-130	0.9 : 131-140	0.7 : 37.6-38.0	0.8 : 38.1-38.5	0.9 : 39.0	0.7 : Happy	-	-

Time			Location			Room Temperature		
0.1 : 00-06	0.2 : 07-08	0.3 : 09-11	0.1 : Area 1	0.2 : Area 2	0.3 : Area 3	0.1 : 0-5	0.2 : 6-9	0.3 : 10-13
0.4 : 12-13	0.5 : 14-16	0.6 : 17-18	0.4 : Area 4	0.5 : Area 5	0.6 : Area 6	0.4 : 14-18	0.5 : 19-22	0.6 : 23-26
0.7 : 19-20	0.8 : 21-22	0.9 : 23-24	0.7 : Area 7	0.8 : Area 8	0.9 : Area 9	0.7 : 27-30	0.8 : 31-33	0.9 : 34-37

Fig. 1. Context Information

Figure 1 shows the six data sets for the context and they are normalized between 0.1 and 0.9. The pulse below 40 and over 180 were eliminated since they represent abnormal human status. The body temperature below 34 and over 40 were eliminated by the same reason. Facial expressions are normalized and categorized as described in [11]. The room temperature is normalized based on the most comfortable temperature which is between 23 and 24 Celsius. The time is normalized based on 24 hours. The location is a user's position in out experimental room.

4 Context-Aware Middleware Framework

Figure 2 shows the overall architecture of the context-aware middleware. The middleware obtains the context data through the context manager. Collected data is fed into the user preference learning module. The user preferable service is automatically provided by the user preference prediction module.

4.1 Context Normalization

As showed in Figure 3, the context manager collects six sets of the contextual data, detects garbage data, normalizes the data, and sends the data to the user preference learning and prediction module. If out of range data is detected, the context manager automatically recollects the same kind of data.

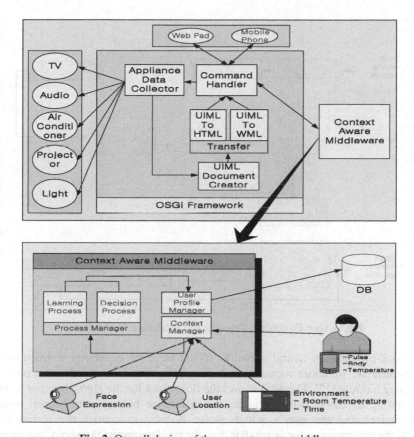

Fig. 2. Overall design of the context aware middleware

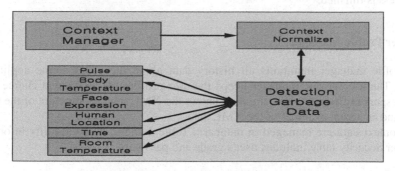

Fig. 3. Context Manager Process

Context data is delivered in machine learning algorithm after pass through normalization process in context manager.

4.2 Learning and Prediction Module

The User preference learning module uses the context data along with the history of past home appliances chosen by the user.

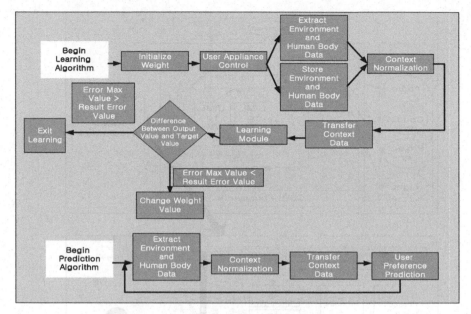

Fig. 4. User Preference Learning and Prediction Modules

Since back-propagation neural networks (NNs) have the capability to learn arbitrary non-linearity, we chose momentum back-propagation NNs for the user preference learning module [12]. Figure 4 shows the flow chart for the algorithm. For training and testing of the learning module, the difference between the trained result (content choice) and the user's actual choice is used to control the connection weights. If the difference value is smaller than the predetermined threshold, the learning process is finished.

4.3 User Profile Manager

User profile manager maintains all history data about the user's home appliance choices. The user profile manager keeps the user profiles of the content choice and the corresponding context data. Figure 5 shows the XML schema structure of the user profile and an actual example of the XML data.

All context data are managed in database. Table 1 shows user's security information. User Security table includes user's grade and password.

Table 1. User Security table

UserID	Password	Grade
como97	********	1

Table 2 shows user service definition. Context aware middleware provides services for 5 appliance. Each appliance is classified by 3 state values (power/channel/sound).

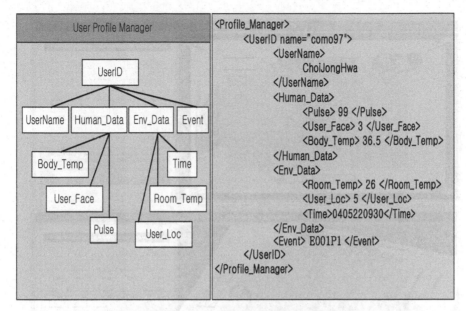

Fig. 5. User Profile Structure and an XML example

Table 2. User Service Definition Table

Service ID	Power	Channel	Sound	Grade
E001(TV)	1	1	1	2
E002(Audio)	1	1	1	2
E003(Air Conditioner)	1	1	0	2
E004(Projector)	1	1	0	1
E005(Light)	1	1	0	3

Table 3 is database table that store 6 context data by user action.

Table 3. User Status Value by User Behavior

UserID	Pulse	User Face	Body Temp	Room Temp	User Loc	Time	Event
como97	89	3	36.5	26	2	0405220930	E001P1
Choi	72	6	37	23	5	0405221052	E002C1

In case of machine learning module loss weight value, user profile manager offers machine learning module all state values that is stored on data base.

5   Visualization of Context Aware Middleware

Figure 6 shows context aware client for context optical representation. Context aware client offers third dimension simulation environment. Context aware client's primary function acquires 6 context data and takes charge of role that transmits data by context aware server. Context aware server transmits sequence to context aware client through learning and prediction module.

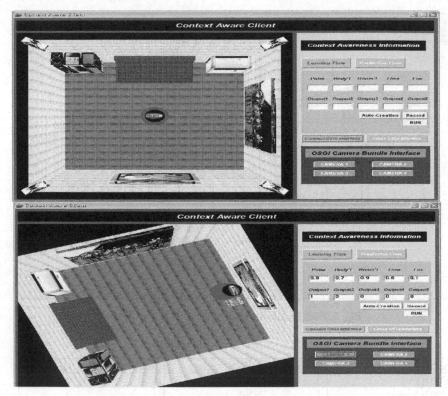

Fig. 6. Context Aware Client for Context Visualization

6 Implementation and Experimental Results

The key modules of the context-aware middleware are the user preference learning and prediction modules. We experimented and evaluated different topologies of NNs. Variation of the topologies of the input layer, the hidden layer and the output layer followed by measuring error signal values by the hidden layer, error signal values by the output layer and the success rates (error signal values means the summation of all error values at a specific layer) (in Table 4). Variation of the number of trainings followed by measuring error signal values by the hidden layer, error signal value by the output layer and the success rates (in Table 5). Variation of the number of neurons at the hidden layer followed by measuring error signal value by hidden layer, error signal value by output layer and the success rate

As shown in Figure 4, the algorithm continues until error signal value by the output layer is small than the predetermined threshold value. In each training experiment, one of the data groups was used to train the NNs, a second group was used for cross-validation [13] during training and the remaining group was used to test the trained NNs. Table 5 shows the definition of the output values produced by NNs.

The experiments show that 6-3-3 topology (6 input units, 3 hidden units, 3 output units) has the best overall results (Table 4). The best number of units for the hidden layer is also three as shown in Table 4.

Table 4. Variation of the topologies

Topology	Success Rate(%)	Cross validation error signal value by output layer	Test error signal value by output value
6-1-1	50	89.032959	91.232523
6-1-3	70	101.343953	102.234234
6-1-5	55	123.254345	124.345234
6-3-1	75	25.172533	26.890749
6-3-3	100	80.8794322	82.3241105
6-3-5	100	160.923144	163.232675
6-5-1	50	142.341646	143.678929
6-5-3	75	136.375948	137.239277
6-5-5	75	128.452788	129.260493

Table 5. Variation of the number of trainings

Learning Count	Success Rate(%)	Cross validation error signal value by output layer	Test error signal value by output layer
10000	100	80.906453	81.185020
20000	100	75.866532	76.109002
30000	100	73.649298	73.979502
40000	100	78.721034	78.957699
50000	100	75.499323	75.668125
60000	100	76.502143	76.601838

Table 6 presents output value in learning and prediction module. Output value is separated into 5 state values.

Table 6. The Definition of the output values

Output Value / Output Layer	TV	Audio	Air Conditioner	Projector	Light
1	0.0	0.25	0.5	0.75	1.0
3	100	110	111	010	001
5	10000	01000	00100	00010	00001

7 Conclusions

This paper described the context-aware middleware providing an automatic home service based on a user's preference at a smart home. The context-aware middleware utilizes 6 basic data for learning and predicting the user's preference of the content: the pulse, the body temperature, the facial expression, the room temperature, the time, the location. The six data sets construct the context model and are used by the context manager module. User profile manager maintains history information for multimedia content chosen by the user. The user-pattern learning and predicting module based on neural network predicts the proper multimedia content for the user.

The testing results show that the pattern of an individual's preference can be effectively evaluated by adopting the proposed context model. Further research will be needed for adopting a different machine learning algorithm such as SVM(support vector machine)[14] and comparing the prediction ratio.

References

1. Weiser, M.: The Computer for the 21st Century. Scientific American, September (1991) 94-104
2. Sherif, M.H.: Intelligent homes: A new Challenge in telecommunications standardization. Communications Magazine, IEEE, Volume 40, Issue 1, Jan. (2002) 8-8
3. Kango, R., Moore, P.R., Pu, J.: Networked smart home appliances - enabling real ubiquitous culture. Proceedings of the 2002 IEEE 5th International Workshop on Networked Appliances, Liverpool, Oct. (2002) 76 – 80
4. Abowd, G.D., Dey, A.K., Brown, P.J. et al.: Towards a better understanding of context and context-awareness. Lecture Notes in Computer Science, Volume 1707, Springer-Verlag (1999)
5. Sun, J.-Z., Sauvola, J.: Towards a conceptual model for context-aware adaptive services. Proceedings of the Fourth International Conference on Parallel and Distributed Computing, Applications and Technologies, Aug. (2003) 90 – 94
6. Anhalt, J., Smailagic, A., Siewiorek, D.P., Gemperle, F.: Towards Context-Aware Computing: Experiences and Lessons. Pervasive Computing (2002)
7. Ranganathan, A., Campbell, R.H.: A Middleware for Context-Aware Agents in Ubiquitous Computing Environments. Lecture Notes in Computer Science, Volume 2672, Springer-Verlag (2003)
8. Capra, L., Emmerich, W., Mascolo, C.: Reflective Middleware Solutions for Context-Aware Applications. Lecture Notes in Computer Science, Volume 2192, Springer-Verlag (2001)
9. Yau, S.S., Karim, F., Wang, Y., Wang, B., Gupta, K.S.: Reconfigurable Context-Sensitive middleware for Pervasive Computing. Pervasive Computing (2002)
10. Crowley, J.L.: Context aware observation of human activities. Proceedings of the 2002 IEEE International Conference on Multimedia and Expo, Volume 1, Aug. (2002) 909-912
11. Charles, D.: The expression of the emotions in man and animals. Electronic Text Center, University of Virginia Library
12. Chen, Z., An, Y., Jia, K., Sun, C.: Intelligent control of alternative current permanent manage servomotor using neural network. Proceedings of the Fifth International Conference on Electrical Machines and Systems, Volume 2, August (2001) 18-20
13. Weiss, S.M., Kulikowski, C.A.: Computer Systems that Learn. Morgan Kaufmann (1991)
14. Burges, C.J.C.: A tutorial on support vector machines for pattern recognition. Data Mining Knowl. Disc., Vol.2, no 2, (1998) 1-47

An Efficient Real-Time Middleware Scheduling Algorithm for Periodic Real-Time Tasks*

Ho-Joon Park and Chang-Hoon Lee

Konkuk University, Seoul 143-701, Korea
{hjpark,chlee}@konkuk.ac.kr

Abstract. For real-time applications, the underlying operating system (OS) should support timely management of real-time tasks. However, most of current operating systems do not provide timely management facilities in an efficient way. There could be two approaches to support timely management facilities for real-time applications: (1) by modifying OS kernel and (2) by providing a middleware without modifying OS. In our approach, we adopted the middleware approach based on the TMO (Time-trigger Message-triggered Object) model which is a well-known real-time object model. The middleware, named TMSOM (TMO Support Middleware) has been implemented on various OSes such as Linux and Windows XP/NT/98. In this paper, we mainly consider TMOSM implemented on Linux (TMOSM/Linux). Although the real-time scheduling algorithm used in current TMOSM/Linux can produce an efficient real-time schedule, it can be improved for periodic real-time tasks by considering several factors. In this paper, we discuss those factors and propose an improved real-time scheduling algorithm for periodic real-time tasks. The proposed algorithm can improve system performance by making the structure of real-time middleware simpler.

1 Introduction

For real-time applications, the underlying operating system should support timely management of real-time tasks. However, most of current operating systems do not provide timely management facilities in an efficient way. There could be two approaches to support timely management facilities for real-time applications: (1) by modifying OS kernel and (2) by providing a middleware without modifying OS. The former approach is to modify OS kernel into a preemptive version. However, it may cause some of OS standard services inoperable. Therefore, we adopted the middleware approach without modifying OS to support timely management of real-time tasks although the middleware approach can support less accurately than the kernel modification approach [1].

Our middleware approach is based on the TMO (Time-trigger Message-triggered Object) model which is a well-known real-time object model. The middleware, named TMOSM (TMO Support Middleware), has been implemented on various OSes such as Linux and Windows XP/NT/98 [2-4]. TMOSM basically provides capabilities for executing TMOs on Linux and Windows XP/NT/98 such as periodic execution of

---

* This work was supported by the Ministry of Information & Communications, Korea, under the University IT Research Center (ITRC) Support Program.

T.G. Kim (Ed.): AIS 2004, LNAI 3397, pp. 304–312, 2005.

real-time tasks, input/output message handling, management of deadline violations, etc. To effectively support these functions, TMOSM contains a real-time middleware scheduler and message handler. The real-time middleware scheduler is activated every time-slice which is defined by the hardware timer interrupt handler. In case a value of time-slice becomes less or a period of a periodic real-time task becomes smaller, the real-time middleware scheduler will be more frequently activated every time-slice. This scheduling algorithm may cause CPU resource to waste and the overhead of a system to increase. Therefore, we propose a new real-time middleware scheduling algorithm which can efficiently handle periodic real-time tasks.

In this paper, we first discuss design issues in TMOSM, mainly TMOSM/Linux, and propose an improved real-time middleware scheduling algorithm for periodic real-time tasks. Additionally, based on our proposed algorithm, we present our experimental results.

The rest of this paper is organized as follows. Section 2 briefly describes the TMO model and TMOSM/Linux. Section 3 discusses some design issues in real-time middleware. Section 4 describes newly proposed real-time middleware structure and scheduling algorithm. Section 5 presents our experimental results. Finally, Section 6 summarizes the paper.

2 Related Works

2.1 TMO Model

TMO is a natural, syntactically minor, and semantically powerful extension of the conventional object(s) [5, 6]. Particularly, TMO is a high-level real-time computing object. Member functions (i.e., methods) are executed within specified time. Timing requirements are specified in natural intuitive forms with no esoteric styles imposed. As depicted in Fig. 1, the basic TMO structure consists of four parts:

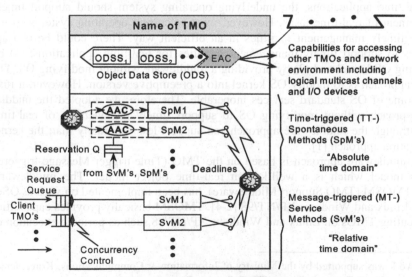

Fig. 1. The Basic structure of TMO (Adapted from [5])

- *Spontaneous Methods (SpM)*: a new type of method, also known as the time-triggered (TT) method. The SpM executions are triggered when the real-time clock reaches specific values determined at design time. An SpM has an AAC (Autonomous Activation Condition), which is a specification of time-windows for execution of the SpM. . An example of an AAC is "for t = from 11am to 11:40am every 20min start-during (t, t+5min) finish-by t+10min" which has the same effect as { "start-during (11am, 11:05am) finish-by 11:10am" and "start-during (11:20am, 11:25am) finish-by 11:30am" }
- *Service Method (SvM)*: conventional service methods. The SvM executions are triggered by service request messages from clients.
- *Object Data Store (ODS)*: the basic unit of storage which can be exclusively accessed by a certain TMO method execution at any given time or shared among concurrent executions of TMO methods (SpMs or SvMs).
- *Environment Access Capability (EAC)*: the list of entry points to remote object methods, logical communication channels, and I/O device interfaces.

There are potential conflictions when SpM's and SvM's access the same data in ODS simultaneously. To avoid such conflictions, a rule named Basic concurrency constraint (BCC) is set up. Under this rule, SvM's cannot disturb the executions of SpM's and the designer's efforts in guaranteeing timely service capabilities of TMO's are greatly simplified. Basically, activation of an SvM triggered by a message from an external client is allowed only when potentially conflicting SpM executions are not in place. An SvM is allowed to execute only if no SpM that accesses the same portion of the Object Data Store (ODS) to be accessed by this SvM has an execution time window that will overlap with the execution time window of this SvM. However, the BCC does not affect either concurrent SpM executions or concurrent SvM executions.

2.2 TMO Support Middleware (TMOSM/Linux)

Figure 2 shows the internal thread structure of TMOSM/Linux. There are two types of threads in TMOSM/Linux, the application thread and the middleware thread (also called system thread). An application thread executes a method (SpM or SvM) of an application TMO as assigned by TMOSM/Linux. Middleware threads are periodic threads (periodically activated by high-precision timer interrupts), each responsible for a major part of the functions of TMOSM/Linux. The middleware threads of TMOSM/Linux are classified by WTMT (Watchdog Timer Management Task), ICT (Incoming Communication Task), and OCT (Outgoing Communication Task). Roles of each middleware thread are as follows:

(1) WTMT : This thread is periodically activated by the timer offered by the underlying OS. The thread schedules other middleware threads and application threads that are assigned to each SpM or SvM of an application. That is, all threads should be controlled by WTMT. WTMT manages updating of system timer, invocation of application threads and deadline. Therefore, WTMT is a core thread.

(2) ICT : This thread manages the distribution of messages coming through the communication network to the destination threads. The computational capacity of ICT is one factor determining the maximum incoming bandwidth that the host node can handle.

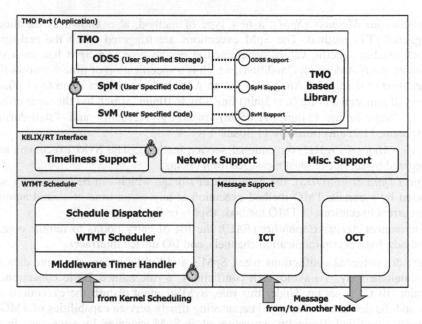

Fig. 2. The Basic Internal Thread Structures of TMOSM/Linux

(3) OCT : This thread manages message transfer from one to another node by the pair of ICT of one node. In TMOSM/Linux, OCT uses UDP socket.

Each middleware thread has its priority and is repeatedly executed in the following order: WTMT, ICT and LIIT. Also these middleware threads are periodically activated to run for a time-slice. Figure 3 shows execution order of tasks in the real-time middleware with 10msec as a time-slice.

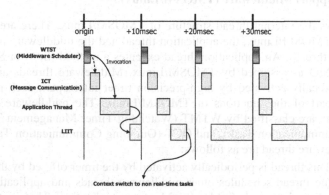

Fig. 3. Execution Order of Tasks

Although both SpM and SvM are registered in similar fashions, TMOSM/Linux handles the two types of methods differently during run-time. In this section, SpM executed by WTMT is described. Figure 4 shows the internal structure of SpM executed by WTMT. WTMT periodically examines the registered SpM's in MCB and

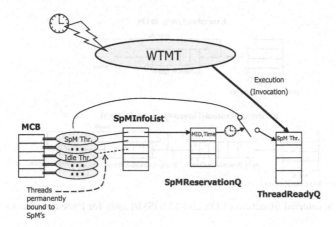

Fig. 4. Internal Structure of SpM Executing by WTMT

identifies the SpM's to be executed in the near future. The identified methods are placed in the SpM-Reservation-Queue for further analysis. Each SpM in SpM-Reservation-Queue is moved into Ready-Application-Thread-Queue later as the time-window for starting the SpM execution arrives. WTMT selects a thread in Ready-Application-Thread-Queue according to the scheduling policy each time a new time-slice opens up.

3   Real-Time Middleware Scheduling Algorithm in TMOSM/Linux

We discuss several design issues in scheduling algorithm of WTMT for the SpM invocation. Those issues are as follows:

(1) From a queue point of view, Figure 5 shows the internal structure of OS and TMOSM/Linux for executing tasks [8]. The Execution-Queue is handled by the OS such as Linux and Windows. Actual activation of a "runnable" task in the Execution-Queue is done by the scheduler of the OS for every time-slice. In general, it is necessary to support queues in the real-time middleware for developing a real-time middleware running on DOS which schedules only single task [9]. However, because the current most popular OS such as Linux and Windows XP/NT/98 can schedule multiple tasks, it is not necessary to support queues in the real-time middleware. Therefore, for TMOSM/Linux to handle the SpM-Reservation-Queue and Ready-Application-Thread-Queue causes CPU resource to waste and the overhead of a system to increase.

(2) Based on time-slice, the timeliness for a real-time system is determined. The reason is that WTMT determines the invocation of each SpM for every time-slice. That is, WTMT is activated every time-slice which is defined by the hardware timer interrupt handler. For example, in case the periodic time of a SpM is 1 sec and the time-slice is 1msec, WTMT should be executed 1,000 times for 1 sec to determine the invocation of SpM. In this case, unnecessary activation of WTMT has happened 999 times whenever SpM is invocated. Therefore, like mentioned above, this scheduling algorithm causes CPU resource to waste and the overhead of a system to increase.

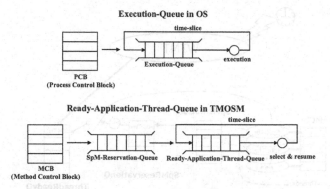

Fig. 5. Internal Structure of OS and TMOSM/Linux for Executing Tasks

(3) Moreover, there is a consideration of a queue structure in the real-time middleware. The basic queue structure is array. The array structure is not an important problem when the number of SpM's on a system is a few. While the overhead of a system or the number of SpM's has been increased, it takes much time to search SpM in a queue [9].

4 Proposed Real-Time Middleware Scheduling Algorithm

4.1 Proposed Structure

As mentioned before, to solve the issues of the real-time middleware scheduling algorithm in TMOSM/Linux, an improved real-time middleware scheduling algorithm for periodic real-time tasks is proposed. Figure 6 shows the proposed internal structure of SpM's invoked by WTMT. WTMT periodically examines the registered SpM's in MCB with our proposed scheduling algorithm. The selected methods are moved into the Execution-Queue and the scheduler of the OS selects a thread in Execution-Queue according to the scheduling policy each time a new time-slice opens up. Without using SpM-Reservation-Queue and Ready-Application-Thread-Queue in this structure, it is not necessary for WTMT to search SpM's in these queues.

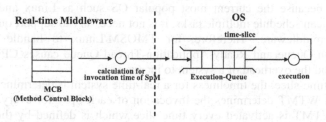

Fig. 6. Proposed Internal Structure of SpM's Invocated by WTMT

4.2 Proposed Scheduling Algorithm

In order to solve the problem of determining the invocation of registered SpM's in MCB each time-slice, the greatest common divisor (G.C.D.) is used in this paper. The following equations show how to calculate G.C.D. based on period of SpM's. With every computed G.C.D. value, WTMT is activated to determine the invocation of SpM's in MCB. That is, the computed G.C.D. value is a periodic time which is used to invoke SpM's in MCB. For example, in case time-slice is 1msec and period of each SpM is 3 sec and 5 sec, 3 sec of SpM corresponds to 3,000 time-slices and 5 sec of SpM corresponds to 5, 000 time-slices. In this paper, 3, 000 and 5, 000 are called conversion period. Therefore, the G.C.D. of 3,000 and 5,000 is 1,000. The 1,000 of G.C.D. is equal to 1,000 times of time-slice. WTMT determines the invocation of SpM's in MCB not each 1 time-slice but each 1,000 time-slice. Therefore, the overhead of a system can be reduced by using the scheme. $T_{GCD-period}$ should be newly calculated whenever new SpM is created or the period of SpM is updated. This scheme is applied in a new proposed real-time middleware scheduling algorithm in this paper.

$$t^1_{conversion-period} = \text{a period of SpM}_1 \text{ / time-slice}$$

$$t^2_{conversion-period} = \text{a period of SpM}_2 \text{ / time-slice}$$

$$\cdots$$

$$t^N_{conversion-period} = \text{a period of SpM}_N \text{ / time-slice}$$

- In case of first execution,

$$T_{GCD-period} = \text{G.C.D} (t^1_{conversion-period}, t^2_{conversion-period}, \cdots, t^N_{conversion-period})$$

- In case of creating new SpM or updating period of SpM,

$$t^{new}_{conversion-period} = \text{a period of SpM}_{new} \text{ / time slice}$$

$$T_{GCD-period} = \text{G.C.D.} (T_{GCD-period}, t^{new}_{conversion-period})$$

As mentioned above, Figure 7 shows an improved real-time middleware scheduling algorithm for invocating SpM's. The following notations are used:

- $T_{SchedulerCounter}$: a value of increasing when each time-slice occurs
- $T_{GCD-period}$: a periodic time for determining the invocation of SpM's in MCB with using G.C.D.
- SpM_{Period} : a periodic time of SpM
- $SpM_{InitTime}$: a initial time of SpM
- $SpM_{InvokeCount}$: times of a SpM invocation

In the improved real-time middleware scheduling algorithm, $T_{SchedulerCounter}$ is increased by the real-time middleware scheduler whenever each time-slice occurs (line 1~2). The real-time middleware scheduler calculates periodic scheduling time of MCB with $T_{SchedulerCounter}$ and computed $T_{GCD-period}$ (line 3). It also calculates the desirable invocation time of SpM, $T_{DesiredInvocation}$ (line 4). If $T_{SchedulerCounter}$ is larger than $T_{DesiredInvocation}$, SpM is invocated (line 5~6). For SpM to be invocated when next period of SpM is occurred, $SpM_{InvokeCount}$ is increased (line 7). Without using SpM-Reservation-Queue and Ready-Application-Thread-Queue in our proposed structure, SpM's are invocated and directly moved into Execution-Queue on the OS by the real-time middleware scheduler. Our proposed scheduling algorithm provides a simple structure of real-time middleware structure and can also reduce the number for SpM's search in MCB. Therefore, CPU resource can be more efficiently used and the over-head of a system can be more reduced. Thus the timeliness for periodic real-time tasks can be more guaranteed.

```
1  for (each time - slice) do {
2      T_SchedulerCounter ← T_SchedulerCounter + 1
3      if (T_SchedulerCounter mod T_GCD-period == 0) {
4          T_DesiredInvocate ← (SpM_InvokeCount * SpM_Period) + SpM_InitTime
5          if (T_DesiredInvocate <= T_SchedulerCounter) {
6              invoke SpM
7              SpM_InvokeCount ← SpM_InvokeCount + 1
8          }
9      }
10 }
```

Fig. 7. Proposed Real-time Middleware Scheduling Algorithm for Periodic Real-time Tasks

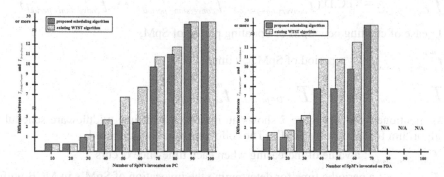

Fig. 8. Simulation Results

5   Experimental Results

In order to test the performance of the proposed scheduling algorithm, the simulation environment is followed. The simulation is performed on a PC with Pentium4 1.5G on X86 LINUX system and a PDA with CPU 206MHz on StrongArm Embedded system. The number of SpM is increased by 10 times from 10 to 100. The periodic time of each SpM's is randomly given from 1 sec to 5 sec. Figure 8 shows simulated result on a PC and a PDA. Whenever SpM is invocated by the real-time middleware scheduler, the difference between $T_{SchedulerCounter}$ and $T_{DesiredInvocate}$ is calculated in the proposed algorithm and the existing algorithm. As can be seen in the Figure 8, our proposed algorithm is better performance than the scheduling algorithm in TMOSM. Without the overload of the system, the difference will be 0 or 1. However, with increasing the number of SpM's, the difference becomes significantly high. In case the number of SpM's is 90 or 100, the difference is very high or is not able to be calculated. The reason is that the performance of PC or PDA is not able to support a lot of SpM's. It is difficult to guarantee timeliness of SpM's as the difference has increased.

6 Conclusions

In this paper, we have examined design issues in real-time middleware scheduling algorithm and proposed a new real-time middleware scheduling algorithm which can determine the invocation time of SpM by G.C.D. method without using queues in the real-time middleware. Furthermore, we showed that our proposed scheduling algorithm is more efficient than the existing scheduling algorithm. The more the number of periodic real-time tasks increase, the more our proposed scheduler is efficient. Moreover, since Linux v2.6 supposes to support preemption in the kernel level, we expect our mechanism to be more suitable for Linux v2.6.

References

1. Park, H.J., and Lee, C.H.: Deadline Handling in a Real-time Middleware on LINUX: IDPT, (2003) pp. 648-651.
2. Kim, K.H., Ishida, M., and Liu, J.: An Efficient Middleware Architecture Supporting Time-Triggered Message-Triggered Objects and an NT-based Implementation: Proc. ISORC'99 (IEEE CS 2ⁿᵈ Int'l Symp. On Object-oriented Real-time distributed Computing), (1999) pp. 54-63.
3. Kim, J.G., Kim, M.H., Min, B.J., and Im, D.B.: A soft Real-Time TMO Platform-WTMOS-and Implementation Techniques: Proc. ISORC'98, Kyoto, Japan (1998) .
4. Kim, M.H. and Kim, J.G.: Linux based TMO execution platform for embedded applications: presented at UKC 2004 (2004) (proceedings will be published in Oct. 2004)
5. Kim, K.H.: APIs for Real-Time Distributed Object Programming: IEEE Computer, Vol. 33, No. 6 (2000) pp.72-80.
6. Kim. K.H.: Real-Time Object-Oriented Distributed Software Engineering and the TMO Scheme: Int'l Jour. Of Software Engineering & Knowledge Engineering, Vol. No.2 (1999) pp. 251-276.
7. Kim, K.H. (Kane) and Kopetz, H.: A Real-Time Object Model RTO.k and an Experimental Investigation of Its Potential: Proc. COMPSAC'94 (IEEE Computer Society's 1994 Int'l Computer Software & Applications Conf.), Taipei (1994) pp. 392-402.
8. Robbins, K.A. and Steven: Practical UNIX Programming: Prentice Hall (1996).
9. Kim, K. H., Subbaraman, C., and Kim, Y.: The DREAM Library Support for PCD and RTO.k programming in C++: Proc. WORDS'96 (IEEE Computer Society 2nd Workshop on Object-oriented Real-Time Dependable Systems), Laguna Beach (1996) pp. 59-68.

Mapping Cooperating GRID Applications by Affinity for Resource Characteristics

Ki-Hyung Kim and Sang-Ryoul Han

Dept. of Computer Eng. Yeungnam University,
214-1 Daedong, Gyungsan, Gyungbuk, Korea
kkim@yu.ac.kr, srman@yumail.ac.kr
http://nclab.yu.ac.kr

Abstract. The Computational Grid, distributed and heterogeneous collections of computers in the Internet, has been considered a promising platform for the deployment of various high-performance computing applications. One of the crucial issues in the Grid is how to discover, select and map possible Grid resources in the Internet for meeting given applications. The general problem of statically mapping tasks to nodes has been shown to be NP-complete. In this paper, we propose a mapping algorithm for cooperating Grid applications by the affinity for the resources, named as *MACA*. The proposed algorithm utilizes the general affinity of Grid applications for certain resource characteristics such as CPU speeds, network bandwidth, and input/output handling capability. To show the effectiveness of the proposed mapping algorithm, we compare the performance of the algorithm with some previous mapping algorithms by simulation. The simulation results show that the algorithm could effectively utilize the affinity of Grid applications and shows good performance.

1 Introduction

Advances in high-speed network technology have made it increasingly feasible to execute even communication-intensive applications on distributed computation and storage resources. Especially the emergence of computational GRID environments [1, 2] has caused much excitement in the high performance computing (HPC) community. Many Grid software systems have been developed and it has become possible to deploy real applications on these systems [3, 4].

A crucial issue for the efficient deployment of HPC applications on the Grid is the resource selection and mapping. There has been much research on this issue [5–9]. Globus [10, 11] and Legion [12] present resource management architectures that support resource discovery, dynamical resource status monitor, resource allocation, and job control. These architectures make it easy to create a high-level scheduler. Legion also provides a simple, generic default scheduler which can easily be outperformed by a scheduler with special knowledge of the application [5, 13].

T.G. Kim (Ed.): AIS 2004, LNAI 3397, pp. 313–322, 2005.

MARS [9] and AppLeS [14] provide application-specific scheduling which determines and actuates a schedule customized for the individual application and the target computational Grid at execution time.

As an approach of a general resource selection and mapping framework instead of relying on application specific scheduling, a *resource selection framework (RSF)* which selects Grid resources by the application's characteristics was proposed [5]. RSF consists of three phases: *selection* of possible resources which form a distributed virtual machine, *configuration*, and *mapping* of application subtasks into the selected virtual machines. These three phases can be interrelated. For the selection of possible resources, RSF defined a set-extended ClassAds Language that allows users to specify aggregate resource properties (e.g., total memory, minimum bandwidth). RSF also proposed an extended set matching matchmaking algorithm that supports one-to-many matching of set-extended ClassAds with resources. The resource selector locates sets of resources that meet user requirements, evaluates them based on specified performance model and mapping strategies, and returns a suitable collection of resources. It also presented a mapping algorithm for the Cactus application [15].

The matching of tasks to machines and scheduling the execution order of these tasks has been referred to as a *mapping*. The general problem of optimally mapping tasks to machines in heterogeneous computing machines has been shown to be NP-complete [7, 9]. Heuristics developed to perform this mapping function are often difficult to compare because of different underlying assumptions in the original study of each heuristic [9].

This paper proposes a mapping algorithm for GRID applications, named as MACA (*M*apping *A*lgorithm for *C*ooperating GRID applications based on the *A*ffinity for GRID resource characteristics) which can be used in RSF. The proposed algorithm utilizes the general affinity of Grid applications for certain resource characteristics such as CPU speeds, network bandwidth, and input/output handling capability. To show the effectiveness of the proposed mapping algorithm, we compare the performance of the algorithm with some previous mapping algorithms by simulation. The simulation results show that the algorithm could effectively utilize the affinity of Grid applications and shows good performance.

The rest of this paper is organized as follows. In Section 2, we briefly describe the resource selection framework which is the basis of our proposed mapping algorithm. In Section 3, we propose MACA. Section 4 presents the performance results of MACA. Finally, we summarize our work in Section 5.

2 Preliminaries

This section describes the preliminary backgrounds of the proposed algorithm. We at first describe the architecture of RSF and a simple mapping algorithm for Cactus application presented in RSF. Then, we show the *Max-min* mapping heuristic which is used for the performance comparison with our proposed algorithm.

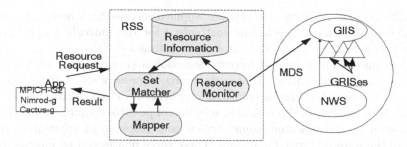

Fig. 1. The architecture of RSF.

2.1   Resource Selection Framework (RSF)

The architecture of the resource selection framework is shown in Fig. 1. Grid information service functionality is provided by the *Monitoring and Discovery Service* (MDS-2) component [16, 17] of the Globus Toolkit [11]. MDS provides a uniform framework for discovering and accessing the system configuration and status information that may be of interest to the schedulers such as server configuration and CPU load. The *Network Weather Service (NWS)* [18] is a distributed monitoring system designed to track periodically and forecast dynamically resource conditions, such as the fraction of CPU available to a newly started process, the amount of memory that is currently unused, and the bandwidth with which data can be sent to a remote host. Grid Index Information Service (GIIS) and Grid Resource Information Service (GRIS) [17] components of MDS provide resource availability and configuration information.

The Resource Selector Service (RSS) comprises three modules. The *resource monitor* acts as a Grid Index Service (GRIS) [17]; it is responsible for querying MDS and NWS to obtain resource information and for caching this information in local memory, refreshing only when associated time-to-live values expire. The *set matcher* uses the set-matching algorithm to match incoming application requests with the best set of available resources. The *mapper* is responsible for deciding the topology of the resources and allocating the workload of application to resources.

For the selection of possible resources, RSF defined a set-extended ClassAds Language that allows users to specify aggregate resource properties (e.g., total memory, minimum bandwidth) [5]. RSF also proposed an extended set matching matchmaking algorithm that supports one-to-many matching of set-extended ClassAds with resources. Both application resource requirements and application performance models are specified declaratively, in the ClassAds language, while mapping strategies, the topic of this paper, can be determined by user-supplied code. The resource selector locates sets of resources that meet user requirements, evaluates them based on specified performance model and mapping strategies, and returns a suitable collection of resources, if any are available.

After selecting the possible resources for a Grid application, the next step is mapping some selected resources for the application. RSF proposed a simple mapping algorithm for Cactus application and showed good performance result with the algorithm. We name it as the simple mapping algorithm of RSF in this paper because the algorithm considers only the bandwidth between nodes while mapping even though the algorithm show good performance for Cactus application. The general step of the algorithm is as follows:

1. Pick the node with the highest CPU speed as the first machine of the line.
2. Find the node that has the highest communication speed with the last node in the line, and add it to the end of the line.
3. Continue Step 2 to extend the line until all nodes are in the line.

2.2   Max-Min Heuristic Mapping Algorithm for Independent Tasks

There has been much research on the mapping issues in heterogeneous computing environments (HCE) [9]. The general problem of optimally mapping tasks to machines in HCE has been shown to be NP-complete [7, 9]. The goal of the heuristic mapping algorithms is to minimize the total execution time, also called as the *makespan*, of the metatask. Heuristics developed to perform this mapping function are often difficult to compare because of different underlying assumptions in the original study of each heuristic [9]. Among them, the *Max-min* heuristic, one of the most typical heuristic mapping algorithms for independent tasks on HCE has shown good performance result.

The Max-min heuristic begins with the set U of all unmapped tasks and the set V of all unmapped nodes in HCE. Then, the set of minimum completion times, $M = min_{0 \le j < \mu}(ct(t_i, m_j))$, for each $t_i \in U$ and $m_j \in V$, is found, where $ct(t, m)$ is the completion time of task t on node m. Next, the task with the overall maximum completion time from M is selected and assigned to the corresponding machine (hence the name Max-min). Lastly, the newly mapped task is removed from U, and the process repeats until all tasks are mapped. Intuitively, Max-min attempts to minimize the penalties incurred from performing tasks with longer execution times.

3   MACA

This section proposes MACA, a mapping algorithm for cooperating Grid applications. The intuition of MACA is to utilize the general affinity of Grid applications for certain resource characteristics such as CPU speeds and network bandwidth. That is, if an application is a group of independent tasks (that is, there is no communication between tasks), MACA considers only the processing power (CPU speed) of nodes during mapping. Conversely, if an application is a group of highly dependent tasks (that is, they communicate frequently with one another), MACA considers the network bandwidth of a node with the first priority during mapping. For this affinity, MACA employs $\alpha(k)$ of a task k as a measure of the tendency of the computation over the total completion time. It ranges from 0 (tasks communicate with one another all the time with no local computation) to 1 (independent tasks).

Fig. 2 shows the pseudo code of MACA. The algorithm takes the sets of tasks (U) and nodes (V) as input, where $\mid U \mid \leq \mid V \mid$. The output is the mapping between all the given tasks in U and some selected nodes out of the given set of nodes (V).

The first step of the algorithm is to order the tasks in the decreasing order of the computation time (line 7 in the algorithm). The algorithm finds the best matching node for each task in this order. The intuition of this ordering of tasks comes from the Max-min algorithm. Intuitively, Max-min attempts to minimize the penalties incurred from performing tasks with longer execution times. Assume, for example, that the Grid application being mapped has many tasks with very short execution times and one task with a very long execution time. Mapping the task with the longer execution time to its best machine first allows this task to be executed concurrently with the remaining tasks with shorter execution times.

The algorithm has an iterative loop until all tasks in U find their own mapped nodes from V. In each step of the iteration, the algorithm finds two candidate nodes for mapping: fc and sc. fc, the first candidate node, is the best matching node from the set of still unmapped nodes while sc, the second candidate node, is for the sake of the next iteration loop of the comparison.

At the first iteration step, the algorithm tries to find fc as the node with the highest CPU speed among the nodes for mapping with the task with the highest computation time (that is, the first one in the ordered task list) (lines 9–11 in the algorithm). The algorithm then moves the mapped task and node from U and V, the sets of input tasks and nodes, to W, the set of mapping candidates.

From the second iteration step, fc becomes a node with the largest value of β, a criteria of the matching from the already mapped node(s). β is defined as follows (line 14 of the algorithm):

$$\beta(y,j) = (CS(j))^{\alpha(k)}/(NL(y,j))^{1-\alpha(k)}, \tag{1}$$

where y is an already mapped node in W, j is the next candidate node of mapping, $\alpha(k)$ is the relative portion of the computation time over the total completion time of task k, $CS(j)$ is the CPU speed of node j, and $NL(y,j)$ is the network latency between node y and j. If α is 1, $\beta(y,j)$ becomes $CS(j)$ which has no relationship with the previously selected node.

The algorithm compares the obtained first candidate node fc with the second candidate node (sc) from the previous step of the iteration and chooses a node with the greater β value for mapping.

For the sake of the next iteration step, the algorithm also finds the node (sc) with the second largest value of β_{sc}. sc will be compared with the first candidate of the next iteration for selecting the next best matching node in the next iteration step.

As an example of the algorithm, we show a typical mapping process for a given tasks and nodes. Fig. 3 and Fig. 4 show an example task and node graphs respectively. Fig. 5 shows the mapping process of the example tasks and nodes. The first step is the calculation of α from the task graph in Fig. 4. Also we order the tasks in the decreasing order of CPU speed for the Max-min fashion

Data: the set U of all unmapped tasks and the set V of all unmapped nodes, where $\mid U \mid \leq \mid V \mid$

Result: the set W of mapping from U to V

```
 1  begin
 2  │   y ← null;
    │   //y is the previously mapped node.
 3  │   fc ← sc ← β_sc ← null;
    │   //fc and sc are the first and second candidate nodes for mapping,
    │     respectively.
 4  │   while U becomes empty do
 5  │   │   For all unmapped tasks i ∈ U and all unmapped nodes j ∈ V, compute
    │   │     ct(i, j), the computation time of task i on node j;
 6  │   │   Get M = {m_i, for all i ∈ U}, where m_i = min_{j∈V} ct(i, j) of task i;
 7  │   │   Select task k such that m_k = max_{i∈U} m_i, where m_i ∈ M;
    │   │   //Select tasks by the Max-min algorithm style.
 8  │   │   if y = null then
 9  │   │   │   pick node n with the highest CPU speed (CS) from all nodes in V;
    │   │   │   fc ← n;
10  │   │   │   map task k into node fc, remove k from U and fc from V, and put
    │   │   │     the mapping between k and fc into W;
11  │   │   │   y ← fc;
12  │   │
13  │   │   else if y ≠ null then
14  │   │   │   fc ← node f with β_f = max_{j∈V} β(y, j), where
    │   │   │     β(y, j) = (CS(j))^{α(k)}/(NL(y, j))^{1−α(k)};
    │   │   │   //α(k) is the relative portion of the computation time over the total
    │   │   │     completion time of task k
    │   │   │   //CS(j) is the CPU speed of node j
    │   │   │   //NL(y, j) is the network latency between node y and j.
    │   │   │   //fc and sc are the first and second candidates for mapping.
15  │   │   │   if β_{fc} < β_{sc} then
16  │   │   │   │   swap fc and sc;
17  │   │   │   │   map task k into node fc, remove k from U and fc from V, and
    │   │   │   │     put the mapping between k and fc into W;
18  │   │   │   │   y ← fc;
19  │   │   │   else
20  │   │   │   │   map task k into node fc, remove k from U and fc from V, and
    │   │   │   │     put the mapping between k and fc into W;
    │   │   │   │   //Determine a second candidate for the next iteration step.
21  │   │   │   │   fc ← node f with β_f = max_{j∈V} β(y, j), where
    │   │   │   │     β(y, j) = (CS(j))^{α(k)}/(NL(y, j))^{1−α(k)};
22  │   │   │   │   if β_{fc} > β_{sc} then
23  │   │   │   │   │   sc ← fc;
24  │   │   │   │   end
25  │   │   │   │   y ← fc;
26  │   │   end
27  │   │
28  │   end
29  end
```

Fig. 2. Pseudo codes of MACA.

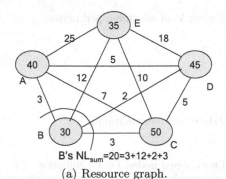

B's NL_{sum}=20=3+12+2+3

(a) Resource graph.

	A	B	C	D	E	NL_{sum}
A		3	7	5	25	40
B	3		3	2	12	20
C	7	3		5	10	25
D	5	2	5		18	30
E	25	12	10	18		65

(b) Network latency matrix.

Fig. 3. An example resource graph.

No de	Task 1			Task 2			Task 3			Task 4		
	Comp. Time	Comm. Time	Total Time	Comp. Time	Comm. Time	Total Time	Comp. Time	Comm. Time	Total Time	Comp. Time	Comm. Time	Total Time
A	55	20	75	50	20	70	65	20	85	60	20	80
B	73	10	83	67	10	77	87	10	97	80	10	90
C	44	13	57	40	13	53	52	13	65	48	13	61
D	49	15	64	44	15	59	58	15	73	53	15	68
E	63	33	95	57	33	90	74	33	107	69	33	101

Fig. 4. An example task graph.

$\alpha(1)=55/75=0.73$ $\alpha(2)=50/70=0.71$ $\alpha(3)=65/85=0.76$ $\alpha(4)=60/80=0.75$

1st loop: CS(A)=40
CS(B)=30
CS(C)=50 ←$fc_{1st-loop}$
CS(B)=45
CS(B)=35

y ← $fc_{1st-loop}$= C
$sc_{1st-loop}$ ← null
W={C-3}

2nd loop: β(C,A)=9.8 ← $sc_{2nd-loop}$
β(C,B)=9.7
β(C,D)=11.6 ← $fc_{2nd-loop}$
β(C,E)=8.1

y←D ←max(β(C,D),0)
$sc_{2nd-loop}$ ← A
W={C-3,D-4}

3rd loop: β(D,A)=9.7 ← $sc_{3rd-loop}$
β(D,B)=10.1 ← $fc_{3rd-loop}$
β(D,E)=6.3

y←B ←max(β(D,B), β(C,A))
$sc_{3rd-loop}$ ←A ← max(β(D,A), β(C,A))

W={C-3,D-4,B-1}

4th loop: β(B,A)=10.2 ← $fc_{4th-loop}$
β(B,E)=5.1 ← $sc_{4th-loop}$

y←A ←max(β(B,A), β(D,A))
$sc_{4th-loop}$ ←E ← max(β(B,E), β(D,A))
W={C-3,D-4,B-1,A-2}

Fig. 5. Mapping process for the example.

mapping: 3, 4, 1, and 2. In the first loop of the iteration, C becomes fc because it has the highest CPU speed among the nodes. In all the other iteration loops, we get two candidate nodes and choose the node with the highest value of β among them.

Task	Min-Max	Simple Mapping of RSF	MACA
1	A, ct(1,A)=75	D, ct(1,D)=64	B, ct(1,B)=83
2	E, ct(2,E)=90	A, ct(2,A)=70	A, ct(2,A)=70
3	C, ct(3,C)=65	C, ct(3,C)=65	C, ct(3,C)=65
4	D, ct(4,D)=68	B, ct(4,B)=90	D, ct(4,D)=68
Makespan	90	90	83

Fig. 6. Comparison of Mapping results.

Simulation Parameters	Range
Network Latency	5 ~ 35
CPU Speed	500 ~ 1050
Task Completion Time	30 ~ 130

Fig. 7. Simulation parameters.

Fig. 6 compares the proposed algorithm with the other two previous algorithms. Max-min considers only the computation time during mapping because it is naturally a mapping algorithm for independent tasks. Simple mapping algorithm of RSF considers only the network latency between nodes during mapping.

4   Performance Evaluation

For evaluating the proposed mapping algorithm, we compared the proposed algorithm with Max-min and the simple mapping algorithm of RSF by simulation. The simulation models of the algorithms are developed by C/C++. The necessary parameters for simulation include the information of a Grid application consisting of several tasks and the node information as shown in Fig. 7. Computation times of tasks are generated randomly. Each application consists of 10 tasks, and each task's computation time is generated by a random number generator. We assume that the computation time of a task depends mainly on the CPU speed even though there are some other aspects such as memory size and the speed of IO. We also assume that the pre-selection procedure screens out unmatched nodes such as nodes with lower memory size than the required minimum memory size. For each value of α, we have performed 10 simulations with different random numbers.

Fig. 8 shows the simulation results for 10 tasks and 15 nodes. The proposed algorithm shows better performance when the value of α becomes greater (that is, the computation time portion of the total completion time becomes greater). When the variation of the CPU speeds of nodes is large, the proposed algorithm also shows better results than the small range. This is because the proposed algorithm considers both the CPU speed and the network latency of resources.

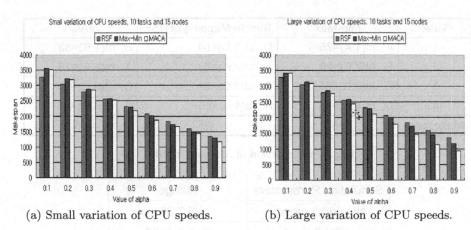

(a) Small variation of CPU speeds.   (b) Large variation of CPU speeds.

Fig. 8. Performance comparison (10 tasks and 15 nodes).

5 Conclusion

The Grid technology enables the aggregation of computational resources connected by high speed networks for high performance computing. A crucial issue for the efficient deployment of GRID applications on the Grid networks is the resource selection and mapping. RSF was proposed as a generic framework for resource selection and mapping. This paper proposes a mapping algorithm for cooperating Grid applications which utilizes the application's affinity for the Grid resource characteristics. Depending upon the communication tendency of the Grid application, the proposed algorithm maps the resources with the affinity. For the performance evaluation of the algorithm, Max-min and the simple algorithm of RSF are compared with the algorithm by simulation. The simulation results show the proposed algorithm shows good performance by utilizing the application's characteristics.

References

1. I. Foster and C. Kesselman. "Computational Grids", Chapter 2 of *The Grid: Blueprint for a New Computing Infrastructure*, Morgan-Kaufman, 1999.
2. I. Foster, C. Kesselman, and S. Tuecke. "The Anatomy of the Grid: Enabling Scalable Virtual Organizations" International J. Supercomputer Applications, vol.15, no.3, pp. 200-222, 2001.
3. G. Allen, D. Angulo, I. Foster, G. Lanfermann, Chuang Liu, T. Radke, E. Seidel and J. Shalf, "The Cactus Worm: Experiments with Dynamic Resource Discovery and Allocation in a Grid Environment," International Journal of High-Performance Computing Applications, vol.15 no.4, 2001.
4. M. Ripeanu, A. Iamnitchi, and I. Foster, "Cactus Application: Performance Predictions in Grid Environments," EuroPar2001, Manchester, UK, August 2001.
5. C.Liu, L.Yang, I.Foster, D.Angulo, "Design and evaluation of a resource selection framework for Grid applications" in proceedings. 11th IEEE International Symposium on High Performance Distributed Computing (HPDC-11), pp. 63–72, 2002.

6. A. Takefusa, S. Matsuoka, H. Casanova, and F. Berman, "A Study of Deadline Scheduling for Client-Server Systems on the Computational Grid," in Proceedings of the 10th IEEE International Symposium on High Performance Distributed Computing (HPDC) pp.406–415, 2001.
7. H. Dail, O. Sievert, F. Berman, H. Casanova, A. YarKhan, S. Vadhiyar, J. Dongarra, C. Liu, L. Yang, D. Angulo, and I. Foster, "Scheduling in the Grid Application Development Software Project," chapter 1 of *Resource Management in the Grid*, Kluwer, 2003,
8. S. Vadhiyar and J. Dongarra, "A Metascheduler For The Grid," HPDC 2002, 11th IEEE International Symposium on High Performance Distributed Computing, Edinburgh, Scotland, IEEE Computer Society, pp. 343–351, 2002.
9. T. Braun, H. Siegel, N. Beck, L. Boloni, M. Maheswaran, A. Reuther, J. Robertson, M. Theys and B. Yao, "A Comparison of Eleven Static Heuristics for Mapping a Class of Independent Tasks onto Heterogeneous Distributed Computing Systems," Journal of Parallel and Distributed Computing 61, pp. 810–837, 2001.
10. K. Czajkowski, et al. "A Resource Management Architecture for Metacomputing Systems," in Proc. IPPS/SPDP '98 Workshop on Job Scheduling Strategies for Parallel Processing. 1998.
11. I. Foster, and C. Kesselman, "Globus: A Toolkit-Based Grid Architecture," In *The Grid: Blueprint for a New Computing Infrastructure*, Morgan Kaufmann, pp. 259–278, 1999.
12. J. Chapin, et al. "Resource Management in Legion," in Proceedings of the 5th Workshop on Job Scheduling Strategies for Parallel Processing (JSSPP '99), 1999.
13. H. Dail, G. Obertelli, and F. Berman. "Application-Aware Scheduling of a Magnetohydrodynamics Applications in the Legion Metasystem," in Proceedings of the 9th Heterogeneous Computing Workshop, 2000.
14. F. Berman and R. Wolski. "The AppLeS project: A Status Report," in Proceedings of the 8th NEC Research Symposium, 1997.
15. G. Allen et al., "Cactus Tools for Grid Applications," Cluster Computing, pp.179–188, 2001.
16. K. Czajkowski, S. Fitzgerald, I. Foster and C. Kesselman, "Grid Information Services for Distributed Resource Sharing," In 10th IEEE International Symposium on High Performance Distributed Computing, IEEE Press, pp. 181–184, 2001.
17. S. Fitzgerald, I. Foster, C. Kesselman, G. Laszewski, W. Smith and S. Tuecke, "A Directory Service for Configuring High-performance Distributed Computations," In Proc. 6th IEEE Symp. on High Performance Distributed Computing, pp. 365–375, 1997.
18. R. Wolski, "Dynamically Forecasting Network Performance Using the Network Weather Service," Journal of Cluster Computing, 1998.

Modeling of Policy-Based Network with SVDB*

Won Young Lee, Hee Suk Seo, and Tae Ho Cho

School of Information and Communications Engineering, Sungkyunkwan University
{sonamu,histone,taecho}@ece.skku.ac.kr

Abstract. There are many security vulnerabilities in computer systems. They can be easily attacked by outsiders or abused by insiders who misuse their rights or who attack the security mechanisms in order to disguise as other users or to detour the security controls. Today's network consists of a large number of routers and servers running a variety of applications. Policy-based network provides a means by which the management process can be simplified and largely automated. This article describes the modeling and simulation of a security system based on a policy-based network that has some merits. We present how the policy rules from vulnerabilities stored in SVDB (Simulation based Vulnerability Data Base) are inducted, and how the policy rules are transformed into PCIM (Policy Core Information Model). In the network security environment, each simulation model is hierarchically designed by DEVS (Discrete EVent system Specification) formalism.

Keywords: Security Policy, PBNM (Policy-based Network Management), network security, DEVS formalism, simulation, Data Mining.

1 Introduction

Present-day networks are large complex systems consisting of a variety of elements, which can come from a variety of vendors. But it is more difficult than before for human administrators to manage more and more new network devices. The challenges facing the enterprise managers include network congestion, network complexity and security. A new wave of multimedia applications now begins to enter into corporate intranets – voice and video can hardly wait to join the bandwidth fray [1]. Every network is made up of a variety of elements, but they must still work together. Because of this heterogeneity and the lack of complete standardization, managing a network with more than a handful of elements can require a significant amount of expertise. The network manger is faced with the difficult task of meeting internal and external security requirements while still providing easy and timely access to network resources to authorized user. The solution to these issues lies in policy-based management. A network manager creates policies to define how resource or services in the network can (or cannot) be used. The policy-based management system transforms these policies into configuration changes and applies those changes to the network. The simplification and automation of the network management process is one of the key applications of the policy framework [2].

Since evaluating the performance of a security system directly in real world requires heavy costs and efforts, an effective alternative solution is using the simulation model [3]. The simulation models that are security model and network component are

---

* This research was supported by University IT Research Center Project.

T.G. Kim (Ed.): AIS 2004, LNAI 3397, pp. 323–332, 2005.
© Springer-Verlag Berlin Heidelberg 2005

constructed based on the DEVS formalism. We construct the SVDB for analyzing the vulnerabilities. SVDB has vulnerabilities of systems and policy information that can be used by security systems. We have simulated to verify the performance of the rule induction using SVDB for DOS (Denial of Service) attack and probing attack.

2 Related Works

2.1 Policy-Based Framework

The general policy-based framework can be considered an adaptation of the IETF policy framework to apply to the area of network provisioning and configuration [4-6]. The policy architecture as defined in the IETF consists of four basic elements as shown in Fig. 1.

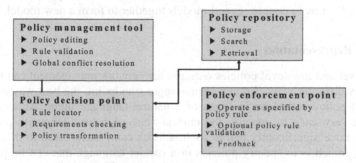

Fig. 1. The IETF policy framework

PMT (Policy Management Tool) is used by an administrator to input the different policies that are active in the network. The PMT takes as input the high-level policies that a user or administrator enters in the network and converts them to a much more detailed and precise low-level policy description that can apply to the various devices in the network. PDP (Policy Decision Point) makes decisions based on policy rules and it is responsible for policy rule interpretation and initiating deployment. Its responsibilities may include trigger detection and handling, rule location and applicability analysis, network and resource-specific rule validation and device adaptation functions. PEP (Policy Enforcement Point) is the network device that actually implements the decisions that the PDP pass to them. The PEP is also responsible for monitoring any statstics or other information relevant to its operation and for reporting it to the appropriate places. Policy Repository is used to store the policies generated by the management tool. Either a directory or a database can store the rules and policies required by the system. In order to ensure interoperability across products from different vendors, information stored in the repository must correspond to an information model specified by the policy framework working group [7].

2.2 DEVS Formalism

The DEVS formalism, developed by Zeigler is a theoretical, well grounded means of expressing hierarchical, modular discrete-event models [8]. In DEVS, a system has a time base, inputs, states, outputs and functions. The system function determines next states and outputs based on the current states and input. In the formalism, a basic model is defined by the structure:

$$M = < X, S, Y, \delta_{int}, \delta_{ext}, \lambda, ta >$$

where X is an external input set, S is a sequential state set, Y is an external output set, δ_{int} is an internal transition function, δ_{ext} is an external transition function, λ is an output function and ta is a time advance function. A coupled model is defined by the structure:

$$DN = < D, \{M_i\}, \{I_i\}, \{Z_{i,j}\}, select >$$

where D is a set of component name, Mi is a component basic model, I_i is a set of influences of I, $Z_{i,j}$ is an output translation, select is a tie-breaking function. Such a coupled model can itself be employed in a larger coupled model. Several basic models can be coupled to build a more complex model, called a coupled model. A coupled model tells how to couple several models together to form a new model.

2.3 Policy Representation

The high-level and low-level policies required for network management can be specified in many different ways. From a human input standpoint, the best way to specify a high-level policy would be in terms of a natural-language input. Although these policies are very easy to specify, the current state of natural-language processing needs to improve significantly before such policies can be expressed in this manner. The next approach is to specify policies in a special language that can be processed and interpreted by a computer. When policies are specified as a computer interpretable program, it is possible to execute them. A simpler approach is to interpret the policy as a sequence of rules, in which each rule is in the form of a simple condition-action pair (in an if-then-else format). The IETF has chosen a rule-based policy representation. IETF Policy Framework WG works especially on the "condition action" part to define Policy Core Information Model [9] for the representation of policy information. The PCIM is the object-oriented information model for representing policy information. This model defines representing policy information and control of policies, and association classes that indicate how instances of the structural classes are related to each other. Policies can either be used in a stand-alone fashion or aggregated into policy groups to perform more elaborate functions. Fig. 2 illustrates the inheritance hierarchy for PCIME (PCIM extensions) [9,10].

2.4 Vulnerability Database

A vulnerability is a condition or weakness in (or absence of) security procedures, technical controls, physical controls, or other controls that could be exploited by a threat [11]. The theme of vulnerabilities analysis is to devise a classification, or set of classifications, that enable the analyst to abstract the information desired from a set of vulnerabilities. This information may be a set of signatures, for intrusion detection; a set of environment conditions necessary for an attacker to exploit the vulnerability; a set of coding characteristics to aid in the scanning of code; or other data.

Government and academic philanthropists, and some companies, offer several widely used and highly valued announcement, alert, and advisory services for free.

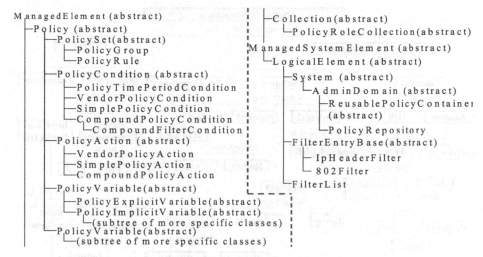

Fig. 2. Class Inheritance Hierarchy for PCIME

Each of those organizations referred to the same vulnerability by a different name. Such confusion made it hard to understand what vulnerabilities you faced and which ones each tool was looking for- or not looking for. The MITRE Corporation began designing a method to sort through the confusion. It involved creating a reference list of unique vulnerability and exposure names and mapping them to appropriate items in each tool and database. We use CVE (Common Vulnerabilities and Exposures) names in a way that lets a security system crosslink its information with other security systems [12].

3   The Structure of Target Network and Simulation Model

The target network has five subnets; subnet_1, subnet_2, subnet_3, subnet4 and subnet_5. The types of component models in the network are Policy Server, IDS, Firewall, Router, Gateway, Vulnerability Scanner model. Each subnet has unix server, linux server, windows NT server and etc. These models are constructed based on the DEVS formalism.

3.1   Network Architecture

The System Entity Structure (SES) [8] is a knowledge representation scheme that combines the decomposition, taxonomic, and coupling relationships. The entities of the SES refer to conceptual components of reality for which models may reside in the model base. Fig. 4 is a system entity structure for the overall target network model.

Network Simulation model is decomposed into Network and EF model. EF Model is again decomposed into Generator model and Transducer model. Generator model generates network input packets. Transducer model measures a performance indexe of interest for the target system. Network model is decomposed into Gateway, Subnet and PolicyCompnent model. PolicyComponent model is decomposed again into PolicyFramework and InterfaceForSVDB model.

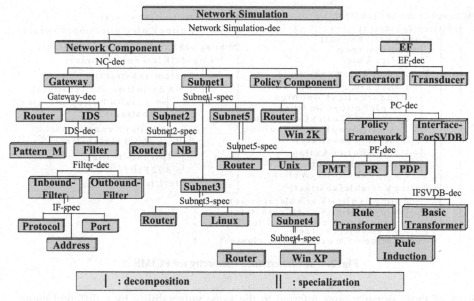

Fig. 3. SES of the target network

3.2 SVDB

SVDB has the specific information that can be used by security agents as well as the common information of vulnerability of system. SVDB has four componets; vulnerability information, packet information, system information and references information. SVDB also has particular parts for accuracy and efficiency of security agents. The payload size is used to test the packet payload size. The offset modifies the starting search position for the pattern match function from the beginning of the packet payload. The payload size and offset have the added advantage of being a much faster way to test for a buffer overflow than a payload content check. URL contents allow search to be matched against only the URL portion of a request. Table 1 shows the table of SVDB for the simulation [13].

Table 1. Table of SVDB

Table	Field
Vulnerability Information	Vulnerability Name(CVE), Summary, Published, Vulnerability Type, Exploitable Range, Loss Type, Vulnerable Software and Versions
Packet Information	IP flags, TTL, Protocol, Source IP, Destination IP, IP options, ICMP code, ICMP type, Source port, Destination port, Sequence number, Acknowledgement number, TCP flag, Offset, Payload size, URL contents, Contents, CVE Name
System Information	Vulnerable Software and Versions, Vendor, Name, Version
References Information	Source, Type, Name, Link

3.3 System Modeling

Fig. 4 shows the structure of PolicyFramework model. PolicyFramework model is divided into PMT model and PDP model. PMT model is composed of Resource Discovery model and Validity Check model. Resource Discovery model determines the topology of the network, the users, and applications operational in the network. In order to generate the configuration for the various devices in the network, the capabilities and topology of the network must be known. Validity Check model consists of various types of checks: Bounds checks, Consistency checks, Feasibility checks.

PDP model is composed of Supervision model, Policy Transformation model and Policy Distributor model. Supervision model receives events from network devices and monitors network usage. The PDP can use this information about the network to invoke new policy-based decisions. Policy Transformation model translates the business-level policies into technology-level policies that can be distributed to the different devices in the network. Policy Distributor model is responsible for ensuring that the technology-level policies are distributed to the various devices in the network.

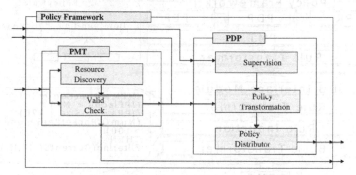

Fig. 4. Structure of Policy Framework Model

Fig. 5 shows the structure of IDS. IDS model is divided into Detector model, Response Generator model and Logger model. Detector model is further decomposed into Pattern Matcher model and Analyzer model. Pattern Matcher model is a rule-based expert system that detects intrusions through pattern matching procedure with packet data and rules. Analyzer model is a statistical detection engine that detects intrusions by analyzing system log and audit. Response Generator model determines a response according to the detection result of Detector model and sends a message. Logger model records all information of detection procedure in the log file.

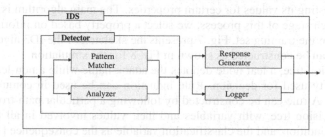

Fig. 5. Structure of IDS Model

4 The Collaboration Between SVDB and Policy-Based Framework

The policy management tool in the common policy-based framework is used by the administrator but we have appended the some module interface for the better automation control. The policy-based framework in the proposed system is accessed by the administrator, SVBD and intrusion detection system. Fig. 6 shows the structure and function of SVDB interface. The main function of each component is as follows:

- Basic Transformer: Basic Transformer provides a connection of DB, data type converting and basic type checking.
- Rule Induction: Rule Induction builds a decision tree from DB using tree induction algorithm. Rule induction serves as a postprocessing of tree induction. A rule is constructed by a particular path from root to a leaf in the decision tree.
- Rule Transformer: Rule transformer provides a transformation between a policy rule and a class of PCIMe.

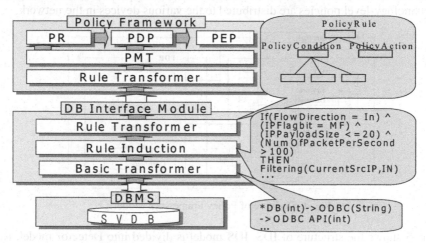

Fig. 6. Structure and function of SVDB interface

4.1 Policy Rule Induction from SVDB

A well-known tree induction algorithm adopted from machine learning ID3, which employs a process of constructing a decision tree in a top-down fashion. A decision tree is a hierarchical representation that can be used to determine the classification of an object by testing its values for certain properties. The main algorithm is a recursive process. At each stage of this process, we select a property based on information gain calculated from the training set. Fig. 7 presents the skeleton of the ID3 algorithm. The decision tree can be constructed as shown in Fig. 8 for the simulation.

In the decision tree, a leaf node denotes the attack name while a non-leaf node denotes a property used for decision. Rule induction can be used in conjunction with tree induction. A rule can be constructed by following a particular path from root to a leaf in the decision tree, with variables and their values involved in all the nonleaf nodes as the condition, and the classification variable as the consequence [10].

Algorithm ID3
Input: a set of example
Output: a decision tree
Method:
```
ID3_tree (examples, properties)
If  all entries in examples are in the same
    category of decision variable
    Return a leaf node labeled with that category
Else
    Calculate information gain;
    Select a property P with highest information
    gain;
    Assign root of the current tree = P;
    Assign properties - P;
    for each value V of P
       Create a branch of the tree labeled with V;
       Assign examples_V = subset of examples
       with values V for property P ;
       Append ID3_tree (example_V, properties) to
       branch V
```

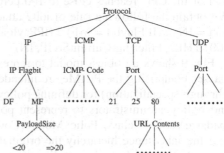

Fig. 7. ID3 algorithm   **Fig. 8.** Construction of decision tree

4.2   Converting a Policy Rule into PCIMe

The policy rule is transformed into PCIMe. Policy conditions, policy actions, a variable and a value are converted into the type of each class in PCIMe. To illustrate the basic idea of converting a policy rule into PCIMe, here we use a simple example.

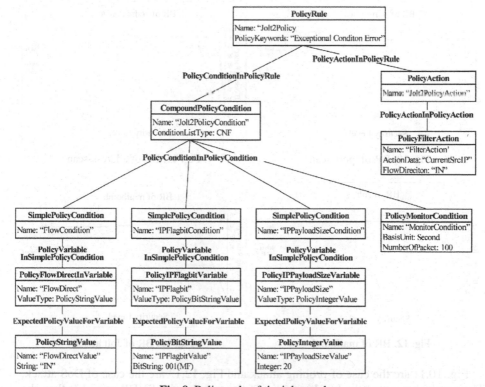

Fig. 9. Policy rule of the jolt attack

Jolt attack is a type of DoS attacks. It slices a IP datagram into small pieces and then transmits the packets of these pieces to the target system. As a result, the utilization of the CPU reaches close to 100 percents and can't handle any other processes. We obtain the following rule of jolt2 attack through rule induction: "IF (FlowDirection = IN) ? (IPFlagbit = MF) ? (IPPayloadSize <= 20) ? (NumOfPacketPerSecond > 100) THEN Filtering(CurrentSrcIP, IN)".

Fig. 9 shows an object model to represent the security policy for jolt attack. The classes comprising the PCIMe are intended to serve as an extensible class hierarchy (through specialization) for defining policy objects that enable network administrators and policy administrators to represent policies of different types. The PolicyIPPayloadsizeVariable class, PolicyMonitorCondition and PolicyFilterAction are inserted into the inheritance hierarchy the original PCIMe classes. This object model is distributed to network devices to be enforced, and is used to map into proper security configurations.

5 Simulation Result

Two main categories of attacks have been simulated, these are DOS (jolt, mailbomb) and Probing (address-scan, port-scan) attack. Blocking ratio (BR) is measured for the performance indexes in the simulation.

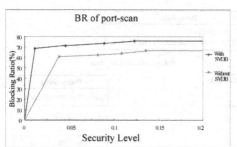

Fig. 10. BR of port-scan

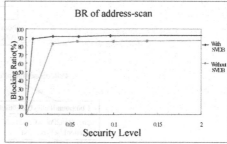

Fig. 11. BR of address-scan

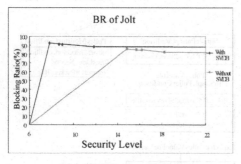

Fig. 12. BR of mailbomb

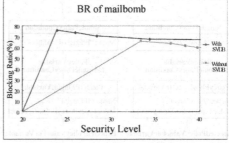

Fig. 13. BR of Jolt attack

Fig. 10,11 are the case of probing attack and Fig. 12,13 are the case of DoS attack. The result shows that the blocking ratio of the system with SVDB is higher than the blocking ratio of system without SVDB. The proposed system detects effectively the attacks through the additional system information in SVDB. Fig. 10 shows that the blocking ratio is increased by strengthening of the security level. Probing attacks have relatively limited variance because they involve making connections to a large number of hosts or ports in a given time. On the other hand, DoS has a wide variety of behavior because they exploit the weaknesses of a large number of different network or system services. As a result, the proposed system provides the added accuracy and efficiency of security systems.

6 Conclusion

We presented a policy-based network simulation environment for the security system. The security system makes a various network situations-the policies should be applied to change the network states. These situations include the response of intrusion detection system and policy change by the firewall, etc. We also proposed the structure for policy rule induction form vulnerabilities stored in SVDB. The security policy rule provides the added accuracy and efficiency in safe guarding the network. The simulation environment can be a good tool to analyze or test a policy, it can help a manager to know that if the applied policies work as expected, and further, it can optimize use of the current network infrastructure.

References

1. Wang Changkun, "Policy-based network management," Communication Technology Proceeding, 2000. WCC-ICCT 2000, International Conference on, Vol. 1. pp. 101-105. Aug. 2000.
2. Verma, D.C., "Simplifying network administration using policy-based management," Network, IEEE, Vol 16, pp 20-26, March-April. 2002.
3. F. Cohen, "Simulating Cyber Attacks, Defences, and Consequences," Computer & Security, Vol.18, pp. 479-518, 1999.
4. Dinesh C. Verna. *Policy-Based Networking: Architecture and Algorithm*, New Rider, 2001.
5. Dave Kosiur. *Understanding Policy-Based Networking*, John Wiley & Sons, Inc. 2001.
6. B. Moore, et al., "Policy Core Information Model-Version 1 Specification," IETF RFC 3060, Feb 2000.
7. E. D. Zwicky, S. Cooper and D. B. Chapman, *Building Internet Firewalls second edition*, O'reilly & Associates, 2000
8. B.P. Zeigler, H. Praehofer, and T.G. Kim, *Theory of modeling and simulation: Integrating discrete event and continuous complex dynamic system*, San Diego: Academic Press, 2000.
9. B. Moore, et al., "Policy Core Information Model (PCIM) Extensions," IETF RFC 3460, Jan 2003.
10. NIST, "An Introduction to Computer Security : The NIST Handbook," Technology Adminstration, U.S.A, 1995.
11. M. Bishop, "Vulnerablities Analysis," Proceedings of the Recent Advances in Intrusion Detection pp. 125-136 Sep. 1999.
12. Robert A. Martin, "Managing Vulnerabilities in Networked Systems," IEEE Computer, Vol.34, No.11, pp. 32-38, Nov. 2001.
13. http://icat.nist.gov, ICAT Metabase Zhengxin Chen, *Data Mining And Uncertain Reasoning: An Integrated Approach*, John Wiley & Sons, 2001.

Timestamp Based Concurrency Control in Broadcast Disks Environment

Sungjun Lim and Haengrae Cho

Department of Computer Engineering, Yeungnam University,
Gyungsan, Gyungbuk 712-749, Republic of Korea
hrcho@yu.ac.kr

Abstract. Broadcast disks are suited for disseminating information to a large number of clients in mobile computing environments. In this paper, we propose a *timestamp based concurrency control* (TCC) to preserve the consistency of read-only client transactions, when the values of broadcast data items are updated at the server. The TCC algorithm is novel in the sense that it can reduce the abort ratio of client transactions with minimal control information to be broadcast from the server. This is achieved by allocating the timestamp of a broadcast data item adaptively so that the client can allow more serializable executions with the timestamp. Using a simulation model of mobile computing environment, we show that the TCC algorithm exhibits substantial performance improvement over the previous algorithms.

Keywords: Mobile computing, broadcast disk, concurrency control, transaction processing, performance evaluation

1   Introduction

Broadcast disks are suited for disseminating information to a large number of clients in mobile computing environments [1, 5]. In broadcast disks, the server continuously and repeatedly broadcasts all the data items in the database to clients without specific requests. The clients monitor the broadcast channel and read data items as they arrive on the broadcast channel. The broadcast channel then becomes a *disk* from which clients can read data items. Due to its capability of involving unlimited number of clients, broadcast disks can support large-scale database applications in a mobile environment like stock trading, auction, electronic tendering, and traffic control information systems [2].

While data items are broadcast, applications at the server may update their values. Then the updated data items are broadcast at the next cycle. This means that the execution results of client applications would be inconsistent if the applications span multiple broadcast cycles. As a result, the broadcast disks require a *concurrency control algorithm* to preserve the consistency and currency of client applications [2, 9, 11]. Note that the traditional concurrency control algorithms such as 2-phase locking and optimistic concurrency control cannot be applied in these environments. This is because they require extensive message passing between the mobile clients to the server [4].

T.G. Kim (Ed.): AIS 2004, LNAI 3397, pp. 333–341, 2005.
© Springer-Verlag Berlin Heidelberg 2005

In this paper, we propose a *timestamp based concurrency control* (TCC) algorithm for read-only client transactions in broadcast disks. While a few concurrency control algorithms have been proposed recently in broadcast disks [8, 9, 12, 11], their primary concern is to reduce unnecessary transaction aborts by using considerable amount of downlink communication from the server to clients for transferring control information. On the other hand, the TCC algorithm requires minimal control information to be broadcast. This comes from the adaptive timestamp allocation strategy for broadcast data items so that each client can allow more serializable executions with the timestamp.

The rest of this paper is organized as follows. Sect. 2 reviews the previous concurrency control algorithms. Sect. 3 presents the proposed algorithm in detail. We have evaluated the performance of the algorithm using a simulation model. Sect. 4 describes the simulation model and analyzes the experiment results. Concluding remarks appear in Sect. 5.

2   Related Work

Note that access to a broadcast data item is strictly *sequential*, since a client transaction needs to wait for the data item to appear on the channel. Furthermore, in most cases, it is not possible to determine the set of data items read by the client transaction before its execution. This means that the client transaction may read data items from different broadcast cycles, which broadcast values from different database states. As a result, the execution of the client transaction can be interleaved with several server update transactions, which might result in inconsistent transaction executions.

Fig. 1 illustrates the inconsistent transaction executions. We will use a term "bcycle" to refer a broadcast cycle. A server transaction ST_1 updates a data item x after a client transaction CT_1 reads x at k^{th} bcycle. This implies that the effective execution order of CT_1 should precede that of ST_1, i.e. $CT_1 \rightarrow ST_1$. Then another server transaction ST_2 updates a data item y using the value of x that was updated by ST_1. As a result, $ST_1 \rightarrow ST_2$ holds. Since ST_1 and ST_2 commit at k^{th} bcycle, their results are broadcast at $k + 1^{th}$ bcycle. Now CT_1 reads y that was updated by ST_2; hence, $ST_2 \rightarrow CT_1$ holds. Then the total execution order becomes a cycle, $CT_1 \rightarrow ST_1 \rightarrow ST_2 \rightarrow CT_1$, and CT_1 should be aborted and restarted.

A problem that makes a concurrency control in broadcast disks be difficult is that clients cannot distinguish the scenario of Fig. 1 from the consistent executions. Suppose that ST_2 does not read y, and thus $ST_1 \rightarrow ST_2$ does not hold. So the resulting transaction execution is consistent. However, CT_1 should also be aborted, since CT_1 cannot inform the difference without expensive uplink communication to the server and CT_1 must assume the worst-case scenario.

To resolve this problem, a few concurrency control algorithms were proposed in broadcast disks [7–9, 12, 11]. In [8], the server broadcasts a transaction dependency graph and data items augmented with the identifier of the last transaction that updated it. In [9], and [12], old versions of data items are broadcast along

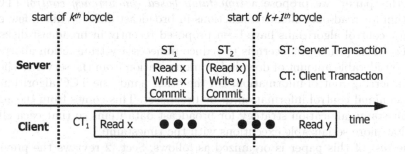

Fig. 1. A scenario of inconsistent transaction execution.

with current values. In [11], the server broadcasts data items along with a control matrix. Note that to reduce the transaction abort ratio and uplink communication, these algorithms require considerable amount of downlink communication for transferring concurrency control information. The control information will use up a substantial percentage of downlink bandwidth especially at large database application, since its size is proportional to the database size or the number of ever executing transactions. As a result, client transactions might suffer from long delay due to waiting the broadcast data items, and thus their response time could be much longer.

The BCC-TI algorithm [6] is closely related to our study in the sense that it requires minimal control information. The BCC-TI dynamically adjusts the position of a client transaction in the current serialization order by recording the timestamp interval associated with the transaction. The server broadcasts a *control information table* (CIT) at the beginning of each bcycle. The CIT includes the timestamps of server transactions committed during the previous bcycle and their write set. When a client transaction reads a data item or CIT, its timestamp interval is adjusted to reflect dependencies between the client transaction and committed server transactions. If the interval becomes invalid (lower bound ≥ upper bound), an inconsistent execution is detected and the transaction is aborted. However, the BCC-TI cannot distinguish the scenario of Fig. 1 from the consistent execution and thus it may reject a consistent one [4].

3   Timestamp Based Concurrency Control

In this section, we propose a *timestamp based concurrency control* (TCC) algorithm. To reduce the number of unnecessary transaction aborts, the TCC algorithm allocates timestamp of a broadcast data item adaptively so that each client can allow more serializable executions.

3.1   Server Algorithm

When a server transaction commits, the transaction and all the data items in its write set are assigned timestamps. Fig. 2 summarizes the procedure of timestamp

A. The following procedure is performed when a server transaction, U, commits during the bcycle B_i.
   1. Assign current timestamp of the server to TS(U).
   2. If U is the first committed transaction at B_i, then $\forall d \in$ WS(U), set TS(d) to TS(U). Copy TS(U) into FIRST.
   3. Else determine whether there exists a dependency between U and any transaction U_c that has already committed at B_i.
      – If {RS(U) $\cup$ WS(U)} $\cap$ WS(U_c) $\neq \emptyset$ or WS(U) $\cap$ RS(U_c) $\neq \emptyset$, then set TS(d) into TS(U), $\forall d \in$ WS(U).
      – Otherwise, set TS(d) into FIRST, $\forall d \in$ WS(U).
   4. Append the pair of TS(U) and WS(U) into CIT.
B. The server broadcasts CIT at the beginning of the next bcycle B_{i+1}.

Fig. 2. Server algorithm of TCC.

allocation. For an update transaction U, WS(U) is a write set of U, RS(U) is its read set. TS(U) is a timestamp of U. TS(d) is a timestamp of a data item, d. The control information table (CIT) includes the timestamp of server transactions committed during the previous bcycle and their write set. The CIT allows each client to determine whether its transaction has introduced a non-serializable execution with respect to the committed server transactions.

The basic idea of Fig. 2 is to distinguish the serialization order of server transactions from their commit order. Note that the timestamp is monotonically increasing and each server transaction has its own timestamp when it commits. This means that the timestamps represent the commit order of server transactions. If we assign the timestamp of each data item to the timestamp of its updating transaction as the BCC-TI does, we could not exploit potential serialization orders those are different from the commit order. To distinguish between two orders, the TCC remembers another timestamp, FIRST. FIRST is the timestamp of a transaction committed first at the current bcycle. After that, when a transaction U commits and it does not depend on other committed transactions of the current bcycle, the timestamps of data items in WS(U) are set to FIRST not TS(U). Then the client may allow more serialization executions when it compares the timestamp of data items as the next section describes.

3.2   Client Algorithm

Similar to the BCC-TI [6], a client transaction is assigned a timestamp interval to detect non-serializable executions. For a client transaction Q, let LB(Q) and UB(Q) denote the lower bound and the upper bound of the timestamp interval of Q, respectively. If the timestamp interval shuts out, i.e. LB(Q) > UB(Q), Q has to be aborted. LB(Q) and UB(Q) are initialized to 0 and ∞, respectively.

When Q reads a data item d, step B of Fig. 3 is performed. The client first changes LB(Q) to the maximum of current LB(Q) and the timestamp of d. Q can proceed as long as the new LB(Q) is still smaller than UB(Q). If LB(Q) becomes larger than UB(Q), then there is a non-serializable execution by Q

A. The following procedure is performed when a client transaction Q is requested.
1. Timestamp interval is set to $(0, \infty)$.
2. $\text{LB}(Q) = 0$, $\text{UB}(Q) = \infty$.

B. The following procedure is performed when Q reads a data item d.
1. $\texttt{ReadSet}(Q) = \texttt{ReadSet}(Q) \cup \{d\}$.
2. $\text{LB}(Q) = \max(\text{LB}(Q), \text{TS}(d))$.
3. If $\text{LB}(Q) = \text{UB}(Q)$ and $d \in \texttt{INV}$, then Q is aborted and restarted.
4. If $\text{LB}(Q) > \text{UB}(Q)$, then Q is aborted and restarted.

C. The following procedure is performed when the client receives the control information table (CIT). $\texttt{INV}$ is initialized to empty before receiving each CIT.
1. For $\forall\, U_i \in \text{CIT}$, do the following steps.
   − If $\text{WS}(U_i) \cap \texttt{ReadSet}(Q) \neq \emptyset$, then $\text{UB}(Q) = \min(\text{TS}(U_i), \text{UB}(Q))$.
   − If $\text{UB}(Q)$ is changed, $\texttt{INV} = \texttt{INV} \cup \text{WS}(U_i)$.
2. If $\text{LB}(Q) > \text{UB}(Q)$, then Q is aborted and restarted.

Fig. 3. Client algorithm of TCC.

and Q has to be aborted. When $\text{LB}(Q)$ is equal to $\text{UB}(Q)$, TCC and BCC-TI behave differently. The BCC-TI has to abort Q whenever $\text{LB}(Q)$ is equal to $\text{UB}(Q)$. However, in the TCC, all of data items updated by independent transactions have the same timestamp ($\texttt{FIRST}$). So the TCC can continue Q even though $\text{LB}(Q)$ is equal to $\text{UB}(Q)$ if the equality comes from independent server transactions. The TCC implements this idea by maintaining an additional variable $\texttt{INV}$ at the client.

When the client receives a CIT at the beginning of the next bcycle, the read set of Q is compared with the write set of every committed server transaction in the CIT. If there is an overlap, the server transaction invalidates Q. So the client changes $\text{UB}(Q)$ to the minimum of current $\text{UB}(Q)$ and the timestamp of the server transaction. If $\text{UB}(Q)$ is changed as a result, the write set of the server transaction is appended to $\texttt{INV}$ (Step C.1). Now coming back to Step B.2, if $\text{LB}(Q)$ becomes equal to $\text{UB}(Q)$ after reading d, the client checks whether d is included in $\texttt{INV}$. If $\texttt{INV}$ does not include d, it means that the updating server transaction of d and the invalidating server transaction of Q are *independent* to each other. Then Q can proceed.

For example, in Fig. 1, we first suppose that ST_2 does not read y and thus the execution is consistent. We assume that $\text{WTS}(x)$ is 1 and current timestamp is 2 at the start of kth bcycle. Then during the kth bcycle $\text{TS}(\text{ST}_1)$ and $\text{TS}(x)$ are set to 2. $\texttt{FIRST}$ is also set to $\text{TS}(\text{ST}_1)$. The $\text{TS}(y)$ is set to $\texttt{FIRST}$ because ST_2 is independent of ST_1. At the client, the timestamp interval of CT_1 is set to $(1, \infty)$ after CT_1 reads x. At the start of $k + 1^{th}$ bcycle, $\text{UB}(\text{CT}_1)$ is updated to 2 and $\texttt{INV}$ becomes $\{x\}$. When CT_1 reads y, the timestamp interval becomes $(2, 2)$. Note that CT_1 can proceed since $\texttt{INV}$ does not include y. If ST_2 reads y and thus there is a dependency between ST_1 and ST_2, $\text{TS}(y)$ becomes $\text{TS}(\text{ST}_2)$ that is greater than 2. As a result, when CT_1 reads y the timestamp interval of CT_1 shuts out, and CT_1 is aborted and restarted.

Table 1. Experiment parameters.

	Server Parameters	
ServerDBsize	Number of data items in database	1000
STlength	Server transaction length	8
NumST	Number of server transactions per bcycle	4 ∼ 36
WriteProb	write operation probability	0.0 ∼ 1.0
	Client Parameters	
CTlength	Client transaction length	1 ∼ 9
OptDelay	Mean inter-operation delay	1
TranDelay	Mean inter-transaction delay	2
TRSizeDev	Deviation of transaction length	0.1

4   Performance Evaluation

We compare the performance of the TCC with the BCC-TI. In this section, we first describe the experiment model. Then we show the experiment results and analyze their meanings.

4.1   Experiment Model

The simulation model consists of a server, a client, and a broadcast disk. We assume that there is only one client because the performance of the concurrency control algorithm for read-only transactions is independent of the number of clients. Transactions access data items uniformly throughout the entire database. Table 1 shows the experiment parameters and their settings. The simulation model was implemented using the CSIM discrete-event simulation package [10].

A server transaction executes *STlength* operations, over which the probability of write operation is *WriteProb*. A client transaction executes read-only operations. The average number of operations per client transaction is determined by a uniform distribution between *CTlength* ± *CTlength* × *TRSizeDev*. For each operation, the client transaction waits for *OptDelay* broadcast units and then makes the next read request. The actual control of client transaction into the system is controlled by the *TranDelay* parameter, which is modeled as an exponential distribution.

The server fills the broadcast disk with data items at the beginning of a bcycle. Each bcycle consists of a broadcast of all the data items in the database along with the associated control information. The control information consists of a table containing the timestamps and write sets of the server transactions committed at the last bcycle and an array of length *ServerDBsize* containing the write timestamps of data items. Each element in the array is broadcast along with the corresponding data items. We do not consider the effects of caching in this performance study. Then if the client misses any data item in the current bcycle, it may have to wait for the next bcycle.

The performance metric used in the experiment are the *abort rate* and *transaction response time*. The abort ratio is the average number of aborts before

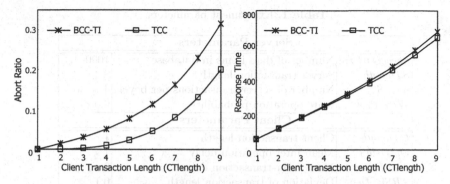

Fig. 4. Varying the length of client transactions.

the client transactions can commit. Transaction response time is measured as the difference between when a client transaction is submitted first and when the client transaction is successfully committed. The time includes any time spent due to restarts.

4.2   Experiment Results

We first compare the performance by varying *CTlength*. *NumST* is set to 16 and *WriteProb* is set to 0.5. Fig. 4 shows the performance results. As *CTlength* increases, both algorithms perform worse. This is because a client transaction has to wait for a larger number of bcycles and the abort ratio increases as a result. This is why the TCC outperforms the BCC-TI when *CTlength* is large. The TCC can reduce the number of unnecessary transaction aborts by allocating timestamp of a broadcast data item adaptively so that the client can allow more serializable executions.

We also compare the performance by varying *WriteProb*. Fig. 5 shows the experiment results where *NumST* is set to 8 and *CTlength* is set to 4 and 5. We see that as *WriteProb* increases, the abort ratio and the response time also increase. Furthermore, the TCC outperforms the BCC-TI at high *WriteProb*. Note that *WriteProb* affects the probability of data conflicts between the client transaction and server transactions. When *WriteProb* is high, the client transaction often conflicts with some server transactions and thus there are more chances to results in false dependency between transactions. This is why the performance improvement of the TCC becomes more distinguished when *CTlength* is set to the high value of 5.

We finally compare the performance by varying *NumST*. Fig. 6 shows the performance results when *WriteProb* is set to 0.5 and *CTlength* is set to 4. By increasing *NumST*, the possibility of data conflicts at the client should also increase. As a result, the abort ratio and the response time increase linearly. Similar to the previous experiment, the TCC performs better that the BCC-TI with respect to both the abort ratio and the response time. The performance difference becomes significant as *NumST* increases. This is because as more server transactions execute per cycle, BCC-TI should suffer from frequent aborts of client transactions due to false dependency with server transactions.

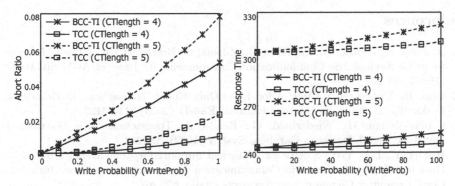

Fig. 5. Varying the write probability of server transactions.

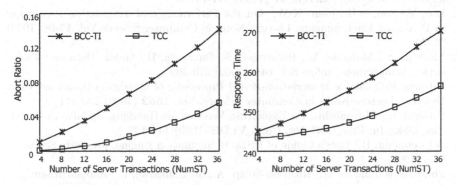

Fig. 6. Varying the number of server transactions (per cycle).

5   Conclusions

In this paper, we have proposed a timestamp based concurrency control (TCC) algorithm for client transactions in broadcast disk environments. To reduce the abort ratio of client transactions, most of previous concurrency control algorithms consume considerable amount of downlink bandwidth for transferring concurrency control information. Instead the TCC allocates the timestamp of a broadcast data item adaptively so that the client can allow more serializable executions with the timestamp. As a result, the TCC can minimize concurrency control information with the reduced abort ratio.

We have compared the performance of the TCC with the BCC-TI algorithm that can also minimize the downlink bandwidth. The TCC performs better than or similar to the BCC-TI throughout the experiments. In particular, the performance improvement is significant (a) when the client transaction is long, (b) when the number of concurrent server transactions is large, or (c) when the server transaction executes a lot of write operations. This result is very encouraging with regard to ever increasing complexity of information system.

References

1. Acharya, S., Alonso, R., Franklin, M., Zdonik, S.: Broadcast Disks: Data Management for Asymmetric Communication Environment. In: Proc. of ACM SIGMOD (1995) 199-210
2. Cho, H.: Concurrency Control for Read-Only Client Transactions in Broadcast Disks. IEICE Trans. on Communications. **E86-B** (2003) 3114-3122
3. Garcia-Molina, H., Wiederhold, G.: Read-Only Transactions in a Distributed Database. ACM Trans. on Database Syst. **7** (1982) 209-234
4. Huang, Y., Lee, Y-H.: STUBcast - Efficient Support for Concurrency Control in Broadcast-based Asymmetric Communication Environment. In: Proc. 10th Int. Conf. Computer Commun. and Networks (2001) 262-267
5. Jing, J., Heral, A., Elmagarmid, A.: Client-Server Computing in Mobile Environments. ACM Comp. Surveys **31** (1999) 117-157
6. Lee, V., Son, S-H., Lam, K-W.: On the Performance of Transaction Processing in Broadcast Environments. Lecture Notes in Computer Science Vol. **1748** (1999) 61-70
7. Madrina, S., Mohania, M., Bhowmick, S., Bhargava, B.: Mobile Data and Transaction Management. Infor. Sci. **141** (2002) 279-309
8. Pitoura, E.: Scalable Invalidation-Based Processing of Queries in Broadcast Push Delivery. Lecture Notes in Computer Science. Vol. **1552** (1999) 230-241
9. Pitoura, E., Chrysanthis, P.: Exploiting Versions for Handling Updates in Broadcast Disks. In: Proc. 25th Int. Conf. VLDB (1999) 114-125
10. Schwetmann, H.: User's Guide of CSIM18 Simulation Engine. Mesquite Software, Inc. (1996)
11. Shanmugasundaram, J., Nithrakashyap, A., Sivasankaran, R., Ramamritham, K.: Efficient Concurrency Control for Broadcast Environments. In: Proc. ACM SIGMOD (1999) 85-96
12. Shigiltchoff, O., Chrysanthis, P., Pitoura, E.: Multiversion Data Broadcast Organizations. Lecture Notes in Computer Science. Vol. **2435** (2002) 135-148

Active Information Based RRK Routing
for Mobile Ad Hoc Network*

Soo-Hyun Park[1], Soo-Young Shin[1], and Gyoo Gun Lim[2]

[1] School of Business IT, Kookmin University, 861-1, Jeongreung-dong,
Sungbuk-ku, SEOUL, 136-701, Korea
shpark21@kookmin.ac.kr, sooyoungshin@korea.com
[2] Department of Business Administration, Sejong University, 98 Kunja-dong,
Kwangjin-gu, Seoul, 143-747, Korea
gglim@sejong.ac.kr

Abstract. In mobile ad hoc network, unlike in wired networks, a path configuration should be in advance of data transmission along a routing path. Frequent movement of mobile nodes, however, makes it difficult to maintain the configured path and requires re-configuration of the path very often. It may also leads to serious problems such as deterioration of QoS in mobile ad-hoc networks. In this paper, we proposed a Reactive Routing Keyword (RRK) routing procedure to solve those problems. Firstly, we noticed it is possible in RRK routing to assign multiple routing paths to the destination node. We applied this feature into active networks and SNMP information based routing by storing unique keywords in cache of mobile nodes corresponding to present and candidate routings in a path configuration procedure. It was shown that the deterioration of QoS, which may be observed in DSR protocol, was greatly mitigated by using the proposed routing technique.

1 Introduction

One of the biggest differences of Mobile Ad-hoc Network (MANET) from backbone networks such as the Internet or mobile communication networks is that a mobile ad-hoc network can configure itself autonomically without fixed mediators.[1][2] In mobile ad-hoc networks, it is required to complete a path configuration in advance of data transmission through the pre-configured routing path. Frequent movement of mobile nodes, however, makes it difficult to maintain the pre-configured path and requires re-configuration of a routing path very often. While mobile nodes are moving, fatal transmission errors may occur to increase the packet drop rate during a new path discovering process and cause consequent loss of messages.[1-4] This can deteriorate the overall QoS of mobile ad-hoc networks. In addition, frequent movement of mobile nodes can cause a waste of resources during the path management process. In this paper, we proposed a Reactive Routing Keyword (RRK) routing procedure to solve these problems. Firstly, we noticed it is possible in RRK routing to configure multiple routing paths to a destination node. We applied this feature into active networks [5][6] and information based routing by storing a unique keyword in cache of

---

* This work was supported by the faculty research program of Kookmin University in Korea.

mobile nodes corresponding to present and candidate routings in a path configuration procedure. It was shown that the deterioration of QoS, which may be observed in DSR protocol, was greatly mitigated by using the proposed routing technique. [7-9]

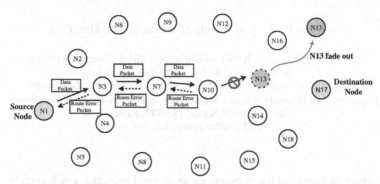

Fig. 1. An example of Route Maintenance Using RRER Packet

2 Dynamic Resource Routing (DSR) Protocol and Its Problems

DSR (Dynamic Source Routing) protocol is one of the various routing protocols that can be applied to mobile ad-hoc networks. DSR protocol has two stages – the route discovery and maintenance stage.

Main problems of DSR protocol, which are described as the followings, arise in a route maintenance procedure and they should be solved or mitigated. The first problem is that if a fatal error occurs by the moving of an intermediate node that is listed in the route records, and then a route discovery procedure should be executed again using the route error packet to deteriorate the overall QoS of the network. For example, in Fig.1, fading out of node N13 causes node N10 not to be able to broadcast packets to the next node N13. Then, N10 will broadcast a route error packet that includes the address of error-occurred nodes, N10 and N13, to a source node SN. Intermediate nodes, including the source node SN, will remove the corresponding links in their cache. At this time, if a source node N1 has another route to a destination node N18 in its cache, the new route is used to broadcast packets. If not, a route discovery procedure has to be executed.

The second problem is that the packet drop rate of sending messages increases and the consequent loss of messages occurs while a new route discovery mechanism is being executed. Furthermore, frequent moving of nodes will cause a waste of resources for route maintenances.[10]

3 Reactive Routing Keyword (RRK) Based Routing

3.1 Proactive Routing Keyword (PRK) Management

To resolve the problems of conventional DSR protocol, we proposed Reactive Routing Keyword (RRK) Management and RRK Routing Procedure based on the fact that multiple routing paths to a destination node are possible to be configured in a route discovery procedure of RRK routing. Keyword is required to be managed uniquely in a sub or local ad hoc network only. Keywords broadcasted in an information based routing should accord along a route between a source and a destination node for accurate transmission. Therefore, we propose a keyup procedure to manage keywords information centrally.

A keyup procedure is a procedure of updating keywords and uses Simple Network Management Protocol (SNMP) of conventional network management system. Therefore, an access to Management Information Base (MIB) is implemented through SNMP PDU (Protocol Data Unit). [11] However, a mirror MIB that contains certain information such as modified MIB is added to the manager for central management of keywords.

3.2 RRK Routing Procedure

3.2.1 Route Discovery Step During RRK Based Routing

In DSR protocol, even though multiple routes exist because RREQs are broadcasted between a source node and intermediate nodes, destination node only replies to the first received RREQ using RREP. Therefore, a path in the source node is unique. In RRK routing program, all messages(RREQ, RRSP) related to routing is configured as active packets to include RRK information. The proposed program executes a similar route discovery procedure as a conventional DSR protocol does. A main difference in the procedure of the proposed program from DSR protocol is that a unique RRK, which was previously generated, is added to RREQ packets. When a destination node received RREQ then the node will reply with RREP, to which keyword confirm fields are added, according to the route records that were stored in the received RREQ. At a source node, only the first arriving RREP is taken and the next ones are discarded. By means of this mechanism, multiple routes from a source to a destination node shall have Reactive Routing Keyword(RRK). Intermediate nodes are consisted of active nodes and have an execution environment that store and process RRK. The following RRK algorithm explains the route discovery mechanism in a RRK based routing.

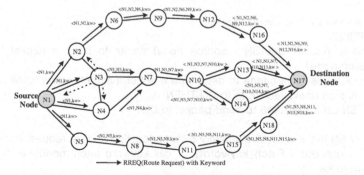

Fig. 2. Building Record Route during Route Discovery with Keyword

Fig. 2 and 3 shows a procedure of RRK-based route discovery. A source node N1 is going to transmit data packets to a destination N17. Firstly, a source node N1 checks whether it has already a routing path. If it has the path then the data will be transmitted through the path or a new route discovery procedure has to be executed.

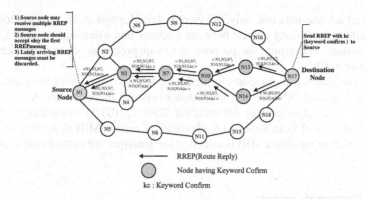

Fig. 3. Propagation of Route Replies with the Route Record & keyword

Followings are the procedure of RRK-based route discovery. Firstly, RREQ active packets are constituted at a source node after a unique keyword is issued from the keyword management system. At this time, this keyword is inserted to the data part of REQ active packet. RREQ active packets are broadcasted to neighbor nodes of N2, N3, N4, N5. Since node N2 that corresponds to HOP 1 in the network is not a destination node, there is not a RREQ's <source node address, unique identification number> in its route discovery table and its address does not exist in the route record of RREQ. So N2 stores its address in the route record of RREQ and broadcast RREQ again to N1, N3 and N6 that are in the range of its propagation. Proceeding as like this scheme, a source node N1 picks the first RREP in various received RREPs and other RREPs are discarded. Through this mechanism, all multiple routes from a source to a destination node get Reactive Routing Keyword(RRK). In the example, N1 is supposed to select <N1, N3, N7, N10, N13, N17> as a route. RRK algorithm is as follows.

■ RRK(Route Record with Keyword) algorithm
Algorithm RRK()
//   Now a mobile node(SN : source node) wants to send a packet to some DN(destination node)
1.   SN consults its route cache to determine whether it already has a route to DN
2.   IF SN has an unexpired route to DN **THEN**
2.1   SN uses this route to send packet to DN
3   **ELSE**
             // SN initiates route discovery by broadcasting a route request packet
3.1   keyword = **Fetch_keyword()**; // call keyword fetch module and get new unique keyword kw
             // call DSR module and bypass new keyword keyword to get novel route record.
3.2   Call **Keyword_SEND_in_SN(keyword)**;
4.   **ENDIF**
5.   Send data packet to DN along with route record
END RRK

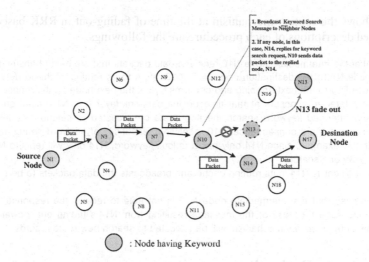

Fig. 4. Routing Mechanism during Fading-out in RRK based Routing

3.2.2 Routing Procedure and Route Maintenance Algorithm

In the stage of route maintenance in Fig. 4, if the path from SN to DN is disconnected by N13's moving to another region, instead of selecting candidate paths and executing routing again or rerouting by sending RERR to a source node, the node at the moment of path disconnection (N10) will broadcast keyword search request messages to neighboring nodes in search of moving nodes with the keyword information. The next step is to go on routing to a neighbor node replying with keyword search response message. Next RMK(Route Maintenance with Keyword) algorithms shows this mechanisms as follows.

■ RMK(Route Maintenance with Keyword) algorithm
Algorithm RMK()
```
//     Now an IN(Intermediate Node) receives data packet going to DN.
1.     IN consults its route cache to determine where is the next hop
2.     IF data link layer encounters a fatal transmission problems THEN
2.1        IN broadcasts keyword_search message with it's keyword to neighbor nodes.
2.2        Wait response_of_keyword_search message.
2.3        IF IN receives response_of_keyword_search message from neighbor node
N THEN
2.3.1          The hop in error is removed from IN's route cache
2.3.2          Forward data packet to neighbor node N
2.4        ELSE
2.4.1          Execute legacy DSR rerouting mechanism
2.4.2          END of Process
2.5        ENDIF
3.     ELSE
3.1        Forward data packet to next hop along route cache
4.     ENDIF
end RMK
```

Fig. 4 shows the routing mechanism at the time of fading-out in RRK-based routing. Detailed descriptions of each procedure are the followings.

[**Step 1**] After the intermediate node N10 received data packets that are being transmitted from the source node N1 to the destination node N17, N10 refers to its routing cache to determine a node corresponding to the next hopping and confirms N13 is the next node on the routing path.

[**Step 2**] If a fatal error occurs at that time on the data link layer (i.e. N13 fades out on the path), keyword–attached keyword_search messages are broadcasted to neighbor nodes in the RRK routing scheme. As a consequence, N10 receives a response_ of_keyword_search message from N14. (In this example, since N14 only has a unique keyword, N14 only will reply to N10 with response_of_keyword_search)

[**Step 3**] N10 discards N14 in its routing cache and broadcasts the data packets to be transmitted to N14.

[**Step 4**] In case that the intermediate node N10 is not able to receive the response_of_keyword_search message (i.e. loss of the message resulting from N14's fading out, power outage and etc.), the route discovery mechanism will be executed to search new route records.

4   Performance Evaluation

4.1   RRK Based Routing Metrics

Moving of mobile nodes is in a series of event of random moving in time domain and obeys Poisson distribution. Supposing t as an arbitrary starting time, the elapsing time of a mobile node I's next moving, T_i, will follows the following probability. [12]

$$P(T_i \succ t) = p_0(t) = e^{-\lambda t} \tag{1}$$

Let's suppose that n mobile nodes in an ad-hoc network are uniformly distributed in a unit area configured as a four-cornered topograph and two arbitrary nodes in a unit area are distributed within a distance of $r(n)$. Then a probability model consisting of the following random graphs can be obtained. Let I_i be an event in case that an arbitrary node i in a given unit area fades out and a wireless link connection is expired as a consequence (i.e. node isolation). Node isolation probability of at least one node can be expressed as $p(\bigcup_{i=1}^{n} I_i)$. Areas, which an arbitrary node is inside, will be at

least $\frac{1}{4}r\, r^2(n)$ (every i, $1\# \ i\# \ n$)

$$P(I_i)\# \left(1 - \frac{1}{4}r\ r^2(n)\right)^{(n-1)} \tag{2}$$

Definition of $p(\bigcup_{i=1}^{n} I_i)$ using the above relation will result in the following probability formula

$$P(\bigcup_{i=1}^{n} I_i) \# \sum_{i=1}^{n} P(I_i)$$

$$\# \ n(1 - 1/4r\ r^2(n)^{(n-1)} = e^{\{\ln n + (n-1)\ln(1 - 1/4r\ r^2(n))\}} \tag{3}$$

$$= e^{\{\ln n - (n-1)\, i\ (1/4r\ r^2(n))\}}$$

$$= e^{\{\ln n (1 - \frac{(n-1)}{\ln n}\, i\ (1/4r\ r^2(n))\}}$$

From the equation (3), we can know that if n converges to an infinite value, then $r(n)$ will converge to zero[2]. $\Theta(n)$ means the number of nodes that are possible to transmit simultaneously in ad-hoc networks. If a distance between two arbitrary nodes in ad-hoc networks, $r(n)$, gets smaller than $\sqrt{\frac{\ln n}{n}}$, the probability of closing of wireless network of an arbitrary node in the network will go to zero(n " 3). To maintain the connection of an arbitrary node (i.e. fading-out probability is zero), therefore, $r(n)$ should be decreased than $\sqrt{\frac{\ln n}{n}}$ gradually. To maintain the connection of two arbitrary nodes in a topograph, $r(n)$ should have the following values in case that n and $c(n)$ goes to an infinite value. And $c(n)$ will converge to an infinite value gradually by arbitrary speed .[13][14]

$$\iota^c \sqrt{\frac{\ln n + c(n)}{n}} \, m \tag{4}$$

4.2 Simulation

To verify the performance of proposed algorithm comparing with the conventional DSR algorithm, a series of computer simulations using NS-2 (Network Simulator 2) was conducted. The number of dropped packets and the throughput were measured in various scenarios with CBR (Constant Bit Rate) traffic on UDP/IP. Simulation parameters are shown in Table 1.

Table 1. Simulation Parameters

Parameter Lists	Value	Parameter Lists	Value
Time	1000 sec	Agent	UDP
Movement #1[a] Start	600 sec	Traffic	CBR
Movement #1 Stop	779 sec	CBR interval	2~8 sec / node
Movement #2[b] Start	780 sec	CBR Pkt size	510 bytes
Movement #2 Stop	999 sec	IF_Queue size	50
Nodes	50	Random	Yes
Max Links	20	Max pkts	10000 / node

[a] Movement #1: All nodes move between Start and Stop.

[b] Movement #2: Arbitrary nodes move intermittently between Start and Stop

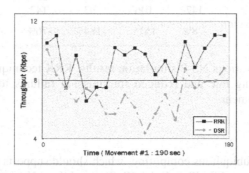

Fig. 5. Throughput in route reconfiguration factor generating section (CBR interval; 8 sec/ node)

Fig. 5 shows the measured throughput in case of the movement #1 scenario. In all situations, RRK shows better throughput then DSR especially in case of lower traffic loads. It is estimated that since RRK algorithm transmits and receives control packets such as key search broadcasting, key response procedure and etc., the consequent extra overheads causes the reduced performance improvement under busier traffic condition. In case that CBR interval is 8 seconds, maximum 32.72 % of performance improvement was measured.

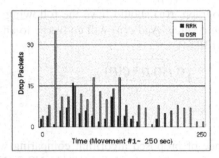

Fig. 6. Number of transmission failures (CBR interval : 8 sec / node)

Number of transmission failures of DSR and RRK algorithm is compared and shown in Fig. 6. The proposed RRK algorithm shows better performance especially in before and behind 600 seconds section, in which the re-configuration of routing paths is executed frequently. Besides, the number of transmission failures in RRK algorithm increases as the traffic load become heavier, which means reduced CBR traffic interval. Especially when moving factors are generated, the transmission failure increased significantly (The worst case is 2 seconds of CBR interval).

Table 2 shows the comparison result of DSR and RRK with respect to drop rates.

Table 2. Drop rate (DSR vs. RRK)

Lists	Entire Path for Packet Tx (250~1000 sec)			Routing Path Reconstruction (590~779 sec)		
CBR interval (sec)	2	4	8	2	4	8
DSR	261	266	264	250	257	252
RRK	147	106	93	142	95	84
Tx Failure rate variation for RRK - DSR(%)	-78%	-151%	-184%	-76%	-171%	-200%

Judging from the various NS-2 simulation results, RRK technique showed the improved data processing rate and reduced transmission failures to show significant performance improvement.

5 Conclusion

Since the concept of ubiquitous computing, which should supports extensive wireless access system such as mobile IP, cellular IP, Personal Area Network (PAN) and etc., is entering the stage recently, MANET begins to be noticed as a framework which can serve wireless access systems in various access systems.

In this paper, we proposed an RRK routing procedure, which is based on active networks and SNMP information-based routing, to solve problems of loss of routing paths and an increasing packet drop rate. By storing a unique keyword in cache of mobile nodes corresponding to present and candidate routings in a path configuration procedures, problems which may observed in DSR protocol, such as QoS deterioration, are expected to be significantly dissolved.

In NS-2 simulations with various scenarios, RRK always showed better performance of throughput comparing with DSR in case of frequent occurrence of mobile node movement after the completion of routing paths configuration. Particularly under light traffic load conditions, RRK showed much better performance than DSR by up to 32.72 % of improvement. At about the time of 600 seconds, in which reconfigurations of routing paths occurs frequently, the proposed RRK technique showed the mitigated performance deterioration and the significant decrease in frequency of transmission failures comparing with DSR. Besides, the frequency of transmission failures in the proposed technique increases as traffic loads being increased- that is, CBR traffic interval being decreased and moving factors being increased. A throughput and frequency of transmission failures is in an inversely proportional relationship. In case of 8 sec/node of CBR interval, it is measured that the throughput was dropped by down to 200%. We have not considered the security in active networks corresponding to keyword transmission. Besides, for future improvement of the proposed routing technique, an additional mechanism for unique keywords management is required.

References

1. Charles E. Perkins, Ad Hoc Networking, Addison-Wesley, 2001.
2. "Chap 5. Performance Analysis of Wireless Ad Hoc Networks", The Handbook of Ad-hoc Wireless Networks", Anurag Kumar and Aditya Karnik, CRC PRESS LLC, 2003
3. IETF MANET Working Group, http://www.ietf.org/html.charters/manet-charter.html
4. Analysis for Electronic Telecommunication Trends, Vol. 18, No. 2, ETRI, 2003
5. D. L.Tennehouse and D. J. Wetherall, "Toward an active network architecture", ACM Computer Communication Review, 26(2):5-18, 1996
6. D. L. Tennenhouse, J. M. Smith, W. D. Sincoskie, D. J. Wetherall, and G. J. Minden, "A Survey of Active Network Research", IEEE Communications Magazine,
7. The Handbook of Ad-hoc Wireless Networks", Edited by Mohammed Ilyads, CRC PRESS LLC, 2003
8. IETF MANET Working Group, http://www.ietf.org/html.charters/manet-charter.html
9. Autoconfiguration Technology for IPv6 Ad-hoc Mobile Wireless Network, IPv6 Forum Korea Technical Doc., 2003-001
10. K.B. Park, "Active Network based Reliable Data Transmission for Mobile Ad hoc Network", Ph.D. Thesis in MyungJi University, Korea, 2002
11. Mark A. Miller, PE, "Inside Secrets SNMP Managing Internetworks", TA Press, 1998
12. Dharma P. Agrawal and Qing-an Zeng, Intorduction to Wirless and Mobile Systems, Thomson Brooks/Cole, 2003
13. P.Gupta and P.R. Kumar, 'Critical Powers for Asymptotic Connectivity in Wireless Networks", A Volume in Honor of W.H. Fleming in Stochastic Analysis, Control, Optimizations and Applications, 1998
14. P. Panchpakesan and D. Manjunath, "On Transmission Range in Dense Ad Hoc Radio Networks", Conference on Signal Processing and Communications SPCOM, IISc, bangalore, 2001

Applying Web Services and Design Patterns to Modeling and Simulating Real-World Systems

Heejung Chang and Kangsun Lee[*]

Department of Computer Engineering, MyongJi University
38-2 San Nam-dong, YongIn, Kyunggi-Do, Korea
ksl@mju.ac.kr

Abstract. Simulation models can be created completely or partly from web services 1) to reduce the development cost, and 2) to allow heterogeneous applications to be integrated more rapidly and easily. In this paper, we present software design patterns useful for modeling and simulation. We illustrate how model federates can be built from web services and design patterns. We also show how the use of software design patterns can greatly help users to build scalable, reliable and robust simulation models with detailed performance analysis.

1 Introduction

A web service is a software application identified by a URI(Uniform Resource Identifier), whose interfaces and binding are capable of being defined, described and discovered by XML artifacts and supports direct interactions with other software applications using XML based messages via internet-based protocols.[1] As the enabling technologies, such as UDDI, WSDL, SOAP, and WSFL [2 – 5] , become standardized, the web services are expected to provide the ideal platform to integrate business resources distributed around different systems.

Web services can be utilized in modeling and simulation in the following ways [6]:

- Whole models as web services: simulation model can be completely created by employing the available web services
- Environmental components as web services: components with infrequent interaction, such as Databases, Spreadsheets, and Visualization Tools, are the best candidates to be separated out as web services. These services may be provided by a third party rather than developed by the simulation community.
- Model Federates as web services: Model federates can be implemented as web services.

Advantages of applying web services to modeling and simulation come from the congruence between the objectives of web services and the objectives of distributed simulation [7]:

- Faster execution times of simulation experiment through distribution of runs and alternatives to banks of available processors
- Geographic distribution of the simulation to allow more convenient collaboration
- Integration of simulations on different hardware devices and operating systems

---

[*] Corresponding author.

T.G. Kim (Ed.): AIS 2004, LNAI 3397, pp. 351–359, 2005.
© Springer-Verlag Berlin Heidelberg 2005

Despite the advantages of web services, there have been limited researches on applying web services to modeling and simulation. Possible reasons are:

- Lack of guidelines on how to architect simulation models using web services
- Lack of empirical studies on how efficiently the QoS(Quality of Service) of simulation models can be improved by applying well-architect web services

Design patterns are about design and interaction of objects as well as providing a communication platform concerning elegant, reusable solutions to commonly encountered programming challenges [8]. Design patterns are the proven best practices that can be safely applied repeatedly to application architecture and system architecture. Our claim is that use of design patterns can greatly help the users to build scalable, reliable, and robust web services and improve the practicality of web services for modeling and simulation. Design patterns can exist at many levels from very low-level specific solutions to broadly generalized system issues, and are now hundreds of patterns available in the literature [8]. We believe that there might be many design patterns applicable to modeling and simulation domains as well.

In this paper, we illustrate how design patterns can be successfully applied to modeling and simulating applications. We also present how the QoS (Quality of Service) of the simulation model can be improved by using web services and design patterns.

The outline of the rest of this paper is as follows. Section 2 presents background study on design patterns and their QoS properties. Section 3.1 illustrates how web services are created by applying design patterns and used as simulation model federates through an example. Section 3.2 discusses how QoS of the simulation model can be improved by utilizing web services and design patterns. Section 4 concludes this paper with summaries and future works.

2   Background Research

In this section, we briefly review the most common software patterns available in object-oriented design and then discuss how the overall QoS can be improved by applying design patterns.

2.1   Overview of Design Patterns

Adapter, Façade and Proxy are the most common software patterns available in object-oriented design. Below is the brief explanation on Adapter, Façade and Proxy patterns.

Adapter Pattern: Adapter pattern provides exposing internal application functionality with a different interface. For example, the client must use an operation named re-quest. An Adaptee object has a specificRequest operation that performs the required function. We create an Adapter object that contains a reference to an Adaptee. The Adapter implements request to call the specificRequest method of the Adaptee [9].

Façade Pattern: Façade pattern provides a unified interface to a set of interfaces in a subsystem. Façade defines a higher-level interface that makes the subsystem easier to use. For example, in Fig 2, ServiceA, ServiceB and ServiceC classes each have operations. Façade class with a doAll operation invokes each of these three operations [9].

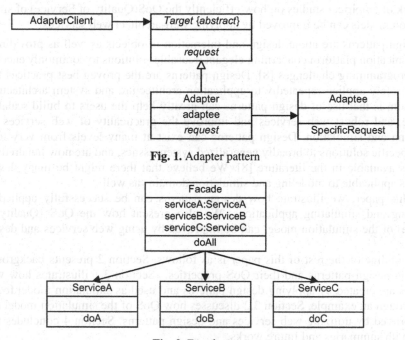

Fig. 1. Adapter pattern

Fig. 2. Façade pattern

Proxy Pattern: Proxy pattern provides a surrogate or placeholder for another object to control access to it. Figure 3 shows an example of Proxy pattern. A RealSubject has a request method that is not always accessible. For example, it may be on host that we cannot reach directly. Suppose we can only reach the RealSubject when the time is an even number. The Proxy class provides a request method that contacts the RealSubject when possible and returns another result otherwise. Both RealSubject and Proxy are subclasses of the abstract Subject Class [9].

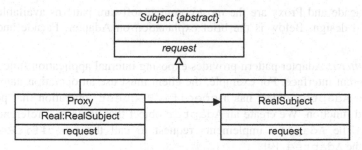

Fig. 3. Proxy pattern

More detailed information on design patterns can be found in References [8-10].

2.2 QoS of Design Patterns

The utility of software design patterns has matched to the design issues of availability, performance, security and interoperability [10].

- Availability: Availability is the property to access a service within a timely manner whenever they may need to do so [11]. Various design patterns provide alternative systems if faults should occur in original system. These patterns help to increase the availability of the original system.
- Performance: Performance can be defined as the ability of a system to deliver the results based on the request of the users, while keeping them satisfied [12]. Various design patterns are used to reduce network traffic and response time for coarse-grained and read-only data.
- Security: A security gives an administrator the power to determine who can access applications and their resources [13]. Many design patterns give limited access rights to an object and can validate the access permissions for that user or systems.
- Interoperability: Interoperability is the ability of a system or a product to work with other systems or products without special effort on the part of the customer [14]. Most design patterns provide approaches making use of a "wrapper" of systems that can convert system's interface into another system's interface on the fly.

In the next section, we will present design patterns useful for web services, and then apply them to simulation models.

3   Simulation Models with Web Services and Patterns

Consider a steam-powered propulsion ship named FULTON. In FULTON, the furnace heats water in boiler. Heat from the furnace is added to the water to form high-pressure steam. The high-pressure steam enters the turbine and performs work by expanding against the turbine blades. After the high-pressure steam is exhausted in the turbine, it enters the condenser and is condensed again into liquid by circulating sea water [15]. At that point, the water can be pumped back to the boiler.

Four components are identified to simulate FULTON: `Boiler`, `Turbine`, `Condenser`, and `Pump`. Section 3.1 illustrates how model federates can utilize design patterns and web services. Section 3.2 discusses how the QoS of the simulation model can be improved by applying web services and design patterns.

3.1   Applying Design Patterns and Web Services

Adapter, Façade and Proxy patterns can be utilized in building FULTON models.

3.1.1   Adapter Pattern

Motivation: An organization will often look to existing software assets to exploit Web services technologies. In some cases, these existing applications utilize languages, protocols, or platforms that are not compatible with the target Web services platform. Managing these incompatible interfaces can be solved through the use of the Adapter patterns. It would typically be used to provide compatibility between multiple platforms or systems. In the design of Web services architectures, the Adapter pattern can be used to expose existing components as Web services. This pattern can be used to wrap any incompatible interfaces with one the Web services platform understands [16].

In FULTON example, assume that `Boiler` and `Pump` are software components in C++, while `Turbine` and `Condenser` are independent software components implemented in Servlet. Figure 4 illustrates how different interfaces and the HTTP protocol can be utilized on a J2EE-based platform [9]. As shown in Figure 4, we can leverage many of our existing software assets. Adapter pattern is one approach to help resolve integration issues.

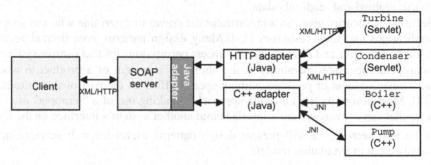

Fig. 4. Applying Adapter pattern to FULTON example

3.1.2 Façade Pattern

Motivation: The Façade pattern can often be applied to build coarse-grained services. The same approach can be leveraged for web services. The idea is to take existing components that are already exposed and encapsulate some of the complexity into high-level, coarse-grained services that meet the specific needs of the client [16].

In FULTON example, assume that `Ship` web service is a coarse-grained service that aggregates the already exposed web services of `Boiler`, `Turbine`, `Condenser`, and `Pump`. Figure 5 shows the relationships between `Ship`'s presentation and business tiers using the Façade pattern [16].

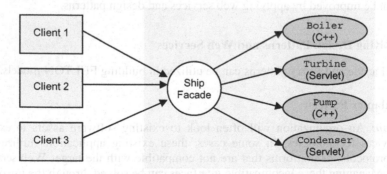

Fig. 5. Applying Facade pattern to FULTON example

3.1.3 Proxy Pattern

Motivation: The Proxy pattern can be used to isolate or move complex processing from the client to a back-end service. Instead of requiring the client to perform the processing, this work can be moved to a proxy server [16].

In FULTON example, assume that clients want to simulate FULTON example with multiple levels of abstraction [15]. For high level of abstraction, clients want to employ the full version of FULTON. However, for a low level of abstraction and possibly faster execution, the developer wouldn't want to interface with the real Pump system. Instead, they would typically write a "dummy" service that would be functionally equivalent to the original Pump. Figure 6 illustrates how mock objects could be used to achieve multiple levels of abstraction.

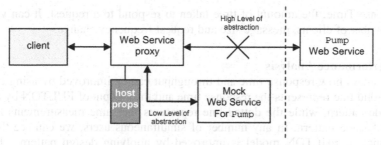

Fig. 6. Applying Proxy pattern to FULTON example

3.2 QoS Improvement

A number of QoS benefits are gained by using the design patterns and Web services to construct simulation models:

- **Interoperability:** As shown in FULTON example, managing incompatible interfaces of web services (e.g., C++, Servelet) can be solved through the use of the Adapter pattern.
- **Security:** When a client needs to interact with a Web service, say across a network, the most preferred way of encapsulating and hiding the interaction mechanism is by using a Proxy object.
- **Availability:** In e-business service, availability is very important issue. This problem can be solved using mirror service that is functionally equivalent to the original. If original Web services are not accessible, mirror service performs instead. Proxy pattern can be applied to achieve the desired availability.
- **Performance:** By using Façade pattern, we can reduce the number of network calls required by the client by reducing the number of client access points. Data is bundled together, typically mapping to a specific service needed by the client. In Section 3.2.1 – 3.2.2, we analyze how the performance of the FULTON model can be improved by applying design patterns, in terms of response time and throughput.

3.2.1 Testing Environment
Table 1 summarizes our testing environment.

Table 1. Performance Testing Environment

Tool	Microsoft Application Center Test
Machine configuration	1.4G CPU, 512 SDRAM, 40GB hard disk drivers.
OS configuration	XP, release version of .NET framework
Network configuration	100Mbps Ethernet LAN

Microsoft Application Center Test (ACT) [17] is designed to carry out stress test on web servers. Our performance measurements are as follows:

- Throughput: the number of requests that can be successfully served by application per unit time (second). It can vary depending on the load (number of users) applied to the web server.

- Response Time: the amount of time taken to respond to a request. It can vary as the number of clients, message size and request frequency change.

3.2.2 Performance Analysis

Figure 7 shows how response time and throughput can be improved by using Façade pattern. Solid line represents the response time and throughput of FULTON by applying Façade pattern, while the dotted line represents the same measurements *not* by applying Façade pattern. In any number of simultaneous users, we can see that the performance of FULTON model is improved by applying design patterns. This is because Façade pattern reduces the interactions between clients and web servers by centralizing the infrastructure.

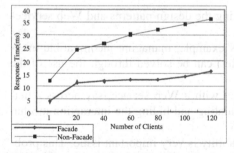 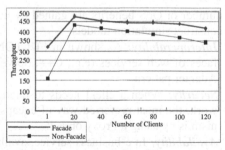

Fig. 7. Response time and throughput with varied number of simultaneous users: Applying Façade pattern improves the performance

Figure 8 shows how the performance of FULTON model can further be improved as the message size increases. Solid line represents the response time and throughput of FULTON by applying Façade pattern, while the dotted line represents the same measurements *not* by applying Façade pattern. In any message size, we can see that the performance of FULTON model is improved by applying design patterns. When message size becomes 100K, we further analyze the response time of FULTON with varying request frequency. More requests cause late response time in overall. However, in any cases, response time is greatly improved by applying design patterns. This is because Façade pattern reduces message exchanges and saves marshalling and unmarshalling overheads.

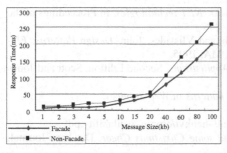 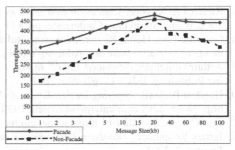

Fig. 8. Response time and throughput with a function of message size: Applying Façade pattern improves the performance

4 Conclusion

This paper presented a number of design patterns useful for modeling and simulation and illustrated possible QoS improvement as the result; Common software design patterns, such as Façade, Proxy and Adapter are successfully applied to building simulation models of FULTON. Design patterns are able to improve QoS of the simulation model in terms of interoperability, security, availability, and performance, as shown in FULTON example. Our experiences indicate that design patterns and web services could help users to build scalable, reliable, and robust simulation models. Design patterns are also good to validate the constructed models by applying the "proven" best practices.

In this paper, we explored only three design patterns (Adapter, Façade, Proxy) to prove their applicability and usefulness to modeling and simulation. We would like to continue our efforts to find other useful design patterns for modeling and simulation and report the benefits of using these patterns. We also would like to study new design patterns specifically for modeling and simulation and apply them to real-world systems.

Acknowledgement

This work was supported by grant No. R05-2004-000-11329-0 from Korea Science & Engineering Foundation.

References

1. W3C, *Web Services Activity*, http://www.w3c.org/2002/ws/
2. *Universal Discovery Description and Integration*, http://www.uddi.org
3. *Web Services Description Language*, http://www.w3.org/TR/wsdl
4. *Simple Object Access Protocol*, http://www.w3.org/TR/soap
5. *Web Services Flow Language*,
   http://www-3.ibm.com/software/solutions/webservices/pdf/WSFL.pdf
6. Senthilanand Chandrasekaran, Gregory Silver, Hoan A. Milller, Jorge Cardoso, and Amit P. Sheth, *Web service technologies and their synergy with simulation*, In Proceedings of the 2002 Winter Simulation Conference, pp. 606 – 615

7. Richard A. Kilgore, *Simulation web services with .NET technologies*, In Proceedings of the 2002 Winter Simulation Conference, pp. 841 - 846
8. Gamma, Eric, Helm, Richard, Johnson, Ralph, and Vlissides, John, *Design Patterns: Elements of Reusable Software,* Addison-Wesley, MA, 1995.
9. Arthur Gittleman, *Advanced Java™ Internet Applications*, Second Edition. Scott and Jones Inc Publishing, ISBN 157676096, 2001.
10. Hewlett-Packard, *Applying Design Issues and patterns in Web Services*, http://www.devx.com/enterprise/Article/10397
11. Eric Cronin, *Availability,* http://www.win.tue.nl/~henkvt/cronin-availability.pdf
12. James Koopmann, *Database Performance and some Christmas Cheer*, Database Journal. January 2, 2003.
13. IBM, *What is security?*, http://www-306.ibm.com/software/webservers/appserv/doc/v40/ae/infocenter/was/0018lite.html
14. Philip Hunter, *Interoperability*, Ariadne. Issue 24, June 2000.
15. Kangsun Lee and Paul A. Fishwick, A methodology for dynamic model abstraction, Transactions of the society for computer simulation international, vol. 13, no 4, 1997, pp. 217-229
16. Chris Peltz, *Applying Design Patterns to Web Services Architectures*, XML Journal, vol. 04, issue 6, 2003.
17. Microsoft. Application Center Test. http://msdn.microsoft.com/library/default.asp?url=/library/en-us/act/htm/actml_main.asp

Ontology Based Integration of Web Databases by Utilizing Web Interfaces

Jeong-Oog Lee, Myeong-Cheol Ko, and Hyun-Kyu Kang

Dept. of Computer Science, Konkuk University,
322 Danwol-dong, Chungju-si, Chungcheongbuk-do, 380-701, Korea
{ljo,cheol,hkkang}@kku.ac.kr

Abstract. The amount of information which we can get from web environments, especially in the web databases, has rapidly grown. And, many questions often can be answered using integrated information than using single web database. Therefore the need for integrating web databases became increasingly. In order to make multiple web databases to interoperate effectively, the integrated system must know where to find the relevant information which is concerned with a user's query on the web databases and which entities in the searched web databases meet the semantics in the user's query. To solve these problems, this paper presents an approach which provides ontology based integration method for multiple web databases. The proposed approach uses the web interfaces through which generally a user can acquire desired information in the web databases.

1 Introduction

One of the advantages of a multidatabase system is that it allows each local DBMS to participate in the system and at the same time it preserves the autonomy of the local DBMSs. However, making it possible for two or more databases to interoperate effectively has many unresolved problems. The most basic and fundamental problem is heterogeneity [1,2,3]. It can be categorized into platform heterogeneity and semantic heterogeneity. The semantic heterogeneity is due to the different database schemas (schema heterogeneity) and/or different logical representation of data (data heterogeneity).

Meanwhile, the amount of information which we can get from web environments, especially in the web databases, has rapidly grown. And, many questions often can be answered using integrated information than using single web database. Therefore the need for integrating web databases became increasingly. In order to make multiple web databases to interoperate effectively, the integrated system must know where to find the relevant information which is concerned with a user's query on the web databases and which entities in the searched web databases meet the semantics in the user's query.

This paper suggests a method to integrate databases, especially web databases. The proposed approach uses the web interfaces through which generally a user can acquire desired information in the web databases.

T.G. Kim (Ed.): AIS 2004, LNAI 3397, pp. 360–369, 2005.

Information in web databases can be provided to the users easily through web interfaces. With analyzing web interfaces, an information integration system can integrate web databases without concerning the database structures.

The rest of this paper is organized as follows. Section 2 shows an integrated retrieval system for web databases using ontology and multi-agents. Section 3 reports results of several experiments about the proposed system. Finally, in section 4, we offer our conclusions.

2   Integrating Heterogeneous Web Databases

2.1   Databases and Web Interfaces

Generally, a web site which provides information via databases is implemented in two phases. First, a database for information of the web site is designed. In next phase, web interface (or template) for retrieving information of the database is constructed. A user's query can be delivered to the query processor of the database through web interface. Figure 1 shows two web interfaces for plant related databases. With analyzing web interfaces, an information integration system can integrate web databases without concerning the database structures. The basic idea is to create virtual database tables for the databases using fields in web interfaces, which provides ease mechanism for integration regardless of database models, DBMSs, and so forth. That is, it needs not to consider platform heterogeneity. Figure 2 shows a sample virtual table for one of the example web databases in figure 1, Shartsmith Herbarium database. The integration system for web databases can integrate web databases easily using these virtual database tables, not needing to know all the actual database schema information.

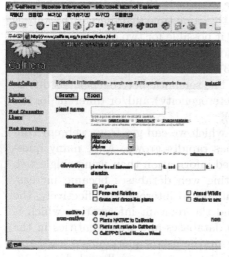

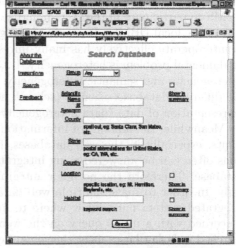

(a) CalFlora database.   (b) Starsmith Herbarium database.

Fig. 1. Sample databases with Web-based interfaces.

Group	Family	Name	Country	State	County	Location	Habitat
Dicots	Asteraceae	Achillea ptarmica L.	Iceland	...	...	...	wetland
Dicots	Ranunculaceae	Ranunculus repens L.	Iceland				wetland
...	...	...	...				...

Fig. 2. A sample virtual database table for Sharsmith Herbarium database.

2.2 SemQL: Semantic Query Language

Seen from a semantic perspective, the process of database design proceeds from the real world to the data. The designer develops his own conceptualization of the real world and turns his conceptualization into a database design. This has led designers to develop different, often incompatible, schemas for the same information. Therefore, users needing to combine information from several heterogeneous databases are faced with the problem of locating and integrating relevant information.

One of the efficient and effective approaches is allowing users to issue queries to a large number of autonomous and heterogeneous databases with his/her own concepts. It frees users from learning schemas. We have proposed SemQL as a semantic query language for users to issue queries using not schema information but concepts that the users know [4,5].

The SemQL is similar to SQL except that it has no FROM clause. The basic form of the SemQL is formed of the two clauses SELECT and WHERE and has the following form:

SELECT < conceptlist >
WHERE < condition >

Here < conceptlist > is a list of concepts whose values are to be retrieved by the query. The < condition > is a conditional search expression that identifies the tuples to be retrieved by the query. As for the entity information as described FROM clause in SQL, it is appended to the query automatically when the query processor processes the query.

The SemQL clauses specify not the entity or attribute names in component database schemas but concepts about what users want. For example, suppose a user wants to find those plants which live in wetland habitats, assuming that the user is familiar with SQL but knows neither of the component database schemas. Then the user might issue a query in SemQL using concepts that he/she knows;

SELECT plant.name
WHERE plant.habitat = "wetland"

Another user might issue a query as follows based on his/her concepts;

SELECT flora.name
WHERE flora.habitat = "wetland"

Though, in above two queries, "*plant*" and "*flora*" are different in the point of word form, the two queries are semantically equivalent.

2.3   The Procedure of Semantic Query Processing

An ontology is an explicit specification of a conceptualization [6,7]. When the knowledge of a domain is represented in a declarative formalism, the set of objects that can be represented is called the universe of discourse. This set of objects, and the describable relationships among them, are reflected in the representational vocabulary with which a knowledge-based program represents knowledge. Ontologies can be applied for inter-operability among systems, communications between human being and systems, increasing system's reusability and so forth. There are several endeavors within the industrial and academic communities that address the problem of implementing ontologies, such as TOP, WordNet, Ontolingua, Cyc, Frame Ontology, PANGLOSS, MikroKosmos, SENSUS, EngMath, PhysSys, TOVE, and CHEMICALS [8,9]. In our approach, ontology has a role of semantic network which is for extracting concepts from user's queries and integrating information of web databases.

Figure 3 shows an integrated retrieval system for Web databases using ontology technology and multiagents [10,11]. The *User Interface Agent* parses the user's query, extracts concepts in the query using ontological concepts and relations, requests for query processing, and displays the processed results to the user. The *Ontology Agent* manages the ontology and cooperates with the *Broker*

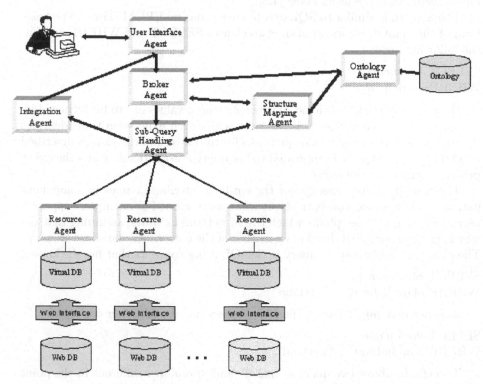

Fig. 3. An information retrieval system for web databases.

Agent and *Structure Mapping Agent*. The *Broker Agent* identifies the relevant Web databases using information received from the *User Interface Agent* and the *Ontology Agent*. The *Structure Mapping Agent* generates the mappings between concepts in the original query and information of the identified Web databases. The *Sub-Query Handling Agent* reformulates the original query into multiple sub-queries for each Web database according to the mappings generated by the *Structure Mapping Agent*. The *Resource Agent* resides in each Web database and executes the sub-query received from the *Sub-Query Handling Agent* according to the schema information of the Web database in which it resides, and then sends the results to the *Integration Agent*. The *Integration Agent*, in turn, manages the intermediate results from various Web databases and presents the integrated results to the users.

3   Experiments and Evaluation

To evaluate the proposed approach for information sharing in multidatabase systems, we have conducted some experiments on several web databases.

General information retrieval systems for web documents require the evaluation of how precise is the answer set. This type of evaluation is referred to as *retrieval performance evaluation*. Such an evaluation is usually based on a test reference collection and on an evaluation measure. The test reference collection consists of a collection of documents, a set of example information requests, and a set of relevant documents provided by specialists for each example information request.

Given a retrieval strategy S, the evaluation measure quantifies, for each example information request, the similarity between the set of documents retrieved by S and the set of relevant documents provided by the specialists. This provides an estimation of the goodness of the retrieval strategy S [12].

Applying the method of the *retrieval performance evaluation* for web documents, the experiments have been conducted in the aspect of retrieval performance which measures how effectively the user's semantic queries are processed through the proposed methods.

3.1   Implementation

As for information sources, we have chosen four Web databases related to plant information listed as follows:

- Component database 1 (CDB_1): CalFlora database at Berkeley University.
- Component database 2 (CDB_2): Sharsmith Herbarium database at San Jose State University.
- Component database 3 (CDB_3): Plants database developed by U.S. Department of Agriculture.
- Component database 4 (CDB_4): Species database constructed by Plants For A Future which is a resource centre for plants.

In Figure 1, web interfaces for CDB_1 and CDB_2 are already shown.

Query1: SELECT plant.name
         WHERE plant.habitat = "wetland"

Query2: SELECT plant.name
         WHERE plant.wetland = "T"

Fig. 4. An example of different query statements requesting the same information.

3.2 Contents and Methods for Experiments

A query may be issued in various forms according to the user's query patterns and provided web interfaces. Figure 4 shows such a case. Though both queries, Query 1 and Query 2, require the same information, they are different in query forms. Of course, Query 1 is a more general query form. However, according to the interfaces of some information sources, queries must be issued like Query 2. A module for query expansion expands the user's original query to all the possible forms of queries.

Experiments have been conducted to determine two things: the query processing time of the query processor, and query processing ability. In both cases, experiments have been conducted with and without query expansion.

Using the constructed system, all the possible queries were prepared for experiments based upon all the concepts in the component database schemas. Afterwards, each query was processed by the query processor, the results were analyzed, and the retrieval performance was evaluated.

In the experiments, given a set of queries, performance was evaluated according to the results of query processing in each component database. The meaning of the notations and equations used in the evaluation are as follows;

- Q: set of total queries.
- R_{CDB_i} (i=1,2,3,4): set of queries related to CDB_i, which must be processed by CDB_i.
- P_{CDB_i} (i=1,2,3,4): set of queries actually processed by CDB_i.
- $R_{CDB_i}, P_{CDB_i} \subseteq Q$.
- $Rp_{CDB_i} = R_{CDB_i} \cap P_{CDB_i}$.
- $|Q|$: the number of queries Q.
- $|R_{CDB_i}|$: the number of queries in R_{CDB_i}.
- $|P_{CDB_i}|$: the number of queries in P_{CDB_i}.
- $|Rp_{CDB_i}|$: the number of queries in Rp_{CDB_i}.
- $Recall_{CDB_i} = \frac{|Rp_{CDB_i}|}{|R_{CDB_i}|}$.
- $Precision_{CDB_i} = \frac{|Rp_{CDB_i}|}{|P_{CDB_i}|}$.
- $Error_{CDB_i} = 1 - Precision_{CDB_i}$.

R_{CDB_i} is extracted from the set of total queries, Q, which must be processed by CDB_i. P_{CDB_i} is a set of queries processed by CDB_i, after each query in Q is processed through the query processor. Rp_{CDB_i} is a set of queries that are included in the intersection set of R_{CDB_i} and P_{CDB_i}.

$Recall_{CDB_i}$ is the ratio of the actually processed queries to the set of queries which must be processed by CDB_i, $Precision_{CDB_i}$ is the ratio of relevant queries to CDB_i to the set of queries which are actually processed by CDB_i. $Error_{CDB_i}$ is the ratio of irrelevant queries to CDB_i to the set of queries which are actually processed by CDB_i, which is the probability of the query processor issuing errors at CDB_i during the query processing procedure.

3.3   Results and Evaluation

The number of total queries involved in the experiments ($|Q|$) is 134. The results are shown below.

- The results of the experiments with query expansion
  $|R_{CDB_1}| = 56, |P_{CDB_1}| = 54, |Rp_{CDB_1}| = 54$
  $|R_{CDB_2}| = 55, |P_{CDB_2}| = 27, |Rp_{CDB_2}| = 27$
  $|R_{CDB_3}| = 23, |P_{CDB_3}| = 23, |Rp_{CDB_3}| = 22$
  $|R_{CDB_4}| = 87, |P_{CDB_4}| = 85, |Rp_{CDB_4}| = 85$
- The results of the experiments without query expansion
  $|R_{CDB_1}| = 56, |P_{CDB_1}| = 31, |Rp_{CDB_1}| = 31$
  $|R_{CDB_2}| = 55, |P_{CDB_2}| = 27, |Rp_{CDB_2}| = 27$
  $|R_{CDB_3}| = 23, |P_{CDB_3}| = 17, |Rp_{CDB_3}| = 16$
  $|R_{CDB_4}| = 87, |P_{CDB_4}| = 35, |Rp_{CDB_4}| = 35$

According to the acquired primary data and the equations in section 3.2, recall rates, precision rates, and error rates for each component database were calculated. Recall rates for each component database are shown in figure 5 in graphical form.

With query expansion, the recall rate for CDB_1 is 96.43% and very excellent, while the recall rate for CDB_2 is 49.09% and is not good compared to that of

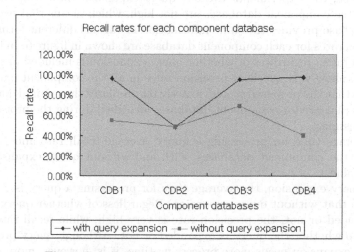

Fig. 5. Recall rates for each component database.

CDB_1. This is because that almost all the input fields of the web interface for CDB_1 are structured such as to enable users to select one of the values for each input field (see figure 1(a)), while all the input fields of the web interface for CDB_2 require users to insert values in the fields (see figure 1(b)).

That is, CDB_1 provides more schema information than CDB_2. If one observes carefully the result of the experiments, one can understand that the web DB which provides more and detailed schema information shows very high recall rate. As almost all the input fields of the interface for CDB_3 are structured such as to enable users to select one of the values for each input field, CDB_3 also shows a very high recall rate. As the same reason, CDB_4 shows a similar level of recall rate.

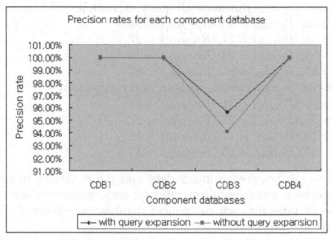

Fig. 6. Precision rates for each component database.

Meanwhile, the experiments without query expansion show that the recall rates of all the component databases are not high, which means that each component database provides the same information in various different forms.

Precision rates for each component database are shown in figure 6. In the case of precision rates, regardless of whether query expansion is included or not, the experiments show very excellent precision rates in all the component databases. This means that the proposed system exactly reformulates the original query into sub-queries that are for component databases identified during the procedure of query processing.

The average time for processing a query, average recall rate and precision rate for all the component databases, with and without query expansion, are shown in Table 1.

With query expansion, the average time for processing a query is 2.7 times slower than that without query expansion. Regardless of whether query expansion is included or not, the precision rate is very high, while recall rate varies according to whether the query is expanded or not. Although query processing with query expansion needs more processing time, it is, however, more efficient in the aspect of recall rate.

Table 1. Average time for processing a query, average recall rate, and average precision rate, with and without query expansion.

	average time for processing a query (sec.)	average recall rate (%)	average precision rate (%)
with query expansion	4.11	84.72	98.91
without query expansion	1.52	53.56	98.53

4 Conclusions

The proposed approach provides common domain knowledge using ontology as knowledge base, so that each information source can be merged into the system without specific domain knowledge of other sources. That is, each information source can describe its domain knowledge and construct its own semantic network independent of other information sources, which enables each information source to be merged into the system easily. This kind of approach guarantees the autonomy of each information source and gives users needing to integrate information an easy and efficient method for information integration. Also, it enables a number of information sources to be merged into the information integration system, so that it gives good extensibility to the system.

As the experiments have been conducted in the level of laboratory with four web databases in the real world, if the proposed approach is applied to an unspecified number of the general databases, both recall rate and precision rate are more or less expected to be decreased. However, if sufficient schema information of each web database is provided, the proposed approach for information sharing among heterogeneous web databases is deserved to be regarded as one of the prominent approaches for information integration.

References

1. R. Hull, "Managing semantic heterogeneity in databases: a theoretical prospective", Proc. 16th ACM SIGACT-SIGMOD-SIGART symposium on Principles of database systems, pp. 51-61, 1997.
2. C. Batini, M. Lenzerini, and S.B. Navathe, "A comparative analysis of methodologies for database schema integration", ACM Computing Surveys, 18(4), 1986.
3. A. M. Ouksel, A. P. Sheth, "Semantic Interoperability in Global Information Systems: A Brief Introduction to the Research Area and the Special Section". SIGMOD Record, 28(1), pp. 5-12, 1999.
4. J. O. Lee, D. K. Baik, "Semantic Integration of Databases using Linguistic Knowledge", Lecture Notes in Artificial Intelligence, LNAI 1747, Springer-Verlag, 1999.
5. J. O. Lee, D. K. Baik, "SemQL: A Semantic Query Language for Multidatabase Systems", Proc. 8th International Conf. on Information and Knowledge Management (CIKM-99), 1999.

6. Mike Uschold and Michael Gruninger, "Ontologies: Principles, Methods and Applications", Knowledge Engineering Review, 1996.
7. Thomas R.Gruber, "Toward Principles for the Design of Ontologies Used for Knowledge Sharing", International Journal of Human-Computer Studies, 1995.
8. Maurizio Panti, Luca Spalazzi, Alberto Giretti, "A Case-Based Approach to Information Integration" , Proceedings of the 26th VLDB conference, 2000.
9. J. Hammer, H. H. Garcia-Molina, K. Ireland, Y. Papakonstantinou, J. Ullman, J. Widom, "Information translation, mediation, and mosaic-based browsing in the tsimmis system", In Proceedings of the ACM SIGMOD International Conference on Management of Data, 1995.
10. Marian Nodine, Jerry Fowler, Brad Perry, "An Overview of Active Information Gathering in InfoSleuth", InfoSlueth Group, 1998.
11. Jeong-Oog Lee,Yo-Han Choi, "Agent-based Layered Intelligent Information Integration System Using Ontology and Metadata Registry", Lecture Notes in Computer Science, LNCS 2690, Springer-Verlag, 2003. 7.
12. R. Baeza-Yates, B. Ribeiro-Neto, Modern Information Retrieval, 1st Ed., pp. 73-97, Addison-Wesley, 1999.

A Web Services-Based Distributed Simulation Architecture for Hierarchical DEVS Models

Ki-Hyung Kim[1] and Won-Seok Kang[2]

[1] Dept. of Computer Eng. Yeungnam University,
214-1 Daedong, Gyungsan, Gyungbuk, Korea
kkim@yu.ac.kr
http://nclab.yu.ac.kr
[2] Advanced Information Technology Research Center (AITrc), KAIST
373-1, Kusung-Dong, Yusong-Gu, Daejon, Korea
wskang@world.kaist.ac.kr

Abstract. The Discrete Event Systems Specification (DEVS) formalism specifies a discrete event system in a hierarchical, modular form. This paper presents a web-services-based distributed simulation architecture for DEVS models, named as *DEVSCluster-WS*. DEVSCluster-WS is actually an enhanced version of DEVSCluster by employing the web services technology, thereby retaining the advantages of the non-hierarchical distributed simulation compared to the previous hierarchical distributed simulations. By employing the web services technologies, it describes models by WSDL and utilizes SOAP and XML for inter-node communication. Due to the standardized nature of the web service technology, DEVSCluster-WS can effectively be embedded in the Internet without adhering to specific vendors and languages. To show the effectiveness of DEVSCluster-WS, we realize it in Visual C++ and SOAPToolkit, and conduct a benchmark simulation for a large-scale logistics system. We compare the performance of DEVSCluster-WS with DEVSCluster-MPI, the MPI-based implementation of DEVSCluster. The performance result shows that the proposed architecture works correctly and could achieve tolerable performance.

1 Introduction

Discrete-event simulation is frequently used to analyze and predict the performance of systems. Simulation of large, complex systems remains a major stumbling block, however, due to the prohibitive computation costs. Distributed discrete-event simulation (or shortly, distributed simulation) offers one approach that can significantly reduce these computation costs.

Conventional distributed simulation algorithms assume rather simple simulation models in that the simulation consists of a collection of logical processes (LPs) that communicate by exchanging timestamped messages or events. The goal of the synchronization mechanism is to ensure that each LP processes events in time-stamp order; this requirement is referred to as the *local causality constraint*. The algorithms can be classified as being either conservative [1] or optimistic [2]. Time Warp is the most well known optimistic method. When an LP

T.G. Kim (Ed.): AIS 2004, LNAI 3397, pp. 370–379, 2005.

receives an event with timestamp smaller than one or more events it has already processed, it rolls back and reprocesses those events in timestamp order.

Since distributed simulation deals with large and complex systems, the following issues should be addressed: model verification and validation, model reusability, and user-transparency of distributed simulation details, etc. The Discrete Event Systems Specification (DEVS) formalism, developed by Zeigler [3], is a formal framework for specifying discrete event models. The DEVS modeling and simulation approach provides an attractive alternative to conventional logical process-based modeling approaches used in distributed simulation by its set-theoretical basis, its independence of any computer language implementation, and its modular, hierarchical modeling methodology.

There have been several research efforts in the distributed simulation of DEVS models. They can broadly be classified into two approaches: *hierarchical* and *non-hierarchical* ones. Adhering to the traditional hierarchical simulation mechanism of DEVS, the hierarchical distributed simulation approach has exploited only specific parallelism inherent in the formalism [3, 4]. The approach can be categorized into two schemes: synchronous and asynchronous ones. Synchronous schemes utilize only the parallelism in simultaneous events, and asynchronous schemes combine both the hierarchical simulation mechanism and distributed synchronization algorithms such as Time Warp [5].

Belonging to the non-hierarchical distributed simulation approach, *DEVS-Cluster* transforms the model structure into a non-hierarchical one during simulation execution [6]. By simplifying the model structure, DEVSCluster can be easily extended to the distributed version. It employs Time Warp as a basic synchronization protocol. It applies distributed object technologies such as CORBA (Common Object Request Broker Architecture) [7] as an underlying communication mechanism which is particularly useful for the DEVS simulation since it is inherently an object oriented modeling formalism.

Meanwhile, the adoption of the web services technology into the scientific distributed computing has been researched actively [8]. The use of the standard protocols, such as SOAP, XML, HTTP, WSDL, and UDDI, easily achieves high interoperability between heterogeneous computing environments such as Grids [13] due to its inherent characteristics: language- and platform-neutral.

In this paper we propose a web-services-based distributed simulation architecture for DEVS models, named as *DEVSCluster-WS*. DEVSCluster-WS is actually an enhanced version of DEVSCluster by employing the web services technology, thereby retaining the advantages of the non-hierarchical distributed simulation compared to the previous hierarchical distributed simulations. By employing the web services technologies, it describes models by WSDL and utilizes SOAP and XML for inter-node communication. Due to the standardized nature of the web service technology, DEVSCluster-WS can effectively be embedded in the Internet without adhering to the specific vendors and languages.

To show the architectural effectiveness of DEVSCluster-WS, we design a model of a large-scale logistics system and perform experiments in a distributed environment. We evaluate the performance of DEVSCluster-WS by comparing the simulation results with DEVSCluster-MPI, the MPI-based implementation of DEVSCluster [6]. The results show that the overhead of the XML processing could be tolerable compared to the efficient MPI-based communication [9].

The rest of this paper is organized as follows: Section 2 describes an overview of the DEVS formalism. Section 3 presents DEVSCluster-WS. Section 4 presents the performance results. Finally, Section 5 concludes the paper.

2   Preliminaries

This section describes the DEVS formalism, the distributed simulation mechanism, and the Web service architecture as the preliminary backgrounds of DEVSCluster-WS.

2.1   DEVS Formalism

The DEVS formalism is a sound formal modeling and simulation (M&S) framework based on generic dynamic systems concepts [3]. DEVS is a mathematical formalism with well-defined concepts of hierarchical and modular model construction, coupling of components, and an object oriented substrate supporting repository reuse. Within the formalism, one must specify (1) the basic models from which larger ones are built, and (2) the way in which these models are connected together in a hierarchical fashion. Top down design resulting in hierarchically constructed models is the basic methodology in constructing models compatible with the multifaceted modeling approach.

A basic model, called the atomic model (or atomic DEVS), specifies the dynamics of the system to be simulated. As with modular specifications in general, we must view the above atomic DEVS model as possessing input and output ports through which all interactions with the external world are mediated. To be more specific, when external input events arrive from another model and are received on its input ports, the model decides how to respond to them by its external transition function. In addition, when no external events arrive until the schedule time, which is specified by the time advance function, the model changes its state by the internal transition function and reveals itself as external output events on the output ports to be transmitted to other models. For the schedule time notice, an internal event (*) is devised as shown in the above definition.

Several atomic models may be coupled in the DEVS formalism to form a multi-component model, also called a coupled model. In addition, closed under coupling, a coupled model can be represented as an equivalent atomic model. Thus, a coupled model can itself be employed as a component in a larger coupled model, thereby giving rise to the construction of complex models in a hierarchical fashion.

Detailed descriptions for the definition of the atomic and coupled DEVS can be found in [3].

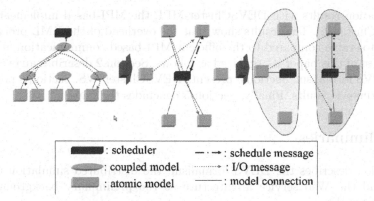

▬ : scheduler	—·→ : schedule message
⬬ : coupled model	·······→ : I/O message
▣ : atomic model	—— : model connection

Fig. 1. Non-hierarchical simulation mechanism for DEVS models.

2.2 DEVSCluster: Non-hierarchical Distributed Simulation Scheme of DEVS Models

DEVSCluster is a non-hierarchical distributed simulation architecture for DEVS models [6]. The basic motivation of DEVSCluster is, as shown in Fig. 1, to transform hierarchically structured DEVS models into a non-hierarchical one before the simulation execution. Based on the flat structured DEVS models, The distributed version of DEVSCluster partitions the models for mapping into the distributed machines, in which simulation engines synchronize each other by Time Warp.

For detailed description of DEVSCluster, refer to [6].

2.3 The Web Service Technology

Web services are self-contained, loosely coupled software components that define and use Internet protocols to describe, publish, discover, and invoke each other [10, 11]. They can dynamically locate and interact with other web services on the Internet to build complex machine-to-machine programmatic services. These services can be advertised and discovered using directories and registries such as the Universal Description, Discovery, and Integration (UDDI) specification [12].

To be useful, each discovered component must be described in a well-structured manner. The Web Services Description Language (WSDL) provides this capability. Using WSDL, an invoking entity can bind to the selected service and start communicating with its externally visible functions via advertised protocols such as SOAP.

SOAP is a simple, lightweight protocol for the exchange of XML based messages between distributed entities. It supports detailed error reporting, application specific XML encodings and is presently transferred over HTTP or SMTP. The integral part of SOAP is the *envelope* root element, which consists of two blocks of XML data called *header* and *body*. While the optional header is used for transferring supplementary information, the body actually contains the main part of XML data – in case of a SOAP-based remote procedure call (SOAP-RPC), parameters and name of the called function.

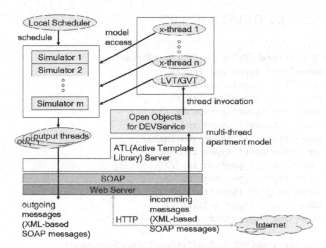

Fig. 2. Block diagram of DEVSCluster-WS.

3   DEVSCluster-WS

In this section, the architecture of the proposed web-services-based distributed simulation architecture, named as DEVSCluster-WS, is presented. Fig. 2 shows the block diagram of DEVSCluster-WS. For inter node communication between distributed models, DEVSCluster-WS utilizes the web services technology, such as SOAP and WSDL, as an underlying distributed object technology. This makes DEVSCluster-WS an adequate simulation engine for heterogeneous network computing environment like Grid [13]. In the figure, output messages are processed in a multi-threaded manner by output threads. Notice that the external input messages are also processed in a multi-threaded manner. This is for avoiding a possible deadlock condition as shown in Fig. 3. Fig. 3 (a) shows the possible deadlock condition when processing external input messages from other nodes. The entire simulators in a node are protected by a global lock in a node, and two entities – a local scheduler for internal event generation and the ATL servants [14] for external message processing – share the critical section. The follow is a possible scenario of deadlock condition; a simulation object in *Node1*, which is now holding a lock of Node1, is now generating output events which should be processed by the ATL servant of *Node2* after acquiring the lock of Node2; the same action happens in Node2 by the chance; thus a deadlock happens.

Fig. 3 (b) shows the solution for deadlock avoidance employed by DEVS-Cluster-WS. It employs the multi-apartment model [14] for multi-threading the execution of external input messages.

The following shows the WSDL codes for DEVSCluster-WS. Through this standard interface description, DEVSCluster-WS can be published in a standardized way.

WSDL Description for DEVSCluster-WS

```
......................
<message name='WebServiceDEVSImp.SendToXMesgThread'>
        <part name='src' type='xsd:int'/>
        <part name='time' type='xsd:double'/>
        <part name='dst' type='xsd:int'/>
        <part name='tN' type='xsd:double'/>
        <part name='simcount' type='xsd:int'/>
        <part name='priority' type='xsd:int'/>
        <part name='dupcount' type='xsd:int'/>
        <part name='type' type='xsd:int'/>
        <part name='sign' type='xsd:int'/>
        <part name='mesgid' type='xsd:int'/>
        <part name='func' type='xsd:int'/>
        <part name='buf' type='xsd:string'/>
</message>
<message name='WebServiceDEVSImp.SendToXMesgThreadResponse'>
</message>
<message name='WebServiceDEVSImp.Cal_Lvt'>
        <part name='lvt' type='xsd:double'/>
</message>
<message name='WebServiceDEVSImp.Cal_LvtResponse'>
        <part name='Result' type='xsd:double'/>
</message>
<message name='WebServiceDEVSImp.SetGVT'>
        <part name='gvt' type='xsd:double'/>
</message>
<message name='WebServiceDEVSImp.SetGVTResponse'>
</message>
.........................
<operation name='SendToXMesgThreadSeq'
        parameterOrder='src time dst tN simcount priority
                dupcount type sign mesgid func buf'>
        <input message='wsdlns:WebServiceDEVSImp.SendToXMesgThread'/>
        <output message='wsdlns:WebServiceDEVSImp.SendToXMesgThreadResponse'/>
</operation>
<operation name='Cal_Lvt' parameterOrder='lvt'>
        <input message='wsdlns:WebServiceDEVSImp.Cal_Lvt'/>
        <output message='wsdlns:WebServiceDEVSImp.Cal_LvtResponse'/>
</operation>
<operation name='SetGVT' parameterOrder='gvt'>
        <input message='wsdlns:WebServiceDEVSImp.SetGVT'/>
        <output message='wsdlns:WebServiceDEVSImp.SetGVTResponse'/>
</operation>
.........................
```

4   Experimental Results

To show the effectiveness of DEVSCluster-WS, we have conducted a benchmark simulation for a large-scale logistics system.

The automatic logistics system becomes more important as the size of the system becomes large and complex. For the help of the decision-making, simula-

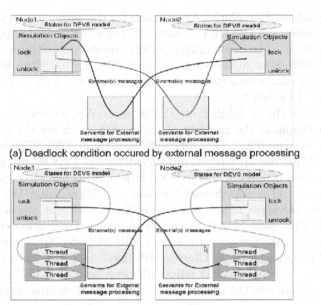

(a) Deadlock condition occured by external message processing

(b) Deadlock avoidance by employing threads for external message processing

Fig. 3. Deadlock prevention mechanism for DEVSCluster-WS.

tion techniques can be used usefully. However, even in the normal sized logistics system, one of the important problems to be tackled is the excessive simulation execution time and the requirement of the large memory. In this case, distributed simulation can help the problem. Fig. 4 shows the basic structure of the logistics system model. In the logistics system, warehouses, stores, and control centers are located over wide areas: stores spend products, warehouses supply the products, and control centers control the logistics cars to chain the stores and warehouses. In this system, the objective is to minimize the number of the required cars while satisfying the orders of all stores. To find a shortest path between warehouses and stores is basically a traveling salesperson problem. We utilized the genetic algorithm for this problem. Also we abstracted the loading of products on cars as a bin packing problem and used the best-fit algorithm for this problem.

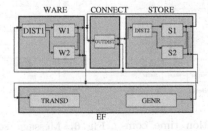

Fig. 4. Basic structure of the logistics system model.

We have conducted simulation experiments for the automatic logistics system using DEVSCluster-WS and DEVSCluster-MPI, the MPI-based implementation of DEVSCluster [6]. The configuration of the simulation platform is four Pentium IV-based Windows 2000 systems connected by the 100 Mbps Ethernet. We implemented the simulation system by Visual C++ 6.0 with SOAPToolkit3.0 [15] and the GUI by JAVA.

For the test of the scalability of DEVSCluster-WS, we conducted the simulation while changing the number of nodes. Fig. 5 shows the result. For 2 and 4 nodes, we compared two different versions of DEVSCluster-WS with DEVSCluster-MPI, the MPI-based implementation of DEVSCluster. We tested a threaded version of DEVSCluster-WS which multi-threads the processing of output messages and a non-threaded version of DEVSCluster-WS. This is to evaluate the effectiveness of the multi-threading for output messages processing especially in the slow SOAP-based inter-node communication environment. The results show that the performance of DEVSCluster-WS could be tolerable compared to efficient DEVSCluster-MPI. Especially the multi-threading of output message processing could reduce the overhead of SOAP processing in DEVSCluster-WS due to the buffering effect of multi-threading, as shown in the Fig. 5 and 6.

Fig. 7 shows the number of rollbacks and GVT (Global Virtual Time) computation time for 4 nodes configuration. The rollback result shows that the number of rollbacks is already in the order of four hundreds and does not critically affect the simulation run time. The GVT computation time is increased compared to DEVSCluster-MPI, and the multi-threading of output messages can reduce this overhead considerably.

5   Conclusion

In this paper, we proposed a web-services-based distributed simulation architecture for the DEVS models, named as *DEVSCluster-WS*. DEVSCluster-WS, an enhanced version of DEVSCluster by employing the web services technol-

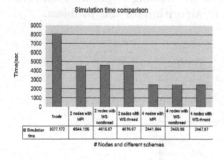

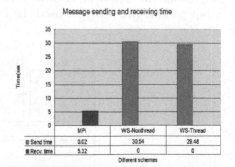

Fig. 5. Simulation execution time comparison of different schemes.

Fig. 6. Message sending and receiving time for different schemes.

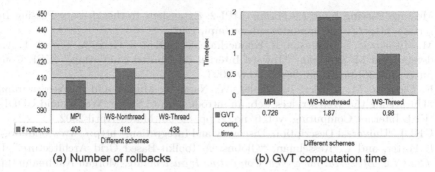

(a) Number of rollbacks   (b) GVT computation time

Fig. 7. Number of rollbacks and GVT computation time for different schemes.

ogy, thereby retaining the advantages of the non-hierarchical distributed simulation compared to the previous hierarchical distributed simulations. Due to the
standardized nature of the web service technology, DEVSCluster-WS can effectively be embedded in the Internet without adhering to the specific vendors and
languages. The performance result of DEVSCluster-WS showed that it works
correctly and could achieve tolerable performance.

References

1. K. Chandy and J. Misra, "Distributed Simulation: A Case Study in Design and
   Verification of Distributed Programs," IEEE Trans. on Software Eng. vol. 5, no.
   5, pp. 440–452, 1978.
2. R. Fujimoto, "Optimistic approaches to parallel discrete event simulation," Transactions of the Society for Computer Simulation International, vol. 7, no. 2, pp.
   153–191, Oct. 1990.
3. B. Zeigler, H. Praehofer, and T. Kim, "Theory of Modeling and Simulation: Integrating Discrete Event and Continuous Complex Dynamic Systems," 2nd Ed.,
   Academic Press, pp. 261–287, 2000.
4. A. Chow, "Parallel DEVS: A parallel, hierarchical, modular modeling framework
   and its distributed simulator," Transactions of the Society for Computer Simulation
   International, vol. 13, no. 2, pp. 55–67, 1996.
5. K. Kim, Y. Seong, T. Kim, and K. Park, "Distributed Simulation of Hierarchical
   DEVS Models: Hierarchical Scheduling Locally and Time Warp Globally," Transactions of the Society for Computer Simulation International, vol. 13. no. 3, pp.
   135–154, 1996.
6. K. Kim and W. Kang, "CORBA-based, Multi-threaded Distributed Simulation of
   Hierarchical DEVS Models: Transforming Model Structure into a Non-Hierarchical
   One," LNCS vol. 3046, pp. 167–176, 2004.
7. Object Management Group: The Common Object Request Broker: Architecture
   and Specification, 2.2 ed., Feb. 1998.
8. K. Chiu, M. Govindaraju, and R. Bramley, "Investigating the limits of SOAP
   performance for scientific computing," in the Proceedings of 11th IEEE International Symposium on High Performance Distributed Computing (HPDC-11), pp.
   246–254, 23-26 July 2002.

9. Message Passing Interface Forum: MPI-2: Extensions to the Message-Passing Interface, http://www-unix.mcs.anl.gov/mpi, 1997.
10. M. Milenkovic, S. Robinson, R. Knauerhase, D. Barkai, S. Garg, A. Tewari, T. Anderson, and M. Bowman, "Toward Internet distributed computing," IEEE Computer, vol. 36, no. 5, pp. 38–46, May 2003.
11. F. Curbera, M. Duftler, R. Khalaf, W. Nagy, N. Mukhi, and S. Weerawarana, "Unraveling the Web services web: an introduction to SOAP, WSDL, and UDDI," IEEE Internet Computing, vol. 6, no. 2, pp. 86–93, March-April 2002.
12. UDDI, "Universal Description, Discovery and Integration," http://www.uddi.org/.
13. I. Foster, and C. Kesselman, "Globus: A Toolkit-Based Grid Architecture," In The Grid: Blueprint for a New Computing Infrastructure, Morgan Kaufmann, pp. 259–278, 1999.
14. "Specifying the Project's Threading Model (ATL)," http://msdn.microsoft.com/.
15. "SOAP Toolkit," http://msdn.microsoft.com/webservices/building/soaptk/.

Automated Cyber-attack Scenario Generation Using the Symbolic Simulation

Jong-Keun Lee[1], Min-Woo Lee[2], Jang-Se Lee[3],
Sung-Do Chi[4], and Syng-Yup Ohn[4]

[1] Cartronics R&D Center, Hyundai MOBIS, Korea
leejkmobis.co.kr
[2] Cooperation of Sena, Korea
momentOmsn.com
[3] The Division of Information Technology Engineering,
Korea Maritime University, Korea
jsleebada.hhu.ac.kr
[4] The Department of Computer Engineering,
Hangkong University, Korea
{sdchi,syohn}hau.ac.kr

Abstract. The major objective of this paper is to propose the automated cyber-attack scenario generation methodology based on the symbolic simulation envi-ronment. Information Assurance is to assure the reliability and availability of information by preventing from attack. Cyber-attack simulation is one of no-ticeable methods for analyzing vulnerabilities in the information assurance field, which requires variety of attack scenarios. To do this, we have adopted the symbolic simulation that has extended a conventional numeric simulation. This study can 1) not only generate conventional cyber-attack scenarios but 2) gen-erate cyber-attack scenarios still unknown, and 3) be applied to establish the appropriate defense strategies by analyzing generated cyber-attack. Simulation test performed on sample network system will illustrate our techniques.

1 Introduction

As we increasingly rely on information infrastructures to support critical operations in defense, banking, telecommunication, transportation, electric power and many other systems, cyber-attacks have become a significant threat to our society with potentially severe consequences [1,2]. Information Assurance is to assure the reliability and availability of information by preventing information and technology that consist, operate and control Information Infrastructure from attack. One of the efforts for in-formation assurance is to set up the standard metrics for design, operation and test, and to develop the proper model and the tool with which information assurance test can be conducted. [3,4]. That is, the deep understanding of system operation and at-tack mechanisms is the foundation of designing and evaluating information assurance activities [2]. Therefore,

T.G. Kim (Ed.): AIS 2004, LNAI 3397, pp. 380–389, 2005.

the advanced modeling and simulation methodology is essential for classifying threats, specifying attack mechanisms, verifying protective mecha-nisms, and evaluating their consequences. Such a methodology may be able to support to find unknown attack behavior if more refined models are allowed.

It is true that Cohen [5], Amoroso [6], Wadlow [7], and Nong Ye [2] have done pioneering work on this field, nonetheless, their studies remain in the conceptual modeling rather than specific simulation trial. Recently, Chi et al [8,9] suggested network security modeling and cyber-attack simulation methodology and developed the SECUSIM that realized the methodology. However, SECUSIM have limits that can only simulate pre-defined cyber-attack scenarios.

In the mean time, if it is possible to predict unknown attack scenarios by adopting the preciseness of modeling and reasoning methodology by using simulation, the effectiveness of simulation is expected to be maximized. To deal with this, we have adopted the symbolic DEVS that has extended a conventional numeric simulation. This study can 1) not only generate conventional cyber-attack scenarios but 2) gener-ate cyber-attack scenarios still unknown, and 3) be applied to establish the appropriate defense strategies by analyzing generated cyber-attack.

2   Brief Descriptions on Symbolic DEVS

Symbolic DEVS, extended Discrete Event System Specification (DEVS) [10,11], expresses discrete event system occurred at symbolic time [12]. The main purpose of symbolic simulation is to track the abnormal actions that may happen between various models rather than acquiring statistic data from conventional simulation. By express-ing the time of event in symbols, symbolic DEVS generates serial trees that show every possible state trajectory reaching special state. This tree can be effectively used to chase the cause of abnormal symptoms (i.e., goal) perceived [11]. That is to say, symbolic DEVS enables symbolic expression of event time by expanding real number for time function of conventional DEVS into linear polynomials [11,12,13,14]. Basically, symbolic DEVS is defined as follows:

$$M = < X, S, Y, \delta_{int}, \delta_{ext}, \lambda, ta > \tag{1}$$

Where X is the set of external (input) event types; S is the sequential state set; Y is the output set; $\delta_{int} : S \times LP + 0, \infty \to S$ is the internal transition function where the $LP + 0, \infty$ is the vector space of linear polynomials over non-negative reals with ∞ adjoined; $\delta_{ext} : Q \times S \to S$ is the external transition function where $Q = \{(s, e) | s \in S, 0 \le e \le ta(s)\}; \lambda : S \to Y$ is the output function; and $ta : S \to 2LP + 0, \infty$ is the time ad-vance function where $ta(s)$ is a finite, non-empty set of linear polynomials [11,13].

The example of unknown attack scenario using symbolic DEVS is shown in Fig. 1. Fig. 1(a) shows the simple state transition diagram of target host, in here, let's say that state transition of solid lines represents known attack, and state transition of dotted line represents unknown attack. So, in this case, known attack scenarios are $\{i_1, i_4\}$ and $\{i_2, i_3, i_4\}$. Meanwhile, unknown scenarios $\{i_2, i_5\}$

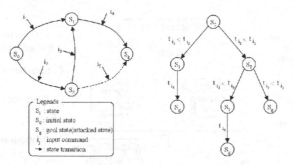

(a) State Transition Diagram of Host model   (b) Generated Simulation Trajectory Tree

Fig. 1. Symbolic DEVS Example.

can be found by using symbolic simulation analysis. That means, in state S_0 it can take one of two external input events i_1 and i_2 which are scheduled to be arrived at symbolic time t_{i1} and t_{i2}, respectively. The external transition corresponding to t_{i1} is $S_0 \rightarrow S_1$; that corresponding to t_{i2} is $S_0 \rightarrow S_2$. That is to say, symbolic time set, t_{i1} and t_{i2} are related to non-deterministic timing, accordingly, by symbolic simulation, simulation trajectory tree shown in Fig. 1(b) is automatically generated so that relationship between every possible state and input can be analyzed effectively.

3   Cyber-attack Scenario Generation Methodology Using the Symbolic DEVS Simulation

Automated cyber-attack generation by symbolic simulation, with which this research is approaching, is achieved by the layered methodology shown in Fig. 2. Here, cyber-attack scenario is defined as command set that transit normal state of victim host to attacked state. First phase is to clearly specify the state of target objective and initial constraints based on information of target network. By defining the goal state to find through simulation in advance, it is possible to reduce simulation scope and to easily analyze the result of simulation. Also, searching space of symbolic simulation can be reduced effectively by reflecting pre-knowledge such as structural constraint of target network and constraints between the states. Second phase is to search or generate the data and models that were already constructed to library or to be constructed, which can be composed of network components like host, attacker's database and symbolic model. By integrating these data and symbolic models, final simulation model in 3rd phase is generated so that simulation proceeds. Symbolic simulation in 4th phase is commenced by start messages input to simulation model. By transferring next components as per coupling and timing relation along with next component, all possible activities are examined up until reaching goal. When non-deterministic timing relation appears, simulation engine automatically generates simulation tree that shows all possible activities [12]. This tree contains all possible attack

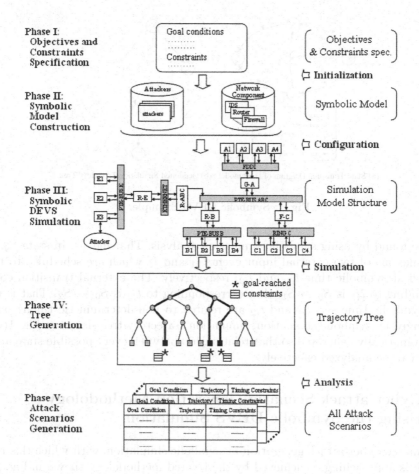

Fig. 2. Automated Cyber Attack Scenario Generation Methodology.

scenarios, and array of the attacks that satisfy goal among them results in table shown in phase 5. This table includes all attack procedures that may successfully attack the target system. What it means is that both known and unknown attack can be found in the process of these attack scenario analysis.

4   Case Study

Simulation test was taken place to examine the validity of suggested methodology. To simplify the test, sample network composed of attacker and victim host omitting net-work between models was assumed as shown in Fig. 3.

4.1   Phase I: Objectives and Constraints

In order to generate the cyber-attack scenario during symbolic simulation, simulation goal is set to let the victim host reach COOLDOWN. COOLDOWN

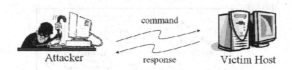

Fig. 3. Sample Network.

mentioned here is one of the 4 steps (COLD, WARM, HOT, and COOLDOWN) Wadlow [7] suggested, it refers the step to conduct destroying activities after successful attack (HOT) (also see Fig. 5). And, the commands that attacker has was limited to 6 general command groups and 1 command group for confirming:

*Group*1: Local host command group: command group that gives command to local host. Commands to access victim host are included.
*Group*2: Telnet command group: it is composed of commands to use telnet service provided from victim host. Setup change and access to other services are possible by this.
*Group*3: FTP command group on telnet: it is composed of commands that can be used when ftp session is open by telnet command.
*Group*4: Authorization command group: it is composed of responses related with ID & password authorization.
*Group*5: FTP command group: it is composed of commands that can be used when ftp service is open
*Group*6: COOLDOWN command group: it is composed of commands group that actually can give damage to victim host.
*Group*7: Check command group: it is composed of commands group that can check the change of victim host condition upon above command groups.

4.2   Phase II: Symbolic Model Construction

Symbolic DEVS model is constructed in this phase. First, Fig. 4 shows the structure of attacker model. As shown in Fig. 4, attacker model is composed of various command models having symbolic time, and each command models are activated upon the response from victim host. Activated command models generate the various commands along with non-deterministic timing value, and generated commands are transferred to victim host. As such, attacker generates various attack scenarios automatically. Symbolic DEVS representation of command model consisting attacker model is summarized as follows:

```
State Variables
   'passive, 'active, 'generating
External Transition Function
   If packet(response) comes from Victim Host and state is 'passive,
   Then hold in 'active during 'symbolic-time
     Else If 'stop comes from other command model and state is 'active,
     Then hold in 'passive during 'INF
Internal Transition Function
   If state is 'active,Then hold in 'generating during 0
   Else If state is 'generating,Then hold in 'passive during 'INF
```

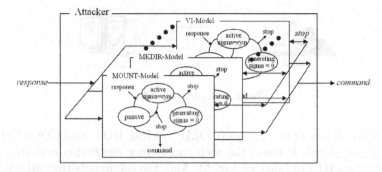

Fig. 4. Attacker model structure.

```
Output Function
   If state is 'active, Then send 'stop to other command models
   Else If state is 'generating,
   Then sent packet(command) to Victim Host
```

State of victim host model [8,9], which receives command from attacker, is defined based on resources of host such as service-type and O/S-type, and it causes various state transitions along with input command. Through these serial state transitions, sometimes, it reaches goal state (i.e., attacked state), accordingly, the set of command changing model's state from initial state to goal state can be defined as attack scenario. Simplified DEVS representation of victim host is summarized as follows:

```
State Variables
   'passive, 'busy, Service-type, H/W-type, O/S-type,
   Registered-user-list, User-size, etc.
External Transition Function
   If packet(command) comes from Attacker and state is 'passive,
   Then change the state variables by executing command
        and hold in 'busy during 'processing-time
Internal Transition Function
   If state is 'busy, Then hold in 'passive during 'INF
Output Function
   If state is 'busy, Then send packet(response) to Attacker
```

4.3   Phase III: Symbolic DEVS Simulation

Symbolic simulation proceeds as per timing relation between attacker and victim host as explained above. Table 1 shows a part of simulation log file. In table 1, 'node' means each condition that may occur by symbolic timing relation, 'children' means next state that may be converted from current state. 'Clock-time' means symbolic time converted into current state, 'imminent' shows the selected command among non-deterministic commands. 'Constraints' mean time restriction resulted from command selection, 'goal' and 'dead-end' shows that whether present node satisfies goal or it is leaf node. For instance, NODE1 is a leaf node and it is no children, and when t_{c1_mount} satisfies the smallest restriction, it generates mount command by MOUNT model being activated in t_{c1_mount} time. Finally, NODE411 is a leaf node that satisfies goal and $t_{c15\_killall1}$ has the smallest restriction condition.

Table 1. Simulation trajectory (partially-shown).

Node	ROOT-NODE
Children	NODE1 NODE2 NODE3 NODE4 NODE5 NODE6 NODE7
Clock-time	0
Imminent	( )
Constraint	( )
Goal	#F
Dead-end	#F
=Node	NODE1
=Children	( )
=Clock-time	$t_{c1\ mount}$
=Imminent	MOUNT
=Constraint	$t_{c1\ mount} \leq \{t_{c1\ echo1},\ t_{c1\ su},\ t_{c1\ mkdir},\ t_{c1\ echo2},\ t_{c1\ cd},\ t_{c1\ rlogin}\}$
=Goal	#F
=Dead-end	#T
	(skipped)
=...=Node	NODE370
=...=Children	NODE411 NODE412 NODE413 NODE414
=...=Clock-time	$t_{c1\ mkdir}+t_{c2\ mount}+t_{c3\ echo1}+t_{c4\ su}+ \dots +t_{c13\ mget}+t_{c14\ vi1}$
=...=Imminent	VI1
=...=Constraint	$t_{c14\ vi1} \leq \{t_{c14\ killall1},\ t_{c14\ killall2},\ t_{c14\ named},\ t_{c14\ vi2}\}$
=...=Goal	#F
=...=Dead-end	#F
=...=Node	NODE411
=...=Children	( )
=...=Clock-time	$t_{c1\ mkdir}+t_{c2\ mount}+t_{c3\ echo1}+t_{c4\ su}+ \dots +t_{c13\ mget}+t_{c14\ vi1}+t_{c15\ killall1}$
=...=Imminent	KILLALL1
=...=Constraint	$t_{c15\ killall1} \leq \{t_{c15\ killall2},\ t_{c15\ named},\ t_{c15\ vi2}\}$
=...=Goal	#T
=...=Dead-end	#T

4.4 Phase IV: Tree Generation

Trajectory tree resulted from symbolic simulation is shown in Fig. 5. Trajectory tree is divided into 4 phases suggested by Wadlow in accordance with degree of attack procedure. In this figure, "○" means normal node, and "●" means leaf node (dead-end). And, "*" marked on leaf shows trajectory that satisfies the goal condition (to have victim host in COOLDOWN condition), bold line presents simulation trajectory that serial attacks have been succeeded. One of trajectory that reached the goal condition in Fig. 5 explains as follows (See the scenario in Table 2):

(1) : mount, echo1, su, mkdir, echo2, cd, and rlogin are the command that is valid in the currnet victim host condition, each command is selected by symbolic time of each command model. In this case, symbolic time of MKDIR model is the shortest, so mkdir command is selected.

(2)–(3), (5)–(6), (8), (10)–(12), (14) : the host condition is changed by executing the selected serial commands, and the next command is selected as per the symbolic time relation of command models that is valid after change of condition.

(4) : this is the case where symbolic time of SU model is the smallest. su command is selected and phase of attack process on victim host becomes WARM by executing su command.

(7) : rlogin is selected as the only command that can be executed in the current host condition.

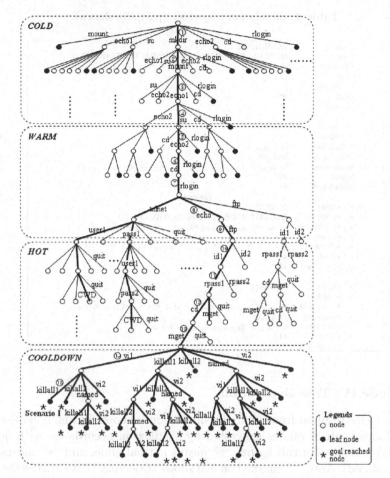

Fig. 5. Generated Simulation Trajectories Tree (obtained from Table 1).

(9) : ftp is selected as the only command that can be executed in the current host condition, and phase of attack process on victim host becomes HOT by executing ftp command.

(13) : mget is selected as the case that symbolic time of MGET model is smaller than that of QUIT model, and phase of attack process on victim host becomes COOLDOWN by executing mget command.

(15) : killall1 is selected as the case that symbolic time of KILLALL1 model is the smallest, and it reaches goal by executing killall1 command.

4.5   Phase V: Attack Scenario Generation

The discovery of new attack scenario is done by analysis of the trajectory tree of Phase IV explained above. Table 2 shows the example of attack scenario obtained from simulation trajectory tree. In this case study, it was confirmed

Table 2. Generated Cyber Attack Scenarios.

Command	Descriptions
mkdir /tmp/mount	Make directory in the local host
mount -nt nfs 172.16.112.20 : home /tmp/mount	Mount directory
echo 'lemin:102:2::/tmp/mount: /bin/csh' >> /etc/passwd	Command of changing the environment to write data to file
su -lemin	Make lemin user
echo '+ +' > lemin/.rhosts	Command of changing the environment to write data to file
cd /	Generate the .rhosts file
rlogin 172.16.112.20	rlogin
Echo "id > /tmp/OUT" > "|sh"	Command of changing the environment to write data to file
ftp localhost	Open ftp session
lemin	Right user id reply
right-pass	Right password reply
cd /	Change directory
mget *	Command of mget
vi inetd.conf	Change the configuration
killall inetd	Restart inetd

that ftp vulnerabilities of system was found and cyber-attack scenario using this was generated, even though the defense of victim host was assumed to be perfect. As such, a unknown cyber-attack scenario can be found by expressing victim host more precisely accord-ing to suggested methodology and defining command set specifically, accordingly.

5 Conclusions

The objective of this paper is to suggest methodology of automated cyber-attack scenario generation that is indispensable in the cyber-attack simulation for information assurance. In order to do this, we defined cyber-attack scenario as command set that transit normal state of victim host to attacked state and modeled the attacker that generates various cyber-attack scenario with all the possible combination of command by adopting the symbolic DEVS. This study can 1) not only generate conventional cyber-attack scenarios but 2) generate cyber-attack scenarios still unknown, and 3) be applied to establish the appropriate corresponding strategies by analyzing generated cyber-attack.

Acknowledgements

This research was supported by IRC(Internet Information Retrieval Research Center) in Hankuk Aviation University. IRC is a Kyounggi-Province Regional Research Center designated by Korea Science and Engineering Foundation and Ministry of Science & Technology.

References

1. T. A. Longstaff et al: Are We Forgetting the Risks of Information Technology, IEEE Computer, pp 43-51, December, 2000
2. Nong Ye and J. Giordano: CACS - A Process Control Approach to Cyber Attack Detection, Communications of the ACM., 1998
3. DoD: Defensive Information Operations, Technical Report #6510.01B, June 1997.
4. A. Jones: The challenge of building survivable information-intensive systems, IEEE Computer, August, pp. 39-43, 2000.
5. Cohen, F.: Simulating Cyber Attacks, Defenses, and Consequences, 1999 IEEE Symposium on Security and Privacy Special 20th Anniversary Program, The Claremont Resort Berkeley, California, May 9-12, 1999.
6. Amoroso, E.: Intrusion Detection, AT&T Laboratory, Intrusion Net Books, January, 1999.
7. Wadlow T. A.: The Process of Network Security, Addison-Wesley, 2000.
8. S.D. Chi et al, "Network Security Modeling and Cyber-attack Simulation Methodology", Lecture Notes on Computer Science series, 6th Australian Conf. on Information Security and Privacy, Sydney, July, 2001.
9. J.S. Park et al, "SECUSIM: A Tool for the Cyber-Attack Simulation", Lecture Notes on Computer Science Series, ICICS 2001 Third International Conference on Information and communications Security Xian, China, 13-16 November, 2001.
10. Zeigler, B.P.: Object-oriented Simulation with Hierarchical, Modular Models: Intelligent Agents and Endomorphic Systems, Academic Press, 1990.
11. Zeigler, B.P., H. Praehofer, and T.G. Kim: Theory of Modeling and Simulation 2ed., Academic Press, 1999.
12. S.D. Chi: Model-Based Reasoning Methodology Using the Symbolic DEVS Simulation, TRANSACTIONS of the Society for Computer Simulation International, Vol. 14, No. 3, pp.141-151, 1996.
13. Zeigler, B.P. and S.D. Chi: Symbolic Discrete Event System Specification, IEEE Trans. on System, Man, and Cybernetics, Vol. 22, No. 6, pp.1428-1443, Nov/Dec., 1992.
14. S.D. Chi, J.O. Lee, Y.K. Kim: Using the SES/MB Framework to Analyze Traffic Flow, Trans. of Computer Simulation International, Vol. 14, No. 4, December, 1997.

A Discrete Event Simulation Study
for Incoming Call Centers
of a Telecommunication Service Company

Yun Bae Kim, Heesang Lee, and Hoo-Gon Choi

Sungkyunkwan University, Suwon, Korea
leehee@skku.edu

Abstract. Call center becomes an important contact point, and an integral part of the majority of corporations. Managing a call center is a diverse challenge due to many complex factors. Improving performance of call centers is critical and valuable for providing better service. In this study we applied forecasting techniques to estimate incoming calls for a couple of call centers of a mobile telecommunication company. We also developed a simulation model to enhance performance of the call centers. The simulation study shows reduction in managing costs, and better customer's satisfaction with the call centers.

1   Introduction

Over the past few years, more and more corporations have become aware of the importance of call centers. Call centers can be used by tele-marketers, collection agencies and fund raising organizations to make contact with customers by phones. In this type of call centers, outgoing calls from the call center are more frequent than incoming calls. Call centers can also be designed and operated to offer after-sales service, software support, or customer inquiry by processing mainly incoming calls.

A typical call center has complex interactions between several "resources" and "entities". Telecommunication trunk lines, private business exchanges (PBX), and agents who call up or answer the customer's calls are the main resources of a call center. The voice response units (VRU), which can replace the human agents for some part of calls or some of calls. Automatic call distributors (ACD) are also standard digital solutions in a PBX, or in a separate system. Entities take the form of calls in a call center. Hence the incoming or outgoing calls navigate call centers and telecommunication networks. In the call centers the calls occupy or share the call center's resources. Hence the objective of call center management is to achieve a high service level for customers and use resources efficiently. The complexity of the management and analysis of a call center depends on several factors such as call types and waiting principles. Call entities that have several types make the analysis difficult. When a call entity is accepted at a call center, and the resources are fully used, the arriving call may be simply lost or hold[1] in

---

[1] This waiting system is theoretically more difficult to analyze than the loss system.

T.G. Kim (Ed.): AIS 2004, LNAI 3397, pp. 390–399, 2005.

an electronic queueing system. A call may also abandon the system during the waiting[2].

To provide enhanced response to the customers and increase customer satisfaction, corporations are providing many resources for call center operations. The design and management of the efficient call center is becoming more complicated due to many factors, such as the fast growing demanding of customer expectations, new computer-telephone integration (CTI) technologies, diverse call routing strategies, and some staffing restrictions.

In this paper we study a nation wide mobile telecommunication company's call center in Korea that has incoming calls as the major entities. The company has about 10 million subscribers and up to 4 million incoming calls to the call centers per month. We focus our research to answer the following questions: "How the call center can be effectively simulated? How can we help the decision makers to make the call center operations efficient?"

The remainder of this paper is organized as follows. In Sect. 2, we present some related research works for the management of call centers. In Sect. 3, we analyze the operation policies and information system resources of the call center. In Sect. 4, we studied the incoming calls for the call center, and used forecasting methods to predict and formulate the entities of the call center. In Sect. 5, we make a discrete event simulation model for the call center. We verify the simulation model with the actual call center. We also suggest some optimal operational strategies for the call center, such as the overflow decision and the agency scheduling. In Sect. 6, we give some concluding remarks and discuss further research topics.

2 Previous Research for Call Center Operations

The most important management decision problem for call center operation is decision about staffing the agents and trunking capacity dimensioning. Analytic models based on Erlang formulas that were designed in 1917 for the queueing system can still be used [1]. However the Erlang based method has too narrow assumptions for today's call center, such as: 1) every call has the same type. 2) a call in a queue never abandons. 3) every call is served by first in first out (FIFO) basis. 4) every agent has the same ability. Many spreadsheet solutions are used for staff scheduling or trunking capacity dimensioning decisions. These methods are based on more realistic assumptions than Erlang models. Software solutions, which are based on spreadsheet or heuristic algorithms, for agent scheduling or dimensioning of the call centers are called Work Force Management (WFM) solutions in the world of call center business.

Many works in operations research have been conducted for the problems related with optimal agent scheduling [2] and references therein. However, these methods have very strong assumptions for randomness and types of entities and resources. [2] presented a deterministic algorithm for the daily agent scheduling problem using dynamic programming and integer programming. They applied

---

[2] The abandoning of calls make the system more realistic but more difficult to analyze.

the algorithm for an emergency call center involving up to several hundred agents and six months. [3] studied a deterministic shift scheduling problem for call centers with an overall service level objective. This problem has a strong assumption that all shifts of agents have the same length without breaks. They showed that with this restriction, a local search algorithm is guaranteed to give a global optimal solution for a well-chosen finite (but very large) neighborhood.

With simple assumptions some stochastic analysis have been reported. [4] studied a stochastic capacity sizing problem of an inbound call center which has a common resource and dedicated servers. For a simpler form, the loss system, where the callers do not wait for service, they suggested an algorithm that can find an optimal capacity. [5] presented a study about an incoming call center that has agents trained to answer different classes of calls. They presented a closed form approximation of the loss probability with a simplifying assumption of the loss system.

There are some discrete event simulation approaches about operations of call centers of telecommunication service companies. [6] discussed how simulation had added value as a decision support tool for the call centers, during a re-engineering initiatives at AT&T. [7] reported that a discrete event simulation study had used for some new problems of the call center of Bell Canada, such as skill-based routing and interactive call handling. However, both studies of AT&T and Bell Canada showed not detailed findings of the simulation works, but some frameworks of the simulation study.

3   System Analysis of the Call Center

In this section we analyze a call center of a Korean mobile telecommunication service company. The company has about ten million subscribed customers for mobile voice calls and wireless internet access. There are two call centers for this company that usually operate incoming subscription-related services such as subscription condition inquiry, service change, software upgrade, and cost rate related services.

We analyzed the flow of the call entities and functional relationships of agents' activities of the call centers of the company, and generated the following system architecture diagram that is a representation of the operational processes of the call centers of the company.

As we see in Fig. 1, the company has two call centers and has two predefined types of customers. The first type of customers is the prestige class and the second type is the ordinary class. The customer classification is based on the customer's business history that is stored in the CTI server. We also classify a separate call class for quality inquiries without considering business history for both classes. Hence, we have three classes of customers, 1) the prestige class without a quality inquiry, 2) the ordinary class without a quality inquiry, and 3) the quality inquiry class.

Two call centers operate independently for the prestige customers or the quality inquiries. However, for an ordinary customer without a quality inquiry,

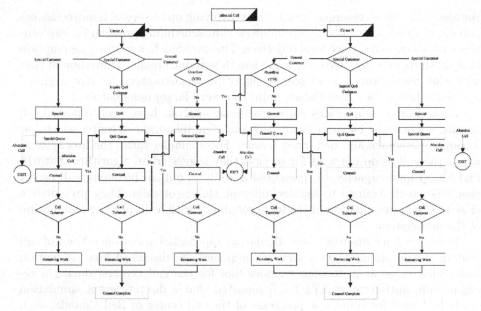

Fig. 1. Process Flows of the Call Center.

when a call center uses full capacities of its resources, and the number of held calls or waiting time of held calls are greater than the predesigned parameter, the other call center can be inundated[3]. Any call that has waited in the call center can abandon waiting for service[4]. Calls can also be transferred to other types of classes after its own service has been rendered. The agent may need some extra work for paper-works and related information processing after the call connection. This process is called the "post-call process".

In our study we surveyed the call center management and concluded that the following data are important performance measures of the operations of the call centers.

 − Service Level
 − Average Waiting Time
 − Reneging Rate
 − Resource Utilization

Note that these performance measures are interrelated. For example if we try to decrease the average waiting time with fixed resources, we may get a worse abandon rate since the pre-queued entities may hold fixed resources more favorably. Using the above performance measure we can find the answers for 1) queueing strategies, 2) load balancing, 3) agent scheduling, and 4) process redesign.

---

[3] As we see in the upcoming section, the decision about how much we allow to overflow is an important problem for an efficient management of the call center operations.

[4] To avoid the overload of incoming calls, the ACD of the call center would abandon a call automatically that experiences three minute waiting.

4   Analysis of Incoming Calls

The arrival of calls is an input of the simulation model. For several months, data regarding arrival rates of the incoming calls for one-day period are collected. We found that the arrival patterns are very different among time periods of a day, among days of a week, and among dates of a month. Fig. 2 shows the arrival patterns of time periods for 15 days. We can see the patterns of Monday through Friday are similar, but patterns of Saturday and Sunday are different from those of other days. From these data we can derive non-homogenous random distributions for 30 minutes period for each day.

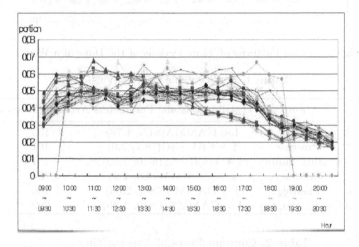

Fig. 2. Arrival Patterns of Time Periods of the Call Center.

Fig. 3 represents incoming call patterns of days of a week for about two months. We can see more incoming calls arrive in Monday and less incoming calls in Saturday and Sunday. Hence, we built a 7 day moving average model for incoming call patterns for the days of a week. The call patterns also changed for the end of a month, and holidays. We also used these special day factor to derive the input call distributions.

Service time data were collected from the CTI server of the call center. We make six service time distributions using ARENA Input Analyzer [8], an input distribution estimation software, for three service classes of two call centers. Table 1 shows each of these distributions.

As we see in Table 1, the service time distributions for quality inquiries are longer than those of other services. After answering a call one telecommunication trunk line the connection with the CTI server are free, but an agent needs a post-call process. Since CTI server does not have this time data for each call, we can calculate only the mean value of post-call process time for each service class. Hence we use constant time of the following mean values as in Table 2.

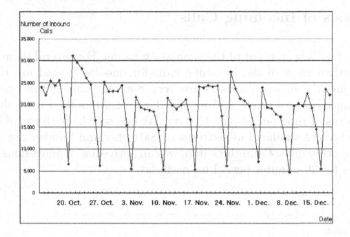

Fig. 3. Arrival Patterns of Time Periods of the Days of a Week.

Table 1. Service Time Distributions.

Center	Group	Distributions and Parameters	Average (Seconds)
Center A	Ordinary	4 + GAMMA(84.5, 1.78)	155
Center A	Prestige	4 + ERLANG(78.7, 2.0)	161
Center A	Quality Inquiry	4 + WEIBUL(195.0, 1.09)	191
Center B	Ordinary	4 + GAMMA(74.8, 2.0)	154
Center B	Prestige	4 + GAMMA(78.9, 2.06)	166
Center B	Quality Inquiry	4 + GAMMA(146.0, 1.36)	202

Table 2. Constant Post-Call Process Times.

Center	Group	Average (Seconds)
Center A	Ordinary	63.10
Center A	Prestige	119.95
Center A	Quality Inquiry	247.26
Center B	Ordinary	52.33
Center B	Prestige	98.88
Center B	Quality Inquiry	218.15

We collected time data for the abandoned calls. As we see in Table 3, calls of the ordinary group and the prestige group can have theoretical probability distributions for abandon time, however for the quality inquiry calls, we should use empirical distributions.

The final data is about transfer rate that is the ratio of transferring from a service to another service. We collected time and ratio data for the transfer and found that the calls from quality inquiry had high (about 10% - 12%) transfer rates. We made random distributions from these data, and used them in the simulation study.

Table 3. Abandon Time Distributions.

Center	Group	Distributions and Parameters	Average (Seconds)
Center A	Ordinary	4 + LOGNORMAL(62.3, 65.1)	32.6
Center A	Prestige	5 + GAMMA(25.8, 2.0)	56.7
Center A	Quality Inquiry	An Empirical Distribution	91.3
Center B	Ordinary	4 + GAMMA(31.4, 1.81)	60.9
Center B	Prestige	4 + GAMMA(32.8, 2.04)	70.9
Center B	Quality Inquiry	An Empirical Distribution	98.5

5 Simulation Model Building, Verifications, and Optimal Call Center Operations

We developed a discrete event simulation model using the process logic which is described in Fig. 1 and input distributions explained in sect. 4. The computer simulation programming is coded by making a Visual Basic program, which has a ProModel as a discrete event simulation engine. The user interface of the developed program is represented in Fig. 4.

We tried to verify if the simulation model portrays the actual call center correctly. We established the following simulation experimental procedure for this verification.

- 1st step: Put a set of input data into the model for some specific time intervals.
- 2nd step: Run a set of computer simulations, and get the output measures. Get statistical confidence intervals from the output measures.

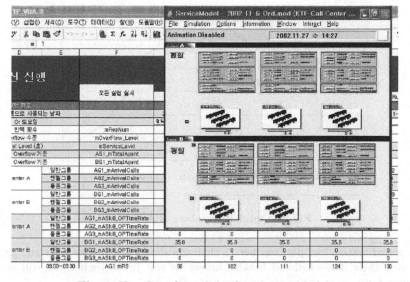

Fig. 4. User Interface of the Simulation Model.

- 3rd step: Compare the actual data, and the confidence intervals from the simulation outputs.
- 4th step: For the significant mismatched data find the causes, and revise the model.

In this verification procedure we learned the following lessons: 1) The recorded data from the CTI servers and actual system operation had some discrepancies; those mainly came from agent break flexibility[5]. We modified the input data for only agents who kept their agent's break schedules, 2) Abilities of the agents are considerably different from each other. We modified the model that had two classes of abilities of agents, 3) The assumption that post-call processing time is constant seems to be incorrect. However, we can not correct the last problem since more precise post-call processing time data , which can not be recorded in the CTI, can not be collected in a limited project schedule.

After performing the verification procedure, we tried to obtain two optimal management policies for the call centers. The first one is to decide the optimal number of agents for each day. The current decision was made by a Work Force Management Tool that was a spreadsheet based on simple heuristic for a given day's call volume forecasting. We already established input call distribution for each day of the week, and each date of a month describe in sect. 4, we needed to find the number of agents that can satisfied the major output performance, the service level. Fig. 5 shows an instance of the optimal numbers of agents for a call center for a given day and date.

The second decision problem for the call centers is about overflow rates between two call centers. Since there are two call centers in the company, we can transfer some incoming calls for a call center to the other call center without

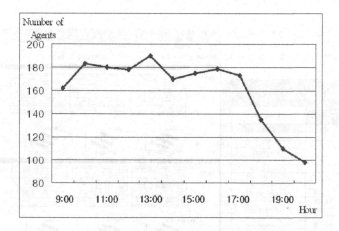

Fig. 5. Optimal Numbers of Agents.

---

[5] The actual system can perform better than the simulation results since the agents can stop their non-scheduled breaks, known as soft breaks, when the call center becomes busy.

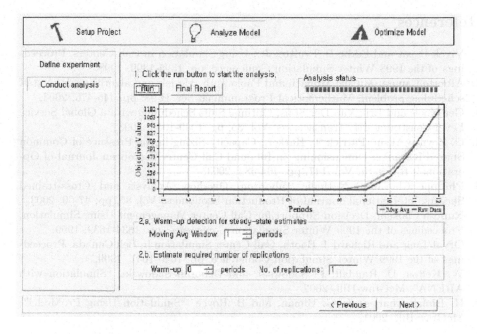

Fig. 6. Optimal Transfer Rates between Call Centers.

increasing the number of the agents for a call center. We can find 10% as the optimal transfer rate of each call center that can satisfy the the service level by using batch runs of the simulation. Fig. 6 shows SimRunner [9], a software that controls the batch runs of the simulation.

6   Conclusions

Call centers become business battle grounds where products and services are purchased, sold, upgraded, and after-serviced. To manage the current complex call centers efficiently, some simple solutions are no longer valid. In this paper, we have shown that discrete simulation is a valid tool to describe and analyze call centers of the corporations. We built a discrete simulation model for the call centers of a large mobile telecommunication service company. We mainly focused on the questions on how the call center can be effectively simulated, and how can we help the decision makers for the efficient call center operations. We can answer the optimal number of agents and optimal transfer rates between two call centers to satisfy some output performances.

In this paper, we did not consider detailed management strategies for the call centers: such as blending each group's agents or process redesigns. One of the further research topics can be the use of the models and results of this paper to answer questions regarding the detailed management strategies, for call centers.

References

1. Vivek Bapat and Eddie B. Pruitte, Jr., Using Simulation in Call Centers, Proceedings of the 1998 Winter Simulation Conference, pp. 1395-1399, 1998.
2. Alberto Capara, Micjele Monaci, and Paolo Toth, Models and algorithms for a staff scheduling problem, Mathematical Programming, Ser. B 98: pp. 445-476, 2003.
3. Ger Koole and Erik Van Der Sluis, Optimal Shift Scheduling with a Global Service Level Constraint, IIE Transactions, Vol. 35, pp. 1049-1055, 2003.
4. O. Zeynep Aksin, Patrick T. Harker., Capacity Sizing in the Presence of Common Shared Resource: Dimensioning an Inbound Call Center, European Journal of Operationals Research, Vol. 147, pp. 464-483, 2003.
5. Philippe Chevalier, Nathalie Tabordon, Overflow Analysis and Cross-trained Servers, International Journal of Production Economics, Vol. 85, pp. 47-60, 2003.
6. Rupesh Chokshi, Decision Support for Call Center Management Using Simulation, Proceedings of the 1999 Winter Simulation Conference, pp. 1634-1639, 1999.
7. Oryal Tanir and Richard J. Booth, Call Center Simulation in Bell Canada, Proceedings of the 1999 Winter Simulation Conference, pp. 1640-1647, 1999.
8. W. Kelton, D. Randall, P. Sadowski, and Deborah Sadowski, "Simulation with ARENA", McGraw-Hill, 2002.
9. H. Biman Charles, K. G. Biman, and B. Royce, "Simulation Using ProModel", McGraw-Hill, 2001.

Requirements Analysis and a Design of Computational Environment for HSE (Human-Sensibility Ergonomics) Simulator

Sugjoon Yoon[1], Jaechun No[2], and Jon Ahn[1]

[1] Department of Aerospace Engineering
Sejong University 98 Gunja-Dong, Gwangjin-Gu, Seoul,
143-747 Republic of Korea
[2] Department of Computer Science & Engineering
Sejong University 98 Gunja-Dong, Gwangjin-Gu, Seoul,
143-747 Republic of Korea

Abstract. Human-Sensibility Ergonomics (HSE) is to apprehend human sensitivity features by measuring human senses and developing index tables related to psychology and physiology. One of the main purposes of HSE may be developing human-centered goods, environment and relevant technologies for an improved life quality. In order to achieve the goal, a test bed or a simulator can be a useful tool in controlling and monitoring a physical environment at will. This paper deals with requirements, design concepts, and specifications of the computing environment for the HSE, which is a part of HSE Technology Development Program sponsored by Korean Ministry of Science and Technology. The integrated computing system is composed of real-time and non-real-time environments. The non-real-time development environment comprises several PC's with Windows NT and their graphical user interfaces coded in Microsoft's Visual C++. Each PC independently controls and monitors a thermal or a light or an audio or a video environment. Each of software and database, developed in the non-real-time environment, is directly ported to the real-time environment through a local-area network. Then the real-time computing system, based on the cPCI bus, controls the integrated HSE environment and collects necessary information. The cPCI computing system is composed of a Pentium CPU board and dedicated I/O boards, whose quantities are determined with expandability considered. The integrated computing environment of the HSE simulator guarantees real-time capability, stability and expandability of the hardware, and to maximize portability, compatibility, maintainability of its software.

1 Introduction

Human-Sensibility Ergonomics (HSE) is a technology measuring and evaluating human sensibility qualitatively or quantitatively in order to develop safe, convenient, and comfortable products and environments. This technology will eventually improve life qualities of mankind. HSE, which is to develop human-centered designs for products and environments, consists of three components [1]. The first component is human sensitivity technology to appreciate sensing mechanism by measuring the five senses responding to outer stimuli. The human sensitivity technology is for developing psychology and physiology indices. The second component is simulation technology. This technology is used to apprehend features of human sensory responses.

T.G. Kim (Ed.): AIS 2004, LNAI 3397, pp. 400–408, 2005.

The HSE simulator is a test-bed to test and evaluate psychological and physiological responses to stimuli from products or environments under intentionally controlled physical conditions. The last component of HSE is application technology, which is to apply the HSE concept to designs of products and environments such as automobiles, home electronics, living or working environments, etc..

This paper presents design concepts and requirements of the computing environment for the HSE simulator developed by KRISS (Korea Research Institute of Standards and Science). The HSE simulator is a testing facility to measure and evaluate human sensitivity factors, such as comfort, satisfaction, movement, spaciousness, and weightiness, processed by a brain and sensory systems including visual, aural, smelling, hearing, and antenna senses as well as an internal ear. For the purpose the simulator sets in a confined space an independent or a composite external environment such as visual scene, sound, smell, taste, temperature, humidity, motion and vibration. When the HSE simulator provides a human with a synthetic environment, the simulator can be used to study human sensitivity technology by measuring human responses to a unit stimulus. The simulator can be also used to develop the application technology, that is, human-friendly goods or environments in the case that composite external environments are furnished. Therefore, the HSE simulator plays a central role among the three HSE components.

In order to design the HSE simulator, experimental facilities in Research Institute of Human Engineering for Quality Life and Life Electronics Research Center in Japan, leading HSE studies in the world, were referenced. Research Institute of Human Engineering for Quality Life in Japan studies living and working environments yielding less stresses as well as product designs reflecting human feelings [2] in the HSMAT (Human Sensory Measurement Application Technology) project. Life Electronics Research Center in Japan researches physiology such as smelling and visual senses and micro electric currents in the body, as well as the subjects in environmental measurement and control [3]. Especially, the HSMAT Project is very similar to the research program "Development of HSE Technologies" sponsored by Korea Ministry of Science and Technology, which supports the study in this paper, in a sense that its research objective is to develop human sensitivity and application technologies. However, the research programs in Japan concentrates on HSE technologies in static environments such as living or working spaces while the Korean research program, whose project year runs through 1995 and 2002, is more interested in dynamic environments including automobiles, airplanes, and railway trains. This may be one of major distinguishing points between the Korean and Japanese research programs.

2   Requirements for the HSE Simulator

As mentioned in the introduction, the subject of this paper is a design of the HSE R&D simulator, more specifically its computing environment. Primary simulation objects of the HSE simulator are dynamic environments such as airplanes and automobiles. Based on physiology and psychology index tables obtained from human sensitivity technology, the HSE simulator effectively measures and evaluates responses of human feelings to composite stimuli in a dynamic environment realized by the simulator. The HSE simulator is composed of a control station, a computing system, environmental controlling and sensing devices. The control station allows a researcher to set HSE environments, such as vision, sound, smell, taste, temperature, humidity, and vibration, at will in a confined space with minimum efforts and time. The computing system comprises system S/W and application S/W whose combination evaluates human responses logically in real-time or non-real-time, based on a relevant HSE database. The controlling and sensing devices are to provide composite stimuli in a simulated dynamic environment and measure responses of human sensitivity factors quantitatively. The system configuration of the HSE simulator developed is illustrated in Fig. 1. In Table 1 are described system requirements relevant to unit testing facilities, each of which generates a different stimulus and measures a relevant human sensitivity factor [1]. Table 1 also shows testing requirements of human sensitivity factors with respect to automobile parts or a whole system. The requirements are actually adopted by the part suppliers and car manufacturers to test their own products. The HSE simulator integrates testing units, realizes their functions seamlessly in a unified test-bed, and provides synthetic composite environments for the HSE studies.

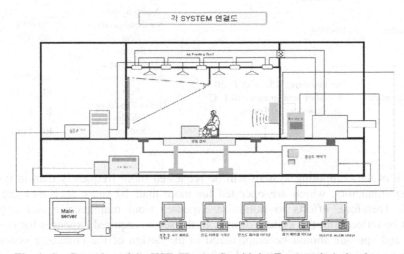

Fig. 1. Configuration of the HSE (Human-Sensitivity Ergonomics) simulator

General requirements applied to the design of the HSE computer system are as follows

- real-time computing capability (update rate: 10hz at minimum)
- stability and durability (operating hours: 8hrs/day, MTBF: 5,000hrs)
- expandability (excessive computing capability and capacity: 50% at minimum)
- S/W portability, compatibility, and maintainability

The above requirements are set based on the experiences and the technical documents [4], [5], [6], [7] in developing airplane, automobile, and railway simulators. Another factor, which has to be considered in the system design, is the operational condition. Research engineers and scientists without any specialties in simulator development or maintenance will operate the HSE simulator. That is, maintenance and

Table 1. Requirement for evaluation and application of human-sensibility factors

Human Sensitivity Factors	Environmental Requirements	Requirements of Car or Parts Manufacturers	
Vision	3D reality of 80% or higher		
Sound and Noise	- air medium: 50~8,000 Hz - solid medium: 40~1,000 Hz - low frequency: 2~50 Hz - when air conditioning stopped: under minimum audible level - when air conditioning operational:: 16dBA	Audio System	100 Hz or less
		Air Conditioner	100 Hz or less
		Seat	100 Hz or less
		Heavy Trucks or Vehicles	10~100 kHz 80 dB or less
		Sedans	125 Hz ~3.2 kHz
Vibration	0.1~50 Hz	Audio System	15~35 Hz
		Air Conditioner	5 kHz or less
		Seat	15~25 kHz
		Sedans	5~400 Hz
Color and Illumination	- lab size: 130~190m^2 variable - transit rate of natural light: 1~5% - brightness of walls and a ceiling: 10~500 cd/m^2		
Humidity	10~95%	Audio System	
		Air Conditioner	10~95%
		Seat	40~95%
		Sedans	
Temperature	- temperature: -15°~+45°C, ±0.5 °C - radiation temperature: ±0.1 °C - air flow rate: ±0.1m/sec	Audio System	
		Air Conditioner	-40°~70°C
		Seat	-30°~150°C
		Heavy Trucks or Vehicles	-40°~100°C
		Sedans	-40°~120°C

upgrade of the computing system will be requested more frequently than in typical training simulators, which are operated for more than 10 years once they are furnished. Therefore, efforts to minimize expenses from maintenance and upgrade should be reflected in the system design. In the sense, required expenses for maintenance and upgrade should be emphasized in the design of the computer system as well as portability, compatibility, and maintainability of the software.

3 Design of the HSE Computing System

The testing units to study human sensitivity factors were already established in the first development stage (1995 – 1998) of the research program "Development of HSE Technologies". The testing units, reflecting the requirements in Table 1, have been used to study human sensitivity factors, and their technical specifications, which should be considered in the design of the new integrated computing system of the HSE simulator, are summarized in Table 2.

The design requirements of the computing system are categorized into general and HSE-specific ones. First, the previously mentioned general requirements are considered to select computer S/W, H/W, and S/W development tools. In order to satisfy the general requirements as much as possible under budget constraints, Pentium PC's and

Table 2. Technical specifications of testing environments for unit HSE factors

		Physiology	Riding Comfort and Fatigue	Sound and Vibration
Computer system		PC Pentium II, 450MHz	PC Pentium II, 450MHz	PC Pentium II, 333MHz
Measurement system	Sensors	EEG, ECG, EMG, PTG, etc.	3-axis accelerometer, me3000p, biopaq, motion capture system	Microphone, Accelerometer
	Output signals	Digital signal	Voltage	Voltage
	Electric power	110 / 220 V	220 V	220V
	A/D Channels	8 bit, 16 ch	10 bit, 30 ch	10 bit, 64 ch
	Update rate	2 kHz	5 kHz (max)	50kHz (max)
Control system	Controllers			Speakers, Hydraulic valve
	Control signals	Digital signal, Voltage	Voltage	Voltage
	Electric power	110 / 220 V		220V
	D/A Channels	8 bit, 24 ch		10 bit, 64 ch
	Update rate	200 kHz		50 kHz (max)
S/W	Operating system	Windows 98	Windows 98	Windows NT 4.0
	Language	Visual C++, Visual Basic Matlab	Visual C++, Matlab	Visual C++, Matlab
	Tool	MS Office, Lab-view Acknowledge	Lab-view	
	Program/data size	1Mb/3Gb	1Mb/3Gb	1Mb/3Gb

Windows NT's are selected as distributed computer platforms and their operating systems. As can be seen in Table 2, present computer systems for testing units to study human sensitivity factors are PC's with either Windows NT or Windows 2000. That is, time and efforts required for system integration will be very effectively and economically used if PC's and Windows NT's are chosen for the integrated system. In addition to this merit, Windows NT guarantees system stability as well as portability, compatibility, and maintainability of the application software. The remaining problem is that Windows NT is not a real-time operating system. The problem can be resolved by either adopting a real-time kernel or adding real features to Windows NT. To achieve real-time capability of the system, it should be also considered how to integrate the distributed computer systems.

Methods integrating unit computer systems to study human sensitivity factors are categorized into LAN (Local Area Network) and bus types. When LAN-type integration is decided, Ethernet is usually considered. The relevant communication protocol will be either TCP/IP or UDP/IP. However, TCP/IP, which has a checking function of communication errors, is not generally recommended in real-time environment, because of following non-real-time characteristics:

- Ethernet collisions
- VME bus spin locks (when the server is of VME bus type)
- dropped packets
- timeouts and retransmissions
- coalescing small packets

In the case of UDP/IP, the protocol does not have error-checking functions, even if it allows faster communications than TCP/IP. When UDP/IP is applied to a real-time system and physical distances of communications are not that long, point-to-point type is often chosen rather than net-type communications. In this case, MAC/LLC (Medium Access Control/Link Level Control) layer just above the physical layer should be edited. But the technology requires profound understanding of UDP/IP protocols, which is not available to the development team of the HSE simulator. The typical speed of Ethernet communication is about 10 Mbps. But it is difficult to say that Ethernet communication is slower than bus types since Ethernet products with 100 Mbps are also available these days in the market. Among bus communications, traditional VME bus and recent cPCI bus are two of the most dominant buses in the market. As can be seen in Table 3, cPCI bus is superior to VME bus in system expandability. Both VME bus and cPCI bus satisfy the real-time requirements of the HSE simulator in communication speeds. VME-based computer systems are also affordable because the configuration comprises not many VME cards in this application, even if cPCI system is less expensive. In conclusion, a cPCI-based system has been selected as an integrated computing environment for the HSE simulator, and the expandability of the system is resolved by using Firewire (P1394) [8].

As mentioned before, Windows NT PC's are chosen for the integrated system. The real-time problem of Windows NT can be resolved by either adopting a real-time kernel or adding real features to Windows NT. QNX [9] and Tornado [10], two of the most popular real-time operating systems, provide real-time kernels, while RTX [11] adds real-time capabilities to the NT kernel. They are compared in Table 4. Tornado is the best in real-time capability, but RTX is superior to the others in functions and S/W compatibility.

All the real-time capabilities of QNX, Tornado, and RTX are acceptable considering the requirements of the unit facilities to test human sensitivity factors in Tables 1 and 2. Therefore, the first criterion in selecting a real-time operating system (executive) should be their available functions, utilities, and maintainability rather than real-time capability. The prices of QNX and RTX are also about the same and lower than that of Tornado. In conclusion, Windows NT with RTX is selected a real-time operating system for the HSE simulator.

The HSE-specific requirements described in Tables 1 and 2 are reflected in the design of the integrated computing system. The H/W configuration is illustrated in Fig. 2 and the S/W configuration in Fig. 3. The following principles are applied to the system design.

Table 3. Comparison of VME and cPCI buses

	VME	cPCI
Slot #	21	8
Bus speed	80 Mbps (VME 64 ext)	128 Mbps (32 bit) 264 Mbs (64 bit)
# of CPU boards	Multi	Single
Processors	68k, PPC, Pentium, i960	Pentium, PPC
hot swap	No	Yes
Connection	Rugged	Rugged
Bus specification	Verified	Verified
Cost	High	50% lower

Table 4. Real-time executives on PC's

		QNX	RTX for Windows NT	Tornado
Operating system		QNX Micro kernel	Windows NT	Wind Micro Kernel
Development tool		Watcom C/C++	Visual C++	GNU C/C++
Debugger			Win DBG/Soft ICE	GNU debugger
Graphic Library		Photon microGUI X Window System for QNX QNX Windows	Any Windows NT compatible graphic libraries	RTX Windows RT GL ZINC
API		QNX dependent API	Win32 API / RTAPI	VxWorks dependent API
RT capa.	Context Switch Time	1.95 s	4.69 s	0.9 s
	Interrupt Latency	4.3 s	8.78 s	1.0 s
	used PC	Pentium, 133MHz	Pentium MMX, 200MHz	Pentium Pro, 150MHz
Cost		< $10,000	$10,000	$22,000

- The HSE simulator is divided into real-time operating and non-real-time development environments.
- The non-real-time environment is composed of PC's with either Windows NT or Windows 98. Their GUI's are programmed in Microsoft's Visual C++. Each PC can independently control the relevant environment such as temperature, humidity, vision, illumination, etc.
- The VME computing system sends on the LAN starting and ending queues of some actions to the VXI system of the motion platform in Figs. 1 and 3, which are the only communication data between two computing systems.
- The VME computing system sends on the bus, starting and ending queues to the physiology measurement system composed of EEG, ECG, EMG, and PTG. The VME computing system intercepts and duplicates the data from the sensors to the processing units of the physiology measurement system.

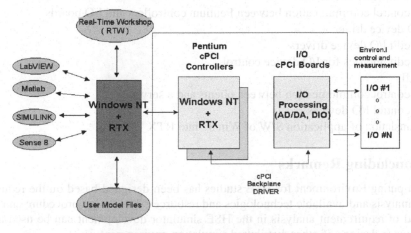

Fig. 2. A software configuration of the HSE real-time computing environment

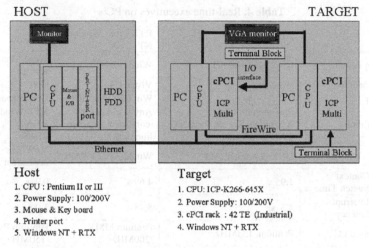

Fig. 3. A hardware configuration of the HSE computing system

4 System S/W Components

In order to realize the computing system for the HSE simulator, a real-time executive, bus communication drivers, I/O drivers, and GUI must be developed. Each of the application S/W for measuring and evaluating human sensitivity factors in the unit systems should be translated or modified in a common language and a unified S/W structure to be a part of the seamless integrating system. The necessary working items are as follows:

- Network Communication
  - control communication between clients and a server
  - edit data received from clients
- cPCI bus driver
  - edit VME bus drivers
  - control communication between Pentium controllers and I/O boards
- I/O device driver
  - edit I/O device drivers
  - edit libraries for I/O device control
- GUI
  - control communication between clients and a server
  - control I/O devices
- Translation of application S/W of Win 32 into RTX

5 Concluding Remarks

A Computing Environment for HSE studies has been designed based on the requirement analysis and available technologies and resources. The design procedure and the method of requirement analysis in the HSE simulator development can be used as a reference to designs of other distributed simulation environments.

Acknowledgement

This research(paper) was performed for the Smart UAV Development Program, one of the 21st Century Frontier R&D Programs funded by the Ministry of Science and Technology of Korea.

References

1. Park S. et al.: Development of Simulator Design Technique for Measurement and Evalua-tion of Human Sensibility, KRISS-98-091-IR, Korea research Institute of Standards and Science (1998)
2. HQL: Catalog of Human Engineering for Quality Life, Japan
3. Life Electronics Research Center: Catalog of Life Electronics research Center, Japan
4. Yoon S.: Development of a FAA Level 7 Flight Training Device, 3rd year Report, Minis-try of Commerce and Trade (1995)
5. Yoon S., Bai M., Choi C., and Kim S.: Design of Real-time Scheduler for ChangGong-91 FTD, Proceedings of 'Making it Real' CEAS Symposium on Simulation technologies, Delft Univ., Nov. (1995)
6. Yoon S., Kim W., and Lee J.: Flight Simulation Efforts in ChangGong-91 Flight Training Device, Proceedings of AIAA Flight Simulation Conference, Aug. (1995)
7. Yoon S.: A Portability Study on Implementation Technologies by Comparing a Railway Simulator and an Aircraft FTD, Proceedings of AIAA Modeling & Simulation Confer-ence, Aug. (1998)
8. http://www.inova-computers.com/
9. http://www.qnx.com/
10. http://www.wrs.com/
11. http://www.vci.com/

Using a Clustering Genetic Algorithm to Support Customer Segmentation for Personalized Recommender Systems

Kyoung-jae Kim[1] and Hyunchul Ahn[2]

[1] Department of Information Systems, Dongguk University, 3-26,
Pil-Dong, Chung-Gu, Seoul 100-715, Korea
kjkim@dongguk.edu
[2] Graduate School of Management, Korea Advanced Institute of Science and Technology,
207-43 Cheongrangri-Dong, Dongdaemun-Gu, Seoul 130-722, Korea
hcahn@kaist.ac.kr

Abstract. This study proposes novel clustering algorithm based on genetic algorithms (GAs) to carry out a segmentation of the online shopping market effectively. In general, GAs are believed to be effective on NP-complete global optimization problems and they can provide good sub-optimal solutions in reasonable time. Thus, we believe that a clustering technique with GA can provide a way of finding the relevant clusters. This paper applies GA-based K-means clustering to the real-world online shopping market segmentation case for personalized recommender systems. In this study, we compare the results of GA-based K-means to those of traditional K-means algorithm and self-organizing maps. The result shows that GA-based K-means clustering may improve segmentation performance in comparison to other typical clustering algorithms.

1 Introduction

As the Internet emerges as a new marketing channel, analyzing and understanding the needs and expectations of their online users or customers are considered as prerequisites to activate the consumer-oriented electronic commerce. Thus, the mass marketing strategy cannot satisfy the needs and expectations of online customers. On the other hand, it is easier to extract knowledge out of the shopping process under the Internet environment. Market segmentation is one of the ways in which such knowledge can be represented and make it new business opportunities. Market segmentation divides the market into customer clusters with similar needs and/or characteristics that are likely to exhibit similar purchasing behaviors [2]. Companies can arrange the right products and services to the right customer cluster and build a close relationship with them with proper market segmentation.

The aim of this study is to carry out an exploratory segmentation of the online shopping market and to find the most effective clustering method for those kinds of data. We adopt some of clustering methods for that purpose. In addition, we compare the performance of each clustering method by using our suggested performance criteria.

The rest of this paper is organized as follows: The next section reviews two traditional clustering algorithms, K-means and self-organizing map (SOM), along with the

T.G. Kim (Ed.): AIS 2004, LNAI 3397, pp. 409–415, 2005.
© Springer-Verlag Berlin Heidelberg 2005

performance criteria. Section 3 proposes the GA approach to optimize the K-means clustering and section 4 describes the data and the experiments. In this section, the empirical results are also summarized and discussed. In the final section, conclusions and the limitations of this study are presented.

2 Clustering Algorithms

Cluster analysis is an effective tool in scientific or managerial inquiry. It groups a set of data in d-dimensional feature space to maximize the similarity within the clusters and minimize the similarity between two different clusters. There exists various clustering methods and they are currently used in wide area. Among them, we apply two popular methods, K-means and SOM, and a novel hybrid method to the market segmentation. Above all, we introduce K-means and SOM in this section. For brief description of each method, we assume that given population is consisted of n elements described by m attributes and it would be partitioned into k clusters. And, $X_i = (x_{i1}, x_{i2}, ..., x_{im})$ represents the vector of the m attributes of element i.

2.1 K-Means Clustering Algorithm

K-means method is a widely used clustering procedure that searches for a nearly optimal partition with a fixed number of clusters. The process of K-means clustering is as follows:

1) The initial seeds with the chosen number of clusters, k, are selected and an initial partition is built by using the seeds as the centroids of the initial clusters.
2) Each record is assigned to the centroid which is nearest, thus forming cluster.
3) Keeping the same number of clusters, the new centriod of each cluster is calculated.
4) Iterate Step 2) and 3) until the clusters stop changing or stop conditions are satisfied.

K-means algorithm has been popular because of its easiness and simplicity for application. However, it also has some of shortcomings. First, it does not deal well with overlapping clusters and the clusters can be pulled of center by outliers. And, the clustering result may depend on the initial seeds but there exists no mechanism to optimize the initial seeds.

2.2 Self-organizing Map

The SOM is the clustering algorithm based on unsupervised neural network model. Since it is suggested by Kohonen [7], it has been applied to many studies because of its good performance. The basic SOM model consists of two layers, an input layer and an output layer. When the training set is presented to the network, the values flow forward through the network to units in the output layer. The neurons in the output layer are arranged in a grid, and the unit in the output layer competes with each other and the one with the highest value wins. The process of SOM is as follows:

1) Initialize the weights as the random small number.
2) An input sample, X_i, is provided into the SOM network and the distances between weight vectors, $W_j = (w_{j1}, w_{j2}, ..., w_{jm})$, and the input sample, X_i, is calculated. Then, select the neuron whose distance with X_i is the shortest. The selected neuron would be called 'winner'.

$$Min \ D(j) = \sum_i (w_{ji} - x_i)^2 \tag{1}$$

3) When α is assumed to be the learning rate, the weights of the winner as well as of its neighbors are updated by following equation.

$$W_j^{NEW} = W_j + \alpha \|X_i - W_j\| \tag{2}$$

4) Iterate Step 2) and 3) until the stop criterion is satisfied.

In spite of several excellent applications, SOM has some limitations that hinder its performance. Especially, just like other neural network based algorithms, SOM has no mechanism to determine the number of clusters, initial weights and stopping conditions.

2.3 Criteria for Performance Comparison of Clustering Algorithms

We compare the performances of the clustering algorithms using 'intraclass inertia'. Intraclass inertia is a measure of how compact each cluster is in the m-dimensional space of numeric attributes. It is the summation of the distances between the means and the observations in each cluster. Eq. (3) represents the equation for the intraclass inertia for a given k clusters [8].

$$F(k) = \frac{1}{n} \sum_K n_K I_K = \frac{1}{n} \sum_K \sum_{i \in C_K} \sum_P (X_P - \overline{X}_{KP}) \tag{3}$$

where n is the number of total observations, C_K is set of the Kth cluster and $\overline{X}_{KP}$ is the mean of the Kth cluster that is $\overline{X}_{KP} = \frac{1}{n_K} \sum_{i \in C_K} X_P$.

3 GA-Based K-Means Clustering Algorithm

As indicated in Section 2.1, K-means algorithm does not have any mechanism to choose appropriate initial seeds. However, selecting different initial seeds may generate huge differences of the clustering results, especially when the target sample contains many outliers. In addition, random selection of initial seeds often causes the clustering quality to fall into local optimization [1]. So, it is very important to select appropriate initial seeds in traditional K-means clustering method. To overcome this critical limitation, we propose GA as the optimization tool of the initial seeds in K-means algorithm.

3.1 Genetic Algorithms

Genetic algorithms are stochastic search techniques that can search large and complicated spaces. It is based on biological backgrounds including natural genetics and
evolutionary principle. In particular, GAs are suitable for parameter optimization
problems with an objective function subject to various hard and soft constraints [10].

The GA works with three operators that are iteratively used. The selection operator
determines which individuals may survive [4]. The crossover operator allows the
search to fan out in diverse directions looking for attractive solutions and permits
chromosomal material from different parents to be combined in a single child. There
are three popular crossover methods: single-point, two-point, and uniform. The single-point crossover makes only one cut in each chromosome and selects two adjacent
genes on the chromosome of a parent. The two-point crossover involves two cuts in
each chromosome. The uniform crossover allows two parent strings to produce two
children. It permits great flexibility in the way strings are combined [6]. In addition,
the mutation operator arbitrarily alters one or more components of a selected chromosome. Mutation randomly changes a gene on a chromosome. It provides the means
for introducing new information into the population. Finally, the GA tends to converge on an optimal or near-optimal solution through these operators [12].

3.2 The Process of GA K-Means Algorithm

GA K-means uses GA to select optimal or sub-optimal initial seeds in K-means clustering. The process of GA K-means clustering is as follows:

1) For the first step, we search the search space to find optimal or near optimal initial seeds. To conduct GA K-means, each observation of the target data should
   have its own ID number and search space consists of ID numbers of the total observations. If we want to find k clusters using GA K-means, there exist k parameters (IDs) to optimize. These parameters to be found should be encoded on a
   chromosome.
2) The second step is the process of K-means clustering which is mentioned in Section 2.1. After the process of K-means, the value of fitness function is updated.
   To find optimal initial seeds, we applied 'intraclass inertia' as a fitness function.
   That is, the objective of our GA K-means is to find the initial seeds that intraclass
   intertia is minimized after K-means clustering finishes.
3) In this stage, the GA operates the process of crossover and mutation on the initial
   chromosome and iterates Step 1) and 2) until the stopping conditions are satisfied.

3.3 Chromosome Encoding

In general, a chromosome can be used to represent a candidate solution to a problem
where each gene in the chromosome represents a parameter of the candidate solution.
In this study, a chromosome is regarded as a set of K initial seeds and each gene is
cluster center. A real-value encoding scheme is used because each customer has
unique customer ID.

For the controlling parameters of the GA search, the population size is set to 100
organisms. The value of the crossover rate is set to 0.7 while the mutation rate is set

to 0.1. This study performs the crossover using a uniform crossover routine. The uniform crossover method is considered better at preserving the schema, and can generate any schema from the two parents, while single-point and two-point crossover methods may bias the search with the irrelevant position of the features. For the mutation method, this study generates a random number between 0 and 1 for each of the features in the organism. If a gene gets a number that is less than or equal to the mutation rate, then that gene is mutated. As the stopping condition, only 5000 trials are permitted.

4 Experimental Design and Results

4.1 Research Data

This study makes use of the public data from the *9th GVU's WWW Users Survey*, part of which concerns with the Internet users' opinions on on-line shopping [5]. Like Velido, Lisboa and Meehan [11], we selected 44 items about general opinion of using WWW for shopping and opinion of Web-based vendors. Total observations were 2180 Internet users.

4.2 Experimental Design

We adopt three clustering algorithms – simple K-means, SOM and GA K-means – to our data. We try to segment the Internet users into 5 clusters. In the case of SOM, we set the learning rate (α) to 0.5.

Simple K-means was conducted by SPSS for Windows 11.0 and SOM by Neuroshell 2 R4.0. GA K-means was developed by using Microsoft Excel 2002 and Palisade Software's Evolver Version 4.06. And, the K-means algorithm for GA K-means was implemented in VBA (Visual Basic for Applications) of Microsoft Excel 2002.

4.3 Experimental Results

As mentioned in Section 2.3, intraclass inertia can be applied to compare the performance of the clustering algorithms. The intraclass inertia of the three clustering algorithms is summarized in Table 1. Table 1 shows that the performance of GA K-means is the best among three comparative methods.

Table 1. Intraclass inertia of each clustering method

Clustering method	K-means	SOM	GA K-means
Intraclass inertia	10850.46	10847.09	10843.70

In addition, Chi-square procedure also provides preliminary information that helps to make the Internet users associated with the segments a bit more identifiable [3]. In this study, Chi-square analysis was used to demographically profile the individuals associated with each of the segments. Table 2 suggests the results of Chi-square analysis. The result reveals that the five segments by all three clustering methods differed significantly with respect to all of demographic variables. So, considering above results, we conclude that GA K-means may offer viable alternative approach.

Table 2. The result of Chi-square analysis

Variables	K-Means		SOM		GA K-Means	
	Chi-square Value	Asy. Sig.	Chi-square Value	Asy. Sig.	Chi-square Value	Asy. Sig.
Age	17.867	0.0222	16.447	0.0364	14.598	0.0670
Gender	51.055	0.0000	48.955	0.0000	52.366	0.0000
Income	148.427	0.0000	134.269	0.0000	129.597	0.0000
Education	68.990	0.0002	72.062	0.0001	78.159	0.0000
Occupation	127.347	0.0000	137.115	0.0000	136.304	0.0000
Years of Web use	197.416	0.0000	203.004	0.0000	206.384	0.0000
Frequency of Web search	528.672	0.0000	535.888	0.0000	559.481	0.0000
Number of purchases over the Web	1027.910	0.0000	1003.725	0.0000	1017.909	0.0000
Experience of Web ordering (yes or no)	662.929	0.0000	654.013	0.0000	654.869	0.0000

5   Concluding Remarks

In this study, we propose new clustering algorithm, GA K-means, and applies it to the market segmentation of electronic commerce. The experimental results show that GA K-means may result in better segmentation than other traditional clustering algorithms including simple K-means and SOM from the perspective of intraclass inertia. However, this study has some limitations. Though this study suggests intraclass inertia as a criterion for performance comparison, it is certain that this is not a complete measure for performance comparison of the clustering algorithms. Consequently, the efforts to develop effective measures to compare clustering algorithms should be provided in the future research. In addition, it is also needed to apply GA K-means to other domains in order to validate and prove its usefulness.

Acknowledgements

This work was supported by Korea Research Foundation Grant(KRF-2003-041-B00181).

References

1. Bradley, P. S., Fayyad, U. M.: Refining Initial Points for K-means Clustering. Proc. of the 15th International Conference on Machine Learning (1998) 91-99
2. Dibb, S., Simkin, L.: The Market Segmentation Workbook: Target Marketing for Marketing Managers, Routledge, London (1995)
3. Gehrt, K. C., Shim, S.: A Shopping Orientation Segmentation of French Consumers: Implications for Catalog Marketing. J. of Interactive Marketing 12(4) (1998) 34-46

4. Hertz, J., Krogh, A., Palmer, R. G.: Introduction to the Theory Neural Computation, Addison-Wesley: Reading, MA (1991)
5. Kehoe, C., Pitkow, J., Rogers, J. D.: Ninth GVU's WWW User Survey. (1998) http://www.gvu.gatech.edu/user_surveys/survey-1998-04/
6. Kim, K.: Toward Global Optimization of Case-Based Reasoning Systems for Financial Forecasting. Applied Intelligence (2004) Forthcoming
7. Kohonen, T.: Self-Organized Formation of Topologically Correct Feature Maps. Biological Cybernetics 43(1) (1982) 59-69
8. Michaud, P.: Clustering Techniques. Future Generation Computer Systems 13 (1997) 135-147
9. Shin, H. W., Sohn, S. Y.: Segmentation of Stock Trading Customers According to Potential Value. Expert Systems with Applications 27(1) (2004) 27-33
10. Shin, K. S., Han, I.: Case-Based Reasoning Supported by Genetic Algorithms for Corporate Bond Rating. Expert Systems with Applications 16 (1999) 85-95
11. Velido, A., Lisboa, P. J. G., Meehan, K.: Segmentation of the On-Line Shopping Market using Neural Networks. Expert Systems with Applications 17 (1999) 303-314
12. Wong, F., Tan, C.: Hybrid Neural, Genetic and Fuzzy Systems. In Deboeck, G. J.: Trading On The Edge. Wiley, New York (1994) 243-261

System Properties of Action Theories

Norman Foo[1] and Pavlos Peppas[2]

[1] School of Computer Science and Engineering, University of New South Wales,
Sydney NSW 2052, Australia
norman@cse.unsw.edu.au

[2] Dept of Business Administration, University of Patras, Patras, 26 500, Greece
ppeppas@otenet.gr

Abstract. Logic-based theories of action and automata-based systems theories
share concerns about state dynamics that are however not reflected by shared insights. As an example of how to remedy this we examine a simple variety of
situation calculus theories from the viewpoint of system-theoretic properties to
reveal relationships between them. These provide insights into relationships between logic-based solution policies, and are suggestive of similar relationships
for more complex versions.

1   Introduction

Reasoning about action in AI has adopted various logical calculi as representational
languages, using classical first-order deduction with a variety of second-order rules or
model selection specifications. A parallel approach, which in fact pre-dates AI logics,
is systems theory. Systems theory unified ad hoc formalisms in different engineering
models of state change by changing the emphasis to systems-theoretic properties such
as causality, realizability and state space minimality. Automata theory was a species of
systems theory. The history of AI logics for reasoning about change has an uncanny
resemblance to that of engineering models. In this paper we begin a program to use the
insights of systems theory to unify diverse strands AI logics. Due to space restrictions a
number of such connections between systems theory and AI logics have been omitted,
but a working paper by the authors is available [Foo and Peppas 2004].

The situation and event calculi (see [Shanahan 1997] for contemporary accounts)
are perhaps the two most popular formalisms for representing dynamic domains in
logic. From the perspective of systems theory we will choose several properties which
are typical of transition systems and automata ([Arbib 1969a]). These properties are
used to classify models of situation calculi, and to examine relationships among them.
We focus on the simplest of situation calculi theories.

2   Situation Trees

To establish notation we first review the situation calculus and automata. The situation
calculus is a multi-sorted first-order theory which admits Herbrand models. The sorts
are: *Sit* (*situations*), *Act* (*actions*), *Flu* (*fluents*). There is a binary function *Result* :
$Act \times Sit \to Sit$, and a binary predicate $Holds : Flu \times Sit$. *Sit* has a distinguished

T.G. Kim (Ed.): AIS 2004, LNAI 3397, pp. 416–427, 2005.
© Springer-Verlag Berlin Heidelberg 2005

member $s0$, called the *initial situation*. The intuition is that Sit is constructed from $s0$ as results of actions from Act, denoting the result of action a on situation s by $Result(a, s)$. Thus Sit is the Herband universe of this theory in which the constants are $s0$ and the action names, and the constructing function is $Result$. The fluents are intuitively the *potentially* observable properties of the situations, as it may be the case that a theory may not specify completely the status of all fluents in all situations. Fluent observation is via the $Holds$ predicate, e.g., $\neg Holds(f, Result(a2, (Result(a1, s0))))$ says that the fluent f does not hold after the actions a1 and a2 are applied (in that order) to the initial situation. A *model* of a situation calculus theory is a Herbrand model. Different names for actions, fluents, and terms denote distinct entities. A contemporary reference for more details is [Shanahan 1997]. A (Moore) automaton $\mathcal{M}$ is a quintuple $\langle I, Q, Y, \delta, \lambda \rangle$ where I is the set of inputs, Y is the set of outputs, Q is the state set, $\delta : Q \times I \rightarrow Q$ is the state transition function, and $\lambda : Q \rightarrow Y$ is the output function. The state set can be infinite. As usual we can represent automata diagrammatically by using labelled circles for states and labelled arcs for transitions. The labels in circles are the outputs for the states, and the arcs are labelled with names of the inputs. Initially, the automata we consider are *loop-free* i.e., there is no path of arcs that lead from one state back to itself. This is to conform to the tree structure of the situations. For our purpose, we will also posit an *initial state* q_0 for the automaton. A standard reference is [Arbib 1969a].

3   Motivating Example

Consider the set Δ of formulas:

$$Holds(f, s0) \wedge Holds(g, s0) \tag{1}$$

$$\neg Holds(f, Result(a, s)) \leftarrow Holds(f, s) \wedge Holds(g, s) \tag{2}$$

$$Holds(g, Result(b, s)) \leftarrow Holds(f, s) \wedge Holds(g, s) \tag{3}$$

This is a simple example of an action specification using the situation calculus. Here $Flu = \{f, g\}$ and $Act = \{a, b\}$. We adopt the diagrammatic convention that $Holds(\neg f, s)$ if and only if $\neg Holds(f, s)$. Here we examine but one way to use the specification – *prediction* – i.e., given the initial situation as above, what can one say about the status of the fluents after the performance of a sequence of actions such as b, a, b? For this example, it is evident that the predictions depend on observation axiom 1 which says what fluents hold in the initial situation $s0$. More generally, such observations can refer to "deeper" situations away from the initial.

Definition 1. *An observation term of depth k is a ground situation term of the form* $Result(a_k, Result(a_{k-1}, (\ldots Result(a_1, s0) \ldots)))$. *An observation sentence of depth k is one in which the only atoms are of the form $Holds(F, S)$ where S is an observation term, and the maximal depth of such terms is at most k.*

It is convenient to define the depth of non-observation terms to be 0, so axioms 2 and 3 are of depth 0.

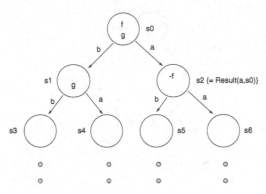

Fig. 1. Minimal information situation tree of $Th(\Delta)$.

Definition 2. *An action specification is predictively-free if all its observation sentences are of depth 0, i.e. only about the initial situation.*

As most specifications are incomplete (for good reason), we may in principle complete them in any consistent manner. Figure 1 is a diagram that displays a fragment of $Th(\Delta)$, the deductive closure of Δ. It can be interpreted as a partial automaton, in which we have identified (in the notation above) Sit with the state set Q, Act with the input set I, the $Result$ function with the state transition function δ, Flu with the output set Y and the $Holds$ predicate with the output function λ. This diagram sets down the minimal consequences of Δ, so that a fluent (output) F (respectively $\neg F$) appears in a situation (state) S if and only if $\Delta \models Holds(F, S)$ (respectively $\Delta \models \neg Holds(F, S)$). The automaton is partial in two senses: (i) it is incomplete for outputs as just explained; (ii) it only suggests that there are situations (states) beyond two actions by means of ellipses. Therefore, completing them amounts to completing both the outputs and extending the depth of the tree ad infinitum. We will initially focus on the former means of completion.

The "filling-in" of the situations in figure 1 is in general constrained by observation formulas. For instance, if it were the case that $\neg Holds(g, Result(a, s0))$ was included in Δ we would be forced to add fluent g to the circle for $s2$ in the figure. But such a possibility is excluded for predictively-free theories since the depth of $Result(a, s0)$ is greater than 0. For the predictively-free Δ as shown, two completions of $Th_2(\Delta)$ are shown as figures 2 and 3, which are of course also fragments of deterministic automata. In figure 2, the choices for completing $s1$ and $s2$ were made before that of higher depth situations $s3$ through $s6$, i.e., commitments were made to chronologically earlier terms. Moreover, the choices were made on the basis of the principle of *inertia* which formalizes the notion that unless a theory forces a fluent change, it should not happen. Thus, fluent f remains true in $s1$, and likewise g remains true in situation $s2$, because no change was forced by $Th(\Delta)$. However, in $s4$, fluent f must change on account of equation 2, but $s3$, $s5$ and $s6$ need not have fluent changes from their preceding states. We define inertia formally later.

In figure 3, fluent f violates inertia in $s2$, where fluent f has changed when it need not done so. This completion has another feature, viz., from $s1$ and $s2$ which have identical fluents holding, the result of applying the same action a to them results in

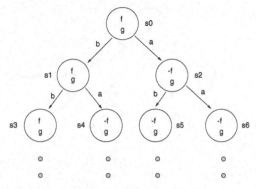

Fig. 2. Chronologically inertial completion of minimal information tree.

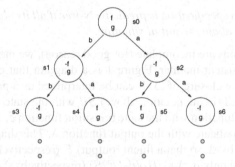

Fig. 3. Another (non-Markovian) completion of minimal information tree.

two situations (states) $s4$ and $s6$ which have different fluents. The condition that situations which agree on fluents make transitions under the same action to situations that also agree on fluents is called *Markovian*. As such, the automaton in figure 3 is non-Markovian. We will define this formally later.

4   Maximal Consistency and Deterministic Automata

Regardless of whether or not an action specification is predictively-free, from the discussion above, it is but a small step to observe that if we are given a situation calculus action theory there is a straightforward correspondence between completions of the theory and the loop-free automata that arise from viewing the situation trees of the theory as a state-transition system. This follows from the well-known fact that every consistent theory can be extended to a maximal consistent theory. Hence, if Γ is a situation calculus theory and Γ_1 is a maximal consistent extension of it, then for every situation s and every fluent f, either $\Gamma_1 \models Holds(f, s)$ or $\Gamma_1 \models \neg Holds(f, s)$. This uniquely determines a deterministic automaton via the translation: (i) the automaton state set Q corresponds to the situations Sit, with initial situation $s0$ identified with initial state q_0; (ii) the automaton input set I corresponds to the actions Act; (iii) and for each fluent f a state (corresponding to) s will output either f or $\neg f$ depending respectively on whether $Holds(f, s)$ or $\neg Holds(f, s)$ is in the maximal extension (thus, the elements of the output set Y are the consistent, complete sets of positive and negative fluents). Since these maximal consistent extensions are also the models of the theory, we may summarize these remarks as:

Observation 1. *The models of a situation calculus action theory may be interpreted as deterministic automata.*

5   Effect Axioms and Constraints

An *effect axiom* is a formula of the form

$$ResultFormula(Result(a, s)) \leftarrow PrecondFormula(s) \qquad (4)$$

where $ResultFormula(Result(a, s))$ is a formula whose literals are only of the form $[\neg]Holds(f, Result(a, s))$, and $PrecondFormula(s)$ is a formula whose literals are only of the form $[\neg]Holds(f, s)$. By $[\neg]$ we mean that the negation may or may not occur. In such a formula, the situation variable s is common between $ResultFormula(Result(a, s))$ and $PrecondFormula(s)$, but fluent variables need not be shared. Intuitively, an effect axiom specifies the *direct* effects of the action a when the pre-condition specified by $PrecondFormula$ is satisfied and the action is performed. Equations 2 and 3 in the motivating example above are effect axioms.

Indirect effects, also called ramifications, can arise via interaction with domain constraints (e.g. [Zhang and Foo 1996]) or causal rules (e.g. [Thielscher 1999]). We confine attention here to constraints, postponing a similar treatment for causal rules to a later report. A *constraint* is a formula whose only situation term is a free s, meaning that it must be satisfied in every model, or in the automaton view, every situation in every possible tree has a fluent set consistent with the constraint.

The incompleteness of an action theory can arise because of one or more of the following reasons: (i) some effect axioms leave the status of some fluents (after an action) unspecified, or (ii) there are non-deterministic effects. The second is complex enough to warrant detailed treatment in a separate paper. In this paper we will concentrate on actions that are definite in the sense defined below.

Definition 3. *Let C be the conjunction of the set of constraints. If for all s, C $\cup$ $ResultFormula(Result(a, s)) \models l_1 \vee \ldots l_n \Rightarrow C \cup ResultFormula(Result(a, s))$ $\models l_i$ for some i, $1 \leq i \leq n$, we say that action a is definite ([Zhang and Foo 1996]).*

If there are no constraints, definiteness simply means that the effect sub-formulas $ResultFormula(Result(a, s))$ are (conjunctions of) literals of the form $[\neg]Holds(f, Result(a, s))$.

Definition 4. *When a is definite, let $Post(a) =$ $\{[\neg]f | C \cup ResultFormula(Result(a, s)) \models [\neg]f\}$.*

The significance of $Post(a)$[1] is that it is exactly the set of fluents determined by action a in the resulting state. Hence for any s, the status of fluents in $Flu - Post(a)$ is not determined in state $Result(a, s)$.

---

[1] It is independent of s.

6   Review of Circumscription

The following *Abnormality Axiom* ([Lifschitz 1994,Shanahan 1997]) is usually introduced to handle the well-known *Frame Problem*.

$$Ab(a, f, s) \leftarrow \neg[Holds(f, Result(a, s)) \leftrightarrow Holds(f, s)] \tag{5}$$

The atom $Ab(a, f, s)$ is meant to signify that fluent f is abnormal with respect to action
a and situation s, i.e., it changes as a result of this action applied to this situation. "Most"
fluents are not abnormal, as intuitively only a small number of fluents are affected by
an action. The hope is that suitable choices for the interpretation of Ab in formula 5
used in conjunction with effect axioms of the form 4 will permit action theories to
be only partially specified, with implicit completions achieved by these choices. It is
also well-known that *Circumscription* [Lifschitz 1994] is a second-order scheme that
enforces such choices, and we shall assume that readers are familiar with it and merely
review notation. $Circ(\Gamma(P); P; Q)$ means the circumscription of the predicate P in
action theory $\Gamma(P)$ in which P appears as a predicate constant, letting the predicate (or
function) Q vary. Let $[\![R]\!]_M$ stand for the extension of predicate R in model M. The
effect of $Circ(\Gamma(P); P; Q)$ is to impose a partial order $\sqsubseteq_P$ on models of $\Gamma(P)$, such
that $M_1 \sqsubseteq_P M_2$ if $[\![P]\!]_{M_1} \subseteq [\![P]\!]_{M_2}$, and the interpretations of all other predicates,
with the exception of Q, agreeing on M_1 and M_2. By saying that Q can vary, we mean
that the extensions (or function interpretations) $[\![Q]\!]_{M_1}$ and $[\![Q]\!]_{M_2}$ can be different.
$Circ(\Gamma(P); P; Q)$ selects those $\Gamma(P)$ models which are $\sqsubseteq_P$-minimal, and it is in this
sense that it mimimizes $[\![P]\!]$.

7   Abnormality, Inertia and the Markov Property

It was once hoped that by minimizing $[\![Ab]\!]$ in formula 5 we can constrain the change of
fluents to the least that is consistent with the specification Γ. This hope was only partially fulfilled, because the models of $Circ(\Gamma(Ab); Ab; Holds)$ sometimes included not
only the ones that were intuitively desired, but also some that achieved the minimization of Ab by trading off its minimization in some places of the situation tree against
those in other places. The prototypical example of this is the infamous Yale Shooting
Problem (YSP) [Hanks and McDermott 1987], and partially represented in figure 4.
    The situation calculus description of the YSP is:

$$Holds(L, Result(Load, s)) \tag{6}$$

$$\neg Holds(A, Result(Shoot, s)) \leftarrow Holds(Loaded, s) \tag{7}$$

$$Holds(A, S0) \tag{8}$$

$$\neg Holds(L, S0) \tag{9}$$

Note that there is no effect axiom for the $Wait$ action. As this is a well-known
example, we recall it merely to cite its second $Circ(\Gamma(Ab); Ab; Holds)$-model (model
II in the figure) as an example of an Ab-minimal automaton that is not inertial in the
sense to be defined in section 8 below.

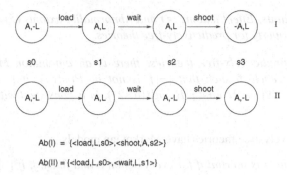

$$Ab(I) = \{<load,L,s0>,<shoot,A,s2>\}$$

$$Ab(II) = \{<load,L,s0>,<wait,L,s1>\}$$

Fig. 4. Partial situation trees for the Yale Shooting Problem.

8 Markovian and Inertial Models

Figures 2 and 3 suggest different ways to complete a theory. In the first figure, because $s0$ and $s1$ have the same fluents, their resulting states also have the same fluents. This is not so in the second figure where, e.g., although $s1$ and $s2$ have the same fluents, their resulting states under action a do not. The predicate $Equiv$ captures the notion that two situations cannot be distinguished using their fluents.

Definition 5. $Equiv(s1, s2) \leftrightarrow \forall f[Holds(f, s1) \leftrightarrow Holds(f, s2)]$

Definition 6. *A automaton M has the Markov property (is Markovian) if for all states s1 and s2, and all actions a,*

$$Equiv(s1, s2) \rightarrow Equiv((Result(a, s1), Result(a, s2)). \tag{10}$$

Figure 2 suggests a Markovian automaton. It is possible (as suggested by figure 3) for an automaton to be deterministic and yet not Markovian. In the next subsections we will argue that standard ways of specifying actions are consistent with the Markov property.

The formula below defines a lower bound on the extension of the predicate *Surprise*. Intuitively, $Surprise(s1, s2)$ holds whenever the states $s1$ and $s2$ violate the Markov property, so if we circumscribe *Surprise* we should minimize this violation, and ideally, we may be able to abolish it altogether, i.e., $[\![Surprise]\!] = \phi$. Since $Equiv$ depends on $Holds$ and $Result$, we could choose to let either of them vary in circumscribing *Surprise*, but in this paper we will let $Holds$ vary. (Varying $Result$, an alternative introduced by Baker [Baker 1991], is treated in a companion paper [Foo, et.al 2000], and also re-examined in [Foo, et.al 2001].

$$Surprise(s1, s2) \leftarrow \neg Equiv(Result(a, s2), Result(a, s2))$$
$$\land Equiv(s1, s2) \tag{11}$$

Observation 2. *An automaton is Markovian if and only if $[\![Surprise]\!] = \phi$.*

The next lemma and observation show that Markovian automata are not overly restrictive as models of predictively-free situation calculus theories. First, by definition Post(a) is independent of s, hence is consistent with the Markov property.

Lemma 1. *Constraints, Effect axioms (4) and Abnormality axioms (5) are consistent with the Markov property for predictively-free theories.*

Corollary 1. *For predictively-free theories, there is an automaton M such that if $Equiv(s1, s2)$, for each f such that $[\neg]f$ is not in $Post(a)$, it is the case that $Holds(f, Result(a, s1)) \leftrightarrow Holds(f, Result(a, s2))$. M is consistent with $[\![Surprise]\!]$ $= \phi$.*

Hence, predictively-free theories have Markovian models.

Definition 7. *A state s is inertial if for every action a and fluent f, if $[\neg]f \notin Post(a)$ then $Holds(f, s) \leftrightarrow Holds(f, Result(a, s))$. An automaton is inertial if every state is inertial.*

Proposition 1. *A predictively-free theory has an inertial automaton. Moreover, the automaton is unique.*

In a trivial way, it is possible for the inertial automaton to satisfy a theory that is not predictively-free. This can occur when all observation sentences of depth $k > 0$ are logically implied by the non-observational part of Δ and the observational sentence about $s0$. Let us call such observational sentences *redundant*, and those that are not so implied *irredundant*. Less trivially, it can simply be the case that each such sentence is consistent with the (theory of) the inertial model even though they may be irredundant. From these simple observations we may conclude that an action theory fails to have an inertial model if it has an irredundant observation sentence that is inconsistent with the inertial model. We investigate this further in a future paper.

9   Relationships

Proposition 2. *The unique inertial automata are Markovian.*

Proposition 3. *Inertial automata are Ab-minimal.*

The dependencies between the three properties of theories can be summarized as follows (see figure 5).

Proposition 4. *1. Markovian $\not\Longrightarrow$ Ab-Min*
 2. Ab-Min $\not\Longrightarrow$ Markovian
 3. Markovian $\not\Longrightarrow$ Inertial
 4. Ab-Min $\not\Longrightarrow$ Inertial
 5. Markovian and Ab-Min $\not\Longrightarrow$ Inertial

These are demonstrated by counterexamples.

For (1), consider figure 6. This represents a theory in which the only action a has no effect axiom, and the only fluent is f. It is clearly Markovian but is not Ab-minimal as it is not inertial. This also shows (3).

For (2), consider figure 7 which is essentially a "gluing together" of the two Ab-minimal models of the YSP (with $Holds$ varying). Its Abnormal triples are displayed.

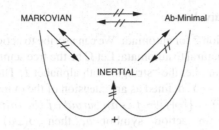

Fig. 5. Dependencies.

There is no effect axiom for action a

Fig. 6. Markovian but not Ab-Min (and not Inertial).

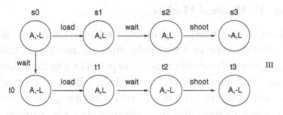

Ab(III) = {<load,L,s0>,<shoot,A,s2>,<load,L,t0>,<wait,L,t1>}

Fig. 7. Ab-Min but not Markovian.

It is not hard to verify (from the fact that the two original models of the YSP are *Ab*-minimal and incomparable) that this is also a *Ab*-minimal model. However, s1 and t1 show that it is not Markovian.

For (4) we use the model II of the YSP as shown in figure 4. It is *Ab*-minimal, but the state s1 is not inertial.

For (5), the model II of the YSP again suffices. It is easy to extend the partial representation in figure 4 to a Markovian model M which remains *Ab*-minimal, but is still not inertial. The basic idea is to specify that in this extension all situations are inertial except for those that are equivalent to $s = \{A, L\}$. Then any smaller interpretation for *Ab* has to eliminate the abnormality due to *wait*, but it then introduces one for the situations equivalent to $Result(wait, s)$, which is not abnormal in M.

10   State Equivalence and Circumscription

In this section we will argue that Markovian models of action theories are in a strong sense "natural". And given this, there is a simple class of automata that can be considered to be natural models of these theories. Moreover, we can reason about the *Ab* predicate using the automata models.

10.1   Input-Output Equivalence

Recall the notation in section 2 for automata. We can ascribe to action theories an *Input-Output (I-O)* view that is natural to automata. Let Ω be the free semigroup over the input symbols I of the automaton, i.e, the "strings" with alphabet I. The I-O function of an automaton is a map $\beta : \Omega \to Y$ defined as an extension of the output map λ as follows. For the empty string ϵ, $\beta(\epsilon) = \{[\neg]f \,|\, [\neg]f \text{ is an output of the initial state}\}$; if ωa is a string ω followed by an action symbol a, then $\beta(\omega a) = \lambda(\beta(\omega), a)$. Intuitively, if ω is the action sequence $a_1 \ldots a_k$, then $\beta(\omega)$ interpreted in the action theory is simply the collection of fluent literals from the formulas $\{[\neg]Holds(f, (Result(a_k, (\ldots (Result(a_1, s0) \ldots) \ldots)\}$.

Definition 8. *Two automata M_1 and M_2 with respective I-O functions β_1 and β_2 are I-O equivalent if $\beta_1 = \beta_2$.*

I-O equivalence is a key concept in systems theory treated in the work of Zeigler and his colleagues [Zeigler, et.al. 2000].

10.2   Markov and Ab-Minimal Models

That the combination of the Markov property and Ab-minimality is insufficient to guarantee Inertia shows that the first two properties are not overly restrictive. In fact, since both the Ab and $Surprise$ predicates occur *positively* in their defining equations, it is possible (see [Lifschitz 1994]) to reduce the *parallel* circumscriptions $Circ(\Gamma(Ab, Surprise); Ab; Holds)$ and $Circ(\Gamma(Ab, Surprise); Surprise; Holds)$ to their separate circumscriptions, and then intersecting their resulting model classes. By corollary 1 earlier, we showed that the latter circumscription will succeed with $[\![Surprise]\!] = \phi$. This permits both the models I and II in figure 4 to survive, but eliminates model III in figure 7.

Suppose we have a model that is Markovian. Then $Equiv$ is not only an equivalence relation on Sit, but it is in fact a right-congruence relation with respect to actions, which indeed is what definition 6 says. Regarded as an automaton M, this Markov property is the basis of a *reduced automaton* M_R ([Arbib 1969a]), which is a "quotient automaton" whose states are the equivalence (congruence) classes $[s]$ of M, and whose state transition function δ_R is defined by $\delta_R([s], a) = [\delta(s, a)]$. The output function is defined by $\lambda_R([s]) = \lambda(s)$. Both of these are well-defined as $Equiv$ is a right-congruence relation. These are classical results in both systems [Arbib 1969b] and automata theory [Arbib 1969a]. It is not hard to see that M and M_R are I-O equivalent in the sense of definition 8. Moreover, the states of M_R all have different output fluent sets. The advantage of viewing Markovian models as automata is due to a standard result, the Nerode finite index theorem ([Arbib 1969b]). It essentially says that $Equiv$ has finite index if and only if M_R is finite state. Hence in any action theory with only a finite number of fluents, its Markovian models are reducible to finite automata. As automata have efficient algorithms for model checking, queries about action theories can be computationally tractable when their models are automata.

The ordering of Markovian models according to the extension of the Ab predicate is preserved in the passage from automaton M to its reduced counterpart M_R. We can make this claim precise. To do so, we make the following observations:

Observation 3. *If $Equiv(s1, s2)$ then in any Markovian automaton of an action theory, $Ab(a, f, s1)$ holds if and only if $Ab(a, f, s2)$ holds.*

Observation 4. *In a reduced Markovian (this is essential) automaton, distinct states have distinct output fluent sets.*

Definition 9. *For two reduced automata M_1 and M_2, $M_1 \sqsubseteq_{Ab} M_2$ if and only if for all actions a, fluents f and (reduced) states s1 and s2 (respectively in M_1 and M_2) which have identical fluent sets, $Ab_1(a, f, s1) \subseteq Ab_2(a, f, s2)$.*

In reduced automata the nodes correspond to states. Note that the notion of abnormality is well-defined even for reduced automata if we treat the (equivalence classes of) states as abstract states in their own right. The next proposition justifies the use of reduced automata as compact representations of those theories of actions that are Markovian, as no Ab-minimal models are lost.

Proposition 5. *If $M1$ and $M2$ are two Markovian models of an action theory and M_1 and M_2 are their reduced counterparts, then $M1 \sqsubseteq_{Ab} M2$ implies $M_1 \sqsubseteq_{Ab} M_2$.*

Non-Markov models are discussed in [Foo and Peppas 2004].

11 Conclusion

This paper is part of a series that investigates relationships among different model restriction and selection policies in logic-based theories of action using a systems-theoretic perspective. We selected the simplest of situation calculus theories so that the basic ideas are not obscured by distracting technicalities. The results are encouraging. For this class of theories we were able to identify the Markov property as a sufficient condition for finite reduced automata to be their models. Moreover, the possible inter-dependencies among Markovian, inertial and abnormality-minimized models were clarified. Nondeterminism is discussed in the full version of the paper ([Foo and Peppas 2004]). Future papers will report similar systems-theoretic insights into theories which have arbitrary observation terms, causal rules and ramifications, and the consequences of letting predicates other than $Holds$ vary. For instance, in [Foo, et.al 2001] we describe how circumscribing Ab in predictively-free theories by letting $Result$ vary guarantees the Markov property. Future work will extend these techniques to event calculus logics.

References

[Arbib 1969a] M. Arbib. *Theories of Abstract Automata*, Prentice-Hall, 1969.

[Arbib 1969b] M. Arbib. Automata Theory: the rapprochement with control theory. In R. Kalman, P. Falb and M. Arbib, *Topics in Mathematical Systems Theory*, McGraw-Hill, 1969.

[Baker 1991] A. Baker. Nonmonotonic Reasoning in the Framework of the Situation Calculus. *Artificial Intelligence*, 49, 1991, 5-23.

[Foo, et.al 2000] N. Foo, P. Peppas, Q.B. Vo, D. Zhang. Circumscriptive Models and Automata. Proceedings NRAC'01 Workshop, 17th International Joint Conference on Artificial Intelligence, IJCAI'01, August 2001, Seattle.

[Foo and Peppas 2004] N. Foo and P. Peppas. System Properties of Action Theories. Working paper downloadable from http://www.cse.unsw.edu.au/ ksg/Pubs/ksgworking.html.

[Foo, et.al 2001] Norman Foo, Abhaya Nayak, Maurice Pagnucco and Dongmo Zhang. State Minimization Revisited. Proceedings of the 14th Australian Joint Conference on Artificial Intelligence, Springer LNAI 2256, ed. M. Stumpner ; D. Corbett ; M. Brooks, pp 153-164, 2001.

[Hanks and McDermott 1987] S. Hanks and D. McDermott. Nonmonotonic Logic and Temporal Projection. *Artificial Intelligence*, 33, 1987, 379-412.

[Lifschitz 1994] V. Lifschitz. Circumscription. In *The Handbook of Logic in Artificial Intelligence and Logic Programming, vol. 3: Nonmonotonic Reasoning and Uncertain Reasoning*, ed. D.M. Gabbay, C.J. Hogger and J.A. Robinson, Oxford University Press, 297-352, 1994.

[Shanahan 1997] M. Shanahan, *Solving the Frame Problem: a mathematical investigation of the commonsense law of inertia*, MIT Press, 1997.

[Thielscher 1999] M. Thielscher. From Situation Calculus to Fluent Calculus. *Artificial Intelligence*, 111, 1999, 277-299.

[Winslett 1988] M. Winslett. Reasoning about actions using a possible models approach. *Proc Seventh National Artificial Intelligence Conference*, AAAI'88, San Mateo, CA., 1988, Morgan Kaufmann Publishers.

[Zeigler, et.al. 2000] B.P. Zeigler, H. Praehofer and T.G. Kim, *Theory of Modeling and Simulation :integrating discrete event and continuous complex dynamic systems*, 2nd ed, Academic Press, San Diego, 2000.

[Zhang and Foo 1996] Y. Zhang and N. Foo, *Deriving Invariants and Constraints from Action Theories* , Fundamenta Informaticea, vol. 30, 23-41, 1996.

Identification of Gene Interaction Networks
Based on Evolutionary Computation

Sung Hoon Jung[1] and Kwang-Hyun Cho[2],*

[1] School of Information Engineering, Hansung University, Seoul, 136-792, Korea
shjung@hansung.ac.kr
http://itsys.hansung.ac.kr/
[2] College of Medicine, Seoul National University, Chongno-gu, Seoul, 110-799, Korea
and Korea Bio-MAX Institute, Seoul National University,
Gwanak-gu, Seoul, 151-818, Korea
Tel.: +82-2-887-2650, Fax : +82-2-887-2692
ckh-sb@snu.ac.kr

Abstract. This paper investigates applying a genetic algorithm and an
evolutionary programming for identification of gene interaction networks
from gene expression data. To this end, we employ recurrent neural networks to model gene interaction networks and make use of an artificial
gene expression data set from literature to validate the proposed approach. We find that the proposed approach using the genetic algorithm
and evolutionary programming can result in better parameter estimates
compared with the other previous approach. We also find that any a
priori knowledge such as zero relations between genes can further help
the identification process whenever it is available.

1 Introduction

With the advent of the age of genomics, an increasing number of researches to
identify gene interaction networks have been reported in [1–13] as a first step of
unraveling their functions. In order to identify gene (interaction) networks (or
genetic regulatory networks), a gene network model should be first built and
then the parameters of the network model should be estimated using gene expression data obtained from experiments such as DNA microarray. Identifying
a gene interaction network can be viewed as a reverse engineering in that the
interactions between genes should be determined from the gene expression data.
Gene networks have been modeled as Boolean networks [3], linear networks [4,
5, 9, 11], Bayesian networks [6], differential equations [14], and recurrent neural networks [1, 12, 13] and the parameters of the networks have been estimated
from gene expression data by simulated annealing, genetic algorithms, gradient
descent methods, and linear regression methods [4, 5, 8, 9, 11]. We can summarize three major problems in the identification of gene network parameters from
gene expression data. First of all, the number of sampling points for the gene expression data obtained from experiments is usually too small compared with the

---

* Corresponding author.

T.G. Kim (Ed.): AIS 2004, LNAI 3397, pp. 428–439, 2005.

large number of parameters of the model, which results in insufficient accuracy of the solutions for the parameter estimates. Secondly, from a viewpoint of reverse engineering, there can be non-unique solutions that all fit the given gene expression data within the same error bound. Third, gene network models themselves can only have limited mathematical structures which might not properly reflect the real gene networks. As a consequence, no approach developed so far provides a satisfactory solution resolving all these problems.

In this paper, we employ recurrent neural network models for the gene interaction networks, and utilize genetic algorithms (GAs) [15–17] and evolutionary programming (EP) [18] to identify the parameters of the network model. Genetic algorithms and evolutionary programming are known as robust and systematic optimization methods which have been successfully applied to many scientific and engineering problems. In the genetic algorithm, a chromosome is generally composed of bit strings. Therefore, the solutions obtained by applying genetic algorithms are discrete and the resolution of the solution is proportional to the number of bits used in formulating the chromosomes. If only one bit is used then the interaction between genes can be represented by either zero (inhibition) or one (activation) which becomes similar to the case of Boolean networks. Evolutionary programming, a kind of evolutionary computation, is another optimization method which encodes a solution to a real value without coding a bit string like GAs. EP uses only the mutation operation without the crossover operation in GAs. In some applications, EP have been reported as showing a better performance than GAs.

Wahde et al. [8] proposed a possible way of identifying the gene interaction networks from the coarse-grained point of view by using a GA to estimate model parameters defined upon a recurrent neural network model. In order to identify model parameters, gene expression data must be available. Here we use artificial gene expression data obtained by a presumed model which was previously used by Wahde et al. [8]. Experimental results are obtained by applying the proposed approach based on the GA and the EP, respectively, and we compare these experimental results with that of Wahde's paper [8]. We further consider using a priori knowledge in building a network model and estimating the parameter values. A notion of zero relations denoting non-interaction between genes is introduced and we investigate how a priori knowledge such as the zero relations can further help the identification process. We also compare the resulting parameter estimates with other ones.

This paper is organized as follows. In section 2, a gene network model is introduced. Section 3 describes the proposed parameter estimation processes based on the GA and the EP. The experimental results and discussions are provided in section 4. Section 5 makes conclusions of this paper.

2   Gene Network Model

Gene interaction networks have been modeled by Boolean networks [3], linear networks [4, 5, 9, 11], Bayesian networks [6], differential equations [14], and recurrent neural networks [1, 12, 13]. In this paper, we use a recurrent neural network model [1, 8, 13] since it is the most descriptive deterministic modeling framework among those, formulated by:

$$\tau_i \frac{dx_i(t)}{dt} = g\left(\sum_{j=1}^{J} W_{i,j} x_j(t) + b_i\right) - x_i(t), \tag{1}$$

where $x_i(t)$ denotes gene expression of a gene i at time instant t, $g : \Re \mapsto \Re$ is an activation function, $W_{i,j}$ is the strength of control of gene j on gene i, b_i is a bias level of gene i, and τ_i is time constant of the ith gene expression product. Sigmoid function $g(z) = (1 + e^{-z})^{-1}$ is generally used as the activation function. Let $z = \sum_{j=1}^{J} W_{i,j} x_j(t) + b_i$, then (1) can be rewritten as

$$\frac{dx_i(t)}{dt} = \frac{1}{\tau_i}\left(\frac{1}{1 + e^{-z}} - x_i(t)\right). \tag{2}$$

This equation can then be further represented by a difference equation as

$$x_i[t+1] = \frac{1}{\tau_i}\left(\frac{1}{1 + e^{-z}} - x_i[t]\right) + x_i[t]. \tag{3}$$

We note that $g(z) \in (0, 1)$ guarantees $(g(z) - x_i[t]) \in (-x_i[t], 1 - x_i[t])$. If the initial value of $x_i[0]$ is in $(0, 1)$, then $(g(z) - x_i[t]) \in (-1, 1)$. If $x_i[t] = 0$ in (3), then $x_i[t+1] = \frac{1}{\tau_i(1+e^{-z})}$, i.e., $x_i[t+1] \in (0, 1/\tau_i)$ If $x_i[t] = 1$ in (3), then $x_i[t+1] \in (1 - 1/\tau_i, 1)$. Therefore, we confirm that $x_i[t+1]$ is bounded between 0 and 1, provided that initial values of $x_i[0]$ lies between 0 and 1, and $\tau_i > 1$. In experiments, we use (3) with all $\tau_i > 1$.

3   Parameter Identification

For the identification of network model parameters, several methods such as simulated annealing, genetic algorithms, gradient descent methods, and linear regression methods have been applied so far [4, 5, 8, 9, 11]. In this paper, we employ genetic algorithms and evolutionary programming to identify the network model parameters since they are most efficient for nonlinear optimization problems.

Wahde et al. [8] also considered making use of the GA to identify model parameters defined upon a recurrent neural network model. They used artificial gene expression data (the patterns of which are very similar to real gene expression data as they showed) obtained from a presumed artificial network. However, we find that the standard deviation of the resulting parameter estimates are too broadly distributed from the average values. On the other hand, we confirm that the proposed approach in this paper reveals better performance in this regard as we can see in the following section. The difference is mainly caused by the structure of GA such as the encoding scheme and crossover methods and by the parameters used in the GA such as the crossover and mutation probabilities and termination conditions. In general, the encoding scheme and crossover methods greatly affect the performances of GA. Algorithm 1. illustrates the operation of the GA used here [16].

Algorithm 1. Genetic Algorithm
   // t : time //
   // P : populations //
1 $t \leftarrow 0$
2 initialize $P(t)$
3 evaluate $P(t)$
4 **while** (**not** termination-condition)
5 **do**
6      $t \leftarrow t + 1$
7   select $P(t)$ from $P(t-1)$
8   recombine $P(t)$
9   evaluate $P(t)$
10 **end**

Chromosomes (in other words, individuals) in the GA are represented by bit strings of a fixed number by a chosen encoding scheme. According to this encoding scheme, a specific number of initial individuals are generated with uniformly distributed random values. Then, the fitness of each individual is assigned by the evaluation operation. To make the next generation, the GA selects parents and then recombines them to generate offsprings with the assignment of their fitness values. This evolution process continues until a solution is obtained within the given specification of an error bound.

We encode the parameters of a gene network model into a number of bit strings. If there are P parameters and the number of bits for one parameter is K, then one chromosome is composed of the $P * K$ number of bits. In this case, we note that the resolution of a solution becomes 2^K. If the K is one, then the recurrent neural network model is similar to the Boolean network in that the parameters can have either of the two values, i.e., one for inhibition and the other for activation between genes. Thus, one of the advantages using a GA is that we are able to control the resolution of a solution as needed. As the P and K increase, the bit strings of a chromosome increase much faster, in which one point crossover is not enough to cover the whole length of chromosomes. Hence we employ in this paper a multi-points crossover scheme. The parameters used in implementing the GA are summarized in Table 1. Note that we use the 40 number of crossover points because the bit strings of an individual is very large (720 bits = 24 parameters * 30 bits).

In order to identify model parameters, we use an artificial gene expression data set obtained by a presumed model which was also used by Wahde *et al.* [8]. The artificial data set is composed of expression levels of 4 genes with 30 time points. We identify total 24 parameters (16 interaction weight parameters ($W_{i,j}$),

Table 1. Parameters for the GA.

Parameters	Values
Crossover probability (p_c)	0.6
Crossover points	40
Normal mutation probability (p_m)	0.01
Population size	20
Individual length	(24 * 30) bits
Termination condition	0.001

4 bias parameters (b_i), and 4 time constant parameters (τ_i)). Each parameter is represented by 30 bits. Each individual contains the set of all encoded parameters to be estimated and the estimates are being updated as generations evolve according to the GA. In order to measure how close an individual j approaches to the solution, the individual j is first decoded into the gene network model. Then, the following mean square error δ_j of the jth individual is obtained by the trajectories from the decoded gene network model.

$$\delta_j = \frac{1}{T * N} \sum_{i=1}^{N} \sum_{t=0}^{T} \left(x_i[t] - x_i^d[t] \right)^2, \tag{4}$$

where N is the number of genes, T is the total number of sampling time points, $x_i[t]$ is the expression level of ith gene under the decoded gene network model, and $x_i^d[t]$ is the desired expression level obtained from the artificial network or an experimental result in real application. The number of parameters identified is $N(N+2)$ when the number of genes is N because the bias b_i and time constants τ_i in addition to the weights $W_{i,j}$ are also identified. For a reasonably accurate estimation of the network parameters, $T > N+2$ is required because the number of data is $T * N$ in (4).

We use a mean square error for the termination condition. If an individual has a mean square error less than the termination condition in a generation, the GA terminates. We note that the mean square error however can not guarantee the quality of the resulting solution since there are a number of parameter sets which can show the similar trajectories as the given data set. This is one of the important difficulties faced in the identification of gene interaction networks.

The mean square error is also used as a measure of fitness for each individual. Since the fitness should be inversely proportional to the mean square error, the fitness of jth individual is given by

$$f_j = \frac{1}{\delta_j + 1}. \tag{5}$$

Even if the GA can be applied to continuous domain problems, the GA is in general more useful when the problem domain is discrete since the individuals are represented as bit strings. Therefore, the resolution of a solution depends on the number of bit strings. The EP on the other hand as another evolutionary

Algorithm 2. Evolutionary Programming

    // $2m$: the number of population, (m parents + m offspring) //

    // X_i : the ith vector of population P //

    // $F(X_i)$: the objective function of vector X_i //

    // J_i : the score of vector X_i, $J_i \leftarrow F(X_i)$ //

    // W_i : the number of winning in competition //

    // σ_i : the perturbation factor of vector X_i, $\sigma_i \propto J_i$ //

1   initialize population $P = [X_0, X_1, \ldots, X_{2m-1}]$

2   **while** (not termination-condition)

3   assign a score J_i to each solution vector X_i

4   compete and count the winning number of X_i with the others $X_j (\neq X_i)$

5   select m parents solutions according to the number order of W_i

6   generate offspring by adding a Gaussian perturbation $N(0, \sigma_i)$ to parents

7   **end while**

computation method, however, does not encode the given problem domain into bit strings and makes use of the real value as they are. This enables the EP to lead to the better resolution of a solution. From this regard, we apply the EP to identify the gene interaction network parameters. Algorithm 2. illustrates the procedure of the EP used in our experiments.

While the main operation in the GA is the crossover and the GA makes each individual evolve towards improved fitness, the main operation in the EP is the mutation (by adding a Gaussian perturbation) and thereby make each individual evolve towards a decreasing score. As shown in the algorithm 2., the more individuals come close to the optimum the smaller Gaussian perturbation is added. Table 2 summarizes the parameter values used for the implementation of the EP. As we already mentioned in the algorithm 2., the σ_i is in general given as proportional to the score J_i; however, we use a constant value here (see Table 2) since the EP sometimes can fall into a local optimum, which is unavoidable if the score at that instant becomes too small.

Table 2. Parameters for the EP.

Parameters	Values
# of population ($2m$)	100
ith vector X_i	the 24 number of real values
score of vector X_j	δ_j at equation (4)
the number of winning in competition	40
σ_i	10.
Termination condition	0.001

4 Experimental Results

In order to identify gene interaction networks, gene expression data from such as DNA microarray must be available. On the other hand, directly using the

Table 3. Parameter values assumed for the artificial data.

$W_{i,j}$				b_i	τ_i
20.0	5.0	0.0	0.0	0.0	10.0
25.0	-5.0	-17.0	0.0	-5.0	5.0
0.0	10.0	20.0	-20.0	-5.0	5.0
0.0	0.0	10.0	-5.0	0.0	15.0

real gene expression data has also a drawback in that the identified results can not be verified until we conduct another experiments to validate the estimated parameter values. Hence we use in this case a set of artificial data obtained from a presumed artificial network model (so we know the exact value of parameters to be estimated and thereby can easily validate the results of the proposed approach). As previously mentioned in Section 2, we use the difference equation model described in (3). We assume a set of gene expression data obtained by using the parameter values shown Table 3 in the model. Figure 1 shows the artificial data with 30 number of time points for T.

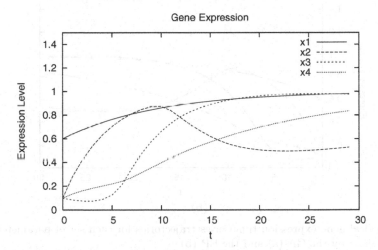

Fig. 1. Artificial data.

We applied the proposed identification approaches based on the GA and the EP to this set of artificial data. Figure 2 illustrates the typical trajectories exhibited by the network model using the identified parameter values. As shown in Figure 2, the trajectories of the EP are more accurate than those of the GA. This is because the EP directly makes use of the real value for encoding while the GA employs bit strings instead. As it is mentioned before, there are other numerous set of parameter values which are in accord with the given expression data within a small error bound. We takes here the average values of 50 runs (as same with that of Wahde *et al.* [8]). Table 4 (a), (b), and (c) compare each experimental result, i.e. the proposed GA, the proposed EP, and the Wahde *et al.*'s GA, respectively. In Table 4, each number denotes the average value of 50 runs and the value within the parenthesis indicates the corresponding standard

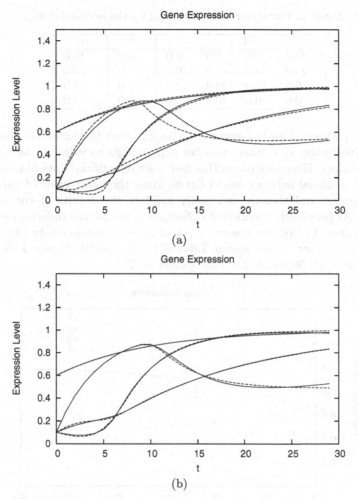

Fig. 2. Typical gene expression time-series trajectories for each set of parameter estimates obtained by the GA (a) and the EP (b).

deviation. We took one of experimental results of Wahde *et al.*'s GA with the same number of time points over the same trajectories. From Table 4, three results look quite similar, but the proposed GA and EP are somewhat better than that of Wahde *et al.*'s GA. We speculate that this result is caused by the difference of the structure and parameter values used for the GA[1]. We need to further investigate which approach can be always better than the others in a certain environments.

In order to examine the effects by the number of simulation runs, we have conducted additional experiment of 50 simulation runs further to the experi-

---

[1] More systematic analysis and discussions for comparison between Wahde *et al.*'s GA and proposed GA are not possible because there is no detailed descriptions about the structure and parameter values of Wahde *et al.*'s GA in their paper [8].

Table 4. Comparison of the experimental results: (a) the proposed GA (b) the proposed EP, and (c) Wahde *et al.*'s GA.

(a)

$W_{i,j}$				b_i	τ_i
17 ($\pm$9.3)	11 ($\pm$13)	6.7 ($\pm$15)	4.0 ($\pm$16)	3.1 ($\pm$5.5)	10 ($\pm$1.1)
20 ($\pm$7.8)	-13 ($\pm$5.3)	-11 ($\pm$10)	-9.4 ($\pm$12)	5.1 ($\pm$4.7)	5.2 ($\pm$0.5)
-12 ($\pm$9.3)	16 ($\pm$7.9)	12 ($\pm$9.5)	13 ($\pm$11)	-4.8 ($\pm$4.6)	5.8 ($\pm$0.6)
-4.5 ($\pm$13)	11 ($\pm$12)	14 ($\pm$11)	1.4 ($\pm$15)	-0.4 ($\pm$5.4)	17 ($\pm$1.8)

(b)

$W_{i,j}$				b_i	τ_i
14 ($\pm$13)	9.1 ($\pm$15)	7.9 ($\pm$14)	7.1 ($\pm$17)	3.7 ($\pm$6.1)	10 ($\pm$1.1)
24 ($\pm$6.1)	-15 ($\pm$4.2)	-18 ($\pm$8.2)	-8.2 ($\pm$10)	8.0 ($\pm$2.9)	5.1 ($\pm$0.3)
-8.7 ($\pm$8.4)	15 ($\pm$8.6)	12 ($\pm$10)	15.2 ($\pm$14)	-8.3 ($\pm$2.7)	5.1 ($\pm$0.6)
6.3 ($\pm$15)	12 ($\pm$14)	11 ($\pm$14)	4.4 ($\pm$15)	-1.3 ($\pm$7.3)	18 ($\pm$1.8)

(c)

$W_{i,j}$				b_i	τ_i
11 ($\pm$11)	7.8 ($\pm$12)	7.5 ($\pm$12)	0.8 ($\pm$13)	3.7 ($\pm$4.8)	11 ($\pm$1.2)
14 ($\pm$7.7)	-16 ($\pm$7.3)	0.45($\pm$13)	-2.4 ($\pm$14)	1.8 ($\pm$4.6)	7.5 ($\pm$1.8)
5.9 ($\pm$14)	11 ($\pm$11)	6.1 ($\pm$13)	3.2 ($\pm$14)	1.9 ($\pm$5.8)	9.1 ($\pm$0.7)
3.0 ($\pm$11)	1.5 ($\pm$13)	5.3 ($\pm$11)	-16 ($\pm$7.0)	1.6 ($\pm$5.7)	16 ($\pm$5.3)

Table 5. Experimental results of our EP at 100 runs.

$W_{i,j}$				b_i	τ_i
15 ($\pm$12)	7.1 ($\pm$15)	7.5 ($\pm$15)	8.4 (+17)	2.6 ($\pm$6.4)	10 ($\pm$1.1)
25 ($\pm$5.7)	-15 ($\pm$4.0)	-18 ($\pm$7.7)	-8.7 ($\pm$10)	8.0 ($\pm$2.7)	5.1 ($\pm$0.4)
-8.6 ($\pm$8.7)	15 ($\pm$8.5)	13 (+10)	13.3 (+14)	-8.4 ($\pm$2.7)	5.1 ($\pm$0.6)
7.5 ($\pm$15)	10 ($\pm$16)	10 ($\pm$13)	5.0 ($\pm$15)	-0.7 ($\pm$7.0)	18 ($\pm$1.8)

ments of Table 4 (b). The result of the total 100 simulation runs of the proposed EP is shown in Table 5. From Table 5, we confirm that there are changes towards the correct values although the changes are very small.

In order to compare the results in a more formal way, we introduce the concept of an identification rate defined as follows. Let $P_n, n = \{1, 2, \ldots, N(N+2)\}$ composed of $W_{i,j}$, b_i, and τ_i, where N is the number of genes in the artificial network that generates the given set of artificial gene expression data. For example, $P_1 = W_{1,1}, P_2 = W_{1,2}, \ldots, P_{N(N+2)} = \tau_N$. Let $P'_n, n = \{1, 2, \ldots, N(N+2)\}$ composed of $W'_{i,j}$, b'_i, and τ'_i the model parameters of the identified network. Let the average value and standard deviation of P_n represented by ρ_{P_n} and σ_{P_n}, respectively. Then the identification rate ψ between the P and P' is given by

$$\psi_{P,P'} = \frac{1}{N(N+2)} \sum_{n=1}^{N(N+2)} \frac{1}{[1 + \eta(\rho_{P_n} - \rho_{P'_n})^2][1 + (1 - \eta)(\sigma_{P_n} - \sigma_{P'_n})^2]}, \quad (6)$$

where η is a ratio how much the average term affects the identification rate compared with the standard deviation.

In (6), $\rho_{P_n} = P_n$ and $\sigma_{P_n} = 0$ since the model parameters in the case of an artificial network have a unique set of real values. If η is one, then the identification rate takes only average value without considering the standard deviation and vice versa. We consider, as an extreme example first, a parameter $W_{i,j}$ and $W'_{i,j}$ have zero standard deviations, respectively and the difference between their average values is 10. Secondly, a parameter $W_{i,j}$ and $W'_{i,j}$ have the same average value and the difference between their standard deviations is 10. Then, the two cases produce a same identification rate for the parameter if η is not considered. Intuitively, the second case should be better than the first case. However, the standard deviation term must not be ignored since even if the average values are same, we still cannot regard it as a good one in the case that the difference of standard deviations is very large. This measure of identification is further used and the results are summarized in Table 7.

If there are certain available a priori knowledge, it can further help the identification process by eliminating infeasible solutions in advance. For instance, we consider a priori knowledge about 'zero relations' (meaning that there exists no interaction between two genes) prior to the proposed EP algorithm. Normally initial individuals are generated by random values within the predefined ranges; however, with this a priori knowledge, we can replace the initial random value with zero at the point already known from the a priori knowledge and similarly for the evolution process of individuals. Table 6 illustrates the experimental result with this a priori knowledge. It is apparent that this result is better than the previous results. However, we note that the weight in the third row and the fourth column $W_{3,4}$ is still positive even though it is set to -20 in the presumed artificial network. This is because the affect of the weight on the overall gene expression data is too small. After all, we find from this experiment that any a priori knowledge about the gene interaction could be of great help in the identification of the network model and the a priori knowledge is in most cases available from the real experimental environments.

Table 6. Experimental results with the zero relations.

$W_{i,j}$				b_i	τ_i
17 (±9.3)	8.3 (±14)	0 (±0)	0 (±0)	0 (±0)	10 (±1.2)
22 (±8.4)	-13.3 (±4.8)	-23 (±6.1)	0 (±0)	7.6 (±4.0)	5.2 (±0.4)
0 (±0)	9.3 (±5.4)	11 (±10)	10(±14)	-9.4 (±0.9)	5.4 (±0.8)
0 (±0)	0 (±0)	12 (±10)	3.2(±14)	0 (±0)	18 (±2.2)

Table 7. Identification rates.

Methods	ψ
Wahde' GA	0.042
Our GA	0.134
Our EP	0.138
Our EP with 100 runs	0.140
Our EP with zero relations	0.458

Table 7 summarizes the comparison of the identification rates of each result under the η is 0.9. We note that the proposed GA and EP approaches show considerably better identification rates than that of Wahde *et al.*'s GA. From the fourth row in Table 7, the more the number of simulation runs increases, the better the identification rates are. As can be easily guessed, using the zero relations increases the identification rates as well.

5   Conclusions

In this paper, we proposed the GA and the EP to identify gene interaction networks by estimating the network parameters. With an artificially generated set of gene expression data, the two approaches have been validated as improving the identification performance compared to the previously reported approach in the literature. We further considered using a priori knowledge such as the zero relations between genes for preprocessing of the proposed EP approach and confirmed that using such a priori knowledge can be of much help in the identification process. We note however that some of the parameter estimates obtained by the proposed approaches can have relatively large errors and it should be further investigated by rigorous analysis.

Acknowledgment

This research was financially supported by Hansung University in the year of 2004 and also partially supported by a grant from the Korea Ministry of Science and Technology (Korean Systems Biology Research Grant, M10309000006-03D5000-00211).

References

1. Reinitz, J., Sharp, D.H.: Mechanism of eve stripe formation. Mechanisms of Development (1995) 133–158
2. Eisen, M.B., Spellman, P.T., Brown, P.O., Botstein, D.: Cluster analysis and display of genome-wide expression patterns. In: Proceedings of the National Academy of Sciences. (1998) 14863–14868
3. Akutsu, T., Miyano, S., Kuhara, S.: Identification of genetic networks from a small number of gene expression patterns under the Boolean network model. In: Pacific Symposium on Biocomputing. (1999) 17–28
4. Weaver, D., Workman, C., Stormo, G.: Modeling regulatory networks with weight matrices. In: Pacific Symposium on Biocomputing. (1999) 112–123
5. D'haeseleer, P., Wen, X., Fuhrman, S., Somogyi, R.: Linear modelling of mRNA expression levels during CNS development and injury. In: Pacific Symposium on Biocomputing. (1999) 41–52
6. Friedman, N., Linial, M., Nachman, I., Pe'er, D.: Using Bayesian networks to analyze expression data. Journal of Computational Biology (2000) 601–620
7. D'haeseleer, P.: Reconstructing Gene Networks from Large Scale Gene Expression Data. PhD thesis, University of New Mexico, Albuquerque (2000)

8. Wahde, M., Hertz, J.: Coarse-grained reverse engineering of genetic regulatory networks. BioSystems (2000) 129–136
9. van Someren, E., Wessels, L., Reinders, M.: Linear modelling of genetic networks from experimental data. In: Proceedings of the Eighth International Conference on Intelligent Systems for Molecular Biology. (2000) 355–366
10. van Someren, E., Wessels, L., Reinders, M.: Genetic network models: A comparative study. In: Proceedings of SPIE, Micro-arrays: Optical Technologies and Informatics. (2001) 236–247
11. Wessels, L., van Someren, E., Reinders, M.: A comparison of genetic network models. In: Pacific Symposium on Biocomputing. (2001) 508–519
12. Wahde, M., Hertz, J.: Modeling genetic regulatory dynamics in neural development. Journal of Computational Biology 8 (2001) 429–442
13. Takane, M.: Inference of Gene Regulatory Networks from Large Scale Gene Expression Data. Master's thesis, McGill University (2003)
14. Chen, T., H. He, G.C.: Modeling gene expression with differential equations. In: Pacific Symposium on Biocomputing. (1999) 29–40
15. Holland, J.: Adaptation in natural and artificial systems. 1st ed.: University of Michigan Press, Ann Arbor; 2nd ed.: 1992, MIT Press, Cambridge (1975)
16. Goldberg, D.: Genetic Algorithms in Search, Optimization and Machine Learning. Addison–Wesley, Reading, MA (1989)
17. Davis, L.: Handbook of Genetic Algorithms. New York: Van Nostrand Reinhold (1991) L. Davis editor.
18. Fogel, D.B.: An Introduction to Simulated Evolutionary Optimization. IEEE Trans. on Neural Networks 5 (1994) 3–14

Modeling Software Component Criticality
Using a Machine Learning Approach

Miyoung Shin[1] and Amrit L. Goel[2]

[1] Bioinformatics Team, Future Technology Research Division,
ETRI, Daejeon 305-350, Korea
shinmy@etri.re.kr
[2] Dept. of Electrical Engineering and Computer Science,
Syracuse University, Syracuse, New York 13244, USA
goel@ecs.syr.edu

Abstract. During software development, early identification of critical components is of much practical significance since it facilitates allocation of adequate resources to these components in a timely fashion and thus enhance the quality of the delivered system. The purpose of this paper is to develop a classification model for evaluating the criticality of software components based on their software characteristics. In particular, we employ the radial basis function machine learning approach for model development where our new, innovative algebraic algorithm is used to determine the model parameters. For experiments, we used the USA-NASA metrics database that contains information about measurable features of software systems at the component level. Using our principled modeling methodology, we obtained parsimonious classification models with impressive performance that involve only design metrics available at earlier stage of software development. Further, the classification modeling approach was non-iterative thus avoiding the usual trial-and-error model development process.

1 Introduction

During the past twenty five years, there have been impressive advances in the theory and practice of software engineering. Yet, software system development continues to encounter cost overruns, schedule slips and poor quality. Due to the lack of an adequate mathematical basis for the discipline of software engineering, empirical modeling approaches are commonly used to deal with the problems of cost, schedule and quality. Two classes of empirical models that have been extensively investigated address software development effort estimation [1], and software component criticality evaluation [2]. However, the applicability and accuracy of most of these models is far from satisfactory. This necessitates further research on these topics. In this paper we focus on the second class of models, viz. models for software component criticality evaluation. This problem can be seen as developing an input-output relationship where the inputs are the software component characteristics and the outputs are their criticality levels. Specifically, a classification model is built from the historical data about the software systems in the database using the known values of some measurable features of software components, called *software metrics*, as inputs and their corresponding criticality levels as outputs. Then, based on metrics data about some previously unseen software component, the classification model is used to assess its class, viz. high or low criticality.

T.G. Kim (Ed.): AIS 2004, LNAI 3397, pp. 440–448, 2005.

A number of studies have been reported in the literature about the use of classification techniques for software components [2,3,4,5,6,7]. A recent case study [2] provides a good review about their application to large software systems. Among the techniques reported in the literature are: logistic regression, case based reasoning (CBR), classification and regression trees (CART), Quinlan's classification trees (C 4.5, SEE 5.0), neural networks, discriminant analysis, principal component regression, neuro-fuzzy logic, etc. Some of the criteria used for selecting, comparing and evaluating classification methods include the time to fit the model, time to use the model, predictive accuracy, robustness, scalability, interpretability, model complexity, and model utility [9, 10]. Obviously, all of these cannot be simultaneously optimized and a compromise is sought based on the application domain. For the current study, i.e., for developing component criticality classifiers, we consider two criteria, viz. predictive accuracy and model utility. Here we evaluate model utility in terms of the number of software characteristics used in the classification model and the software development stage at which they become available for use in the model. A model that uses only a few metrics that become available early has a higher utility than the one that uses several metrics, some of which become available late in the software development life-cycle. Predictive accuracy is measured here by classification error on the test set.

As a classification model, we consider a machine learning approach called Gaussian radial basis functions (GRBF) because it possesses strong mathematical properties of universal and best approximation. This model performs a nonlinear mapping of the input data into a higher dimensional space for classification tasks. Justification for this mapping is found in Cover's theorem. Qualitatively, this theorem states that a complex classification problem mapped in a higher dimensional space nonlinearly is more likely to be linearly separable than in a low-dimensional space. Thus, we investigate the GRBF models for software component criticality evaluation in this paper. Particularly, we are interested in (1) developing good models for classifying the criticality of software components based on their design metrics and (2) evaluating the efficacy of these models in comparison to the models constructed with the coding metrics and the combined design and coding metrics, respectively.

This paper is organized as follows. In Section 2, we provide a brief description of the software data set employed in the study. An outline of our classification methodology using GRBF is given in Section 3. The main results of the empirical study are discussed in Section 4. Finally, some concluding remarks are presented in Section 5.

2 Software Data Description

The software data employed in this study are obtained from the publicly available NASA software metrics database. It provides various project and process measures for many space related software systems and their subsystems. We took the data for eight systems used for tracking launched objects in space, consisting of a total of 796 components. The number of components in a system varies from 45 to 166 with an average of about 100 components per system. Among the various measurable features of software components, we only considered three design metrics and three coding metrics, which are shown in Table 1. It should be noted that the design metrics become available to software engineers much earlier in the software development life cycle than the coding metrics. Therefore, the classification or predictive models based on design metrics have higher utility, i.e., are more useful, than those based on, say, coding metrics that become available much later in the development life cycle. Typical values of the six metrics for five components in the database are shown in Table 2, along with the component criticality level where 1 refers to high criticality and 0 to low criticality. These levels are determined according to the scheme described below. Prior to modeling, all the metrics data were normalized to lie in the range 0 to 1.

Table 1. List of metrics

Metrics	Variable	Description
Design Metrics	X_1	Function calls from this component
	X_2	Function calls to this component
	X_3	Input/Output component parameters
Coding Metrics	X_4	Size in lines
	X_5	Comment lines
	X_6	Number of decisions

Table 2. Values of selected metrics for five components from the database

Component Number	X_1	X_2	X_3	X_4	X_5	X_6	Criticality levels
1	16	39	1	689	388	25	1
2	0	4	0	361	198	25	0
3	5	0	3	276	222	2	0
4	4	45	1	635	305	32	0
5	38	41	22	891	407	130	1

The classification of software component into high criticality (class 1) and low criticality (class 0) was determined on the basis of the actual number of faults detected in a component, by taking the philosophy conveyed in the well known Pareto principle which has been found to be applicable to fault distribution among software components. According to this principle, in our context, approximately 80% of the faults are found in 20% of the components. After analyzing the fault distribution for our data set, components with 5 or less faults were labeled as class 0 (i.e., low criticality) and others as class 1 (i.e., high criticality). As a result, our software data consists of about 25% of the high critical components and about 75% of low critical components.

For model development, the data set of 796 components was arbitrarily divided into three subsets, 398 components for training, 199 components for validation, and 199 components for testing. Further, 10 random permutations of the original data sets were analyzed to minimize the effect of random partitioning of the data into three subsets on classification performance.

3 Modeling Methodology

The modeling task in this study is to construct a model that captures the unknown input-output mapping pattern between the software metrics of a component and its criticality level. It is to be accomplished on the basis of limited evidence about its nature. The evidence available is a set of labeled historic data (training data) denoted as

$$D = \{(\mathbf{x}_i, y_i): \mathbf{x}_i \in R^d, y_i \in R, i = 1,..,398\}.$$

Here the inputs ($\mathbf{x}_i$) represent the metrics data of software components and their corresponding outputs (y_i) represent the class label, which is either zero or one for our problem. In the above setup, d is the dimensionality of the input data.

The GRBF model is a nonlinear model where the estimated output $\hat{y}$ for input vector $\mathbf{x}$ is represented in the following functional form:

$$\hat{y} = f(\mathbf{x}) = \sum_{j=1}^{m} w_j \phi_j(\mathbf{x}) = \sum_{j=1}^{m} w_j \phi (\| \mathbf{x} - \mu_j \| / \sigma_j) = \sum_{j=1}^{m} w_j \exp(-\| \mathbf{x} - \mu_j \|^2 / 2\sigma_j^2)$$

where $\phi_j(.)$ is called a basis function and μ_j and σ_j are called the center and the width of jth basis function, respectively. Also, w_j is the weight associated with the jth basis function output and m is the number of basis functions. Thus, the GRBF model is fully determined by the parameter set $P = (m, \mu, \sigma, \mathbf{w})$.

From a modeling point of view, we seek a classification model that is neither too simple nor too complex. A model that is too simple will suffer from under fitting because it does not learn enough from the data and hence provides a poor fit. On the other hand, a model that is too complicated would learn too many details of data including even noise and thus suffers from over-fitting, without providing good generalization on unseen data. Hence, it is an important issue how to determine an appropriate number of basis functions that represents a good compromise between the above conflicting goals, as well as good selection of basis function parameters. For this purpose, we employed our recently developed algorithm [1, 11] for determining GRBF model parameters. It consists of the following four steps.

Step 1: Select a range of values for width σ and a desired complexity measure δ, Heuristically, we take $0 \le \sigma \le \sqrt{d/2}$ where d is the number of input variables.

Step 2: Determine a value of m that satisfies the δ criterion. This step involves singular value decomposition of the interpolation matrix computed from the ($n \times d$) input data matrix.

Step 3: Determine the centers of the m basis functions such that these centers would maximize structural stability provided by the selected model complexity, m. This step involves the use of QR factorization.

Step 4: Compute weight using pseudo inverse and estimate output values.

It should be noted that the desired complexity measure δ needs to be specified by the user at the beginning, along with a global width σ. For our analyses, we used a default value of $\delta=0.05$ as our desired complexity; generally its value is taken to be in the range [0.01, 0.1]. The value of the global width σ is heuristically determined by considering the dimensionality of the input data, as indicated above. Once these parameters are specified, the algorithm automatically finds a good choice of the parameters $P= (m, \mu, \sigma, \mathbf{w})$ for the GRBF model. The mathematical and computational details of the algorithm can be found in [1,11].

For each model developed from the training data, its classification error on validation data is computed as a fraction of the total components that are incorrectly classified. The model with the smallest validation error is chosen as our classification
model. Finally, to predict its generalization performance, we compute classification
error for test data so as to get an estimate of the classification error rate for future
components of which criticality levels are not known yet.

4 Experiment Results

In this section we perform three experiments of developing classification models for
software component criticality evaluation with (1) design metrics, (2) coding metrics,
and (3) combined design and coding metrics. In each case, we develop GRBF classifiers according to the methodology described above. Since the model development
algorithm used in this study provides consistent models without any statistical variability, we do not need to perform multiple runs for a given data set. In order to assess model utility, we perform comparative assessment for the classification models
that are obtained with three different sets of metrics, viz. design, coding and combined design and coding. These results are discussed below.

4.1 Results with Design Metrics

The training, validation, and test error rates of the classification models that were
developed based on design metrics (X_4, X_5, X_6) are shown in Table 3. We note that
the training error varies from 21.6% to 27.1% with an average of 24.3%. The validation error varies from 21.1% to 29.2% with an average of 25.9%, and the test error
varies from 21.6% to 28.1% with an average of 25.0%. Clearly, there is a noticeable
variation among the results from different permutations. To illustrate this variability
graphically, a plot of the three sets of error rates is given in Figure 1.

Table 3. Classification results for design metrics

Permutation	Classification Error (%)		
	Training	Validation	Test
1	27.1	29.2	21.6
2	25.2	23.6	24.6
3	25.6	21.1	26.1
4	24.9	26.6	22.6
5	21.6	27.6	28.1
6	24.1	25.1	24.6
7	22.6	26.6	24.6
8	24.4	25.6	26.1
9	24.4	28.6	24.1
10	23.1	24.6	27.1
Average	24.3	25.9	24.95
SD	1.58	2.42	1.97

In order to find statistical significance of the above results, we compute confidence
bounds for classification errors based on the results from the ten permutations. It

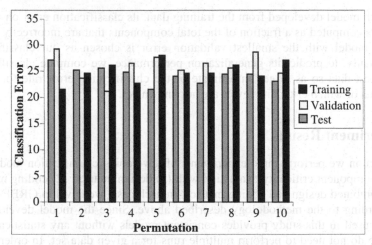

Fig. 1. Plots of classification errors for various data permutations: design metrics

should be noted that the confidence bounds given here reflect the variation in classification errors occurred by random assignment of high and low criticality components to the three subsets in different permutations. Using the well known t-statistic, 95 % confidence bounds for true classification error that is unknown are obtained as shown below, where $t(9;0.025)$ is the upper 2.5 percentile of the t-distribution with nine degrees of freedom.

$$\{\text{Average of Test Errors} \pm t(9; 0.025)* (\text{SD of Test Errors}) / \sqrt{10} \}$$
$$= \{24.95 \pm 2.26 \ (1.97)/\sqrt{10}\}$$
$$= \{23.60, 26.40\}$$

The above values are interpreted as follows. Based on the results for test errors presented in Table 3, we are 95 % confident that the true classification error is between 23.6 % and 26.4 %. Similarly, 90% confidence bounds for test error are {23.81, 26.05}. As confidence value increases, the confidence intervals get narrower while they get wider as confidence value decreases.

4.2 Results with Coding and Combined Metrics

Since our primary interest here is the test error, we now present in Table 4 only the test errors for coding metrics (X_4, X_5, X_6) and for combined (design plus coding) metrics (X_1 to X_6). Standard deviations and confidence bounds for these cases were obtained as above. A summary of test error results for the three cases is given in Table 5.

4.3 Discussion

The results in Table 5 provide insights into the issues addressed in this paper. A measure of the variability in classification performance caused by random assignment of component data to training, validation and test sets is indicated by the interval widths of confidence bounds. For both the 90% and 95% bounds, the interval width is quite reasonable in light of the fact that software engineering data tends to be quite noisy. Further, the variability in classification performance obtained by coding metrics alone is much higher than those for the other two cases. On the other hand, in terms of classification accuracy, it appears that the classification models based on the design metrics alone is comparable to those based on the coding and combined metrics. Overall, it appears that a test error of, say, 24% is a reasonable summary for the data analyzed in this study since this value of 24% is easily contained in all the confidence bounds listed in Table 5.

Table 4. Test errors for coding metrics and combined metrics, respectively

Permutation	Coding Metrics Test Error (%)	Combined Metrics Test Error (%)
1	14.6	20.6
2	26.1	26.1
3	24.1	26.1
4	18.6	21.1
5	26.1	27.6
6	25.1	24.6
7	23.6	22.6
8	23.6	24.6
9	23.6	22.6
10	24.6	27.6
Average	23.0	24.35
SD	3.63	2.54

Table 5. Summary of test error results

Metrics	90% Confidence Bounds	Interval width	95% Confidence Bounds	Interval width
Design Metrics	{23.81, 26.05}	2.24	{23.60, 26.40}	2.80
Coding Metrics	{20.89, 25.11}	4.22	{21.40, 25.80}	4.40
Combined Metrics	{22.89, 25.81}	2.92	{22.55, 26.15}	3.60

Generally, the observed error rate for an application will be determined by various factors such as the noise in the data, the choice of the classifier type (GRBF here), the choice of the algorithm, and etc. In software engineering modeling studies, error rates tend to be rather high. Even though it is not feasible to provide exact comparisons of our results in the above with other studies, we can say that classification error rates for "similar" studies, e.g. [12], which are the error rates observed based on the use of several different classification techniques, tend to be about thirty percent. Thus, it can be said that our modeling methodology produces good classification models that exhibit impressive performance on benchmark data.

From the aspect of model utility, it is not intuitively clear that design metrics alone should perform statistically as well as coding and combined metrics. Nevertheless, our experiment results support the case here. Further, taking into account the confidence bounds presented above, this observation is also confirmable. Therefore, based on this empirical study, we can say that the design metrics alone are enough to provide as good classifiers in a statistical sense as those that employ coding metrics or combined metrics at least in terms of classification accuracy on unseen data.

5 Concluding Remarks

In this paper we investigated an important issue in software engineering of identifying critical components using their software characteristics as predictor variables. A software criticality evaluation model, if it is parsimonious and employs the metrics easily available at earlier stage, can be an effective analytical tool to reduce the likelihood of operational failures. However, the efficacy of such a tool depends on the mathematical properties of the classifier and its design algorithm. In this study we employed Gaussian radial basis function classification models that possess the powerful properties of best and universal approximation. Further, our new design algorithm enables us to produce consistent and reliable models in a systematic way. For model evaluation, we analyzed the performance of the models obtained from ten different permutation of a given data set and considered their averages, standard deviations and confidence bounds for the test error. A comparison of confidence bounds based on design, coding, and combined metrics indicated that test error rates in the three cases are almost equal in a statistical sense. We believe that owing to our modeling methodology using the radial basis function model, we were able to establish effective predictors of potentially critical components only based on design metrics, which are easily available early in the software development cycle. Early identification of critical components and additional resources allocated to them can be instrumental in minimizing operational failures and thus improving system quality. It may be interesting to pursue such studies further using additional datasets and alternative classification techniques.

In conclusion, we believe that in this empirical study the use of a mathematically powerful machine learning technique, pursued in a principled manner, was essential in achieving quite impressive results that are potentially of great practical benefit to the discipline of software engineering.

References

1. Shin, M and Goel, A.L.: A. Empirical Data Modeling in Software Engineering using Radial Basis Functions. IEEE Transactions on Software Engineering, (28).(2002) 567_576.
2. Khoshgoftaar, T., and Seliya, N. Comparative Assessment of Empirical Software Quality Classification Techniques: An Empirical Case study. Software Engineering (2004), Vol. 9, 229-257.
3. Khoshgoftaar, T., Yuan, X., Aleen, E. B., Jones, W. D., and Hudepohl, J. P.: Uncertain Classification of Fault-Prone Software Modules. Empirical Software Engineering (2002), Vol. 7, 297-318.
4. Lanubile, F., Lonigro, A., and Vissagio, G.: Comparing Models for identifying Fault-Prone Software Components, 7th International Conference on Software Engineering and Knowledge Engineering, 312-319, Rockville, Maryland, June 1995.
5. Pedrycz, W.: Computational Intelligence as an Emerging Paradigm of Software Engineering. Fourteenth International Conference on Software Engineering and Knowledge Engineering, Ischia, Itlay, July 2002, 7-14.

6. Pighin, M. And Zamolo, R.: A Predictive Metric based on Discriminant Statistical Analysis, International Conference on Software Engineering, Boston, MA, 1997, 262-270.
7. Zhang, D. And Tsai, J. J. P.: Machine Learning and Software Engineering. Software Quality journal (2003), Vol. 11, 87-119.
8. Shull, F., Mendonce, M. G., Basili V., Carver J., Maldonado, J. C., Fabbri, S., Travassos, G. H., and Ferreira, M. C.: Knowledge-Sharing Issues in Experimental Software Engineering, Empirical Software Engineering (2004), Vol. 9, 111-137.
9. Goel, A.L.and Shin, M.: Tutorial on Software Models and Metrics. International Conference on Software Engineering (1997), Boston, MA.
10. Han, J. and Kamber, M.: Data Mining Morgan Kauffman, 2001.
11. Shin. M. and Goel, A.L.: Design and Evaluation of RBF Models based on RC Criterion, Technical Report (2003), Syracuse University.
12. Khoshgoftaar, T. and et.al.: Uncertain Classification of Fault Prone Software Modules, Empirical Software Engineering, (2002), Vol.7, 297-318.

Component Architecture Redesigning Approach Using Component Metrics

Byungsun Ko and Jainyun Park

Major in Computer Science, Division of Information Science,
Sookmyung Women's University
Yongsan-ku, Seoul, 140-742, Korea
{kobs,jnpark}@sookmyung.ac.kr

Abstract. Components are reusable software building blocks that can be quickly and easily assembled into new systems. Many people think the primary objective of components is reuse. The best reuse is reuse of the design rather than implementation. So, it is necessary to study the component metrics that can be applied in the stage of the component analysis and design. In this paper, we propose component architecture redesigning approach using the component metrics. The proposed component metrics reflect the keynotes of component technology, base on the similarity information about behavior patterns of operations to offer the component's service. Also, we propose the component architecture redesigning approach. That uses the clustering principle, makes the component design as the independent functional unit having the high-level reusability and cohesion, low level complexity and coupling.

1 Introduction

Component-based software development (CBD) has become widely accepted as a cost-effective approach to software development. CBD is capable of reducing the developmental costs using the well-designed components. Therefore, it is capable of improving the reliability of an entire software system, and emphasizes the design and the construction of software systems using reusable components [1,2]. Many want to design something once and use it over and over again in different context, thereby realizing the large productivity gains, taking advantage of the best solutions, the consequent improved quality, and so forth. The quality is importance in every software development, but particularly that is so in CBD because of its emphasis on reuse. The best reuse is reuse of the design rather than implementation.

In this paper, we propose component architecture redesigning approach. That is based on the component metrics and the clustering principle.

The component is an individual functional unit, made of interfaces and classes. Conceptually, interfaces define the service of the component, and specifically, that is a set of operations that can be invoked by other components or clients. The invoked operations offer the component's service by interacting with the inside and outside classes of the component. The interaction information is extracted in the analysis & design stage of the component-based system, and

T.G. Kim (Ed.): AIS 2004, LNAI 3397, pp. 449–459, 2005.
© Springer-Verlag Berlin Heidelberg 2005

represents behavioral types of operations. So, we define the component metrics based on the interaction information. We expect to develop the high quality component design model by measuring and redesigning of the component architecture.

The rest of this paper is organized as follows. Section 2 explains related works. Section 3 describes abstract representation models for the components and defines component metrics. Section 4 shows the component architecture redesigning process. Section 5 shows experiments on the case study. Our conclusions are presented in section 6.

2   Related Works

2.1   The Keynote of Components

The component based systems adhere to the principle of divide and conquer for managing complexity − break a large problem down into smaller pieces and solve those smaller pieces, then build up more elaborate solutions from simpler and smaller foundations. The goal of CBD is high abstraction and reuse. Therefore, it is necessary to measure component's quality for the more abstraction and reuse [3,4,5].

To measure component's quality, it is necessary for us to define the terms previously. So, we will investigate the definition and keynote about component technology [1,2,5].

- Component is an executable unit of independent production, acquisition, and development that can be composed into a functioning system.
  Component is a unit of composition with contractually specified interfaces, and can be developed independently and is subject to composition by third parties.
- Interface represents a service that the component offers, is the component's access point.
  The interface as an access point allows clients to access and connect the services provided by the component. Namely, the interface plays dual roles as it serves both providers of the interface and clients using the interface. Technically, the interface is a set of operations that can be invoked by clients.
- Component follows the object principle of combining functions and related data into a single unit.
  At this time, the classes belong to the role of data and the interfaces belong to the role of function about the component. So, we can consider the component as the group of classes and operations.

2.2   Quality Models

The structural properties of a piece of software, such as cohesion, coupling and complexity, can serve as indicator of important software quality attributes. For instance, highly complex components are more likely to contain faults, and therefore may negatively impact on system reliability and maintainability [3].

Quality modeling based on the structural properties of the artifacts is particularly valuable, because it can provide easily feedback on the quality of components. So, it is necessary component metrics. Other advantages of quality modeling based on the measurement of structural properties are the measurement process is objective, does not rely on subjective judgment, and can be automated [6].

3   Component Quality Measurement

3.1   Abstract Representation Models

We believe that a good abstract representation model objectively depicts the relationships among the members of a component, and it is the precondition of a well-defined component quality measurement. If the component quality measurement can abstract the relationship among the members of a component properly, then there is little questionable. Therefore, an abstract representation model is important.

In this section, we will propose good abstract representation models. They stand for operations, interfaces, classes, components, communication and relationship among them. That is, the abstract representation models stand for the necessary information to derive component metrics.

Definition 1. *Component Architecture Organization Graph (CAOG)*

CAOG represents the organization of the component-based system. Generally speaking, the component architecture about the system made of components and the component made of operations and classes. CAOG shows the members of components and the interactions among them, is defined to be a directed graph.

$$CAOG = (V_{Co},\ E_{Co})$$

where,

- V_{Co} is a set of vertex, $V_{Co} = V_o \cup V_c$. V_o is a set of operations of an interface in a component and V_c is a set of classes in a component.
- E_{Co} is a set of edges such that $E_{Co} = \{\langle x,y \rangle | x \in V_o, y \in V_c \text{ such that } x \text{ uses } y\}$.

The notations are like this. A circle or a rectangle represents each vertex. The circle represents operations, and the rectangle represents classes. An arrow represents each edge with the direction from operation to class, which shows the relationships between operations and classes to deliver the service of the component. Each component is represented by a dotted rectangle. It is clear that the CAOG about the design model of the component-based system consists of two types of nodes, operations and classes, and one type of edges, that represents use-dependence between operations and classes, respectively.

Definition 2. *Operation Use Matrix (OUM)*

For any operation in the interface, the OUM shows the amount of classes that any pairs of operations use in common to deliver the service of a component.

The OUM is a square matrix. Its dimension equals to the total number of operations. Each element of OUM is calculated from the CAOG. The value of the OUM is the number of classes, each pair of operations use in common to deliver the very component's functionality. So, the value is a positive number. The diagonal elements of OUM represent the number of classes that each operation uses one's own operation, and the off-diagonal elements represent the number of classes that a pair of operations uses in common to deliver the component's service.

Each row in the OUM is a row vector, which denotes the behavioral type of that operation. That is, it represents how much the operation interacts with other operations to offer the service.

$$OUM(p,q) = Common\_Use(p,q)$$

$$Common\_Use(p,q) = \begin{cases} Common\_Use(p), & \text{if } p = q \\ Common\_Use(p,q), & \text{if } p \neq q \end{cases}$$

where,

- $Common\_Use(p)$ means the number of classes which operation p uses by itself, $Common\_Use(p)$ is the same as $Common\_Use(q)$.
- $Common\_Use(p,q)$ means the number of classes which operation p and operation q use in common.
- p and q are the number, represents the order of the operations, is used in row-number or column-number of matrix.

Definition 3. *Operation Similarity Matrix (OSM)*

For any operation in the interface of the component, the OSM shows the degree of the similarity in the behavioral type between operations to deliver the service of the component.

The OSM is a square matrix. Its dimension equals to the total number of operations, and is the same as OUM. The OSM is calculated from the OUM. To deliver the service of the component, the operations interact with inside classes or outside classes of the component. We can find out the behavioral type of operations according to the amount of the operation's communication. The similarity about the behavioral type between operations is expressed as the cosine value. The cosine value is the ratio of the length of the side adjacent to an acute angle to the length of the hypotenuse in a right triangle. The abscissa at the endpoint of an arc of a unit circle centered at the origin of a Cartesian coordinate system, the arc being of length x and measured counterclockwise from the point $(1, 0)$ if x is positive or clockwise if x is negative. Here, we calculate the cosine value about the angle between row vectors from the OUM. That means the degree of the similarity between the row vectors to deliver the service of the component.

The OSM is a symmetric matrix. Because the diagonal elements of OSM represent the similarity of the same operations, its value is 1. And the off-diagonal elements represent the similarity about the interaction between the operations, its value is between 0 and 1, and two values are inclusive. The cosine value between the two row-vectors is derived from the gap of angles. The narrower the gap of an angle is, the closer to 1 the value is. Therefore, we can find out the degree of the similarity about the behavioral type between the row-vectors to deliver the component's service.

$$OSM(p,q) = Type\_Similarity(p,q)$$

$$Type\_Similarity(p,q) = \frac{ObliqueSideV}{\sqrt{BaseSideV}}$$

$$ObliqueSideV = \sum_{i=1}^{t} Common\_Use(p,i) \times Common\_Use(q,i)$$

$$BaseSideV = \sum_{i=1}^{t} \left(Common\_Use(p,i) \times Common\_Use(q,i)\right)^2$$

Where,

- t is the total operation number of all the components in component architecture.
- p and q are the number which represents the order of the operations. So, let $1 \leq p, q \leq t$.

3.2   Component Metrics

In this section, we will define component cohesion and coupling measurement based on the CAOG. Then, we use briand's criterion to validate our metrics and show that our metrics have the properties what a well-defined cohesion and coupling measurement should have.

Cohesion is a measurement about the degree of the dependency among the internal elements of the component. The component serves as a unit of encapsulation and independence, consists of data and the functions that process those data. The unification of the internal elements of component, improves the cohesion. Therefore, we define the component cohesion metric based on the similarity about the behavioral type of operations in the component.

Coupling is the degree to which various entities in the software connected. These connections cause dependencies between the entities that have an impact on various external system quality attributes [7,8,9]. Component coupling is a measurement about the degree of dependency among components. It implies how much independent the component is potentially [1,7]. Therefore, we define the component coupling metric based on the similarity about the behavioral type of operations in interactions among the different components.

Definition 4. *Cohesion Metric by Operation Type (COHOT)*

The cohesion measure $COHOT(Co_i)$ of a component Co_i is defined out of the OSM, which denotes the similarity about the behavioral type of operations.

Cohesion represents the degree of the dependency among the internal elements of the component. If the behavioral type of operations is similar to each other, they have much dependency each other internally. Therefore, if the operations with the similarity close to each other are included into the same component, cohesion of the component will be stronger.

$$COHOT(Co_i) = \frac{SumCHV1 - SumCHV2}{TotalCH}$$

$$SumCHV1 = \sum_{p=1}^{n_i} \sum_{q=1}^{n_i} Type\_Similarity(p, q)$$

$$SumCHV2 = \sum_{p=1}^{n_i} \sum_{q=p+1}^{n_i} Type\_Similarity(p, q)$$

$$TotalCH = \frac{n_i(n_i+1)}{2}$$

Where,

- i means the order of components in the component architecture, it is called as a sequential number.
- n_i is the total number of operations in the interface of component Co_i

We include the similarity only in case of $p \geq q$ in the calculation. So, we consider the similarity only in case of $1 \leq p \leq n_i$ and $1 \leq q \leq p$. That reason is to avoid the duplication in the process of calculation, because OSM is the symmetric matrix.

Now, we will investigate the theoretic validation of component cohesion metric. A well-defined component cohesion measure should have these properties, otherwise the usefulness of that measure is questionable. Although these properties are not sufficient conditions of a well-defined measure, they can be used to check the problems in a measure and can be used as a guide to develop a new measure. Briand stated that a well-defined class cohesion measure should have the following four properties [10,11]. And they make the metrics stricter and less experimental. Table 1 shows the properties and meaning about cohesion measure.

Definition 5. *Coupling Metric by Operation Type (COPOT)*

The coupling measure $COPOT(Co_i)$ of a component Co_i is defined out of the OSM, which denotes the similarity of the behavioral type of operations.

Coupling represents the degree of the dependency between components. If the behavioral type of operations in one component is similar to that in the other component, components have much dependency each other. Therefore, if the operations of the different components have the similarity close to each other, the coupling of the component will be stronger.

Table 1. The properties of cohesion measurement.

Property	Contents
Cohesion 1	Non-negativity and Normalization $(\exists Co)(Cohesion(Co) \in [0,1])$
Cohesion 2	Null Value $(\exists Co)(R_c = \emptyset \Rightarrow Cohesion(Co) = 0)$
Cohesion 3	Monotonicity $(\exists Co)(R_{Co} \leq R'_{Co} \Rightarrow Cohesion(Co) \leq Cohesion(Co'))$
Cohesion 4	Cohesive components $(\exists Co_1, \exists Co_2)(max\{Cohesion(Co_1), Cohesion(Co_2)\} \geq Cohesion(Co_3))$

$$COPOT(Co_i) = \frac{SumCPV1 - SumCPV2}{TotalCP}$$

$$SumCPV1 = \sum_{p=1}^{n_i} \sum_{q=1}^{t} Type\_Similarity(p, q)$$

$$SumCPV2 = \sum_{p=1}^{n_i} \sum_{q=1}^{n_i} Type\_Similarity(p, q)$$

$$TotalCP = n_i \times (t - n_i)$$

Now, we will investigate the theoretic validation of component coupling metric. Table shows the properties and meaning about coupling measure. And they are satisfied.

Table 2. The properties of coupling measurement.

Property	Contents
Coupling 1	Non-negativity $(\exists Co)(Coupling(Co) \geq 0)$
Coupling 2	Null Value $(\exists Co)(R_c = \emptyset \Rightarrow Coupling(Co) = 0)$
Coupling 3	Monotonicity $(\exists Co)(R_c o \leq R_C o' \Rightarrow Coupling(Co) \leq Coupling(Co'))$
Coupling 4	Merging of components $(\exists Co_1, \exists Co_2)(\Sigma\{Coupling(Co_1), Coupling(Co_2)\} \geq Coupling(Co_3))$
Coupling 5	Disjoint component additivity $(\exists Co_1, \exists Co_2)(\Sigma\{Coupling(Co_1), Coupling(Co_2)\} = Coupling(Co_3))$

4   Component Architecture Redesigning Approach

In this section, we will define the component architecture redesigning process. The objective of that is to enhance component's independent functionality. To evaluate an adequateness of the component design model, we use cohesion and coupling metrics altogether. For the independency of the components, components should have as high-cohesion and low-coupling as possible. That is a general and abstract standard for the cohesion and coupling. But there is not concrete standard to find out the appropriate of the component design model. To use cohesion and coupling metrics collectively, we measure the independence of a component. The independence is calculated by subtracting the coupling from the cohesion. So, the independence of a component is normalized from −1 to 1, two values is inclusive. If the independence of a component is a positive number close to 1, we can judge a component to be an independent functional unit. And if the independence of a component is a negative number close to -1, it is opposed to that. Consequently, we can easily evaluate a component design model with the independence metric.

If components have low-independency, we should make components have high-independency by the component redesigning process. The process makes use of the clustering principle as grouping a similar thing. The process is as follows and is showed in Figure 1, in the form of the Nassi-Shneiderman chart.

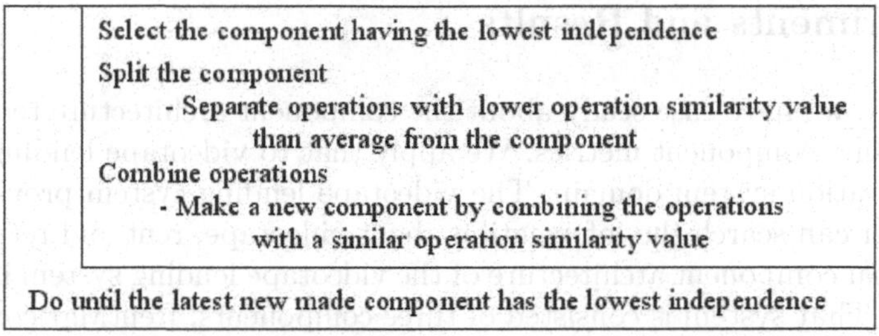

```
Select the component having the lowest independence
Split the component
        - Separate operations with lower operation similarity value
                than average from the component
Combine operations
        - Make a new component by combining the operations
                with a similar operation similarity value

Do until the latest new made component has the lowest independence
```

Fig. 1. The component redesigning process.

Step 1. Selecting the component having the lowest independence, and then splitting that.

First, select the component having the lowest independence. For the selected component, calculate the operation similarity average. And then, partition the operations having the lower operation similarity than average from that component. Therefore, that component has been divided into several components. The component cohesion, coupling and independence are calculated from the operation similarity. If the operations having the operation similarity below the average are partitioned from the component, the cohesion will be stronger and the coupling will be weaker. Therefore, the independence of the component will be stronger.

Step 2. Combining components that are alike in the operation similarity for creating new components.

The component is an independent functional unit. Hence, we can regard a component as a group of operations. The operation similarity value about the component takes an average the operation similarity of all the operations in that component. If components having the alikeness in the operation similarity are combined all together, the cohesion of a new made component will be high. When combining components, we use the clustering principle. Consequently, the new made components will have high-independence.

Step 3. Repeating the component redesigning process, until the new combined component has the lowest independence.

The component redesigning process is repetitious. That is the consecutive process, partition and combination according to the operation similarity. To improve the cohesion of components, we separate the operations having the operation similarity below the average. And then, we make the new component by combining components having the similar operation similarity. We can find out the component design model is whether good or not, by measuring the independence of a component. If that component has the lowest independence among all the components, we should split that component again. In this case, the same step of component redesigning process is executed repeatedly, separation and combination. So, we should stop the process of component redesigning.

5   Experiments and Results

In this study, we have case study about the component architecture redesigning approach using component metrics. We apply that to videotape lending system in the reservation system domain. The videotape lending system provides that the customer can search the information about videotape, rent and return that.

The initial component architecture of the videotape lending system is showed in Figure 2. That system is consisted of three components, RentMgr component managing the information about rent and return the videotape, CustomerMgr component managing the information about customer, and TapeMgr component managing the information about videotape or rent. The service offered by each component is IRentMgt interface, IcustomeMgt interface and ITapeMgt interface.

Figure 3 is CAOG of the initial component architecture. Now, we adapt the component redesigning process to the initial component architecture. Figure 4 is the component architecture after adapting the component redesigning process continuously. About the before and after component redesigning process, the result of the measurement is like Figure 5. In conclusion, we can find out the component redesigning process can upgrade the quality of the component design model. Therefore, we can find out the usefulness of component metrics and component redesigning process.

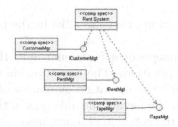

Fig. 2. The initial component architecture.

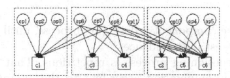

Fig. 3. CAOG of the initial component architecture.

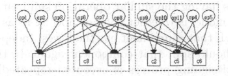

Fig. 4. CAOG of the component architecture after redesigning process.

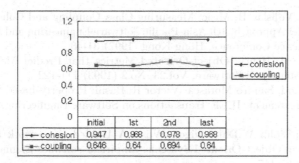

Fig. 5. Improvement of component quality with the redesigning process.

6   Conclusion

In this study, we have studied the component architecture redesigning approach using component metrics. And we have empirically evaluated the component redesigning approach with a case study. In accordance with the measurement result, we can improve the quality of the component design model the analysis and design phase.

We expect several things through this study. The component metrics can obtain the merits which reflect the measured and revaluated results of the component's quality in the component development more quickly. And, these component metrics can be the method of forecasting the functional degree of independence, the facility of maintenance and the reusability of the component. In addition to that, the suggested component redesigning process can make the component design model have the high level reusability and cohesion, low level complexity.

References

1. Clemens Szyperski, Dominik Gruntz, Stephan Murer: Component Software: Beyond Object-Oriented Programming. 2nd Edition, Addison-Wesley (2002)
2. Desmond F. D'Souza, Alan C. Wills: Object, Component and Framework with UML: The Catalysis Approach. Addison Wesley (1999)
3. John Cheesman, John Daniels: Uml Components: a Simple Process for Specifying Component-Based Software. Addison Wesley (2001)
4. Paul Allen: Realizing e-Business with Components. Addison-Wesley (2001)
5. George T. Heineman, William T. Councill: Component-Based Software Engineering: Putting the Pieces Together. Addison-Wesley (2001)
6. Stephen H. Kan: Metrics and Models in Software Quality Engineering. Addison-Wesley (1995)
7. Colin Atkinson, Joachim Bayer, Christian Bunse, Erik Kamsties, Oliver Laitenberger, Roland. Laqua, Dirk Muthig, Barbara Peach, Jurgen Wust, Jorg Zettel: Component-Based Product Line Engineering with UML. Addison-Wesley (2002) 372-408

8. Simon Moser, Vojislav B. Misic: Measuring Class Coupling and Cohesion: A Formal Metamodel Approach. 4th Asia-Pacific Software Engineering and International Computer Science Conference, Hong Kong (1997) 31-40
9. Wei Li, Sallie M. Henry: Object-Oriented Metrics that Predict Maintainability. Journal of Systems and Software, Vol.23, No.2 (1993) 111-122
10. Lionel C. Briand, Sandro Morasca, Victor R. Basili: Property-based Software Engineering Measurement. IEEE Transactions on Software Engineering, Vol.22, No.1 (1996) 68-86
11. Lionel Briand, John W.Daly, Jügen Wüst: A Unified Framework for Cohesion Measurement in Object-Oriented Systems. Empirical Software Engineering, Vol.3, Issue 1 (1998) 65-117

A Workflow Variability Design Technique for Dynamic Component Integration

Chul Jin Kim[1] and Eun Sook Cho[2]

[1] School of Computer Science and Information Engineering, Catholic University
43-1 Yeokgok 2-dong, Wonmi-gu, Bucheon-si, Gyeonggi-do 420-743, Korea
cjkim@otlab.ssu.ac.kr
[2] College of Computer and Information Science, Dongduk Women's University
23-1 Wolgok-dong, Sungbuk-ku, Seoul 136-714, Korea
escho@dongduk.ac.kr

Abstract. Software development by component integration is the mainstream for Time-to-Market and is the solution for overcoming the short lifecycle of software. Therefore, the effective techniques for component integration have been working. But, the systematic and practical technique has not been proposed. Main issues for component integration are a specification for integration and the component architecture for operating the specification. In this paper, we propose a workflow variability design technique for component integration. This technique focuses on designing the connection contract based on the component architecture. The connection contract is designed to use the provided interface of component and the architecture can assemble and customize components through the connection contract dynamically.

1 Introduction

The current enterprise system has evolved rapidly. Therefore, the business model has frequently been modified. Because of these frequent modifications and the increment of software complexity and maintenance costs, enterprises require high quality reusable components[1,2,3]. Component technology is widely accepted in both academia and industry. Since object-oriented technology could not improve software development innovatively, component technology came to be applied to the development of software. One of the main reasons was that the object could not reduce the burden of development because what it covers is pretty limited. The component significantly decreases the overhead of development, because it has workflow and is offered as the function unit that objects are composed of. The component evolved software development into an assembly concept, which can rapidly and easily satisfy the requirements of domains. Components must be made more diverse to satisfy various domain requirements using component interfaces. But, it is not easy to design variability to provide the diversity, and the procedural technique for variability design has not been studied. Currently, reusable business components for satisfying a variety of domains have not been satisfactorily provided to developers. The reason is that the component is developed to focus on only the business logic and the reusability of the component is not considered.

This paper proposes variability design techniques for developing high quality reusable components(or complex component). It aims to improve the reusability of com-

T.G. Kim (Ed.): AIS 2004, LNAI 3397, pp. 460–469, 2005.

ponents, to provide variability to components to satisfy a variety of domains. Also, as the components should be included all variable logics to satisfy a variety of requirements, it happens the problem that the size of the components is increased.

In this paper, we propose processes for designing the workflow variability between components. The design techniques for workflow variability consist of techniques for designing variable message flows between components based on the component architecture. We propose the connector technique, which plays the role of coordinator, for operating message flows in the component architecture.

This paper is organized as follows: we present the limitations of the existing component integration researches and related researches, describe the concept of workflow variability in Section 2, and propose workflow variability design systematically in Section 3. In Section 4, we also present the comparison of the proposed technique with the existing method using the results of case study experiments. Finally, we describe conclusion and future works.

2 Related Works and Foundation

2.1 Component Integration Researches

There are two researches named [4] and [5] covering the way of customizing the software through the composition of components, which is compared to this paper. [4] is the technique to customize the software through dynamically defining the required interface for an unexpected modification. [5] represents through a structural description and a behavior specification as well as the provided interface and the required interface for assembling components. The structural description represents using the communication channel which works as the access point for services of component interfaces. The behavior specification represents the connections and collaborations between components to use Petri net.

These researches do not cover designing the workflow through the connection between components. Also, they do cover only the composition of in-house components.

2.2 C2 Architecture Style

C2 Architecture Style was designed as the architectural language to develop a GUI software in UCI(University of California in Irvine)[6]. C2 invokes the entered message, or creates the output message via the top and bottom port of a component. C2 interacts with the message exchange and the messaging between components is exchanged by the connector. Components represent a message to their ports and transfer a message to other components through connectors indirectly.

To design the workflow clearly using C2 Architecture Style is dissatisfactory because it represents the flow of a message through the composition of components, but, cannot represent the flow of a data.

2.3 OCL(Object Constraints Language)

OCL is a formal language to express constraints, such as invariants, pre conditions, and post conditions[7]. Components can specify constraints of interface through OCL and component users can understand the behaviors of component clearly through specified constrains. OCL can be used for the different purposes to specify invariants on classes and types in the class model, to specify type invariant for Stereotypes, to describe pre- and post conditions on operations and methods, to describe guards, and specify constraints on operations[7].

But, since OCL represents constraint of only a behavior, it is short of expressing the interaction between components.

2.4 Workflow Variability

A product family is a set of products with many commonalities in the same domain[8]. A member is a product that is included in the product family. A product family has commonality and variability. The commonality is the reusable area and the variability is the difference between products in the product family. For example, the variations in a TV product family may be the kind of output device, the size of the output device, the broadcasting standards that it supports, the quality of picture and sound, and so on. These differences are the variability factors that can organize the members of a TV product family.

Workflow variability is the difference among message flows that process the same functionality among members of the product family. Elements of the workflow variability are message flows of the operation among component interfaces[9]. The flow between objects within the component is designed with the unit of the component interface, and the workflow variability is designed to be adapted it in a variety of domains when the workflow is used.

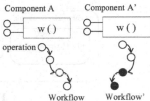

Fig. 1. Workflow Variability

3 Workflow Variability Design

Workflow variability is designed within the component architecture using only the provided interface of components. As shown in Figure 2, connector manages the sequential message invocation, and plays as the mediator between application and component architecture.

Connector manages a semantic(configuration) of workflow and the semantic is used to call service of components when applications are developed or customized services. Generally, the semantic can be configured using XML.

3.1 Workflow Variability Design Process

As shown in Figure 3, workflow variability design process consists of steps to design the connection contract, workflow and dataflow.

The first step defines connection contracts as the interaction standard[10]. The second step defines workflow through the mapping rule for contracts. Finally, it defines the data flow for input data of operation. Following is the detailed description for each step.

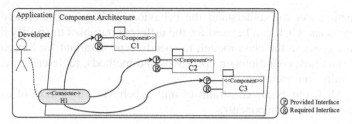

Fig. 2. Workflow based on Component Architecture

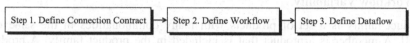

Fig. 3. Design Process

3.1.1 Define Connection Contract

This step defines the connection contract that works as the interaction standard between components. The connection contract defines three contracts for operations of the provided interface consisting of the workflow(Figure 4).

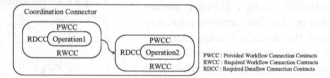

Fig. 4. Connection Contracts

PWCC is the information for the current operation and RWCC is the contract for the one invoked following the current one to compose the workflow within the coordination connector. RDCC is the contract to provide input parameters to the current operation. Thus, each operation within coordination connector defines three contracts and the workflow design through these connection contracts. Also, the workflow can be customized through changing contracts within the coordination connector.

PWCC is the contract that provides information of registered operation within the coordination connector. PWCC is called as the reference for other operations by RWCC. As shown in Figure 5, PWCC defines the contract information of the component, the interface, the operation, the input parameters, and the output parameter providing the current operation.

RWCC is the contract for the one operation invoked following the current operation. RWCC consists of the dynamic connection referring to the PWCC of other operations. PWCC and RWCC become the factors of workflow design and customization. As shown in Figure 5, RWCC defines the contract information of the component, the interface, the operation, the input parameter, and the output parameter required following the current operation.

RDCC is the contract for input parameters of operations composing workflow. To gain data of input parameters define the information of required operations. As shown in Figure 5, RDCC defines the contract by the input parameter of the operation defined in PWCC. RDCC defines the component, the interface, and the operation required for each input parameter.

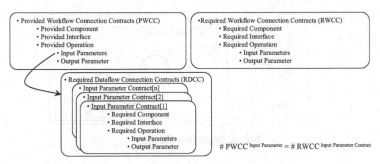

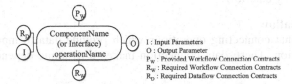

Fig. 5. PWCC / RWCC / RDCC

To design workflow within coordination connector, we propose the operation notation like as Figure 6.

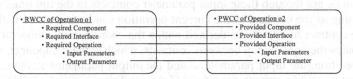

I : Input Parameters
O : Output Parameter
P_W : Provided Workflow Connection Contracts
R_W : Required Workflow Connection Contracts
R_D : Required Dataflow Connection Contracts

Fig. 6. Operation Notation for Workflow Design

This notation specifies the component or interface and operation name within box and notes input and output parameter and contracts in outside of box.

3.1.2 Define Workflow

This step designs the workflow through mapping between the connection contracts. That is to say, designing the workflow is the mapping PWCC to RWCC. As shown in Figure 7, the mapping between contracts is accomplished only between two operations because workflow is the flow of sequential operation. But, in the case of concurrency processing can be accomplished the mapping among several operations.

Fig. 7. Contracts Mapping for Workflow Design

RWCC of current operation is compared with PWCC of other operations within coordination connector. It iteratively is checked consistency between contracts, which is the component, the interface, the operation, input parameters and output parameter as the number of operations within coordination connector. Input parameters are checked the additional contract information, such as the number of parameters, the data type, and the order of parameters. Output parameter is checked only data type. If condition is satisfied, RWCC of the current operation is connected with PWCC of the other operation.

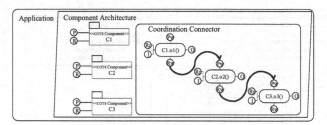

Fig. 8. Workflow Design

If the contracts between two operations are satisfied each other, workflow is designed as Figure 8. Since RWCC of operation "o1" requires operation "o2", two contracts is connected. The workflow is not related with dependency between components. Because workflow manages only sequential operation invocation, it cannot know dependency by dataflow.

3.1.3 Define Dataflow

Dataflow means that connecting operations which provide data to input parameters within one operation. These data connection makes a dynamic dependency among components.

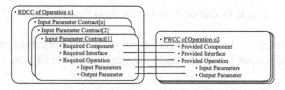

Fig. 9. Contracts Mapping for Dataflow Design

Dataflow is designed to map RDCC using as the contract of the current operation with PWCC using as the one of the other operation. The mapping between factors of RDCC and PWCC is same as Figure 9. The input parameter contracts of RDCC define as the number of input parameters of the current operation. The input value of operation can be got through these input parameter contracts in the run time.

RDCC using as the contract of the current operation can be compared with PWCC using as the contract of operations defined within the coordination connector. It iteratively compares the consistency between contracts, which is the component, the interface, the operation, the input parameters, and the output parameter as the number of input parameters within current operation. If the condition is satisfied, RDCC of the current operation is connected with PWCC of the other operation.

If the contracts between two operations are satisfied each other, dataflow is designed as Figure 10. Since RDCC of operation "o3" agrees with PWCC of operation "o1" and "o2", two input parameters of operation "o3" takes the output of operation "o1" and "o2".

3.2 Component Architecture for Integrating Components

In this chapter, we propose the architecture of coordination connector that can realize the proposed workflow variability design and customization techniques.

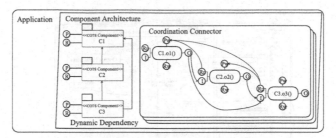

Fig. 10. Dataflow Design

As shown in Figure 11, the coordination connector is based on the RTTI(Reflection) [11] that can invoke component dynamically. Connection Contracts designed by Workflow Design Tool are parsed by Contract Parser(XML parser) and Connector invokes components to use these parsed contracts(PWCC, RWCC). Also, the component invocation for dataflow is operated to the same mechanism.

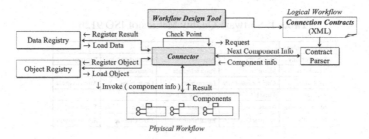

Fig. 11. Component Architecture for Component Integration

4   Case Study

In this chapter, we evaluate the reusability of component through variability design techniques. First, we extracted workflow variability surveying customization cases. The extracted variability is statistical results by the surveyed customization cases. Next, we designed variability using the proposed variability design techniques in this paper. And, we redeveloped component by the designed artifacts. Finally, we evaluated the reusability of component that is applied variability.

4.1   Design and Apply Variability

Generally, the connection contracts can be defined to XML and can be generated contracts(XML) automatically using the Workflow Design Tool as shown in Figure 12. Coordination connector can invoke components sequentially following the generated contracts. This tool can operate Java class or EJB, based on J2EE.

As shown in Figure 12, the workflow is designed to connect among component methods and the dataflow is designed to connect among parameters. The dataflow is connected from output parameter to input parameter as the mapping rule between RDCC and PWCC. This tool generates the connection contract through these design process.

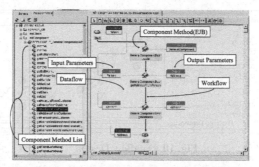

Fig. 12. Workflow(Component Integration) Design Tool

4.2 Survey Applied Variability

This step measures the quality of a component which is applied the proposed variability design techniques in this paper, in the reusable point of view.

Table 1. Software Quality Metrics of ISO 9126

Quality attribute	Sub-Characteristic			
Functionality	Suitability	Accuracy	Inter-operability	Security
Reliability	Maturity	Fault tolerance	Recoverability	
Usability	*Understandability*	Learnability	Operability	Attractiveness
Efficiency	Time behavior	Resource utilization		
Maintainability	Analyzability	*Changeability*	Stability	Testability
Portability	Adaptability	Installability	Co-existence	*Replaceability*

As shown in Table 1, the measurement metrics are based on the software quality measurement metrics of ISO 9126[12], [13], and [14]. We defined the Understandability, Changeability, Replaceability, Extensibility, and Generality as metrics for measuring the reusability of component.

```
<Understandability>
U1.1 When the variability of component is designed using this technique, can you understand this technique easily ?  Yes ( )  No ( )
U1.2 When the variability of component is designed using existing method, can you understand existing method easily ? Yes ( )  No ( )
U2.1 When the component is customized using this technique, can you understand the variability ? Yes ( )  No ( )
U2.2 When the component is customized using the existing method, can you understand the variability ? Yes ( )  No ( )
U3.1 When the component is customized using this technique, do exist the required data for variability ? Yes ( )  No ( )
U3.2 When the component is customized using the existing method, do exist the required data for variability ? Yes ( )  No ( )
U4.1 When the component is customized using this technique, can you understand input and output ? Yes ( )  No ( )
U4.2 When the component is customized using the existing method, can you understand input and output ? Yes ( )  No ( )
```

Fig. 13. Understandability Survey List

Figure 13 is the survey list for measuring the Understandability of variability when the component is reused. Also, the survey list for Changeability, Replaceability, Extensibility, and Generality metric are based on ISO 9126.

Figure 14 (1) represents the measured results of Understandability, Changeability, Replaceability, Extensibility, and Generality metrics for workflow variability.

The graph of Figure 14 (2) represents the difference between this technique and the existing method. Through these results, we can see that the reusability of a component, which is applied using this technique, is improved. The Figure 14 (3) presents the average value of the measured results from five metrics. Nearing a measured value of one means that the component is of high quality from a reusable point of view. In significance level, if the measured value is larger than 0.5, Null Hypothesis assumes that the reusability of a component is high quality. And, if the measured value is smaller than 0.5, Alternative Hypothesis assumes that the reusability of a component is not high quality. As the measured results by proposed technique is larger than 0.5 mostly, the reusability of a component applied this technique is high quality as Null Hypothesis. This technique can improve the reusable quality more than twice that of the existing method.

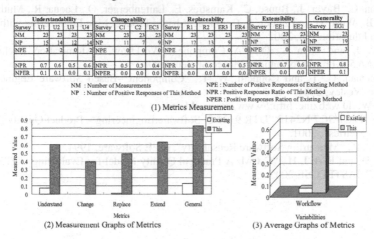

Understandability				Changeability				Replaceability					Extensibility			Generality	
Survey	U1	U2	U3	U4	Survey	C1	C2	EC3	Survey	R1	R2	ER3	ER4	Survey	EE1	EE2	Survey	EG1
NM	23	23	23	23	NM	23	23	23	NM	23	23	23	23	NM	23	23	NM	23
NP	15	14	12	14	NP	11	7	9	NP	12	13	9	11	NP	15	14	NP	19
NPE	3	2	0	2	NPE	0	0	0	NPE	1	0	0	0	NPE	0	0	NPE	3
NPR	0.7	0.6	0.5	0.6	NPR	0.5	0.3	0.4	NPR	0.5	0.6	0.4	0.5	NPR	0.7	0.6	NPR	0.8
NPER	0.1	0.1	0.0	0.1	NPER	0.0	0.0	0.0	NPER	0.0	0.0	0.0	0.0	NPER	0.0	0.0	NPER	0.1

NM : Number of Measurements   NPE : Number of Positive Responses of Existing Method
NP : Number of Positive Responses of This Method   NPR : Positive Responses Ratio of This Method
                                                                                     NPER : Positive Responses Ration of Existing Method

(1) Metrics Measurement

(2) Measurement Graphs of Metrics   (3) Average Graphs of Metrics

Fig. 14. Comparison with This and Existing Method

5   Conclusion Remarks

In this paper, we proposed how to design the workflow variably through the detailed research for the interaction among components. To design the variability of the workflow, we proposed the design techniques based on connector within the component architecture. The connector is defined as the interaction standard for interacting between components. The interaction standard consists of contracts for workflow and for dataflow. Also, it showed that workflow is generated and is customized dynamically using workflow design tool.

In the future, we will study a technique that can support the interaction among components having the different component model standard, such as COM, EJB, and CCM.

References

1. Felix Bachman, Technical Report, CMU/SEI-2000-TR-008, ESC-TR-2000-007.
2. Kang K: Issues in Component-Based Software Engineering, 1999 International Workshop on Component-Based Software Engineering.
3. D'Souza, D. F., and Wills, A. C., Objects, Components, and Frameworks with UML, Addison Wesley Longman, Inc., 1999.

4. I. Sora, P. Verbaeten, and Y. Berbers, "Using Component Composition for Self-Customizable Systems", Workshop on CBSE, ECBS 2002, April 8-11, 2002, Lund, Sweden.
5. R. P. e Silva, et al., "Component Interface Pattern", Procs. Pattern Languages of Program, 1999.
6. The C2 Sytle, http://www.ics.uci.edu/pub/arch /C2.html, Information and Computer Science, University of California, Irvine.
7. Object Management Group, "Object Constraint Language Specification", Version 1.3, 1999.
8. Atkinson, C., Bayer, J., Bunse, C., Kamsties, E., Laitenberger, O., Laqua, R., Muthig, D., Paech, B., Wust, J., and Zettel, J., Component-based Product Line Engineering with UML, Pearson Education Ltd, 2002.
9. Kim C. J. and Kim S. D., "A Component Workflow Customization Technique", Vol.27,No.5, Korea Information Science Society, 2000.
10. 10.Heineman, G. T. and Councill, W. T., "Component –Based Software Engineering", Addison-Wesley, 2001.
11. 11.JavaWorld webzine, http://www.javaworld.com.
12. ISO/IEC JTC1/SC7 N2419 "DTR 9126-2: Software Engineering – Product Quality Part 2 – External Metrics", 2001.
13. Jeffrey S. P., "Measuring Software Reusability", IEEE Software, 1994.
14. Kim S. D. and Park J. H., "C-QM: A Practical Quality Model for Evaluating COTS Components", IASTED, SE 2003.

Measuring Semantic Similarity
Based on Weighting Attributes of Edge Counting

JuHum Kwon[1], Chang-Joo Moon[1], Soo-Hyun Park[2], and Doo-Kwon Baik[1]

[1] 225-Ho, Asan Science Building, 1, 5-ka, Anam-dong, Sungbuk-ku, 136-701, Seoul, Korea
{jkweon,mcjmh,baik}@software.korea.ac.kr
[2] Kookmin Univ., 861-1, Chongnung-Dong, Songbuk-Gu, Seoul 136-702 Korea
shpark21@kookmin.ac.kr

Abstract. Semantic similarity measurement can be applied in many different
fields and has variety of ways to measure it. As a foundation paper for semantic
similarity, we explored the edge counting method for measuring semantic similarity by considering the weighting attributes from where they affect an edge's
strength. We considered the attributes of scaling depth effect and semantic relation type extensively. Further, we showed how the existing edge counting
method could be improved by considering virtual connection. Finally, we compared the performance of the proposed method with a benchmark set of human
judgment of similarity. The results of proposed measure were encouraging
compared with other combined approaches.

1 Introduction

Semantic similarity measurement gives a good alternative when certain symbolically
different concept needs to be substituted with the semantically very close one. That
means substitution of one for the other never changes the meaning of a sentence in
which the substitution was made. That is a fundamental issue for natural language
processing and information retrieval for many years [1], [2]. A variety of applications
can benefit from the advantage of incorporating the similarity measurement. Examples include word sense disambiguation [3], automatic hypertext linking [4], and
information retrieval [5], [6], [7]. Broadly, the similarity measurement has been based
on four different categories: dictionary-based approaches [8], [9], thesaurus-based
approaches [10], [11], semantic network-based approaches [12], [13], [14], and integrated approaches [15], [16]. For extensive comparisons of some representative similarity measures, please refer to [17]. Single common method among aforementioned
approaches is edge counting using lexical semantic network such as WordNet [13],
[17], [18]. Edge counting calculates the shortest path as a distance between concepts.
This method has been considered to have relatively low performance compared with
other measurements. Despite of its low performance, edge counting has an advantage
in terms of calculation cost. Most of the methods, which have relatively high rate of
similarity, use various information sources. For example, in [3], [24], [28], other
methods measure the similarity using large corpus of statistical information so that
the semantic distance between nodes in the semantic space constructed by the
taxonomy can be better quantified with the computational evidence derived
from a distributional analysis of corpus data. However, this kind of approach needs

T.G. Kim (Ed.): AIS 2004, LNAI 3397, pp. 470–480, 2005.
© Springer-Verlag Berlin Heidelberg 2005

high computational cost to analyze the large volume of corpus data. Ideally, the best would be if one could get similarity results from the taxonomy as high as those of the combined approaches. Several methods have tried to improve the similarity rate based on edge counting. They gave weights on the edges according to the weighting attributes. Weighting attributes are the factors that affect an edge's strength in edge counting. The commonly used weighting attributes are the *scaling depth effect, local semantic density, and semantic relation type*. However, the edge counting as well as other combined methods has been considered not to be satisfying in terms of performance aspect although some of the approaches used those different weighting attributes. According to our investigation, the main reasons are as follows:

- First, traditional edge counting method accounted the *IS-A* type only [14]. This proves to be absolutely wrong for certain domains. For example, in the figure 1, the concept pair <*"computer", "platform"*> has 4 edges between them using pure *IS-A* relation. However, if the *Has-part* relation type be used then, the distance is just 1.
- Second, although, some researchers used different semantic relation types such as Part-of, Has-part, etc [22], [25], they did not give weights on each edge, but instead they gave weight on each source and destination node. For example, based on human judgment, we know that the concept pair <*"computer", "platform"*> pair is more similar than the concept pair <*"computer", "machine"*> in term of the context of computer domain [19], [20], [21], [22], [23]. However, both concept pairs have the same distance of 1 if we use both of the relation types *IS-A* and *Has-Part*. That means, particular care should be taken when we give weights according to different semantic relation types.
- Third, in traditional edge counting based on WordNet, researchers gave extraordinary long distance between concepts from where they located in different domains. For example, as will be discussed in the evaluation section, the WordNet consists of many different sub-domains and some of those sub-domains are connected by the concept *entity* virtually on the top level of taxonomy. This means that the concepts cannot be directly connected. In that case, the existing methods would give maximum distance for the concept pairs. However, according to our investigation, this decreases the overall correlation from the human judgment of similarity.

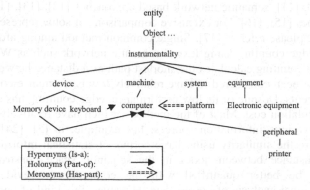

Fig. 1. Lexical concept taxonomy generated from WordNet

In this paper, we measure the similarity using only lexical semantic network to help in considering the weighting of the attributes and virtual connections. Our similarity model uses the WordNet [13], [17], [18] semantic network due to its wide spread use and growing popularity [3], [6], [24]. Originally, we considered three weighting attributes as the main components that affect edge strength such as *scaling depth effect, semantic relation type, and local semantic density*. However, we dropped the attribute of local density effect because the notion of local semantic density comes from the fact that different domains have different depth of nodes. For example, the biology domain tends to be much denser than other domains. Theoretically this is correct but, according to our observation, if we exclude the small number of those exceptional domains, the average depth of other domain is quite similar. In WordNet, most of the domains do not exceed the maximum depth of 15 nodes. Furthermore, the effect of local semantic density can be diluted quite a lot if we give proper weight on the calculation of depth effect. That is because we see that the information content of each edge in WordNet varies according to its depth level. In particular, the upper part of the edges connects the concepts more loosely than the connection in the lower part of edges. This directly reflects the effect of local semantic density. For example, the particular sub-domain such as biology domain is located in the lower part of the taxonomy. Thus, if we give more weight on the low parts of edges then, the local semantic density effect can be reflected in the calculation of depth effect. The novel contribution of our approach is that it generates quite viable similarity results using only lexical semantic network compared with other combined approaches.

The organization of this paper is as follows: the next section examines the weighting attributes and formulates them to adjust our similarity model. Section 3 evaluates our method by comparing the similarity results of proposed method with the human similarity judgment. Finally, in section 4, we give a conclusion and address future work.

2 Edge Counting Using Weighting Attributes

In this section, we investigate the weighting attributes and formulate the similarity model according to each weighting attribute. Before we introduce the calculation of similarity, we need to clarify the meanings between semantic distance and semantic similarity, in short similarity. In this paper, we differentiate the terms of semantic distance and semantic similarity. In our approach, we see the relation between semantic distance and similarity is inversely proportional such that if the semantic distance in the knowledge taxonomy is long then the semantic similarity is low. For example, the similarity between synonymous words is 1 while the distance is 0.

2.1 Formulation of Similarity Based on Edge Counting

In order to provide formulations of the intuitive concept of similarity, we first clarify our intuitions about similarity.

Intuition 1: In edge-based measure, if the number of edges (i.e., the distance) between concepts is short then, the two concepts are more similar than those of concepts having higher number of edges.

In order to formulate the above intuition, we need to represent each edge as a metric. We set the distance for each edge as 1 for the base value, and then we put more weight on them according to the value of weighting attributes. Values of the weighting attributes may cover a large range up to infinity, while the interval of similarity should be finite with extremes of "exactly the same" or "not similar at all". In edge-based measure, if we assign "exactly the same" with a value of 1 and no similarity as 0, then the interval of similarity is [0, 1]. On the same token, the similarity for self-distance is 0, which means that the two concepts are exactly the same semantically. For example, the similarity between car and car, the same Synset in the WordNet such as Car, automobile, etc are all set to 0. Based on this intuition, we justify the similarity range for an edge as follows:

$$S_{range} = 1 / (1 + Weight_{attr}).$$ (1)

From the similarity range of (1), we can assumes that if the weight derived from weighting attributes is high, then the similarity range should be decreased. If the weight is low, then the similarity range should be increased. That is, if the weight is very high, say, the weight goes extream ∞ (i.e., the self distance or the same Synset in the WordNet) then the similarity range become 0. On the other hand, if the weight goes very close to 0, then the value becomes 1. In our model, we add up each similarity range value to form a distance measure. The distance dist(c1, cn) between concepts, c1 and cn, is then the sum of each similarity range value.

$$dist(c_1, c_n) = \Sigma_{i=1, n} S_{range\ i}.$$ (2)

In addition to knowing that an edge's similarity range is affected by the weighting attributes, an edge's strength is the sum of values of the each weighting attribute. Thus, each edge can be represented as weighted sum of weighting attributes. Let e be an edge, then the edge strength $e_{i, j}$ {$e_{i, j}$ | $1 \leq i \leq n$, $1 \leq j \leq m$} be the jth weighting attribute in the ith edge, where n and m are integers. The value of an edge's strength can be represented as:

$$Weight_{attr}(e_i) = \Sigma_{j=1, m} Max\ [\ edge\_strength(e_{i, j})\ ].$$ (3)

Where, i represents edge(s) between concepts, c_1 and c_n, and j represents attributes, which affect the edge strength. Finally, the similarity in edge counting for the concepts, c_1 and c_n, can be expressed as the minimum path distance.

$$S(c_1, c_n) = (30 - Min\ [\ dist(c_1, c_n)\ ]) / MAX\_SIM.$$ (4)

Where, the number 30 comes from the fact that the maximum depth of WordNet hierarchy is 15. Therefore, the sum of depths for both sides cannot exceed more than 30. And the MAX_SIM is a constant. The value of MAX_SIM depends on the maximum similarity value. For example, if an application system sets the maximum similarity for two concepts as 1, then the MAX_SIM should be 30.

2.2 Edge Strength in Terms of Weighting Attributes

It is necessary to introduce some constraints to the development of similarity measure before proceeding to the presentation of calculation of weighting attributes. As mentioned previously, the similarity between words is context-dependent and asymmetric [16], [17]. That is, people may give different ratings when asked to judge the similarity. For example, as Rada stated in [14], student may consider the statement, "*Lectures are like sleeping pills,*" to be more true than the statement, "*Sleeping pills are like lecture.*" Although similarity may be asymmetric, we do not consider asymmetric factor. Because experimental results [16] investigating the ratings for a word pair is less than 5 percent. For the weighting factors, as discussed in the previous section, we assume that edge strength is affected by the *semantic relation type and scaling depth effect* (we ignore the effect of local semantic density). Therefore, we propose that edge strength between each node (concept) be a sum of functions of attributes as follows:

$$edge\_Strength(e_i) = \alpha.f(t) + \beta.f(h) . \tag{5}$$

Where f(t) and f(h) represent semantic relation type, scaling depth effect respectively. And α and β are constant. The assumption in (5) comes from the following considerations: semantic relation type is derived from a lexical database WordNet directly, while the scaling depth effect is computed from relative depth length. Those derivations are detailed in the following sub-sections. The independence assumption in (5) enables us to investigate the contribution of individual attributes to overall similarity by combining them. Again, the intuition beyond the edge counting is that the interval of similarity should be finite with extremes of exactly the same or nothing similar at all of [0, 1].

2.2.1 Semantic Relation Type

Many edge-based models consider only the IS-A link hierarchy [14]. In fact, other link relation types, such as Meronym/Holonym (Part-of, Substance-of), should also be considered, as they would have different effects in calculating an edge's weight. The WordNet has eight different semantic relation types. In [25], Sussna also considered the weight for semantic relation type. However, our approach is different from Sussna's model in two aspects. First, although Sussna also accounted the scaling depth effect in the formula, the depth effect computation is quite intuitive because Sussna took only the deeper node of the two nodes for depth calculation. That means the depth effect of the other node is not accounted for. As in (6, 7), we deal with the depth and relation type separately for weighting attributes. Second, Sussna considered the asymmetric aspect in her model. However, we ignore it [26]. Rather, our model is simple. Because, we set the weight for *IS-A* type to be 0 as a base line and the *Part-of* and *Has-part* types to be 1 (In this paper, we only consider the *IS-A*, *Part-of* and *Has-part* types). Proposed weights are realistic because normally, as shown in figure 1, the *Part-of* and *Has-part* types are more close to the specific context. For example, if we consider the concept *printer* in the terms of the context of computer domain, it should be dealt more closely to the concept *computer* rather than to the concepts of *machine* or *device*, which both have the distance 1 and 3 in *IS-A* type based calculation respectively, although the distance between *printer* and *computer* is 6 in an *IS-A* type. Thus, the following formula (6) justifies our intuition on semantic relation type in edge based distance calculation.

$$f(t) = weight_{type} (A \rightarrow_t B) = \begin{cases} 0, \text{ for IS-A relation type} \\ 1, \text{ for Part-of, has-part relation type.} \end{cases} \tag{6}$$

Where →_t is a semantic relation of type t. The range for weights varied from 0 to 1.

2.2.2 Scaling Depth Effect

According to the information theory [27], the upper level concepts have relatively low information content than the concepts in the lower level in the taxonomy. That means the upper layer concepts are less informative than the concepts in lower layer [3]. Same holds for edges also. The edges in the lower part of the taxonomy link the concepts more informatively than the connection in the upper level. In our model, we divided the level of information content according to the depth. As the depth moves up to the taxonomy, the connection is less informative. As for measuring an edge's informative ness, we leveled each edge according to its relative depth from top to bottom as shown in (7). For a semantic network organized hierarchically, such as WordNet, the edge strength between two concepts, c_l and c_n, can be determined from the following cases in terms of scaling depth effect:

1. c_l and c_n are in the same concept. That is, the distance is 0.
2. If both for the concepts, c_l and c_n, are located in deeper place from the top, then the edge strength should be higher than the concepts in the upper link.
3. The edge weight should be increased exponentially as it moves from top to bottom in the node-based information content [22].

Case 1 implies that c_l and c_n is the same concept. As Rada stated in [14], this is the case of reflective metric in edge counting. Consequently, that means that the maximum weight should be given on the concepts regardless of the level of depth. Case 2 and case 3 indicate that the edge strength should be increased exponentially as the edge moves far from top node. Taking the above considerations into account, we set the scaling depth effect $f(h)$ be a exponentially increasing function of i:

$$f(h) = e^{i/N}. \tag{7}$$

where N is the highest level of depth in the WordNet, i is the current level of depth (i.e., the edge between the concepts entity and artifact is the first level, then i is 1). Therefore i/N represents the ith level of edge from the top edge.

3 Evaluation

The best way to evaluate the performance of machine measurements of semantic similarity between concepts is to compare them with human ratings on the same data set. Then, the correlation between human judgment and machine calculations should be examined [29]. Therefore, in the following sub-section, we evaluate our similarity model using WordNet data set and compare the similarity results with human judgment.

3.1 The Benchmark Data Set

The work reported here used WordNet's taxonomy of concepts represented by nouns in English. We used the noun taxonomy from WordNet version 1.7.1. The best way to show the quality of computational method for calculating word similarity can be established by investing its performance against human common sense. The most commonly used word set is from an experiment by Miller and Charles [30]. An experiment by Miller and Charles provided appropriate human subject for the task. In

Table 1. Semantic similarity of human rating vs. similarity with weighting attributes

Word Pair	R & G	M & C	Num. of edges (IS-A type only)	Num. of edges with Virtual connection	Shortest path with different types	Scaling depth effect	Weight according to the types
Rooster-voyage	0.04	0.08	30	**17**	17	22.921425	22.921425
Noon-string	0.04	0.08	30	**12**	12	24.710528	24.710528
Car-journey	1.55	1.16	30	**12**	12	24.710528	24.710528
Food-fruit	2.69	3.08	3	8	**3**	28.698679	**28.821611**
Furnace-stove	3.11	3.11	2	7	**2**	29.213282	**29.3283**
Cord-smile	0.02	0.13	12	12	12	24.758218	24.758218
Glass-magician	0.44	0.11	9	9	9	25.947089	25.947089
Monk-slave	0.57	0.55	4	4	4	28.378423	28.378423
Coast-forest	0.85	0.42	6	6	6	27.315081	27.315081
Monk-oracle	0.91	1.1	7	7	7	27.221965	27.221965
Lad-wizard	0.99	0.42	4	4	4	28.362516	28.362516
Forest-graveyard	1	0.84	9	9	9	25.947089	25.947089
Food-rooster	1.09	0.89	13	13	13	24.624059	24.624059
Coast-hill	1.26	0.87	4	4	4	28.297689	28.297689
Crane-implement	2.37	1.68	4	4	4	28.313807	28.313807
Brother-lad	2.41	1.66	4	4	4	28.362516	28.362516
Bird-crane	2.63	2.97	3	3	3	28.936673	28.936673
Bird-cock	2.63	3.05	1	1	1	29.63026	29.63026
Brother-monk	2.74	2.82	1	1	1	29.63026	29.63026
Asylum-madhouse	3.04	3.61	1	1	1	29.614594	29.614594
Magician-wizard	3.21	3.5	0	0	0	30	30
Journey-voyage	3.58	3.84	1	1	1	29.614594	29.614594
Coast-shore	3.6	3.7	1	1	1	29.58257	29.58257
Implement-tool	3.66	2.95	1	1	1	29.58257	29.58257
Boy-lad	3.82	3.76	1	1	1	29.614594	29.614594
Automobile-car	3.92	3.92	0	0	0	30	30
Midday-noon	3.94	3.42	0	0	0	30	30
Gem-jewel	3.94	3.84	0	0	0	30	30

their study, 38 undergraduate subjects were given 30 pairs of nouns that were chosen to cover high, intermediate, and low levels of similarity (as determined using a previous study [31], Rubenstein & Goodenough, 1965), and those subjects were asked to rate "similarity of meaning" for each pair on a scale from 0 (no similarity) to 4 (perfect synonymy). The average rating for each pair thus represents a good estimate of how similar the two words are. However, our method on similarity is slightly different from Miller and Charles. Because they set the scale from 0 to 4, but on the contrary, we set each interval of an edge be 0 (perfect synonym) to 1 (no similarity). And the sum of those edges is the similarity measure. Note that, the experiment from Miller and Charles used only 28 word pairs of the Rubenstein and Goodenough's original 65 words set [31]. We also used the Miller and Charles' word pairs. The words pairs are listed in table 1.

3.2 Experiments According to the Weighting Attributes

We assumed that the similarity based on edge counting in the taxonomy is affected by weighting attributes. We tested the impact of weighting attributes separately to show the effect of each attribute. Our test consists of several steps in order to investigate the effectiveness of proposed similarity model. After each test step, we measured the performance of the similarity in terms of the calculation of correlation coefficient between the computed semantic similarity values and Miller and Charles's values.

Step 1. Considering virtual connection. As shown in table 1, the existing edge counting only accounted the IS-A relation type, and it also neglected the connection between entity and other top-level concepts. For example, in Wordnet taxonomy, each domain is connected with the concept entity virtually. That means if we try to find the super-ordinate concept from a concept of sub-domain, some domains cannot be reached to the concept entity directly. The concept pairs in the first 3 rows from table 1 such as <Rooster-voyage>, <Noon-string>, and <Car-journey>, cannot be connected to each other directly, because the concept rooster is in the entity domain while the concept voyage is in the act domain. Those two domains have no physical connection and in this paper, we regard them to be linked virtually. Usually, the existing researches counted the distance between concepts in the virtual connection as the maximum 30. This is obviously wrong. For example, the direct super-concept of voyage is journey in the act domain. And the concept Rooster and Car are in the entity domain. Then, the distance between <Rooster-voyage> and <Car-Journey> should be treated different but they were all treated as 30 because the domains of act and entity are different in the top level. Further, they have no connection for similarity calculation. In our model, we counted the distance between concepts in the virtual connection. However, we do not count them as a distance as the name virtual connection imply. For example, the distance entity and act should be 0. This increases the correlation coefficient dramatically as of the original value from 0.6335 to 0.7667.

Step 2. Shortest path with different relation types. Our similarity model used not only for the relation type IS-A but also for the other relation types such as Part-of and Has-Part. Therefore, our similarity model tries to link the concepts with the shortest path as long as they are linked with IS-A, Part-of, and Has-Part relation type.

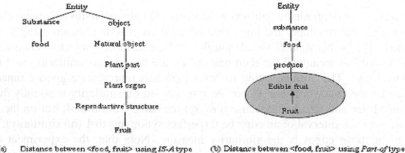

Fig. 2. A fragment of WordNet taxonomy with different relation types

The figure 2 illustrates well this calculation. In our 28 pairs of sample data, we have two concept pairs, <food, fruit> and <furnace, stove>, which use part-of or has-part relation as a shortest path relation type. According to figure 2 (a), the original distance using IS-A type is 8. But in (b), the concept pair is connected with part-of relation type. And their distance is 3. Using this shortest path with relation type Part-of, we get 0.8234 for the correlation coefficient.

Step 3. Scaling depth effect. As discussed previously, the information content in the WordNet increases as it moves down to the lower level of depth in the taxonomy. Furthermore, the increment is changed exponentially, which was given in (7). We measured the similarity based on the edge calculation of step 2. The correlation coefficient between the scaling depth effect and human rating was 0.82748. This also shows the increment of correlation coefficient from the previous steps.

Step 4. Weight according to relation types. As discussed in step 2, we have two concepts pairs, <food, fruit> and <furnace, stove>, where they are connected with Part-of and Has-Part types. In step 2, we just connected them with shortest path. However, in this step, according to (6), we put weight on each edge between them. Results obtained are also encouraging with the 0.8288 of correlation coefficient value.

Table 2 summarizes the increment of correlation value according to the each step. Among them, step 1 that used virtual connection, and step 2 that used the shortest path with different relation type, showed the sharp increment of correlation values. We also measured the correlation between our approach and Rubenstein & Goodenough' approach from which we had higher correlation values than the comparison with Miller & Charles. Although, the upper value of human judgment between Rubenstein & Goodenough and Miller & Charles is 0.9015 [15], which is much higher than our results, our model produced high correlation value compared with other machine-calculated approaches.

Table 2. Correlation coefficient values according to each step

Similarity measure according to the steps	Correlation with M & C	Correlation with R & G
Num. of edges with IS-A type	0.6335	0.6611
Step 1 (Num. of edges with the virtual connection)	0.7667	0.7999
Step 2 (step 1+Shortest path with different relation types)	0.8234	0.8441
Step 3 (Step 2+Scaling depth effect)	0.8274	0.8454
Step 4 (step 3 +weight according to relation types)	0.8288	0.84630

Table 3 shows the comparison between different approaches. From the table 3, we can assume that although the value from Jing & Conrath's approach is quite similar with our approach, our approach is much better in terms of cost effectiveness, since most of other combined approaches calculate the similarity from various information sources before it can be applied to semantic network. Further, other combined approaches calculated similarity against large corpus data beforehand. However, our approach used only the semantic network.

Table 3. Summary of experimental results of different approaches

Similarity Method	Correlation (r)
Human Judgment	0.9583
Node Based (information content)	0.745
Edge Counting (IS-A only)	0.6335
Combined Approach (Jing & Conrath)	0.8282
Edge Based (Proposed approach)	0.8288

4 Conclusion

This paper presented semantic similarity based on edge counting. We gave weights on each edge to reflect the edge strength. For weighting attributes, scaling depth effect and semantic relation type were employed to increase the correlation coefficient values from human judgment of similarity. Further, we found that the similarity based on edge counting was sharply affected by consideration on the virtual connection. We also found that some of the classifications in the WordNet do not exactly fit the similarity measure. That might come from their specific classification scheme on the hierarchy. Although our similarity model used only semantic network *WordNet*, the results are encouraging. Compared with the other combined approaches, if we consider the fact that they used high computing cost and the resulting maximum value of correlation coefficient was 0.8282, our similarity model is quite viable with the correlation coefficient value of 0.8288. For the future work, we are considering to mix our pure edge based model with the information content approach, as in the combined approach.

References

1. Tversky, A.: Features of similarity. Psychological Rev., Vol. 84, (1977) pp. 327-352
2. McGill et al.,: An evaluation of factors affecting document ranking by information retrieval systems. Project report, Syracuse University School of Information Studies, (1979)
3. Resnik, P.: Semantic Similarity in a Taxonomy: An Information-based Measure and its Application to Problems of Ambiguity in Natural Language. Journal of Artificial Intelligence Research Vol. 11, (1999)
4. 4 Green, S.J: Building Hypertext Links by Computing Semantic Similarity. IEEE Trans. Knowledge and Data Eng., Vol. 11, No. 5, (1999) pp. 713-730
5. Lee, J. H., Kim, M. H.: Information retrieval based on conceptual distance in is-a hierarchies. Journal of Documentation, Vol. 49, No. 2, (1989)
6. Guarino, N., Masolo, C., and Vetere, G.: OntoSeek: Content-BasedAccess to the Web. IEEE Intelligent Systems, Vol. 14, No. 3, (1999)
7. Voorhees, E.: Using WordNet for Text Retrieval. WordNet: an electronic Lexical Database, C. Fellbaum, ed., Cambridge, Mass.: The MIT Press, (1998) pp. 285-303
8. Hideki Kozima and Teiji Furugori: Similarity between words computed by spreading activation on English dictionary. In Proc. of 6th conference of the European chapter of the association for computational linguistics (EACL-93), Utrecht, (1993) pp 232-239
9. Hideki Kozima and Akira Ito: Context-sensitive measurement of word distance by adaptive scaling of a semantic space. In Ruslan Mitkov and Nicolas Nicolov, editors, Recent Advances in Natural Language Processing: Selected Papers from RANLP'95, Vol 136 of Amsterdam Sudies in the Theory and History of Linguistic Science: Current Issues in Linguistic Theory, chapter 2, (1997) pp 111-124
10. Jane Morris and Graeme Hirst: Lexical cohesion computed by thesaural relations as an indicator of the structure of text. Computational Linguistics, Vol. 17, No. 1, (1991)
11. Manabu Okumura and Takeo Honda: Word sense disambiguation and text segmentation based on lexical cohesion. In Proc. of the Fifteenth International Conference on Computational Linguistics (COLING-94), Vol. 2, Kyoto, Japan, (1994) pp. 755-761
12. Joon Ho Lee, Myong Ho Kim, and Yoon Joon Lee: Information retrieval based on conceptual distance in IS-A hierarchies. Journal of Documentation, Vol. 49, No. 2, (1993) pp. 188-207

13. Miller, George A., Richard Beckwith, Christiane Fellbaum, Derek Gross, and Katherine Miller: Five papers on WordNet. CSL Report 43, Princeton University, (1990), revised August (1993)
14. Roy Rada, Hafedh Mili, Ellen Bicknell, and Maria Blettner: Development and application of a metric on semantic nets. IEEE Trans. On systems, Man, and Cybernetics, Vol. 19, No. 1, (1989) pp. 17-30
15. Philip Resnik: Using information content to evaluate semantic similarity. In Proc. of the 14th International Joint conference on Artificial Intelligence, Montreal, Canada, August (1995) pp. 448-453
16. Alexander Budanitsky: Lexical Semantic Relatedness and Its Application in Natural Language Processing. Technical Report CSRG-390, Computer Systems Research Group, University of Toronto, August (1999)
17. Miller, G. A.: *WordNet: A lexical database for English*. Communications of the ACM, Vol. 38, No. 11, (1995)
18. Miller, G.: Nouns in WordNet. WordNet: An Electronic Lexical Database, C. Fellbaum, ed., Cambridge, Mass.: The MIT press, (1998) pp. 23-46
19. Anna Formica and Michele Missikoff: Concept Similarity in *SymOntos*: An Enterprise Ontology Management Tool. The Computer Journal, Vol. 45, No. 6, (2002) pp. 583-594
20. Kasahara, K., Matsuzawa, K., Ishikawa, T. and Kawaoka, T.: *Viewpoint-based measurement of semantic similarity between words*. In Proc. Int. Workshop on Artificial Intelligence and Statistics, (1995)
21. Kozima, H., Ito, A.: A Context-sensitive Measurement of Semantic Word Distance. Journal of Information Processing, Vol. 38, No. 03–011, (2001)
22. Li, Yuhua, Zuhair, A. Bandar, David McLean: An Approach for Measuring Semantic Similarity between Words Using Multiple Information Sources. IEEE Trans. on Data and Knowledge engineering, Vol. 15, No. 4, July/August (2003) pp. 871-882
23. Finkelstein L., Gabrilovich E., Matias Y., Rivlin E., Solan Z., Wolfman G., and Ruppin E.: *Placing search in context: the concept revisited*. ACM Trans. on Information Systems, Vol. 20, No. 1, January (2002) pp 116-131
24. Jiang, J. and Conrath, D.: *Semantic similarity based on corpus statistics and lexical taxonomy*. In Proc. on International Conference on Research in Computational Linguistics, Taiwan, (1997)
25. Sussna, M.: Word sense disambiguation for free-text indexing using a massive semantic network. In Proceedings of the Seoul International Conference on Information and Knowledge management (CIKM-93) Arlington, Virginia, 1993.
26. Medin, D.L., Goldstone, R.L. and Gentner, D.: Respects for Similarity. Psychological Rev., Vol. 100, No. 2, (1993) pp. 254-278
27. Dekang Lin: An Information-Theoretic Definition of Similarity. Proc. 15th International Conf. on Machine Learning, (1998)
28. Richardson, R., Smeation, A. F., Murphy, J.: Using WordNet as a Knowledge Base for Measuring Semantic Similarity between Words. Technical Report CA-1294, Dublin City University, School of Computer Applications, (1994)
29. Edwards, A.L.: An introduction to linear regression and correlation. San Francisco: W.H. Freeman, (1976)
30. Miller, G. and Charles, W.G.: Contextual correlates of semantic similarity.. Language and Cognitive Process, Vol. 6, No. 1, (1991) pp. 1-28
31. Rubenstein, H. and Goodenough, J.B.: Contextual correlates of Synonymy. Comm. ACM, Vol. 8, pp. 627-633, 1965.

3D Watermarking Shape Recognition System Using Normal Vector Distribution Modelling

Ki-Ryong Kwon[1], Seong-Geun Kwon[2], and Suk-Hwan Lee[3]

[1] Division of Digital Information Eng., Pusan Univ. of Foreign Studies,
55-1 Uam-dong, Nam-gu, Pusan, 608-738, Republic of Korea
krkwon@pufs.ac.kr
[2] R&D Group 3, Mobile Communication Division,
Information & Communication Business,
Samsung Electronics Co., LTD, Republic of Korea
seonggeun.kwon@samsung.com
[3] School of Electrical Engineering and Computer Science, Kyungpook National Univ.,
1370, Sankyuk-dong, Buk-gu, Daegu, 702-701, Republic of Korea
skylee@m80.knu.ac.kr

Abstract. We developed the shape recognition system with 3D watermarking using normal vector distribution. The 3D shape recognition system consists of laser beam generator, linear CCD imaging system, and digital signal processing hardware and software. 3D Watermark algorithm is embedded by 3D mesh model using each patch EGI distribution. The proposed algorithm divides a 3D mesh model into 4 patches to have the robustness against the partial geometric deformation. Plus, it uses EGI distributions as the consistent factor that has the robustness against the topological deformation. To satisfy both geometric and topological deformation, the same watermark bits for each subdivided patch are embedded by changing the mesh normal vectors. Moreover, the proposed algorithm does not need the original mesh model and the resampling process to extract the watermark. Experimental results verify that the proposed algorithm is imperceptible and robust against geometrical and topological attacks.

1 Introduction

Many digital watermarking schemes have recently been proposed for copyright protection and other application due to the rapid growing demand for multimedia data distribution. Watermark designing issues include detection robustness, detection reliability, imperceptibility, and capacity. Recently, 3D geometric models, such as 3D geometric CAD data, MPEG-4, and VRML, have been receiving a lot of attention, such as, various 3D watermarking algorithms have also been proposed to protect the copyright of 3D geometric data [1–5]. Ohbuchi et al. [6,7] proposed an algorithm that adds a watermark to a 3D polygonal mesh in the mesh spectral domain. However, this algorithm is not robust against attacks that alter the connectivity of meshes, such as mesh simplification and remeshing.

T.G. Kim (Ed.): AIS 2004, LNAI 3397, pp. 481–489, 2005.

Beneden *et al.* [8] also proposed an algorithm that adds a watermark by modifying the normal distribution of the model geometry. Although this algorithm is robust against the randomization of points, mesh altering, and polygon simplification, it is not robust against cropping attacks, as the normal distribution is calculated for the entire model. Kanai *et al.* [9] proposed a watermarking algorithm for 3D polygons using multiresolution wavelet decomposition. Yet, the application is restricted to a certain topological class of mesh, as the wavelet transform can only be applied to 4-to-1 subdivision connectivity schemes. However, this algorithm requires a complex resampling process of a suspect mesh in order to obtain a mesh with the same geometry, yet with a given connectivity.

In this paper, we proposed the 3D watermarking schemes of 3D shape recognition system using the normal vector distribution. This 3D watermarking system implements hardware and software of digital signal processing parts of 3D shape recognition system. 3D watermark algorithm is embedded by 3D mesh model using each patch EGI distribution. The proposed algorithm divides a 3D mesh model into 4 patches to have the robustness against the partial geometric attacks and topological deformation. To satisfy both geometric and topological attacks, the same watermark bits for each subdivided patch are embedded by changing the mesh normal vectors. Moreover, the proposed algorithm does not need both the original mesh model and the resampling process to extract the watermark. Experimental results verify that the proposed algorithm is imperceptible and robust against geometrical and topological attacks.

2   The Proposed 3D Watermarking Algorithm

A block diagram of the proposed 3D watermark embedding algorithm of 3D shape system is shown in Fig. 1. The vertices and connectivity of the 3D mesh are entered from the scanned 3D shape system, then the coordinate values of the vertices are changed by the watermark algorithm. The center points of the patches and order information of the patch EGIs are then needed to extract the watermark.

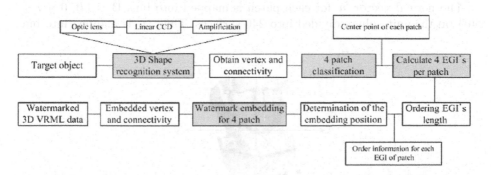

Fig. 1. The block diagram of the proposed embedding watermark algorithm.

2.1   Watermark Embedding

Patch Division of 3D Model. The 3D mesh model is divided into 4 patches $P_{i\in[1,4]}$ using a distance measure. Above all, the initial center points $I_{i\in[1,4]}$ are specified as the points with the maximum distance in the direction of 4 unit vectors $\pm\mathbf{x}$, $\pm\mathbf{y}$, and $\pm\mathbf{z}$ from the origin and considered as the respective center points of 4 patches. Then, all the vertices v are clustered into one patch P_i

$$P_i = \{v : d(v, I_i) < d(v, I_j), all\, j \neq i, 1 \leq i,j \leq 4\} \tag{1}$$
$$where\ d(v, I_i) = ||v - I||^2$$

which has the minimum distance among the initial center points $\mathbf{I}$. After calculating the center of the clustered vertices for each patch, $\mathbf{I}$ are updated to the new center point. Then, the clustering and updating process is iterated until the condition $|D_{m-1} - D_m|/D_m \leq 10^{-5}$ is satisfied, where m is the iteration number and D_m is expressed as $D_m = \sum_{i=1}^{4}\sum_{v\in P_i}||I_i - v||^2$. Fig. 2 shows the patches into which the face model was divided. $P_{i\in[1,4]}$ represent the patches in the direction of the 4 unit vectors $\pm\mathbf{x}$, $\pm\mathbf{y}$, $\pm\mathbf{z}$ respectively. The meshes are then divided into the patches that include their vertices. In the case where meshes have vertices that are divided into different patches at the boundary of each patch, these meshes are not excluded from the patch EGI.

Patch EGI. The normal distribution of each patch is represented by an orientation histogram, called an Extended Gaussian Image (EGI). In the current paper, the unit sphere is divided into 240 surfaces that have the same area as shown in Fig. 3.

The respective patch EGIs $P_{i\in[1,4]}$ are obtained by mapping the mesh normal vectors in $P_{i\in[1,4]}$ into the surface that has the closest direction among the 240 surfaces. To calculate a normal vector with a consistent direction, the vector of the vertex at the patch center point $\mathbf{I}$ is used without referring to the origin of the entire model. Thus, the unit normal vector $\vec{n}_{ij}$ and area A_{ij} for the j-th mesh surface in are calculated as $\vec{n}_{ij} = (\mathbf{I}_i\mathbf{v}_1 - \mathbf{I}_i\mathbf{v}_2) \times (\mathbf{I}_i\mathbf{v}_1 - \mathbf{I}_i\mathbf{v}_3)/||n_{ij}||^2$, $A_{ij} = (\mathbf{I}_i\mathbf{v}_1 - \mathbf{I}_i\mathbf{v}_2) \times (\mathbf{I}_i\mathbf{v}_1 - \mathbf{I}_i\mathbf{v}_3)/2$.

The normal vector $\vec{n}$ for each patch is mapped into bins $\mathbf{B} = \{B_i|0 \leq i \leq 240\}$ on the unit sphere divided into 240 surfaces. Thus, $\vec{n}$ is mapped into bin

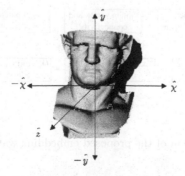

Fig. 2. The divided mesh model into 4 patches.

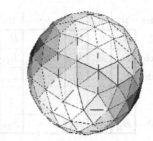

Fig. 3. Unit sphere discretized into 240 cells with the same area.

B_i with the smallest angle among the angles between $\overrightarrow{n}$ and the unit normal vector of bin $\overrightarrow{BC}$;

$$B_i = \{\overrightarrow{n} : ||\overrightarrow{BC}_i \cdot \overrightarrow{n} < ||\overrightarrow{BC}_j \cdot \overrightarrow{n} \ all \ j \neq i, 1 \leq i,j \leq 240\} \tag{2}$$

The maximum value θ_m of $\overrightarrow{BC}_i \cdot \overrightarrow{n}$ is half the angle between the $\overrightarrow{BC}$s of the neighborhood bins as shown in Fig. 4. As $\overrightarrow{n}$ around θ_m can be mapped into a different bin after being attacked, it is excluded from the EGI. Thus, only $\overrightarrow{n}$ within a certain range is selected as follows;

$$0 \leq cos^{-1}(\overrightarrow{BC}_i \cdot \overrightarrow{n}) \leq \theta_{max}, \ \theta_{max} < \theta_m \tag{3}$$

The length of B_i is the sum of area A for all the meshes that are mapped into B_i. There are 240 bins $\mathbf{B}_j = \{B_{ji}|0 \leq i \leq 240\}$ in the j-th patch EGI

The watermark, a 1-bit random sequence, is embedded in the ordered B of **PE**. When the 1-bit watermark sequence of n length is embedded, the normal vectors for all the meshes that are mapped into $4n$ bins are changed according to the watermark bit allocated to each bin. The j-th watermark w_j $(1 \leq j \leq N)$ is embedded into the j-th ordered B_{ijs} $(1 \leq j \leq 4)$ in each **PE**, as shown in Fig. 5. The extraction of the watermark requires the ordered information of **PE**.

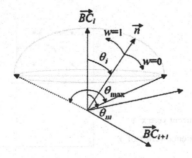

Fig. 4. Angle range where mesh normal vector $\overrightarrow{n}$ can be mapped into i-th bin B_i.

watermark	w_1	w_2	w_3	$\cdots$	w_N
1st. patch	B_{11}	B_{12}	B_{13}	$\cdots$	B_{1N}
2nd. patch	B_{21}	B_{22}	B_{23}	$\cdots$	B_{2N}
3rd. patch	B_{31}	B_{32}	B_{33}	$\cdots$	B_{3N}
	$\vdots$	$\vdots$	$\vdots$	$\cdots$	$\vdots$
4th. patch	B_{41}	B_{42}	B_{43}	$\cdots$	B_{4N}

Fig. 5. Embedding watermark bit into the high ranked cells per patch EGI **PE**.

Embedding of Watermark. All the normal vectors in the selected bin are changed according to the watermark bit. To change a normal vector, the position of the vertex needs to be changed. Yet, if the position of a vertex is changed, the normal vectors of all the meshes connected to this vertex will also be changed. Thus, it is important to identify the position of a vertex, as the normal vectors of all the meshes connected to this vertex will be changed according to the watermark bit of the bin that they are mapped into. Fig. 6 (a) shows 4 meshes, $S_{i\in[1,4]}$ and 4 vertices, $valence(v)$ connected to vertex v. After calculating the patch P_i that each mesh S_i is divided into and the bin into which the normal vector $\vec{n}$ of its mesh is mapped in PE_i of P_i, the watermark bit allocated in its bin is matched. Then, vertex v is changed to the optimum position that entirely satisfies the watermark bits of each mesh $S_{i\in[1,4]}$ within the search region of its vertex.

The optimum position is identified for all the vertices that satisfy the cost measure for the condition of watermark embedding within the search region. To be invisible, the search region for changing the position of a vertex must be below the value of each coordinate in all the vertices connected to the current vertex. Thus, the search regions for each coordinate of the current vertex $v = (x, y, z)$ are $x \pm \triangle x$, $y \pm \triangle y$ and $z \pm \triangle z$; $\triangle x = 0.5 \times min|x - v_x^i|_{v^i \in valence(v)}$, $\triangle y = 0.5 \times min|y - v_y^i|_{v^i \in valence(v)}$, $\triangle z = 0.5 \times min|z - v_z^i|_{v^i \in valence(v)}$. $valence(v)$

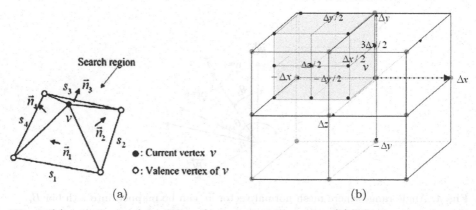

(a)   (b)

Fig. 6. (a) 4 meshes and 4 vertices that connected in vertex (b) 2 Step process in step algorithm to identify optimum vertex.

represents all the vertices that are connected to the current vertex v and v_x, v_y, and v_z are the x, y, and z coordinate values of $valence(v)$, respectively. The step algorithm is used to identify the optimum vertex within the search region. In the 1-th step, 27 positions for the vertex $v = \{(x, y, z) | x \in \{x - \triangle x, x, x - \triangle x\}, y \in \{y - \triangle y, y, y - \triangle y\}, z \in \{z - \triangle z, z, z - \triangle z\}$ are selected for the initial search. This means that the normal vector direction of all the meshes mapped into B_i are either moved toward the $\overrightarrow{BC}_i$ ($\theta_{w_i} = 0$) or further away from the $\overrightarrow{BC}_i$ ($\theta_{w_i} = \theta_{max}$), as shown in Fig. 4. Then, $\triangle x$, $\triangle y$, and $\triangle z$ are decreased by half as shown in Fig. 6 (b). In the next step, the vertex v' is updated through the above process, which is performed in 3 iterations.

2.2   Extraction

The watermark can still be extracted from a watermarked model that has been attacked by connectivity altering, such as remeshing and simplification. First, the normal vector and its EGI distribution are calculated after dividing the original model into 4 patches based on the known center point I of each patch. Using the location information obtained from the EGI distribution of each patch with watermark embedding, the watermark can be extracted based on the average difference of the angle between all the normal vectors mapped into each bin and the bin center point. The watermark w_{ij} of the B_{ij} bin with a j-th length for the i-th patch is extracted as 1, if $0 \leq \theta_{ij} \leq \theta_{th}$, otherwise, it is extracted as 0. θ_{ij} is represented as $\theta_{ij} = 1/N \sum_{k=0}^{N} cos^{-1} |\overrightarrow{BC}_i \cdot \overrightarrow{n}_k|$ where N equals the number of all the mesh normal vectors projected into the B_{ij} bin. A watermark decision of 1 bit is performed based on the w_{ij} obtained from the 4 patches. Namely, w_j is expressed as $w_j = INT(1/4 \sum_{i=1}^{4} w_{ij} + 0.5)$.

3   Experimental Results

The practical picture of 3D shape recognition system and the operating software menu are shown in Fig. 7. VRML face model extracted in 3D shape recognition system has 128,000 vertices and 254,562 faces. The 3D VRML data of the face model was used to evaluate the robustness of the proposed algorithm against geometrical and topological attacks. The experiment used a 1-bit watermark with a 50 length $W_{N=50}$ generated by a Gaussian random sequence. Therefore, each watermark was embedded into 50 bins per patch (total 200 bins in a model). θ_m, half the angle between two bins in the neighborhood, was calculated as 10.27 degrees. The maximum angle θ_{max} that the normal vector could be mapped into a bin was experimentally determined as 8.6 degrees, while the threshold θ_{th} to extract the watermark was experimentally determined as 4.8 degrees. The original model and watermarked model are shown in Fig. 8 (a) and (b), clearly demonstrating the invisibility of the watermark.

To evaluate the robustness of the proposed algorithm, the watermarked model was attacked by mesh simplification, cropping, and additive random noise. The results are shown in Table 1, where the numbers of bin bit errors indicate the

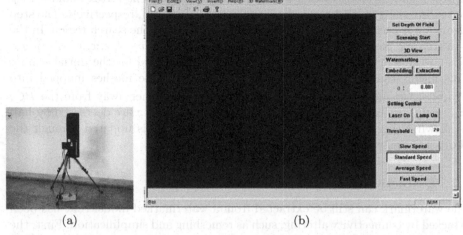

(a)   (b)

Fig. 7. (a) The 3D shape recognition system and (b) The operating program menu.

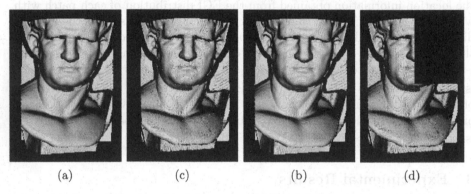

(a)   (c)   (b)   (d)

Fig. 8. (a) Original face model, (b) watermarked model, (c) mesh simplified to 50%, and (d) cropped (c) model.

number of bins with a bit error among the 200 watermarked bins of all patches. Plus, the BEP of a bin bit indicates the percentage of the number of bin bit errors in 200 bits. Although bit errors due to attacks did occur in some watermarked bins, the watermark still remained in each patch.

The watermarked model was attacked with remeshing and simplification using MeshToSS [10]. However, the watermark remained until the model was simplified to 10% of the original model vertex. In addition, 68% of the watermark remained until the model was simplified to 70% of the vertex. Randomization of a vertex was performed, where vertex sampled randomly was added to $v\alpha \times uniform()$. The modulation factor α was 0.008 and $uniform()$ was a uniformly random function of [-0.5 0.5]. In table, 50% and 100% of random noise indicate the percentage of the number of vertices that were randomly sampled to add random noise. In both cases, all watermark remained. In the case of crop-

Table 1. Experimental results of robustness against various attacks.

Attacks		Proposed algorithm			
		# of bin bit error	BEP of bin bit [%]	# of watermark bit error	BEP of watermark bit [%]
Simplify	115,200 vertex (10%)	0	0	0	0
	89,600 vertex (30%)	19	95	6	12
	64,000 vertex (50%)	34	17	9	18
	38,400 vertex (70%)	68	34.5	16	32
Random noise 50%		0	0	0	0
Random noise 100%		2	1	0	0
Remeshing		0	0	0	0
Cropping		25	12.5	3	6
Simplify (50%) + Cropping		54	27	21	42

ping, although there were some patches that had no vertices, all the watermarks could be extracted from the other patches. Plus, in a cropped model with 50% of the vertices, 64% of the watermark still remained.

4   Conclusions

We proposed the 3D watermarking schemes of 3D shape recognition system using the normal vector distribution. A 3D watermarking algorithm was presented that embeds a watermark into the normal vector distributions of each patch. In the proposed algorithm, the same watermark bit sequence is embedded in 4 patch EGIs of a 3D model, thereby making the watermark robust to geometrical and topological deformation. Invisibility is also improved, as the watermark is embedded based on identifying the optimum position of a vertex using a step-searching algorithm within the search region of each vertex. Moreover, the proposed algorithm does not need the original model to extract the watermark or a resampling process. Experimental results confirmed that the proposed algorithm is imperceptible as well as robust to geometrical and topological attacks. Also, as this result, this paper presented possibility of watermark embedding in 3D shape cognition system.

Acknowledgements

This work was supported by grant No. (R01-2002-000-00589-0) from the Basic Research Program of Korea Science & Engineering Foundation.

References

1. S. H. Lee, T. S. Kim, K. R. Kwon, and K. I. Lee.: 3D Polygonal Meshes Watermarking Using Normal Vector Distributions. International Conference on Multimedia & Expo(ICME2003), Vol. 3. (2003) pp. 105-108
2. R. Kwon, S. H. Lee, T. S. Kim, K. S. G. Kwon, and K. I. Lee.: Watermarking for 3D polygonal meshes using normal vector distributions of each patch. International Conference on Image Processing(ICIP2003). (2003)
3. E. Praun, H. Hoppe, and A. Finkelstein.: Robust Mesh Watermarking. SIGGRAPH 99, (1999) pp. 49-56
4. M. G. Wagner: Robust Watermarking of Polygonal Meshes. Geometric Modeling & Processing 2000, pp. 201-208, Apr. 2000.
5. B-L. Yeo and M. M. Yeung: Watermarking 3D Objects for Verification. IEEE CG&G, pp. 36-45, Jan./Feb. 1999.
6. R. Ohbuchi, H. Masuda, and M. Aono.: Watermarking Three-Dimensional Polygonal Models Through Geometric and Topological Modification. IEEE JSAC, (1998) pp. 551-560
7. R. Ohbuchi, S. Takahashi, T. Miyazawa, and A. Mukaiyama.: Watermarking 3D Polygonal Meshes in the Mesh Spectral Domain. Graphics Interface 2001, (2001) pp. 9-17
8. O. Benedens.: Geometry-Based Watermarking of 3D Models. IEEE CG&A, (1999) pp. 46-55, Third Information Hiding Workshop. (1999)
9. S. Kanai, H. Date, and T. Kishinami.: Digital Watermarking for 3D Polygons using Multiresolution Wavelet Decomposition. Sixth IFIP WG 5.2 GEO-6, (1998) pp. 296-307
10. T. Kanai, MeshToSS Version 1.0.1, http://graphics.sfc.keio.ac.jp/MeshToSS/indexE.html

DWT-Based Image Watermarking
for Copyright Protection

Ho Seok Moon[1], Myung Ho Sohn[2], and Dong Sik Jang[1,*]

[1] Industrial Systems and Information Engineering, Korea University
1, 5-ka, Anam-Dong, Sungbuk-Ku, Seoul 136-701, South Korea
{bawooi,jang}@korea.ac.kr
[2] Business Administration, Myongji College
356-1, Hongeun 3-Dong, Seodaemun-Ku, Seoul 120-776, South Korea
totalsol@mail.mjc.ac.kr

Abstract. A discrete wavelet transform (DWT)-based image watermarking algorithm is proposed, where the original image is not required for watermark extracting, is embedded watermarks into the DC area while preserving good fidelity and is achieved by inserting the watermark in subimages obtained through subsampling. Experimental results demonstrate that the proposed watermarking is robust to various attacks.

1 Introduction

A digital watermark has been proposed as a valid solution to the problem of copyright protection for multimedia data in a network environment [1]. Enforcement of digital copyright protection is an important issue, because digital contents that have been used in a network environment are easy to be duplicated. The digital watermark is a digital code embedded in the original data, which is unremovably, robustly, imperceptibly, and typically contains information about ownership [2]. To achieve maximum protection, the watermark should be: 1) undeletable by hackers; 2) perceptually invisible; 3) statistically undetectable; 4) resistant to lossy data compression; 5) resistant to common image processing operations.

In general, watermarking in the frequency domain is more robust than watermarking in the spatial domain, because watermark information can be spread out to the entire image. In the viewpoint of frequency domain, high frequency area should be avoided for robustness while low frequency area should be avoided for fidelity. The DC area has been excluded from the consideration for watermark embedding, even though it can give the best robustness [1, 3, 4, 5].

Image watermarking can be viewed as superimposing a weak signal (watermark) onto a strong background signal (image). The superimposed signal can be detected by human visual system (HVS) only if they exceed the detection threshold of HVS. According to Weber's law, the detection threshold of visibility for an embedded signal is proportional to the magnitude of the background signal [6, 7]. In other words, compared with AC components, although DC components cannot be changed by a larger percentage in order to avoid block artifacts under the constraint of invisibility, they can be modified by a much larger quantity due to their huge peak in the magni-

---

* Corresponding author.

T.G. Kim (Ed.): AIS 2004, LNAI 3397, pp. 490–497, 2005.

tude distribution. This indicates that DC components have much larger perceptual capacity than AC components [7, 8].

Watermarking techniques can be alternatively split into two distinct categories depending on whether the original image is necessary for watermark extraction or not. Recently, the pursuit of a scheme that does not need the original image during watermark recovery has become a topic of intense research [9]. This is partly due to practical issues, like the fact that the recovery process can be simplified without comparison with the original image. Also, in many instances, release of original material for any purpose is not desirable or even prohibited.

In this paper, our embedding strategy is based on a discrete wavelet transform (DWT). We use a visually recognizable watermark, because the invisible watermarks like ID number are not good to characterize owner [3, 4, 10]. The proposed technique can embed watermarks into the DC area while preserving good fidelity and does not require the original image for watermark extraction. We introduce here an alternate method that is based on comparing the DWT coefficients of two subimages among the four DWT subimages obtained by subsampling the image. It is shown that it is possible to recover the watermark without comparison with the original image by employing modification to the DWT coefficients pertaining to different subimages.

This paper is organized as follows. The embedding method is described in Section 2. Section 3 describes the extraction method. In Section 4, the experimental results are shown. Section 5 concludes this paper.

2 The Embedding Method

In the proposed watermarking, the original image is a gray-level image of size N_1 x N_2, and the digital watermark is a binary image of size M_1 x M_2. The resolution of a watermark image is assumed to be eighth of that of the original image (Fig. 2). Given the image, $X[n_1, n_2]$, $n_1 = 0, ..., N_1 - 1$, $n_2 = 0, ..., N_2 - 1$, then

$$X_1[k_1, k_2] = X[2n_1, 2n_2], X_2[k_1, k_2] = X[2n_1 + 1, 2n_2],$$
$$X_3[k_1, k_2] = X[2n_1, 2n_2 + 1],$$
$$X_4[k_1, k_2] = X[2n_1 + 1, 2n_2 + 1]$$
(1)

for $k_1 = 0, ..., N_1 / 2 - 1$, $k_2 = 0, ..., N_2 / 2 - 1$ are the subimages obtained by subsampling. These are transformed via DWT to obtain sets of coefficients $Y_i[k_1, k_2]$, $i = 1, 2, 3, 4$. Since the subimages X_i are highly correlated, it is expected that $Y_i \approx Y_j$, for $i \neq j$. This is indeed the case in practice for many images of interest.

In the DWT, an image is first decomposed into four subbands, LL_1, LH_1, HL_1, and HH_1 by cascading horizontal and vertical two-channel critically subsampled filter banks. Each entry in subbands LH_1, HL_1, and HH_1 represents the finest scale wavelet coefficients. To obtain next coarser scale of wavelet coefficients, the subband LL_1 is further decomposed and critically subsampled. The process continues until some final scale is reached. Fig. 1 shows that the image is decomposed into seven subbands for two scales. It is noted that the lowest frequency subband is at the top of left corner and the highest frequency subband is at the bottom of right corner. Thus, in our implementation, we modified the wavelet coefficients selected from LL_2 subband.

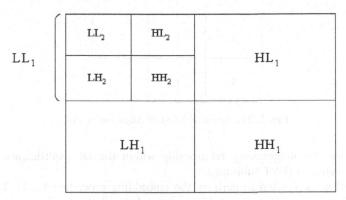

Fig. 1. Two-Level Wavelet Decomposition.

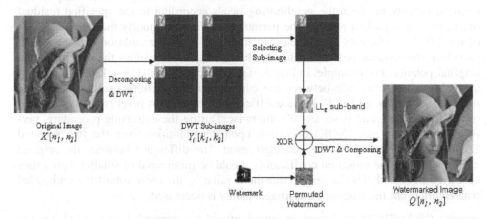

Fig. 2. Watermark Embedding Steps.

We proposed embedding steps in Fig. 2. The algorithm works according to the following steps.

DWT of the subimage generated by decomposing the original image: The original image is decomposed according to equation (1). The subimages are transformed via DWT. We take two DWT subimages among the DWT subimages that are examined in pair to verify whether they are appropriate for insertion. If two coefficients are very different in amplitudes, they will not be modified; this is to avoid causing excessive distortion. And we embed the watermark into one of the two selected DWT subimages.

Pixel-based pseudo random permutation of the watermark: in order to disperse the spatial relationship of the binary pattern, a pseudo random permutation is performed as follows. 1) Number each pixel from zero to $M_1 \times M_2$. 2) Generate each number in random order. 3) Generate the coordinate pairs by mapping the random sequence number into a 2-D sequence.

Modification of DC coefficients in the transformed subimage: there are various embedding techniques to achieve this work. In the proposed method, the watermark is

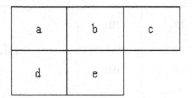

Fig. 3. The Residual Mask of Adjacent Pixels.

embedded into the neighboring relationship within the DC coefficients (LL_2 sub-band) of the selected DWT subimage.

A residual mask is used to perform the embedding procedure [3, 4]. That is, the watermark is not embedded as an additive noise. Instead, the watermark is embedded into the "neighboring relationship" within the transformed image. First, compute the residual polarity between the neighboring pixels according to the specified residual mask. For each marked pixel of the permuted watermark, modify the relevant pixels of the DWT coefficients of the selected subimage by either addition or subtraction, such that the residual polarity of the modified sequence becomes the reverse of the original polarity. For example, in Fig. 3, if a = b = c = 0, d = -1, e = 1, then the residual value is the difference between the current and the previous pixels. That is, the residual polarity is set as "1" if the coefficient of the current pixel is larger than that of the former one, and is set as "0" otherwise. During the extracting procedure, perform the exclusive-or (XOR) operation upon the polarities from the two selected DWT subimages to obtain the extracted result. The difference between the original coefficients and the modified coefficients should be minimized of smaller than a user specified threshold. The larger the threshold value is, the more robust the embedded watermark is but the watermarked image quality is decreased.

Inverse DWT (IDWT) and compose image: after the watermark is embedded into the selected subimage, the DWT subimages are IDWT transformed. And we compose the watermarked image with the IDWT subimages.

3 The Extracting Method

Whereas most of methods for watermark extraction require the original image, the proposed method does not. Figure 4 shows the proposed extracting steps.

The extraction steps are as follows:

Decomposing watermarked image and DWT: the watermarked image is decomposed according to equation (1). The subimages that are decomposed from the watermarked image are transformed via DWT.

Select two DWT subimages and take LL_2 sub-bands: we take the pairs of coefficients that are selected in embedding steps.

Generation of polarity pattern: one pair of coefficients (LL_2 subbands) from two different subimages situated in the same DWT domain location is used to extract a watermark. We make use of the residual mask applied during the embedding steps and generate the polarity patterns for LL_2 subbands.

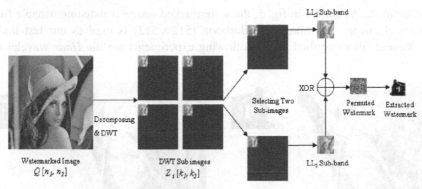

Fig. 4. Watermark Extracting Steps.

Extract the permuted watermark: perform the XOR operation on the polarity patterns to obtain the permuted binary data.

Reverse the permutation: reverse the pseudorandom permutation according to the predefined pseudo random order.

Similarity measurement: the extracted watermark is a visually recognizable pattern. The viewer can compare the result with the referenced watermark subjectively. However, the subjective measurement is dependent on factors such as the expertise of the viewers, the experimental conditions, etc. Therefore, we use a similarity measurement of the extracted watermark, $W^*[i, j]$ and the referenced watermark, $W[i, j]$ given in equation (2). $M_1 \times M_2$ is the size of the watermark image.

$$NC = \frac{\sum_{i=0}^{M_1-1} \sum_{j=0}^{M_2-1} W[i, j] W^*[i, j]}{\sum_{i=0}^{M_1-1} \sum_{j=0}^{M_2-1} W[i, j] W[i, j]} \tag{2}$$

Quantitative measure of imperceptibility: to establish a more quantitative measure of imperceptibility, we make use of the peak signal-to-noise ratio (PSNR) metric. Although this measure is generally not very accurate, it serves as a good rule of thumb of the invisibility of the watermark. The PSNR is defined as equation (3) in unit of dB, where $X[i, j]$ is the original image, $Q[i, j]$ is the watermarked image, and $N_1 \times N_2$ is the size of the original image. In general, if the PSNR value is greater than 35 dB then the perceptual quality is acceptable, i.e. the watermark is almost invisible to human eyes [2].

$$PSNR = 10 \log \frac{255 \times 255}{\frac{1}{N_1 N_2} \sum_{i=0}^{N_1-1} \sum_{j=0}^{N_2-1} |X[i, j] - Q[i, j]|^2} \tag{3}$$

4   Experimental Results

Figure 5 shows an example of embedding and extracting results. The image of "Lena" (512 x 512) is used as a test image and the binary image of "house" is used as a watermark. As shown in fig. 5, the watermarked image is indistinguishable from the original image. The image of "Barboon"(512 x 512) is used as our test image also. Wavelet filters applied to out following experiments are the *Haar* wavelet filters.

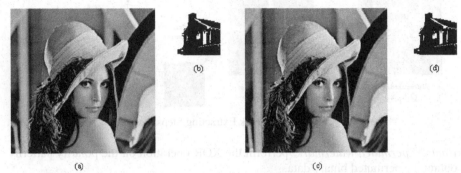

Fig. 5. Example of the Proposed Watermarking Approach. (a) Test Image Lena. (b) Watermark. (c) Watermarked Image (with PSNR = 40.24 dB). (d) Extracted Watermark with NC ≈ 1.0.

Image processing operations: Figure 6 shows a blurred version of the watermarked image, the extracted result still with high NC value (0.82), a contrast-enhanced version of the watermarked image and the extracted result still with high NC value (0.91) also.

JPEG lossy compression: Figure 7 shows the extracted watermarks from JPEG compressed version with compression ratio from 3.22, 4.82, 5.92, 6.66, 8.15 to 9.22, and corresponding NC values of the extracted watermarks of 0.89, 0.86, 0.82, 0.77, 0.73, and 0.67, respectively. As the compression ratio increases, the NC value decreases accordingly in Fig. 7. Therefore, as the compression ratio is high enough to quantize DCT coefficients very coarsely, the watermark will be destroyed and become indiscernible. However, in this situation, the quality of the JPEG compressed image (without being watermarked) will be degraded severely so that the processes of digital watermarking become less meaningful.

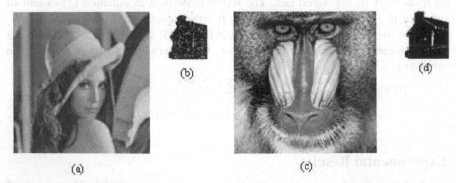

Fig. 6. (a) Blurred Version. (b) Extracted Watermark with NC = 0.82. (c) Image Enhanced Version. (d) Extracted Watermark with NC = 0.91.

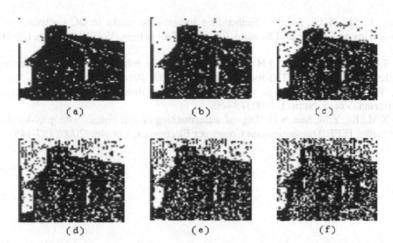

Fig. 7. The Extracted Watermarks of JPEG Compressed Version of the Watermarked Image. The Extracted Watermarks of JPEG Compressed Version of Fig. 5(c), where (a) with compression ratio 3.22 and NC = 0.891, (b) with compression ratio 4.82 and NC = 0.865, (c) with compression ratio 5.92 and NC = 0.829, (d) with compression ratio 6.66 and NC = 0.776, (e) with compression ratio 8.15 and NC = 0.736, and (f) with compression ratio 9.22 and NC = 0.676.

5 Conclusion

We introduce here an alternate method that is based on comparing the DWT coefficients of two subimages among the four DWT subimages obtained by subsampling the image. Main motivation is the development of a watermarking scheme where the presence of the original image is not required for watermark recovery. And the watermark is embedded into the neighboring relationship within the DC coefficients (LL_2 subband). Experimental results demonstrate that the proposed method is resistant to some image processing operations and JPEG lossy compression that unauthorized users may use as illegal methods. The watermarked image quality is also good.

References

1. Lin, S.D., Chen, C.F.: A robust DCT-based watermarking for copyright protection. IEEE Transactions on Consumer Electronics, Vol. 46, Num. 3. (2000) 415-421
2. Chen, L.H., Lin, J.J.: Mean quantization based image watermarking. Image and Vision Computing , Vol. 21. (2003) 717-727
3. Hsu, C.T., Wu, J.L.: Multiresolution watermarking for digital images. IEEE Transactions on Circuits and Systems, Vol. 45, Num. 8. (1998) 1097-1101
4. Hsu, C.T., Wu, J.L.: Hidden digital watermarks in images. IEEE Transactions on Image Processing, Vol. 8, Num. 8. (1999) 58-68
5. Miu, X.M., Lu, Z.M., Sun, S.H.: Digital watermarking of still images with gray-level digital watermarks. IEEE Transactions on Consumer Electronics, Vol. 46. (2000) 137-145
6. Shi, Y.Q, Sun, H.: Image and video compression for multimedia engineering. Fundamentals, Algorithms and Standards, Boca Raton, FL, CRC, (1999)

7. Huang, J., Shi, Y.Q., Shi, Y.: Embedding image watermarks in DC components. IEEE Transactions on Circuits and Systems for Video Technology, Vol. 10, Num. 6. (2000) 974-979
8. Joo, S.H., Suh, Y.H., Shin, J.H., Kikuchi, H.: A new robust watermark embedding into wavelet DC component. ETRI Journal, Vol. 24. (2002) 401-404
9. Chu, W.C.: DCT-based image watermarking using subsampling. IEEE Transactions on Multimedia, Vol. 5, Num. 1. (2003) 34-38
10. Niu, X.M., Lu, Z.M, Sun, S.H.: Digital watermarking of still images with gray-level digital watermarks. IEEE Transactions on Consumer Electronics, Vol. 46. (2000) 137-145

Cropping, Rotation and Scaling Invariant LBX Interleaved Voice-in-Image Watermarking

Sung Shik Koh[1] and Chung Hwa Kim[2]

[1] Dept. of Information and Communication Engineering, Osaka City University,
3-3-138, Sugimoto, Sumiyoshiku, Osaka, 558-8585, Japan
phdkss@sys.info.eng.osaka-cu.ac.jp
[2] Dept. of Electronics, Chosun University,
375, Seosuk-dong, Dong-gu, Gwangju, 501-759, Korea
jhdkim@chosun.ac.kr

Abstract. The rapid development of digital media and communication network urgently has highlighted the need of data certification technology to protect IPR (Intellectual property rights). This paper proposed a new watermarking method for embedding the owner's voice signal using our LBX (Linear Bit-eXpansion) interleaving. This method uses a voice signal as a watermark to be embedded, which makes it quite useful in claiming ownership, and has the advantage of restoring a voice signal that has been modified and removed by image removing attacks by applying our LBX interleaving. Three basic stages of this watermarking include: 1) Encode the analogue owner's voice signal by PCM and create new digital voice watermark; 2) Interleave a voice watermark by LBX; and 3) Embed the interleaved voice watermark in the low frequency band on DHWT (Discrete Haar Wavelet Transform) of the blue and red channels of the color image. Therefore the resulting model can be used to maximize the robustness against attacks for removing a part of image such as cropping, rotation and scaling, because de-interleaver can correct the modified watermark information.

1 Introduction

The rapid growth of digital media and communication network has highlighted the need for IPR protection technology for digital multimedia. Watermarking can be used to identify the owners, license information, or other information related to the digital object carrying the watermark. Watermarks may also provide the mechanisms for determining if a particular work has been tampered with or copied illegally. There are several approaches to invisibly embedding a watermark into an image. Spatial domain methods usually modify the least-significant bits (LSB) of the image contents [1], but it is generally not sufficiently robust for operations such as low-pass filtering. A great deal of work has also been carried out to modify the data in the transform domain which include the either Fourier [2], DCT [3]-[5], or the wavelet transform [6], [7]. However, these are not sufficiently robust against the image operations such as image cropping or image rotation. In Reference [8] the watermark was added in the blue channel of a color image because the human visual system is less sensitive to this component, but it is not sufficiently robust against image removal attacks.

T.G. Kim (Ed.): AIS 2004, LNAI 3397, pp. 498–507, 2005.

Many proposed watermarking techniques are sensitive to image transformations such as cropping, rotation and scaling, etc. In order to deal with this problem, this paper proposes a robust color image watermarking method that offers protection against image removal attacks such as cropping, rotation and scaling.

2 Voice Watermark and LBX Interleaving

2.1 Payload of Voice Signal

An amount of the owner's voice signal to be embedded is determined according to the size of the original color image, the level of the wavelet transform, and the size of the marking space on the DHWT domain. Accordingly, the voice signal payload proposed in this paper is calculated by

$$Voice\ signal\ payload = \left(\frac{2^n \times 2^n}{4^{l+1}} \right) \times 8\ [bits]. \tag{1}$$

where $2^n \times 2^n$ indicates the size of the x and y axis of the original color image, l indicates the level of the DHWT, and in the l-level DHWT, $1/4^{l+1}$ indicates the size of the marking space equivalent to the size of the low frequency band. For example, if the size of the original image is 512x512 and a voice signal is embedded in the low frequency band of the 1-level DHWT domain, the voice signal payload to be embedded is found as $(512 \times 512/4^2) \times 8 = 131,072[bits]$.

2.2 Transform of Voice Signal into Voice Watermark

This section explains the method for transforming an owner's voice signal obtained through a voice input device into a voice watermark.

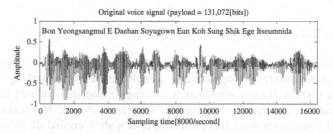

Fig. 1. Wave of owner's voice signal

The voice signal used in this paper is an analogue sound of the Korean pronunciation: "*Bon Yeongsangmul E Daehan Soyugown Eun Koh Sung Shik Ege Itseumnida*" ("The ownership of this image contents is held by Koh Sung Shik" in English). The wave of the voice signal is shown in Fig. 1. In order to create a voice watermark, the digital voice codes obtained by PCM are placed in the new position as shown in Fig. 2.

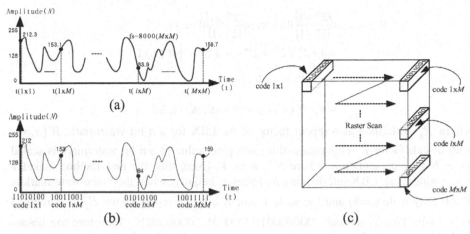

Fig. 2. Transform sequence for creating a voice watermark. (a) Analogue voice signal and a PAM pulse. (b) 8-bit quantized PCM pulse and PCM code. (c) Voice watermark

The transform sequence of a voice watermark consists of the following stages:

Stage 1. Fix payload of the analogue voice signal to be embedded using Equation (1).

Stage 2. Take the sample values of the analogue voice signal as 8KHz sampling frequency.

Stage 3. Quantize with 8-bit steps per sample.

Stage 4. Encode by 8 bits per quantized sample signal.

Stage 5. Arrange the bit strings obtained through the above stages in the new space via a Raster Scan and then obtain the voice watermark.

2.3 LBX Interleaving of Voice Watermark

Interleaving is the process by which the natural transmission order of the bit sequence is deliberately altered using a data communication technique. When the interleaved bits experience an unauthorized external attack, a burst of bit errors may occur during the attacking period. The burst of bit errors, which are consecutive in an altered order, are in fact separated from each other when the bit sequence is de-interleaved to its original natural order. The proposed LBX interleaving also has a better chance of correcting the bit errors. Therefore, if it is applied to a watermarking system, the error correction efficiency can be enhanced even though a burst error occurs.

In order to derive the interleaving, the linear bit-expansion expression in Chae and Manjunath's watermarking method [8] was applied as follows

$$W_N(x, y) = (2^N - 1)\left(\frac{W(x, y) - W_{min}}{W_{max} - W_{min}}\right). \tag{2}$$

where $W_N(x, y)$ is a set of linearly N-bit expanded watermarks, and W_{max} and W_{min} are the maximum and minimum values of watermark $W(x, y)$. In order to obtain the proposed LBX interleaving equation, the value of W_{max} was set to $2^N - 1$ and W_{min} was set to zero respectively. Therefore, Equation (2) is induced to

$$W_N(x,y) = \frac{\left(2^N - 1\right)}{\left(2^n - 1\right)} W(x,y) = \frac{\left(2^{nK} - 1\right)}{\left(2^n - 1\right)} W(x,y)$$

$$= \left[1 + 2^n + 2^{2n} + 2^{3n} + \cdots + 2^{(K-1)n}\right] \cdot W(x,y)$$

$$= \left[\sum_{i=1}^{K} 2^{(i-1)n}\right] \cdot W(x,y) \tag{3}$$

$$= R_K^n \cdot W(x,y), \text{ for } N = nK, \; K = 1,2,3 \cdots.$$

where R_K^n is defined as a repeat factor of the LBX for a n-bit watermark, $W(x,y)$, with a scale factor K. This means that each pixel value of a n-bit watermark is scaled to an N-bit representation, where N is n by K. Using this, the new method makes a voice watermark LBX interleaving as follows: Initially, when the voice watermark is a 256 gray-scale ($n=8$) and the scale factor K is 8, the repeat factor R_8^8 is computed as a 64-bit binary number "0000000100000001...00000001". Therefore the linear-scaled bit string is then described on the repeating of 8-bit strings 8 times. The eight 8-bit strings established from MSB to LSB are then relocated to the new space respectively as shown in Fig. 3.

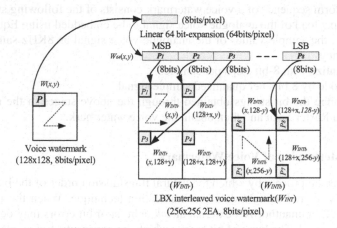

Fig. 3. LBX interleaving of voice watermark

Therefore, the size of the interleaved voice watermark becomes four times larger than that of the original watermark, and it has the same dimensions as the marking space of the *LL* band on a 1-level DHWT.

3 Voice Watermark Embedding and Extracting

3.1 Embedding of Interleaved Voice Watermark

Visual perception ranges from approximately 700nm (red) to 400nm (violet). However, the highest sensitivity in RGB mode is to the green range with a middle value of approximately 546nm. Therefore, the strongest response is to green and the weakest response is to red and blue. Accordingly, the marking space of this method uses the *LL* band on the DHWT domain of the red and blue channels in order to gain non-visual processing. In order to embed the interleaved voice watermark in the marking space, when the size of original image is $2^2 \times 2^n$, the fused coefficient can be then calculated as

$$\hat{X}_{LL}(i,j) = X_{LLg}(i,j) + \hat{X}_{LLr}(i,j) + \hat{X}_{LLb}(i,j)$$
$$\hat{X}_{LLr}(i,j) = X_{LLr}(i,j) + \alpha_r \cdot W_{INTr}\left(2^{n-2} \cdot (p-1) + x, \, 2^{n-2} \cdot (q-1) + y\right)$$
$$\hat{X}_{LLb}(i,j) = X_{LLb}(i,j) + \alpha_b \cdot W_{INTb}\left(2^{n-2}(p-1) + x, \, 2^{n-1} \cdot q - y\right),$$
$$\text{for } i,j = 1,2,\cdots,2^{n-1}, \quad x,y = 1,2,\cdots,2^{n-2} \text{ and } p,q = 1,2.$$

$$(4)$$

Where α_r and α_b are the perceptual parameters, $X_{LLr}(x,y)$ and $X_{LLb}(x,y)$ are the marking spaces of the *LL* bands respectively on the 1-level DHWT domain of the red and blue channels, and $W_{INTr}(x,y)$ and $W_{INTb}(x,y)$ are the LBX-interleaved voice watermarks.

Finally, the watermarked color image is computed with $\hat{X}_{LL}(x,y)$ and the other tree bands (*LH, HL,* and *HH*) using the inverse DHWT. Fig. 4 illustrates the LBX interleaving process used in the watermarking procedure.

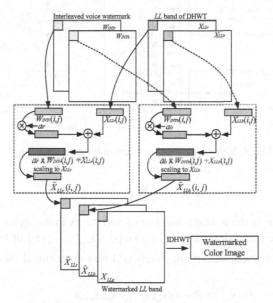

Fig. 4. Voice watermark insertion sequence

3.2 De-interleaving for Extracting Voice Watermark

De-interleaving is used to improve the burst error correction capability. Therefore, it is processed to an interleaving order in reverse. That is, using Equation (5), the interleaved voice watermark can be extracted from two *LL* bands on the DHWT domain of both the watermarked and original image.

$$W_{INTr}(i,j) = \frac{\hat{X}_{LLr}(i,j) - X_{LLr}(i,j)}{\alpha_r}, \quad W_{INTb}(i,j) = \frac{\hat{X}_{LLb}(i,j) - X_{LLb}(i,j)}{\alpha_b}.$$

$$(5)$$

Fig. 5 shows the procedure for selecting an optimal watermark using our de-interleaver.

In order to obtain the optimal voice watermark from $W_{deINT}(x,y) = \{P_1, P_2, P_3, \cdots, P_8\}$ rearranged by the de-interleaver, it is important to set an optimal threshold value. A normal distribution curve was used for $W_{deINT}(x,y)$ as follows. Initially, the mean $m(x,y)$ of the rearranged watermark, $W_{deINT}(x,y)$, was calculated using the following Equation (6):

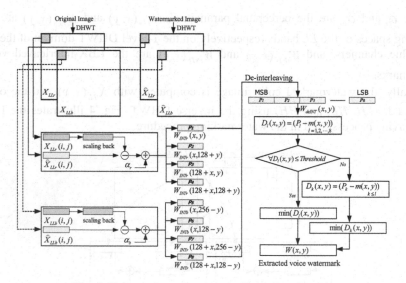

Fig. 5. Voice watermark extraction by de-interleaving

$$m(x,y) = \frac{1}{8}\sum_{l=1}^{8} P_l. \tag{6}$$

The mean deviation is the average number of variations in the items in a distribution from the mean. The mean deviation $D_l(x,y) = \{D_1, D_2, D_3, \cdots, D_8\}$ of between the mean $m(x,y)$ and the extracted watermark pixels (P_i) was determined using the following Equation (7):

$$D_l(x,y) = P_l - m(x,y), \quad l = 1,2,3,\ldots,8. \tag{7}$$

In Equation (8), the standard deviation, σ, of the probability distribution is defined as the square root of Equation (7).

$$\sigma(x,y) = \sqrt{\frac{1}{8}\sum_{l=1}^{8} D_l(x,y)^2}. \tag{8}$$

In a normal distribution curve, approximately 68.27% of the standard deviation cases are included between $m-\sigma$ and $m+\sigma$, approximately 95.45% between $m-2\sigma$ and $m-2\sigma$. In order to select the threshold value needed to estimate whether or not the occurred error is present, the range of 68.27% within one standard deviation of between $m-\sigma$ and $m+\sigma$ was used as the threshold.

Using this, an optimal watermark data, $P_i \mid \min\{D_i(x,y)\}$, was chosen for the minimum mean deviation, $\min\{D_i(x,y)\}$, when all the mean deviations $D_i(x,y)$ were less than or equal to the threshold value $m \pm \sigma$. On the other hand, when there is any mean deviation more than the threshold value, the watermark data is considered to have information-lossy by common geometric distortion such as cropping, rotation, etc., and that data is removed as part of the standard processing. After the mean deviation, $D_k(x,y)$, is again calculated using P_k, an optimal watermark data, $P_k \mid \min\{D_k(x,y)\}$, is then chosen for the minimum mean deviation, $\min\{D_k(x,y) \mid k < i\}$. Therefore, an optimal voice watermark $W(x,y)$ can be selected as a pixel with a minimum error by the above de-interleaving procedure. When it is decoded in inverse of Fig. 2, the content owner's voice signal can be obtained.

4 Simulation Results of Test Image

This section verifies the robustness of the watermarking proposed in this paper through simulation. In order to demonstrate the propriety of the proposed method, the robustness against image removal attacks such as cropping, rotation and scaling, which are representative benchmark, was evaluated.

The effectiveness of our watermarking was evaluated by calculating the PSNR between the original image and the watermarked image and the SNR between the original voice signal and the extracted voice signal. In order to measure the distortion of the watermarked image, the PSNR was calculated using

$$PSNR = 10\log\left\{ \sum_{i=0}^{2^n-1} \sum_{j=0}^{2^n-1} \frac{255^2}{\left[X_{rgb}(i,j) - Y_{rgb}(x,y)\right]^2} \right\} \text{ [dB].} \tag{9}$$

Where $X_{rgb}(i,j)$ is an original color image of 8 bits per pixel and size $2^n \times 2^n$, and $Y_{rgb}(i,j)$ an watermarked color image. In order to measure the similarity of the extracted voice signal, the SNR was defined as

$$SNR = 10\log\left\{ \frac{\frac{1}{M}\sum_{n=1}^{M} S(n)^2}{MSE} \right\} \text{ [dB], } MSE = \frac{1}{M}\sum_{n=1}^{M}\left[y(n) - s(n)\right]^2. \tag{10}$$

where $s(n)$ is a sample of the original voice signal, $y(n)$ is a sample of the extracted voice signal, and M is the total number of samples.

Fig. 6a is the "Lenna" test image 512x512 in size and Fig. 6b is the result of embedding a voice watermark in the test image when the perceptual parameter, α_r and α_b, are 0.035 and 0.07, respectively. This shows that there is very little visual image distortion for maintaining the PSNR 40.12dB quality compared with the original image.

(a)   (b)

Fig. 6. The test image (512×512) and watermarked image. (a) "Lenna" test image. (b) Image embedded with a voice watermark (α_r =0.035, α_b =0.07 and *PSNR* 40.12dB)

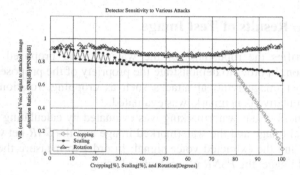

Fig. 7. VIR comparison to efficiency against image removing attacks

Fig. 7 shows a graph of VIR ("*Voice to Image Ratio*" which we define as the extracted voice signal to the attacked image distortion ratio) comparison to the efficiency against image removing attacks such as cropping, rotation and scaling.

There are two clear differences between the three waves. The first feature to notice about a cropping attack is the fact that the cropping within 75% shows the graph of the infinite values. The reason is that a de-interleaver can reconstruct watermark information even although a part of the image has been removed. The next feature of significance is the fact that the graph against a rotation attack, for all the attack periods, is better than the scaling, showing a higher level for the test image. This means that the method is the most efficient, particularly to cropping, of all image-removing attacks.

Fig. 8 to 10 show the simulation results of the "Lenna" test image. Fig. 8a shows the distorted image of the *PSNR* 6.67dB against a cropping 70% attack and Fig. 8b shows the voice signal of the infinite *SNR* extracted from Fig. 8a. Fig. 9a shows the distorted image of the *PSNR* 14.62dB against a rotation 20° attack where a part of the image is removed and the pixel values are modified, and Fig. 9b shows the voice signal of the *SNR* 14.90dB extracted from Fig. 9a. Fig. 10a shows the distorted image of the *PSNR* 30.23dB against a scaling 20% and Fig. 10b shows the voice signal of the *SNR* 12.39dB extracted from Fig. 10a. Therefore, our watermarking system is robust against an attack involving the removal of a part of an image.

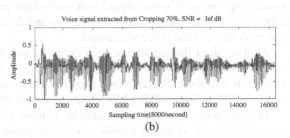

(a)   (b)

Fig. 8. Cropping attack and extracted voice signal. (a) Against cropping 70% (*PSNR* 6.67dB). (b) Extracted voice signal (*SNR* inf. dB)

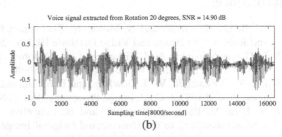

(a)   (b)

Fig. 9. Rotation attack and extracted voice signal. (a) Against rotation 20° (*PSNR* 14.62dB). (b) Extracted voice signal (*SNR* 14.90dB)

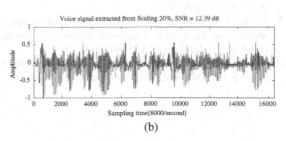

(a)   (b)

Fig. 10. Scaling attack and extracted voice signal. (a) Against scaling 20% (*PSNR* 30.23dB). (b) Extracted voice signal (*SNR* 12.39dB)

5   Conclusions

In this paper, a cropping, rotation and scaling resistant watermarking method was proposed for color images where it can embed the owner's voice signal by LBX interleaving. In order to embed a watermark using human visual system, the red and blue channels of the RGB color image were used as the marking space. In order to claim ownership, owner's voice signal was embedded in the low frequency band on the DHWT of a color image. LBX interleaving and de-interleaving was used to maximize the robustness of the voice watermark.

As a result of the simulation, in a cropping attack which removes a part of an image, when the image removal rate is less than 75%, 100% of the voice signal could be extracted, and the extracted voice signal could be kept over *SNR* 14dB quality up to rotation 20° and over *SNR* 12dB quality against a scaling ratio up to 20%.

In conclusion, it was demonstrated by computer simulations that the proposed watermarking method could provide excellent protection even when the images are removed as in cropping or modified by scaling and rotation. In future, a voice perception evaluation will be applied to more reasonably claim ownership using a voice perception algorithm in an extracted voice signal.

References

1. W. Bender, D. Gruhl, and N. Morimoto, Techniques for Data Hiding, Proc. SPIE, Storage and Retrieval for Image and Video Database III, Vol. 2420 San Jose (1995) 164-173
2. R. G. van Schyndel, A. Z. Tirkel and C. F. Osborne, A Digital Watermark, Proc. IEEE Int. Conf. Image Processing, Vol. 2 Austin (1994) 86-90
3. J. J. K. O Ruanaidh, W. J. Dowling, and F. M. Boland, Phase Watermarking of Digital Images, Proc. IEEE Int. Conf. Image Processing, Vol. 3, Switzerland (1996) 239-242
4. A. Piva, M. Barni, F. Bartolini, and B. Cappellini, DCT-based Watermark Recovering Without Resorting to the Uncorrupted Original Image, Proc. IEEE Int. Conf. Image Processing, Santa Barbara CA (1997) 520-523
5. M. Barni, F. Bartolini, V. Cappellini, A. Lippi. And A. Piva, A DWT-based Technique for Spatio-frequency Masking of Digital Signatures, Proc. SPIE/IS&T Int. Conf. Security Watermarking Multimedia Contents, Vol. 3657 San Jose CA (1999) 31-39
6. D. Kundur and D. Hatzinakos, Digital Watermarking using Multiresolution Wavelet Decomposition, Proc. IEEE Int. Conf. Acoustics, Voice, Signal Processing, Vol. 5, Seattle, WA (1998) 2969-2972
7. M. Kutter, F Jordan, and F. Bossen, Digital Signature of Color Images Using Amplitude Modulation, Proc. SPIE Electronic Imaging97, Storage and Retrieval for Image and Video Databases V, San Jose CA (1997) 518-526
8. Chae J. J., and Manjunath B. S., A Robust Embedded Data from Wavelet Coefficients, Proc. SPIE EI'98, Vol. 3312 San Jose (1998) 308-317

Data Aggregation for Wireless Sensor Networks Using Self-organizing Map*

SangHak Lee[1,2] and TaeChoong Chung[2]

[1] Ubiquitous Computing Research Center, Korea Electronics Technology Institute,
270-2, SeoHyun-Dong, PunDang-Gu, SungNam-Si, KyungGi-Do, 463-771, Korea
shlee@keti.re.kr
[2] Department of Computer Engineering, KyungHee University,
1, SeChen-Ri, GiHeung-Eup, YongIn-Si, KyongGi-Do, 449-701, Korea
tcchung@khu.ac.kr

Abstract. Sensor Networks have recently emerged as a ubiquitous computing platform. However, the energy constrained and limited computing resources of the sensor nodes present major challenges in gathering data. In this work, we propose a self-organizing method for aggregating data in ad-hoc wireless sensor networks. We present new network architecture, CODA (Cluster-based self-Organizing Data Aggregation), based on the Kohonen Self-Organizing Map to aggregate sensor data in cluster. Before deploying the network, we train the nodes to have the ability to classify the sensor data. Thus, it increases the quality of data and reduces data traffic as well as energy-conserving. Our simulation results show that CODA increases the accuracy of data than traditional aggregation of database system. Finally, we show a real-world platform, TIP, on that we will implement the idea.

1 Introduction

With the advancement of low-cost processor, memory, and radio technologies, it becomes possible to build inexpensive wireless sensor nodes. Although these nodes are not so powerful, it is possible to build a high-quality, fault-tolerant sensor network by using hundreds or thousands of them. Wireless sensor networks are assumed to be consists of tens or hundreds of thousand of energy constrained sensor nodes. Many researches have proved that wireless communication is more energy consuming in transmitting than computation of data [1]. The key challenge in such data gathering and routing is conserving the sensor's energies, so as to maximize their lifetime.

The difference between sensor network and standard temperature sensor is the ability to interconnect nodes intelligently in cluster and to aggregate data collectively. Instead of improving the quality of the individual sensor(s), the quantity is increased in distributed sensing. Benefits of this approach have been

* This research has been partially supported by Ubiquitous Autonomic Computing and Network Project,the Ministry of Science and Technology (MOST) 21st Century Frontier R&D Program in Korea.

T.G. Kim (Ed.): AIS 2004, LNAI 3397, pp. 508–517, 2005.
© Springer-Verlag Berlin Heidelberg 2005

mentioned early on in sensor fusion literature: (1) redundancy in sensors leads to a more robust system since faulty sensors have little effect on the output, (2) distributed sensors have a higher chance to capture relevant aspects because of their spatial spreading, and (3) the cost of producing many sensor modules that sense concurrently is considered to be smaller. These sensors can be produced in smaller sizes because high precision is not a major factor for this application. However, distributed self-clustering is a difficult challenge. We will concentrate in this paper on clustering data originated from set of sensors, concurrently operating in the same environment. A prime requirement is that the clustering should be based on data and done in a decentralized way since it is being implemented on a hardware platform based on microcontrollers with limited memory.

Some studies have been done on topics related to method for building clustering and finding aggregation function. However, they did not use data but geometrical location and distance between nodes. In this paper, we present a new efficient method for clustering node. The key idea in our algorithm is first to cluster nodes using unsupervised learning, the Kohonen Self-Organizing Map (SOM) and then to combine or partition the clusters. Our proposed algorithm increases the quality of data. The rest of the paper is organized as follows: in section II, related works, we survey the previous works related data clustering and aggregation. In section III, we assume the model of network and define the problem. Section IV present the CODA (Cluster-based self-Organizing Data Aggregation) protocol and argues if it satisfies its goals. Section V discusses detailed experimental results. Finally, Section VI gives concluding remarks and directions for future work.

2   Related Works

In this section, we discuss related works from both sensor network and unsupervised learning communities. There are already a lot of works related to cluster for vector quantization. We review sensor network routing, aggregation, and data classification algorithm to apply the Kohonen SOM to wireless sensor network.

Since data generated in a sensor network is too much for an end-user to process and data from close sensors is highly correlated, methods for combining data into a small set of meaningful information is highly required. Several studies have been made on aggregation or clustering. Most of the previous work [3–6] in the related area focus at reducing the energy expended by the sensors during the process of data gathering. LEACH [4] analyzes the performance of cluster-based routing mechanism with in-network data compression. Their network architecture is shown in Fig. 1 (c). LEACH leverages balancing the energy load among sensor s using randomized rotation of cluster heads. Even though they achieve to prolong the lifetime of network, they do not specify how flows and opportunities for aggregation would be activated. In PEGASIS [5], sensors form chains so that each node transmits and receives from a nearby neighbor. Gathered data moves from node to node, gets aggregated and is eventually transmitted to the base station. In [6], the authors propose a hierarchical scheme based on PEGASIS that reduces the average energy and delay incurred in gathering the sensed data. In

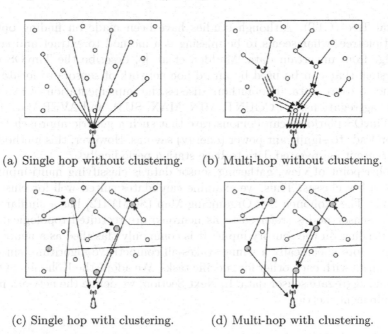

(a) Single hop without clustering.   (b) Multi-hop without clustering.

(c) Single hop with clustering.   (d) Multi-hop with clustering.

Fig. 1. This shows a figure of different types of network architecture.

HEED [9], they shows scheme of data aggregation (Clustering) and single hop vs. multi-hop routing in wireless sensor network. Our method, CODA, is based on Fig. 1 (c) or (d).

Fig. 1 shows that since data gathering in sensor network without clustering may result in many packets, thus increasing the network traffic, clustering mechanism is needed.

Directed diffusion [4] is based on a network of nodes that can coordinate to perform distributed sensing of an environmental phenomenon. Such an approach achieves significant energy savings when intermediate nodes aggregate responses to queries. The SPIN protocol [5] uses meta-data negotiations between sensors to eliminate the transmission of redundant data through the network.

In related work, Bhardwaj et al. [7] derive upper bounds on the lifetime of a sensor network that collects data from a specified region using some energy-constrained nodes. Krishnamachari et al. [2] already showed that gains due to data aggregation are obtained theoretically. Equation (1) shows the number of data-centric transmission to address-centric transmission ratio.

$$\lim_{d \to \infty} \frac{N_D}{N_A} = \frac{1}{k} \tag{1}$$

N_D is the total number of transmissions required for the optimal Data-Centric protocol; N_A is the total number of transmissions required for the optimal Address-Centric protocol; d is the distance of the shortest path from source to the sink; k is the number of sources. They examined three suboptimal schemes; Center at Nearest Source (CNS), Shortest Paths Tree (SPT) and Greedy Incremental Tree (GIT). Although studies have been made on finding optimal aggregation tree, what seems to be missing is a method to extract and encode knowledge from uncertain data. Madden et al. [8] describe the TinyOS operating system that can be used by an ad-hoc network of sensors to locate each other and to route data. The authors discuss the implementation of five basic database aggregations, i.e. COUNT, MIN, MAX, SUM, and AVERAGE, based on the TinyOS platform can demonstrate that such a generic approach for aggregation leads to significant power (energy) savings. However, this method also has to transmit aggregated data to base station periodically.

In other point of view, gathering sensor data is classifying multi-input signals into a few classes. Thus, we examine candidates suited well for clustering algorithms. The Kohonen Self-Organizing Map (SOM) [12] has a similar characteristic: sensor nodes (referred to as neurons) are recruited topologically for tasks depending on the sensory input. It is commonly classified as a neural network, and more specifically a winner-takes-all competitive algorithm, since the units compete with each other for specific tasks. We adopt the Kohonen SOM to cluster and aggregate sensor data. In Next Section, we define the network model and requirement metrics.

3   Problem Statement

In this section, we assume the network model and propose problem to be addressed. Our goal of research is to find an algorithm to cluster sensor network for transmitting data efficiently without degrading the quality of data.

There are various models for sensor networks. In this work we mainly consider a sensor network environment where:

- Each node periodically senses its nearby environment and would like to send its data to a base station located at a fixed point.
- Sensor nodes are homogeneous and energy constrained.
- Sensor nodes know their location.
- Sensor nodes are quasi-stationary while base station has fixed location.
- Sensor nodes control communication radio range dynamically, but maximum range is limited.
- Data fusion or aggregation is used to reduce the number of messages in the network. We assume that combining n identical packets of size k results in one packet of size k instead of size nk.

The aim is to efficiently transmit of all the data to the base station so that the lifetime of the network is maximized in terms of rounds, where a round is defined as the process of gathering all the data from sensor nodes to the base station regardless of how much time it takes. Also, it requires also no degrading quality of data.

In TEEN [10], they propose a formal classification of sensor network on the basis of their mode of functioning and the type of target application. We will mention their definition to achieve a better understanding of the process of sensor network.

Proactive Networks: The nodes in this network periodically switch on their sensors and transmitters, sense the environment and transmit the data of interest. Thus, they provide a snapshot of the relevant parameters at regular intervals. They are well suited for applications requiring periodic data monitoring.

Reactive Networks: In this scheme the nodes react immediately to sudden and drastic changes in the value of a sensed attribute. As such, they are well suited for time critical applications.

Most of sensor network applications may have both properties. Since nodes are deployed to sense overlapped region, closer sensor may get more correlated data. Sensor network requires method to detect not each node's data, but aggregated data accurately in any case. There are two kinds of aggregation method; (1) centralized aggregation (2) localized aggregation. In centralized aggregation, all nodes transmit data to sink, and then sink aggregates. On the other hand, localized aggregation has sensors in local area that collaborate to cluster data, and send aggregated data to sink. Since the centralized aggregation scheme requires all of the data, thus increasing traffic, localized aggregation seems to be more energy efficient than centralized aggregation. However, there may be a possibility of in degrading quality of information.

Thus, we present an algorithm, CODA. In this algorithm, we regard a node as neuron, or agent, since each node has a small but sufficient microcontroller, memory, and wireless transceiver. Thus a node can either be a cluster head or a source. And they can process signal cooperatively. Finally, each node can memorize the learning algorithm and the results. The algorithm will be able to achieve three primary goals: (1) clustering sensor network efficiently based on data, (2) prolonging network lifetime by reducing the traffic, and (3) increasing accuracy of aggregated data by using unsupervised learning.

4 CODA (Cluster-Based Self-organizing Data Aggregation)

In this section, we propose a protocol which train sensor node using SOM (Self-Organizing Map), re-cluster network based on data, and transmit only aggregated data to sink node.

First, we will take a look at the Kohonen Self-Organizing Map. In general, SOM are unsupervised learning systems employed to map high dimension inputs to a lower dimension output where similar inputs are mapped near each other. The essential constituents of feature maps are as follows [11]:

- an array of neurons that compute simple output functions of incoming inputs of arbitrary dimensionality
- a mechanism for selecting the neuron with the largest output
- an adaptive mechanism that updates the weights of the selected neuron and its neighbors

The training algorithm proposed by Kohonen for forming a feature map is summarized as follows. Each unit j has its own prototype vector w_j, being a local storage for one particular kind of input vector that has been introduced to the system.

Step 1) *Initialization:* Choose random values for the initial weights $w_j(0)$

Step 2) *Winner Finding:* Find the winning unit j^* at time t, using the minimum-distance Euclidean criterion

$$j^* = \arg\min_j \|x_j(t) - w_j\|, j = 1, ..., N \qquad (2)$$

where $x_j(t)$ represents the input pattern, N is the total number of unit, and $\|\cdot\|$ indicates the Euclidean norm.

Step 3) *Weights Updating:* Adjust the weights of the winner and its neighbors, using the following rule:

$$w_j(t+1) = w_j(t) + \alpha N_{j^*}(t)(x_j(t) - w_j(t)) \qquad (3)$$

where α is a positive constant and $N_{j^*}(t)$ is the topological neighborhood function of the winner unit j^* at time t. The neighborhood function is traditionally implemented as a Gaussian (bell-shaped) function:

$$N_{j^*}(t) = \frac{1}{\sqrt{2\pi}\sigma} e^{-(j^*-j)^2/2\sigma^2} \qquad (4)$$

with σ a parameter indicating the width of the function, and thus the radius in which the neighbors of the winning unit are allowed to update their prototype vectors significantly. It should be emphasized that the success of the map formation is critically dependent on how the values of the main parameters (i.e., α and $N_{j*}(t)$), initial values of weight vectors, and the number of iterations are prespecified. The Kohonen SOM mainly has implementations based on a single-processor, centralized method. We implement the Kohonen SOM simulator on PC. Fig. 2 shows the graphical network map before training (a) and after training (b).

There are some different requirements that we have to modify from the original Kohonen SOM. Since sensor network may consist of hundreds or thousands of nodes, it is difficult to implement in centralized method. Although inputs for

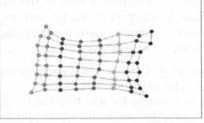

(a) The initial map.   (b) After 1,000 epoch.

Fig. 2. (a) The map based on initial random weight vector and (b) the map after 1,000 iterations running.

each of the units may contains a small magnitude of difference since data are collected from the same region, it is important for us to recognize the difference in the sensed data.

In a traditional SOM, inputs for all units are exactly the same. This situation is already addressed in [12]. They used a fixed number of units to percept distributedly the sensed environment. Thus, we implement the Kohonen SOM algorithm and weight value on each sensor platform. Our idea is to classify the sensed data as well as to re-cluster the network using the Kohonen SOM. We choose a cluster head which manages the communication schedule, merges and partitions the clusters. There are geographical clusters immediately after deploying sensor network and then clusters are merged and partitioned based on data. Data-centric cluster scheme can achieve the accuracy of aggregated data and thus make user to become aware the situation of sensor field.

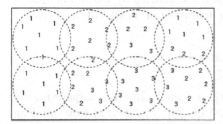 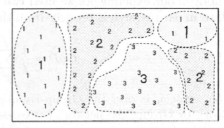

(a) Geographical Clustering and Data.   (b) Data-Centric Re-Clustering and Data.

Fig. 3. (a) geographical clustering and (b) re-clustering based on data classified by the SOM.

In original SOM, each unit starts out with initial random weight w_j. In contrast, our strategy is to make node to learn enough, and then deploy the node with learned weight w'_j. Sensor node may regard as an agent from the viewpoint of artificial intelligence. How well the node is trained off-line will decide the quality of data, the lifetime of network, adaptability of node and the level of collaboration on-line.

We design two different method of clustering: merging and partitioning. Initial cluster head node schedules local data gathering period. Each non-cluster head node transmits data to head node in their scheduled time. Then, head node identifies the winner, update prototype vector, and calculates errors. Partitioning algorithm is carried out after the time for sensor to aggregate values has expired. Merging algorithm between neighbor clusters which have similar data is activated.

Partitioning Algorithm: Partitioning is carried out basically based on error value. Continuous higher error means two cases: one is abrupt change in the environment. Another is that there are different types of data in some units. Partitioning algorithm is activated in the second case. Since cluster head node knows each node's location and error value, it can partition the nodes based on error value.

Merging Algorithm: After partitioning internal cluster, merging algorithm between close neighbor clusters is executed. Each cluster head broadcast their local aggregated sensor value, and then the two clusters are merged in which difference between their data is under certain threshold.

The quality of the network is based on the quality of the aggregated data set, so the requirement of correlated level between clusters is varied in applications and sensors. Thus, we don't fix the threshold value and remain it as a design factor.

5   Performance Evaluation

In this section, we evaluate the performance of the CODA algorithm via simulations. We use a packet-level simulation to explore the performance of proposed method. In our experiments we used two metrics: *quality of data*. The quality of data in our research is brought from the definition of TiNA [13] as follows:

$$QoD = \frac{1}{T} \sum_{t=1}^{T} 100 - \delta_t \qquad (5)$$

In (5), error δ_t is over the Group-By query at time t. In case of querying the network by Group, the network is expressed by $\{g_1, ..., g_n\}$, and exact and aggregated values over measure attribute M in the group g_i are m_i and m_i'. Then, error in group g_i is $\epsilon_i = \frac{|m_i - m_i'|}{m_i} \times 100$, the overall error is $\delta = \frac{1}{n} \sum_{i=i}^{n} \epsilon_i$.

In this work, we inspect whether data-centric adaptive clustering scheme affect the quality of data. Thus we first take a look at the quality of data in a cluster. There are five units in a cluster where they aggregate data using different methods: MAX, MIN, AVERAGE and SOM. They sense different value from the same event. We measured the difference between original values and aggregated using light sensor. Fig. 4 shows difference between AVERAGE and

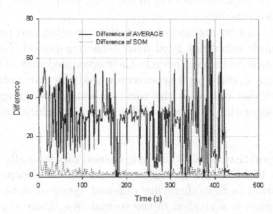

Fig. 4. Comparison the original data with the aggregated data in two cases of AVERAGE and SOM.

SOM aggregation. We exclude the MIN, MAX aggregation, since difference is too high in case of that. Since the aggregation methods used in database system have no knowledge in deciding if the heterogeneous data should be included in one data set, it's hard to partition the dataset. These simulation results verify that clustering based on data is effective at aggregating data.

In order to run the CODA on real-world platform, we have made a small, embedded sensor network node platform, called TIP (Tiny Interface for Physical World). The core of sensor module is an ATMEGA 128L microcontroller clocked at 8 MHz, which offers 128Kbytes of flash memory and 4Kbyte of SRAM/EEPROM. Fig. 5 represents the picture of Tiny Interface for Physical World (TIP) from Korea Electronics Technology Institute (KETI).

Fig. 5. A TIP from KETI.

An RF stack provides wireless communication, at a maximum rate of 56 kbit/s. Two serial connectors are available for connecting the sensor module to a PC. One of them is RS-232 serial port for data and another is USB for ISP or JTAG. Current sensor options include: temperature, humidity, and light. We use I2C connectors to get data from temperature and humidity, and read the light sensor through ADC (Analog Digital Converter). We do not have yet implemented the CODA on the TIP completely.

6   Conclusions and Future Work

In this paper, we present a self-organizing clustering method based on data classified by competitive learning neural network. Sensor networks have properties in which geographically close nodes may have correlated data. Hence, clustering network and aggregating data are positively necessary in wireless sensor network. There is a trade-off between clustering and the quality of data. Although clustering reduces the traffic and prolong the lifetime of network, it may degrade the accuracy of data. Sensor network needs an efficient data aggregating method which doesn't degrade the quality of data.

We use the unsupervised self-organizing neural network, Kohonen SOM, to map sensor data to context information. Our proposed algorithm, the CODA clusters the network based on data after sensing the environment in certain period of time and re-cluster the cluster by using the value aggregated. It works well in increasing the quality of data. We will have to implement the CODA on the real-world platform, TIP, and verify the results in real environment using various sensors.

References

1. Dasgupta, K., Kalpakis, K., Namjoshi, P.: An Efficient Clustering-based Heuristic for Data Gathering and Aggregation in Sensor Networks. Wireless Communications and Networking (2003) 1948–1953
2. Krishnamachari, L., Estrin, D., Wicker, S.: The Impact of Data Aggregation in Wireless Sensor Networks. Proc. of 22nd Int. Conference on Distributed Computing Systems Workshops (2000) 575–578
3. Intanagonwiwat, C., Govindan, R., Estrin, D.: Directed diffusion: A scalable and robust communication paradigm for sensor networks. Proc. of 6th ACM/IEEE Mobicom Conference. Boston, Massachusetts, United States (2000) 56–67
4. Heinzelman, W.R., Kulik, J., Balakrishnan, H.: Adaptive Protocols for Information Dissemination in Wireless Sensor Networks. Proc. of 5th ACM/IEEE Mobicom Conference. Seattle, Washington, United States (1999) 174–185
5. Lindsey, S., Raghavendra, C. S.: PEGASIS: Power Efficient Gathering in Sensor Information Systems. Proc. of IEEE Aerospace Conference, Vol. 3. (2002) 1125–1130
6. Lindsey, S., Raghavendra, C. S., Sivalingam, K.: Data Gathering in Sensor Networks using the Energy*Delay Metric. Proc. of IPDPS Workshop on Issues in Wireless Networks and Mobile Computing (2001) 2001–2008
7. Bhardwaj, M., Garnett, T., Chandrakasan, A.P.: Upper Bounds on the Lifetime of Sensor Networks. Proc. of International Conference on Communications (2001) 785–790
8. Madden, S., Szewczyk, R., Franklin, M. J., Culler, D.: Supporting Aggregate Queries over Ad-Hoc Wireless Sensor Networks. Proc. of 4th IEEE Workshop on Mobile Computing and Systems Applications (2002) 49–58
9. Younis, O., Fahmy, S.: Distributed Clustering in Ad-hoc Sensor Networks: A Hybrid, Energy-Efficient Approach. IEEE INFOCOM 2004 (2004)
10. Manjeshwar, A., Agrawal, D.: APTEEN: A Hybrid Protocol for Efficient Routing and Comprehensive information Retrieval in Wireless Sensor Networks. Int. Parallel and Distributed Processing Symposium: IPDPS 2002 Workshops (2002)
11. Kohonen, T.: The self-organizing map. Proc. of the IEEE (1990) 1464–1480
12. Catterall, E., Laerhoven, K., Strohbach, M.: Self-organization in ad hoc sensor networks: an empirical study. Proc. of the eighth international conference on Artificial life (2002) 260–263
13. Sharaf, M., Beaver, J., Labrinidis, A. Chrysanthis, P. K.: TiNA: A Scheme for Temporal Coherency-Aware in-Network Aggregation. Proc. of the 3rd ACM international workshop on Data engineering for wireless and mobile access (2003) 69–76

Feasibility and Performance Study of a Shared Disks Cluster for Real-Time Processing

Sangho Lee, Kyungoh Ohn, and Haengrae Cho

Department of Computer Engineering, Yeungnam University,
Gyungsan, Gyungbuk 712-749, Republic of Korea
hrcho@yu.ac.kr

Abstract. A great deal of research indicates that the shared disks (SD) cluster is suitable to high performance transaction processing, but the aggregation of SD cluster with real-time processing has not been investigated at all. By adopting cluster technology, the real-time services will be highly available and can exploit inter-node parallelism. In this paper, we investigate the feasibility of real-time processing in the SD cluster. Specifically, we evaluate the cross effect of real-time transaction processing algorithms and SD cluster algorithms with the simulation model of an SD-based real-time database system (SD-RTDBS).

Keywords: Performance evaluation, cluster computing, real-time processing, cache coherency, transaction processing

1 Introduction

There has been an increasing growth of real-time transaction processing applications, such as telecommunication system, stock trading, electronic commerce, and so on. A real-time transaction has not only ACID properties of traditional transactions but also time constraints of completing its execution before deadline [8]. The major performance metric for real-time processing is the percentage of input transactions missing their deadlines.

A cluster is a collection of interconnected computing nodes that collaborate on executing an application and presents itself as one unified computing resource. Depending on the nature of disk access, there are two primary flavors of cluster architecture designs: *shared disks* (SD) and *shared nothing* (SN) [14]. The SD cluster allows each node to have direct access to all disks. In the SN cluster, however, each node has its own set of private disks and only the node can directly read and write its disks. The SD cluster offers many advantages compared to the SN cluster, such as dynamic load balancing and seamless integration, that make it attractive for high performance transaction processing. Furthermore, the rapidly emerging technology of *storage area networks* (SAN) makes SD clusters the preferred choice for reasons of higher system availability and flexible data access. The recent parallel database systems using the SD cluster include IBM DB2 Parallel Edition [5] and Oracle Real Application Cluster [13].

T.G. Kim (Ed.): AIS 2004, LNAI 3397, pp. 518–527, 2005.

Although there has been a great deal of independent research in real-time processing and SD cluster, their aggregation has not been investigated at all. By adopting cluster technology, it is possible to support highly available real-time database services, which are the core of many telecommunication services. Furthermore, the cluster can achieve high performance real-time transaction processing by exploiting inter-node parallelism and reducing the amount of disk I/O with judicious data caching.

In this paper, we investigate the cross effect of real-time transaction processing algorithms and SD cluster algorithms with the simulation model of an SD-based real-time database system (SD-RTDBS). The emphasis of this paper is not to develop a new algorithm for SD-RTDBS, but to understand the feasibility of SD-RTDBS over a wide range of real-time workload and system parameters, and to identify the workload characteristics best suited for each algorithm. Specifically, we explore the following issues about the relationship of real-time transaction processing and SD cluster.

- Is the SD cluster feasible to the real-time transaction processing?
- How much of the SD cluster algorithm performs differently at real-time applications?
- How does the SD cluster affect the performance of real-time transaction processing algorithms, especially concurrency control algorithm?

The remainder of this paper is organized as follows. Sect. 2 presents real-time transaction processing algorithms on SD-RTDBS considered in this paper. Sect. 3 describes a simulation model of SD-RTDBS and Sect. 4 analyzes the experiment results. Concluding remarks appear in Sect. 5.

2   Real-Time Transaction Processing on SD-RTDBS

This section presents three classes of real-time transaction processing algorithms to be considered for SD-RTDBS. They include cache coherency algorithm, transaction routing algorithm, and concurrency control algorithm.

2.1   Cache Coherency Algorithm

Each node in the SD cluster has its own buffer pool and caches database pages in the buffer. Caching may substantially reduce the number of disk I/O operations by utilizing the locality of reference. However, since a particular page may be simultaneously cached in different nodes, modification of the page in any buffer invalidates copies of that page in other nodes. This necessitates the use of a *cache coherency algorithm* so that the nodes always see the most recent version of database pages [2–4, 9].

The complexity of cache coherency scheme critically depends on the locking protocols. In case of the page locking, the cache coherency scheme is rather simple. This is because different transactions cannot access a page with conflicting modes simultaneously. On the other hand, if the record locking is used, the cache coherency scheme becomes very complicated. This is due to the fact that different records of the same page could be concurrently updated at different nodes. As a result, each node's page copy is only partially up-to-date and the page copy on disk may contain none of the updates (at first).

The record locking can reduce the lock conflict ratio; hence, it can support well the traditional transaction processing applications where there are many concurrent transactions. This is why most commercial products of SD cluster implement the record locking [5, 13]. However, large amounts of message and page transfer could occur in the record locking due to frequent lock requests and maintaining cache coherency. Excessive page transfer also causes frequent disk I/O for flushing log records [9]. This shortcoming of the record locking could be particularly problematic in some real-time applications where the degree of concurrency would not be so high. As a result, we have to analyze the effect of lock granularity on the cache coherency algorithm under a wide variety of real-time workloads.

2.2   Transaction Routing Algorithm

The key idea of the successful SD-RTDBS is to maximize inter-node parallelism with minimal inter-node synchronization overhead. A font-end transaction router may implement this idea with an efficient transaction routing. In the SD cluster, if transactions referencing similar data are clustered together to be executed on the same node (*affinity node*), then the buffer hit ratio should increase and the level of interference among nodes due to buffer invalidation will be reduced. This concept is referred to as *affinity-based routing* [11, 15, 16].

We consider two affinity-based routing algorithms: pure affinity-based routing (PAR) [15, 16] and dynamic affinity-based cluster allocation (DACA) [11]. PAR is a static algorithm in the sense that the affinity relationship between transaction classes and nodes are fixed. While PAR is simple and easy to implement, the deadline miss ratio of some transaction class must increase dramatically if transactions of the class are congested. DACA can avoid overloading individual node by making an optimal balance between the affinity-based routing and indiscriminate sharing of load in the SD cluster. Specifically, DACA allocates more affinity nodes for a congested transaction class and returns to the original state if the congestion is resolved. Note that the affinity relationship is determined dynamically. Furthermore, DACA can reduce the frequency of buffer invalidations by trying to limit the number of affinity nodes allocated to a congested transaction class if the load deviation of each node is not significant.

In traditional transaction applications, we have shown that DACA performs better than PAR and other dynamic transaction routing algorithms [11]. In this paper, we will investigate whether this result is still valid under the real-time workloads where the performance metric is a deadline miss ratio and a system applies a real-time concurrency control described in the next section.

2.3   Concurrency Control Algorithm

A concurrency control algorithm is required to maintain the database consistency among concurrent transactions accessing the same data [10]. In real-time appli-

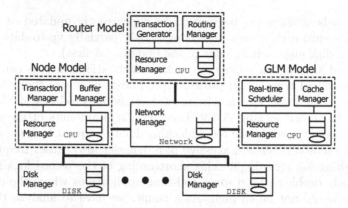

Fig. 1. A simulation model of an SD-RTDBS.

cations, the algorithm also has to handle the *priority inversion* problem that lower priority transactions block the execution of higher priority transactions. It might degrade performance because urgent transactions cannot be executed until blocking transactions of low priority complete. Authors in [1] proposed two real-time concurrency control algorithms, *wait promote locking* (WP) and *high priority locking* (HP). WP prevents the priority inversion by increasing the priority of blocking transaction to be as high as that of blocked transactions whenever a priority inversion occurs. On the other hand, HP aborts blocking transactions of low priority if they cause the priority inversion.

Authors in [1] show that WP outperforms HP in real-time applications with soft deadline. In firm deadline applications, authors in [6] show that HP performs better than WP. However, both results are obtained from the centralized database system model and they might not be hold in the SD cluster. The reason is that (a) each node of the SD cluster can execute transactions concurrently and the data conflict ratio would increase as a result, and (b) the performance of concurrency control algorithm is affected by the underlying cache coherency algorithm and transaction routing algorithm.

3   Simulation Model

To evaluate the feasibility of real-time processing in SD cluster, we have developed a simulation model of an SD-RTDBS using CSIM [12] discrete-event simulation package. Fig. 1 shows the simulation model.

We model the SD cluster consisting of a single router and a global lock manager (GLM) plus a varying number of nodes, all of which are connected via a local area network. The router model consists of a *transaction generator* and a *routing manager*. The transaction generator has a role to generate transactions, each of which is modeled as a sequence of database operations. The routing manager captures the semantics of a given transaction routing algorithm.

The model for each node consists of a *buffer manager*, which manages the node buffer pool using an LRU policy, and a *resource manager*, which models

Table 1. Simulation parameters.

System Parameters		
CPUSpeed	Instruction rate of CPU	1 GIPS
NetBandwidth	Network bandwidth	100 Mbps
NumNode	Number of computing nodes	1 ~ 16
NumDisk	Number of shared disks	20
MinDiskTime	Minimum disk access time	0.01 sec
MaxDiskTime	Maximum disk access time	0.03 sec
PageSize	Size of a page	4096 bytes
RecPerPage	Number of records per page	10
ClusterSize	Number of pages in a cluster	10000
HotSize	Size of hot set in a cluster	2000 pages
DBSize	Number of clusters in database	8
BufSize	Per-node buffer size	10000 pages
FixedMsgInst	Number of instructions per messaging	20000
LockInst	Number of instructions per locking	2000
PerIOInst	Number of instructions per disk I/O	5000
PerObjInst	Number of instructions for a DB call	15000
LogIOTime	I/O time for writing a log record	0.005 sec
Transaction Parameters		
TrxSize	Transaction size (# of records)	10
SizeDev	Transaction size deviation	0.2
UpdatePr	Probability of updating a record	0.0 ~ 0.8
MPL	Number of concurrent transactions	40 ~ 400
ClassNum	Number of transaction classes	8
Locality	Probability of accessing local cluster	0.8
HotPr	Probability of accessing hot set	0.8

CPU activity and provides access to the shared disks and the network. For each transaction, the *transaction manager* forwards lock request messages and commit messages to the GLM. The disks are shared by every node.

The GLM has a role to perform the real-time concurrency control and the cache coherency control. The *real-time scheduler* implements WP or HP. The lock granularity is defined as a record or a page. Transactions wait on any conflicting lock requests. They are aborted in case of either deadlock, priority inversion in HP, or missing their deadlines. The *cache manager* implements ARIES/SD algorithm [9], which is a representative cache coherency algorithm in the SD cluster.

Table 1 shows the simulation parameters for specifying the resources and overheads of the system. Many of the parameter values are adopted from [3, 16]. The *network manager* is implemented as a FIFO server with 100 Mbps bandwidth. The number of shared disks is set to 20, and each disk has a priority queue of I/O requests. Disk access time is drawn from a uniform distribution between 10 milliseconds to 30 milliseconds.

We model that the database is logically partitioned into several *clusters*. Each database cluster has 10000 pages (40 Mbytes), and it is affiliated to a specific transaction class. The number of classes (*ClassNum*) is set to 8. The transaction parameter of *Locality* determines the probability that a transaction operation accesses a data item in its affiliated database cluster. The *HotPr* parameter models

"80-20 rule", where 80% of the references to the affiliated database cluster go to the 20% of the database cluster (*HotSize*). We refer the 20% of the database cluster as *hot set*, and the remaining part as *cold set*. The average number of records accessed by a transaction is determined by a uniform distribution between *TrxSize* ± *TrxSize* × *SizeDev*. The parameter *UpdatePr* represents the probability of updating a record. The processing associated with each record, *PerObjInst*, is assumed to be 15000 instructions.

For a transaction, T, we determine its deadline (D_T) as follows [6]: $D_T = A_T + SF \times E_T$, where A_T and E_T are the arrival time and estimated execution time of T, respectively. SF is a *slack factor* that provides control over the tightness of deadlines. We set the value of SF to 4 like [6]. E_T is computed as follows: $E_T = NumRead_T \times (PerObjInst + MaxDiskTime) + NumWrite_T \times PerObjInst$, where $NumRead_T$ and $NumWrite_T$ are the number of read and write operations of T, respectively. The disk time for writing updated pages is not included in E_T since these writes occur after the transaction has committed [6].

The performance metric used in the experiments is a *deadline miss ratio*, which is the percentage of input transactions that the system is unable to complete before their deadlines. We also use an additional performance metric, *buffer hit ratio*, which gives the probability of finding the requested pages at buffers.

4   Experiment Results

We compare the performance of the algorithms under the following three categories: (a) transaction routing – PAR and DACA, (b) lock granularity – page and record, and (c) concurrency control – WP and HP. By combining each category, we implement eight algorithm sets for the SD-RTDBS.

4.1   Experiment 1: Feasibility of Real-Time Processing in SD Cluster

To verify the feasibility of real-time processing in SD cluster, we first compare the performance of SD cluster by varying *MPL* and *NumNode*. Fig. 2(a) shows the experiment results where every input transaction is assumed to be same class (*ClassNum* = 1). HP with record locking is used as a concurrency control, and DACA is used for transaction routing. *UpdatePr* is set to 0.2. When MPL is low, every system configuration exploits similar performance behavior. In this case, even a single node can process all the input transactions within their deadlines. As MPL increases, systems with large number of nodes exploit significant performance improvement. This is due to *load balancing* and *high buffer hit ratio*. If the number of nodes is large, each node may take over a small portion of system load. The buffer hit ratio also increases due to large aggregate buffer size as Fig. 2(b) shows. As a result, the benefit of load balancing and high buffer hit ratio dominate the negative effect of synchronization overhead between nodes.

We also compare the performance of single node system and SD clusters with different transaction routing algorithms. Fig. 2(c) shows the result. Both

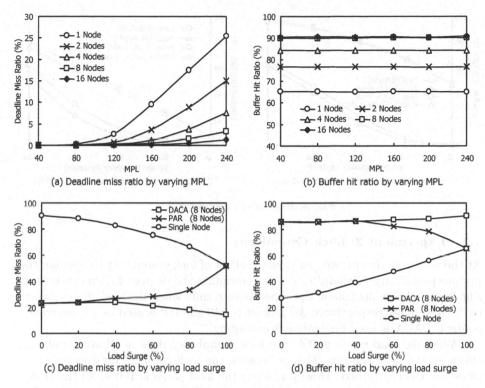

(a) Deadline miss ratio by varying MPL

(b) Buffer hit ratio by varying MPL

(c) Deadline miss ratio by varying load surge

(d) Buffer hit ratio by varying load surge

Fig. 2. Results of experiment 1.

NumNode and *ClassNum* are set to 8. *MPL* is set to 400, and thus the steady state load per each transaction class is 50 transactions. The *load surge* is expressed as a fraction of its steady state load. For example, a load surge of 20% implies that the load of each non-surge class decreases about 20% (10 transactions) and the total sum of additional load (70 transactions) goes to the single surge class.

When the transaction load is evenly distributed (load surge = 0%), the single node system performs worst. This is because the single node cannot cache all the pages accessed by different transaction classes. Many buffer misses must increase the number of transactions missing their deadlines. On the other hand, both PAR and DACA can allocate each transaction class to its affinity node and can achieve very high buffer hit ratio as Fig. 2(d) shows. As the load surge increases, PAR performs worse because PAR does not distribute the extra load of the surged transaction class to other nodes, and thus it suffers from lower buffer hit ratio and limited computing facility. When the load surge is 100%, PAR performs exactly like to the single node system. DACA performs better as the load surge increases. Allocating more nodes to the surged transaction class can achieve load balancing between nodes. Furthermore, the aggregate buffer space for the surged transaction class increases also. This is why the buffer hit ratio of DACA increases as the load surge increases.

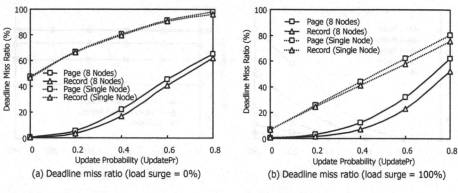

Fig. 3. Results of experiment 2.

4.2 Experiment 2: Lock Granularity

At the next experiment, we compare the effect of lock granularity by varying the update probability ($UpdatePr$) which determines the degree of data contention. Fig. 3(a) and Fig. 3(b) show the experiment results when the load surges are set to 0% and 100%, respectively. MPL is set to 240 and HP is used as a concurrency control. DACA is used for transaction routing.

When the load surge is 0%, the lock granularity does not have significant effect on the performance. This is because the lock conflicts between transactions are inherently rare. However, when the load surge is 100%, all the transactions belong to the same transaction class and they access the same portion of database. As a result, there are many lock conflicts between transactions. In this case, the record locking performs better than the page locking as $UpdatePr$ increases. The page locking causes many transactions being blocked or being aborted due to priority inversion. As a result, the number of transactions missing their deadlines increases. The record locking does not suffer from its cache coherency overhead seriously. Though not shown in this paper, the page locking outperformed slightly the record locking when a transaction accesses several records for each page and $UpdatePr$ is low. In this case, the record locking results in several message communications to check the consistency of a cached page, while the page locking requires only one message for the page.

Another observation is that the SD cluster is more sensitive to the setting of $UpdatePr$ than the single node system. This is particularly true when the load surge is high as Fig. 3(b) shows. The reason is due to the cache coherency overhead. When $UpdatePr$ and the load surge are high, the probability of internode buffer invalidation increases and more pages has to be transferred between nodes. As a result, the transaction execution time becomes longer.

4.3 Experiment 3: Concurrency Control Algorithms

The last experiment compares the performance of real-time concurrency control algorithms, HP and WP, by varying $UpdatePr$. Fig. 4(a) and Fig. 4(b) show the experiment results when the load surges are set to 0% and 100%, respectively.

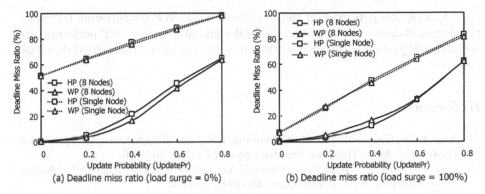

(a) Deadline miss ratio (load surge = 0%)   (b) Deadline miss ratio (load surge = 100%)

Fig. 4. Results of experiment 3.

MPL is set to 240 and DACA is used for transaction routing. We set the lock granularity as a page to create a situation where lock conflicts are more frequent.

When the load surge is 0%, WP outperforms HP in the SD cluster with eight nodes. This is because (a) HP would produce unnecessary transaction aborts, and (b) the priority inheritance of WP allows the disk manager to reduce I/O queue waiting time for the high priority transactions. Note that in this workload minimizing I/O queue waiting time is essential since the buffer hit ratio is relatively low. On the other hand, HP performs better than WP when the load surge is 100%. The buffer hit ratio of this workload is rather high, so handling data contention efficiently is more important to meet transaction deadlines. It is well known that WP performs worse under high data contention workload [6]. WP may block high priority transactions repeatedly due to low priority transactions. Furthermore, the priority inheritance mechanism causes many transactions executing at the same priority. This means high priority transactions effectively receive little or no preferential treatment in WP.

5   Concluding Remarks

Although there has been a great deal of independent research in real-time processing and SD cluster, their aggregation has not been investigated at all. By adopting cluster technology, the real-time services will be highly available and can exploit inter-node parallelism. In this paper, we investigate the cross effect of real-time transaction processing algorithms and SD cluster algorithms with the simulation model of an SD-based real-time database system (SD-RTDBS).

The basic results obtained from the experiments can be summarized as follows. First, the SD cluster can process real-time transactions better when the number of concurrent transactions is large, there are many nodes, and efficient transaction routing algorithm is implemented. This result implies that the SD cluster is an effective and scalable alternative to the centralized database system for real-time transaction processing. Next, the cache coherency overhead of record locking is not significant. As a result, the record locking outperforms the page locking through most of the workloads. Last, WP outperforms HP when the degree of data contention and the buffer hit ratio are low. HP performs better under high data contention workload as in case of the centralized database system.

References

1. Abbott, R., Garcia-Molina, H.: Scheduling Real-Time Transactions: A Performance Evaluation. ACM Trans. on Database Syst. **17** (1992) 513-560
2. Cho, H., Park, J.: Maintaining Cache Coherency in a Multisystem Data Sharing Environment. J. Syst. Architecture **45** (1998) 285-303
3. Cho, H.: Cache Coherency and Concurrency Control in a Multisystem Data Sharing Environment. IEICE Trans. on Infor. and Syst. **E82-D** (1999) 1042-1050
4. Cho, H.: Database Recovery using Incomplete Page Versions in a Multisystem Data Sharing Environment. Infor. Processing Letters **83** (2002) 49-55
5. DB2 Universal Database for OS/390 and z/OS – Data Sharing: Planning and Administration. IBM SC26-9935-01 (2001)
6. Harita, J., Carey, M., Livny, M.: Data Access Scheduling in Firm Real-Time Database Systems. J. Real-Time Syst. **4** (1994) 203-241
7. Kanitkar, V., Delis, A.: Real-Time Processing in Client-Server Databases. IEEE Trans. on Computers **51** (2002) 269-288
8. Lam, K-Y., Kuo, T-W. (ed.): Real-Time Database Systems: Architecture and Techniques. Kluwer Academic Publishers (2000)
9. Mohan, C., Narang, I.: Recovery and Coherency-Control Protocols for Fast Intersystem Page Transfer and Fine-Granularity Locking in a Shared Disks Transaction Environment. In: Proc. 17th VLDB Conf. (1991) 193-207
10. Moon, A., Cho, H.: Global Concurrency Control using Message Ordering of Group Communication in Multidatabase Systems. Lecture Notes in Computer Science, Vol. **3045**. (2004) 696-705
11. Ohn, K., Cho, H.: Cache Conscious Dynamic Transaction Routing in a Shared Disks Cluster. Lecture Notes in Computer Science, Vol. **3045**. (2004) 548-557
12. Schwetmann, H.: User's Guide of CSIM18 Simulation Engine. Mesquite Software, Inc. (1996)
13. Vallath, M.: Oracle Real Application Clusters. Elsevier Digital Press (2004)
14. Yousif, M.: Shared-Storage Clusters. Cluster Comp. **2** (1999) 249-257
15. Yu, P., Dan, A.: Performance Analysis of Affinity Clustering on Transaction Processing Coupling Architecture. IEEE Trans. on Knowledge and Data Eng. **6** (1994) 764-786
16. Yu, P., Dan, A.: Performance Evaluation of Transaction Processing Coupling Architectures for Handling System Dynamics. IEEE Trans. on Parallel and Distributed Syst. **5** (1994) 139-153

A Web Cluster Simulator
for Performance Analysis
of the ALBM Cluster System*

Eunmi Choi[1] and Dugki Min[2]

[1] School of Business IT, Kookmin University,
Chongnung-dong, Songbuk-gu, Seoul, 136-702, Korea
emchoi@kookmin.ac.kr
[2] School of Computer Science and Engineering, Konkuk University,
Hwayang-dong, Kwangjin-gu, Seoul, 133-701, Korea
dkmin@konkuk.ac.kr

Abstract. A distributed server cluster system is a cost-effective solution to provide scalable and reliable Internet services. In order to achieve high qualities in service, it is necessary to tune the system varying configurable parameters and employed algorithms that significantly affect system performance. In this purpose, we develop a simulator for performance analysis of a traffic-distribution cluster system, called the ALBM (Adaptive Load Balancing and Management) cluster. In this paper, we introduce the architecture of the proposed simulator. Major design considerations are given to the flexible structures that can be easily expanded for adding new features, such as new workloads and scheduling algorithms. With this simulator, we perform two simple cases of performance analysis: one is to find appropriate overload and underload thresholds and the other is to find the suitable scheduling algorithm for a given workload.

1 Introduction

A distributed server cluster system is a cost-effective solution to provide scalable and reliable internet services [1]. Distributed server cluster systems for Internet services, so called Web cluster systems, haven been developed in various forms. The most popular form is based on hardware L4 switches [2,3] as traffic managers that would direct IP traffics to the highly appropriate server in a cluster. The Linux Virtual Server (LVS) is a well-known software load balancer [4,5] and an open software solution, so that it can be customized to collaborate with other software tools or extended by add-ing an improved scheduling algorithm. The third class is in a hybrid form; that is, a hybrid-type web cluster system employs both S/W L4 network switch appliances and middleware services on server nodes

* This work was supported by the Korea Science and Engineering Foundation (KOSEF) under Grant No. R04-2003-000-10213-0. This work was also supported by research program 2004 of Kookmin University in Korea.

for proactive and adaptive management. The ALBM (Adaptive Load Balancing and Management) cluster, proposed in our previous research work [6], is in this class.

In this paper, we present a web cluster simulator that can be used to solve some decision problems in a web cluster system. The web cluster simulator aims to simulate the ALBM cluster that employs the L4 traffic manager and middleware management services on server nodes. This paper introduces the architecture of the proposed simulator. Major design considerations are given to the flexible structures that can be easily expanded for adding new features, such as new workloads and scheduling algorithms. With this simulator, we perform two simple cases of performance analysis. The first case is to get an insight of how to find appropriate overload and underload thresholds for a given system environment. This simulation results can be used to tune configurable parameters of an adaptive scheduling mechanism in a web cluster system for the better performance. The second is to help to find the appropriate scheduling algorithm for a given workload characteristic. Since the workload assignment problem is NP-Complete, it cannot be solved in a polynomial time. With the help of simulation, we provide the approximation approach to find the proper assignment algorithm of workloads.

This paper is organized as follows. Section 2 introduces the ALBM cluster system. Section 3 introduces the design architecture of the simulator. Section 4 and 5 show the simulation results with scenarios and analyze them. We summarize in Section 6.

2   The ALBM Active Cluster Architecture

Before showing our simulator of a cluster system, we introduce the real architecture of ALBM (Adaptive Load Balancing and Management) cluster system [6] that serves Web applications to a huge number of clients. As shown in Figure 1, the ALBM cluster system is composed of active switches, application servers, and the management station. Although the cluster consists of tens of application servers, called 'nodes', it is published with one site name and one virtual IP address that are assigned to one or more active switches, called 'Traffic Managers' (TMs). The TMs interface the rest of cluster nodes with the Internet, making the distributed nature of the architecture transparent to Internet clients. All inbound packets to the system are received and forwarded to application servers by the TMs. It provides network-level traffic distribution services, balancing the servers' loads. On each of application servers, a middleware service, called 'Node Agent' (NA), is running. The NAs are indicated by Circle 'A's on the bottom of each node in Figure 1. The NA makes the underlying server be a member node of the ALBM system. It is in charge of management and operation of the ALBM cluster, such as cluster formation, membership management, and adaptive management and load balancing.

The management station, called 'M-Station', is a management center of the entire ALBM cluster system, working together with the management console

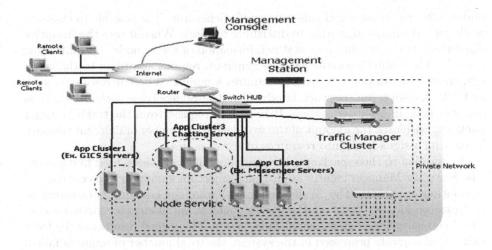

Fig. 1. The Architecture of the ALBM Active Cluster System.

through Web-based GUIs. All administrators' commands are received into and executed through the M-Station. Communicating with NAs, the M-Station performs all kinds of cluster management operations, such as cluster creation/removal and node join/leave. The M-Station also collects the current configurations and states of the entire system, and provides information to other authenticated components in the system. Using the server state information provided by NAs, it performs proactive management actions according to the predefined policies. Besides, the M-Station checks the system state in the service-level, and carries out some actions when values monitored from service-level quality are significantly far behind the service-level QoS objectives.

The adaptive scheduling algorithms in the ALBM active cluster system adjust their schedules, taking into accounts of dynamic state information of servers and ap-plications collected from servers. By collecting appropriate information of server states, the NAs customize and store the data depending on the application architecture, machine types, and expectation of a system manager. Each NA decides if the current state is overloaded or underloaded by using upper or lower thresholds of resource utilization determined by system configuration and load balancing policies of cluster management. After collecting state data of all NAs in a cluster, the Master NA reports the state changes to the TM. By using the state information reported by Master NAs, the TM adjusts traffics of incoming requests properly to balanced server allocation. The scheduling algorithms are applied to the TM through the control of M-Station.

3   The Architecture of the Simulator

The simulator consists of three packages and a number of common classes. The three packages are trafficManager, trafficGenerator, and nodeService packages. The trafficManager package is in charge of delivery of traffic requests into server nodes, after receiving workloads from trafficGenerator. It is possible to choose a workload-scheduling algorithm to distribute requests. When it uses the Adaptive algorithm, it receives and uses status information of server nodes by controlling periods. The trafficGenerator package generates request according to the setup information of workload. A request requires a number different resources, such as CPU, memory, and network bandwidth. The nodeService package works as a server node in a cluster system. It receives a request from the trafficManager package, calculates the amount of the workload to a number of different resource loads, and assigns loads into resources of server nodes.

In addition to those packages, there are a number of classes used in common. The SimulatorManager creates an instance object for simulator and controls the simulation operation. The TrafficGeneratorParameter contains parameters to generate workloads. The NodeParameter contains parameters for virtual nodes. The Log analyzes and stores simulation results. The logged results are the total number of requests processed in the system, the total number of requests failed, the number of requests processed in a server node, the overload rate in a server node, etc. The entire results are stored with the timestamps under a specific excel file name and, at the same time, resource usages of each server node during simulation are stored separately. The Clock keeps the logical time of simulation, while keeping the globalTime as a static attribute. The logical time is used to report timing of resource assignment and release, and the total simulation time.

3.1   The Structure of the trafficManager Package

The trafficManager package has the design structure shown in Figure 2. In order to easily add an additional workload-scheduling algorithm, the factory design pattern [8] is employed. When the Adaptive workload-scheduling algorithm is used, each server node declares overloaded status by evaluating its current status with the threshold value, and reports the status to the trafficManager. Thus, the trafficManager stops further distribution of incoming requests to the overloaded node.

In Figure 2, there are a number of classes: TMManager, TMFactory, AbstractTM, and various concrete TMs. The TMManager generates a TM through the TMFactory in the beginning of simulation. During simulation, the TMManager receives workload from trafficGenerator and pass the workload to the TM generated. The TMFactory is a factory that creates proper TMs required. The AbstractTM is an abstract class, which contains abstract methods that are overridden by concrete TM classes. Also the AbstractTM calculates the releasing time of workload that are assigned into the TM, and reports into the workload, checks states of server nodes, and passes workload into server nodes via a workload scheduling algorithm applied. The abstract getNextNode() method is overridden by subclasses to implement their specific workload scheduling algorithms. The RRTM, LCTM, RTTM, ARRTM, ALCTM, ARTTM are concrete classes for the AbstractTM, by applying the Round-Robin (RR), the Least Connection (LC), the Response Time (RT), the Adaptive-RR (ARR), the Adaptive-LC (ALC), the Adaptive-RT (ART) scheduling algorithms.

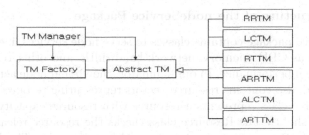

Fig. 2. The Structure of the trafficManager package.

3.2 The Structure of the trafficGenerator Package

The trafficGenerator package is in charge of generating workloads that are logically composed of service requests. Each service request requires some amounts of resources in a server node. In a real system, each request acquires a number of resources at the same time. Thus, one workload is composed of a number of loads. A load becomes CPU, memory, and network system resources with some statistical requirement. The structure of the trafficGenerator package is shown in Figure 3. In Figure 3, there are a number of classes to build the trafficGenerator package. The TrafficGenerator provides a workload by gathering a number of loads from resource generators. The Load shows a resource load, such as amount of resource usage, resource holding time, arrival time, etc. The Workload contains one or more loads in a form of Vector. The ResourceGeneratorManager creates resource generators and pass the references to the TrafficGenerator. The AbstractResourceGenerator is an abstract class that generates statistical values according to the users input parameters. The statistical distributions that the AbstractResourceGenerator generates contain Normal distribution, Exponential distribution, and Uniform distribution. While producing distribution values, any random numbers can be generated and pseudo random numbers from a specific seed number can be generated. There are a number of concrete subclasses from the AbstractResourceGenerator: CPUGenerator, MemoryGenerator, and NetworkGenerator to generate CPU load, memory load, and network load, respectively.

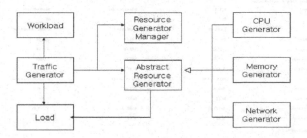

Fig. 3. The Structure of the trafficGenerator package.

3.3   The Structure of the nodeService Package

The nodeService package contains classes to serve node services, including node resources such as CPU, memory, network bandwidth, and other resources that affect the system performance. In our simulator, the virtual node acts as a server node agent and also contains resource vectors representing resource loads.

The Resource class represents a resource with resource capacity, and upper and lower thresholds. The Resource class checks the resource release, recovery, and status after the resource loads release the assignment. The VirtualNode class contains a resource vector and works as a server node to provide a service according to client's request. The VirtualNode class contains the node status, overload information, node ID, etc. It also assigns loads from the workload into the proper resources, and evaluates resource states to define the current node status.

4   The Implementation of the Simulator

In this section, we show the GUIs of the simulator according to the processing steps of the simulation. In order to generate workloads, the user can set up the input parameters of the load amount, the duration time of load in a resource, and statistical parameter values as shown in Figure 4. The execution environment is set up before starting simulation. It includes environment parameters, arrival time parameters, and concurrent load parameters. The user can set up the number of virtual server nodes, simulation time, the number of resources in a node, and workload scheduling algorithm as shown in Figure 5 (a). The parameters related to the arrival time are set up by controlling the interval time as shown in Figure5 (b). The user adjusts the time interval between any two workload requests by applying statistical distribution. A large number of workloads generated by the trafficGenerator package may produce requests concurrently to the trafficManager. Thus, this concurrency rate can be controlled by the input parameters with statistical distribution as shown in Figure 5 (c).

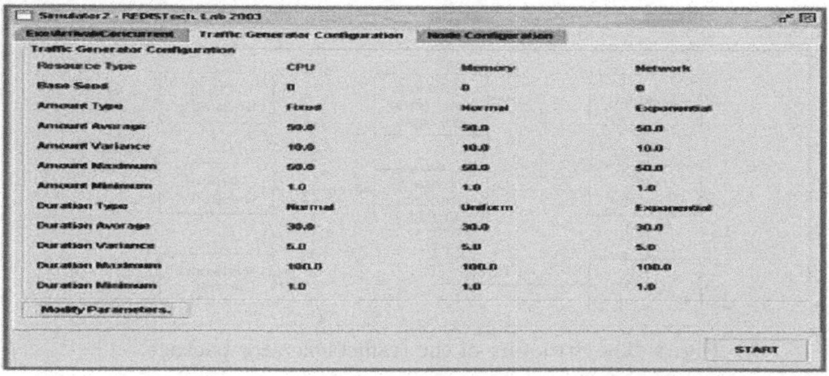

Fig. 4. The Workload Parameters Setup.

534   Eunmi Choi and Dugki Min

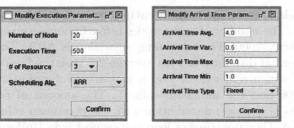

(a) Environment Setup. (b) Arrival Time Setup. (c) Concurrent Load Setup.

Fig. 5. Setup GUIs of the Simulator.

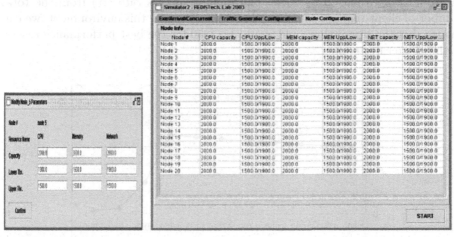

(a) Virtual Node Setup GUI. (b) Virtual Node Setup Information.

Fig. 6. Virtual Node Setup.

The resource capacities of each virtual node and upper & lower thresholds are set up as in Figure 6 (a). According to simulation environmental setup, the user can set up a cluster of homogeneous system or a cluster of heterogeneous system. Figure 6 (b) shows the resource capacities of each node and the threshold values.

5   Simulation Scenarios

In this section, we show two simulation scenarios of generating workloads in a server cluster system and analyzing the performance results.

5.1   Finding the Optimal Thresholds

By using the simulator we propose, we can find the optimal threshold values in a virtual cluster system with Adaptive scheduling algorithm. The optimal threshold points result in the efficient workload scheduling and distribution, yielding the load-balanced server nodes. The experimental environment of finding the optimal thresholds is follows. The number of server nodes is five in homogeneous form with the resource capacity 2000. The scheduling algorithm is ARR. The arrival time internal is in Normal distribution with average 20, variance 5, and the maximum limit 20. The concurrent rate of load is in Uniform distribution with the maximum value 100 and the minimum value 1. The amount of resource required per request is in Uniform distribution with the maximum value 200 and the minimum value 1. Duration time of load per request is in Uniform distribution with the maximum value 500 and the minimum value 1.

Figure 7 shows the success rate of client requests as the threshold value changes. We predefined the lower threshold as 1500, and changed the upper threshold from 1500 up to 2000. The highest success rate (%) from the total number of client requests is achieved at 1850. Thus, in this environment, we can ensure that the upper threshold would be 1850 for the best performance results in the cluster system.

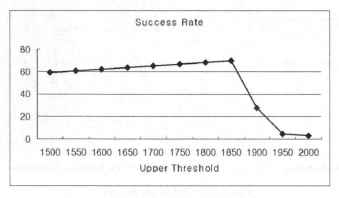

Fig. 7. Success Rates based on the Threshold Values.

5.2   Comparing Workload Scheduling Algorithms

The second simulation scenario is to test workload-scheduling algorithms with a fixed form of workload and a random form of workload, respectively. The fixed form of workload provides the same size and resource duration time. The random form of workload generates any size of workload and arbitrary duration time. The fixed form of workload has the fixed arrival time interval (5), the fixed concurrent load (15), the fixed amount of resource per request (4.4), and the fixed duration time per request (750). The random form of workload has the arrival time interval in Normal distribution with average 6.4, variance 0.5, and the maximum limit 50, the concurrent load in Normal distribution with average 15, variance 0.5, and the maximum limit 50, the amount of resource per request in Uniform distribution with the maximum limit 400 and the minimum limit 1, and duration time per request in Uniform distribution with the maximum limit 200 and the minimum limit 1.

Table 1. Simulation Testing Results from Workload Scheduling Algorithms (Unit: Success rate (%) of the total number of requests).

Algorithm	Fixed Workload	Random Workload
RR	37.08%	17.23%
LC	37.08%	44.87%
RT	37.08%	44.99%
ARR	93.63%	97.01%
ALC	93.63%	96.99%
ART	93.63%	96.97%

When generating heavy workloads as much as the server is down with the fixed form of workload and the random form of workload, respectively. Table 1 shows worse results from RR, LC, and RT workload scheduling algorithms. The unit of resulting values is the success rate (%) of the total number of client requests that are served without problem by the cluster system. As in the Table 1, the RR case with the random form of workload achieves only 17% success rate. However, with the same workload sets, the performance results are much better when applying Adaptive scheduling algorithm by considering server status and assigning the incoming requests into the server nodes. Another interesting point is that when considering the results from ARR, ALC, and ART, the ARR is a little better than other cases. It means that if we can apply the proper adaptive mechanism for scheduling workloads, the ARR that contains first-fit property can achieve the better performance than the ALC or ART that contain the best-fit property.

6   Conclusion

In this paper, we designed and developed a simulator to simulate and test a distributed server cluster system by generating workloads and scheduling workloads according to the user's input parameter and statistical requirement. The simulation results can be compared and analyzed in terms of the effects of workloads on performances of a cluster system. We can control workload environment and requirement statistically, apply various workload-scheduling algorithms to traffic distribution, and change various parameter values for resource requests of server nodes as the input parameters. Our simulator can simulate a given workload environment for a cluster system, so that we can predict the performance results by changing the workload scheduling algorithms, and other system parameters. In this paper, we also provide two simulation scenarios to find the optimal point of thresholds for the Adaptive-scheduling algorithm, and to compare the performance results from various workload-scheduling algorithms.

References

1. Andrew S. Tanenbaum, Maarten van Steen: "Distributed Systems: principles and Paradigms", Prentice Hall, 2002

2. Jeffray S. Chase: Server switching: yesterday and tomorrow. Internet Applications (2001) 114-123
3. Ronald P. Doyle, Jeffrey S, Chase, Syam Gadde, Amin M Vahdat: The trickle-down effect: Web caching and server request distribution. In Proceedings of the Sixth International Workshop of Web Caching and Content Distribution, 2001
4. Wensong Zhang, Shiyao Jin, Quanyuan Wu: Scaling Internet Service by LinuxDirector. High Performance Computing in the Asia-Pacific Region, 2000. Proceedings. The Fourth International Conference/Exhibition, Volume: 1, (2000) 176 -183
5. Wensong Zhang: Linux Virtual Server for Scalable Network Services. Linux Symposium 2000, July (2000)
6. Eunmi Choi, Dugki Min: A Proactive Management Framework in Active Clusters. LNCS on IWAN, December 2003
7. Valeria Cardellini, Emiliano Casaliccho, Michele Colajanni, Philip S. Yu: The State of the Art in Locally Distributed Web-server Systems. IBM Research Report, RC22209(W0110-048) October (2001) 1-54
8. Mark Grand, "Patterns in Java, Volume1, A Catalog of Reusable Design Patterns Illustrated with UML", John Wiley, 1998

Dynamic Load Balancing Scheme
Based on Resource Reservation for Migration
of Agent in the Pure P2P Network Environment*

Gu Su Kim, Kyoung-in Kim, and Young Ik Eom

School of Information and Communication Engineering,
Sungkyunkwan University, 300 Chunchun-dong, Jangan-gu, Suwon,
Kyunggi-do, 440-746, Korea
{gusukim,kimki95,yieom}@ece.skku.ac.kr

Abstract. Mobile agents are defined as processes which can be autonomously delegated or transferred among the hosts in a network in order to perform some computations on behalf of the user and co-operate with other agents. Currently, mobile agents are used in various fields, such as electronic commerce, mobile communication, parallel processing, search of information, recovery, and so on. In pure P2P network environment, if mobile agents that require computing resources rashly migrate to another peers without the consideration on the peer's capacity of resources, the peer may have a problem that the performance can be degraded, due to the lack of resources. To solve this problem, we propose resource reservation based load balancing scheme of using RMA (Resource Management Agent) that monitors workload information of the peers and decides migrating agents and destination peers. In mobile agent migration procedure, if the resource of specific peer is already reserved, our resource reservation scheme prevents other mobile agents from allocating the resource.

1   Introduction

A mobile agent is a process that represents a user in a network environment and is capable of migrating from one node to another, performing computations on behalf of the user [1]. By deploying the mobile agent environments, we can get several advantages such as reduction of network traffic, asynchronous and autonomous activities of the mobile agents, dynamic adaptation capability, and the robustness and fault-tolerance [2]. But, if mobile agents migrate to a specific node without consideration of the available resource of the node, it decreases the efficiency of the corresponding system.

In this paper, we propose a load balancing scheme that increases the efficiency of the system by reserving resources of the peer that have enough resources in the pure P2P environment. If the workload of the peer that executes the mobile agent becomes higher than the pre-defined threshold value, the peer gathers the

---

* This paper was supported by Samsung Research Fund, SKKU 2002.

T.G. Kim (Ed.): AIS 2004, LNAI 3397, pp. 538–546, 2005.

information on the amount of available resources and the information about the workloads of its neighbor peers, and decides whether to migrate the mobile agent or not. This scheme prevents mobile agents from being centralized in a specific peer, and increases the efficiency of the resource usage.

The rest of the paper is organized as follows. In section 2, we describe some related works on mobile agent systems, the P2P networks, and the load balancing schemes using mobile agent. Section 3 describes our system environment, message formats, and algorithms. Section 4 shows the performance evaluation. Finally, Section 5 concludes our scheme.

2   Related Works

In this section, we describe P2P networks, mobile agent concept, and existing load balancing schemes that uses mobile agents.

2.1   P2P Network

In general, each peer in a P2P network performs both roles of client and server. To overcome the concentration of traffic in the existing centralized network architectures, each peer tries to maximize the performance by distributing its jobs, cooperating with other peers, and sharing resources. The type of P2P network can be classified into two categories according to whether the servers exist or not: one is the pure P2P network and the other is the hybrid P2P network. Fig. 1 shows these two P2P network architectures [3].

Pure P2P network does not require any server for system operation and service. In this architecture, all the peers have both roles of client and server. This architecture has high reliability and scalability because any peer can get a service it wants from any other peer that provides the service. This architecture is used in Gnutella [4].

Hybrid P2P network provides services by using some servers. In this environment, after connecting to a server in the network, each peer sends the information on the shared files or services that he has. The server then maintains the shared file list information for each peer that is connected to him and acts as a mediator between service requester and service provider. This architecture is used in Napster [5].

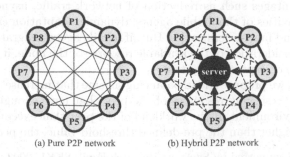

(a) Pure P2P network   (b) Hybrid P2P network

Fig. 1. P2P network architecture.

2.2   Mobile Agent System

Mobile agent is a program that is capable of migrating autonomously in the heterogeneous distributed environment. It helps network traffic reduction, asynchronous interaction, load balancing, dispersion of service, and concurrent processing [6]. Several academic and industrial research groups are currently investigating and building mobile agent systems such as Aglets [7], Ajanta [8, 9], Voyager [10], Tacoma [11], and So on.

2.3   Load Balancing Using Mobile Agents

The aim of the load-balancing strategy is to adapt to the utilization and performance requirements of the machines available and to the requirements of the agents. The load balancing schemes can be classified into two categories: static load balancing schemes and dynamic load balancing schemes [12, 13]. Static load balancing scheme uses only information about the average behavior of the system and ignores the current state of the system. In static load balancing, generally, a task cannot be migrated elsewhere once it was launched on a host. Dynamic load balancing scheme, on the other hand, reacts to the system state that changes dynamically, and, to distribute dynamically the tasks, it may migrate the tasks many times to increase the performance of the system.

The Comet mobile agent system [14] has the load balancing mechanism that uses mobile agents. The Comet has one central host and several compute hosts. The role of the central host is to initiate system startup, suspension, and termination. It is also responsible for providing an interface for query of the current system load and agent distribution. The central host is also a decision-maker about whether there is a need for agent migration or not, and also the chief commander of the selection and location policies. Other than the central host, there are a number of compute hosts. These are the actual hosts that perform the computation and each agent starts and runs on the compute hosts. Migration of agents may occur as needed in-between the compute hosts. Central Agent residing at the central host gathers all the information submitted by Comm Agent at each compute host, detects whether it is suitable to initiate an agent migration, and determines which agent should be migrated to which compute host. But, Comet has several problems. If central host exits, agents cannot migrate to other compute hosts. Also, if the number of compute hosts increases, it takes more time for Central Agent to gather workload information.

3   Proposed Scheme

In this section, we describe our system architecture, message format, and communication algorithm for mobile agent based load balancing in pure P2P network environments.

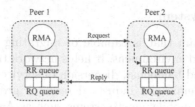

Fig. 2. System architecture.

3.1 System Architecture

Our proposed scheme is based on pure P2P network environment and restricts resources of each peer to CPU and main memory. Fig. 2 shows our simple system architecture.

Each peer has a RMA (Resource Management Agent) and message queues. RMA is a static agent residing at the peer and monitors available resource amount of the peer. RMA is also responsible for gathering the resource information of other peers and selecting the agent to migrate and the target peer to receive the agent. RMA deals with resource reservation message and revoke message received from other peers. Each peer has two message queues: RQ (Resource reQuest) queue and RR (Resource Reservation) queue. RQ queue saves the resource request messages, which includes information on the resource type and amount requested from other peer. Reply messages received from the peers having enough resources are saved into RR queue.

3.2 Message Formats

There are three message types: RREQ (Resource REQuest message), RREP (Resource REPly message), and RREV (Resource REVoke) message. RREQ message is the resource request message that the peer sends to the neighbor peers when he lacks resources. RREP message is the reply message for RREQ message. RREV message is the revocation message that cancels the resource reservation. Fig. 3 shows the content of these messages.

In Fig. 3, *Type* field distinguishes the type of message and *Seq* field is used to prevent duplicate messages. *S_Peer* field specifies the ID of the peer that requests the resource and *R_Type* and *R_Size* fields specify the resource type and amount wanted. *R_Peer* field in RREP message is the field having variable

Type	Seq	S_Peer	R_Type	R_Size	TTL	R_Peer

(a) RREQ(Resource REQuest) message

Type	Seq	S_Peer	R_Peer	R_Type	R_Size

(b) RREP(Resource REPly) message

Type	Seq	S_Peer	E_Peer	TTL

(c) RREV(Resouce REVoke) message

Fig. 3. Message format.

length, where IDs of the peers that received the corresponding message is added. Consequently, the transfer path of the message is stored into the *R_Peer* field of RREP message. The ID of the target peer to receive the mobile agent is saved into the *E_Peer* field of RREV message.

3.3 Algorithms

When a peer lacks resources to complete its jobs, RMA in the peer sends RREQ message including resource type and amount to all its neighbor peers. Fig. 4 shows the resource request algorithm.

```
input : RT (Resource Type), RA (Resource Amount)
output:reserved peer
{
   msg = Type(RREQ)+Seq(Seq.number)+S_Peer(myPeerID)+R_Type(RT)
       +R_Size(RS)+TTL(MH)+R_Peer(NULL);
   send msg to the connected peers;
   TimeOut = C1×2×TT×MH+C2;
   sleep(TimeOut);
   select the most appropriate entry in RQ queue;
   make RREV message and send the message to the connected peers except selected one;
}
```

Fig. 4. The resource request algorithm.

In Fig. 4, MH (Max Hop) is the maximum hop count of message transfer and TT (Transfer Time) is the mean message transfer time of one hop. We assume that the timeout to receive the reply message is set to $2 \times TT \times MH$ (ms).

The RMA which monitors the available resource sends the RREQ message including the required resource type and amount to its neighbor peers. The RMA sending RREQ message waits for RREP messages during the specified period ($C1 \times 2 \times TT \times MH+C2$ ms, $C1$, $C2$ are constants), and decides the optimal peer to migrate a mobile agent by investigating information in the received RREP messages.

The peer received RREQ message creates a QUEUE_ENTRY and inserts this entry into RR queue. Table 1 shows the fields of the QUEUE_ENTRY.

Initially, *ResFlag* is 0, and, if available resource exists when RREQ message is received, the *ResFlag* becomes 1. Fig. 5 shows the algorithm for RREQ message processing.

The peer that received the RREQ message creates a QUEUE_ENTRY and inserts the entry into the RR queue. If the peer has enough available resource, he reserves the resource and sends RREP message to the peer that initially requested the resource reservation. Otherwise, the peer only stores RREQ message into the RR queue and waits until the resource is revoked.

Fig. 6. shows the algorithm for RREP message processing.

If a RREP message arrives at the peer that initially requested the resource, it is saved into RQ queue in the peer. Eventually, RMA finds out an optimal

Table 1. QUEUE_ENTRY.

Field	Description
S_Peer	The ID of the peer that requested the resource allocation.
Seq	Sequence Number
RA(ResourceAmout)	The amount of the resource requested
RT(ResourceType)	The type of the resource requested
ReservedTime	The time that received the reservation request
ResFlag	The flag whether the resource is reserved or not (default: 0)

```
input : msg(RREQ message received)
output : none
{
    if (duplicate message) return;
    record the Seq number of the msg and S_Peer;
    make a QUEUE_ENTRY;
    insert the QUEUE_ENTRY into the RR queue;
    if ( GetResourceSize(msg.R_Type) >= msg.R_Size ) {
        Reserve the requested resource;
        msg2 = Type(RREP)+Seq(msg.Seq.number)+S_Peer(msg.S_Peer)
            +R_Peer(msg.R_Peer+myPeerID)+R_Type(msg.R_Type)+R_Size(msg.R_Size);
        send msg2 to the peer that sent msg;
    }
    msg.TTL--;
    if ( msg.TTL > 0 ) {
        msg.R_Peer += myPeerID;
        send msg to the connected peers except the peer that sent msg;
    }
}
```

Fig. 5. The algorithm for RREQ message processing.

```
input : msg(RREP message)
output : Reserved peer
{
    if ( S_Peer field of msg != myPeerID ) {
        send msg to the previous peer within R_Peer field of msg;
        return;
    }
    insert msg into RQ queue;
}
```

Fig. 6. The algorithm for RREP message processing.

peer from its RQ queue information, migrates the mobile agent to the selected peer, and sends RREV message to the other peers in its RQ queue except the selected peer. Fig. 7 shows the algorithm for RREV message processing.

```
input : msg (RREV message)
output : none
{
    msg.TTL−−;
    if ( msg.TTL < 0 ) send msg to the connected peers except the peer that sent msg;
    if ( E_Peer field of msg == myPeerID ) return;
    remove the corresponding entry in RR queue;
}
```

Fig. 7. The algorithm for RREV message processing.

4   Performance Evaluation

In this section, we evaluate the performance of our scheme through simulation and analyze the result. We have simulated our load balancing scheme in pure P2P network using Visual C++ 6.0 on Window 2000 platform, and measured the agent response ratio and turnaround time of our proposed scheme. Also, we compared the performance of our scheme with non-migration scheme. The agent response ratio can be computed as follows:

agent response ratio = service time / (sleep time + service time)

In our simulation, we assumed that interarrival time of mobile agents at each peer follows exponential distribution with mean 100ms, and the average execution time of mobile agents follows follows exponential distribution with mean 5000ms. We also assumed that message transfer time takes 100ms and the maximum hop count is set to 3 hops.

Fig. 8 shows the agent response ratio according to the number of mobile agent creations per second. As the number of the agents created increases, the agent response ratio of the two models decreases. But, the agent response ratio of our proposed scheme is higher than non-migration scheme that does not consider workload of each peer.

Fig. 9 shows the agent response ratio according to the agent execution time. In this simulation, 10 mobile agents are created per second. When the execution

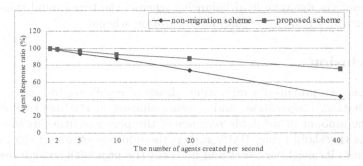

Fig. 8. Agent response ratio according to the number of agents created.

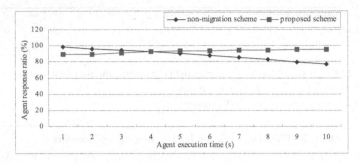

Fig. 9. Agent response ratio according to the agent execution time.

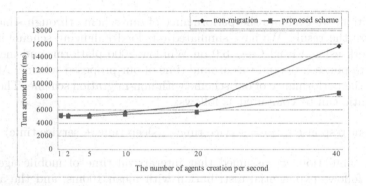

Fig. 10. Turnaround time according to the number of agents created.

time of the agent is smaller than 3 seconds, non-migration scheme is more efficient than our proposed scheme because the agents in non-migration environments seldom sleep. But, the agent response ratio of our proposed scheme becomes higher than that of non-migration scheme when the execution time of mobile agent increases. The reason is that, in non-migration scheme, the time an agent waits for resource increases continuously, but agents in the environment of our proposed scheme can migrate to the peer with enough available resources.

Fig. 10 shows the turnaround time according to the number of mobile agent creations per second. Because the proposed scheme decrease the sleep time of mobile agents via dynamic load balancing scheme, the turnaround time of the mobile agents slowly increase than non-migration scheme.

5   Conclusion

In this paper, we proposed a migration scheme of mobile agents that supports load distribution, cooperation, and resource sharing in P2P network environment. If the amount of available resource of the peer executing mobile agents is under the threshold value, RMA gathers the information about the available resources from its neighbor peers and finds out the peer with enough resources. The peer can migrate its mobile agents to the selected peers. RMA, which resides in each peer, takes charge of gathering the information about available resources in neighbor peers. To increase the reliability and efficiency, if a mobile agent reserves a resource of the specific peer, the peer prevents other mobile agent from migrating into him. Therefore, our scheme can increase the agent response ratio and decrease the turnaround time in normal condition.

References

1. C. Harrison, D. Chess, and A. Kershenbaum, Mobile Agents: Are They a Good Idea?, *Research Report 1987*, IBM Research Division, 1994.
2. D. B. Lange and M. Ohima, Programming And Deploying Java Mobile Agents with Aglets, Addison Wesley, 1998.
3. D. Barkai, An Introduction to Peer-to-Peer Computing, *Developer Update Magazine*, Intel Corporation, Feb. 2000.
4. CLIP2, The Gnutella Protocol Specification v0.4, *Technique Report*, http://www.clip2.com
5. A. Oram, Peer-To-Peer, O'Reilly, Mar. 2001.
6. J. Baumann, et. al., Communication Concepts for Mobile Agent Systems, *Lecture Notes in Computer Science*, Vol. 1219, Springer-Verlag, 1997.
7. A. Gopalan, S. Saleem, and D. Andresen, Bablets: Adding Hierarchical Scheduling to Aglets, *The 8th IEEE International Symposium on High Performance Distributed Computing*, Redondo Beach, California, Aug. 1000.
8. N. M. Karnik and A. R. Tripathi, Security in the Ajanta mobile agent system, *Technical Report*, University of Minnesota, Minneapolis, MN 55455, U. S. A, May 1999.
9. N. Karnik and A. Tripathi, Agent Server Architecture for the Ajanta Mobile Agent System, *Proceedings of the 1998 International Conference on Parallel and Distributed Processing Techniques and Applications (PDPTA'98)*, pages 66-73, Jul. 1998.
10. Glass G., Voyager Core Package Technical Overview, *White Paper*, ObjectSpace, 1999.
11. D. Johansen, R. van Renesse, and F. B. Schneider, An Introduction to the TACOMA Distributed System, *Technical Report*, Department of Computer Science University of Tromso, Jun. 1995.
12. J. Gomoluch and M. Schroeder. Information Agents on the Move: A Survey on Load-Balancing with Mobile Agents, *Software Focus*, Vol. 2, No. 2, Wiley, 2001.
13. D. Gupta and P. Bepari, Load Sharing in Distributed Systems, *Proc. National Workshop on Distributed Computing*, Jadavpur University, Calcutta, Jan. 1999.
14. K. P. Chow and Y. K. Kwok, On Load Balancing for Distributed Multiagent Computing, *IEEE Transaction on Parallel and Distributed System*, Vol. 13, No. 8, Aug. 2002.

Application of Feedforward Neural Network for the Deblocking of Low Bit Rate Coded Images

Kee-Koo Kwon[1], Man-Seok Yang[1], Jin-Suk Ma[1],
Sung-Ho Im[1], and Dong-Sun Lim[1]

Embedded S/W Technology Center, ETRI
161 Gajeong-dong, Yuseong-gu, Daejeon 305-350, Korea
{kwonkk,msyang,jsma,shim,dslim}@etri.re.kr

Abstract. In this paper, we propose a novel post-filtering algorithm to reduce the blocking artifacts in block-based coded images using block classification and feedforward neural network. This algorithm exploited the nonlinearity property of the neural network learning algorithm to reduce the blocking artifacts more accurately. At first, each block is classified into four classes; smooth, horizontal edge, vertical edge, and complex blocks, based on the characteristic of their discrete cosine transform (DCT) coefficients. Thereafter, according to the class information of the neighborhood block, adaptive feedforward neural network is then applied to the horizontal and vertical block boundaries. That is, for each class a different multi-layer perceptron (MLP) is used to remove the blocking artifacts. Experimental results show that the proposed algorithm produced better results than those of the conventional algorithms both subjective and objective viewpoints.

1 Introduction

Transform coding technique has been widely used in many image compression applications [1]-[8]. And block DCT (BDCT)-based coding technique has been adopted in many international standards, including the still and moving image coding standards such as JPEG, H.263, MPEG, and others [1], [2]. However, such techniques produce noticeable blocking artifacts along blocks boundaries in decompressed images at a low bit rate, because the coefficients for each are processed and quantized independently [2]-[8]. Moreover, the discontinuity effect between adjacent blocks in decompressed images is more serious for highly compressed images. Consequently, an efficient blocking artifacts reduction scheme is essential for preserving the visual quality of decompressed images.

A variety of blocking artifacts reduction algorithms have been proposed to improve the visual quality of decoded image in the decoder, such as adaptive filtering methods in the spatial domain [3]-[5], the projections onto convex sets (POCS)-based method [6], estimating the lost DCT coefficients in the transform domain [7], and a filtering method in the wavelet transform domain [8]. The POCS-based method has produced good results, however, since it is based on an iterative approach, it is computationally expensive and time consuming. The

T.G. Kim (Ed.): AIS 2004, LNAI 3397, pp. 547–555, 2005.
© Springer-Verlag Berlin Heidelberg 2005

filtering method in the wavelet transform domain is also unsuitable for images containing a large portion of texture. In contrast, the spatial domain filtering methods have the advantage of simplicity and easy hardware implementation.

Among the spatial domain filtering methods, Ramamurthi's algorithm [3] classifies each block as either monotone or an edge area, then a two-dimensional (2-D) filter is performed to remove grid noise from the monotone areas, then a one-dimensional (1-D) filter is performed to remove staircase noise from the edge areas. However, this algorithm is unable to accurately classify monotone and edge blocks.

The algorithm proposed by Kim et al. [4] classifies an image into a smooth region mode and default mode using the pixel difference in the block boundary. A 1-D LPF is applied to the smooth region mode, while a LPF based on the frequency components in the block boundary is applied to the default mode. Although this algorithm can conserve complex regions, it is still unable to eliminate the blocking artifacts in edge regions.

Qui's algorithm [5] used a feedforward neural network. In this algorithm, the useful features extracted from the decompressed image are used as the input to a feedforward neural network. Yet, since the neural network is applied to all block boundaries, the edge components are blurred. And since this algorithm processes only two pixels near the block boundary, other blocking artifacts are occurred to the inner regions of the block.

Accordingly, we propose a new blocking artifacts reduction algorithm that can reduce the blocking artifacts in block-based coded image using block classification and feedforward neural network. In the proposed algorithm, each block is classified into one of the four classes; smooth, horizontal edge, vertical edge, and complex blocks, based on the characteristic of their DCT coefficients. Thereafter, according to the class information of the neighborhood blocks, adaptive MLP then applied to the horizontal and vertical block boundaries.

Experimental results show that the proposed algorithm improved the PSNR and visual quality of JPEG compressed images and produced better results than those of the conventional algorithms.

2   Proposed Post-filtering Algorithm

2.1   Structure of MLP

A multi-layer feedforward neural network consists of a set of sensory units that constitute the input layer, one or more hidden layers of computation nodes, and an output layer of computation nodes. The input signal propagates through the network in a forward direction, on a layer-by-layer basis. MLP has already been successfully ap-plied to a variety of difficult and diverse problems based on supervised learning using an error back-propagation learning algorithm [9].

The error back-propagation (EBP) learning algorithm consists of two passes through the different layers of the network; a forward pass and backward pass. In a forward pass, an input vector is applied to the sensory nodes of the network, and its effect propagates through the network layer by layer. Finally, a set of

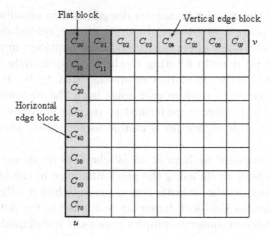

Fig. 1. Block classification using 8×8 DCT coefficients C_{uv} distribution.

outputs is produced as the actual response of the network. During a forward pass, the synaptic weights of the networks are all fixed. In contrast, during a backward pass, the synaptic weights are all adjusted in accordance with an error-correction rule. Specifically, the actual response of the network is subtracted from the desired response to produce an error signal. This error signal is then propagated backward through the network, against the direction of the synaptic connections. The synaptic weights are adjusted to make the actual response of the network move closer to the desired response in a statistical sense.

The current study uses MLP with an EBP learning algorithm to reduce the blocking artifacts. That is, we use neural network filters rather than adaptive linear filters be-cause compression and decompression are highly nonlinear and complex and neural network filters have been proven effective in finding nonlinear mappings between inputs and outputs. After training the MLP, the final synaptic weights of the MLP are used as the processing filters to reduce the blocking artifacts.

2.2 Block Classification

Each block is classified into one of four categories, that is, smooth block, horizontal edge block, vertical edge block, and complex block, according to the distribution of an 8×8 DCT coefficients C_{uv}, in which u and v represent the horizontal and vertical frequency coordinates, respectively. That is, for block classification, we use the fact that the distribution of the DCT coefficients is important measures.

The block classification process is as follows. First, calculate the mean value $m_{C_{uv}}$ of the DCT coefficients C_{uv} for each block. Then, calculate

$$\hat{C}_{uv} = Round(\frac{|C_{uv}|}{m_{c_{uv}}}) \tag{1}$$

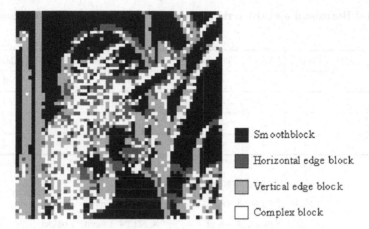

Smoothblock

Horizontal edge block

Vertical edge block

Complex block

Fig. 2. Result of block classification on a JPEG decoded LENA image with a bit rate of 0.271 bpp.

where $Round()$ is the rounding to the nearest integer and $|\cdot|$ is an absolute value operator. Because the DCT coefficients C_{uv} have a floating-point value, using these DCT values for block classification, the considering coefficients are a lot, so increase the computational complexity. Thus in the proposed method, normalize the coefficients using the mean value $m_{C_{uv}}$ and then transpose the integer value. So the small DCT coefficients have zero value.

Thereafter, as shown in Fig. 1, a block is classified into one of four categories as follows:

1) Class 0 (smooth block): if i) more than one thing among $\hat{C}_{00}, \hat{C}_{01}, \hat{C}_{10}, \hat{C}_{11} \neq 0$ and ii) $\hat{C}_{uv} = 0$ for $u > 1, v > 1$
2) Class 1 (horizontal edge block): if i) more than one thing among $\hat{C}_{20} \sim \hat{C}_{70} \neq 0$ and ii) all $\hat{C}_{uv}$ except for $\hat{C}_{20} \sim \hat{C}_{70}, \hat{C}_{00}, \hat{C}_{01}, \hat{C}_{10}, \hat{C}_{11}$ are zero values
3) Class 2 (vertical edge block): if i) more than one thing among $\hat{C}_{02} \sim \hat{C}_{07} \neq 0$ and ii) all $\hat{C}_{uv}$ except for $\hat{C}_{02} \sim \hat{C}_{07}, \hat{C}_{00}, \hat{C}_{01}, \hat{C}_{10}, \hat{C}_{11}$ are zero values
4) Class 3 (complex block): all blocks except for class 0, class 1, and class 2.

Fig. 2 shows the result of the proposed block classification algorithm on a JPEG decoded LENA image with a bit rate of 0.271 bpp. This classification scheme is used to determine the class of the current processing block and then select the corresponding MLP to adaptively reduce the blocking artifacts.

2.3   Adaptive Inter-block Filtering

Based on the above classification scheme, adaptive inter-block filtering is processed the horizontal and vertical block boundaries. The horizontal inter-block filtering is performed in a horizontal direction at a horizontal block boundary.

Table 1. (a) Horizontal and (b) vertical inter-block filtering methods using adaptive MLP.

R \ N	Class 0	Class 1	Class 2	Class 3
Class 0	1	2	3	4
Class 1	2	2	3	4
Class 2	3	3	3	4
Class 3	4	4	4	4

(a)

R \ N	Class 0	Class 2	Class 1	Class 3
Class 0	1	2	3	4
Class 2	2	2	3	4
Class 1	3	3	3	4
Class 3	4	4	4	4

(b)

R: Current block, N: Neighborhood block

1: MLP1 5 input, 6 output   2: MLP2 3 input, 4 output

3: MLP3 3 input, 4 output   4: MLP4 1 input, 2 output

Likewise, the vertical inter-block filtering is performed in a vertical direction at a vertical block boundary.

Horizontal and vertical inter-block filtering is performed adaptively to the class information of the current block and the horizontal or vertical neighborhood block, as shown in Table 1.

Because human visual system is sensitive to blocking artifacts in smooth region more than complex region, the proposed method designs the MLP which is adaptive to the characteristic of the block. That is, as shown in Fig. 3, a strong 1-D 5-tap filtering is applied to smooth blocks, whereas a weak 1-D 3-tap filtering is applied to horizontal and vertical edge blocks. For complex block, inter-block filtering is only applied two block boundary pixels so as to preserve the image details. The filtering of the horizontal and vertical directions is processed separately using a different MLP.

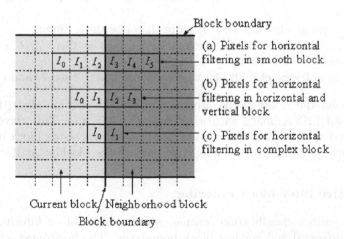

Fig. 3. Adaptive filtering methods for smooth block, horizontal and vertical edge block, and complex block.

As shown in Fig. 3, to process the horizontal blocks, the input vectors for the MLP are formed as follows:

$$D_n = |I_{n+1} - I_n|, \quad n = 0, 1, \cdots, m \tag{2}$$

where I_n and m denote the intensity value and number of input nodes in the network, respectively. The corresponding desired output vectors are formed based on the difference between the original image and the decompressed image as

$$\begin{aligned} t_n &= I(i+n, j) - \tilde{I}(i+n, j), && n = 0, 1, \cdots, m+1, \\ i &= 7, 15, \cdots, M-8, j = 0, 1, \cdots, N-1 \end{aligned} \tag{3}$$

where $I(i,j)$, $\tilde{I}(i,j)$, M, and N denote the original image, decompressed image, horizontal image size, and vertical image size, respectively. We use the pair of training samples in (2) and (3) to train the MLP until it converges. Once the MLP is trained, its weights are saved. That is, the proposed MLP produce the difference between the original image and the decompressed image. So the blocking artifacts are removed by adding the outputs of the MLP to the decompressed image. That is, the postprocessed image $\hat{I}(i,j)$ is formed as follows:

$$\begin{aligned} \hat{I}(i+n, j) &= \tilde{I}(i+n, j) + y_n, && n = 0, 1, \cdots, m+1, \\ i &= 7, 15, \cdots, M-8, && j = 0, 1, \cdots, N-1 \end{aligned} \tag{4}$$

where y_n denotes the outputs of the MLP.

After the horizontal blocks are processed, the vertical blocks are processed in a similar manner with a different MLP of the same size.

3   Experimental Results

To evaluate the performance of the proposed algorithm, computer simulations were performed using JPEG decoded images. The MLP were trained using three training images ('BABOON', 'BARBARA', and 'BANK') due to the generalization capability of the MLP. Another three images ('LENA', 'BOAT', and 'PEPPERS') were then used as the test images. Each image was 512×512 in size with 256 gray levels and compressed by JPEG at various bit rates. The peak signal-to-noise ratio (PSNR) was used to measure the performance of the postprocessing algorithms. For $M \times N$ images with [0, 255] gray level range, PSNR can be defined as

$$PSNR = 10 log_{10}(\frac{255^2}{MSE}) \tag{5}$$

$$MSE = \frac{1}{M \times N} \sum_{i=0}^{M-1} \sum_{j=0}^{N-1} (I(i,j) - \hat{I}(i,j))^2 \tag{6}$$

where, $I(i,j)$ and $\hat{I}(i,j)$ denote the original image and postprocessed image, respectively.

Table 2. Experimental results for JPEG decoded images.

Test images	Bit rate [bpp]	PSNR [dB]					
		JPEG	Yang	Kim [8]	Kim [4]	Qui	Proposed
LENA	0.208	30.41	31.05	31.15	30.72	30.94	31.08
	0.271	31.95	32.45	32.49	32.15	32.36	32.48
	0.324	32.96	33.34	33.33	33.03	33.23	33.41
BOAT	0.258	28.13	28.57	28.64	28.32	28.57	28.68
	0.350	29.53	29.88	29.90	29.66	29.87	29.90
	0.420	30.49	30.76	30.74	30.56	30.73	30.87
PEPPERS	0.212	30.13	30.62	30.87	30.49	30.55	30.87
	0.272	31.53	31.88	32.05	31.77	31.80	31.98
	0.325	32.43	32.66	32.82	32.54	32.58	32.76

The structures of the horizontal and vertical MLP were the same. The number of hidden neurons is decided through experiment and seven, five, five, and two hidden neurons were used for class 0, class 1, class 2, and class 3, respectively. The initial synaptic weights were initialized with a floating point interval [0., 0.00001], and after training the MLP the final synaptic weights of the MLP are used as the processing filters to reduce the blocking artifacts. To compare the proposed algorithm with other conventional algorithms, the PSNR performances of the proposed algorithm and those of the four conventional algorithms [4]-[6], [8] are presented in Table 2. The proposed algorithm improved the PSNR by 0.10 dB to 0.74 dB in the JPEG decoded images, which was roughly the same or better than the performances of the conventional algorithms. Yet, Yang's algorithm [6] and Kim's algorithm [8] have good results, but these algorithms have a high computational complexity, because both algorithms include an iterative or wavelet transform.

A magnified portion of the LENA image decoded by JPEG with a bit rate of 0.271 bpp, and the postprocessed images are shown in Fig. 4. The proposed algorithm effectively reduced the blocking artifacts and preserved the original high-frequency components, such as edges. And we showed that the postprocessed image by using Kim's method [4] has the most subjective quality among the conventional methods. As shown in Fig. 4, the proposed method effectively reduced the blocking artifacts in smooth and complex regions and the mosquito noise near the edges. Kim's method [4] reduced the blocking artifacts in smooth regions, but the blocking artifacts in complex regions and the mosquito noise near the edges are still remained.

4   Conclusions

A new blocking artifacts reduction algorithm to reduce the blocking artifacts in block-based coded images was proposed using block classification and adaptive MLP. In this algorithm, each block is classified into one of the four classes based on the characteristics of their DCT coefficients. Thereafter, according to the class information of neighborhood blocks, adaptive MLP is applied to the horizontal and vertical block boundaries. As such, different MLP are applied to remove

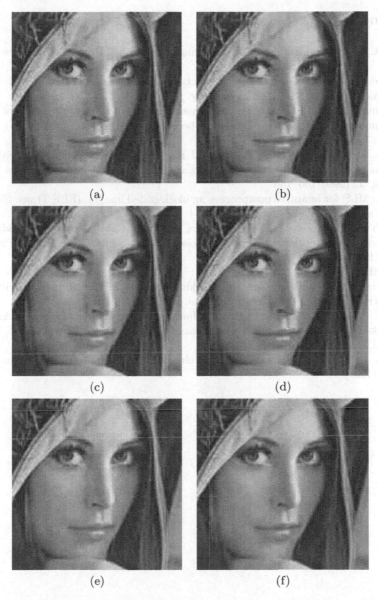

(a)   (b)

(c)   (d)

(e)   (f)

Fig. 4. A magnified portions of (a) JPEG decoded image with a bit rate of 0.271 bpp, and postprocessed images using (b) Yang's method, (c) Kim's method [8], (d) Kim's method [4], (e) Qui's method, and (f) proposed method.

the blocking artifacts. Experimental results showed that the proposed algorithm improved the PSNR from 0.10 dB to 0.74 dB in JPEG decoded images, which was roughly same or the better results than the performances of the conventional algorithms. That is, the proposed algorithm effectively reduced the blocking artifacts and mosquito noise, preserving the original high-frequency components, including the edges.

References

1. Jain, A. K.: Fundamentals of Digital Image Processing. Prentice Hall, New York (1990)
2. Wallace, G. K.: The The JPEG still picture compression standard. IEEE Trans. Consumer Electron. 38 (1992) xviii-xxxiv
3. Ramamurthi, B. and Gersho, A.: Nonlinear space-variant postprocessing of block coded images. IEEE Trans. Acoustics, Speech, Signal Processing, 34 (1986) 1258–1268
4. Kim, S. D., Yi, J. Y., Kim, H. M., and Ra, J. B.: A deblocking filter with two separate modes in block-based video coding. IEEE Trans. Circuits Systems Video Technol. 9 (1999) 156–160
5. Qui, G.: MLP for adaptive postprocessing block-coded images. IEEE Trans. Circuits Syst. Video Technol. 10 (2000) 1450–1454
6. Yang, Y., Galatsanos, N., and Katsagelos, A.: Projection-based spatially adaptive reconstruction of block-transform compressed images. IEEE Trans. Image Processing. 4 (1995) 896–908
7. Paek, H., Kim, R. C., and Lee, S. U.: A DCT-based spatially adaptive postprocessing technique to reduce the blocking artifacts in transform coded images. IEEE Trans. Circuits Syst. Video Technol. 10 (2000) 36–41
8. Kim, N. C., Jang, I. H., Kim, D. H., and Hong, W. H.: Reduction of blocking Artifact in block-coded images using wavelet transform. IEEE Trans. Circuits Syst. Video Technol. 8 (1998) 253–257
9. Haykin, H.: Neural Networks: A Comprehensive Foundation. Tom Robbins, New Jersey (1999)

A Dynamic Bandwidth Allocation Algorithm
with Supporting QoS for EPON

Min-Suk Jung[1], Jong-hoon Eom[2], Sang-Ryul Ryu[3], and Sung-Ho Kim[1]

[1] Dept. of Computer Engineering, Kyungpook National University, Daegu, Korea
msjung@mmlab.knu.ac.kr, shkim@bh.knu.ac.kr
[2] Telecommunication Network Laboratory, KT, Daejeon, Korea
jheom@kt.co.kr
[3] Dept. of Computer Science, Chungwoon University, Chungwoon, Korea
rsr@mail.chungwoon.ac.kr

Abstract. An Ethernet PON (Passive Optical Network) is an economical and efficient access network that has received significant research attention in recent years. A MAC (Media Access Control) protocol of the PON, the next generation access network, is based primarily on TDMA (Time Division Multiple Access). In this paper, we addressed the problem of a dynamic bandwidth allocation in QoS (Quality-of-Service) based Ethernet PONs. We augmented the bandwidth allocation algorithms to support QoS in a differentiated service framework. Our differentiated bandwidth guarantee allocation (DBGA) algorithm allocates effectively and fairly the bandwidths between end-users. Moreover, we showed that a DBGA algorithm that perform weighted bandwidth allocation for high priority packets results in a better performance in terms of average and maximum packet delay, as well as the network throughput compared with some other dynamic allocation algorithms. We used the simulation to study the performance and validate the effectiveness of the proposed algorithm.

1 Introduction

In the last decade, the bandwidth in the backbone networks has significantly increased up to several Tbps by using WDM technology. The bandwidth in the local area networks also has increased up to 10Gbps based on Ethernet. Though access networks have improved their bandwidth using xDSL, Cable modem, etc., their bandwidth has been limited to several Mbps. Their limited bandwidth causes the bottleneck phenomenon in the access networks. Many service providers have already attempted to construct high speed access networks with the ultimate goal of FTTH [1-3]. The xDSL is currently used as a high-speed subscription access network. Although this type of network model is effective for an area where the demand is high, it is ineffective when the demand is only sporadic, plus its transmission capacity is inadequate for the latest multimedia services. Accordingly, a PON is a new technology for subscription access networks that supports the ability to connect subscribers using FTTx over the distances of up to 20km. In addition, a PON has the capacity to simultaneously connect the maximum number of 64 ONUs through an optical splitter using an optical cable. Consequently, a PON has the ability to support the transmission of large amounts of data and is a very economical way to construct networks [4-

T.G. Kim (Ed.): AIS 2004, LNAI 3397, pp. 556–564, 2005.
© Springer-Verlag Berlin Heidelberg 2005

6]. Among the various methods of transmitting data based on a PON, the first developed method was an APON. Yet, when constructing a new network, using Ethernet has two advantages: Competitive price and home networks are constructed using the Ethernet. As such, the IEEE 802.3 EFM SG (Ethernet in the First Mile Study Group) was established with the goal of standardizing EPONs.

In this paper, we propose a method that controls the ratio of high-priority, medium-priority, and low priority without the additional bandwidth for high-priority traffic to each ONU, so as to minimize the delay of high-priority traffic. In addition, we design a simulation model whereby our proposed method is applied to the Ethernet PON and verify our proposed method is verified by analyzing the end-to-end delay and the throughput. The rest of the paper is organized as follow. Section 2 describes the fashion of the study in Ethernet PON. In Section 3, we propose an efficient dynamic bandwidth allocation method. Section 4 describes Control message scheduling method. We will simulate our model and analyze its results in Section 5. Section 6 offers the conclusion and describes future research.

2 Ethernet PON

In this section, we describe differences between the Ethernet PON and ATM PON. We also explain the fundamentals of the operation in Ethernet PON. Figure 1 shows the EPON system structure, as suggested by the IEEE 802.3 EFM SG. The OLT and the ONU are located at the End Point of a PSS (Passive Star Splitter), whereby they are connected by an optical fiber. The PON is either distributed into several identical optical signals or is united into one signal according to the transfer direction of the optical signal. PSSs are economical as they are easy to construct, require little maintenance and are inexpensive to repair cost. Also, since a PSS is a passive component, it does not require any extra power supply. In addition, since the OLT and the ONU are connected by a Point-to-Multipoint form, the installment cost of the optical fiber is lower than that of a Point-to-Point form.

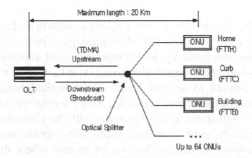

Fig. 1. The structure of the Ethernet PON is proposed in IEEE 802.ah

To compare the Ethernet PON with an ATM PON, a very different point is the transmission data type is required. Ethernet PON is able to transmit the variable length IP packets but the ATM PON only transmits the fixed length (53bytes) cells [8]. Moreover, the Ethernet PON is able to expand easily the channel speed up to 10 Gbps, but the ATM PON is limited in 622Mbps.

3 The Bandwidth Allocation Algorithm

In Figure 2, we present the bandwidth allocation procedure. The bandwidth allocation procedure is as follows: First, an ONU notifies its buffer status using a REPORT message. After the OLT receives a REPORT message, it executes the bandwidth allocation algorithm and grants their allocated bandwidth using GATE messages to each ONU. The bandwidth management plays an important role in guaranteeing QoS. The proposed bandwidth allocation algorithms in [7] are fixed, limited, credit, and gated. Here, they applied a strict priority scheduling mechanism (defined 802.1D) to the limited algorithm which is estimated higher than the other bandwidth allocation algorithms in [7]. Low-priority class traffic, however is not able to guarantee a sufficient bandwidth at the high load. As a result, low-priority class traffic may suffer from a starvation. In [9] they propose a DBA2 algorithm that combines the predicted credit based method with a priority-based scheduling mechanism.

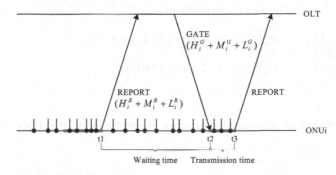

Fig. 2. The Bandwidth Allocation Procedure in Ethernet PON

The method that adds the predicted credit to the request time for high-priority traffic may reduce the packet delay in this cycle, but the extended cycle time influences the next cycle time. Finally, it will accumulate so that it increases the queuing delay. In addition, if the predicted credit is not used, the throughput will decrease. Therefore, we propose a DBGA (Differentiated Bandwidth Guarantee Allocation) algorithm that is fair and does not increase the queuing delay for high-priority class traffic.

B_i^R represents the requested bandwidth from ONUi and consists of high-priority, medium-priority, and low-priority traffic. It is as follows:

$$B_i^R = H_i^R + M_i^R + L_i^R \tag{1}$$

If we represent B_i^R as the granted bandwidth to ONUi, as below:

$$B_i^G = (H_i^R + L_i^R(1-\omega)) + M_i^R + L_i^R\omega \quad for \quad W_i^{MIN} \le B_i^G \le W_i^{MAX} \tag{2}$$

ω represents the yield ratio whereby the low-priority class gives the high-priority class a part of itself. If the OLT allows the transmission of all packets in each ONU's buffer. Then, the ONU that has the most packets will monopolize the upstream bandwidth. To avoid this situation the OLT limits the maximum transmission window size W_i^{MAX} and the minimum transmission window size W_i^{MIN}. So, W_i^{MAX} guarantees the maximum transmission delay and W_i^{MIN} guarantees the minimum bandwidth, thereby it is able to satisfy SLA (Service Level Agreement). To reduce the control packet scheduling time, the OLT allocates the bandwidth by dividing the total number of ONUs into two parts using an interleaving method. The algorithm is the following:

```
DBGA_Algorithm (H_R[], M_R[], L_R[], i)
{
  if (i == |N/2|)
    for (i=1; i<|N/2); i++)
    {
      H_G[i] = H_R[i] + L_R[i](1-w);
      M_G[i] = M_R[i];
      L_G[i] = L_R[i] - L_R[i]*w;
    }
  if (i == N)
    for (i=|N/2|+1; i<N; i++)
    {
      H_G[i] = H_R[i] + L_R[i](1-w);
      M_G[i] = M_R[i];
      L_G[i] = L_R[i] - L_R[i]*w;
    }
}
```

The detailed control messages scheduling the problem are described in the next section.

4 Control Message Scheduling

Another important factor concerned with the bandwidth allocation algorithm is the generation of the grant table. The GATE message and the REPORT message generation are classified into two methods, the polling method whereby the data transmission is started by the OLT timer and the method whereby the data transmission is started by the ONU timer. As the polling method is considered in various characteristics of the Ethernet PON system, it is beneficial in scheduling the upstreaming traffic and has many advantages.

In Figure 3 (a), receiving the REPORT message of the ONUs, the OLT in the simple bandwidth allocation performs the bandwidth allocation algorithm, so that it generates the GRANT table. This method, however, needs the DBA_TIME in order to perform the algorithm and generate the GRANT table. So the IDLE time which does not use the channels is wasted. The IDLE time occupies DBA_TIME + RTT. Figure 3 (b) shows the method which does not waste the IDLE time. After receiving the REPORT message in the last ONU in the n-1 cycle, the method generates the GRANT table. The generated table is used in sending the GATE message in n+1 cycle. Therefore, because the requested bandwidth timeslot in the n-1 cycle is allocated in the n+1 cycle, the delay time is increased. In order to resolve this problem, the DBA2 method in [9] sends the GRANT for which the ONU that requested the small bandwidth uses the channels in advance of the IDLE time. During this time, it performs the algorithm, and then generates the remainder of the GRANT table. This

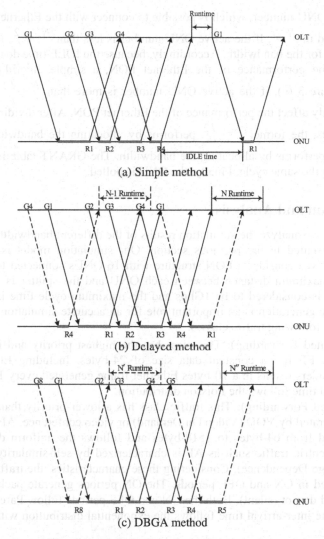

Fig. 3. Bandwidth allocation performance time

method functions well, when the number of the ONU which is $B_i^R < B_i^{MAX}$ is many. In this case, because the bandwidth is sufficient, the DBA performance time is not important. When the traffic is not heavy in the Ethernet PON system, the situation that has many nodes occurs. The problem is that all nodes have to wait to complete the GRANT table, because there is no ONU which is $B_i^R < B_i^{MIN}$, which sends the GRANT in advance. In fact, the case whereby the bandwidth is used efficiently, is when the traffic is high. This causes wasting the IDLE time. This paper proposes a method to minimize the delay and to function efficiently operate in the busy traffic of all ONUs.

The whole ONU number, which is possible to connect with the Ethernet PON system, is declared in N_{total}. If the active ONU number is less than $N_{total}/2$, it is considered sufficient for the bandwidth. Accordingly, because the IDLE time does not affect significantly the performance of the Ethernet PON, a simple algorithm is used. Shown in Figure 3 (c), if the active ONU number is more than $N_{total}/2$, the IDLE time may greatly affect the performance of the Ethernet PON. After dividing T_{cycle} into two equal parts, the former $N_{total}/2$ performs by allocating the bandwidth and the latter $N_{total}/2$ performs by allocating the bandwidth. The GRANT table is generated alternatively in the same cycle. Finally, the ONU is polled.

5 Simulation and Analysis

In this section, we analyze the simulation results of the different bandwidth allocation algorithms presented in the previous section. Our simulation model is developed using OPNET. We consider a PON structure with 16 ONUs connected to a tree topology. The maximum distance between each ONU and the splitter is 20km. The channel speed is considered to be 1Gbps and the maximum cycle time is 2ms. The practical traffic generation is an important role for an accurate simulation. Therefore we support the following traffic classes:

1. EF (Expedited Forwarding): This class has the highest priority and is a typical voice data. EF has a constant data size of 24-bytes. Including Ethernet and UDP/IP headers results in a 70-bytes Ethernet frame generated every 125 μs. The inter-arrival time follows the Poisson distribution.
2. AF (Assured Forwarding): This traffic class has a lower priority than EF traffic and is generated by VOD (Video On Demand) or video conference. AF frame size is increased from 64-bytes to 1518-bytes and follows the uniform distribution. The data-centric traffic such as AF is characterized by self-similarity and LRD (Long Range Dependence). Considering these characteristics, the traffic model is implemented in ON and OFF periods. The ON periods generate packets but the OFF period do not (silent). Both the ON and OFF periods follow Pareto distribution, and the inter-arrival time follows the exponential distribution within the ON periods.
3. BE (Best Effort): This class has the lowest priority and is the fully data-centric traffic. This class includes http, ftp, e-mail, etc. The traffic has properties the same as AF (self-similarity and LRD).

We simulated and analyzed the results of fixed, limited, DBA2 and the proposed DBGA allocation method about the three kinds of the data traffic. The experiments consider the current Internet traffic. Since a real-time service like VOD on the web occupies 70% of the total traffic, we generated a 1:1:2 traffic rate on the EF traffic, AF traffic and BE traffic. With that situation, we wanted to determine how the proposed algorithm worked. In the implemented experiment model, we applied a strict priority scheduling mechanism to the fixed allocation algorithm and a priority-based scheduling mechanism to the limited, DBA2 and DBGA allocation method.

Figure 4 shows (a) the maximum delay and (b) the average delay which impact high-priority traffic on each allocation algorithm. As shown in (a) and (b) of Figure 4, high-priority traffic has the end-to-end delay in the order of DBGA, Limited, DBA2 and Fixed allocation methods in the load which is below 0.7. Though DBA2 and DBGA are based on the limited allocation method, DBGA is what better than DBA2, because DBA2 method is based on credit, and the ONU requests more data than what is queuing currently. So, the total cycle time becomes longer.

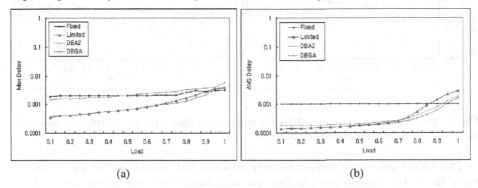

Fig. 4. The end-to-end delay of the high-priority class for each algorithm: (a) the maximum end-to-end delay and (b) the average end-to-end delay

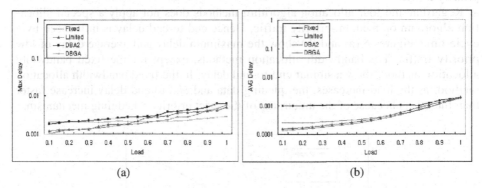

Fig. 5. The end-to-end delay of medium-priority traffic class for each algorithm: (a) the maximum end-to-end delay and (b) the average end-to-end delay

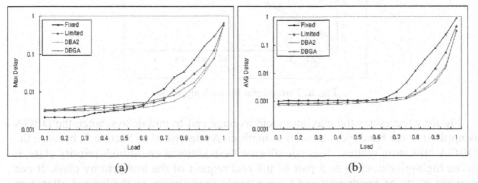

Fig. 6. The end-to-end delay of low-priority traffic class for each algorithm: (a) the maximum end-to-end delay and (b) the average end-to-end delay

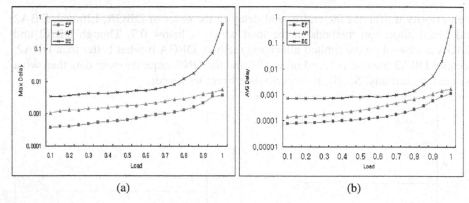

(a)   (b)

Fig. 7. The end-to-end delay of DBGA for each traffic class : (a) the maximum end-to-end delay and (b) the average end-to-end delay

The DBGA allocation method linearly allocates additional data that are borrowed from BE to EF. Accordingly, it does not have much influence on the entire cycle time. Figures 5 (a) and (b) show the maximum delay and average delay in medium-priority traffic.

Regarding the four allocation algorithm methods does not apply a special allocation algorithm on medium-priority traffic. Hence end-to-end delay is influenced by a cycle time. Figures 6 (a) and (b) show the maximum delay and average delay of low priority traffic. The bandwidth allocation methods, except for the fixed bandwidth allocation method, show a similar end-to-end delay. In the fixed bandwidth allocation method, as the load increases, the queuing data and end-to-end delay increase faster than the others because of the unfairness of the strict priority scheduling mechanism.

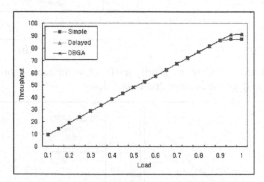

Fig. 8. Throughput for each algorithm

Figure 7 represents the maximum and average end-to-end delays, when the DBGA is applied to the bandwidth allocation method. As described in Section 3, because the DBGA allocation method does not allocate the additional credit high-priority class, it gives high-priority class to a part of the real request of the low-priority class. It can prevent waste of bandwidth and keep a cycle time similar to the limited allocation method. It can guarantee the end-to-end delay of high priority traffic. As shown in

Figure 8, they are similar to each of the bandwidth allocation algorithms except for fixed bandwidth allocation algorithm, That is, to guarantee QoS the cycle time must be reduced as much as possible.

6 Conclusion

This paper dealt with the dynamic bandwidth allocation problem in the Ethernet PON. In particular, we proposed dynamic allocation algorithm supporting QoS in the differentiated service. The bandwidth allocation method is based on strict priority scheduling mechanism results in increasing delays for a specific traffic class. The method that allocates predicted credit to the high-priority service shows that it didn't improve the performance because of the increasing total cycle time. In order to correct this drawback, we proposed a DBGA that adds a part of the timeslot for low-priority packets to the timeslot for high-priority packets.

When the proposed dynamic bandwidth allocation algorithm is compared with other algorithms, it shows that the proposed allocation algorithm is much greater than the average and maximum packet delay. Furthermore, the performance of the bandwidth allocation algorithm is dependant on the cycle time as well as the timeslot size. So, the proposed algorithm that represented the shorter-cycle time in the appropriate range is a better performer than the long-cycle time. To verify the proposed algorithm, we used the OPNET simulation tool. Further study is required to verify about the performance in Ethernet PON, when the bandwidth is at 10Gbs. Also, we plan to analyze the performance in the DBGA algorithm, which is adapted in the system model using a control channel of the fixing transceiver in the WDM PON, and a data channel of the tunable transceiver.

References

1. A. Cook and J. Stern, "Optical fiber access Perspectives toward the 21st century," IEEE Commun. Mag., pp. 78-86, Feb. 1994.
2. Y. Takigawa, S. Aoyagi, and E. Maekawa, "ATM based passive double star system offering B-ISDN, N-ISDN, and POTS," Proc. of GLOBECOM'93, pp. 14-18, Nov. 1993.
3. S. S. Kang, H. J. Kim, Y. Y. Chun, and M. S. Lee, "An Experimental Field Trial of PON Based Digital CATV Network," IEICE Trans. Commun., Vol. E79-B, No. 7, pp. 904-908, July 1996.
4. G. Pesavento and M. Kelsey, "PONs for the Broadband Local loop," Lightwave, Vol. 16, No. 10, pp. 68-74, Sep. 1999.
5. B. Lung, "PON Architecture Future proof FTTH," Lightwave, Vol. 16, No. 10, pp. 104-107, Sep. 1999.
6. G. Kramer B. Mukherjee, and G. Pesavento, "Ethernet PON(Ethernet PON): Design and Analysis of an Optical Access Network," Photo. Net. Commun., Vol. 3, No. 3, pp. 307-319, July 2001.
7. G. Kramer and B. Mukherjee, "IPACT: A Dynamic Protocol for an Ethernet Passive O- ptical Network (Ethernet PON)," IEEE Commun., Mag., pp. 74-80, Feb. 2002.
8. ITU-T Recommendation G.983.1, "Broadband Optical Access Systems Based on PON," Geneva, Oct. 1998.
9. Chadi M. Assi, Y. Ye, Sudhir Dixit, Mohamed A. Ali, "Dynamic Bandwidth Allocation for Quality-of-Service Over Ethernet PONs" Journal of Sel. Area in Communication Vol. 21 No 9, Nov, 2003.

A Layered Scripting Language Technique for Avatar Behavior Representation and Control*

Jae-Kyung Kim[1], Won-Sung Sohn[2], Beom-Joon Cho[3],
Soon-Bum Lim[4], and Yoon-Chul Choy[1]

[1] Dept of Computer Science, Yonsei University,
Shinchon-dong, Seodaemun-gu, 120-749, Seoul, Korea
{ki187cm,ycchoy}@rainbow.yonsei.ac.kr
[2] Computational Design Program,
School of Architecture Carnegie Mellon University,
Pittsburgh, PA15213, USA
sohnws@u.washington.edu
[3] Department of Computer Engineering,
Chosun University Kwangju, Korea
bjcho@chosun.ac.kr
[4] Department of Multimedia Science, Sookmyung Women's University,
Chungpa-dong, Yongsan-gu, 140-742, Seoul, Korea
sblim@sookmyung.ac.kr

Abstract. The paper proposes a layered scripting language technique for representation and control of avatar behavior for simpler avatar control in various domain environments. We suggest three layered architecture which is consisted of task-level behavior, high-level motion, and primitive motion script language. These layers brides gap between application domain and implementation environments, so that end user can control the avatar through easy and simple task-level scripting language without concerning low-level animation and the script can be applied various implementations regardless of application domain types. Our goal is to support flexible and extensible representation and control of avatar behavior by layered approach separating application domains and implementation tools.

1 Introduction

With computer becoming an important part of our lives, many of our activities are achieved in virtual environment. Now, computers are not just a machine doing simple jobs but one of the interfaces to our social activities [1]. Thus, interactive user inter-face techniques between human and computer which can induce interests is becoming more and more important. The avatar is a representative example of the interface techniques. Avatar techniques have rapidly progressed over the recent years. Gartner group [2] selected the avatar application as the

---

* This work was supported by Ministry of Commerce, Industry and Energy.

T.G. Kim (Ed.): AIS 2004, LNAI 3397, pp. 565–573, 2005.

major communication technique among ten branch information which were paid attention to in 21st century.

Today, there has been an active research on the application of avatar along with representation and control of avatar behavior. The control of avatar especially through the control of web based language XML along with behavior of XML based behavior script made standard representation and control of avatar behavior feasible. Thus, various script languages that can control avatar in various purposes and environment are being researched.

In this paper, we defined XML based avatar behavior expression and control script through hierarchical approach method to provide users simple avatar behavior control interface in various domain environment and implementation environment.

The proposed script is made up of task-level behavior, high-level and primitive motions. Each layer of script operates independently and avatar expression and control is made possible in formal way in various domains and implementation environment.

2   Related Works

2.1   Task-Level Behavior

Task is avatar's behavior with specific purpose or compensation [11]. For example, avatar behavior such as 'walk to table' or 'jump here to there' is simply not walk, jump motions itself but they have the purpose to go specific location. Task is defined differently depending on the domain, but in the case of cyber class domain, task can be explanation and query answer like 'explain', 'answer', 'query' and in shopping mall, task is 'introducing' and 'selling' a product.

The existing research using this kind of task based behaviors are STEVE [10], CPSL [6], and Wizlow [9]. In the task based avatar system, user does not have to ap-point complicated avatar's behavior, and simply assign necessary tasks which are corresponding to domain.

However, in most of the systems, task-level behaviors are dependent to animations which avatar engine offers. Therefore, task-level behavior can be dependent on specific system's low-level motion, and avatar behavior's extensibility or reusability, efficiency can be deteriorated.

2.2   High-Level Motion

For general users to control avatar motion by using each joint rotational angle motion expression like lower level motion expression is a very difficult task. Therefore, users should be able to control avatar without concerning low-level animation data. The representative research of avatar high-level motion are AML [3], CML [4], VHML [5], STEP [8], TVML [7]. All of these are XML based and express avatar motion independently from specific implementation environment.

The purposes and scopes of the scripts are somewhat different. In the case of AML, it controls through parameter by calling low-level motion data from fixed motion library like MPEG4 from the script. In the case of CML, primitive motion is defined as tag and avatar motion is expressed as an independent script from task domain and avatar tool. STEP is designed to control avatar's body through logic language and CPSL supports avatar motions for cyber teaching environment. However, these script-ing languages represent motions by complicated parameters, common users experience difficulty in controlling avatar motion.

3 Avatar Behavior Hierarchical Representation and Control

In this paper, behavior expression and control language is defined to control hierarchical avatar behavior, and to gain high extensibility, convertibility, and reusability of avatar high-level motion. We suggest three layered architecture: task-level behavior, high-level motion, and primitive motion. The task-level behavior represents task oriented avatar behaviors used in various task domains so that end user can control the avatar through easy and simple task-level user interface. High-level motion provides abstract and parameterized actions independent from task domains and implementation.

Also, these script languages are translated into another level of motion. The changes in the script were to extract and analyze object information of avatar in virtual environment and achieve appropriate interaction in virtual environment.

Table 1. Characteristics of proposed scripting language.

Component	Characteristic
Task-Level Behavior	- required set of behaviors to perform tasks for specific domain such as shopping mall, and game as cyber class - domain dependent - implementation such as character motion engine or library independent
High-Level Motion	- parameterized representation of general avatar motions - abstract representation and control - domain independent - implementation such as character motion engine or library independent
Primitive Motion	- primitive motions supported by the specific character motion engine - physical representation and control - domain independent - implementation dependent

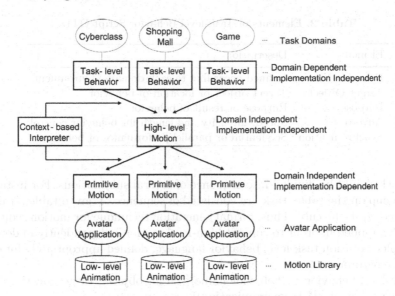

Fig. 1. Overall Procedure of the Proposed System.

Therefore avatar's task is properly performed regardless of dynamically changing physical system of virtual environment.

The overall organizational system is shown in fig.+1 and in the next paragraph. The role of each layer's language and characteristics will be explained.

3.1   Avatar Task-Level Behavior

Avatar's task-level behavior is to accomplish assigned task. For example, in domain such as cyber education, there are demands for explanation of the contents of the lecture, query and annotation of task-level behavior, and in case of shopping mall, presentation of product and product manufacturing behavior is needed. In this paper, the following design concept, avatar task-level behavior to be used in various task domains is defined.

Simplicity : Easy to write by human scripter.
Abstraction : Completely abstracted from low-level concepts.
Readability : Human-readable, also machine readable.
Extensibility : Usable and extensible for various domain task environments.
Parameterization : Behaviors controlled by various parameters.
Synchronization : Sequential and parallel control of behaviors.

According to this concept, task-level script language is defined in XML DTD. Since XML is independent and has superior extensibility, defined task-level behavior DTD can be applied to various domain environments, and when necessary extensibility of behavior is easily achieved. DTD elements were designed as shown in table 2.

Table 2. Elements of Task-level Behavior Script DTD.

Element	Description
Behavior Name	Defined according to each domain environment
Target Object	Target object or location of behavior
Purpose	Purpose or result of behavior
Adverb	Speed, intensity and length for behavior control
Synchronization	Sequential or parallel performance of behavior

A task should be divided into sequence of several sub motions. For instance, 'get the cup on the table' task are consisted of sequence of 'go to table', 'raise a hand' and 'grab the cub'. Thus, formal generation technique for motion sequence should be applied in order to convert high-level motion independently in domain and implementation task-level behavior language, defined appropriately for each domain environment.

Task-level behavior = ⟨behavior name⟩[⟨target object⟩|⟨target location⟩] [⟨purpose⟩] [⟨adverb⟩]* [⟨synchronization⟩]+

According to the following process of model, motion is created to translate the task behavior into a standardized high-level motion.

Formal Task Translation Model

- Identification of target
  - find location of target object (L_o)
- Locomotive motion (M_l)
  - identification of present avatar location (L_a)
  - spatial distance between L_a and L_o
  - generate L_m, if L_a L_o
- Manipulative motion (M_m)
  - generate M_m for L_o
- Verbal information (V_i)
  - generate verbal speech for avatar behavior if available
- Speed and intensity parameters
  - parameterize M_m and M_m for speed, intensity and duration

Depending on the proposed conversion model, task behavior is divided into and composed of locomotive and manipulative motion. These are transformed into lower stage, high-level motion expression.

3.2 High-Level Motion

Avatar high-level motion is not dependent on any specific domain or implementation environment, and generally has abstract expression motion. High-level motion ex-presses and controls avatar motion with its parameters like velocity, intensity, and direction.

In the proposed method, for the definition of avatar high-level motion, we analyzed the existing parameters and property in the existing high-level motion script to draw factors for avatar motion expression.

The high-level motion expresses avatar's name, role, and gender and contains at least one of the motions lists. The motion list is composed of motion elements such as motion name and parameters like space, time, intensity, and verbal elements.

First, spatial element is divided into destination property appointing target with hand and feet gestures and target property expressing avatar's direction. Time element is avatar's speed, continuance duration of motion, repetition of motion, and sequential and parallel motion. Intensity element is element that stresses or changes the intensity of the avatar motion property. Finally, verbal element is to express speech information for output of avatar's voice and sound effect.

These parameters are converted to lower hierarchy, primitive motion, which is de-pendent to implementation and express physical information of virtual world. High-level motion expresses avatar motion independently and abstractly from task domain or implementation environment. In the proposed script (fig. 2), the above parameters were expressed after the defined XML based high-level motion DTD.

3.3   Primitive Motion

Primitive motion expresses avatar motion, which is provided by avatar engine or motion library. As we mentioned before, high-level motion expresses avatar motion in general and abstract concept. On the other hand, in primitive motion, its format and expression changes depending on the specific avatar engine and motion library. Expression also becomes subordinate of implementation environment and forms a physical shape. For example, in high-level motion, motions like ⟨go to="table"⟩ is expressed ⟨walk to x="100" y="12" z="-2"⟩ in primitive motion layer.

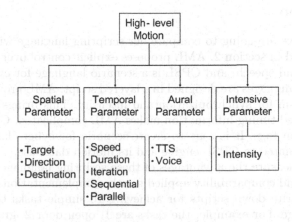

Fig. 2. High-level Motion Parameters.

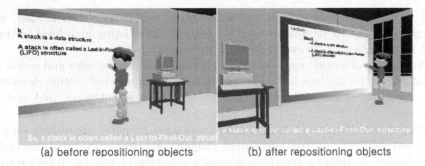

(a) before repositioning objects   (b) after repositioning objects

Fig. 3. The avatar is teaching an algorithm in cyber classroom.

4   Implementation Results

Our world represents an ordinary classroom with some components: a lecture, a blackboard, a computer, a table, a door, walls and several lecture-related objects. In task-level scripting language, author command the lecturer by combination of these components and behavior. After the example script is translated to a high-level script language by the suggested formal translator, the high-level script will be loaded to the system and converted to primitive motions and the avatar performs its tasks in fig. 3(a). As we mentioned before, the primitive motion contains the physical information of the system, the same task-level script can be properly applied even though physical structure of the virtual world is reorganized as shown in fig. 3(b).

In addition, in different implementation environment, for example 2D environment like web, task script can be applied. Fig. 4 is an example of task level script that was prepared using MS agent in web environment avatar control system. Because our architecture takes layered approach, application tools and script languages are explicitly decoupled. This makes easier to apply script language to various implementations.

5   Discussion

In this section, we are going to compare our scripting language with AML and CPSL, mentioned in section 2. AML proposes explicit control over synchronization of motion and speech, and CPSL is a scenario language for cyber teaching assistant. Both languages are designed in a layered script. AML provides detailed control mechanism for synchronization and intensity. This brings flexibility to AML, but detailed timing information makes AML complicated. CPSL has predefined animation tags. It is easy to create scenario for cyber class, in return, CPSL is subordinated to CPSL engine and its motion database.

In order to measure the advantages of these scripting languages on creating scenario script and comparability applied to other implementations. First, users were asked to write down scripts for achieving 10 simple tasks by using each scripting language. For example, the tasks are 1. open door 2. grab the cup 3. wave a hand. Second, the scripts user made are applied to avatar implementation.

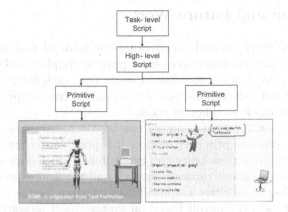

Fig. 4. The avatar performs his tasks in the 3D and 2D web environments.

Because of no standard for avatar engine, we selected MS Agent engine with basic motion library. Then, check the successful task achieved by the agent using each scripting languages.

First test uses scales from 1 (the least satisfaction) to 10 (the most satisfaction) to compare the difficulty of creating scenario script and spent time. Second test also uses the same scale to compare the number of correctly accomplished tasks. A group of 20 subjects consisted of 14 male and 6 female graduate students participated in the experiment. An ANOVA with repeated measures was used to analyze the result. Fig. 5 shows the results.

As shown in fig 4, the users rated that the results of the proposed system were easier and required less time to create the script. Significant effects were found between the spent time for the proposed script and the other script ($F_{(2,57)}$ = 8.98, $P < 0.05$). Most of the users rated that the scripting language by the proposed system were easier to create than those in the other system. In particular, higher ratings were given for the proposed system with AML script which support complicated synchronization control. Also, the proposed system got more number of achieved tasks. Note that CPSL has the low score because of its implementation dependency.

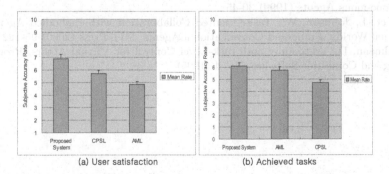

Fig. 5. Subjective evaluation of user satisfaction and achieved tasks.

6 Conclusion and Future Works

We suggested the three layered architecture consisted of task-level behavior, high-level motion and primitive motion to provide simple interface for avatar control at various application domains to users. Also each layer interacts independently. Task-level behavior is explicitly separated from implementations so that it could be reusable at different tools.

Using this approach, avatar behavior can be controlled more easily in task-level and in high level and primitive motion, it is possible to control and express avatar motion with great reusability and extensibility which does not depend on implementation environment through abstract and physical expression.

In future works, an intuitive graphical user interface for the input of avatar tasks, and avatar motion controls based on avatar-object interaction technique are required for providing more efficient interface to users.

References

1. Prendinger, H.: Life-like Characters. Life-like characters book, Springer-Verlag (2003) 3-17
2. Woo, S.: Virtual Human Trends. Journal of Korea Multimedia Society, Vol. 6. No. 4 (2000)
3. Kshirsagar, S., Thalmann, D., Kamyab, K.: Avatar Markup Language. Proceeding of the workshop on Virtual environments (2002) 169-177
4. Arafa, Y., Mamdani, E.: Scripting embodied agents behaviour with CML. Proceeding of Intelligent User Interfaces (2003) 313-315
5. Marriott, A., Stallo, J.: VHML- Uncertainties and Problems A discussion. Proceeding of Embodied conversational agents for AAMAS2002, Bologna, Italy (2002)
6. Yoshiaki, S., Matsuda, H.: Design and Implementation of Scenario Language for Cyber Teaching Assistant. International conference on Computers in Education (2001)
7. Hayashi, M.: TVML. ACM SIGGRAPH 98 Conference on applications (1998) 292-297
8. Huang, Z., Eliens, A., Visser, C.: Implementation of a scripting language for VRML/X3D-based embodied agents. Proceeding of web technology (2003) 91-100
9. Lester, C., Zettlemoyer, S., Gregoire, P., Bares, H.: Explanatory Lifelike Avatars. Autono-mous Agents (1999) 30-45
10. Ricket, J., Johnson, W.: Task-Oriented Collaboration with Embodied Agents in Virtual Worlds. Embodied Conversational Agents, MIT Press (2000) 95-122
11. Thalmann, D.: Autonomy and Task-Level Control for Virtual Actors. Programming and Computer Software, No. 4 (1995)

An Integrated Environment Blending Dynamic and Geometry Models

Minho Park and Paul Fishwick

Department of Computer and Information Science and Engineering,
Bldg CSE, Room 301, University of Florida, Gainesville, Florida 32611, USA
{mhpark,fishwick}@cise.ufl.edu

Abstract. Modeling techniques tend to be found in isolated communities: geometry models in CAD and Computer Graphics and dynamic models in Computer Simulation. When models are included within the same digital environment, the ways of connecting them together seamlessly and visually are not well known even though elements from each model have many commonalities. We attempt to address this deficiency by studying specific ways in which models can be interconnected within the same 3D space through effective ontology construction and human interaction techniques. Our work to date has resulted in an environment built using the open source 3D Blender package, where we include a Python-based interface, as well as scene, geometry, and dynamics models.

1 Introduction

A model is a simplified representation of the real entity in order to increase our understanding of that entity. Modeling is the process of making the model. In the modeling process, we can express the model with one of the model types, such as a geometry model, a dynamic model, or an information model. On the other hand, a multimodel [1] is a model that contains multiple heterogeneous models. Multimodeling is the process that creates a human interface environment for allowing a user to alternate and juxtapose the representations of the heterogeneous models. Ideally, we can explore and execute these models within a 3D scene to unify our diverse modeling efforts. To accomplish multimodeling, we started by posing four general questions: 1) "Is there any way we can include different model types within the same environment?"; If so, 2) "How can we connect them together seamlessly, visually and effectively?"; 3) "How can we overcome semantic heterogeneity between the models?"; and 4) "How can we simulate the models under the multimodel environment?" We found that the ability of creating customized 3D models, effective ontology construction, and Human-Computer Interaction (HCI) techniques should help us to blend and stitch together different model types.

We present a new modeling and simulation process called *integrative multimodeling* [1]. The purpose of integrative multimodeling is to provide a Human-Computer Interaction environment that allows components of different model types to be linked to one another – most notably dynamic models used in simulation to geometry models for the phenomena being modeled. To support integrative multimodeling, the OWL Web Ontology Language [2] is employed to bridge semantic gaps between the

T.G. Kim (Ed.): AIS 2004, LNAI 3397, pp. 574–584, 2005.

different models, facilitate mapping processes between the components of the different models, and construct a *Model Explorer*, which will be discussed in section 4. The Human-Computer Interaction techniques are needed to provide effective interaction methods inherent within the process and application of modeling. This process involves designing, manipulating, and executing the model, while leveraging model presentation and interface metaphors.

In this paper, we provide a methodology for integrative multimodeling using an example to be discussed in section 4. Related work is addressed in section 2. The technologies used for supporting integrative multimodeling are discussed in section 3, and section 5 concludes the paper by discussing future research. The work we describe in this paper is an extension to the preliminary research described in [3].

2   Related Work

In the area of modeling and simulation, *Dynamic* and *Static* models can be defined as one of the model classifications [4, 5, 6]. A dynamic model, which is compared with a static model, represents a system as model variables in the system evolve over time. We can represent the dynamic behaviors as one of the model types, such as Functional Block Model (FBM), Finite State Model (FSM), Equation Model, System Dynamics Model, or Queuing Model [4]. On the other hand, a variety of geometric model types are classified in a Graphics and CAD community. For example, Boundary Representations, Constructive Solid Geometry (CSG), and wireframe models are used to represent geometric models [7, 8].

Many techniques are being used for visualization, interaction, and navigation in the Virtual Environment (VE). Campbell and his colleagues [9] have developed a virtual Geographical Information System (GIS) using GeoVRML and Java 3D software development packages. They employ a menu bar and toolbars for ease of use because most users immediately understand how to use the menu bar and toolbars. Kirner and Martins [10] developed the infoVIS system, which combined the Information Visualization and the Virtual Reality techniques to generate graphical representations of the information, and Cubaud and Topol [11] present a VRML-based user interface for a virtual library, which applied 2D Widows-based interface concepts to a 3D world. They allow users to move, minimize, maximize, or close windows by dragging and dropping them or by pushing a button, which is usually provided in a traditional Windows system environment. Lin and Loftin [12] provide a functional Virtual Reality application for geoscience visualization. They employ Virtual Button and bounding box concepts to interact with geoscience data. If interaction is needed, all the control buttons on the frame can be visible; otherwise, they are set to be invisible so that the frame simply acts as the reference outside the bounding box. Hendricks, Marsden, and Blake [13] present a VR authoring system. The system provides three main modules, graphics, scripting, and events modules, for supporting interactions. The important point is that they support a scripting-based interaction method for non-computer expert users.

The Semantic Web [14] technologies, such as ontology languages involving RDF [15], RDF-S [16], and OWL, are employed in a variety of communities to share or exchange the information, as well as dealing with semantic gaps between different domains. In an information systems community, including a database systems community, ontologies are used to achieve semantic interoperability in heterogeneous information systems by overcoming structural heterogeneity and semantic heterogeneity between the information systems [17, 18, 19, 20]. In a simulation and modeling community, Miller, Sheth, and Fishwick propose the Discrete-event Modeling Ontology (DeMO) [21], to facilitate modeling and simulation [22]. To represent core concepts in the discrete-event modeling domain, they define four main abstract classes in the ontology: *Demodel*, *ModelConcepts*, *ModelComponents* and *ModelMechanisms*. Liang and Paredis [23] define an OWL-based *port* ontology to capture both syntactic and semantic information for allowing modelers to reason about the system configuration and corresponding simulation models.

3 Infrastructure

An ontology describes meaning of terms and their interrelationships used in a particular domain. Ontologies consist of three general elements: classes, properties, and the relationships between classes and properties. OWL can be used to represent ontologies so that the information contained in documents can be processed by applications. OWL has more power to express meaning and semantics than RDF, which is a general-purpose language for representing information on the Web. RDF-S describes how to use RDF to describe RDF vocabularies, since OWL is built on RDF and extends a vocabulary of RDF-S.

Blender [24] is a free and fully functional 3D modeling/rendering/animation/game package for Linux/Unix/Windows systems. OpenGL-based Blender is fast becoming the standard 3D design application, especially within the Linux community. Blender has developed into a suite of technologies enabling the creation and replay of real-time, interactive 3D content. The software offers a versatile animation system, contemporary modeling principles, advanced rendering tools, and an editor for postproduction.

Virtual Reality Modeling Language (VRML) represents the standard 3D language for the Web [25]. To be precise, VRML is not a modeling language but an effective 3D file interchange format, which is a 3D analog to HTML. The 3D objects and worlds are described in VRML files using a hierarchical scene graph, which is composed of entities called nodes. Nodes can contain other nodes, and this scene graph structure makes it easy to create complex, hierarchical systems from subparts. VRML also defines an event or message-passing mechanism by which nodes in the scene graph can communicate with each other and Script nodes can be inserted between event generators and event receivers.

The next-generation X3D (eXtensible 3D) Graphics specification was designed and implemented by the X3D Task Group of the Web3D Consortium [26], who expressed the geometry and behavior capabilities of the VRML 97 in XML (eXtensible Markup Language). We can think of X3D as an XML version of VRML with several added functionalities.

RUBE [27, 28], which is developed within the Graphics, Modeling and Arts (GMA) laboratory at the University of Florida, is an XML-based Modeling and Simulation framework and application, which permits users to specify and simulate a dynamic model, with an ability to customize a model presentation using 2D or 3D visualizations. Two XML-based languages, MXL (Multimodel eXchange Language) and DXL (Dynamic eXchange Language) [27], are designed to capture the semantic content for the dynamic model. The MXL file describes the behavior of the model to represent the model file. That file describes a heterogeneous multimodel in an abstract level, such as Functional Block Model (FBM) and Finite State Machine (FSM). The DXL is a lower-level modeling language, which can be described with homogeneous simple block diagrams. It is translated into an executable programming code for the model simulation.

4 Integrative Multimodeling

We use Blender as a 3D authoring tool for supporting integrative multimodeling and the RUBE framework, and we use the Python scripting language for implementing an interface, which can allow users to build their customized or personalized dynamic models such as FBM, FSM, and Queuing Model in Blender. In addition, the interface, which is called *Blender Interface*, contains other functionalities: *object import/export* [29], *VRML exporter* [30], and *ontology instance creation*. The object import/export capability is provided for users to create their customized or personalized objects. The VRML exporter, which converts Blender into VRML objects, can be used for VRML users. For supporting model reusability and constructing model database, which will be shown in the next stage of *Blender Interface*, the ontology instance creation functionality is included. This will help us to build dynamic and geometry models conveniently. Because all instances in an ontology can have the location information of geometry and dynamic objects as properties, we can make the two models directly by just clicking a proper instance button provided by Model Explorer, which is similar to a Windows Explorer concept. We are now in a position to discuss integrative multimodeling and the RUBE framework using the example of the combat scene, as shown in Figure 7. Recall that the purpose of integrative multimodeling is to provide a Human-Computer Interaction environment that allows components of different model types to be linked to one another.

To do integrative multimodeling with simulation for the scene in 3D space, we need the following components:

- Ontology describing the combat scene explicitly
- Geometry model for the scene
- Dynamic model for the scene
- Interaction model for Human-Computer Interactions
- RUBE framework for XML-based modeling and simulation

We will explain each part with the example in the following subsections.

4.1 A Military Modeling Example

We introduce the combat scene as an example of how to achieve interaction multimodeling and use the RUBE framework. In this particular example, three geometry objects represent one Unmanned Aerial Vehicle (UAV), one F15, and one Joint Surveillance Target Attack Radar System (JSTARS), respectively. The objects are flying around and communicating with each other. To avoid simulation complexity, we simplify a simulation scenario, which means the UAV passes target information to the JSTARS, and then the JSTARS commands the F15 to attack the target. We can represent the scenario as a Functional Block Model (FBM). Figure 1 shows the 2D diagram representation for the FBM which can be modeled as a 3D dynamic model in *Blender Interface*. The UAV block randomly generates boolean value (0/1) and sends it to the JSTARS block. If the JSTARS block receives "True," it conveys the value to the F15 block. The F15 block counts the "True" value as the number of received attack messages.

Fig. 1. 2D Dynamic FBM representation of the combat scene

4.2 Ontology: A Framework for Encoding Modeling Knowledge

An ontology for the combat scene is necessary to define the scene domain and formalize and justify connections between geometry and dynamic model components for integrative multimodeling. To create the ontology, we employ the Protégé ontology editor [31] and visualize the ontology using ezOWL [32]. Figure 2 shows the graphical representation of the ontology for the combat scene created by ezOWL. In Figure 2, two properties for each object, *has_geometry* and *has_dynamic*, can be found. We can interpret it as all objects can have geometry and dynamic model components and be interchangeable between the components in the scene domain. In addition, we define two dynamic models, *MXLforUAV_Activity* for the combat scene and *MXLforIM* (Interaction Model) for user interactions, which can be taken out of the dynamic properties of the ontology and wrapped up into the models. MXL describes dynamic model behaviors using XML syntax to represent the dynamic model. The details will be discussed in subsection 4.4. Any instance, which will be generated in the modeling process within the *Blender Interface* environment, is inserted into the ontology so that we can implement Model Explorer and Model Database.

4.3 The Integrative 3D Modeling Process

Geometry and Dynamic Models. FOr the example, we can create or append geometry objects with Blender and generate dynamic objects with *Blender Interface*. To avoid model complexity, we use different layers of the scene for each model. In the first layer, we place geometry objects. The dynamic model for the scene is generated in the second layer using *Blender Interface*.

To use *Blender Interface*, RUBE must be installed. RUBE has four folders: *predefined primitive*, *predefined theme*, *user-defined theme*, and *rube_utility* folders. As the names imply, the predefined primitive folder has primitive blender objects, such as cube and sphere, with the corresponding MXL file for each model type, such as FSM, FBM, or Queuing Model. The predefined theme folder contains customized and personalized blender objects like architecture. If users want their own model

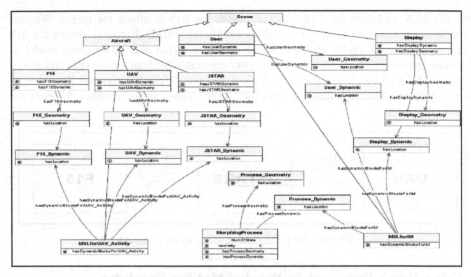

Fig. 2. Scene Ontology

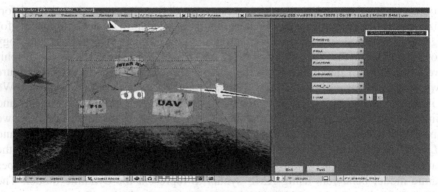

Fig. 3. A snapshot of *Blender Interface*

representations, they can create an object and store it into a proper model type folder under the user-defined theme folder using export function provide by *Blender Interface*. After installing RUBE, the user loads "Blender_Interface.py" from rube_utility folder on the machine and runs it. Figure 3 shows the snapshot of *Blender Interface*. The user then chooses one of the model presentation types. In this example, we select "primitive" because we represent a dynamic model as a primitive type such as cube. Then FBM is selected as a model type. Because FBM consists of two model elements, such as Function and Transition, we have to choose one of the model elements and then select any proper function or transition types. To represent the first block in Figure 1, the Boolean random generator is selected as Function. If we press the "import" button, we can see the corresponding object in the 3D window. To represent the first arrow between the first and second blocks in Figure 1, we choose Transition type and select "Arrow," then we press the "import" button. As with the first block, the arrow object appears in the scene. All blocks and transitions in the 2D diagram can be generated with the same manner as previously explained. *Blender Interface* utilizes a

Windows Explorer concept to provide a relatively easy way for modeling to users since most users are familiar with Windows Explorer systems.

Interaction Model. *Blender Interface* provides two simulation environments, VRML-based and Blender-based. To achieve integrative multimodeling within the different environments, we need different approaches for making an interaction model. An interaction model is simply a dynamic model for users who are driving the system. We can make an interaction model as dynamic model. Figure 4 shows a 2D representation of an interaction model for our example. If we construe the interaction model (IM) verbally, the following interpretation is possible:

- The user wants to see and know the scene by touching a sensor
- The scene provides two kinds of model types, geometry and dynamic models
- The two models are interchangeable
- The user can choose a model type
- The conversion process is achieved through morphing

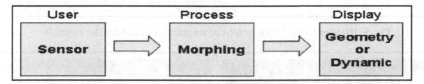

Fig. 4. 2D Dynamic Model (FBM) representation of interaction model

In VRML, this interaction model is really the routing graph and a kind of dynamic model. We can therefore build the interaction model like the scene dynamic model in the third layer of Blender using *Blender Interface* since RUBE provides sensor and transformation functions along with arithmetic functions in libraries. In Blender, the interaction model can be built by the Blender Game Engine without making the model explicitly. In the Blender Game Engine, there are Sensors (i.e., VRML/X3D sensors), Controllers, and Actuators. These are used to create "Logic Brick" graphs, which drive the interaction. We can consider the Logic Brick graphs as dynamic models as well. Sensors are FBM blocks with no inputs (they drive or generate the data), Controllers are FBM blocks with inputs and outputs, and Actuators are FBM blocks with no outputs. The Game Engine can cover most interaction behaviors since it has well-defined built-in Sensors and Actuators. If a user, however, wants more complicated interactions, the user can handle the interactions by Python scripting language. After creating a Python file, the user puts it in one of the Controllers and connects the Controller with any proper Actuator. Figure 5 represents the Blender Logic Brick graphs of the interaction model for the example.

RUBE Framework for XML-Based Modeling and Simulation. After finishing the 3D modeling processes, we can generate a simulation code for simulating the target system from RUBE and given dynamic models. The simulation code must be JavaScript or Python because *Blender Interface* provides two different simulation environments. A user can choose his simulation environment by pressing *RUBE for VRML*, or *RUBE for Blender* buttons in *Blender Interface*. In the VRML case, the process for 3D scene dynamic model includes five steps:

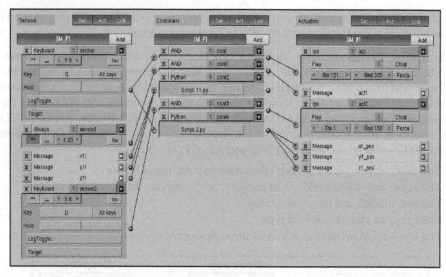

Fig. 5. Blender Logic Brick graphs of interaction model for the example

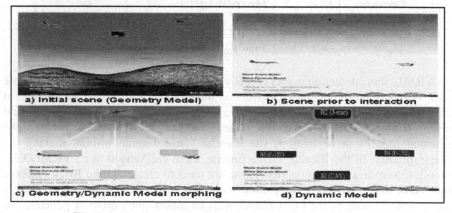

Fig. 6. Integrative multimodeling under the VRML environment

1) *BlenderToMXL*: First, we create an MXL file for a given scene dynamic model by gathering all segmented MXL files for the scene dynamic objects from the corresponding object libraries in RUBE since the RUBE library system contains a Blender object file and the corresponding MXL file in pairs.
2) *MXLToDXL*: Using XSLT, the MXL file is converted to a low-level MXL language called DXL.
3) *DXLToJavaScript*: JavaScript code for simulation is generated from the DXL by using DOM.
4) *BlenderToVRML*: Blender is transformed into VRML scene.
5) *VRMLToX3D*: Using NIST Translator and XSLT, an X3D file is created from the JavaScript code and the VRML file.

On the other hand, the process for an interaction model performs the first three steps of the previous five steps and then the JavaScript code is merged with the final

X3D file. Currently RUBE supports only the process for the 3D scene dynamic model. We are still working on the process for interaction model. The VRML-based integrative multimodeling environment of our example is created manually, as shown in Figure 6. Two button objects, show dynamic model and show geometry model, with touch sensors are created for user interactions. We change the transparencies of objects to obtain the desired effect, morphing. In the Blender case, we need three steps to generate the Python code for the scene simulation, BlenderToMXL, MXLToDXL, and DXLToPythonCode. All steps are almost identical with the proc-ess for the VRML case except the last one, DXLToPythonCode, which produces a Python code instead of a JavaScript code. In the example, the generated Python code is placed within the Controller of a dummy object so that the Actuator of the object can distribute current simulation values to each Sensor of the dynamic blocks for simulation. For interaction model of the example in the Blender environment, we create two Keyboard sensors, "D" and "G" keys, two Python files for tracking object positions as Controllers, and two IPO Actuators for the morphing effect between a geometry model and a dynamic model as shown in Figure 5. Consider Figure 7, which represents a conversion process with simulation from a geometry model to a dynamic model applied to one F15, one JSTARS, and one UAV. The number in the figure implies the result of the F15, that is, the number of received attack messages from the JSTARS.

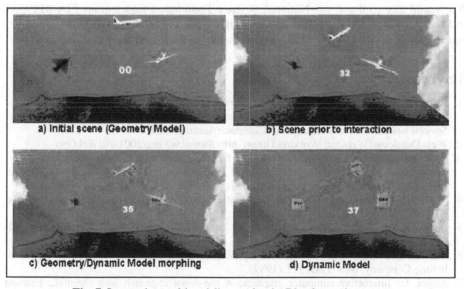

a) Initial scene (Geometry Model) b) Scene prior to interaction

c) Geometry/Dynamic Model morphing d) Dynamic Model

Fig. 7. Integrative multimodeling under the Blender environment

5   Conclusions and Future Research

We have presented a strategy for integrative multimodeling using the example of the combat scene, and we have discussed and explained our modeling and simulation approaches for integrative multimodeling under the VRML and Blender environ-ments. In addition, we have shown that to perform integrative multimodeling, we need a scene ontology, geometry model, dynamic model, interaction model, and RUBE framework. We have learned that effective ontology construction and Human-Computer Interaction techniques are essential factors to support integrative multi-modeling. HCI techniques play an especially important role in integrative multimodeling to connect different model types together visually and seamlessly within the same digital environment. One of the distinct points in integrative multimodeling is that we applied the Blender Game Engine concepts to the interaction model. In the Blender Game Engine, as we mentioned in section 4, Logic Brick graphs consist of Sensors, Controllers, and Actuators. We consider these elements as FBM blocks: Sensors are FBM blocks with no inputs, Controllers are FBM blocks with inputs and outputs, and Actuators are FBM blocks with no outputs. Thus we adopt Game Engine schemes for implementing an interaction model.

The ontology and Blender Interface, which are used for the example, are intended to be a starting point for integrative multimodeling. For the next stage of integrative multimodeling, we will add an advanced interface technology, the Model Explorer, to our current approach, as well as considering Human-Computer Interaction requirements, such as usability, emotion, immersion, and customization [1].

Acknowledgements

We would like to thank the National Science Foundation under grant EIA-0119532 and the Air Force Research Laboratory under grant F30602-01-1-05920119532. We thank Jinho Lee and Hyunju Shim for their technical help and fruitful discussions.

References

1. Fishwick, P.: Toward an Integrative Multimodeling Interface: A Human-Computer Interface Approach to Interrelating Model Structures. Accepted for SCS Trans. on Simulation, Special Issue in Grand Challenges in Computer Simulation (2004)
2. Web-Ontology (WebOnt) Working Group. http://www.w3.org/2001/sw/WebOnt/
3. Park, M. and Fishwick, P.: A methodology for integrative multimodeling: Connecting dynamic and geometry models, Proceedings of Enabling Technology for Simulation Science, Part of SPIE Aerosense '04 Conference (2004)
4. Fishwick, P.: Simulation Model Design and Execution. Prentice-Hall, Englewood Cliffs, NJ (1995)
5. Law, A. and Kelton. W.: SIMULATION MODELING & ANALYSIS. McGraw-Hill (1991)
6. Hoover, S. and Perry. R.: Simulation: a problem-solving approach. Addison-Wesley Publishing Company, New York, NY (1989)
7. Angel. E.: Interactive computer graphics: A top-down approach with OpenGL. Pearson Education (2003)
8. Chasen, S.: Geometric Principles and Procedures for Computer Graphic Applications. Prentice-Hall, Englewood Cliffs, NJ (1978)
9. Campbell, B., Collins, P., Hadaway, H., Hedley, N. and Stoermer, M.: Web3D in ocean science learning environments: virtual big beef creek. Proceedings of the seventh International Conference on 3D Web Technology (2002) 85-91
10. Kirner, G. and Martins, V.: Development of an Information Visualization Tool Using Virtual Reality. Proceedings of the 2000 ACM symposium on Applied computing (2000) 604-606

11. Cubaud, P. and Topol, A.: A VRML-based user interface for an online digitalized antiquarian collection. Proceedings of the sixth International Conference on 3D Web Technology (2001) 51-59
12. Lin, C., and Loftin, R.: Application of virtual reality in the interpretation of geoscience data. Proceedings of the ACM Symposium on Virtual Reality Software and Technology (1998) 187-194
13. Hendricks, Z., Marsden, G and Blake, E.: A meta-authoring tool for specifying interactions in virtual reality environments. Proceedings of the 2nd International Conference on Computer Graphics, Virtual Reality, Visualisation and Interaction in Africa (2003) 171-180
14. Berners-Lee, T., Hendler, J. and Lassila, O.:
    http://www.sciam.com/article.cfm?articleID=00048144-10D2-1C70-84A9809EC588EF21
15. Resource Description Framework (RDF). http://www.w3.org/RDF/
16. RDF Vocabulary Description Language 1.0: RDF Schema.
    http://www.w3.org/TR/rdf- schema/
17. Bouquet, P., Dona, A., Serafini, L. and S. Zanobini: ConTeXtualized Local Ontologies Specification via CTXML. Proceedings of the AAAI Workshop on Meaning Negotiation (2002) 64-71
18. Kashyap, V. and Sheth, A.: Semantic and schematic similarities between database objects: A context-based approach. The VLDB Journal: The International Journal on Very Large Data Bases (1996) 276-304
19. Maedche, A., Motik, B., Silva, N. and Volz, R.: MAFRA - A MApping FRAmework for Distributed Ontologies. Proceedings of the 13th International Conference on Knowledge Engineering and Knowledge Management. Ontologies and the Semantic Web (2002) 235-250
20. Wache, H., Voegele, T., Visser, U., Stuckenschmidt, H., Schuster, G., Neumann, H. and Huebner, S.: Ontology-Based Integration of Information: A Survey of Existing approaches. Proceedings of the IJCAI-01 Workshop on Ontologies and Information Sharing (2001) 108-118
21. Miller, J., Baramidze, G., Fishwick, P. and Sheth, A.: Investigating Ontologies for Simulation Modeling ," Proceedings of the 37th Annual Simulation Symposium (2004)
22. Fishwick, P. and Miller, J.: ONTOLOGIES FOR MODELING AND SIMULATION: ISSUES AND APPROACHES. Winter Simulation Conference (2004)
23. Liang, V. and Paredis. C.: A Port Ontology for Automated Model Composition. Winter Simulation Conference (2003) 613-622
24. Blender. http://www.blender.org/
25. Carey, R. and Bell, G.: The annotated VRML 2.0 Reference Manual, Addison-Wesley, New York, NY (1997)
26. X3D Graphics Working Group. http://www.web3d.org/x3d.html
27. Kim, T., Lee, J. and Fishwick, P.: A two-stage modeling and simulation process for web-based modeling and simulation. ACM Transactions on Modeling and Computer Simulation (2002) 230-248
28. Kim, T., and Fishwick, P.: A 3D XML-based customized framework for dynamic models. Proceeding of the seventh International Conference on 3D Web Technology (2002) 103-109
29. Lynch, C.: obj_io_modif228. http://jmsoler.free.fr/util/blenderfile/py/
30. Kimball, R.: wrl2export. http://kimballsoftware.com/
31. Protégé ontology editor. http://protege.stanford.edu/
32. ezOWL. http://iweb.etri.re.kr/ezowl/index.html

Linux-Based System Modelling
for Cyber-attack Simulation

Jang-Se Lee[1], Jung-Rae Jung[2], Jong-Sou Park[3], and Sung-Do Chi[3]

[1] The Division of Information Technology Engineering,
Korea Maritime University, Korea
jslee@bada.hhu.ac.kr

[2] Cooperation of appeal telecom, Korea
dayfly77@freechal.com

[3] The Department of Computer Engineering, Hangkong University, Korea
{jspark,sdchi}@mail.hangkong.ac.kr

Abstract. The major objective of this paper is to describe modeling on the linux-based system for simulation of cyber attacks. To do this, we have analyzed the Linux system from a security viewpoint and proposed the Linux-based system model using the DEVS modeling and simulation environment. Unlike conventional researches, we are able to i) reproduce the detail behavior of cyber-attack, ii) analyze concrete changes of system resource according to a cyber-attack and iii) expect that this would be a cornerstone for more practical application researches (generating cyber-attack scenarios, analyzing vulnerability and examining countermeasures, etc.) of security simulation. Several simulation tests performed on sample network system will illustrate our techniques.

1 Introduction

The growth of information technology and easy access to computers has enabled hackers and would-be terrorists to attack information systems and critical infrastructures in the world [1]. A computer and network system must be protected to assure security goals such as availability, confidentiality and integrity. Namely, the complete understanding of system operation and attack mechanisms is the foundation of designing and integrating information protection activities [2]. We need to establish the advanced simulation methodology for analyzing the vulnerability, survivability, etc. of a given infrastructure as well as the expected consequences of successful attacks and the effect of the defense policy [3].

Cohen [3], who was a pioneer in the field of network security modeling and simulation, interestingly suggested a simple network security model composed of network model represented by node and link, cause-effect model, characteristic functions, and pseudo-random number generator. However, cyber attack and defense representation based on cause-effect model is so simple that practical difficulty in application comes about. Amoroso suggested that the intrusion model [4] should be represented by sequence of actions, however, the computer simulation approach was not considered clearly. Wadlow [5] suggested an intrusion model with four classified states such as COOL, WARM, HOT, and

T.G. Kim (Ed.): AIS 2004, LNAI 3397, pp. 585–596, 2005.

COOLDOWN, but it failed to go beyond the conceptual modeling level. Nong Ye [2] noticeably proposed a layer-based approach to a complex security system, but failed to provide a practical modeling and simulation technique of the relevant layers.

To deal with this, we have analyzed Linux-based system and performed Linux system modeling using the DEVS (Discrete Event System Specification) which is the hierarchical and modular modeling and simulation environment [6,7,8]. Unlike conventional research for security simulation that has still been at the conceptual level or focused on statistical method, by adopting a discrete event modeling and simulation methodology, it is possible to i) reproduce the detailed behaviors of cyber-attack, ii) analyze concrete changes of system resource according to cyber-attack and iii) expect that this would be a cornerstone for more practical application researches (generating cyber-attack scenario, analyzing vulnerability and examining countermeasures, etc.) of security simulation.

2   Brief Descriptions on DEVS Modelling and Simulation Environment

The DEVS formalism is a theoretically well-grounded means of expressing modular discrete event simulation models developed by Zeigler [6,7]. A DEVS is a structure:

$$M =< X, S, Y, \delta_{int}, \delta_{ext}, \lambda, ta >$$

X means the set of events that occur outside the system. Y means the set of output variables. S means the cross product of definition areas of state variables and s ($s \in S$) means the sequential snap shot of system according to time progress. $ta(s)$ is defined as the time allowed to be at the state s unless system doesn't get external events. δ_{int} is defined as the function that explains the change of the state of model according to time progress when there are no external events. δ_{ext} is defined as the function that represents the change of the state of model by the events occurred in the outside of the system. λ is defined as the output of the system in the state s. The DEVS environment supports building models in a hierarchical and modular manner, in which the term "modular" means the description of a model in such a way that is has recognized input and output ports through which all interaction with the external world is mediated [6,7,8]. This property enables hierarchical construction of models so that the complex network security models can be easily developed.

3   Security Modelling for Linux-Based System

In the ever-changing world of global data communications, inexpensive Internet connections, and fast-paced software development, security is constantly being compromised. Linux system being similar to Unix in mechanism is open-source. That means that the source code of the operating system is available – anyone can view the source code and examine it, modify it, and suggest and make changes to it. So, many hacks, exploits and network security tools are written on Linux because it's readily available [9]. Therefore, in the modeling on information systems based on various operating systems, the modeling of Linux-based systems can be effectively applied to analyzing hacking mechanism through cyber-attack simulation, analyzing vulnerability and studying countermeasures.

3.1 Linux-Based System Model Design

We have tried modeling through the abstraction and detailing of components and functions of Linux system from the security viewpoint on the basis of DEVS framework. Also, we have performed modeling of command level to analyze the detailed behavior of the system against cyber-attack.

Fig. 1 shows the security model structure for Linux. As shown in Fig. 1, a Linux system model that can become coupled with Attacker Model is divided largely into Kernel Model that represents Operating System and Process Models that represent application services. Kernel Model again consists of Network Manager Model, File System Manager Model and Process Manager Model. Process Models are subdivided according to the provided functions into various service models including the FTP Model, the Telnet and the Http Model, etc. Like this, hierarchical modeling of Linux system has an advantage of low complexity for modeling and has a merit that can consider concretely security factors in every model.

The explanation of general flow of the proposed Linux-based System is as follows. First, if Linux-based System Model receives the packet from Attacker

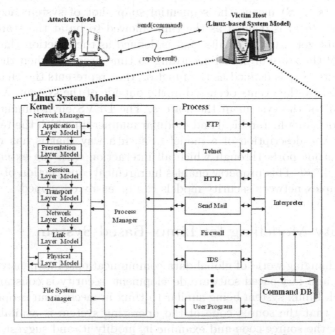

Fig. 1. Security Model Structure for Linux-based System.

Model, it transmits the packet to the Network Manager Model in Kernel Model. Network Manager Model performs network connection. If network connection is normally done in Network Manager Model, the packet is transmitted to Process Manager Model. Process Manager Model confirms that the packet is from an already-connected user, and if not, it creates a new process and saves that information in Process Map. Like this, after the allocation of process is completed, Process Manager Model confirms an appropriate service and transmits the packet to an appropriate Process Model. Then Process Model confirms whether the command included in it is right, transmits the packet to Interpreter Model, and receives pre/post condition according to commands from Interpreter Model, which transmits the result of query of Command DB to Process Model. If pre/post condition received from Interpreter Model needs to access a file system, Process Model transmits that to Process Manager Model. And Process Manager Model transmits the packet to File System Manager Model. File System Manager Model performs various functions related to file including searching requested file, destroying file, etc. Finally, processed result is transmitted to outside in opposite order.

3.2   Network Manager Model

Network Manager Model consists of models coming under OSI 7 layer. This performs functions like creating packets and transmitting them outside or inside, and processing them. Models composing Network Manager Model are as follows.

3.2.1 Link Layer Model

Link Layer Model examines physical address (MAC Address). If that is normal, this model transmits the packet to Network Layer Model that is the upper layer. The DEVS representation of Link Layer Model is as follows.

```
State Variables
  'passive, 'active, 'b-active, MAC Address
External Transition Function
  If packet comes from physical layer model and state is 'passive
  Then check packet.MAC address and
       set state as 'active during processing-time
  Else if packet comes from network layer model and state is 'passive
  Then set state as 'b-active during processing-time
Internal Transition Function
  If state is 'active, Then change state to 'passive
  Else if state is 'b-active, Then change state to 'passive
Output Function
  If state 'active, Then send packet to network layer model,
  Else if state is 'b-active, Then send packet to physical layer model
```

Let's explain above pseudo code according to DEVS formalism. 'passive, 'active, and 'b-active of state variable come under S that shows the state set of models. Packets inputted from the physical layer model and network layer model come under input event set X. The external transition function that represents δ_{ext} examines if the MAC address of input packet is right and maintains the state of 'active during processing-time defined as $ta(s)$ when state is 'passive and packets come from physical layer model. And, if external input comes from the network layer model, it maintains the state of 'b-active. The internal transition function that represents δ_{int} changes the state into 'passive when the state is 'active or 'b-active. The output function, λ transmits packets to the network layer model when the state is 'active and it transmits them to the physical layer model when the state is 'b-active. Here, packets that are transmitted to outside through output function come under output event set Y.

3.2.2 Network Layer Model
This analyzes the content of packet transmitted from Link layer Model with IP lists of permitted users, examines the source IP of the packet, and decides whether or not the user is a permitted. If not, an error message is sent. The DEVS representation of Network Layer Model is as follows.

```
State Variables
    'passive, 'active, 'b-active, IP Address List
External Transition Function
    If packet comes from link layer model and state is 'passive,
    Then check packet.sourceIP with IP Address List and
        set state as 'active during processing-time
    Else if packet come from transport layer model and state is 'passive
    Then set state as 'b-active during processing-time
Internal Transition Function
    If state is 'active, Then change state to 'passive
    Else if state is 'b-active, Then change state to 'passive
output function
    If state is 'active, Then send packet to transport layer model
    Else if state is 'b-active, Then send packet to link layer Model
    Else send error packet to link layer model
```

3.2.3 Transport Layer Model
This confirms the requested service with port number. If the requested service is offered, this transmits packet to Application Layer Model. If not, error message is sent. The DEVS representation of Transport Layer Model is as follows.

```
State Variables
    'passive, 'active, 'b-active, Open Port List
External Transition Function
    If packet comes from network layer model and state is 'passive,
    Then check packet.destPort with Open Port List and
        set state as 'active during 'processing time
    Else if packet comes from session layer model and state is 'passive
    Then set state as 'b-active during 'processing-time
Internal Transition Function
    If state is 'active, Then change state to 'passive
    Else if state is 'b-active, Then change state to 'passive
Output Function
    If state is 'active, Then send packet to session layer model
    Else if state is 'b-active, Then send packet to network layer model
    Else send error packet to network layer model
```

3.2.4 Physical Layer Model and the Remaining Higher Layer Models

These are abstraction models of basic function and play a middleman role between the high layer and the low layer. The DEVS representation of these models is as follows.

```
State Variables
  'passive, 'active, 'b-active
External Transition Function
  If packet comes from lower layer model and state is 'passive,
  Then set state as 'active during processing-time
  Else if packet comes from higher layer model and state is 'passive
  Then set state as 'b-active during processing-time
Internal Transition Function
  If state is 'active, Then change state to 'passive
  Else if state is 'b-active, Then change state to 'passive
Output Function
  If state is 'active ,Then send packet to higher layer model
  Else if state is 'b-active, Then send packet to lower layer model
```

3.3   File System Manager Model

File System Manager Model has the information about important files and directory related to security in Linux system. And it decides the process of requests like as read, write and delete according to file permission and sends the result. Each file has file name, file size, file type, file permission, file ownership, group ID, saved location, and current state which is working or idle, etc. and it is saved on Inode-Table. The file requested from external is processed according to various environment variables and file information saved on Inode-Table. The DEVS representation of File system Manager Model is as follows.

```
State Variables
  'passive, 'active, Inode-Table, Env-attr
External Transition Function
  If packet comes from process manager model and state is 'passive,
  Then modify Inode-Table according Env-attr and
    set state as 'active during 'processing-time
Internal Transition Function
  If state is 'active, Then change state to 'passive
Output Function
  If state is 'active, Then send packet to process manager model
```

3.4   Process Manager Model

Process Manager Model saves, manages and schedules information about process. Process Manager Model has individual process information including process ID(PID), current user ID, provided service name and process state in process map so that it connects with an appropriate process properly whenever requests enter. The DEVS representation of Process Manager Model is as follows.

```
State Variables
  'passive, 's-active, 'r-active, 'f-active, 'waiting, process map
External Transition Function
  Update process map
  If packet comes from network manager model,
  Then set state as 's-active during processing-time
  Else If packet comes from process models,
          If file access is needed,
          Then set state as 'waiting during processing-time
          Else set state as 'r-active during 'processing-time
  Else If packet comes file system manager model,
  Then set state as 'f-active during processing-time
Internal Transition Function
  If state is 's-active or 'r-active or 'f-active or 'waiting,
  Then change state to 'passive
Output Function
  If state is 's-active, Then send packet to process model
  Else If state is 'r-active or 'f-active,
  Then send packet to network manager model
  Else if state is 'waiting,
  Then send packet to file system manager model
```

3.5   Process Models

Process Models consist of various service models performed according to user's service requests. For example, Process Models can consist of Telnet Model that offers remote connection service, FTP Model that offers file transmission service between two ends, Http Model that offers web service, Send-Mail Model that offers electronic mail service, Router Model that passes packet to other network or objective component model, Firewall Model that blocks illegal packets, and IDS Model that detects illegal intrusion from outside, etc. Among Process Models, the DEVS representation of Telnet Model is as follows.

```
State Variables
  'passive, 'waiting, 'parsing, 'busy
External Transition Function
  If packet comes from kernel and state is 'passive,
  Then parse command in the packet and
       set state as 'parsing during processing-time
  Else if packet comes from interpreter model and state is 'waiting,
  Then process pre/post-condition and
       set state as 'busy during processing-time
Internal Transition Function
  If state is 'parsing, Then change state to 'waiting
  Else if state is 'busy, Then change state to 'passive
Output Function
  If state is 'parsing, Then send packet to interpreter model
  Else if state is 'busy, Then send packet to kernel model
```

3.6   Interpreter Model

Interpreter Model analyzes the command being in the input packet and obtains pre-condition and post-condition through Command DB. Command DB can be

Table 1. Pre/post-condition representation of Linux commands (partially-shown).

Command	Pre-condition (current states)	Output	Post-condition (next states)
rmdir	Check the emptiness of the directory	Remove directory entries	Change directory attributes
cd	Check the file existence	Change working directory	Change directory attributes
chmod	Check the file existence	Change the permission mode	Change permission attributes

created by command-level modeling, which is accomplished by grouping and characterizing of commands that are used in various services. Table 1 shows an example of command-level modeling using pre/post-condition representation in Linux. Here pre-condition represents the condition for executing the command, output represents the results by command execution, and post-condition represents the changed properties after command execution [10].

The DEVS representation of Interpreter Model is as follows.

```
State Variables
  'passive, 'busy
External Transition Function
  If packet comes from process model and state is 'passive,
  Then get pre/post-condition from command-DB and
      set state as 'busy during processing-time
Internal Transition Function
  If state is 'busy, Then change state to 'passive
Output Function
  If state is 'busy, Then send pre/post-condition to process-model
```

3.7   Attacker Model

Attacker Model performs attacks by changing resources of the target system according to various cyber-attack scenarios. Cyber-attack scenario is expressed by command set, which is the sequential list of continuous commands using appropriate vulnerabilities. Command set of attack scenario is shown in Table 2 [11].

Table 2. An example of cyber-attack scenario.

(1) telnet %Dest-IP	(5) cat > ls /bin/sh
(2) login %ID %PW	(6) chmod 755 ls
(3) ls -al	(7) Export PATH= .
(4) cd tmp	(8) /home/prog1

Attacker Model transmits each command being composed of a cyber-attack scenario to target system sequentially and receives processed results and the state of target system. The DEVS representation of Attacker Model is as follows.

```
State Variables
  'passive, 'active, scenario-type, target-host
External Transition Function
  If packet comes from target-host and state is 'passive,
  Then assign next command from scenario-table and
       set state as 'active during processing-time
Internal Transition Function
  If state is 'active, Then change state to 'passive
Output Function
  If state is 'active, Then send packet to target-host
```

4   Case Study

We have built models of a simple sample network to analyze concrete changes of Linux system against cyber-attack and performed many simulations about this. Fig. 1 shows a sample network and it consists of previously explained Linux System Model and Attacker Model. Network between two models is omitted. As shown in Fig. 1, Attacker Mode sends a packet including the command of cyber-attack to Victim Host and receives the result of processing from the victim host.

Table 3 illustrates the simulation result using the cyber-attack scenario in Table 2 that gains root authorization by using files that set Setuid due to the error of configuration. Here, 'Clock' means the simulation time and 'Model' means the model changed at that time. 'What' means commands processed in Model, and 'Remark' means the supplementary explanation about 'What'. Let's take a look at simulation processing briefly. First of all, Attacker performs login by telnet connecting to Victim Host. If the command packets of telnet and login come in Victim Host, the packets are transmitted to Process Manager Model through Network Manager Model. And Process Manager Model creates and saves a new process in Process Map that saves and manages process information. Let's suppose there is a file called *prog*1 that performs automatically "ls -al" which is one of telnet commands by searching for files having setuid permission for attack. Attacker moves to *tmp* directory and transmits "cat > ls /bin/sh" command packet to Victim Host to create an executable file, ls, being able to start /bin/sh. At this time, *ls* file is created with "rw−r−−r−−" access right according to the default setting. Attacker sets the executing right of *ls* file through "chmod 755 ls" and changes the value of PATH into a value that starts *ls* file in *tmp* directory. Finally, Attacker performs *prog*1 and */tmp/ls* set in PATH, that is, *ls* make by Attacker is performed. Then */bin/sh* is started and Attacker gains shell of root-authorization.

Through analyzing cyber-attack simulation results, we can know the mechanism of cyber-attack that gains root-authorization by using the file having setuid permission and executing the file of malicious code from the attacker. Also, as a defense strategy, we can prevent cyber-attack by removing the file having setuid permission and limiting execution of user's file. Meanwhile, vulnerability can be considered the state of system resource. For example, when 'Clock' is 20.3 in Table 3 there is vulnerability that a file of malicious code can be executed by

Table 3. A simulation trajectory: "Obtaining root-authorization by using setuid file".

Clock	Model	What	Remarks
0.3	Attacker (203.253.146.215)	telnet 203.253.150.2	Connect to Victim host (203.253.150.2)
0.8	Network Manager Model of Victim host	telnet 203.253.150.2	Checking MAC Address, IP Address, and Open Port List at each layer
1.4	Process Manager Model of Victim host	telnet 203.253.150.2	Creating process: addition [203.253.146.215 / Telnet / Idle / Anybody] to Process Map
...	...	...	...
2.8	Attacker	login dayfly 1230	Try to log in 203.253.150.2
...	...	...	...
4.2	Process Manager Model of Victim host	login dayfly 1230	Altering Process Map:[203.253.146. 215 / Telnet / Waiting / dayfly]
...	...	...	...
5.6	Attacker	ls -al	Finding setuid file
...	...	...	...
8.4	File System Manager Model of Victim host	ls -al	Suppose that prog1 is setuid file: Inode-Table[0] = [Name = "prog1" / Dir = "/home" / Owner = dayfly77 / Permit = "rwsr-xr-x" / Content = "ls -al" / State = idle]
...	...	...	...
9.8	Attacker	cd /tmp	Change Directory
...	...	...	...
11.3	File System Manager Model of Victim host	cd /tmp	Changing Working Directory: Work-DIR = /tmp, PATH = /bin/sh/
...	...	...	...
13.7	Attacker	cat > ls /bin/sh/	Create the "ls" file
...	...	...	...
16.3	File System Manager Model of Victim host	cat > ls /bin/sh/	Altering Inode-Table: Inode-Table[1] = [Name = "ls" / Dir = "/tmp" / Owner = dayfly77 / Permit = "rw−r−−r−−" / Content = "/bin/sh" / State = idle]
...	...	...	...
18.8	Attacker	chmod 755 ls	Change Permission
...	...	...	...
20.3	File System Manager Model of Victim host	chmod 755 ls	Altering Inode-Table: Inode-Table[1] = [Name = "ls" / Dir = "/tmp" / Owner = dayfly77 / Permit = "rwxr-xr-x" / Content = "/bin/sh" / State = idle]
...	...	...	...
22.5	Attacker	Export PATH = .	Change Path for execution
...	...	...	...
24.8	File System Manager Model of Victim host	Export PATH = .	Changing Path for execution: Work-DIR = /tmp, PATH = /tmp
...	...	...	...
26.0	Attacker	/home/prog1	Execute the setuid file
...	...	...	...
28.3	Process Manager Model of Victim host	/home/prog1	Altering Process Map: [203.253.146. 215 / Telnet / Running / dayfly77]

Table 3. (Continued).

28.9	File System Manager Model of Victim host	/home/prog1	Executing the prog1 file: Inode-Table[0] = [Name = "prog1" / Dir = "/home" / Owner = dayfly77 / Permit = "rwsr-xr-x" / Content = "ls -al" / State = using] Inode-Table[1] = [Name = "ls" / Dir = "/tmp" / Owner = dayfly77 / Permit = "rwxr-xr-x" / Content = "/bin/sh" / State = using]
29.2	Process Manager Model of Victim host	/home/prog1	Creating Process Map: [203.253.146. 215 / Telnet / Running / root]
...	...	...	...

changing permission. Namely, it is possible to analyze vulnerability quantitatively by defining the degree of resource's change related to system vulnerability [11]. Also, cyber-attack can be defined as command set that can transform the normal state of target system into the state that has vulnerability. Therefore, unknown cyber-attacks can be explored by performing simulations of all possible occasions on the proposed Linux-based System Model [12].

5 Conclusions

This study have discussed on the modeling on the linux-based system from a security viewpoint. To do this, we have analyzed Linux system and designed the Linux-based system model in hierarchical fashion using the DEVS modeling and simulation environment. Unlike conventional research for security simulation that has still been at the conceptual level or focused on statistical method, by applying a discrete event modeling and simulation methodology, it is possible to i) reproduce the detailed behavior of cyber-attack, ii) analyze concrete changes of system resource according to cyber-attack and iii) expect that this would be a cornerstone for more practical application researches (generating cyber-attack scenario, analyzing vulnerability and examining countermeasures, etc) of security simulation. In the foreseeable future, research for the vulnerability analysis of various information systems will have to be continued by detailed modeling of systems based on different operating systems.

References

1. T. A. Longstaff, C. Chittister, R. Pethia, and Y. Y. Haimes, "Are We Forgetting the Risks of Information Technology", IEEE Computer, pp 43-51, December, 2000
2. N. Ye and J. Giordano, "CACS – A Process Control Approach to Cyber Attack Detection", Communications of the ACM., 1998
3. Cohen, F., "Simulating Cyber Attacks, Defenses, and Consequences", 1999 IEEE Symposium on Security and Privacy Special 20th Anniversary Program, The Claremont Resort Berkeley, California, May 9-12, 1999.

4. Amoroso, E., Intrusion Detection, ATandT Laboratory, Intrusion Net Books, January, 1999.
5. Wadlow T. A., The Process of Network Security, Addison-Wesley, 2000.
6. B.P. Zeigler, Multifacetted Modeling and Discrete Event Simulation, Academic Press, 1984.
7. Zeigler, B.P., H. Praehofer, and T.G. Kim. Theory of Modeling and Simulation 2ed. Academic Press, 1999.
8. B.P. Zeigler: Object-Oriented Simulation with Hierarchical, Modular Models. Academic Press, 1990.
9. Brian Hatch, James Lee, George Kurtz, Hacking Linux Exposed: Linux Security Secrets and Solutions, Mc Graw Hill, 2001.
10. S.D. Chi et al, "Network Security Modeling and Cyber Attack Simulation Methodology", Lecture Notes on Computer Science series, 6 th Australian Conf. on Information Security and Privacy, Sydney, July, 2001.
11. J.S. Lee et al, "Simulation-based Vulnerability Analysis", Computer Systems Science and Engineering (Submitted)
12. M.W. Lee, "Automated Cyber-attack Scenario Generation Using the Symbolic Simulation", Master Thesis, Hangkong University, 2002.

A Rule Based Approach to Network Fault and Security Diagnosis with Agent Collaboration

Siheung Kim[1], Seong jin Ahn[2], Jinwok Chung[1], Ilsung Hwang[3],
Sunghe Kim[3], Minki No[3], and Seungchung Sin[4]

[1] Sungkyunkwan University, 440-746 Suwon, Korea
{shkim,jwchung}@songgang.skku.ac.kr
[2] Sungkyunkwan University, 110-745 Seoul, Korea
sjahn@comedu.skku.ac.kr
[3] Korea Institute of Science and Technology Information, 305-806 Taejon, Korea
{shil,shkim,mkno}@kisti.re.kr
[4] Hansei University, 435-742, Kunpo, Korea
expersin@hansei.ac.kr

Abstract. This paper introduces rule-based reasoning (RBR) Expert System for network fault and security diagnosis and a mechanism for optimization. In this system, we use agent collaboration mechanism which is the process that the system gathers network environment data from distributed agent. Collaboration mechanism make accurate diagnosis by making inferences based on the results of various tests and measurements. Additionally, we consider optimization of the system. In some comprehensive systems like network fault diagnosis system or system using decisive loops reasoning, so much more studies on its reasoning time are necessary. So we consider reasoning time and rule probability to optimize the system. For our purpose, rule reasoning time estimation algorithm will be used, with the simulation, where a comparison with the previous rule-based fault diagnosis system is possible.

1 Introduction

The Internet appeared to make a noticeable advancement, based on recent computer telecommunication technologies and data telecommunication technologies. Besides, a variety of network equipments organizing the internet and systems linked to the equipments have increased in complexity and scale. With the extension of internet, it is extremely difficult to manage the network faults. Further more the cost by network fault emerges critical problem in business area. To cope with above problem, rule based approach to network fault is previously proposed [1, 4, 9–12]. The LODES system is one of the systems that are using the RBR to control the network errors. LODES system is capable of examining the error, but it does not have the ability to control and recover the errors [1]. Cisco proposes TCP/IP network diagnosis tool using ping, traceroute and packet debugging. And Lucent suggests network fault management software, which is real time network fault detection tool. But these systems are not enough to detect and recovery network fault problem accurately, especially network security, because of

T.G. Kim (Ed.): AIS 2004, LNAI 3397, pp. 597–606, 2005.

their methodological model. Therefore, we suggest rule based network fault and security diagnosis expert system with agent collaboration and its optimization algorithm that can be used other rule based system.

Chapter 2 will cover the collaborative process among agents of the RBR Expert System. In Chapter 3, rule reasoning time estimation algorithm is described. Finally, there will be a comparison of the agent collaboration model with existing model and a proof of rule reasoning time estimation algorithm through simulation.

2   RBR Based Network Fault Diagnosis and Recovery System

2.1   RBR Fault Management Algorithm Through Collaboration Among Agents

The RBR Expert System introduced in this study analyzes the causes of some possible faults and recover them through collaboration among agents shown in Fig. 1. This would heighten the reliability for an analysis of network fault. For an application of rules based on collaboration as mentioned above, the following introduces the four types of agents. The first is the T-Agent which is detect network fault and is suppose to solve network problems. It analyzes network fault through rule base hypotheses which may require another agents to send information about their own network information. The second is the F-Agent that resides on the same network in which the T-Agent is. The F-agent processes rule requested by the T-agent and then return the result. The third is the R-Agent which is a portal agent for agents in another network. The R-agent plays a role in gathering test result from another network. A example can be shown in Figure 2. When one host can't have access to destination host through

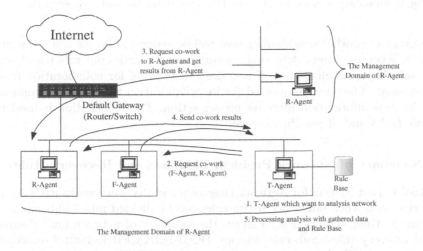

Fig. 1. Rule processing using agent collaboration.

Diagnosis Rules	Check items	Next Job
S11	Title : Default Connectivity Configuration Fault Detection(DCCFD) NIC, NIC driver, cable connection, subnet mask, broadcast setup and IP address(comparing with backup files)	S12 (DCCFR)
S21	Title : Network Environment Fault Detection(NEFD) Whether the fault is one of interior network or exterior network If the result of pinging test to default gateway is **unknown host** then go to S31, if not, go to S41	S22 (Report)
S31	Title : Routing Configuration Fault Detection(RCFD) Whether static routing or dynamic. Whether system can reach next 1-hop using tracing route. Whether **default** or **0.0.0.0** entry exists in routing table. The state of default-router If IP address of default-router exists in default-router file. Whether routing demon processes exist. Whether Operating file of routing demon processes exists	S32 (RCFR)
S41	Title : Name Service Configuration Fault Detection(NSCFD) Whether the system is a resolver or a name server. Discriminating between using file sources and using DNS for name service. Whether the IP address and the host name in host file exist. Whether the IP address and DNS server in resolv.conf file exist. Whether the response of DNS server is normal. Whether the IP address of current system is equals to the IP address in recolv.conf file.	S42 (NSCFR)
S51	Title : Internet Demon Process Configuration Fault Detection(IDPCFD) Whether the entry for operating internet demon process in inetsvc script exists. Whether internet demon process is operating normally. Compare current inetd.conf file to backup inetd.conf file	S52 (IDPCFR)
S61	Title : Application Demon Configuration Fault Detection(ADCFD) Whether the entry for operating demon process in inetd.conf file or etc/services file exist. Whether the demon process operates normally. Whether the demon exist.	S62 (ADCFR)
S71	Title : Security Week Point Detection(SWPD) Whether packet rate or CPU rate or DNS query rate is too high based on information from R-Agent. Whether abnormal port open.	S72 (ADCFR)

Fig. 2. Summary of network fault and security diagnosis and recovery rules.

a given router, another neighboring host will be requested for its test to secure accuracy. When one agent detects a network fault, it carries out rule-based network fault diagnosis algorithm, if necessary, requesting for collaboration from another agent. The agent requested for its collaboration transmits information about its own information after its proper action. The information is used to diagnose faults and, if possible, recover them.

2.2   Summary of Network Problem Diagnosis and Recovery Rules

The RBR Expert System for Network Diagnosis and Recovery works itself based on its rule base when faults occur. Its rules can be divided into 7 sub-rule sets according to their actions. Fig. 3 shows the summary of network fault diagnosis and recovery rules. Sub-rule sets are DCCFD,DCCFR(Default Connectivity Configuration Fault Detection and Recovery), NEFD(Network Environment

Fault Detection), RCFD,RCFR (Routing Con-figuration Fault Detection and Recovery), NSCFD,NSCFD (Name Service Con-figuration Fault Detection and Recovery), IDPCFD,IDPCFR (Internet Demon Process Configuration Fault Detection and Recover) and ADCFD,ADCFR (Application Demon Configuration Fault Detection Recovery). The sub-sets except for NEFD can be classified into diagnosis rules (Rule x1xx) and recovery rules (Rule x2xx). NEFD contains only diagnosis rules. DCCFD,DCCFR is a Rule-set that diagnoses the status of NIC, the cable, IP addresses and subnet mask including the basic connection configuration of a system. SWPD is a rule-set that find week point of host based on security policy. For example, by checking DNS query rate or CPU rate, agent can find the possibility of warm virus or attack. The whole rules can be found in [2, 3, 5].

3   Efficiency Enhancement of Rule-Based Expert System

3.1   The Model of Improved Rule-Based Expert System

In this study, To implement rule base, we adopt the tree structure similar to B-tree which supports various basic dynamic set operations including Search, Predecessor, Successor, Minimum, Maximum, Insert, and Delete in time proportional to the height of the tree [6, 7]. Additionally, we devise the object, called controller, which manipulates the rule base. In Fig. 4, there is the controller that manages the rule set. The controller has some factor of each rule set, and based on this factor rule's application order is decided. Each rule set is applied according to the rule application order, in which the controller has defined and the result is reported to the controller. If the relevant fault is recovered, it adds the number of recovery that it has made itself. Whereas not, the switch applies the next rule set. If we define the rule's application order through this method, the frequency of each fault is recorded whenever a fault occurs so the time used to diagnose and recover a specific frequently occurred fault could be reduced. Keeping record of the frequency of the fault is similar to the learning effect of this system. A performance improvement could be expected by peripherally apply-ing the diagnosis rules that are fit to the characteristics of the relevant fault through an estimation of the characteristics of fault within the system or the network, in which the fault diagnosis and recovery system is installed. Figure 5 shows the actual implementation of the rule and each diagnosis rule module through c-like code. Each rule object stores some information to be used in reordering. Information about the rules could refer to its probability and reasoning time. The time taken to test the hypotheses of the rules varies among them, by means of which their applied order is determined, affecting the optimization of the whole system [8].

3.2   Reasoning Time Estimation Algorithm

To find network faults, the system needs to check the rules in rule-base one by one. From the aspect of optimization, the system can be improved by reducing

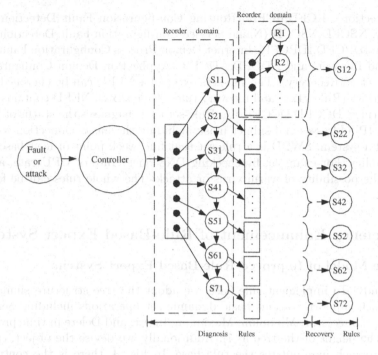

Fig. 3. Data structure of Rule-Base.

```
System_working()
{
        while(true){
        if (check_network_fault = true)
        Controller_run();
        }
}

Controller_run()
{
        Rule_set *rule_set;
        Rule *rule;

        for(int i=0;i<num_of_ruleset;i++){
        rule_set = Controller->rule_set[i];

        for(int j=1;i< num_of_rule_in_ruleset; j++){
        rule = rule_set->rule[i];

        if ( rule->execute() = hit){
                rule->frequency++;
                rule_set->frequency++;
                rule->update_avg_RPT();
                rule_set->update_avg_RPT();
                Controller->reorder_ruletree();
                break this function;
        }
        }
        }
}
```

Fig. 4. Controller function.

the process-ing time and rapidly search for an appropriate rule to solve network problems. But the processing time is under the influence of the network environment which cannot be controlled by the expert system. Therefore, in the RBR Expert System, the fast search for the diagnosis rule helps optimization of the system, for which the reasoning time and the probability of a specific fault happen within the networks(rule probability) serve as a factor to be considered for the re-ordering. To measure reasoning time of each rule, we must closely define the procedure of the rule. In our network fault management system, there are three different kinds of rules. That distinction depends on that the rule requires network information gathered by agent collaboration. They are the rule which doesn't need network information from other agents, the rule which needs F-Agents information and the rule which needs F-Agents and R-Agent information. For now, on a basis of the message-exchange processes described above (Fig. 1), the rule processes time will be covered. Processing the rules without helps from neighboring agents will require as much as the time taken to process rules within a system. When it comes to the rules requiring helps from other Agents, the whole rule reasoning time is that the time of the rule reasoning time in each agent plus the time taken for messages to be transmitted to agents. Thus, the whole rule reasoning time could be formulated as follows:

$$\tau_i(T) \leq \tau_i \leq \tau_i(T) + Max\ [0, (rtt_{r\ to\ t} + \tau_i(F)),$$
$$(rtt_{r\ to\ t} + rtt_{r\ to\ t'} + \tau_i(R'))] \tag{1}$$

- $\tau_i(T), \tau_i(F), \tau_i(R')$: rule reasoning time in T,F,R'-Agent
- $rtt_{r\ to\ t}, rtt_{R\ to\ R'}$: round trip time from R-Agent to T-Agent, R-Agent to R'-Agent
- R'-Agent : R-Agent in another network

Where $\tau_i(T)$ refers to the time taken to process rules by T-agents and rtt is the round trip time which means the time taken to transmit messages among agents. After rule reasoning time Estimation, the controller records the whole rule reasoning time to the controller in Fig. 4, which is used to determine the applied order of rules after net-work fault diagnosis and recovery. Followings are rule reasoning time estimation algorithm which was derived from [7,8].

Assume that there are s pieces of rules in rule base. The reasoning time of each rule is $\tau_1, \tau_2, \cdots, \tau_3$.

$$Rule\ Set\ = \{R_1, R_2, \cdots, R_s\} \tag{2}$$

Assuming that the initial probability of each rule during the reasoning process is $\{P_1, R_2, \cdots, P_s\}$ and the followings are conditional probability of rules.

$$P = \begin{pmatrix} p_{11} & \cdots & p_{1n} \\ \vdots & \ddots & \vdots \\ p_{m1} & \cdots & p_{mn} \end{pmatrix} \tag{3}$$

Blow we define τ_{ij}.

$$\tau_{ij} = \begin{cases} \tau_{R_s} + \tau_{R_j} + \tau_{R_i} & i < j \\ \tau_{R_j} + \tau_{R_i} & i > j \end{cases} \tag{4}$$

Initial step, We can estimate that expected reasoning time of ith rule is

$$T_i[1] = p_i \cdot \tau_{R_i} \tag{5}$$

Second step, We can estimate that expected reasoning time of ith rule is

$$T_i[2] = p_i \cdot \tau_{R_i} + \sum_{j=1}^{s} p_i \cdot p_{ij}\tau_{ij}$$

$$= p_i \cdot \tau_{R_i} + \sum_{j=1}^{i} p_i \cdot p_{ij} \cdot (\tau_{R_s} + \tau_{R_j} - \tau_{R_i}) + \sum_{j=1}^{s} p_i \cdot p_{ij} \cdot (\tau_{R_j} - \tau_{R_i}) \tag{6}$$

$$= T_i[1] + \sum_{j=1}^{i} p_i \cdot p_{ij} \cdot \tau_{R_s} + \sum_{j=1}^{s} p_i \cdot p_{ij} \cdot (\tau_{R_j} - \tau_{R_i})$$

Every time network fault occurred, controller estimates rule reasoning time by above process and reorder the sequence of rules.

4   Experiments and Study

In First experiment, there was a comparison of network fault detection and recovery ability of the proposed model with that of the other model. In next experiment, we will show how much improved with reasoning time estimation model.

4.1   Experiments 1

Table 1 shows comparison of proposed model and other models. Network configuration and network device problem can be easily detected by above models because the problem occurred in host itself. But Invisible Network node failure can be detected by only proposed model because it needs not only information of test host data but also information of other agents. In case Service Level Detect, assume that default router has different policy to each host. By comparing packet rate in other agents, it is possible to detect service level. With this data, the system can conclude that the host has low data rate because of default router policy.

4.2   Experiments 2

The followings are the experimental results that indicate the improvement of the system. Assume that there are 100 rules in the rule base used for the experiment. The 100 rules were divided into 100/6 sub-sets of rules. In addition to that, the

Table 1. A Comparison of the agent collaboration model with existing model

Case Description	Proposed Model Fault Detect	Other Model Fault Detect	Proposed Model Fault Recovery	Other Model Fault Recovery
Network Configuration Problem	A	A	P	P
Network Device Problem	A	A	N	N
Default Router Failure	A	A	N	N
Router Filtering	A	P	N	N
Network Node Failure	A	P	N	N
DNS failure	A	A	N	N
Service Level Detect	A	P	N	N

A : All, P : Partial, N : Not

whole rule reasoning time was randomly set to fit the sub-sets. For example, In the case of the rule sets for routing configuration fault diagnosis, their whole rule reasoning time was engineered to be 20m seconds on the average and 2 in variance. On the condition that network faults can be always diagnosed by diagnosis rules, their types were set to range from type 1 to type 100 where they were made to happen in accordance with normal distribution of 50 on the average and 5 in variance. Followings are the assumption used in experiments.

- There are 100 types of the network faults.
- Network fault occurred at the rate of the normal distribution, (average = 50, variance = 5)
- All network faults can be solved by the rule in the rule-base.
- All network faults match each rule one-by-one.
- Rule processing times followed by the rule set definition are distributed 0 to 20 randomly.

Fig. 5 shows a part of data where the number of the network faults amounts to 100 faults. In the case that normal system after 100 times of the network faults, the average was 0.46 seconds, and where system with rule reasoning time estimation algorithm, the average refers to 0.28 seconds.

5 Conclusion

This study proposed the model that diagnoses and recovers the network fault and network security problem through the collaboration of agents and its optimiz-

reasoning time(s)

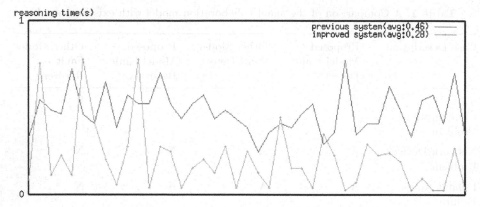

Fig. 5. Rule processing using agent collaboration.

ing algorithm. To make accurate network fault analysis, we devise collaboration mechanism which is based on distributed agents. The agents are divided into T-Agent, F-Agent, and R-Agent according to the location of the network, and diagnose and recover the network fault occurred in host through the process collaborating in the form of query and answer. Moreover, it reports concerning the network configuration model based on RBR. As well as fault management model, we mainly discuss how to optimize the system. For such purpose, collaborations among agents within networks, and the methodology of re-ordering rules of rule base with controller between the drive part of the system and the rule base were utilized. Such factors as the probability and the rule reasoning time of the specific types of faults within networks were used to determine the order of the applied rules, and the data in comparison those of the existing system were brought forth as a proof of optimization. Finally, the future study should pursue combining the RBR Expert System with other reasoning systems and extend rule base so that the former could heighten the ability of network fault diagnosis, and with NMS(Network Management System) so as to strengthen the aspects of network management.

References

1. T. Sugawara, Acooperative LAN diagnosistic and observation expert system, computers and communications, Proceeding of the Ninth Annual International Phoenix Conference, 1990, pp. 667-674.
2. Yunseok jang, Seongjin Ahn, Jin Wook Chung, RBR Based Network Fault Detection and Recovery System using Agent Collaboration, ICOIN, 2003
3. Kwang Jong Cho, Seongjin Ahn, Jin Wook Chung, RBR Based Network Configuration Fault Diagnosis Algorithm using Agent Collaboration, Thesis of master's course, Sung-kyunkwan Univ.
4. Kang Hong Cho, Seongjin Ahn, Jin Wook Chung, Rule-based Agent system for Fault Detection and Location on LAN, KIPS, vol. 7-7 pp.2169-2178, 2000

5. Taein Hwang, Seongjin Ahn, Jin Wook Chung, Design and Implementation of Rule-Based Network Configuration Fault Management System, Thesis of master's course, Sungkyunk-wan Univ.
6. Zheng, S.Q.; Sun,M.; Constructing Optimal Search Trees in Optimal Time; Computers, IEEE Transactions on , Volume: 48 , Issue: 7 , July 1999, Pages:738-748
7. Kato, K., Persistently Cashed B-trees, IEEE Transactions on, Volume: 15, Issue: 3, May-June 2003, Pages 706-720.
8. Pan Aihuam, Lku Chuntu, Real Time Analysis on Expert Systems, Proceeding of the 3rd World Congress on Intelligent Control and Automation, 2000
9. J.-M. Yun,, S.-J. Ahn, J.-W. Chung, Web Server Fault Diagnosis and Recovery Mechanism Using INBANCA, 2000, PP. 2467-2504.
10. K. Ohta, T.Mori, N.Kato, II.Sone, G.Mansfield, Y.Nemoto, Divideand Conquer Technique for Network Fault Management, Proceedings of ISINM97, 1997
11. Kohei Ohta, Takumi Mori, Nei Kato, Hideaki Sone, Glenn Mansfield, Yoshiaki Nemoto, Divide and Conguer Technique for Network Fault Management, Proceedings of ISINM97, 1997
12. Taein Hwang, Seongjin Ahn, Jin Wook Chung, A study on the rules and algorithm for the diagnosis and recovery of routing configuration, SCI 2000, World Multiconference on Sys-temics, Cybernetics and Informatics 4(2000) 137-141

Transient Time Analysis
of Network Security Survivability Using DEVS

Jong Sou Park and Khin Mi Mi Aung

Dept. of Computer Engineering,
Hankuk Aviation University, Seoul, Republic of Korea
{jspark,maung}@hau.ac.kr

Abstract. We propose the use of discrete event system specification
(DEVS) in the transient time analysis of network security survivability.
By doing the analysis with the time element, it is possible to predict
the exact behaviors of attacked system with respect to time. In this
work, we create a conceptual model with DEVS and simulate based on
time and state variables change when events occur. Our purpose is to
illustrate the unknown attacks' features where a variety of discrete events
with various attacks occur at diverse times. Subsequently, the response
methods would be performed more precisely and this is a way to increase
the survivability level in the target environment. It can also be used
to predict the survivability attributes of complex systems while they
are under development and preventing costly vulnerabilities before the
system is built.

1 Introduction

Survivable Information System is intended to tolerate the mission critical functions in face of known and future attacks. It is very critical to simplify design,
model and analysis of such a system. This work emphasizes the transient time
analysis of survivability using Discrete Event System specification (DEVS) to
predict the exact behavior in a system with respect to time. We have studied
the rejuvenation as a proactive approach of survivability [1, 2]. To perform the
rejuvenation, the timing is incredibly important. Our approach is dynamic system where a variety of discrete events with various attacks occur at diverse times
and our system will respond to those events. Survivability system with unknown
attacks' features cannot be precisely illustrated by a mathematical model that
can be estimated analytically. DEVS can be used as the core formalism, in order to enable proof of correctness. EASEL [5, 6] is currently a discrete event
simulation language with limited support for continuous variables. Moitra et.al.,
[7] were modeling survivability of networked systems. They developed a model
to evaluate the tradeoffs between the cost of defense mechanisms for networked
systems and resulting expected survivability after a network attack. They mentioned a major need in survivability management is for more data collection so
that managers can better assess survivability and security and can make better
decisions regarding the costs and benefits of alternative defense strategies for

T.G. Kim (Ed.): AIS 2004, LNAI 3397, pp. 607–616, 2005.

their systems. Chung et.al., [3] had simulated intrusions in sequential and parallelized forms and they presented an algorithm for automatically transforming a sequential intrusive script into a set of parallel intrusive scripts, which simulate a concurrent intrusion. But parallelizing an intrusive script is difficult in practice because of the rich set of shell-level commands and various constructs supported by the script language. F.Cohen [4] presented cause-effect model. He discussed the limitations on modeling and simulation such as limits on accuracy of the models, limits on the accuracy of the data upon which the simulation is based and the ability to explore the simulation space through the use of multiple runs of the simulator through the space. Based on those related and relevant approach, our goals in this work are

- to create a conceptual model that emphasizes events and their relations to one another with DEVS,
- to simulate based on time and state variables change when events occur rather than the way normal time changes continuously,
- to evaluate the survivability of systems and services, as well as the impact of any proposed changes on the overall survivability of systems, and
- to simulate the effects of attacks, accidents, and failures, and can be used to predict the survivability attributes of complex systems while they are under development, preventing costly vulnerabilities before the system is built.

In this paper, first we describe our proposed model and then we give a brief introduction to DEVS. In section four, we denote DEVS Simulation and Network Security Survivability.

2   A Proposed Model

We have studied the fusion method for survivability (Fig. 1) and we have implemented some modules in our previous work [1]. We had solved the second portion of our work, transient state analysis, using semi-Markov process and

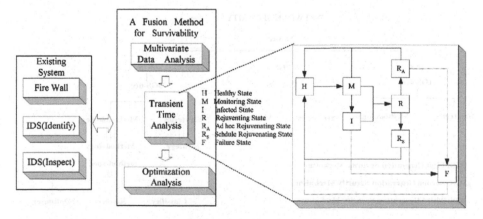

Fig. 1. A Proposed Model.

Weibull distribution function for survivability in our previous work [2]. In this work, we address the transient time analysis using DEVS and our experiences show that DEVS can be used as the core formalism, in order to enable proof of correctness rather than illustrating by a mathematical model that can be estimated analytically.

We consider five states such as healthy state, monitoring state, infected state, rejuvenating state (schedule and ad hoc) and failure state. As shown in (Fig. 1), the system would get back to healthy state, even after infected, by means of three ways (solid lines). And the system would be totally failed in three ways (dotted lines). We analyze these two situations with two models, named, Healthy Monitoring Rejuvenation (H-M-R) model and Healthy Monitoring Infected Rejuvenation (H-M-I-R) model.

2.1 DEVS Formalism

DEVS is the formalism that allows a modeler to specify a hierarchically decomposable system as a discrete event model that can be later simulated by a simulation engine. In DEVS, we have to specify the atomic models and the coupled models. Detail descriptions of the atomic model, coupled model and DEVS formalism can be found in [8, 9].

2.2 DEVS Simulation and Network Security Survivability

The most important purpose of DEVS formalism is that it provides a formal way of how discrete event languages specify their discrete event system parameters. By doing the analysis with the time element involved in it, it is possible to see the exact behavior of an event in time. Therefore we utilize DEVS to analyze the transient time in network security survivability.

The most important security approach for the survivable system is to provide third generation security (3GS) mechanisms. We have constructed our 3GS strategy and methodologies and we illustrate it with the system entity structures, a

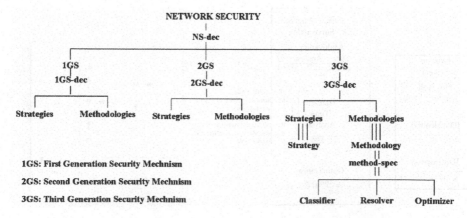

Fig. 2. System Entity Structures for Network Security.

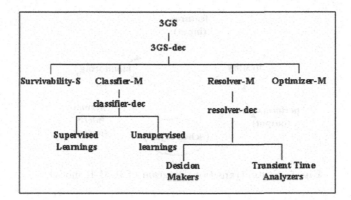

Fig. 3. Pruned Entity Structure of 3 GS.

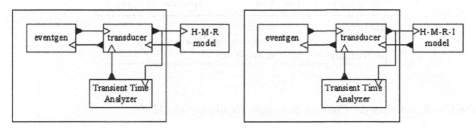

Fig. 4. DEVS models for Transient Time Analysis.

structured knowledge representation scheme (Fig. 2). The root entity network-security has an aspect node that implies the descendants (1GS, 2GS and 3GS) can be coupled.

Pruning is required to create a pure SES, no specialization and at most one aspect under each entity. (Fig. 3) shows a Pruned Entity Structure (PES) of 3 GS. It specifies a hierarchical discrete-event model, which is reduced structure by pruning the SES according to the objectives of our study. In this work, we address transient time analysis with DEVS. (Fig. 4) shows an overall system model of H-M-R and H-M-I-R.

Our response methods for network security survivability are schedule rejuvenation and ad hoc rejuvenation. In the next section, we model and simulate these behavior atomic models.

2.3 Transient Time Analysis on DEVS-Based Schedule Rejuvenation Model

We have collected the features from the event list and it has prepared by using classifiers [1]. In the simulation cycle, we insert the features to monitor the system. And there has an international transaction and it will output as alerts. Since the output is normal-alerts then the simulation will end by performing schedule rejuvenation. And we evaluate the performance for this model.

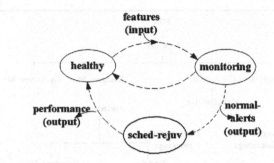

Fig. 5. State Transition Diagram of H-M-R model.

Table 1. Schedule rejuvenation model events and states.

External transition	State	Internal transition
features	healthy	
	monitoring	normal-alerts
performance	sched-rejuv	

DEVS-Based Pseudo Code of Schedule Rejuvenation Model

```
State variables
    'healthy, 'monitoring, 'sched-rejuv

External transition function
    if current state is 'healthy,
    then features inserts from event list
    and set state as 'monitoring during processing-time

Internal transition function
    if current state is 'monitoring and output is normal-alerts
    then set state as 'sched-rejuv during processing-time
    else if current state is 'sched-rejuv
    and output is performance
    then change state to 'healthy

Input function
    if current state is 'healthy
    then insert features from event list

Output function
    if current state is 'monitoring
    then check features and send alerts
    if current state is 'sched-rejuv
    then send performance
```

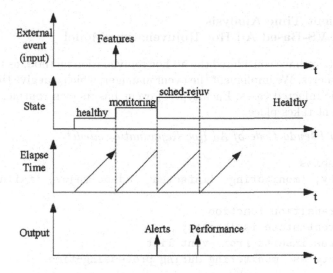

Fig. 6. Time-line analysis of H-M-R model.

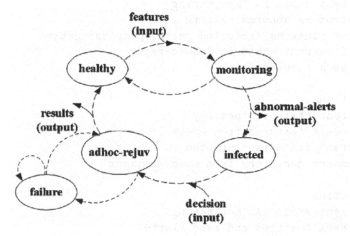

Fig. 7. State Transition Diagram of H-M-I-R model.

Table 2. Ad hoc rejuvenation model events and states.

External transition	State	Internal transition
features	healthy	
	monitoring	abnormal-alerts
decision	infected	
	adhoc-rejuv	Results

2.4   Transient Time Analysis on DEVS-Based Ad Hoc Rejuvenation Model

In this model, we have considered the ad hoc rejuvenation method, as the output is abnormal-alerts. We implement the event manager, which can give the decision in the specific infected cases. Each type of event has its own routine, to be run when the event takes place.

DEVS-Based Pseudo Code of Ad Hoc Rejuvenation Model

```
State variables
    'healthy, 'monitoring, 'infected, 'adhoc-rejuv, 'failure

External transition function
    if current state is 'healthy
then features inserts from event list
and set state as 'monitoring during processing-time

Internal transition function
    if current state is 'monitoring
    and output is abnormal-alerts
    then set state as 'infected during processing-time
    else if current state is 'adhoc-rejuv
    then check results

Input function
    if current state is 'healthy
    then insert features from event list
    if current state is 'infected
    then insert decisions from event manager

Output function
    if current state is 'monitoring
    then check features and send alerts
    if current state is 'adhoc-rejuv
    then check decisions and send results
```

3   Simulation Results for Network Security Survivability

The presented experiment demonstrates the network security survivability with H-M-R and H-M-I-R models. 41 Collected features [1] are inserted from the event list and the simulation cycle will be in processing time while it was monitoring. Depending upon the output alerts the elapsed time will be varied case by case.

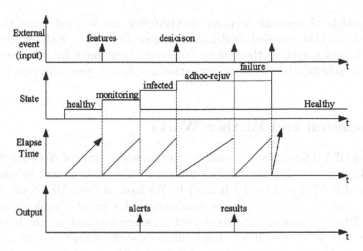

Fig. 8. Time-line analysis H-M-I-R model.

Simulation Results for Network Security Survivability H-M-R Model

```
Simulation Results for Network Security Survivability
after 25 Simulation Runs:
- - - - - - - - - - - - - - - - - - - - - - - - - - -
Total number of features = 41
Mean Monitoring Time              = 0.2231 min
Mean Monitoring Time of Phase1 (H-->M)  = 0.2231 min

Duration = 0.2231 min


H-M-R Model
    Simulation Results for Network Security Survivability
after 25 Simulation Runs:
- - - - - - - - - - - - - - - - - - - - - - - - - - -
Total number of features = 41
Mean Monitoring Time              = 0.2231 min
Mean Monitoring Time of Phase1 (H-->M)  = 0.2231 min
Processing Time of Phase3 (M-->RS)  = 15.229 min

Duration = 15.4521 min


H-M-I-R Model
    Simulation Results for Network Security Survivability
after 25 Simulation Runs:
- - - - - - - - - - - - - - - - - - - - - - - - - - -
Total number of features = 41
Mean Monitoring Time              = 0.2231 min
Mean Monitoring Time of Phase1 (H-->M)  = 0.2231 min
Processing Time of Phase3 (M-->I)   = 5.349 min
Processing Time of Phase4 (I-->RA)  = 18.115 min

Duration = 23.6871 min
```

Two models of network security survivability are verified using the time-line analysis and the coupled model information. From this analysis, we get the useful information such as the system behaviors, how much time to wait before applying the schedule or ad hoc rejuvenation and how a system responds to each event.

4 Conclusion and Further Works

We applied DEVS formalism to analyze the transient time of Network Security Survivability and developed an associated simulation environment by employing execu-tion of H-M-R and H-M-I-R models. We have utilized DEVS not only for predicting performance but also for analyzing behavior of complex system. In Section 4, DEVS time-line analysis is used to design and analyze the behavior of the models. We used the DEVS couple model specification to create a coupled model. Using DEVS, we are able to have as a feature of time in the design and analysis. It helps us how a system responds to an event, and it also guides us the behaviors of the system. We can fully analyze the transient time before performing our response methods to each attack. This work is a paradigm of property-based types. And our models are discrete event simulation with limited support for continuous variables in arbitrary time. These models use arbitrary time units. While the time units are thought to be relatively consistent, further work is required for validated real-world values. We are implementing the real time simulation that will represent the network security survivability to handle non-deterministic event time models.

Acknowledgement

This research has been supported by Information Technology Research Center (ITRC) Project of Ministry of Information and Communication, Republic of Korea.

References

1. Aung, K. M. M., Park, J. S.: Software Rejuvenation Approach to Security Engineer-ing, ICCSA 2004, Special Issue of Lecture Notes in Computer Science, Vol. 3046, Perugia, Italy (2004) 574–583
2. Aung, K. M. M.: The Optimum Time to Perform Software Rejuvenation for Sur-vivability, IASTED, SE2004, Int. Conf., Innsbruck, Austria (2004) 292–296
3. Chung M., Mukherjee B., Olsson R.A., Puketza N.: Simulating Concurrent Intru-sions for Testing Intrusion Detection Systems: Parallelizing Intrusions, Proc. of the 18th NISSC, 1995
4. Cohen F.: Simulating Cyber Attacks, Defenses, and Consequences, IEEE Sympo-sium on Security and Privacy, Berkeley, CA, 1999
5. Fisher, D. A.: Design and Implementation of EASEL-A Language for Simulating Highly Distributed Systems. Proceedings of MacHack 14, the 14th Annual Confer-ence for Leading Edge Developers. Deerborn, MI, June 24-26, 1999

6. Fisher, D. A.: Survivability and Simulation. Third Information Survivability Workshop (ISW-2000). Boston, MA, October 23-26, 2000
7. Moitra S.D., Konda S.L: Simulating Cyber Attacks, Defenses, and Consequences, IEEE Symposium on Security and Privacy, Berkeley, CA, 1999
8. Zeigler, B.P.: Multifacetted Modelling and Discrete Event Simulation, Academic Press, London, UK and Orlando, FL, 1984
9. Zeigler, B.P.: Object-Oriented Simulation with Hierarchical Modular Models: Intelligent Agents and Endomorphic Systems, Academic Press, Boston, MA, 1990

A Harmful Content Protection in Peer-to-Peer Networks

Taekyong Nam[1], Ho Gyun Lee[1], Chi Yon Jeong [1], and Chimoon Han[2]

[1] Privacy Protection Research Team, Electronics and Telecommunications Research Institute,
161 Gajeong-dong, Yseong-gu, Daejeon, 305-350, Korea
{tynam,hglee,iamready}@etri.re.kr
[2] Dept. of Electronics and Information Engineering, Hankuk University of Foreign Studies,
89, Wangsan-ri, Mohyun-myen, Yngin-si, kyenggi-do, 449-791, Korea
cmhan@hufs.ac.kr

Abstract. A variety of activities like commerce, education, voting is taking place in cyber-space these days. As the use of the Internet gradually increases, many side effects have appeared. Therefore, it is time to address the many troubling side effects associated with the Internet. In this paper, we propose a method of prevention of adult video and text in Peer-to-Peer networks and we evaluate text categorization algorithms for Peer-to-Peer networks. We also suggest a new image categorization algorithm for detecting adult images and compare its performance with existing commercial software.

1 Introduction

The rapid distribution of high-speed Internet and the development of Internet technology made it possible for us to engage in many cyber-space activities. As the results, the total sales of e-commerce now exceed than those of Television-home shopping. Moreover there has been dramatic decrease in traditional mail correspondence due to the ubiquity of email. Also Internet messenger and video chatting have become new communication alternatives. As the Internet becomes a necessary part of many daily lives, many side effects of the Internet have appeared. For example, spam mail affects nearly everyone who uses e-mail. There are more serious examples such as the exposure of children to pornography and the propagation of anti-social information such as suicide and terrorism. Thus, It is increasingly important to take innovative measures in order to mitigate deleterious side effects [1, 2].

Until now, most computer security systems have concentrated on the defense of the computer itself, for example defense against computer viruses, Distributed Denial of Service (DDOS) attacks or on the encryption of information for a secure communication. However, in order to prevent the exposure of harmful content from reaching children information needs to be detected and prevented automatically. Many companies have already operated the monitoring system in their intranet for precaution of leakage of secret information. But it is illegal from the viewpoint of the law to detect private information. So our system implementation for the purpose of detecting and preventing harmful information is automatically done on PC only when user agrees.

In this paper we implemented a system that identifies adult video and text contents in Peer-to-Peer (P2P) networks and isolates them automatically. This system detects and blocks the harmful information in P2P networks on personal computers by text, image and video categorization.

T.G. Kim (Ed.): AIS 2004, LNAI 3397, pp. 617–626, 2005.
© Springer-Verlag Berlin Heidelberg 2005

The rest of this paper is organized as follows. In Section 2, we discuss related work regarding the harmful information filtering and the information categorization. Section 3 presents the behavior of our system and its components. The results of our study appear in Section 4, and the conclusion and references to future work in is discussed in section 5.

2 Related Works

Harmful information filtering technology has been used in Internet filtering software like parental control. The technology is divided into two ways: the pre-processing method and the post-processing method. The pre-processing method uses a harmful site Uniform Resource Locator (URL) database. When a user tries to access any site, the site URL is compared with URL database to decide whether the site is harmful or not. If the database replies that the site is harmful, the system blocks access to that site.

The post-processing method monitors incoming and outgoing traffic in real time and inspects the traffic contents like text, image and video. If the traffic has any harmful content, that session is closed. The merit of pre-processing lies in the accuracy and the speed of detection. Because harmful site databases are automatically categorized by machine learning algorithms such as SVM (Support Vector Machine), MLP, and kNN and verified by an administrator, the pre-processing method has greater accuracy than post-processing in general. The weakness in pre-processing is that the databases may not have all the harmful site information and it is possible for information about harmful sites to be inaccurate because site contents may have changed. The merit of post-processing is that it does not need a database server or the assistance of humans. The system can decide the harmfulness of site by itself. The defect of post processing is low correctness and slowness. Because post processing decides the harmfulness of contents in real time by automatic categorization, it is less accurate and it inspects real time traffic contents by hooking the network traffic, so user may feel that it is slower.

The method of making databases of harmful sites describes in detail [3]. In [3], system uses text, images, and the URL name for feature vector to make harmful site databases.

In this paper, we use a new post-processing method to prevent harmful contents in P2P networks. P2P networks are a new way to distribute harmful content like pornography, violence, and illegal software [1,2]. Preventing harmful content in P2P networks differs from blocking the harmful site in the technical viewpoint. In this paper, we discuss these differences.

The current goal of harmful information filtering is to improve the accuracy of automatic categorization of contents. The automatic categorization of contents can be divided into a text and an image categorization [3]. The former has been studied in information categorization and retrieval in many ways. And the previous methods in text categorization field can be used to achieve good performance [4,5,6]. Specially, the text categorization for a harmful information filtering can achieve greater performance. But we cannot expect good performance of the previous methods in P2P networks because we can use only several words that are the result of file search query and are composed of file name, type and size. In this paper, we find optimal solution for text categorization algorithm in P2P networks.

Recently, methods of categorizing harmful images based on its contents have been studied in many ways. There are two main ways. The one is finding adult images using image retrieval feature in Content Based Image Retrieval (CBIR). The other is finding harmful image by extracting a skin region and using feature vector in the skin region. The CBIR methods take a long time and the using skin region method has low accuracy. In this paper we use appearance information to categorize harmful image correctly. By this method we can detect and block harmful image effectively in P2P networks.

In existing parental control products, they focus on still images, but the most harmful information in P2P networks is video file. So to address this situation, in this paper we propose a mechanism of intercepting video file that is being transmitted in P2P networks, slicing this file to key frame and deciding whether it is harmful or not. Also we show performance of our harmful image detection engine by experiment.

3 Frameworks

Harmful Information Protection System consists of P2P traffic sensor, text categorization, image categorization, and video categorization.

First, P2P sensor is a prerequisite function to block the transmission of harmful information through P2P. On the client side, it can be checking whether the P2P program is running or not. There are mainly two approaches: One is to scan all known port numbers for P2P and check against currently allocated port numbers, which is similar to firewall. But this method has some weaknesses. For example, the port number can be changed and new P2P program use unknown port number. The other is to learn the P2P traffic pattern and distinguish the P2P traffic as suggested in [9]. Surely, it is feasible to learn P2P traffic pattern as in web traffic and ftp traffic by using protocol pattern, traffic size, port number, and the characteristics of peer.

Second, text categorization mechanism is the first step to categorize the P2P contents. Text categorization can be done these ways: dictionary-based checking, dictionary-based threshold comparison, and learning. Text categorization includes morphological analysis and harmful words dictionary. Therefore, if keywords and file names contain harmful words, the session is closed [3]. Text categorization with learning is depicted in figure 1. Text categorization with learning has functions such as morphological analysis, feature selection, and text rate classification. Morphological analyzer outputs nouns and adjectives in the text and feature weight computation is done in feature selection. Finally, text rate is determined after learning model is being compared with features.

Text categorization algorithm is employed in determining the harmful rate of search keyword and file name. Search keyword and file names are usually 10byte ~ 128byte long. We analyzed the words used in P2P and created harmful words dictionary especially for P2P after analyzing the characteristics of words, which frequently appear in P2P networks.

Third, video categorization mechanism is the special function for categorization of P2P contents. In the case of eDonkey, file is fragmented and sent to the recipient.

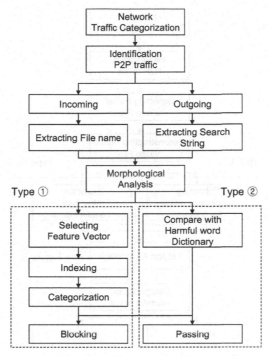

Fig. 1. Text Categorization Algorithms

Hence, the file cannot be executed on receiving and must be reassembled before running. As a result, real time processing is not important aspect in video categorization. The most fundamental function is retrieving still picture from video. Open source like VirtualDub implements this function.

Like the character set problem in text categorization, video processing needs several CODEC. We are not sure of how many still images need to be retrieved to determine the harmful rate. If general video contains harmful scenes, which are 2 or 3 minute long, they can be classified as harmful video. But, in this paper, we target video files that contain harmful scenes through the film such as commercial adult films.

There are two mechanisms in retrieving still images. One is the key frame extraction and the other is extracting at the regular interval. Key frame extraction can avoid retrieving the same image, but it takes longer time. Extracting at the regular interval might retrieve the same image, but it takes less to operate. Figure 2 describes the process to determine a harmful rate.

Last, still image is analyzed and the harmful rate is determined. In this paper, the appearance information of image is used. The overall process is shown in figure 3. Using the uniform characteristics of skin area, skin area is extracted and image feature vector is created. If the skin area is relatively small, the image is classified into harmlessness. Image feature vector contains both skin color information and appearance information, and is of size 40x40. Image feature vector is given to SVM to determine whether the image is harmful or not [8].

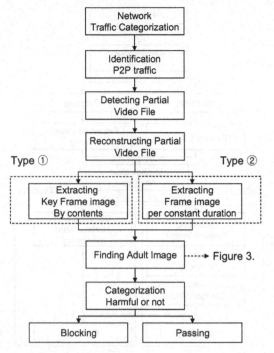

Fig. 2. Video Categorization Algorithms

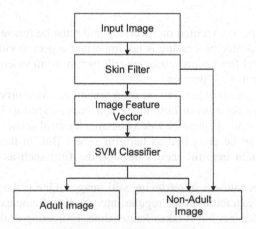

Fig. 3. Image Categorization Algorithm

Figure 4 describes the relationship among the components. If the search keywords in outbound traffic are found to be harmful, the session will be closed. When the file names in inbound traffic are found to be harmful, transmission of the file is also suspended. In the case of image and video, image rate categorization engine is employed and proper action is taken [7,8].

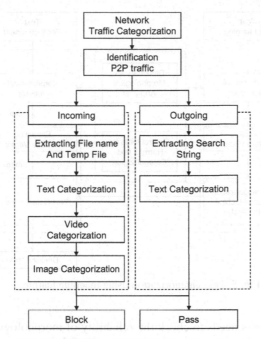

Fig. 4. P2P Harmful Information Prevention

4   Performance Evaluations

We will illustrate the results of text and image filtering which are the components of the P2P harmful information prevention system in this section. First, we test the file name-based filtering method, one of the methods described in figure 4, for incoming P2P traffics. We compare the performance of the following three methods.

- Pattern matching-based method: First, the system compares the input text with the harmful word dictionary and then counts the number of harmful words in input text. If the number of harmful words is more than one, input text is classified as harmful information.
- Threshold-based method: First, the system calculates the weight for each word in harmful word dictionary and then sums the weight of the harmful words in input text. If the sum of the weights is more than threshold, input text is classified as harmful information.
- Machine Learning based method: First, the system creates the classification model using the machine learning algorithm and natural language processing technique and then input text is classified by the classification model. This method has two parts: One is the training part that generates the classification model and the other is the classification part classifies the input text using the classification model. Experimental process is shown in figure 5.

(1) Text Rate Classification Model Learning Process
   The system performs the morphological analysis from the sample filename data-set (harmful file name: 1000, non-harmful file name: 1000) and extracts the

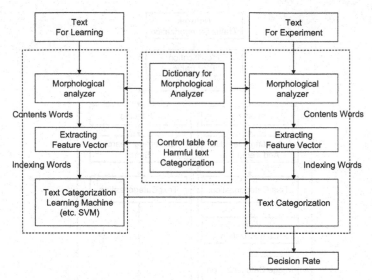

Fig. 5. Text categorization using machine learning

nouns. This process needs to check the reliability of morphological analyzer. The system calculates feature vectors from the result of the noun extraction to classify the harmful filename and trains the classification model using the feature vectors. This process needs to check the processing time. Then the system passes the classification model to the text rate classification process.

(2) Text Rate Classification Process

The system performs the morphological analysis from the input filename and extracts the nouns and calculates feature vector from the result of the noun extraction. A feature vector is used as an input vector to classification model and the classification model classifies the rate of the input filename.

4.1 Experiment Dataset

We collect the harmful and non-harmful filenames using the eDonkey. The training set contains 1000 harmful filenames and 1000 non-harmful filenames and the test set also contain 1000 harmful filenames and 1000 non-harmful filenames. The words in harmful word dictionary are used as keywords to collect harmful filenames. The words, which are generally used in the economics, sports, TV, movies and so on, are used as keywords to collect non-harmful filenames. The collected filenames are classified as harmful or non-harmful manually and the average number of the words in filenames is 8.

4.2 Experiment 1

In the case of the web document containing many words, machine learning-based text filtering method has a good performance than the pattern matching or threshold-based method.

However, when the size of the filename on P2P system is less than 256 bytes, we cannot be sure of that. So in this experiment, we inspect the effectiveness of the machine learning-based text-filtering method on P2P by comparing the other methods.

Training data: Harmful filenames (1000), Non-harmful filenames (1000)
Test data: Harmful filenames (1000), Non-harmful filenames (1000)

Since the filenames on P2P system do not contain many words, the number of extracted nouns from morphological analysis result is not big, either. According to the result of the noun extraction, harmful filenames are usually composed of harmful nouns only and non-harmful filenames are usually composed of the non-harmful nouns only. The experimental results show that machine learning-based text-filtering method and pattern matching-based method have a similar performance. Because the difference between harmful filenames and non-harmful filenames is clear, we cannot expect the effect of machine learning algorithm.

To improve the performance of the filename filtering system on P2P, we need a more elaborate morphological analyzer that can deal with the compound nouns or the mistyped words.

Table 1. Pattern matching based method

Accuracy	0.935000
Precision	0.995000
Recall	0.888393

Table 2. Threshold based method

Accuracy	0.950500
Precision	0.995000
Recall	0.913682

Table 3. Machine learning based method

Accuracy	0.941500
Precision	0.995000
Recall	0.898826

Table 4. Image categorization for performance test

Scan Level	Harmful Image Block Rate (100)	Normal Image				
		Human1 Mistake Rate (100)	Human 2 Mistake Rate (100)	Human 3 Mistake Rate (100)	Scene 1 Mistake Rate (100)	Scene 2 Mistake Rate (100)
2	37%	5%	20%	10%	5%	9%
5	75%	18%	38%	22%	18%	16%
8	96%	44%	59%	42%	42%	3%
ANID	93%	18%	16%	11%	2%	1%

4.3   Experiment 2

In experiment 2, we compare the performance of the harmful image detection methods in table 4. Test data contain the 100 harmful images and the 500 non-harmful images. In non-harmful datasets, Human 1 and 3 dataset contain general human images, and Human 2 dataset contains many sexy images that have many skin regions. Scene 1 dataset contains many traditional Korean images that have many yellow color regions similar to skin regions. Scene 2 dataset contains a lot of modern city scenes.

In this experiment, we compare the performance between the proposed harmful image detection method and the commercial software. Since the commercial software can change the detection level (0 to 10) of the harmful image, we use the three levels 2, 5, and 8 for experiments. The high detection level means that the threshold for harmful image detection is low. ANID [8] is an appearance-based nude algorithm that shape information is used to classify the nude images, and detect small nude images in a large background image. It finds skin regions using texture characteristics of the human skin, which then generates the skin likelihood image. Since the skin likelihood image contains shape information as well as skin color information, ANID used the skin likelihood image as a high level feature to classify the nude images. The image feature vector is used as an input to a nonlinear-SVM.

The experimental results of the harmful image detection are shown in table 4. In table 4, the proposed method has the similar detection rate to commercial software at detection level 8 and the similar false detection rate to commercial software at detection level 5 or less. Experimental results show that the proposed harmful image detection method is superior to the commercial software.

5 Concluding Remark

In this paper, we implemented the P2P harmful information prevention system, inspecting the harmfulness of the transferring P2P traffic and blocking the harmful P2P traffic. Since P2P is the new network paradigm and one of the important network applications, it will be a good strategy to block the P2P traffics by its content than to block all P2P traffics. We proposed the harmful text and image categorization algorithm for harmful information prevention.

Experimental results show that the proposed image categorization algorithm has an excellent performance in preventing P2P harmful information and that text categorization can perform well by improving morphological analyzer.

The proposed method can be used to prevent the distributions of illegal software, advertisements and harmful messages on the Internet as well as to prevent the P2P harmful information.

References

1. Baldonado Henry A. Waxman, Steve Largent, Children's Access to Pornography Through Internet File-Sharing Programs;' United St ates House of Representatives Committee on Government Reform Stafff Report, July 2001, www.reform.house.gov
2. Tom Davis, Henry A. Waxman, "Children's Exposure to Pornography on Peer-to-Peer Newworks", United States House of Representatives Committee on Government Reform Stafff Report, March 2003, www.reform.house.gov

3. Mohamed Hammami, Oussef Chahir, and   Liming Chen, "WebGuard : Web based Adult Content Detection and Filtering System", IEEE/WIC International Conference on Web Intelligence, 2003.
4. H.C.M. de Kroon, T.M. Mitchell, and E.J.H Kerchhoffs, "Improving Learning Accuracy in Information Filtering", In International Conference on Machine Learning – Workshop on Marchine Learning Meet HCI (ICML-96), 1996.
5. Thorsten Joachims, "A Probabilistic Analysis of the Rocchio Algorithm with TFIDF for Text Categorization", In Proc. ICML-97, 143-146, 1997.
6. Douglas W. Oard, Nicholas DeClaris, Bonnie J. Dorr, and Christos Faloutsos, "On Automatic Filtering of Multilingual Texts", In Conference Proceeding, 1994 IEEE International Conference on Systems, Man, and Cybernetics, volume 2, pages 1645-1650, October 1994, at http:/www.glue.umd.edu/oard/research.html
7. D.Forsyth and M.Fleck, "Finding naked people", Proceedings of European Conference of Computer Vision. Berlin, Germany Springer-Verlag, 593-602, 1996.
8. Chi-Won Jeong, Jong-Sung Kim, Ki-Sang   Hong, "Appearance-Based Nude Image Detection", In 17th International Conference on Pattern Recognition (ICPR 2004), August 2004.
9. Jeff Sedayao, "World Wide Web Network Traffic Patterns", In Proceedings of the COMPCON'95 Conference, 1995.

Security Agent Model
Using Interactive Authentication Database

Jae-Woo Lee

Software System Lab., Dept. of Computer Science & Engineering,
Korea University, 1, 5-ka, Anam-dong, SungBuk-ku, 136-701, Seoul, Korea
It21c@korea.ac.kr

Abstract. Authentication agent enables an authorized user to gain authority in the Internet or distributed computing systems. It is one of the most important problems that application server systems can identify many clients authorized or not. To protect many resources of web server systems or any other our computer systems, we should perform client authentication process in the Internet or distributed client server systems. Generally, a user can gain authority using the user's ID and password. But using client's password is not always secure because of various security attacks of many opponents. In this paper, we propose an authentication agent system model using an interactive authentication database. Our proposed agent system provides secure client authentication that add interactive authentication process to current systems based on user's ID and password. Before a user requests a transaction for information processing to distributed application servers, the user should send a authentication key acquired from authentication database. The agent system requests an authentication key from the user to identify authorized user. The proposed authentication agent system can provide high quality of computer security using the two passwords, user's own password and authentication key or password. The second authentication password can be acquired by authentication database in every request transaction without user's input because of storing to client's database when the user gets authority first. For more secure authentication, the agent system can modify the authentication database. Using the interactive database, the proposed agent system can detect intrusion during unauthorized client's transaction using the authentication key because we can know immediately through stored the authentication password when a hackers attack out network or computer systems.

1 Introduction

Authentication agent enables an authorized user to gain authority in the Internet or distributed computing systems. Nowadays, there is no place that information systems do not reach on our society whole by information technology. In this information society, many information processing request generally are performed by the Internet environment that a client request information to a web server systems in network. In such computing environments, it is one of the most important problems that application server systems can identify many clients authorized or not. There have been many risks that many unauthorized users attack our network and computers for acquiring many information or destroying our resources. At that case, an authentication

T.G. Kim (Ed.): AIS 2004, LNAI 3397, pp. 627–634, 2005.

agents can act as intelligent decision-makers, active resource monitors, wrappers to encapsulate legacy software, mediators, or as simple control functions [1].

To protect many resources of web server systems or any other our computer systems, we should perform client authentication process in the Internet or distributed client server systems. Generally, a user can gain authority using the user's ID and password. But using client's password is not always secure because of various security attacks of many opponents. Therefore, various security services should be needed to protect our network and computers during security attacks of opponents, such as authentication, encryption, etc. There are several types of security attacks in network and computer security, such as interruption, interception, modification and fabrication. Interruption is that resources of systems is destroyed or becomes unavailable by interrupting transmission between sender and recipient, interception is that an unauthorized opponent gains many information in a network. Modification is that transmission between sender and recipient is modified by opponent's attacks, such as changing values or altering data file. Fabrication is that unauthorized user access network or systems as authorized client [2, 3].

Security service is an information technology that enhances the security of resources of computer systems. Several security services are needed to protect our network or computers for preventing these various security attacks, such as authentication, access control, confidentiality, integrity and non-repudiation. Authentication service is to assuring that a client is authentic to request resources of server by authorized user. Access control is preventing unauthorized users from accessing network, database and computers by wrong action. Confidentiality is for protection of traffic flow or observing information source by using encryption or flow control. And then the security attack is not useful to opponent. Integrity is for modification of security attack created by unauthorized user. Non-repudiation prevents sender or receiver from denying transmission of information for proving transaction between sender and receiver[2, 3].

In the Internet or distributed client server systems, many clients access various server for acquire many information. And then, first of all, server systems check whether the client is authorized or not. It is one of the most important problems that we protect many resources of systems using authorized client authentication in Internet. It is general process of client authentication that a user can gain authority by the user's ID and password. But using only client's password is not always secure because there are many security attacks in a network occurred by many opponents. Many unauthorized users always try to know ID and password of authorized clients and hackers know easily authorized user's password using security attacks, like replay.

In this paper, we propose an authentication agent system model using an interactive authentication database. Our proposed agent system provides secure client authentication that add interactive authentication process to current systems based on user's ID and password. Before a user requests a transaction for information processing to distributed application servers, the user should send a authentication key acquired from authentication database. The agent system requests an authentication key from the user to identify authorized user. The proposed authentication agent system can provide high quality of computer security using the two passwords, user's own password and authentication key or password. The second authentication password can be acquired by authentication database in every request transaction without user's input because of storing to client's database when the user gets authority first. For more secure authentication, the agent system can modify the authentication database. Using the interactive database, the proposed agent system can detect intrusion during unauthorized client's transaction using the authentication key because we can know immediately through stored the authentication password when a hackers attack out network or computer systems.

This paper is as following. In section 2, we introduce briefly about security attacks and various security services, and define an authentication agent system model using an interactive authentication database. In section 3, we define our proposed agent's procedure about interactive authentication database for client authentication and explain propose client authentication algorithm. In section 4, we describe characteristic of our agent system for security service, and evaluate our proposed authentication agent. Finally, in conclusion we establish more secure client authentication agent systems and plan future works.

2 Client Authentication and Proposed Authentication Agent Model

2.1 Security Service for Authentication

Authentication service is to assuring whether a client is authentic or not, by using user's ID, password or internet address, etc. In the Internet including distributed computing systems, a server system requires a user's ID and password for preventing unauthorized users from using resources of the server [2]. There are several types of security attacks to networks or computer systems. Those are interruption, interception, modification and fabrication. Interruption is concerned with availability in the Internet or client server systems, interception is an attack on confidentiality. Modification is concerned with integrity and fabrication is an attack on authentication [2, 6].

To protect our networks or computing systems, it is very important to identify authorized users during many security attacks. Several security attacks types are shown in figure 2.1. Interruption is a security attack that destroy or make unavailable resources of systems or network by interrupting transmission between sender and recipient, interception is illegal action that an unauthorized opponent gains many information in a network for information stealing. And then thus stolen information

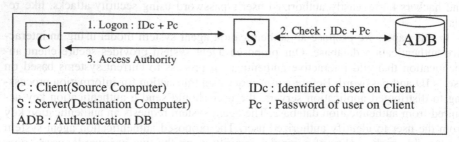

C : Client(Source Computer)   IDc : Identifier of user on Client
S : Server(Destination Computer)   Pc : Password of user on Client
ADB : Authentication DB

Fig. 2.1. Security Attacks in the Internet or Distributed Computing Systems

will be used in accessing, copying, spreading for others. Modification is that transmission between sender and recipient is modified by opponent's attacks, such as changing values or altering data file. And those are transmitted modified in network, an authorized user is misunderstood by them. Fabrication is that unauthorized user access network or systems for transmitting illegal transaction as authorized client [2, 3, 6, 8, 9].

2.2 Our Proposed Authentication Agent Model

As above mentioned in section 2.1, authentication service enables an authorized user to gain authority in the Internet or distributed computing systems. To protect many resources of web server systems or any other our computer systems, our proposed authentication agent system model uses an interactive authentication database. Our proposed agent system provides secure client authentication that add interactive authentication process to current systems based on user's ID and password. Before a user requests a transaction for information processing to distributed application servers, the user should send a authentication key acquired from authentication database. The agent system requests an authentication key from the user to identify authorized user. The proposed authentication agent system can provide high quality of computer security using the two passwords, user's own password and authentication key or password. The second authentication password can be acquired by authentication database in every request transaction without user's input because of storing to client's database when the user gets authority first.

In this authentication agent system model, the interactive authentication database is used for identifying user's authority at every transaction processing. The authentication database is shared between the agent system and users. That is, for more secure authentication, the agent system requests authentication key from the user besides the user's id and password. Thus our authentication agent model is shown in Figure 2.2.

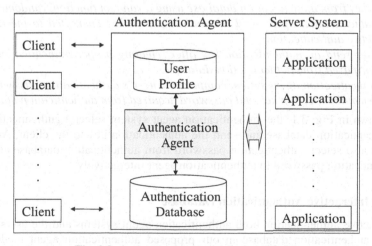

Fig. 2.2. Authentication Agent System Model

3   Authentication Agent Model Using Authentication Database

3.1   Interactive Authentication Agent

In our proposed authentication agent model, authentication database is used for secure user authentication. In section 2.1, we have described various security attacks and client authentication services. In this paper, we consider that password for transaction and authentication is needed more complex. So, we propose an interactive authentication database between authentication agent and users.

Before a user requests a transaction for information processing to distributed application servers, the user send his or her own ID and password to authentication agent system. And then the authentication agent system checks authority of the user using user profile. And the authentication agent selects authentication key from authentication database using a function, random number, randomize(), and requests authentication password corresponding the authentication key to the user. The user should send an authentication key after selecting proper authentication password acquired from authentication database. And then the authentication agent system compare the user's authentication password with it's own interactive authentication database. If the procedure is correct or normal, the transaction can be processed in application server. The proposed authentication agent system can provide high quality of computer security using the two passwords, user's own password and authentication key or password. The second authentication password can be acquired by interactive authentication database in every request transaction.

As mentioned above, the authentication agent sends a authentication key to user and the user returns the authentication password interactively. The our proposed authentication agent's procedure is summarized as follows:

1. *A client inputs it's own ID and password for log-in and send the message for authentication to authentication agent system.*
2. *The authentication agent check the client's ID and password, and select authentication key from authentication database using a random function, randomize().*
3. *The authentication agent send the authentication key encrypted to the user for interactive authentication.*
4. *The user send an authentication key after selecting proper authentication password acquired from authentication database.*
5. *The authentication agent system compare the user's authentication password with it's own authentication key and password acquired from authentication database.*

As shown in Fig. 3.1, the authentication agent system select a authentication key from authentication database and send the authentication key to the client. And then the client also select authentication password from authentication database and send the authentication password to authentication agent interactively.

3.2   The Interactive Authentication Database

For interactive authentication both authentication agent systems and the users should maintain authentication database in our proposed authentication agent model. The authentication database is used for secure user authentication. The database has au-

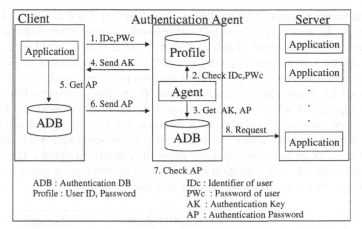

Fig. 3.1. An Interactive Authentication Procedures

thentication keys and passwords. Using the authentication key they can search authentication password for interactive authentication.

Interactive authentication database table has two fields, key(ak_i) and password(ap_i). The layout of authentication table is shown in table 3.1, ak_1, ak_2, ..ak_i, ..,ak_n are set of authentication key. And authentication password are defined as ap_1, ap_2, ..ap_i, ..,ap_n . When authentication agent select a authentication key, the agent create a random number for authentication key using function, randomize(). And agent queries to authentication database.

Table 3.1. Authentication Database Table

Authentication Key	Authentication Password
ak_1	ap_1
ak_2	ap_2
...	...
ak_i	ap_i
ak_n	ap_n

4 Secure Transaction Processing

In our proposed authentication agent systems, client authentication is performed by interactive authentication between the client and authentication agent. So, the proposed authentication agent system can provide high quality of computer security using the two passwords, user's own password and authentication key or password. The second authentication password can be acquired by authentication database in every request transaction without user's input because of storing to client's database when the user gets authority first. For more secure authentication, the agent system can modify the authentication database. Using the interactive database, the proposed agent system can detect intrusion during unauthorized client's transaction using the authentication key because we can know immediately through stored the authentication password when a hackers attack out network or computer systems.

If several security attacks like modification or fabrication are conducted by unauthorized users, the hackers can't intrude our networks or computers without the proposed authentication password. When an authorized client has received authentication key from authentication agent system, the authentication password is stored only authorized client's database. And the client use two passwords including authentication password for request to application server. Every time a request from a client reaches to application server, the authentication agent system checks the two password after select interactive authentication database. So, even though a hacker gains a password through security attack, such as replay or masquerade, when the hacker access our network or computer systems using the authorized client's ID and password, the agent checks authentication database with authentication password. Therefore the hacker cannot access the systems or data file.

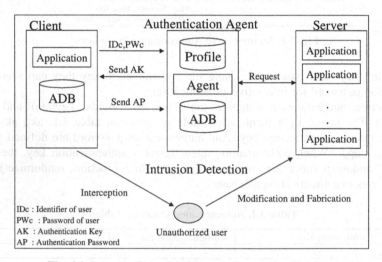

Fig. 4.1. Intrusion Detection Using Authentication Database

In our proposed authentication agent system, we can detect those security attacks by checking the authentication password in the authentication database. As shown in figure 4.1, an unauthorized user intercept the message created by an authorized client and the hacker send a false message with fabrication or modification. When the authentication agent system receives the false message from a hacker, agent system can detect a security attack. The authentication key and password is unique key in the authentication database and randomly generated, so when agent system receives the message with other password, the agent can checks the authentication password. Our proposed authentication agent system model is very simple and interactive authentication provides secure transaction processing easily.

5 Conclusion

Authentication agent enables an authorized user to gain authority in the Internet or distributed computing systems. It is one of the most important problems that application server systems can identify many clients authorized or not. To protect many resources of web server systems or any other our computer systems, we should perform client authentication process in the Internet or distributed client server systems. Generally, a user can gain authority using the user's ID and password. But using client's password is not always secure because of various security attacks of many opponents. We have shown general concepts of security and authentication and propose an authentication agent system model using interactive authentication database.

In this paper, we propose an authentication agent system model using an interactive authentication database. Our proposed agent system provides secure client authentication that add interactive authentication process to current systems based on user's ID and password. Before a user requests a transaction for information processing to distributed application servers, the user should send a authentication key acquired from authentication database. The agent system requests an authentication key from the user to identify authorized user. The proposed authentication agent system can provide high quality of computer security using the two passwords, user's own password and authentication key or password. The second authentication password can be acquired by authentication database in every request transaction without user's input because of storing to client's database when the user gets authority first. For more secure authentication, the agent system can modify the authentication database. Using the interactive database, the proposed agent system can detect intrusion during unauthorized client's transaction using the authentication key because we can know immediately through stored the authentication password when a hackers attack out network or computer systems.

In the future, we will further research to prevent various security attacks and detect intrusion using various authentication agents. It will be very important that we define security attack in detail and more secure client authentication model.

References

1. Alex L. G. Hayzelden and Rachel Bourne, Agent Technology for Communications Infrastructure, Chichester John Wiley & Sons, Ltd., 2001
2. William Stallings, Network Security Essentials : Application and Standards, Prentice Hall, 1999
3. Charlie Kaufman, Radia Perlman, Mike Speciner, Network Security : Private Communication in a Public World, Prentice Hall, 1995
4. William Stallings, Cryptography and Network Security : Principles and Practice, Prentice Hall, 1999
5. B.C. Neuman, Theodore Ts'o. Kerberos, "An Authentication Service for Computer Networks," IEEE Communications, 32(9):33-38, September, 1994
6. S. M. Bellovin and M. Merritt, "An Attack on the Interlock Protocol When Used for Authentication," IEEE Transactions on Information Theory, 40(1):273-275, Jan., 1994
7. Ravi Sandhu and Pierangela Samarati, "Authentication, Access Control, and Audit," ACM Computing Surveys, 28(1):241-243, March, 1996
8. Soichi Furuya, "Slide Attacks with a Known-Plaintext Cryptanalysis," Lecture Notes in Computer Science, Vol. 2288:214-225, Springer-Verlag, 2001
9. James Giles, Reiner Sailer, Dinesh Verma, and Suresh Chari, "Authentication for Distributed Web Caches," Lecture Notes in Computer Science, Vol. 2502:126-145, Springer-Verlag, 2002

Discrete-Event Semantics for Tools
for Business Process Modeling in Web-Service Era

Ryo Sato

Institute of Policy and Planning Sciences, University of Tsukuba
Tsukuba, Ibaraki 305-8573, Japan
rsato@sk.tsukuba.ac.jp

Abstract. A business process itself is artificial, and is a discrete-event system. This paper uses the business transaction system, which is a multicomponent DEVS proposed by Sato and Praehofer, as a device to design business processes in which web-service like software components are part of the processes. The static model of business transaction system, called activity interaction diagram (AID, for short), plays a fundamental role in developing a unified framework for universal modeling language (UML), architecture of integrated information systems (ARIS) and event-driven process chain (EPC), and data flow diagram (DFD). The framework provides us with both the meaning of those tools and possibility of automatic transformation of the models described by any of them. Furthermore, it is a guideline for a modeling method to have a validated business process.

1 Introduction

From 1974 on, the electric data interchange has been developed in Japan. Yet web-EDI is not widely used, the movement of web service and enterprise application integration envisions connected business processes that beyond organizational boundaries.

Many modeling tools for business process/architecture have been proposed to analysis and design business processes with such information technology.

A business process itself is artificial, and is a discrete-event system. This paper uses the business transaction system, which is a multicomponent DEVS[17,18] proposed by Sato and Praehofer[8], so that Web-service like software components will be incorporated in the design of business processes. Section 2 develops a concise explanation of business transaction systems. The static model of business transaction system, called activity interaction diagram (AID, for short), plays a fundamental role in developing a unified framework for universal modeling language (UML)[2,16], architecture of integrated information systems (ARIS)[13,14] and event-driven process chain (EPC), and data flow diagram (DFD). They are considered in Sections 3, 4, and 5, respectively. The framework of business transaction system provides us with both the meaning of those tools and possibility of automatic transformation of the models described by any of them. Furthermore, it gives a guideline for a modeling method to have a validated business process. Though security issues in process formation are of importance[7], it is beyond the paper.

T.G. Kim (Ed.): AIS 2004, LNAI 3397, pp. 635–644, 2005.

2 Business Transaction Systems

A business transaction system is a discrete-event model of business process with information system [8]. It has intrinsic structure that can be drawn with a diagrammatic tool defined below, and also has precise state transition mechanism.

Definition. Activity Interaction Diagram (AID) [8]
An activity interaction diagram is a diagram that has three kinds of components. They are (1) activities, (2) queues, and (3) connecting arrows. Activities should be connected with queues, and vice versa. That is, in the graph theoretic sense, an AID is a directed graph.

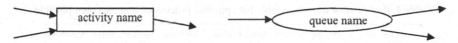

Fig. 1. Components of activity interaction diagrams

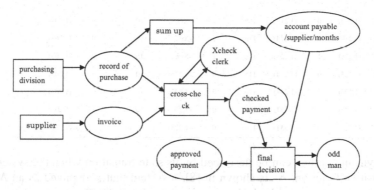

Fig. 2. Approval of trade account payable

There are several symbols to draw an AID as shown in Figure 1. For example, a business process for approval of trade account payable is depicted in Figure 2. The intrinsic meaning of Figure 2 is as follows.

(1) The purchasing division records the transactions for purchase.
(2) Suppliers send us invoices based on past purchase.
(3) The record and invoice are cross-checked by a responsible clerk.
(4) After cross checking, payment is requested for final decision.
(5) The odd man decides the validity of the payment and suitable bank account for the payment.
(6) After the decision, approved payments are sent to treasurer who will remit the amount or make bills due in certain days/months (that is, sight).

The time evolution of the business transaction system depicted in an AID follows the flowchart of Figure 3. An activity with incoming queues is called an internal transaction, while one with no incoming queues is called an external transaction. In Figure 2, "supplier" and "purchasing division" are external. The other activities are internal.

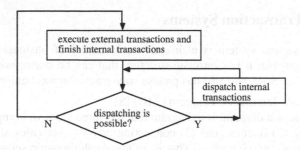

Fig. 3. Flowchart for the processing of a dynamic structure

Table 1. Part of state transition table for approval process of trade account payable

time	supplier	invo	PurcDiv	recPurch	XchkClk	Xcheck	chkdPymnt	oddMan	finDec	apprvPymnt
ta	15		7			10			29	
220	(1,210)	0	(1,217)	115	1	-Xck-	10	0	(1,203)	7
224	(1,210)	0	(1,224)	116	1	-Xck-	10	0	(1,203)	7
225	(1,225)	1	(1,224)	116	1	-Xck-	10	0	(1,203)	7
225	(1,225)	0	(1,224)	115	0	(1,225)	10	0	(1,203)	7
231	(1,225)	0	(1,231)	116	0	(1,225)	10	0	(1,203)	7
232	(1,225)	0	(1,231)	116	0	(1,225)	10	1	-finDec-	8
232	(1,225)	0	(1,231)	116	0	(1,225)	9	0	(1,232)	8

Any dynamic system can be described by its state transition when the system has a state transition function. As is shown in [8], a system that is depicted as an AID is a discrete-event system. Table 1 shows a time evolution of the approval process of trade account payable, starting from time zero with arbitrary fixed initial condition. Table 1 is called a state transition table. The first row of Table 1 shows the following: "time" shows clock. The "supplier" shows data for supplier activity. The "invo" is used as an abbreviation of invoice to show the invoices currently accumulated in the system. The "PurcDiv" is purchase division. The "recPurch" is an abbreviation of record of purchase. The "XchkClk" shows the available number of the clerks for cross check, and "Xcheck" the activity of cross check. The "chkdPymnt" is an abbreviation of checked payments. The "oddMan" is the number of available odd man who is responsible for checked payment to be approved payment. Odd man's activity is called final decision that is shown as "finDec" in Table 1. The second line shows several numbers, corresponding to each on activities. For example, 15 for supplier means that it takes 15 time units for suppliers to issue a new invoice. In the same way, it takes 7 time units for the purchase division to issue a new record of purchase, 10 time units to cross check a payment, and 29 time units to make final decision on a checked payment.

At time 220 the value for supplier attribute is (1, 210). It means that a supplier had started preparing one invoice at time 210. Since 15 time units are needed to complete the activity, that invoice will be input into the business process at 225. Both of the values (1, 217) for "PurcDiv" and (1, 203) for "finDec" have the same meaning. The

Table 2. A state of approval process of trade account payable

supplier invo	PurcDiv	recPuech	XchkClk Xcheck	chkdPymnt	oddMan	finDec	apprvPymnt
(1,210) 0	(1,217)	115	1 -Xck-	10	0	(1,203)	7

value "-Xck-" for "Xcheck" means that there exists no cross check activity. The value 0 for "invo" means that there is zero invoices at 220, while 115 for "recPurch" shows that there are 115 records of purchase waiting for cross checking. According to the flow chart in Figure 3, the values at the line for 220 changes to the next line at time = 224.

Notice that the contents of connecting queues of an AID are modeled as simple objects. That is, in the approval process of trade account payable, only the number of accumulated invoice are concerned and processed in turn. In general, suppliers supplies different materials with various amount, preparation and processing of them needs various length time, and also objects in invoice queue should be described by some attributes to characterize each invoice. A discrete-event system with AID and a data model is called a business transaction system [8].

A business transaction system is a discrete-event system with AID and a data model that describes the logical structure of data that is implemented as a set of tables like Figure 4 in a database management system.

Though the mathematical definition of the state transition function of a business transaction system is provided in Sato and Praehofer [8], the state evolution is nothing but the flow chart in Figure 3, and, if applied, produces the state transition table. So, if we provide a suitable and correct data model for Figure 2, then the approval business process in Figure 2 becomes a business transaction system. Since a data

time=220	supplier	purchDivision	XcheckClerk	oddMan
	(#1282981, 210)	(#9328178, 217)	-	(#100186, 203)

(A spplier is preparing No.1282981 slip. PurchaseDivision started preparing to a record of No.9328179.)
(The odd man started No. 100186 at t= 203)

paymentOverview			
pymnt #	yearMont	payTo #	issuedSlip #
83258	20XX/02	2031	342032
83243	20XX/02	14	2,250,000

purchaseOverview			
purchase #	supplier #	puchDate	puchAmnt
9328145	546	20XX/02/(	231,000

purchaseItem		
purchase #	Item #	puchAmnt
9328145	010	200,000
9328145	020	31,000

checkedPymntOverview			
chkdPymnt #	Pymnt #	PymntDat	reconciltn
100185	83243	20XX/02/	2,250,000
100186	83258	20XX/02/	320,000

checkedPymntLine		
chkdPymnt #	Line #	purchase #
100185	010	9328139

approvedPymnt		
appPymnt #	Emp #	Bank #
100183	1022	1458176

accountPayable/supplier/month			
payTo #	yearMonth	carryOverEOM	sumTotal4Month
2031	20XX/01	33,500	220,525
1095	20XX/01	23,500	908,000

t=224	supplier	purchDivision	XcheckClerk	oddMan
	(#1282981, 210)	(#9328179, 224)	-	(#100186, 203)

Data for No.9328178 was added. PurchaseDivision is now preparing to a record of No.9328179.

purchaseOverview			
purchase #	supplier #	puchDate	puchAmnt
9328145	546	20XX/02/(	231,000
9328178	72	20XX/01/1	145,700

purchaseItem		
purchase #	Item #	puchAmnt
9328145	010	200,000
9328145	020	31,000
9328178	010	30,150

Fig. 4. State transition of business transaction system

model can be realized in tables and then form a file system as a whole, the state transition table is analyzed in the same way as Table 1. As being depicted in Figure 4, the state transition of the business transaction system has the same structure as Table 1. Notice that the file system can describe arbitrary variety of possible invoices or payments. (It is still assumed that the processing time of activities is the same as Table 1 for comparison purpose.)

By observing this state transition in Figure 4, we can conclude that if a data model is suitable and correct, then the whole business transaction system works well.

3 Sequence Diagram and Collaboration Diagram in UML

UML (Unified Modeling Language) is a set of tools such as use case diagrams/ analysis, class diagrams, se-quence diagrams, collaboration diagrams, state diagrams, activity diagrams, component diagrams, and deployment diagrams[16].

In order to describe dynamic aspect of business proc-esses, sequence and collaboration diagrams are important, both of which describe the same from different view points.

In this section the meaning of sequence diagram is con-sidered. Figure 5 shows the components of a sequence diagram. Based on a use case diagram and analysis that describe how necessary functions are provides for users, a sequence diagram diagrams shows required classes and objects and the order of interaction among them through messages. The messages are methods of the objects. Time axis is set vertically from top to bottom using a line for each object, which is called life line because an object is on (alive) when its method is called (or, used) by other objects.

Figure 6 is a sequence diagram for inquiry process of clothes from customer [5]. The process is a combination of a web-site for customer order, called webOrder system, and web services of apparel manufacturers. If we forget the adjective, "apparel", in the figure, the same mechanism will be used for most of businesses that deal with configurable products. The point is that inquiry of a configurable product needs some interaction processes among customers and related organizations. A web service is a mechanism of remote procedure call (RPC) that is invoked from web client or web server [7]. Thus, a chain of web services is formed to automate complex business process.

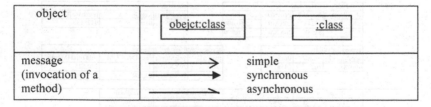

Fig. 5. Components of sequence diagram

In Figure 6, the process proceeds as follows:

(1) A customer specifies a manufacturer, and then sees products on a browser.
(2) The customer selects a product, and then sees the possible customization.

(3) The customer specifies configuration and order quantity, and push send button on the browser.

(4) The customer sees the quotation.

The sequence diagram shows mutual interaction among human actors, graphical user interfaces (GUI), messages, web services, and databases.

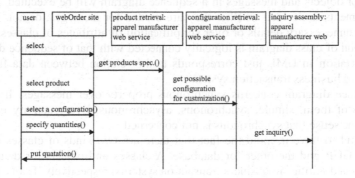

Fig. 6. Sequence diagram: Inquiry process through web services

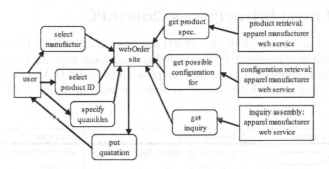

Fig. 7. Activity interaction diagram: Inquiry process through web services

Since the process in Figure 6 is a business process, and should have a structure of business transaction system, the corresponding AID should exist. Table 3 is the transformation rule and Figure 7 is the resultant AID for the sequence diagram in Figure 6.

In the transformation of sequence diagrams into AID, the directions of arrows for "get something" messages are reversed. The resultant data flows in an AID do not have any names. It means that the content of data flow is ambiguous until the attributes of data stores are precisely defined.

Table 3. Transformation rule from sequence diagram to AID

sequence diagram	activity interaction diagram
object and class	external process and data store
method	activity
message-arrow	connecting arrow

Guideline for Modeling Business Processes Through Sequence Diagrams

Messages in a sequence diagram are activities in the corresponding activity interaction diagram (AID). Any activity needs necessary data to start its processing in a business transaction system with AID. Since a sequence diagram shows a skeleton of business process, it does not concern data flows. Therefore, when elaboration of data elements for objects and messages in a sequence diagram will be executed, selection of data elements is decided whether they are needed for messages, or not. Then we can expect unnecessary details or over simplification of attributes of classes. That is, the validation of class diagram is logically connected with that of sequence diagrams, and their relation in UML just corresponds to the relation between data flows and activities in a business transaction system.

A sequence diagram concerns synchronous property over messages. It provides three kinds of them: simple, synchronous, asynchronous. Every activity in AID is simple, in the sense that synchronous is not concerned.

It is useful to have in mind the fact that there are two kinds of classes in UML. One is for GUI, and the other for database. A classes and objects for database are entity types and entities in business transaction systems, respectively. Entity types are implemented as tables and entities are realized as rows in tables in some DBMS.

4 ARIS and Event-Driven Process Chain (EPC)

Event-driven process chains (EPC, for short) had been used in SAP R/3 that is the most popular enterprise re-source planning package software, describe business process. EPC is the diagrammatic tool of an enterprise architecture (EA), called ARIS – architecture of integrated information systems, which is originated by A.W. Sheer[6,13,14].

Name	Symbol	Description	Example
event		time with object that requires processing	arrival of order; release of order
function (activity)		transformation of input to output	check an order; assembly
connective		OR; AND; XOR	
control flow		order of processing	
external process	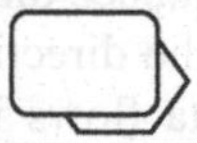	inbound flow; outbound flow	planning and control system; customer
organizational unit	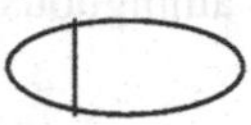	department; machine; human resource	sales division; assembly line
information object		data; table	customer order; employee
information flow		creation; read; addition; deletion	

Fig. 8. Components of event-driven process chain (EPC)

The components of EPC are in Figure 8. The intrinsic meaning of respective components is self-explanatory.

Figure 9 is an EPC of costing process for a customer order [13]. Arrival of customer order happens as an event, and cost calculation for an order ends as an event. Events in EPC have more information than time epoch. For example, the event of arrival of customer order contains information about the order. Thus, an event in EPC should be taken as a request for some processing. In other words, events are input of succeeding activities.

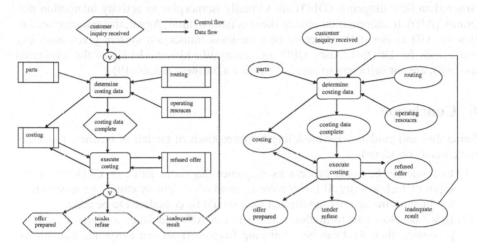

Fig. 9. Event-driven process chain: Costing process for a customer order

Fig. 10. Activity interaction diagram: Costing process for a customer order

The transformation rule from EPC to AID is shown in Table 4. Both events and information objects in EPC are data stores in AID, which determines the condition for commencement of activities.

Table 4. Transformation from EPC to AID

event-driven process chain (EPC)	activity interaction diagram (AID)
event	data store
information objects	data store
function (activity)	activity
control or information arrow	connecting arrow
connectives (AND, OR, XOR)	(disappear)

Guideline for Modeling Business Processes Through Event-Driven Process Chain

EPC does not presuppose the structure of business transaction systems, or discrete-event nature of business process. As in similar way as sequence diagrams, the transformed AID provide us with a check list for validity of the EPC model.

For each function in EPC, if any data is lacking to determine the function can commence, then additional information or event should be added.

Furthermore, in some stage of elaboration of model, data should be modeled so that the whole consistency can be checked for two issues: (1) Are necessary data and event specified as input of a function in EPC? (2) Is any information assigned to functions as output? Otherwise, that information is supposed to be produced from vacuum.

5 Data Flow Diagrams

Since data flow diagrams (DFD) are virtually isomorphic to activity interaction diagrams (AID), transformation among them is quite natural. Again, the dynamic semantics of AID as the static structure of a business transaction system gives modeling guidelines for DFD[8]. Also, DFD and entity life history (ELH) of the structured analysis, are not sufficient to precisely define a business process [9].

6 Conclusion

Semantics and guideline for modeling for three kinds of models of business processes have been considered.

(1) In modeling business process as sequence diagrams, EPC, or DFD, check the input of each activity. If the commencement of an activity cannot be specified by the input of the activity, additional data should be considered to be added.
(2) If we develop transformation rules between AID and other models of business processes, then AID can be a unifying framework so that automatic transformation of models are possible. Such facility supports modeling process.
(3) Diagrammatic tools considered in this paper do not provide a design method for some performance measure, such as minimal inventories of parts and materials, shorter lead times, or maximal through put. Since no comprehensive theory is developed for that direction yet, this situation is not responsible for modeling methods. Development of design of dynamic properties of business processes has been called on [1,3,10,11,12,15].

Acknowledgements

The author is grateful to SAP Japan for their donation of R/3 to the university.

References

1. Baccelli, F.L., Cohen, G., Olsder, G.J., and Quadrat, J.P. Synchronization and Linearity – An algebra for discrete event systems. John Wiley (1992)
2. Eriksson, H-E. and Penker M., Business Modeling with UML - Business Patterns at Work, John Weiley & Sons (2000)
3. Hopp, W. and Spearman, M.L.: Factory Physics, 2nd edition, McGraw Hill (2000)
4. Hotaka, R., Database systems and data model, ohom-sya (in Japanese) (1989)
5. Ida, A., Design and Implementation of "Customized Clothes" Business Process on Web Service Network, Graduation Thesis of University of Tsukuba (in Japanese) (2003)
6. Keller & Teufel: SAP R/3 process oriented implementation, Addison-Wesley(1998)

7. Nishizawa, H. et al., Web Service: A Guide for Analysis and Design, Soft Bank Publishing (in Japanese) (2002)
8. Sato, R. and Praehofer, H. "A discrete event model of business system - A Systems Theoretic for Information Systems Analysis: Part 1." IEEE Trans. Systems, Man, and Cybernetics, Volume 27-1 (1997) 1-10
9. Sato, R.: Meaning of Dataflow Diagram and Entity Life History - A Systems Theoretic Foundation for Information Systems Analysis: Part 2, IEEE Transactions on Systems, Man, and Cybernetics, 27-1 (1997) 11-22
10. Sato, R. and Tsai,T.L.: An Agile Production Planning and Control System with Advance Notification to Change Schedule, Int. J. Production Research, 42-2 (2004) 321-336
11. Sato, Ryo: "Modeling and Simulation of Business-Logistics with Business Process Equation", Proceedings for 2000 AI, Simulation and Planning in High Autonomy Systems, The Society for Computer Simulation International, pp259--264 (2000)
12. Sato, R. "On control mechanism for business processes," Journal of Japan Society for Management Information. 8 - 1, pp17 - 28 (in Japanese) (1999)
13. Scheer, A.W.: Business Process Engineering(2 edtion), Springer (1994)
14. Scheer, A.W.: ARIS-Business Process Framework, 2nd edition, Springer (1998)
15. Tsai,T.L. and R. Sato: A UML model of agile production planning and control system, Computers in Industry, 53 (2004) 133-152
16. Yoshida, Y. et al, , Guide for Object-Oriented Development Using UML, Gijutsu Hyoronsha. (in Japanese) (1999)
17. Zeigler, B. P. Theory of modelling and simulation. John Wiley (1976)
18. Zeigler, B. P. Multifaceted modeling and discrete event simulation. Academic (1984)

An Architecture Modelling
of a Workflow Management System[*]

Dugki Min[1] and Eunmi Choi[2]

[1] School of Computer Science and Engineering, Konkuk University,
Hwayang-dong, Kwangjin-gu, Seoul, 133-701, Korea
dkmin@konkuk.ac.kr
[2] School of Business IT, Kookmin University,
Chongnung-dong, Songbuk-gu, Seoul, 136-702, Korea
emchoi@kookmin.ac.kr

Abstract. This paper presents the design and implementation of a
workflow management system by which one can determine a work process
at runtime. Our workflow management system consists of a build-time
part and a run-time part. The build-time part employs a process definition model that can support various types of work processes. A process
definition is described in XML and converted to objects at runtime. In
order to extend the function of activity, such as condition checking and
invoked applications, we allow inserting real java code in XML process
definition. The core of run-time part is a workflow engine that schedules
and operates tasks of a work process according to a given process definition. Activities and transitions are designed as objects that have various
execution modes, including simple manual or automatic modes.

1 Introduction

Dynamics of business process will be more popular in the age of so-called *realtime business*. According to the Gartner's prediction [13], in 2012 the service
development cycle will be shorten within a day based on the Web Service technologies [11,12]. That is, when an e-Business idea comes in mind, the e-Business
service should be started within a day unless the business chance will be lost (i.e.
realtime business). Therefore, the realtime business systems should be developed
fast by integrating existing Web Services, instead of developing from the scratch.
In order to support this kind of realtime business, the e-Business systems should
be developed so as to add, delete, and change easily business processes that are
created in a day.

The workflow management system [9,10] gives a solution to the problem of
business dynamics. The workflow management system separates business process logic from its execution environment. There have been three standardization

---

[*] This work was supported by the Korea Science and Engineering Foundation
(KOSEF) under Grant No. R04-2003-000-10213-0. This work was also supported
by research program 2004 of Kookmin University in Korea.

T.G. Kim (Ed.): AIS 2004, LNAI 3397, pp. 645–654, 2005.

specifications related to workflow management. The first one is the workflow reference model that had been proposed by the WfMC [1,2,3]. The workflow reference model introduces a workflow management framework with sex standard interfaces that make its core components interop-erable [4]. This framework has been used as the basis of the following workflow management systems. Our workflow management system gets many design idea from the WfMC's. The second one is the Workflow Facility of OMG [7], which is a part of object management architecture (OMA). The Workflow Facility proposes an object-oriented architecture of workflow management system that is based on the WfMC's workflow reference model. The third one is the SWAP of IETF [5,6]. The SWAP (Simple Workflow Access Protocol) sends an execution request encoded in XML on top of HTTP. The concept is very similar to SOAP of the Web Service technology. In order to support the workflow mechanism, the SWAP specifies a number of standard methods in XML.

This paper presents the design and implementation of a workflow management system by which one can determine a work process at runtime. Our workflow management system consists of a build-time part and a run-time part. The build-time part employs a process definition model that can support various types of work processes. A process definition is described in XML [8] and converted to objects at runtime. In order to extend the function of activity, such as condition checking and invoked applications, we allow inserting real java code in XML process definition. The core of run-time part is a workflow engine that schedules and operates tasks of a work process according to a given process definition. Activities and transitions are designed as objects that have various execution modes, including simple manual or automatic modes. The engine supports various transition types and relevant data types. In addition, our system provides monitoring and notification services for auditing and managing the progress of business processes.

Section 2 presents the structure of our workflow management system. In Section 3, we present the architecture of build-time system. The architecture of run-time system is described in Section 4 that is the encore of the workflow management system. We summarize in Section 5.

2   The Structure of Workflow Management System

Our workflow management system consists of three parts: the build-time management part that is used to define a business process, the run-time management part that executes instance of business process and the human interaction part that controls IT invoked application for processing various activity steps. The build-time management part is used to define a business process (or workflow process) that consists of a number of activities. In our build-time management system, a process definition is an instance of process definition model that can define a various types of business processes. A process definition is described in XML and stored in a repository in build time. A process definition describes a workflow of business process in terms of activities, transitions, conditions, invoked applications and roles of persons. When a process instance is created by the run-time workflow engine, the corresponding XML process definition is used as an input of the run-time engine. Inside the run-time management part, i.e. workflow engine, the XML process definition is converted to process definition objects and used in an object-oriented fashion.

The run-time management part is the major part of workflow management system that creates process instances and executes them according to the process definition defined in build time. When we use the term 'workflow engine' in a broad sense, it means the run-time management part as a whole. In a narrow sense, a workflow engine represents a WfEngine component that will be described in Section 4. When a request of process creation is arrived, the run-time workflow engine parses the process definition and manages a process instance according to the process definition. While executing a workflow process, it sequences activities and manages the scheduled activities, arranging work items to user's work lists and invoking related applications that are needed for activity execution.

The runtime engine interacts with the external environment in order to execute an activity. The external environment might be a person or an external application that can be invoked by the workflow engine. For human interaction, the workflow engine creates work items and arranges them to the person who is in charge. When a work item is processed by a person through a worklist handler, an external application might be invoked. Related to the interaction with the external environment, our workflow management framework provides mechanisms to invoke external application, to pass the relevant data to the invoked applications and to check the execution state of invoked applications.

In this research, we design and implement a workflow management system under a distributed environment. The design of the workflow management is designed in component-based, considering extensibility and scalability. Since most of B2B applications usually execute on different computers, its implementation is in Java for achieving platform independent portability. Also for dealing with the network failure problem that may occur while executing a long-lived business process, our system is developed based on our reliable UDP communication infrastructure.

3   The Architecture of Build-Time System

Figure 1 shows the workflow process model that is used in our workflow management system. The workflow process model consists of three major parts: the business process graph part, the application part, and the organization part. The *Business Process Graph Part* presents the structure of a workflow process in a graph. A node of the graph indicates an activity that should be performed in each step within a work process. An activity can be any type of task that is a transactional task, a human interaction web task, or an automatic execution task. Also, an activity can be a sub-process that is composed of a set of subtasks. A link of the graph indicates a transition between activities. The *Application Part* is to decide what to do in specific in executing an activity and how to handle

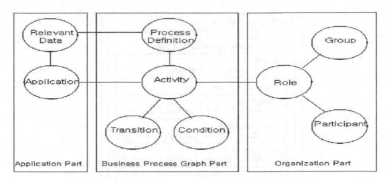

Fig. 1. Definition Model of Workflow Process.

the results. An application object presents the actual application program or a module to be executed. The Relevant Data which is in the Application Part contains the data that are needed to execute a task. The *Organization Part* is to define who is going to execute activities. It is composed of using participators, groups, and roles and used for authentication and authorization management.

We use XML [10] for presenting the workflow process model in a computer. Figure 2 shows the XML structure of workflow process model presented in Figure 1. Tags in the XML format are as follows. The *Add_Process_Definition* tag lets the workflow engine add more process definitions. As its attributes, it contains the Process_Definition tag. The *Process_Definition* tag is the top most tag of a workflow process definition. It defines a workflow process. It includes a number of workflow features, control flows of workflow, and tasks. The *Activity tag* defines each activity in workflow process. Each activity connects the person who performs the activity task to the application that is to be invoked for the task. The *Automation_Mode tag* of an activity is used to select an operation mode, in which the activity is executed automatically by a workflow engine or manually through human interactions. The sub tags of the *Automation_Mode tag* define the necessary operations before executing activities (*Start_Mode tag*), while executing activities (*Execution_Mode tag*), and when to terminate activities (*Finish_Mode tag*). For automatic execution, the condition code in Java is periodically checked to examine the current status. When the condition becomes true, the next operation is automatically executed.

The *Transition tag* defines activity transitions. According to the type of transition, a transition can have a number of input types and output types. The input types are 'single', 'join xor', 'join and', 'join or', 'join replicated xor', 'join replicated and', and 'join replicated or'. The output types are 'single', 'split xor', 'split and', 'split or', and 'split replicated'. Any type of workflow can be constructed as a concatenated combination of various input and output types. The *Input_Activity tag* and *Output_Activity tag* define inputs and outputs of transitions. The number of inputs and outputs can be more than one. As the properties of *Input/Output_activity tag*, there are the activity IDs and the transition conditions to/from an activity.

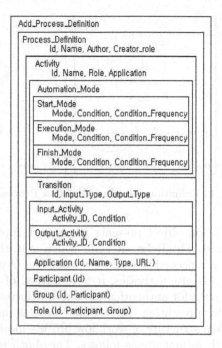

Fig. 2. The XML Structure of Process Definition.

The *Participant tag* and the *Group tag* define the participants and the groups that participate in the workflow process. The *Role tag* defines the role of participants or groups in an activity execution of workflow process. The *Application tag* defines applications invoked for activity executions. There are several types of an invoked application: local process calls, shell scripts, ORB calls, remote execution calls, message passing, transactions, component calls, web applications, and workflow embedded codes.

4   The Architecture of Runtime System

This section introduces the architecture of workflow engine that is the core part of workflow management system. The workflow engine manages the progress of workflow process. It parses the process definition and manages a process instance according to the process definition. While executing a workflow process, it sequences activities and schedules work items to user's work lists. It also has an ability to invoke related applications for executing a specific activity.

4.1   The Structure of Workflow Engine

Figure 3 presents the architecture of our workflow engine. The workflow engine consists of main six modules: Definition Repository Module, Process Definition Module, Enactment Module, Organization Manager Module, Monitor Agent

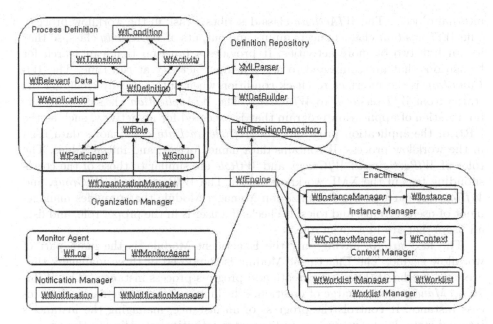

Fig. 3. Structure of Workflow Engine Classes.

Module, and Notification Manager Module. These six modules can be classified into three groups according to their functionalities. The Definition Repository Module, the Process Definition Module, and the Organization Manager Module are in one group. The Enactment Module is another group. The Monitor Agent Module and the Notification Manager Module are in the last group.

The first group, including the Definition Repository Module and the Process Definition Module, takes a role of creating process definition objects. When the workflow engine initially starts at build time, the *WfDefinitionRepository* reads workflow definition XML, creates *WfDefinition* objects with the help of *WfDefBuilder*, and manages the *WfDefinition* objects, so that the *WfDefinitionRepository* can manage workflow process definition objects. When creating *WfDefinition* objects, the *WfDefinitionRepository* delegates the creation task to the *WfDefBuilder* and the *WfDefBuilder* delegates again the XML analyzing tasks to the *XMLParser*. Thus, the *WfDefinitionRepository* brings the *WfDefinition* objects with the helps of the *WfDefBuilder* and the *XMLParser*. We utilize the dynamic class loader in order to redefine the process definition objects dynamically. Whenever the *XMLParser* finds Java codes in the process, the *XMLParser* uses dynamic class loader in Java. When compiling the code, it uses Java compiler information written in the *WfConfiguration*. The workflow reference model defined in XML is converted to a number of classes whose template classes are defined in the Process Definition Module.

The classes in the Process Definition Module are mapped to tags of the XML workflow process model described in Figure 2. The classes provide templates that can be use to create process definition objects by the *WfDefBuilder*. The *WfDefinition* class presents workflow process definition and manages other internal objects. The *WfActivity* class describes a task in the workflow process. The *WfTransition* class connects inputs and outputs of *WfActivity* class, so that it can link two or more activities. It provides a function that can search for transitions that are connected to a given input activity at run time. The *WfCondition* is the interface to check condition when automatically changing the status from *WfTransition* to *WfActivity*. The *WfApplication* class contains the information of application program that is executed for an activity, such as the URL of the application program. The *WfRelevantData* class shows data used in the workflow process. It contains data name, format, and initial values. The roles of *WfParticipant*, *WfGroup*, and *WfRole* are similar to those of the corresponding tags in the XML workflow model. The *WfParticipant*, *WfGroup*, and *WfRole* are parts of the Organization Manager Module that provides management of users, groups, and roles. It checks if a user is in the proper role, and lists all users that are in a given role.

The next group, having only the Enactment Module, is the main part of workflow engine. The Enactment Module is composed of three managers that are used by the *WfEngine* to create and progress process instances. The *WfInstanceManager* is in charge of creation, scheduling, and managements of a process instance. It controls the progress of an instance, managing the instance's internal state. It also sequences and executes activities according to the process definition, invoking related applications. The state of activity is also managed by the *WfInstanceManager*. How to manage the states of process instances and activities is explained in detail in the next section. When the *WfEngine* creates a process instance through the *WfInstanceManager*, it also creates a context through the *WfContextManager*. A context holds the relevant information that should be shared by the consecutive activities of a process instance. A context includes a security context, a transaction context, or data to be transmitted between consecutive activities. For executing interactive activities, the *WfEngine* calls the *WfWorklistManager* to create work items that are sent to a person or a group of persons who have the role of taking the interactive activities. The *WfWorklistManager* manages the progress of each work item so that the work item could be done completely. A workitem can be implemented in various formats, such as an email or a process call, depending on the application domain and the system environment.

The last group contains the modules that provide additional services for the workflow engine. Included are the Monitor Agent Module and the Notification Manager Module. The *WfMonitorAgent* collects the states of workflow engine periodically or on-demand. The collected state logs are used to analyze the correctness or speed of the on-going work processes. It can also report the current state of workflow engine according to a predefined schedule. The *WfNotificationManager* is used by the workflow engine to notify some event messages generated by the process instance. The generated events are distributed to the event listeners registered to the *WfNotificationManager*. It is applied to send messages to associated users when having special event occurrence, such as terminating business processes.

4.2   Scheduling of Workflow Process

One of the important tasks performed by a workflow engine is workflow process scheduling. The *WfInstanceManager* of the Enactment Module is in charge of this scheduling task. When a client application requests to start a business process, the *WfEngine* (i.e. workflow engine) delegates the task to the *WfInstance-Manager*. Receiving the request message of starting process, the *WfInstance-Manager* creates a process instance and manages the lifecycle of the process instance, controlling its process scheduling. The main task of workflow process scheduling is to manage the states of a process instance and its activities. When the *WfInstanceManager* invokes the instance state transition method, it checks the state of workflow process. Based on the current instance state, the *WfInstanceManager* makes the process instance moves to the next step, adds/deletes workitems or executes invoked applications.

There exist two states inside the workflow instance. One is the state of process instance itself, and the other is the state of activities that are executing for the process instance. The state of instance becomes the *Initiated* state when it is activated as shown in Figure 4 (a). In the Initiate state, the instance waits for transition to the *Running* state. The state transition from the *Initiated* state to the *Running* state can occur automatically or human-interactively. During the *Running* state, the process instance progresses its task. According to the process definition, it schedules the next activities and executes the scheduled activities. Detailed description of the state changes of activities is in the next paragraph A process instance goes to the *Suspended* state, when it is temporarily paused waiting for an occurrence of a specific event. When a process instance is abnormally terminated, it goes to the *Terminated* state, and when successfully done, the *Complete* state. After the *Terminated* state or *Complete* state, the instance goes to the *Dead* state, where it releases references and memory back to the system and finishes the instance.

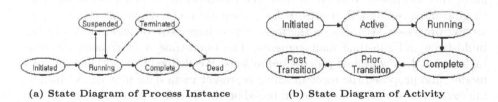

(a) State Diagram of Process Instance   (b) State Diagram of Activity

Fig. 4. Scheduling of Workflow Process.

An activity in execution has six states as in Figure 4 (b). The *Initiated* state is the state where an activity is newly created, but not in execution yet. Depending on the Start mode of the activity, which is described in process definition XML, there are two ways to move to the Active state. If the Start mode is set to *'interactive'*, the activity waits for a human response. If the Start mode is set to *'automatic'*, the activity checks the starting condition and becomes *Active* if satisfied. The *Active* state is another pre-stage before executing an activity. If

653 An Architecture Modelling of a Workflow Management System

the Execution Mode of activity is automatic, the activity executes the invoked application and goes to the *Running* state. If the Execution Mode is interactive, the activity arranges a workitem and goes to the *Running* state. The *Running* state is the state where the activity is in execution. In the *Running* state, the activity checks the finishing condition. If the finishing condition is satisfied, the state of activity changes to the *Complete* state. The *Complete* state checks the Finish mode. If the Finish mode is interactive, all the created workitems are all deleted. If the Finish mode is automatic, the *Complete* state moves to become the *Transition* state. It is necessary to stay in the *Transition* state before activating the next activity, since the complete activity needs to be satisfied with specific conditions to transfer and also the next activity needs to be satisfied with the precondition to be activated. Thus, to consider the *Transition* state, there are more extra kinds of states: *Prior* and *Post Transition* states. The *Prior Transition* state checks the current activity and finds the proper next transition that accepts this activity as the input activity. By observing the input condition of transition, the *Prior Transition* decides to transfer the current working activity. According to the given various input values, types, and the expected conditions, it is necessary to check the prior conditions to enter the next activity. If the condition is satisfied, it applies to the *Post Transition* state. The *Post Transition* state checks output conditions of transition. As done for input cases, the conditions need to be checked according to the given various output values and types. If the output conditions are satisfied, the current *Transition* state moves to the next step and activates another activity, and the next activity becomes *Initiated* state.

5   Conclusion

In this paper, we designed and implemented a workflow management system to allow the workflow of tasks to be changed dynamically as well as statically, and to guarantee to support processes for automatic and asynchronous executions among processes. We focus on the workflow management system in terms of the build-time and run-time managements. The build-time management concerns tasks related to workflow process definitions. We defined the workflow process model, and proposed the mechanism to represent meta data in XML. Considering execution of workflow operation based on the workflow process definition, we handled to generate the processing objects from the XML models. Also we designed process definition to cooperate with the workflow engine by using XML that becomes the Web information standard over Web. In terms of run-time management, we presented the structure of workflow engine that is the encore of run-time workflow management. The engine proposed consists of parsing and processing workflow process definition, managing process instances, sequencing activities, and progressing workflow schedules. Regarding schedules, we handled workflow process and activity state managements. Our workflow management system considers reusability of distributed components, and integrity of information systems for various heterogeneous information handling. Also it is possible to support various workflow services of B2B on Web, for example, parallel structure, internal container structure, and cycle structure. We implemented our workflow management system in Java, so that the workflow engine is also able to achieve platform-independent and portable workflow services.

References

1. Workflow Management Coalition: Workflow Standard Interoperability Abstract Specification, WfMC-TC-1012, Oct. (1996) (http://www.wfmc.org/standards/docs/if4-a.pdf)
2. Workflow Management Coalition: Workflow Reference Model. WfMC-TC-1003, 1.1, 19 Jan (1995) (http://www.wfmc.org/standards/docs/tc003v11-16.pdf)
3. Workflow Management Coalition: Workflow Standard Interoperability. Wf-XML Binding, WFMC-TC-1023, May (2000) (http://www.aiim.org/wfmc/standards/docs/ Wf-XML-1.0.pdf)
4. David Holliingsworth: workflow Management Coalition: The Workflow Reference Model Jan (1995) (http://www.aiim.org/wfmc)
5. Internet Engineering Task Force(IETF): Requirements for Simple Workflow Access Protocol, August (1998) (http://www.ietf.org)
6. Keith Swenson: Simple Workflow Access Protocol (SWAP). Internet-Draft, 7 Aug. 1998, http://www.ics.uci.edu/ ietfswap
7. Object Management Group(OMG): Workflow Management Facilit, Revised submission, 4 July (1998) (ftp://ftp.omg.org/pub/docs/bom/98-06-07.pdf)
8. W3C: Extensible Markup Language (XML). REC-xml-19980210, 1.0, 10 Feb. (1998)
9. John A. Miller, Devanand Palaniswami, Amit P. Sheth, Krys J. Kochut and Harvinder Singh: WebWork: METHOR2's Web-Based Workflow.
10. John A. Miller, Amit P. Sheth, Krys J. Kochut and Devanand Palaniswami: The Future of Web-Based Workflows, The University of Georgia (1998)
11. T. Sollazzo, S. Handschuh, S. Staab, M. Frank: Semantic Web Service Architecture-Evolving Web Service Standards toward the Semantic Web. Proc. of the 15th International FLAIRS Conference. Pensacola, Florida. May 16-18, (2002) AAAI Press.
12. Guenter Orth: The Web Services Framework: A Survey of WSDL, SOAP and UDDI. Mater's Thesis, Information System Institute, Vienna University of Technology. May (2002)
13. Gartner's Report: Enterprise IT Architecture and the Real-Time Enterprise (http://www.tibco.com/mk/geitaw.jsp)

Client Authentication Model
Using Duplicated Authentication Server Systems

Jae-Woo Lee

Software System Lab., Dept. of Computer Science & Engineering, Korea University,
1, 5-ka, Anam-dong, SungBuk-ku, 136-701, Seoul, Korea
It21c@korea.ac.kr

Abstract. In the Internet and distributed systems, we can always access many application servers for gaining many information or electronic business processing, etc. Despite of those advantages of information technology, there have been also many security problems that many unauthorized users attack our network and computer systems for acquiring many information or destroying our resources. In this paper, we propose a client authentication model that uses two authentication server systems, duplicated authentication. Before a client requests information processing to application web servers, the user acquire session password from two authentication servers. The proposed client authentication model can be used making high quality of computer security using the two authentication procedures, user's password and authentication password. The second password by two authentication servers is used in every request transaction without user's input because of storing to client's disc cache when a session is opened first. For more secure authentication we can close session between client and server if a request transaction is not created during a time interval. And then user will acquire authentication password again using logon to the authentication servers for requesting information processing. The client authentication procedure is needed to protect systems during user's transaction by using duplicated password system. And we can detect intrusion during authorized client's transaction using our two client authentication passwords because we can know immediately through stored client authentication password when a hackers attack our network or computer systems.

1 Introduction

In the Internet and distributed systems, we can always access many application servers for gaining many information or electronic business processing, etc. The Internet is comes from interconnecting worldwide servers in networks. That is enables us to live in information society. There is no place that information systems do not reach on our society whole by information technology. Relating to distributed systems, there are various information processing type in addition to the Internet such as Intranet and Extranet. Intranets are networks based on Internet standards and exist within an organization for the benefit of its employees. An Extranet is a private wide area network (WAN) using Internet protocols. An example would be a manufacturer that uses an Extranet to exchange information with its suppliers [1, 2]. In this information society, we can always gain various information easily that we need without restriction of location of the information.

T.G. Kim (Ed.): AIS 2004, LNAI 3397, pp. 655–662, 2005.
© Springer-Verlag Berlin Heidelberg 2005

Despite of those various advantages in the internetworking environments, there have been also many security problems that many unauthorized users attack our network and computer systems for acquiring many information or destroying our resources. Anybody in the Internet can access any server systems in anywhere. Therefore, various security services should be needed to protect our network and computers during security attacks of opponents, such as authentication, encryption, etc. There are several types of security attacks in network and computer security, such as interruption, interception, modification and fabrication. Interruption is that resources of systems is destroyed or becomes unavailable by interrupting transmission between sender and recipient, interception is that an unauthorized opponent gains many information in a network. Modification is that transmission between sender and recipient is modified by opponent's attacks, such as changing values or altering data file. Fabrication is that unauthorized user access network or systems as authorized client[3, 4].

In the Internet or distributed systems, many clients access application web server for gaining many information or business processing. And then, first of all, server systems check whether the client is authorized or not. It is one of the most important problems that we protect many resources of systems using authorized client authentication in the Internet or distributed systems. It is general process of client authentication that a user can get authority by the user's ID and password. But using only client's password is not always secure because there are many security attacks in a network occurred by many opponents, so called hackers. Many unauthorized users always try to know ID and password of authorized clients and hackers know easily authorized user's password using security attacks, like replay [2, 3].

In this paper, we propose a client authentication model that uses two authentication server systems, duplicated authentication. The client authentication model is composed of the user's ID and password and two authentication keys from the duplicated server systems. Before a client requests information processing to application web servers, the user acquire session password from two authentication servers. The proposed client authentication model can be used making high quality of computer security using the two authentication procedures, user's password and authentication password. The second password by two authentication servers is used in every request transaction without user's input because of storing to client's disc cache when a session is opened first. For more secure authentication we can close session between client and server if a request transaction is not created during a time interval. And then user will acquire authentication password again using logon to the authentication servers for requesting information processing. The client authentication procedure is needed to protect systems during user's transaction by using duplicated password system. And we can detect intrusion during authorized client's transaction using our two client authentication passwords because we can know immediately through stored client authentication password when a hackers attack our network or computer systems.

This paper is as following. In section 2, we introduce briefly about security attacks and various security including client authentication. In section 3, we propose client authentication model using two authentication servers. In section 4, explain characteristics of our proposed client authentication model and evaluate our proposed authentication, intrusion detection. Finally, in conclusion we establish more secure client authentication and plan future works.

2 Security Attacks and Authentication

2.1 Various Security Problems

In the Internet or distributed systems, there are several types of security attacks such as interruption, interception, modification and fabrication. Interruption is concerned with availability in web server systems, interception is an attack on confidentiality. Modification is concerned with integrity, and fabrication is an attack on authentication [3, 7].

Interruption is a security attack that resources of systems or network are destroyed or becomes unavailable by interrupting transmission between sender and recipient, interception is illegal action that an unauthorized opponent gains many information in a network for information stealing. And then thus stolen information will be used in accessing, copying, spreading for others. Modification is that transmission between sender and recipient is modified by opponent's attacks, such as changing values or altering data files. And those are transmitted in the network, and an authorized user misunderstood the information by them. Fabrication is that unauthorized user accesses networks or systems for transmitting illegal transaction as an authorized client [3, 4, 7, 9, 10].

2.2 Security Policies

Against the various security attacks many security technology or policies should be needed in our networks or server system. Security service is a protection technology and policy that enhances the security of resources of computer systems. Various security services are applied to computer systems for protecting our network or server systems such as authentication, access control, confidentiality, integrity and non-repudiation. Authentication is to protect server systems by using authorized client authentication in the Internet or distributed systems [3, 4]. Authentication service is to assuring whether a client is authentic or not, by using user's ID, password or internet address, etc.

In the Internet or distributed systems, a server system requires a user's ID and password for preventing unauthorized users from using resources of the server. Authentication is focus on fabrication attack. In addition to the authentication, it is also very important all of messages in a network should be encrypted by protocol of sender and receiver. And then even though hackers gain many information about our network or server systems, the stolen information or message is no useful to them [3, 6].

2.3 Current Client Authentication

Client authentication is process for identifying user's authority by using identity of user or user's own characteristics, etc [3, 4]. In the Internet or distributed server systems, generally we use user's ID, password or internet address, etc. for identifying authorized client. Also, the messages related user's information should be encrypted for protecting contents[4, 5]. The client sends it's own ID and password to destination server. And then, first of all, after selecting authentication database, server systems check whether the client is authorized or not. If the user's id and password is correct,

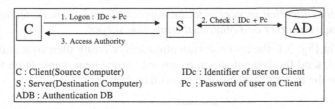

Fig. 2.1. A Current General Authentication in Distributed Systems

the server systems send a message for authority to requested client. Those procedures are shown in figure 2.1.

Many hackers in a network pretend to be an authorized client. Masquerade and replay usually occurred by unauthorized users. They repeat and access until know a password or messages. So, using only client's password does not always provide security because of many times security attack, like replay [8].

3 A New Client Authentication Model

3.1 Duplicated Session Authentication

As above mentioned in section 2, using only client's password is not always secure because there are many security attacks in a network occurred by many opponents. In this paper, we consider that user's transaction is secure and password is needed more complex. We propose a client authentication model that uses two authentication server systems, duplicated authentication. The client authentication model is composed of the user's ID and password and two authentication keys from the duplicated server systems. Before a client requests information processing to application web servers, the user acquire session password from two authentication servers. The proposed client authentication model can be used making high quality of computer security using the two authentication procedures, user's password and authentication password. The second password by two authentication servers is used in every request transaction without user's input because of storing to client's disc cache when a session is opened first. For more secure authentication we can close session between client and server if a request transaction is not created during a time interval. And then user will acquire authentication password again using logon to the authentication servers for requesting information processing. The client authentication procedure is needed to protect systems during user's transaction by using duplicated password system. The proposed duplicated authentication procedure is shown in figure 3.1. The two authentication server create session password using function, randomize(). And the client and application web server store the session password in it's own local disc or database respectively. As result of the duplicated session authentication, the client gains second password, authentication password. The our proposed client authentication procedure is summarized as follows:

1. *A client inputs it's own ID and password for log-in and Send the message for authentication to two authentication server, respectively*
2. *The two authentication servers check the client's ID and password, and create authentication key using a function, randomize()*

3. The two authentication servers send the authentication key encrypted to the client, application web server and other authentication server

As shown in Fig. 3.1, the two authentication server create their own authentication key, password, send the authentication password to the client, application web server, and they share the two authentication password each other.

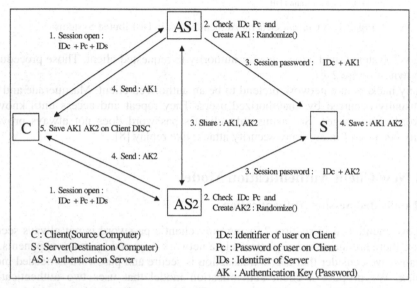

Fig. 3.1. Duplicated Session Authentication Procedures

3.2 Secure Client Authentication

Using the above duplicated authentication, the client acquires two session authentication passwords, and then the client uses two authentication passwords for transaction processing to application web server. First, the client sends its messages to two authentication servers, respectively. And then the two authentication servers check the client's ID, password and authentication password. If the passwords are correct, the client's messages are transferred to application web server. The application web server checks client's ID, password and authentication password and compare the two transferred messages from authentication servers. Using those procedures the web server can detect intrusion during authorized client's transaction because the web server can know intrusion immediately through comparing the two messages and stored session authentication password. As shown in figure 3.2, a client sends two messages including authentication password, Pc and AK_1, AK_2 after session authentication.

1. *A client sends messages to two authentication servers for transaction processing*
2. *The two authentication servers check the client's ID and password, and transfer the client's messages to application web server*
3. *The application web server checks the passwords and compare the two messages from authentication servers*

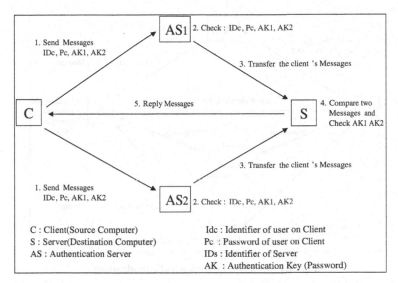

Fig. 3.2. Client Authentication Using Duplicated Authentication Server Systems

4   Characteristics of Our Client Authentication Model

4.1   Secure Client Authentication

Modification is one of the security attacks that a hacker intercepts a message from an authorized user for modifying the contents or data file. The above our proposed client authentication model is focus on modification. At that cases, authentication server or application server can detect the intrusion with comparing authentication password stored database of the server or comparing the two messages from authentication servers.

When an authorized client has received authentication passwords from two authentication servers, the passwords is stored in server's database until session close. And the client uses two passwords including authentication password for request to application web server. Every time a request from a client reaches to application server, the server checks two passwords after comparing the two messages. So, even though a hacker gains a password through security attack, such as replay or masquerade, when the hacker access our network or computer systems using the authorized client's ID and password, server check authentication passwords and two messages transferred from authentication servers. Therefore the hacker cannot access the systems or data file.

4.2   Intrusion Detection and Broadcasting

Intrusion detection is defined as a system detect unauthorized accessing its resources of computer by an unauthorized opponent. When an authorized client send a message to server, a hacker intercept the message and send a false message to the server after modifying the message. And then we can detect the security attacks but that is not easy or impossible without intrusion detection facility or technology.

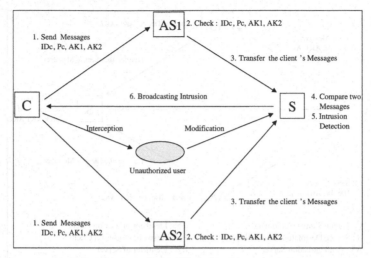

Fig. 4.1. Intrusion Detection and Broadcasting

As shown in figure 4.1, an unauthorized user intercept the message created by a authorized client and the hacker send a false message with fabrication or modification. When server receives the false message from a hacker, server can detect a security attack. The session authentication password is unique key in the server system database, so when server receives the message with any other password the server broadcasts to related client with intrusion, and write the logging in a file. Our proposed authentication model is very simple and authentication database in a server can be maintained easily.

5   Conclusion

In the Internet and distributed systems, we can always access many application servers for gaining many information or electronic business processing, etc. Despite of those advantages of information technology, there have been also many security problems that many unauthorized users attack our network and computer systems for acquiring many information or destroying our resources. It is general process of client authentication that a user can gain authority by the user's ID and password. But using only client's password is not always secure because there are many security attacks in a network occurred by many opponents.

In this paper, we have shown general concepts of security and security attack. And we have described various security services for preventing the security attack. In order to secure user's password we propose a client authentication model that uses two authentication server systems, duplicated authentication. Before a client requests information processing to application web servers, the user acquire session password from two authentication servers. The proposed client authentication model can be used making high quality of computer security using the two authentication procedures, user's password and authentication password. The second password by two authentication servers is used in every request transaction without user's input because of storing to client's disc cache when a session is opened first. For more secure authentication we can close session between client and server if a request transaction is not created during a time interval. And then user will acquire authentication password again using logon to the authentication servers for requesting information processing. The client authentication procedure is needed to protect systems during user's transaction by using duplicated password system. And we can detect intrusion during authorized client's transaction using our two client authentication passwords because we can know immediately through stored client authentication password when a hackers attack our network or computer systems.

In the future, we will further research to prevent various security attacks and detect intrusion. And more secure client authentication model should be needed. It will be very important that we define security attack in detail and various security services for secure client authentication.

References

1. Gurpreet Dhillon, Information Security Management : Global Challenges in the New Millennium, Hershey, Pa. Idea Group Publishing, 2001
2. Rolf Oppliger, Security Technologies for the World Wide Web, Boston, MA Artech House, Inc., 2000
3. William Stallings, Network Security Essentials : Application and Standards, Prentice Hall, 1999
4. Charlie Kaufman, Radia Perlman, Mike Specincr, Network Security : Private Communication in a Public World, Prentice Hall, 1995
5. William Stallings, Cryptography and Network Security : Principles and Practice, Prentice Hall, 1999
6. W. Diffie and M. Hellman, "New Directions in Cryptography," IEEE Transactions on Information Theory, IT-22(6):644-654, 1976
7. S. M. Bellovin and M. Merritt, "An Attack on the Interlock Protocol When Used for Authentication," IEEE Transactions on Information Theory, 40(1):273-275, Jan., 1994
8. Shai Halevi and Hugo Krawczyk, "Public-key Cryptography and Password Protocols," ACM Transactions on Information and System Security, 2(3):230-268, August, 1999
9. Soichi Furuya, "Slide Attacks with a Known-Plaintext Cryptanalysis," Lecture Notes in Computer Science, Vol. 2288:214-225, Springer-Verlag, 2001
10. James Giles, Reiner Sailer, Dinesh Verma, and Suresh Chari, "Authentication for Distributed Web Caches," Lecture Notes in Computer Science, Vol. 2502:126-145, Springer-Verlag, 2002

Dynamic Visualization of Signal Transduction Pathways from Database Information*

Donghoon Lee, Byoung-Hyun Ju, and Kyungsook Han**

School of Computer Science and Engineering Inha University, Inchon 402-751, Korea
khan@inha.ac.kr

Abstract. The automatic generation of signal transduction pathways is challenging because it often yields complicated, non-planar diagrams with a large number of intersections. Most signal transduction pathways available in public databases are static images and thus cannot be refined or changed to reflect updated data. We have developed an algorithm for visualizing signal transduction pathways dynamically as three-dimensional layered digraphs. Experimental results show that the algorithm generates clear and aesthetically pleasing representations of large-scale signal-transduction pathways.

1 Introduction

Advances in biological technology have produced a rapidly expanding volume of molecular interaction data. Consequently, the visualization of biological networks is becoming an important challenge for analyzing interaction data. There are several types of biological networks, such as signal transduction pathways, protein interaction networks, metabolic pathways, and gene regulatory networks. Different types of network represent different biological relationships, and are visualized in different formats in order to convey their biological meaning clearly. The primary focus of this paper is the representation of signal transduction pathways.

A *signal transduction pathway* is a set of chemical reactions in a cell that occurs when a molecule, such as a hormone, attaches to a receptor on the cell membrane. The pathway is a process by which molecules inside the cell can be altered by molecules on the outside [1]. A large amount of data on signal transduction pathways is available in databases, including diagrams of signal transduction pathways [2, 3, 4]. However, most of these are static images that cannot be refined or changed to reflect updated data. It is increasingly important to visualize signal transduction pathway data from databases dynamically.

From the standpoint of creating diagrams, signal transduction pathways are different from protein-protein interaction networks [9] because (1) protein-protein interaction networks are non-directional graphs, whereas signal transduction pathways are directional graphs (digraphs for short) in which a node represents a molecule and an edge between two nodes represents a biological relation between them, (2) protein-protein interaction networks have proteins and edges of uniform types, whereas signal

---

* This study was supported by a grant of the Ministry of Health & Welfare of Korea under grant 03-PJ1-PG3-20700-0040.
** To whom correspondence should be addressed.

T.G. Kim (Ed.): AIS 2004, LNAI 3397, pp. 663–672, 2005.

transduction pathways contain nodes and edges of various types, and (3) signal transduction pathways impose more restrictions on edge flows and node positions than
protein-protein interaction networks.

Fig. 1A shows the mammalian mitogen-activated protein kinase signaling pathway
(http://kinase.uhnres.utoronto.ca/pages/maps.html), represented by a force-directed
layout algorithm. The force-directed layout is a general graph layout, often used for
drawing protein-protein interaction networks. The drawing shown in Fig. 1A follows
several drawing rules faithfully, including no edge crossing and no overlapping

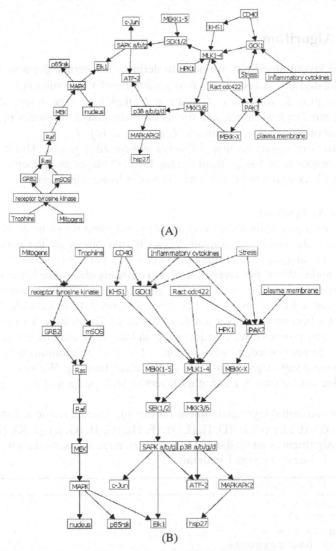

Fig. 1. (A) The mammalian mitogen-activated protein kinase (MAPK) pathway, represented by
a force-directed layout algorithm. (B) The same pathway represented as a layered graph by our
algorithm

among edges and vertices. However, Fig. 1A does not convey its meaning as clearly as Fig. 1B, which represents the same pathway but visualized as a layered graph.

Our experience is that signal transduction pathways convey their meaning best when they are represented as layered digraphs with uniform edge flows, typically either downward or to the right. The Sugiyama algorithm [7, 8] or its variants is most widely used for drawing layered graphs, but the algorithm produces a diagram with various types of edge flow even for an acyclic graph. In the present paper we introduce a new algorithm for automatically representing signal transduction pathways as layered digraphs with uniform edge flows and no edge crossing.

2 Layout Algorithm

To discuss the layout algorithm, we need to define a few terms. Suppose that $G=(V, E)$ is an acyclic digraph. A layering of G is a partition of V into subsets $L_1, L_2, \dots, L_h$, such that if $(u, v) \in E$, where $u \in L_i$ and $v \in L_j$, then $i < j$. The *height* of a layered digraph is the number h of layers, and the *width* is the number of nodes in the largest layer. The *span* of an edge (u, v) with $u \in L_i$ and $v \in L_j$ is $j - i$.

We visualize signal transduction pathways as layered digraphs. The visualization algorithm is composed of 4 steps from the top level: (1) layer assignment, (2) dummy node creation, (3) crossing reduction, and (4) z-coordinate adjustment.

Step 1: Layer Assignment
This step assigns a y-coordinate to every node by assigning it to a layer. Nodes in the same layer have the same y-coordinate value. It first places all the nodes with no parent in layer L_1, and then each remaining node n in layer L_p+1, where L_p is the layer of n's parent node. When the layer of a node is already determined, the node is assigned the larger layer value. For example, node L in the middle graph of Fig. 2 is assigned to layer 4 from path (A, E, I, L), but to layer 3 from path (B, G, L). The larger value of 4 becomes the layer number of node L. The drawback of this layering is that the digraph produced may be too wide and the edge may have a span greater than one. The number of edges whose span > 1 should be minimized because they slow down subsequent steps (steps 2-4) of the algorithm [5]. We place the source node of an edge whose span > 1 in higher layers so that the span of the edge becomes one.

Fig. 2 shows an initial layer assignment for the input signal transduction data: {(A, E), (B, F), (B, G), (C, F), (C, I), (D, I), (E, I), (F, H), (G, J), (G, L), (I, K), (I, L)}. The initial layer assignment is later adjusted to minimize edge spans, as shown in Fig. 3.

Algorithms 1-3 describe step 1 in detail.

Algorithm 1 AssignLevel ()

```
for each node n ∈ V
   n.nodeLayerLvl = -1;
for each node n ∈ V
   if(not (n has parent))
     AssignNodeLevel(n, 0);
AlignBottom();
```

Algorithm 2 `AssignNodeLevel (node, nodeLvl)`

```
if(node.nodeLayerLvl < nodeLvl) {
   node.nodeLayerLvl = nodeLvl;
   for each downward edge e of node
   AssignNodeLevel(child (node), nodeLvl+1);
}
```

Algorithm 3 `AlignBottom ()`

```
for each node n ∈ V {
   if(n.nodeLayerLvl != n.minChildLvl-1)
   n.nodeLayerLvl = n.minChildLvl-1;
}
```

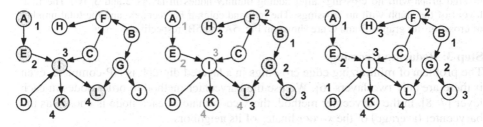

Fig. 2. An example of assignment of nodes to layers. The layer numbers in gray refer to the previously assigned numbers

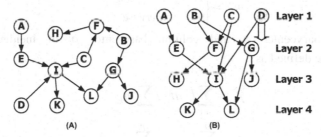

Fig. 3. (A) Initial digraph. (B) Layered digraph. Node D is moved to layer 2 since the span of the edge (D, I) is 2

Step 2: Dummy Node Creation

When every edge is represented by a straight line segment it may cross other edges and/or nodes. In order to avoid this we add a dummy node along an edge and bend the edge at the dummy node. A dummy node is inserted into a layer through which an edge with span > 1 passes (see Fig. 3 and Algorithm 4). Dummy nodes are treated as general nodes when computing their location but are not shown in the final diagrams.

Algorithm 4 `AddDummy()`

```
for each node n ∈ V {
   for each downward edge e of n {
      if (n.nodeLayerLvl+1 < child(n).nodeLayerLvl+1)
      CreateDummyNodes;
   }
}
```

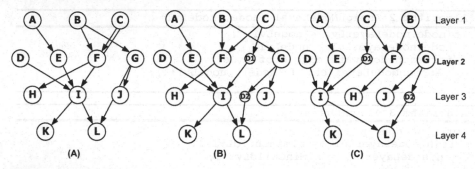

Fig. 4. (A) Initial layered graph with crossings caused by edges (C, I) and (G, L). (B) Modified layered graph with no crossings after adding dummy nodes in layers 2 and 3. (C) The final layout of the graph with no crossings. The row and column barycenters, and the total number of crossings for graphs B and C are shown in Fig. 5A and 5B, respectively

Step 3: Reducing Crossing

The problem of minimizing edge crossings in a layered digraph is NP-complete, even if there are only two layers [6]. We use the barycenter method to order nodes in each layer [7, 8]. In the barycenter method, the x-coordinate of each node is chosen as the barycenter (average) of the x-coordinates of its neighbors.

Suppose that the element $m_{kl}^{(i)}$ of incidence matrix $M(i)$ is given by

$$m_{kl}^{(i)} = \begin{cases} 1: & if (v_k, v_l) \in E \\ 0: & otherwise \end{cases} \tag{1}$$

The row barycenter γ_k and column barycenter ρ_l of incidence matrix $M = (m_{kl})$ are defined as

$$\gamma_k = \sum_{l=1}^{q} l \cdot m_{kl} / \sum_{l=1}^{q} m_{kl} \tag{2}$$

$$\rho_l = \sum_{k=1}^{p} k \cdot m_{kl} / \sum_{k=1}^{p} m_{kl} \tag{3}$$

Then, the number C of crossings of the edge between v_k and v_l is given by

$$C(v_k, v_l) = \sum_{\alpha=k+1}^{p} \sum_{\beta=1}^{l-1} m_{\alpha\beta} \cdot m_{kl} \tag{4}$$

When rearranging the order of rows (columns), the row (column) barycenters are computed and arranged in increasing order with the order of columns (rows) fixed. By repeatedly alternating row and column barycenter ordering, the total number of crossings is reduced. The algorithm for reducing the total number of crossings is given in Algorithms 5-7, and Fig. 5A gives an example of computing the initial row and column barycenters, and the total number of crossings for the graph in Fig. 4B. Fig 5B displays the final row and column barycenters, and the total number of crossings after applying Algorithms 5-7, and the final layout obtained is shown in Fig. 4C. Fig. 6 shows an example of reduction of edge crossings by Algorithms 5-7.

Algorithm 5 `baryCenter(sourceLayer, targetLayer)`

```
for each node n ∈ sourceLayer
  n.calculateBCValue_source(targetLayer);
for each node n ∈ targetLayer
  n.calculateBCValue_target(sourceLayer);
if(isNeedSort) {
  sourceLayer.sortbyBCValue;
  return true;
} else return false;
```

Algorithm 6 `calculateBCValue_target(targetLayer)`

```
BCValue = 0; edge_cntSub = 0;
  for each node n ∈ targetLayer
    if(n has upward edge) {
      BCValue += n.rowIdx;
      edge_cntSub++;
    }
  }
BCValue /= edge_cntSub;
```

Algorithm 7 `calculateBCValue_source (sourceLayer)`

```
BCValue = 0; node_layer_cntSub = 0;
  for each node n ∈ sourceLayer {
    if(n has downward edge) {
      BCValue += n.colIdx;
      edge_cntSub++;
    }
  }
BCValue /= edge_cntSub;
```

Step 4: Z-Coordinate Adjustment

We visualize simple signal transduction pathways as two-dimensional (2D) diagrams. However, we visualize complicated signal transduction pathways as three-dimensional (3D) diagrams for two reasons. First, the lengths of the edges can be more easily made uniform in 3D diagrams than in 2D diagrams. Second, most complicated signal transduction pathways contain a large number of crossings that cannot be removed in a two-dimensional diagram. When there is no edge crossing in the graph after step 3, the side view of the graph will look like Fig. 7A, and the graph is sufficiently clear as a 2D diagram (that is, all nodes have the same z-coordinate value). For graphs with edge crossing even after step 3 (Fig. 7B, for example), we divide the nodes involved in the edge crossing into groups and adjust the z-coordinate values of the groups so that the edge crossing is removed (Fig. 7C). Examples of the side views of graphs involving groups with adjusted z-coordinate values are shown in Figs. 7C and 7D.

Step 4 first computes the number of groups by calculating the maximum number of edge crossings from target (child) layer to source (parent) layer (Algorithm 8) and then partitions the nodes into groups if there is an edge crossing (Algorithm 9). The total number of groups required in a given layer is one more than the maximum number of crossings in that layer.

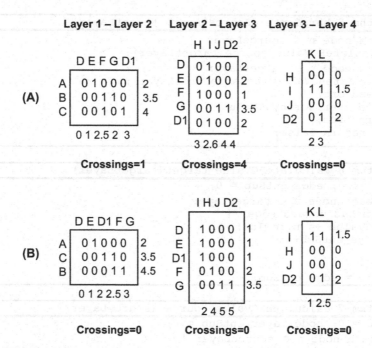

Fig. 5. (A) The initial row and column barycenters, and the total number of crossings for the graph in Fig. 4B. (B) The final row and column barycenters, and the total number of crossings of the graph in Fig. 4C

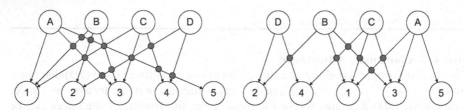

Fig. 6. *Left*: The number of edge crossings is initially 17. The number of crossings among 3 edges is counted 3 instead of 1. *Right*: The number of edge crossings is 6 after reducing crossings

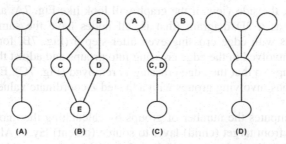

Fig. 7. (A) Side view of a planar graph. (B) Example of a non-planar graph with edge crossings. (C) Side view of the graph shown in (B) after varying the z-coordinate values of nodes A and B. (D) Side view of a graph with 3 groups of nodes in layer 1

Algorithm 8 `getGroupCnt ()`

```
groupCnt=0;
for each group g in targetLayer
  for each node n ∈ g
    for each upward edge e of n {
      if(leftCrossing(e)>groupCnt) groupCnt=leftCrossing(e);
      if(rightCrossing(e)>groupCnt)groupCnt=rightCrossing(e);
    }
groupCnt++;
```

Algorithm 9 `AssignGroupCnt ()`

```
for each upward node n_ij connected to group g_i of targetLayer
{
  Assign node n_i1 to group g1;//n_ij: j-th upward node of g_i
  for each node n_ij {
  Assign node n_ij to group g_x+y, where
  x = group number to which (j-1)th node is assigned.
  y = 1 if there is an edge crossing between (j-1)th node
and j-th node.
  y = 0 otherwise.
  if (x+y > total number of groups)
       The group number of the j-th node is 1;
  }
}
```

The leftCrossing and rightCrossing in Algorithm 8 are the number of edge crossings associated with the left and right nodes, computed from equations 5 and 6, respectively.

$$\text{leftCrossing}(e) = \sum_{\alpha=k+1}^{p} f(\alpha,l), \text{where } f(\alpha,l) = \begin{cases} 0 & if \sum_{\beta=1}^{l-1} m_{\alpha\beta} = 0 \\ 1 & otherwise \end{cases} \quad (5)$$

$$\text{rightCrossing}(e) = \sum_{\alpha=1}^{k-1} f(\alpha,l), \text{where } f(\alpha,l) = \begin{cases} 0 & if \sum_{\beta=l+1}^{q} m_{\alpha\beta} = 0 \\ 1 & otherwise \end{cases} \quad (6)$$

3 Results and Discussion

We implemented the algorithms in Microsoft C#. The program runs on any PC with Windows 2000/XP/Me/98/NT 4.0 as its operating system. The program accepts the input signal transduction data in several formats:

- **pnm:** a pair of names for the source node and target node, separated by a tab, in each line (the input data of the graph in Fig. 2).
- **pid:** a pair of node indices for the source node and target node, separated by a tab, in each line. Additional information is provided by pid_Label and pid_Pos files. Pid_label contains node labels and pid_Pos contains the x-, y-, and z-coordinates of each node.
- **stp:** consists of 2 sections. The node section contains node ID, x-, y-, and z-coordinates of the node, node type and node label. The edge section contains the source node ID, target node ID, and edge type.

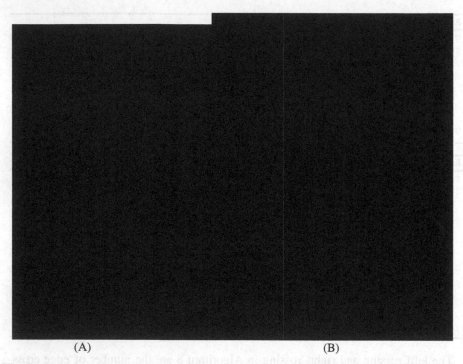

(A)   (B)

Fig. 8. (A) The signal transduction pathway of a tyrosine kinase-linked receptor and G-protein-coupled receptors, displayed by our algorithm. Red arrows indicate that the source nodes activate the sink nodes, while blue lines ending with filled circles indicate that the source nodes inhibit the sink nodes. Black, dotted arrows refer to signals transferred between sub-cellular locations. (B) Proteins involved in a metabolic pathway of *E. coli*. Edge bends at dummy nodes are removed by transforming the bended edges into spline curves

- **mdb:** data from a Microsoft Access database, which can provide all the information supported by the pnm and stp formats.

Fig. 8A shows the signal transduction pathway of a tyrosine kinase-linked receptor and G-protein-coupled receptors (http://www.hippron.com), visualized with our algorithm. The pathway contains exclusively straight-line segments and edges of span=1, but the diagram contains no crossings. Fig. 8B shows a collection of proteins involved in a metabolic pathway of *E. coli* (http://maine.ebi.ac.uk:8000/services/biolayout/examples/metabolic_pathways), visualized by a new version of the algorithm. Edge bends at dummy nodes are removed by transforming the bended edges into spline curves. The drawing appears to have edge crossings, but it actually contains no edge crossing when it is visualized as a three-dimensional drawing on a video monitor. The program allows the user to explore the three-dimensional drawing by rotating or by zooming in or out of it. The size of a node can also be changed in proportional to its Z-coordinate value.

In summary, most drawings of signal transduction pathways in databases are static images and thus cannot be refined, or changed later, to reflect new data. We have developed an algorithm for automatically representing signal transduction pathways from databases or text files. Unique features of the algorithm include (1) it does not place all sink nodes (nodes with no parent) in layer 1 but moves them to a lower layer so that edge spans can be minimized; (2) edge bends occur only at dummy nodes and the number of edge bends is minimized; (3) edge crossings are removed by adjusting the z-coordinates of nodes. We are currently extending the program to overlay various types of additional information onto the signal transduction pathways.

References

1. Hirsch, E.D., Kett, J.F., Trefilet J. (ed.): The New Dictionary of Cultural Literacy. Boston (2002)
2. Kanehisa, M., Goto, S., Kawashima, S., Nakaya, A.: The KEGG databases at GenomeNet. Nucleic Acids Research 30 (2002) 42-46
3. Hippron Physiomics, Dynamic Signaling Maps. http://www.hippron.com/products.htm
4. BioCarta. http://www.biocarta.com
5. Gansner, E.R., Koutsofios, E., North, S.C., Vo, K.-P.: A technique for drawing directed graphs. IEEE Transactions on Software Engineering 19 (1993) 214-230
6. Garey, M.R., Johnson, D.S.: Crossing Number is NP-Complete. SIAM J. Algebraic Discrete Methods 4 (1983) 312-316
7. Sugiyama, K., Tagawa, S., Toda, M.: Method for visual understanding of hierarchical system structures, IEEE Transaction on Systems, Man, and Cybernetics SMC-11 (1981) 109-125
8. Sugiyama, K.: Graph Drawing and Applications for Software and Knowledge Engineering. Singapore (2002)
9. Han, K., Ju, B. H.: A fast layout algorithm for protein interaction networks. Bioinformatics 19 (2003) 1882-1887

Integrated Term Weighting, Visualization, and User Interface Development for Bioinformation Retrieval

Min Hong*, Anis Karimpour-Fard*, Steve Russell*, and Lawrence Hunter

Bioinformatics, University of Colorado Health Sciences Center,
4200 E. 9th Avenue Campus Box C-245, Denver, CO 80262, USA
{Min.Hong,Anis.Karimpour-Fard,Steve_Russell,Larry.Hunter}
@UCHSC.edu

Abstract. This project implements an integrated biological information website that classifies technical documents, learns about users' interests, and offers intuitive interactive visualization to navigate vast information spaces. The effective use of modern software engineering principles, system environments, and development approaches is demonstrated. Straightforward yet powerful document characterization strategies are illustrated, helpful visualization for effective knowledge transfer is shown, and current user interface methodologies are applied. A specific success of note is the collaboration of disparately skilled specialists to deliver a flexible integrated prototype in a rapid manner that meets user acceptance and performance goals. The domain chosen for the demonstration is breast cancer, using a corpus of abstracts from publications obtained online from Medline. The terms in the abstracts are extracted by word stemming and a stop list, and are encoded in vectors. A TF-IDF technique is implemented to calculate similarity scores between a set of documents and a query. Polysemy and synonyms are explicitly addressed. Groups of related and useful documents are identified using interactive visual displays such as a spiral graph that represents of the overall similarity of documents. K-means clustering of the similarities among a document set is used to display a 3-D relationship map. User identities are established and updated by observing the patterns of terms used in their queries, and from login site locations. Explicit considerations of changing user category profiles, site stakeholders, information modeling, and networked technologies are pointed out.

1 Introduction

The volume of biological and medical information has been growing rapidly, leading to increased interest and research in information retrieval and presentation, knowledge extraction, and the enhanced discovery of new ideas. The text indexing effort here examines terms used in a database of Medline papers to determine internal consistencies and patterns. Text ranking applies the frequencies of the constituent text terms in order to characterize documents. Different user types have varied interests in documents and in the distinctive terms in these documents. Similarities in users and documents are used here as indicators of the technical arenas, concepts, timeframes, and research directions with the highest value to the reader.

---

* Contributed equally.

T.G. Kim (Ed.): AIS 2004, LNAI 3397, pp. 673–682, 2005.
© Springer-Verlag Berlin Heidelberg 2005

The frequencies of words in each document and in the user's query are used to derive a measure of information pertinence. Documents closest to a query which is also similar to the user's overall word preferences, are ranked and returned in order. The ratings based on key terms are also visualized in a spiral display. The care that is expended in biomedical information site implementation (as in commercial websites) is directly related to the ultimate success and utility of the site.

2   Methods

The objective here is to illustrate effective document characterization, helpful information visualization, and adaptive awareness to user categories and individual interests.

2.1   Text Indexing and Analysis

A vector space model is implemented to process document terms. Similarity of publications is represented using the term frequency vector for each document. The test domain here is Medline abstracts for articles on breast cancer. The word characterization method is based on stemming text terms, using a stop list to exclude common terms, and on creating a term-frequency vector for each document. Such vector space models are widely used based on TF-IDF (term frequency – inverse document frequency) rules for calculating similarity between documents. There has been growing research in biological text processing focusing on different areas such as clustering [11], categorization using predefined categories [22], detection of keywords [8, 16], and the detection and extraction of relations [4, 26, 27]. These advances are applied for the tasks here, using text terms and user profiling, and in the next phases it is anticipated that categorization of terms and direct query modifications will also be implemented.

There are many biological terms that have multiple synonyms, and there is a lack of uniformity in the set of terms used by researchers. One consequence is that identical queries from dissimilar users can result in their getting the identical set of returned documents, which is rarely the best set of publications for their separate needs. Documents and queries are represented as bags of terms and statistical values. Each document vector shows the presence or absence of a term, or the weight of each term within the document. The construction of the document-term vector space can be divided into three different stages: document indexing, weighting of index terms, and document ranking. In the document indexing stage, non-significant terms and words that do not describe context are removed. Word stemming reduces terms to a root form. Such feature reduction can improve efficiency as well as accuracy in document clustering and classification [28, 29].

Different weighting schemes can be used to discriminate one document from the other. Experimentally it has been shown that TF-IDF discrimination factors lead to more effective retrieval [25]. Terms that occur in only few documents are more valuable for characterization than the terms that repeatedly appear in many documents (IDF = Inverse document frequency). On the other hand, the more often that a term occurs within a certain document, the more likely it is that that term is important to that document (TF = term frequency). The formula that is used here is [18]:

$$IDF (t) = \log (N / N_t)$$
$$N = \text{total number of documents in the set}$$
$$N_t = \text{number of documents in which the term t occurs}$$

The weighting of the ith term in a document is given from its frequency f_i by:

$$W_i = \text{weight of term } i = f_i * IDF$$

One of the methods of adding concepts to augment the vector space approach is latent semantic indexing (LSI). This approach creates a matrix of terms that is analyzed by singular value decomposition (SVD) for the most different and predictive items. This is similar to clustering in that it solves the synonym problem but not polysemy [3, 7]. Other methods such as clustering or categorization of documents can be applied after documents vectors are created. Manual categorization is becoming impractical [14, 15] due to the volume of documents. So, clustering is most often accomplished using an unsupervised learning algorithm. Categorization generally achieves better results, using human supervision.

A simple approach to add conceptual information is the method used by the Semantic And Probabilistic Heuristic Information Retrieval Environment (SAPHIRE). SAPHIRE predefines a set of allowable terms [10], and maps the words and phrases to a set of canonical terms. Another approach is Northern Light, which classifies the documents into pre-defined subjects [34]. Another example using added concepts is the DYNOCAT system, which uses a knowledge server [14, 15]. It maps words and phrases to canonical terms, but it does not weight the term vector with frequency information [2]. This dynamic categorization uses a semantic base representation of terms, from the semantic type of each term. The most traditional way for adding conceptual information is natural language processing. Using this approach after pre-processing, tagging is applied. Many systems have been implemented using this approach [8, 17, 32]. Vector space modeling such as the SMART project does not place restrictions on the subject matter. SMART uses relevance feedback to tighten its search, but its user interface is quite unfriendly [19, 30]. Another vector system is Fox, where concepts other than normal content terms are used. An extended vector is assembled with classes of information about documents. Subvectors represent different concept classes, and similarity between extended vectors is calculated as a linear combination of corresponding subvectors [6].

Here, vectors are created from Medline abstracts to see how eliminating common words and infrequent terms affect document classification. The steps are:

1. Eliminate common English words:
   a. A "stop- list" containing 582 words is identified, ignoring words with little information content. Different lists are gathered from various sources and are merged together and sorted.
   b. The vector length is reduced by selecting a subset of key words. Then, a refinement is performed to find biomedical terms.
   c. The abstracts are parsed, stemming is applied, and a frequency count is tallied for each term. The number of documents in which each term appears is determined and the terms that occur most are identified. The importance of each term is assumed to be inversely proportional to the number of documents that contains that term. This IDF weighing scheme specifies that terms with low document membership are less discriminating [28].

2. Abstracts are chosen if they contain "breast", "cancer", "brca", or "gene".
3. Terms are normalized to lower case; non-alphanumeric terms are removed.
4. Common biomedical terms are removed.
5. For terms with the length six or more [11] forms (like cancer, cancerous) are re-placed with their stem (cancer), using the Porter algorithm [12, 13].
6. Less frequent terms are deleted using a cut-off threshold, reducing the dimension-ality of the document vectors, since infrequent terms have little resolving power (Zipf's Law) [1, 25, 28].

If the document collection contains n unique terms, then each document is repre-sented as a vector of length n: $(t_1, t_2, t_3, t_4, \ldots t_n)$ where t_i is proportional to the fre-quency of term i. The vector space is also represented in a simplified form by using 0s and 1s if the term is absent or present [1, 21- 25, 28]. The jth document d_j is thus written as an n-tuple $d_j = <w_{1j}, w_{2j}, w_{3j}, w_{4j}, \ldots, w_{nj}>$, and the ith keyword ti is written as an m-tuple $t_i = <w_{i1}, w_{i2}, w_{i3}, w_{i4}, \ldots, w_{im}>$.

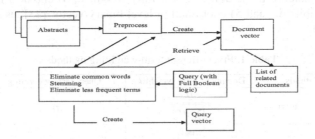

Fig. 1. Creating query vectors from abstracts

2.2 Query Process

The words in a user query are analyzed in a manner similar to that performed on the document abstracts. In addition, there are methods for improving information re-trieval that involve adjusting the user query. For instance, the query can be expanded by adding all the terms that occur in the same category or cluster as each of the query terms [9, 31]. The list of returned documents is determined by Boolean operands "and" or "or". Synonyms are entered manually, but in further versions of the system this will done automatically. The steps in processing a query are 1) eliminate com-mon English words, 2) apply stemming, 3) create a query vector, 4) compare to the space of document term vectors, and 5) return the most related documents.

2.3 Document Ranking Using User Personalization

The relative importance-weighting levels of terms for a particular user (or group) are obtained and applied at query time. If the user query contains a given term, the slot for that term is set to one – otherwise the slot is set to zero. Profile information relates to terms that are important to a user. Medline users could be, say, medical doctors (MDs), college students, college teachers, PhD researchers, or even information-browsing patients. A list of terms is assembled that reflects those words commonly found in queries from such users. User evaluations of the relative importance of terms give a scale from zero to ten.

TERM	Overall Occurrence	Importance	MD	PhD Lab	PhD Teach	Student	Browser
adult	1		7	5	6	5	8
advanc	10		5	4	5	4	6
advantag	3		7	7	6	7	7
advens	1		4	4	4	4	4
advers	4		7	4	4	4	8
advis	3		7	4	4	4	7
advoc	1		7	4	4	4	9
affect	20		7	7		7	8
afford	1		7	4	4	4	9
african	4		6	4	4	6	6
age	58		9	7		6	9

Fig. 2. Term importance for user type

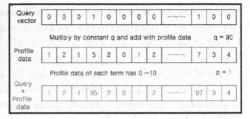

Fig. 3. Query vector and profile data

The vector for the query is combined with the preference-profile vector for that type of user. The influence of the profile is initially set so that 90% of the importance in the final query+profile vector is derived from the actual query, and the other 10% is derived from the profile vector. The system offers five variations of term-weighting: 1) standard tf (default), 2) best fully weighted, 3) classical tf-idf, 4) best weighted probabilistic, and 5) coordination level binary vector [23], as shown in the following table.

Table 1. Five different term-weighting methods

Methods	Document term weighting	Query term weighting
txc/txx	$\dfrac{tf}{\sqrt{\sum (tf_i)^2}}$	tf
tfc/nfx	$\dfrac{tf\log\frac{N}{n}}{\sqrt{\sum \left(tf\log\frac{N}{n_i}\right)^2}}$	$\left(0.5+\dfrac{0.5tf}{\max tf}\right)\log\dfrac{N}{n}$
bxx/bxx	1 (if term is present)	1 (if term is present)
tfx/tfx	$tf\log\dfrac{N}{n}$	$tf\log\dfrac{N}{n}$
nxx/bpx	$0.5+\dfrac{0.5tf}{\max tf}$	$\log\dfrac{N-n}{n}$

Documents are ranked and presented in an order that matches the query and the user's inferred interests. There are two well-known methods for such purposes, cosine similarity and distance similarity. The cosine vector similarity (inner product or dot product) is used here [33]. Each document vector is compared with the query-profile vector to give a -1 to +1 value. The similarity is computed as:

$$sim(Q,D_i) = \frac{\sum_{j=1}^{t} w_{q_j} \cdot w_{d_{ij}}}{\sqrt{\sum_{j=1}^{t} (w_{q_j})^2 \cdot \sum_{j=1}^{t} (w_{d_{ij}})^2}}$$

Fig. 4. Similarity between documents

When the documents are well-aligned, the angle between them is small and the cosine is near positive one.

2.4 Interactive Display and Intuitive Refinement

A spiral ranking graph is adopted [35], where each document is shown by a small dot, starting with the highest ranked (similar) document in the center. The user can see if the scores are distributed evenly or whether just a few documents are good matches. Documents that cluster together have a similar interest value to the user.

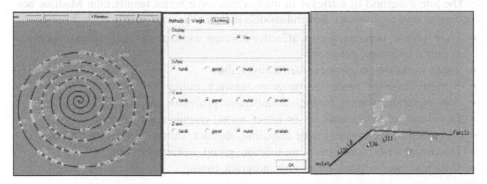

Fig. 5. Spiral ranking graph (left), clustering menu (middle), 3D vector visualization (right)

The interface includes sliders that enable refinement of the query term-weighting and group-profiling. The graph then dynamically changes the corresponding arrangement of the dots. Visualization of high dimensional term factors is narrowed through a manual selection of three selected term-dimensions. A k-means clustering algorithm is applied to distinctively color groups of similar documents.

2.5 Web Site Design

Website development should be based on principles, engineered to be flexible, and grounded in measurability. Commercial sites generally involve teams with special skills, integrating their efforts during the site's design, construction, evaluation, and support. Among the key areas are usability, User Experience Design (UXD), and personalization. Specific personas are walked through mainstream and side-path visits, and storyboards are used to explicitly trace out the key page-clicks and navigation. Novice web site designers often suffer from "developer's fallacy" in assuming that all users perceive the site like they do. Unfortunately, foolproofing a site is quite frustrating. Unlike commercial sites, academic sites are often in transitional situations over time, without a sustained vision or a way of measuring value and choosing the most important upgrades.

Several popular and important biologically-related Internet sites were reviewed, with a starting point of usability features and aids. Only a subset of all review approaches and metrics could be considered, so there were several informal interviews with various users of the sites. A walkthrough session by a typical user is common in commercial website evaluations, where the person makes selections and follows typical paths with good and not so good services from the site. Generally, bio-sites were found to be indifferent to user needs, differences, and difficulties.

The development here was undertaken to expose ideas and clarify alternatives related to the above objectives. In achieving modest site-development and information-access goals, a more central system coding purpose was also achieved - that of following explicit examples of design choices, time tradeoffs, and user awareness - as a contribution to the partnership of informatics and the pursuit of biological understanding.

The site's method of retrieval in many cases gave better results than Medline according to user evaluations. Several models underlie the user interface for the site.

- User: profile, persona, group, affective concern, motivation, goal, value, expertise, country, language, pet peeve
- Task: workflow, component, difficulty, failure
- System: database, processor, program, network, policy
- Site owner: institution, finder, administrator, professor
- Developer: coder, consultant, architect, tester, graphic artist, student
- Document: page layout, image map, link, abstract, summary, title, rating, size
- Document writer: profession, motivation, group, institution, country
- User interface: session, page, visit, program, timing

2.6 User Acceptance and Term-Associated User Profile Vectors

Another issue addressed explicitly is that of the measurement of success. The site was pre-determined to have a standard of satisfactory capability if a subset of documents could be accessed in an individual manner by disparate users.

3 Results

188 documents were used to evaluate the system, 29 of which did not contain any abstract. Only a small number of the most important words within a document provide the key information for classification, so a next phase of this project will address the elimination of terms that provide little semantic content or significance. A query ("breast" & "cancer" & "brca" & "gene") was entered as a test of the system's re-

Table 2. Identification and pruning of key terms

Step	# of terms
1.a Create stop-list	582 terms in stop-list
1.b Identify common terms within corpus (weight threshold = .15)	3 terms added to 1a
2 Terms within the corpus of documents	34832
3 After eliminating numeric and non-alphanumeric terms	30875
4 After eliminating common words	17884
5 After stemming and eliminating redundant terms	2334
6 After eliminating less frequent terms (threshold = .01)	1649

trieval. In this case, 106 out of the 159 documents were successfully retrieved. 34 out of the 53 documents that were not retrieved did not contain the term "gene". Two of the documents did not contain any of the exact terms. The system here missed some documents that Medline retrieved, since the Medline search also uses concept enhancements rather than just the query or document terms.

It is possible for the user to adjust the weightings to assess the sensitivity of the with respect to key terms. The sensitivity related to the term changed can be calculated from the number of positions in the returned ranking that each document changes. The prototype site here is adaptable in that it encourages user self-identification for improved query handling. The user can also tell the site what terms were favored or to be excluded. The site was also set up to be automatically adaptive, with adjustments based on user activities, in the form of web visit logs. Different term-weighting methods give different ranking results.

4 Discussion

The speed of computer processors has grown at a rate of an 18 month doubling (Moore's Law). The amount of data has also grown, but it is poorly understood that this rate exponentially exceeds processing power ("Russell's Law") and that the ability to transport data is also lagging behind. The increasing percentage of unused data is a concern that makes encyclopedic understanding ever less possible in areas like bio-tech and medical research. There is an increasing reliance on search engines, yet queries often return thousands of documents in less than optimal orderings. There is a need to provide different users with focused information to reduce the complexity of details.

In this paper a system is shown that incorporates adaptive term-frequency alignment with user profile data. The system also provides interactive visualization to give an immediate overview of the retrieval space. Query-term clustering indicates the relationships between documents, so the user can readily find similar documents.

Improvements to the system here could involve adding concepts and semantic types to the term processing by using UMLS [5]. In addition, categorizing results from queries using the MeSH hierarchy (Medical Subject Headings) could be of value.

References

1. V. Anh, O. Kretser, and A. Moffat, Vector-space ranking with effective early termination, Proceedings of the 24th annual international ACM SIGIR conference on Research and development in information retrieval, p.35-42, September 2001, New Orleans, Louisiana, United States
2. D. Berrios, R. Cucina, P. Sutphin, L. Fagan, Methods for Semi-Automated Indexing For High Precision Information Retrieval. 2002
3. Berry, M., Dumais, S., and Letsche, T., Computational methods for intelligent information access, Proc. of Supercomputing '95, San Diego, CA: USA, 1995.
4. C. Blaschke, M. Andrade, C. Ouzounis, and A. Valencia, Automatic extraction of biological information from scientific text: protein-protein interactions. Intelligent Systems for Molecular Biology, Heidelberg, p. 60, 1999.

5. K. Campbell, D. Oliver, and E. Shortliffe, The Unified Medical Language System: Toward a Collaborative Approach For Solving Terminologic Problems Submitted to a Special Issue of the Journal of the American Medical Informatics Association
6. C. Crouch, S. Apte, and H. A. Bapat. An IR approach to XML retrieval based on the extended vector model, http://www.dumn.edu/~ccrouch/pubs/XML_Paper_final.pdf
7. Deerwester, S., Dumais, S., Furnas, G., Landauer, T., & Harshman, R. Indexing by latent semantic analysis. Journal of the American Society for Information Science, 41, 391-407, 1990
8. K. Fukuda, A. Tamura, T. Tsunoda, and T. Takagi, Toward information extraction: identifying protein names from biological papers, Pac. Symp. Biocomput, p. 707, 1998
9. A. H. Griffiths, Claire Luckhurst, and Peter Willett, P Using inter-document similarity information in document retrieval systems, Journal of the American Society for Information Science, vol. 37, 3-11, 1986.
10. W. Hersh, and R. Greenes, SAPHIRE – An information retrieval system featuring concept matching, automatic indexing, probabilistic retrieval, and hierarchical relationships. Computers and Biomedical Research 23: 410-425, 1990
11. Iliopoulos, A. J. Enright, C. A. Ouzounis; TEXTQUEST: Document Clustering of Medical Abstracts for Discovery in Molecular Biology; Proceedings of the Sixth Annual Pacific Symposium on Biocomputing (PSB 01), 384-395, 2001
12. B. Lovins, Development of a stemming algorithm. Mechanical Translation and Computational Linguistics, 11, 22-31, 1968
13. M.E. Porter, An algorithm for suffix stripping program, 14(3), 130-137, 1990
14. Pratt, L. Fagan, The Usefulness of Dynamically Categorizing Search Results, Journal of the American Medical Informatics Association (JAMIA), 7(6), 605-617, 2000
15. W. Pratt, H. Wasserman, QueryCat: Automatic Categorization of MEDLINE Queries. Proceedings of the American Medical Informatics Association (AMIA) Fall Symposium 2000
16. D. Proux, F. Rechenmann, L. Julliard, V. Pillet, and B. Jacq, Detecting gene symbols and names in biological texts: a first step toward pertinent information extraction, Genome Informatics Workshop, Tokyo, p. 72., 1998
17. TC Rindflesch L. Tanabe, JN Weinstein, L. Hunter, EDGAR: extraction of drugs, genes and relations from the biomedical literature, Pac Symp Biocomput, 517-28.
18. Robertson, S., Walker, S., Some simple effective approximations to the 2-Poisson model for probabilistic weighted retrieval, Proceedings of the 17th annual international ACM SIGIR conference on Research and development in information retrieval, 232-241, July 3-6, 1994
19. J. Rocchio, The SMART retrieval system - experiments in automated document processing, Relevance feedback information retrieval, ed. G. Salton, p. 313, Prentice-Hall, Englewood Cliffs NJ, 1971.
20. S. Russell, Knowledge Liquidity, Proceedings of Knowledge World, July, 1999.
21. G. Salton, Automatic content analysis in information retrieval, University of Pennsylvania, PA, 1968.
22. G. Salton, Developments in automatic text retrieval, Science 253, 974, 1991.
23. G. Salton, Term-weighting approaches in automatic text retrieval, Information Processing & Management, Vol 24, 5, 513-523, 1988
24. G. Salton., Automatic Text Processing: the transformation, analysis, and retrieval of information by computer, (Addison-Wesley, Reading MA, 1989)
25. G. Salton., A. Wang, and C. Yang, A vector space model for information retrieval. In Journal of the American Society for Information Science, vol 18, 613-620, 1975
26. T. Sekimizu, H. Park, and J. Tsujii, Identifying the interaction between genes and gene products based on frequently seen verbs in Medline abstracts, Genome Informatics Workshop, Tokyo, p. 62, 1998

27. J. Thomas, D. Milward, C. Ouzounis, S. Pulman, and M. Carroll, Automatic extraction of protein interactions from scientific abstracts, Pac. Symp. Biocomput, p. 538, 2000

28. T. Tokunaga, and M. Iwayama, Text categorization based on weighted inverse document frequency. Technical Report 94 TR0001, Department of Computer Science, Tokyo Institute of Technology, Tokyo, Japan, 1994.

29. S. Uger, and S. Gaucho, Feature reduction for document clustering and classification. Technical report, Computing Department, Imperial College, London, UK, 2000.

30. E.M. Voorhees, in WordNet: an electronic lexical database, Using WordNet for text retrieval, ed. C. Fellbaum, p. 285, MIT Press, Cambridge MA, 1998.

31. P. Willett, An algorithm for the calculation of exact term discrimination values, Information Processing and Management: an International Journal, v.21 n.3, 225-232, 1985

32. Yoshida, K. Fukuda, T. Takagi, PNAD-CSS: a workbench for constructing a protein name abbreviation dictionary, Bioinformatics, 16:169-75, Feb 2000.

33. Information Retrieval lecture note (University of Massachusetts Amherst) http://ciir.cs.umass.edu/cmpsci646

34. Northernlight, http://www.northernlight.com/

35. Interactive 3D Visualization for Document Retrieval, http://zing.ncsl.nist.gov/~cugini/uicd/viz.html

CONDOCS: A Concept-Based
Document Categorization System
Using Concept-Probability Vector with Thesaurus

Hyun-Kyu Kang[1], Jeong-Oog Lee[1], Heung Seok Jeon[1], Myeong-Cheol Ko[1],
Doo Hyun Kim[2], Ryum-Duck Oh[3], and Wonseog Kang[4]

[1] Dept. of Computer Science, Konkuk University,
322 Danwol-dong, Chungju-city, Chungbuk, 380-701, Korea
{hkkang,ljo,hsjeon,cheol}@kku.ac.kr
[2] Dept. of Internet & Multimedia, Konkuk University,
1 Hwayang-dong, Gwangjin-gu, Seoul, 143-701, Korea
doohyun@konkuk.ac.kr
[3] Dept. of Computer Science, Chungju National University,
123 Goemdan-ri, Eryu-meon, Chungju-city, Chungbuk, 380-702, Korea
rdoh@gukwon.chungju.ac.kr
[4] Dept. of Computer Engineering Education, Andong National University,
388 SongChun Dong, Andong, Kyungpook, 760-749, Korea
wskang@anu.andong.ac.kr

Abstract. Traditional approaches in document categorization use the term-based classification techniques to classify the documents. The techniques, for enormous terms, are not effective to the applications that need speedy response or not much space. This paper presents an effective concept-based document categorization system, which can efficiently classify Korean documents through the thesaurus tool. The thesaurus tool is the information extractor that acquires the meanings of document terms from the thesaurus. It supports effective document categorization with the acquired meanings. The system uses the concept-probability vector to represent the meanings of the terms. Because the category of the document depends on the meanings than the terms, even though the size of the vector is small, the system can classify the document without degradation of the performance. The system uses the small concept-probability vector so that it can save the time and space for document categorization. The experimental results suggest that the presented system with the thesaurus tool can effectively classify the documents. The results show that even though the system uses the contracted vector for document categorization, the performance of the system is not degraded.

1 Introduction

The majority of document classification systems use term-based techniques to classify documents. However, for the term-based document categorization system [1,3], it is difficult to effectively classify the documents for enormous terms. If we decrease the number of terms to get effective classification, the classification quality of the system will be lowered. So, we need the system to get effective result without declining the performance. This paper presents such an effective system with the thesaurus

T.G. Kim (Ed.): AIS 2004, LNAI 3397, pp. 683–691, 2005.
© Springer-Verlag Berlin Heidelberg 2005

tool. It is a concept-based document categorization system (CONDOCS) that classifies Korean documents using the thesaurus tool. Even though there have been researches on the thesaurus tool [13], there is no practical thesaurus tool for Korean. Because the semantics of Korean is different from other languages, the previous thesaurus tools are not adequate to Korean document categorization. In this paper, we present a thesaurus tool for Korean to use in Korean document categorization.

In order to look for the categories of the documents, we must firstly represent the documents. The thesaurus tool provides the meanings and upper meanings of the document words so that the system can represent the document with not the terms but the concepts. In general, the concepts are more closely associated with the category of the document than words. For this reason, we can do the good document categorization even if we represent the document with a small number of concepts. Then, we present a concept-based document categorization system that classifies the documents with the thesaurus tool so that it can be useful to application that need speedy response or not much space.

Document categorization has been researched in two approaches; the man-made approach and the automatic approach. In the man-made approach [4], the condition-action rules to classify documents are designed by the human. Many systems use this approach, however, the man-made approach is very expensive in the time and effort to define good classification rules and expand the constructed rule system. In addition, it is impossible to learn the feedback knowledge acquired during classification process.

To supplement the man-made approach, there have been researches to the machine-generated approach that automates the rule construction [3,5,11,12,14]. This approach infers classification rules from training data. According as what kind of inference method is used, it can be also divided into three methods; the probability-based method, the neural network method, and MBR method. The probability-based method classifies the documents with the probability acquired from training data. [11] determine the category of the document by inferring the Bayes association estimate between keywords and category. [2,15] classify the category of the document by measuring the similarity between the probability of the document and categories. Here, the probability means the frequency of keywords in the document. In this model, as the knowledge acquired in classifying the documents is added to the system, it is possible to incrementally learn. The neural network method constructs the neural network to classify the document from training data. It determines the category of the document with the constructed network. This method is also possible to incrementally learn, however, it needs much learning time. MBR(Memory Based Reasoning) method [1,5] estimates the similarity between the input document and training documents, so the system determines the category of the input document with the category of the most related document. This method has the similarity estimate overhead between the input document and every training document.

Therefore, in this paper, we adopt the probability-based method that has no similarity measure overhead to all training documents for effective system. This system is easy to expand for incremental learning capability. In the following section, we briefly explain CONDOCS for effective document categorization. In section 3, we describe the experiments and results of this system. Finally, in section 4, we describe the conclusion.

2   Effective Concept-Based Document Categorization System (CONDOCS)

The presented effective concept-based document categorization system (CONDOCS) has the structure like Fig. 1. CONDOCS consists of the Korean morphological analyzer, the thesaurus tool, and the similarity measurer. The Korean morphological analyzer transforms the input document into the list of terms. The thesaurus tool gets the terms from the Korean morphological analyzer and draws the concepts included in the terms. Finally, the similarity measurer finds the category code by measuring the similarity between the input document and the categories.

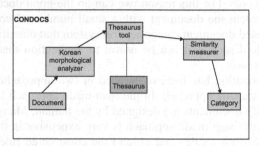

Fig. 1. Structure of **CONDOCS**

To classify the document, we must firstly do a morphological analysis on Korean sentences in the document. In this paper, we briefly explain the analyzer as the analyzer is designed as the preprocessor of the thesaurus tool. In general, document classification has been researched with a basis on the assumption that the category of the document is closely connected with the terms of noun grammatical category [7]. For this reason, we focus on the extraction of the terms of noun grammatical category. This analyzer should divide the word phrases in the sentences into morphemes to extract the noun terms from input sentences. The word phrases means a sequence of morphemes between two spaces. The analyzer splits postfix, prefix, postposition, and ending from a word phrase so it can easily get the noun terms [10]. The analyzer transfers the analyzed noun terms to the thesaurus tool so that it gets the concepts of the terms.

2.1   Thesaurus Tool

Since this system classifies the document with the concepts provided by the thesaurus tool, the performance of the system depends on the thesaurus tool. Even though there have been researches on the thesaurus tool [13], they are not adequate to represent the concepts of Korean terms. So, we make the thesaurus tool to provide the concepts of Korean terms.

To get the good document categorization, the thesaurus tool should have the correctness and generality. The correctness means whether the tool acquires the correct meanings to a given word, and the generality means whether the tool provides the general concepts that can be applicable to every domain. It is not easy to design the correct and general thesaurus. To get correctness, the construction of thesaurus should be complete and objective. In this system, in order to construct as correctly as possible, we repeated the validation process in constructing the thesaurus. To authenticate to be correctness, we will review the experimental results of the system using the tool. To get the generality, we define the concepts of the thesaurus that can be applied to every domain by considering the previous researches [8,9] on concept taxonomy. The number of the defined concepts is 150. The concepts may have the IS-A relation between each other.

The thesaurus generally contains the meaning of words with the noun, verb, adjective, and adverb, but the thesaurus in CONDOCS includes only the concepts on the noun important on document classification. However, this thesaurus has the event concept since the concept on noun also includes the action or event concept. The structure of the concept taxonomy is the tree structure. In the structure, the higher the concept is, the more abstract it is.

CONDOCS uses the 150 concepts to classify the document. To get the better classification quality, we may use the more many concepts than 150. If we use too many concepts, the system may not be efficiently operated. As there is usually a tradeoff between the number of the concepts and effectiveness of the system, the concepts should be carefully selected. In this system, we examine and review the thesaurus with the 150 concepts. In the meantime, the thesaurus tool should get the concepts on all words appeared in the document. However, it is time-consuming and expensive to develop such thesaurus tool. In this paper, we firstly construct the thesaurus tool that can get the concepts on the 3990 words that are frequently used in the documents. In future, we will increase the number of the words what the thesaurus tool can get the concepts.

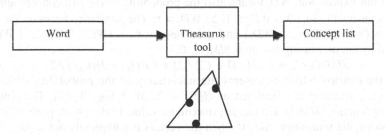

Fig. 2. Structure of the thesaurus tool

In Fig. 2, we depict the structure of the thesaurus tool. The tool finds the concepts to the input word from thesaurus. The tool also finds the upper concepts of the concepts. Table 1 shows the processing result of the tool. If the word has two more meanings, it provides the all concepts on the word. To resolve the multiple meanings, we must design the disambiguation procedure. In this paper, we exclude the problem.

2.2   Similarity Measurer

This system adopts the probability-based method to categorize the document. As this system uses the probability of the concepts, we should represent the document with the concept-probability vector. The similarity measurer compares the concept-probability vectors of an input document and categories.

Table 1. Concept list to example words

Words	Concept lists
가객(a guy)	animal animate-thing art domain human music physical-thing thing
가게(a shop)	abstract-thing inanimate-thing man-made-thing organization physical-thing structure thing
가격(price)	domain economics feature measurement society
가계(livelihood)	abstract-thing domain domestic industry organization thing
가곡(a song)	abstract-thing art domain intellectual-thing music thing word
가곡집(a collection of songs)	abstract-thing art domain information intellectual-thing music thing
가공(manufacturing)	composition domain event industry
......	......

Firstly, we construct the concept-probability vectors from training documents to every category. Then, the thesaurus tool transforms the terms of the documents into concepts so that we can represent the documents with the concept-probability vectors. In this system, we use the 76 category codes and 12 category codes to identify the category. If we use the 76 category codes, the number of the concept-probability vectors for each category will be 76. Secondly, we transform the input document into the concept-probability vector for input document with the thesaurus tool. Finally, the similarity measurer compares the input vector with the vectors for categories. In the following, we describe the estimate function of the similarity measurer.

We represent the category C_i with the concept-probability vector $C_i = (WC_{i1}, WC_{i2}, \dots , WC_{in})$. In this expression, WC_{ij} means the probability of the j-th concept appeared in the document set on the category C_i. So, $0 \leq WC_{ij} \leq 1$, $\Sigma j\ WC_{ij} = 1$. The input document is represented with the concept-probability vector $D = (WD_1, WD_2, \dots, WD_n)$. In this expression, WD_j means also the probability of the j-th concept appeared in the document D. So, $0 \leq WD_j \leq 1$, $\Sigma j\ WD_j = 1$. The similarity between the document $D = (WD_1, WD_2, \dots , WD_n)$ and the category $C_i = (WC_{i1}, WC_{i2}, \dots , WC_{in})$ is measured by the measuring function $SIM(D, C_i)$.

$$SIM(D, C_i) = 1 - H[(D + C_i) / 2] + [\ H(D) + H(C_i)\ ] / 2$$

$H(P)$ is the entropy which represents the uncertainty of the probability vector $P = (W_1, W_2, \dots , W_n)$ and is calculated as $H(P) = -\Sigma i\ W_i * log_2\ W_i$ [2]. The similarity measuring function $SIM(D, C_i)$ has the maximum value 1 when both probability vectors are same, the minimum value 0 when the vectors is completely not same.

3 Experiments and Results of CONDOCS

We examined CONDOCS with the factors that effect the performance of the system. They are the number of training documents, the size of the concept vector, and the number of words that the thesaurus tool can get the concepts. The training documents are extracted in the Kemongsa's encyclopedia [6]. The total number of training documents is 21558. Firstly, we describe the experimental results according to the number of training documents.

3.1 Experiment of CONDOCS According
to the Number of Training Documents

We constructed the systems with 3593 training documents, 7186 documents, 10779 documents, 14372 documents, and 17965 documents. We called the systems as

CONDOCS-1, CONDOCS-2, CONDOCS-3, CONDOCS-4, and CONDOCS-5. Table 2 describes the success rates on the five systems on the test documents. The number of the test documents is 3593. The first column of the table 2 depicts the kinds of systems. The second and third column represents the best success rates on two category sets. The two category sets are as Table 3. In the fourth and fifth column, we express the success rates summed up success rates of the first and second candidate categories. Because the category of the document may be ambiguous, we include the success rates to both the best and secondary results.

Table 2. Success rates on the five systems

Kinds of system	Best success rate		Secondary success rate	
	category 76	Category 12	Category 76	Category 12
CONDOCS-1	0.524	0.645	0.674	0.841
CONDOCS-2	0.535	0.647	0.686	0.837
CONDOCS-3	0.543	0.652	0.689	0.839
CONDOCS-4	0.543	0.650	0.694	0.840
CONDOCS-5	0.547	0.650	0.692	0.840

Table 3. Two category sets in Kemongsa encyclodepia

Abstract category set	specific category set
0100. philsophy	0101.philsophy, 0102.logics, 0103.psychology, 0104.ethics, 0199.Related names
0200. religion	0201.general, 0202.Buddhism, 0203.Christ, 0204.shamanism, 0205.Etc religion, 0299.Related names
0300. society	0301.general, 0302.customs, 0303.politics, 0304.economics, 0305.sociology, 0306.law, 0307.administration, 0308.education, 0309.military, 0310.press, 0399.Related names
0400. science	0401.general, 0402.mathematics, 0403.physics, 0404.chemistry, 0405.astronomy, 0406.geology, 0499.Related names
....	....

In Fig. 3, the experimental result shows that the more training documents are increased, the better the success rate is. However, the improvement of the success rate is very tiny. Though we construct the system with the more many training documents, we can not get the good success rate. Hence, we experiment to the concept-probability vector and the thesaurus tool in the following section.

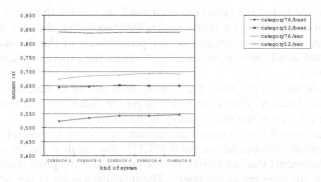

Fig. 3. Experiment results per each system

3.2 Experiment of CONDOCS to Each Concept Vector

CONDOCS uses the concept-probability vector to represent the document. As the size of the concept-probability vector is smaller than term-probability vector, it can support effective concept-based document classification. Since the performance of the system depends on the designed concepts, we carefully designed the concepts. In this paper, we use the 87 concept vector, 150 concept vector, and 768 concept vector. The former two vectors are carefully designed, but the third 768 vector is defined by adding 5-6 words related per 150 concepts for experiment of the vectors. Fig. 4 shows the experimental result on these vectors.

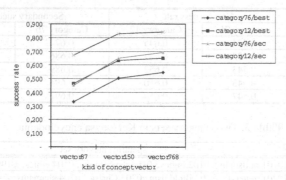

Fig. 4. Experiment result per each concept vector

The result shows that the more the size of vector is increased, the higher the success rate becomes. But the improvement of success rate to the vector 768 is not good. It seems the vector 768 is not carefully designed. If we design the vector 768 with deep study, we can expect the better success rate. If improvement rate is linear, the success rate to the carefully designed 768 vector will be 65%(category 76) and 80%(category 12). As stated above, when we design the concept vector, we must think tradeoff between the number of the concepts and effectiveness of the system.

3.3 Experiment of CONDOCS to Each Word Set
That the Thesaurus Tool Can Get the Concept

In this section, we tested the system to each word set that the thesaurus tool can get the concept. In this system, we firstly select 3990 words that the thesaurus tool can get the concept. For the experiment on the performance of the thesaurus tool, we divided the 3990 words into three sets. Fig. 5 shows the experimental results on the defined word sets.

The result shows that the more the size of the word set is increased, the success rate is increased linearly. Hence, in order to get the better success rate, we should firstly increase the size of word set used in the thesaurus tool.

The performance of the document categorization system, in general, is measured by recall, precision, and breakeven point [11,12]. The recall is the proportion of all the relevant documents that are retrieved. The precision is the proportion of retrieved documents that are relevant to the query. The breakeven point is the point at which

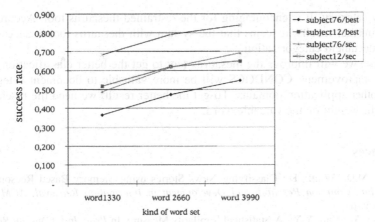

Fig. 5. Experiment result per each word set

recall equals precision. In this system, we evaluate the system with the thesaurus tool of each word set. We can get the 0.65 breakeven point with the final thesaurus tool. The result shows that the more the size of the word set is increased, the more the breakeven point is increased. Then, this result says that we can effectively classify the documents with the thesaurus tool to expanded word set.

3.4 Review on the Experiment Results

In above sections, we reviewed the experiments on CONDOCS. As a result of the review, we must firstly improve the word set used in the thesaurus tool. In next, we have to redefine the 768 concept vector to get the correct and general concept set. We carefully reviewed the failed cases for the test documents and found the following reasons.

(1) Low performance of the thesaurus tool. As stated above experiments, in order to get the better performance, we must improve the thesaurus tool.
(2) Frequent appearance of unrelated words to category of the document. It needs the definition of the weight to each word in the document. To do so, we must use the linguistic analysis technique.
(3) Polysemy problem. To resolve this problem, we must disambiguate the meaning of the words.
(4) Ambiguous category of the document. The problem is due to selecting the best category. If necessary, we can select two or more categories.

4   Conclusion

In this paper, we designed and constructed the effective concept-based document categorization system CONDOCS. It uses the thesaurus tool so that we can effectively classify the document. Then, the thesaurus tool finds the concepts of terms in the document to represent the document with the concept-probability vector. CONDOCS got 65% success rate for the best category and 84% rate including the secondary category. It is encouraging for the restrained thesaurus tool. According to the experimental results, it seems that the system with thesaurus tool is necessary in effective document categorization.

In future, we will improve the thesaurus tool to get the better classification result. After the improvement, CONDOCS will be more valuable to document categorization and other application domains. To get the better result, we need the research on defining the weight on the valuable word.

References

1. Linoff, M.D., Waltz, D.: Classifying News Stories using Memory Based Reasoning, In *Proc. Intl. Conf. on Research and Development in Information Retrieval, ACM SIGIR* (1992) 59-65
2. Wong, K.M., Yao, Y.Y.: A Statistical Similarity Measure, In *Proc. Intl. Conf. on Research and Development in Information Retrieval, ACM SIGIR* (1987) 3-12
3. Kweon, O.W.: Optimizing for Text Categorization Using Probability Vector and Meta Category, M.S. Thesis, KAIST Computer Science Dept. (1995)
4. Hayes, J.: Intelligent High-Volume Text Processing Using Shallow, Domain-Specific Technique, In Paul S. Jacobs, editor, *Text-Based Intelligent Systems: Current Research and Practice in Information Extraction and Retrieval*, Hillsdale, New Jersey (1992) 227-241
5. Yang, Y.: Expert Network: Effective and Efficient Learning from Human Decision in Text Categorization and Retrieval, In *Proc. Intl. Conf. on Research and Development in Information Retrieval, ACM SIGIR* (1994) 13-22
6. ETRI Natural Language Processing Lab.: ETRIKEMONG SET, ETRI (1997)
7. Lee, H.A., Lee, J.H., Lee, G.B.: Concept-based Noun Phrase Indexing Method Using Syntactic Analysis and Cooccurence Information, Proc. Of the 7th Hanguel and Korean Information Processing Conference (1996)
8. EDR Technical Report: Concept Dictionary, Japan Electronic Dictionary Research Institute (1988)
9. Kang, W.S.: Semantic Analysis of Prepositional Phrases in English-to-Korean Machine Translation, KAIST Ph.D. Thesis (1995)
10. Kim, S.Y.: Morphological Analyzer using Tabular Parsing Method and Concatenation Information, KAIST M.S. Thesis (1987)
11. Lewis, D.D.: An Evaluation of Phrasal and Clustered Representations on a Text Categorization Task, ACM SIGIR'92 (1992)
12. Apte, C., Famerau, F., Weiss, S.M.: Automated Learning of Decision Rules for Text Categorization, ACM Tr. on Information Systems, Vol.12, No.3 (1994)
13. Miller, G.A., Beckwith, R., Fellbaum, C., Gross, D., Miller, K.: Introduction to WordNet:An On-line Lexical Database, Report of WordNet, Princeton University (1990)
14. Sebstiani, F.: Machine Learning in Automated Text Categorization, ACM Computing Surveys, Vol.34, No.1, (2002) 1-47
15. Yang, Y., Zhang, J., Kisiel, B.: A Scalability Analysis of Classifiers in Text Categorization, In Proceedings of SIGIR-03, 26th ACM International Conference (2003) 96–103

Using DEVS for Modeling and Simulation of Human Behaviour

Mamadou Seck, Claudia Frydman, and Norbert Giambiasi

LSIS, Domaine Universitaire de Saint Jérôme, Avenue Escadrille Normandie-Niemen,
13397 Marseille Cedex 20, France
{Mamadou.seck,Claudia.frydman,Norbert.giambiasi}@lsis.org

Abstract. Throughout the last decade, research in cognitive sciences has proven emotions to be playing a prominent role in human behaviour.
The military is being more and more interested in modelling human behaviour for simulation and training purposes.
The aim of our work is to, from models coming from physiology and psychology, give to such models an operative semantics in order to simulate the human behaviour. We propose a DEVS model of stress states as well as a model of physical tiredness.The latter models interact with the behavioural model within an architecture that we also present.

1 Introduction

Throughout the last decade, research in cognitive sciences has proven emotions to be playing a prominent role in human behaviour.

The military is being more and more interested in modelling human behaviour for simulation and training purposes.

The work presented here utilises existing physiology theories to design novel behavioural models that take into account factors such as stress and tiredness. We propose a DEVS [6] model of stress states as well as a model of physical tiredness. The latter models interact with the behavioural model within an architecture that we also present.

2 Stress

The understanding of the processes underlying stress has evolved recently, yet leaving numerous uncertainties.

A stressor is any event or situation requiring a change in adaptation or behaviour of an individual. It represents a threat to their welfare or survival. It can have positive or negative effects.

A stressor implies physiological reflexes that prepare to tackle a situation or to flee it.

2.1 Types of Stressors

We can distinguish between physical and mental stressors. A physical stressor is one which has a direct effect on the body. This may be an external environmental condi-

T.G. Kim (Ed.): AIS 2004, LNAI 3397, pp. 692–698, 2005.

tion or the internal physical/physiologic demands of the human body. A mental stressor is one in which only information reaches the brain with no direct physical impact on the body. This information may place demands on either the cognitive systems (thought processes) or the emotional system (feeling responses, such as anger or fear) in the brain. Often, reactions are evoked from both the cognitive and the emotional systems.

Table 1. Types of stressors [1]

Physical stressors	Mental stressors
ENVIRONMENTAL	**COGNITIVE**
Heat, cold	Information : too much or too little
Vibrations, noise, blast	Isolation
Hypoxia ...	Hard judgement ...
PHYSIOLOGICAL	**EMOTIONAL**
Lack of sleep	Threats
Dehydratation	Frustration
Muscular fatigue ...	Boredom/inactivity ...

2.2 Stress / Performance Relationship

Stress is a process that promotes survival in certain contexts. An optimal level of stress is useful for task performance. When stress is too low, it implies distraction and tasks are done haphazardly. An intense level of stress may cause poor motor coordination or even choking.

Many models try to link stress to performance. One of the most cited is the Yerkes-Dodson law [2], also referred to as the inverted-U hypothesis. As displayed in figure 2, it states that at low stress, performance is low; it improves until its highest point corresponding to the optimal level of stress or arousal. Then, performance level decreases when stress intensifies, leading to complete disorganization when stress level reaches panic.

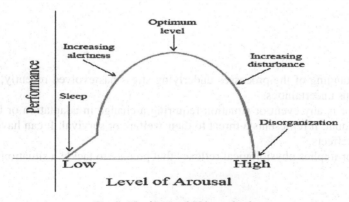

Fig. 1. The inverted-U hypothesis

The inverted-U certainly does not capture the whole complexity of the stress/performance relationship, but appears to be a largely accepted principle. The model has even been perfected by Hancock [3]. His extended-U model relates stress level to physiological and psychological adaptability. He defines various zones of the curve corresponding to physical and mental states depending on stress level.

3 The DEVS Formalism

DEVS is a modular formalism for deterministic and causal systems' modelling. It allows for behavioural description of systems. A DEVS model may contain two kinds of DEVS components: Atomic DEVS and Coupled DEVS.

3.1 Formal Definition of an Atomic DEVS

AtomicDEVS = <S, ta, δint, X, δext, Y, λ>

- The time base is continuous and not explicitly mentionned : T= |R
- S represents the set of sequential states : The dynamics consist in an ordered sequence of states in S.
- ta(s) is the lifetime function of a state in the model.
- δ_{int} is internal transition function, allowing the system to go from one state to another autonomously.
- Y is the set of outputs of the model.
- X is the set of (external) inputs of the model. They interrupt its autonomous behaviour by the activation of the external transition function δ_{ext}.

The system's reaction to an external event depends on its current state, the input and the elapsed time.

3.2 Formal Definition of a Coupled DEVS

The coupled DEVS formalism describes a discrete events system in terms of a network of coupled components.

$$\text{CoupledDEVS} = < \text{Xself, Yself, D, } \{Mi\}, \{Ii\}, \{Zi,j\}, \text{ select} >$$

Self stands for the model itself.

- X*self* is the set of possible inputs of the coupled model.
- Y*self* is the set of possible inputs of the coupled model.
- D is a set of names associated to the model's components, self is not in D.
- {Mi} is the set of the coupled model's components, with i being in D. These components are either atomic DEVS or coupled DEVS.
- {Ii} is the set of l'ensemble des influencees of a component. That is what defines the coupling structure.
- {Zi,j} defines the model's behaviour, transforming a component's output into another componant's input within the coupled model.

As concurrent components can be coupled, many state transitions can have to occure at the same simulation time. A selection mecanism then becomes necessary, in order to choose which transition is to be executed first. So is the role of the "select" function.

4 The DEVS Stress Model

Based on those theories, we propose a model of stress in the DEVS formalism. It consists in a representation of discrete states of stress. Depending on their nature, external events make the model evolve from one state to another. Those events are sorted according to their seriousness or intensity by an appraisal function.

Inspired by the inverted – U hypothesis, our model consists of 5 states corresponding to points chosen on the curve. We sort events in 4 types, each type defining a certain evolution of the stress model.

During a mission consisting in reaching a distant point in enemy zone, such a typology of events can be made:

TE1: suspect noises, blasts …;
TE2: shots, losses in our camp
TE3: loss of a pal, wounds
TE4: positive events

Such event types will also have to be generated to define a given situation as a stressor. For example, as lack of information is a stressor (it isolates the soldier [1]), a rule can be activated to generate a "T1 event" if no information has been received by the soldier in a given period of time. These features will be developed in the appraisal component of our architecture. The model may also evolve autonomously as a way of representing the natural efforts of the individual to recover equilibrium.

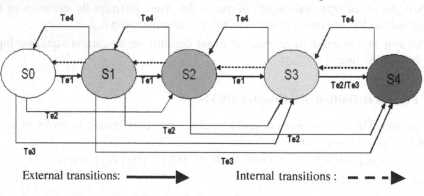

External transitions: ➡   Internal transitions : – – –▶

Fig. 2. DEVS model of stress

These stress states have the following behavioural features:

S0: Sleep, inactivity
S1: Slow reactions, average precision
S2: Optimal level
S3: Risk taking, bad choices
S4: Bad precision, choking, panic

5 Modelling Tiredness

Following [5], transitions of a DEVS model can be conditioned by thresholds of a continuous function.

We can model tiredness by a polynomial function of task durations multiplied the task's constant of difficulty.

The rate of efficiency of the task can be conditioned by the computed value of tiredness. The possibility for an agent to start or complete a physically demanding task can also be conditioned by such a function.

The values of thresholds and of the constants of difficulty will be defined and refined during phases of test.

Figure 3 shows the evolution of the tiredness function while the agent is executing various tasks alternately:

T1: average physical difficulty
T2: low physical difficulty
T3: high physical difficulty
T4: Rest.

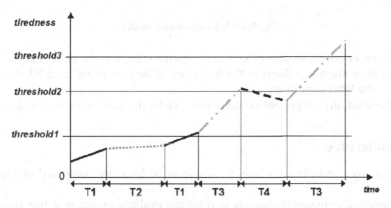

Fig. 3. Continuous function of physical tiredness

6   DEVS Behavioural Model

Now that we have seen how to represent stress and tiredness, let us present our behavioural model.

Our DEVS behavioural model represents the current task of the agent by a state. Events that induce a change in behaviour activate transitions that can be conditioned by values of variables representing stress, tiredness or any other environmental information.

Let us model the behaviour of a soldier in a mission consisting in reaching a distant point in enemy zone:

- The soldier's initial state is stop.
- He can receive the order of reaching the objective point and then starts walking.
- While walking, if he detects an enemy, or if he is shot upon, while his stress level is S1 or S2, he retreats to cover. If he were in stress level S3, he would shoot back directly or worse, if he were in stress level S4, he would freeze, as a result of panic.

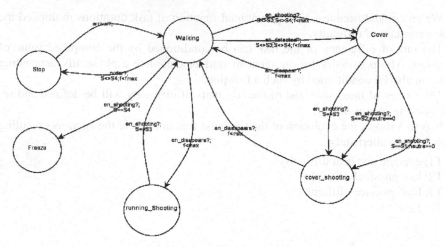

Fig. 4. DEVS behavioural model

- While under cover, he shoots back at the enemy if he is in the stress state S1 or S2 and if there are no civilians in the trajectory. If he is in stress level S3, he would not for the latter condition.
- All these actions are possible if tiredness is under the maximum threshold.

7 Architecture

The previously cited elements interact within an architecture composed of four elements.

The appraisal component consists in rules that evaluate events to define their seriousness and how stressful they are. Such rules will be developed in the spirit of the goal based appraisal described by Gratch in [4]. Events and situations are appraised in regard with the goals of the mission.

The output of that component is stressor types TE1, TE2, TE3 and TE4.

The DEVS stress model receives inputs from the appraisal component. The events appraised as stressors make the stress level evolve.

In a parallel manner, tiredness level is computed as a continuous function of task durations and task's constant of difficulty.

Both those models and external events are used to make the behavioural model. In that way, our architecture takes into account human factors such as stress and tiredness and other environmental factors to create a behavioural model.

8 Discussion

The architecture we present here allows an explicit modelling of human behaviour under constrains such as stress and physical tiredness.

We wanted a framework close to the theory, but still easily usable for military simulation purposes.

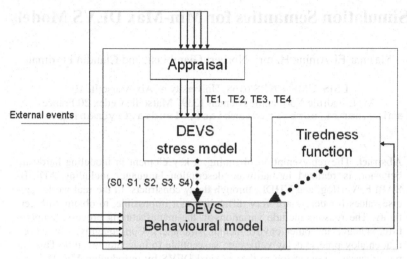

Fig. 5. Architecture

DEVS offers a direct manipulation of events, states and their evolutions, thus allowing us to build robust and reliable models.

Many improvements will complement this framework.

First of all, our modelling of physical tiredness is at a very early level. Future investigations will help us better handle the complexity involved in such a process. Parameters like weather or terrain will be added to the model to make it more accurate.

Another planned evolution is the implementation of interpersonal relation. DEVS coupled models seem to be an interesting feature for that kind of application [6].

We will also add the possibility of modelling personality differences as far as susceptibility to stress is concerned.

Our architecture certainly has interesting features and will develop new ones in order to achieve high standard simulation of human activity.

References

1. www.globalsecurity.org/military/library/policy/army/fm/22-51/22-51_toc.htm. Field manual n° 22-51, Leader's manual for combat stress control, Chapter 2 : Stress and combat performance. Headquaters, Department of the army. Washington DC 29 september 1994
2. Lyle E. Bourne, Jr. and Rita A. Yaroush. Stress and cognition: A cognitive psychological perspective NASA. Grant Number NAG2-1561 February 1, 2003
3. Hancock, P.A., & Warm, J.S. (1989). A dynamic model of stress and sustained attention. Human Factors, 31, 519-537.
4. Jonathan Gratch. Why you should buy an emotional planner, Proceedings of the Agents'99 Workshop on Emotion-based Agent Architectures (EBAA'99) 1999.
5. Praehofer, Herbert and Pree, D. Visual Modeling of DEVS-based Multiformalism Systems Based on Higraphs. Winter Simulation Conf., pp.595-603, Soc.Comp. Simulation, LA CA 1993.
6. Zeigler, B. Theory of Modeling and Simulation. Krieger Publishing Company. 2nd Edition, 1984.

Simulation Semantics for Min-Max DEVS Models

Maâmar El-Amine Hamri, Norbert Giambiasi, and Claudia Frydman

LSIS, UMR-CNRS 6168, University of Aix-Marseille III
Av. Escadrille Normandie Niemen 13397 Marseille Cedex 20 France
{amine.hamri,norbert.giambiasi,claudia.frydman}@lsis.org

Abstract. The representation of timing, a key element in modeling hardware behavior, is realized in hardware description languages including ADLIB-SABLE, Verilog, and VHDL, through delay constructs. In the real world, precise values for delays are very difficult, if not impossible, to obtain with certainty. The reasons include variations in the manufacturing process, temperature, voltage, and other environmental parameters. Consequently, simulations that employ precise delay values are susceptible to inaccurate results. This paper proposes an extension to the classical DEVS by introducing Min-Max delays. In the augmented formalism, termed Min-Max DEVS, the state of a hardware model may, in some time interval, become unknown and is represented by the symbol, ϕ. The occurrence of ϕ implies greater accuracy of the results, not lack of information. Min-Max DEVS offers a unique advantage, namely, the execution of a single simulation pass utilizing Min-Max delays is equivalent to multiple simulation passes, each corresponding to a set of precise delay values selected from the interval. This, in turn, poses a key challenge – efficient execution of the Min-Max DEVS simulator.

1 Introduction

Hardware Description Languages (VHDL, Verilog, ADLIB, ...) and logic gate simulators use the concept of delay [1], [2] to represent timed behaviors. Different classes of delay models have been introduced [1], [3-6] to allow different kinds of temporal analysis.

In the DEVS formalism, Zeigler introduced the concept of the lifetime of a state (called time advance function), which corresponds to the concept of precise delay values in digital circuits. However, in a design or an analysis process of a real system, the lifetime of transitory states is generally unknown and, as for digital circuits, the user defines a mean value. For a more realistic modeling, we propose to apply the concept of Min-Max delays to the specification of discrete event models. We proposed an extension of the DEVS formalism, called Min-Max DEVS [7], and the purpose of this paper is to present its simulation semantics by the way of the conceptual Min-Max DEVS simulator.

The aim of Min-Max DEVS is to model and simulate real systems for which the delays values are not known with accuracy (the scattering of parameters due to the manufacturing process, for example) and not as in [8] to do real-time simulations. For a best understanding of the meaning attach to Min-Max DEVS, we recall the meaning of ambiguous delay in the domain of digital circuits, the minimal delay corresponds to the faster circuit and the maximum delay corresponds to the slower one. All the transitions of states, in the possible real circuits, are between these two limits.

T.G. Kim (Ed.): AIS 2004, LNAI 3397, pp. 699–708, 2005.

After a brief recall of the DEVS formalism [9], [10] and of classical delay models, we present the Min-Max DEVS [7] formalism and its operational semantics. The most difficult problem to solve is to obtain a fast simulation technique, knowing that with Min-Max delays, all the possible precise models are to be simulated during only one stage of simulation; the presentation of the conceptual Min-Max DEVS simulator is detailed in section 4. Section 5 illustrates our approach by using an example; then we present the simulation results and their interpretation.

2   Recall

The simplest type of delay used in the digital circuit field is the pure transport delay. This delay is defined by one value, which represents a mean value of real delays for the considered category of gates. A rise-fall delay is defined by two precise values corresponding to a mean value for each possible state change (0 to 1 and 1 to 0), the delay depends on the state change. In [3], [11], [12] the simulation algorithm of this delay gives us its semantics that is a preemptive semantics.

The problem with this kind of delays is that we assume we know a precise value of the delays of the real gates, which is evidently impossible because it depends on the environment and on the manufacturing process. For a more precise representation of real gates, probabilistic and fuzzy models have been introduced [6], [13]. These categories of delays have not widely used because the simulation algorithms are complex and heavy. Another category of delay models uses a time interval to define possible values of the delay. In this time interval, each point corresponds to a possible value of a real delay. These models are widely used because they represent the imprecise knowledge about real delays and they have very fast simulation techniques. These delay models are called ambiguous delays or Min-Max delays [1], [3].

An ambiguous transport delay is defined by two values [tmin, tmax], which represent the distribution of the possible real delays of the corresponding gate. In this case, the variables of a model have three possible values [0, 1, Φ]. Φ represents an unknown value due to the imprecision on the delay values. In fact, when a gate model is in the state Φ, the interpretation is that some possible real gates are in the state 0 and some other are in the state 1 depending on the real value of the delay.

In fact, with this modeling we represent in one model all the possible behaviors for the different delay values. The simulation process consists in using 4 events: (0 to Φ), (Φ to 0), (Φ to 1) and (1 to Φ). The occurrence of the event (0 to Φ) represents the faster possible signal (faster circuit corresponding to a real circuit with the minimum delays for all its logic gates), the event (Φ to 0) represents the slower circuit.

3   Min-Max DEVS Formalism

In [7] we have proposed a Min-Max DEVS formalism to model systems with lifetime of transitory states represented by time intervals and not an exact value. That is to say that the faster event must be interpreted as defining the beginning of a temporal window in which a real system can send out the considered output event, the slower event defines the end of this window. At the occurrence time of the slower event all the possible real systems have had the considered state change.

We define first an external Min-Max specification, which corresponds to a user specification. An internal specification will be associated to it in order to define the operational semantics of the formalism. The external specification of Min-Max DEVS is identical to RT-DEVS [7], [12], but the interpretation and the operational semantics are different.

A Min-Max discrete event specification is a structure:

$$M= <X, S, Y, \delta_{int}, \delta_{ext}, \lambda, D>,$$

Where:

- X, S and Y are the same sets than in classical DEVS
- δ_{int}, δ_{ext} and λ are the same functions than in classical DEVS
- the lifetime function $D: S \rightarrow R^+ x R^+$

  $D(si) = [dmin, dmax]$, where
  - dmin is the minimum time the model may remain in the state si, dmin corresponds to the faster real system.
  - dmax represents the maximum time the model may remain in the state si, dmax corresponds to the slower real system.

As in the digital circuit field, we introduce a Min-Max definition for the lifetime of transitory states because in general:

- in an analysis process, the length of the state cannot be measured with precision, only a window of possible values can be defined
- in a design process, there are several possible realizations of the designed system and the manufacturing process is never perfected (all the products have slightly different features).

In other words, we consider that it is impossible to obtain a precise value for the length of the transitory states of a real system and then, we define these lengths with a minimal value and a maximal value. For a more detailed interpretation, we define now the internal model with its operational semantics.

Given a user specification of a Min-Max DEVS:

$$EM= < X, S, Y, \delta_{int}, \delta_{ext}, \lambda, D>,$$

We associate with this specification, an internal model as follow:

$$IM= < XI, SI, YI, \delta I_{int}, \delta I_{ext}, \lambda I, DI>,$$

Where:

- $XI = X x \{fast, slow\}$

The event $(xj, fast)$ is interpreted as an event with the value xj, this event represents (as we will show in the next sections) the faster signal in the real world, its occurrence time defines the beginning of a temporal window in which, the event xj can occur at each time (depending on the real value of the delay).

The event $(xj, slow)$ is interpreted as an event with the value xj, this event represents (as we will see in the following) the slower signal in the real world, ending the window.

- $SI = (S U \Phi) x (S U \Phi) x A$

With: $A = \{fast, slow, autonomous, passive\}$,

Φ is an unknown value.

The interpretation of a model state (si, sj, A) is:

If $si = sj$, the external model is in the state si and all the possible real systems are in the state si.

If $si \neq sj$, the external model is in unknown state, because some fast real systems are in the state sj and the slower systems are still in si.

The interpretation of a state (Φ, Φ, A) is that the model is in a completely unknown state, representing the fact that there is, at this time, *more than two possible situations in the real world*. The interpretation of an unknown value is:

- some real systems have already had a state change,
- some other real systems don't have even a state change.

In fact, the simulation of a model with Min-Max delays corresponds to the simulation of all the possible real systems with only one stage of simulation.

- $YI = Y \times \{fast, slow\}$, where:

    $((yj, fast), tj)$ is an output event with the occurrence date tj, representing the response of the faster real system,

    $((yj, slow), tk)$ is an output event with the occurrence date tk, representing the response of the slower real system.

In the temporal window $[tj, tk]$, the external model is in an unknown state, this window represents all possible occurrence times of the event yj. With these definitions, we can now define the transition functions of the internal model.

Internal Transition Function δI_{int}.

δI_{int}: $(S \cup \Phi) \times (S \cup \Phi) \times A \rightarrow (S \cup \Phi) \times (S \cup \Phi) \times A$

with the following rules:

$\delta I_{int} (si, si, autonomous) = (si, \delta_{int} (si), autonomous)$

with: $DI(si, si, autonomous) = Min (D(si))$.

The second element of the triplet $(si, si, autonomous)$ represents the faster real system, and we apply the internal transition function to this element after the min. delay. The internal model is now in a state $(si, sj, autonomous)$, which is associated with an unknown state for the external model. The length of this unknown state is $Max(D(si))-Min(D(si))$ time units, after this time all the real systems have had the corresponding state change.

$\delta I_{int} ((si, sj, autonomous))$

$= (sj, sj, passive)$ if sj is a passive state

$= (sj, sj, autonomous)$ if sj is an active state

with: $DI (si, \delta_{int} (si), autonomous) = Max (D(si))-Min (D((si))$.

To understand these rules, we have to consider that we model simultaneously the state changes of all the possible real systems. To obtain the previous rule, we consider the faster and the slower real systems. The model state $(si, sj, autonomous)$ represents that some real systems (the faster systems) have had a state change and the other are still in the previous state (the slower systems). For a more detailed presentation of the internal transition function see [7].

Output Function λI. As in classical DEVS, an output event is emitted before the activation of the internal transition function: $\lambda I(si, si, A) = (\lambda(si), fast)$

$\lambda I(si, sj, A) = (\lambda(si), slow)$

The interpretation of the output event $(\lambda(si), \textbf{fast})$ is:

- this event occurs when the *faster possible real system* has a state change
- and if no input event occurs, the next output event will be emitted at the time of the state change *for the slower system* corresponding to the output event $(\lambda(si), \textbf{slow})$.

That is to say that the faster event must be interpreted as defining the beginning of a temporal window in which a real system can send out the considered output event, the slower event defines the end of this window: at the occurrence time of the slower event all the possible real systems have had the considered state change.

External Transition Function δI_{ext}. To define the external transition function of the internal model we have to analyze the different possible cases. However, due to the fact we represent all the possible real systems, for an internal state (si, sj, A) the slow event must be apply to the slower system, that is to say that the state si, and the fast event must be apply to the fast system, that is to say that the state sj. For the description of all the possible cases see [7].

In addition, we introduce the totally unknown state $(\Phi, \Phi, \text{passive})$ to model situation in which more than two evolutions are possible depending on the different lifetimes of the states and the occurrences time of the input events. This state is passive, and:

$$\delta I_{ext}((\Phi, \Phi, \text{passive}), (xj, (\text{fast or slow}))) = (\Phi, \Phi, \text{passive}).$$ The model remains definitely in the unknown state.

4 Min-Max DEVS Conceptual Simulator

The simulation semantics of a Min-Max DEVS is given by the Min-Max DEVS-simulator. As in classical DEVS simulation, the simulator receives messages from the coordinator to execute the model functions and sends in the form of messages the output function results.

The simulator employs four time variables tl, tf, ts and tn. The first variable holds the simulation time when the last event occurs. tf and ts indicate the occurrence time of the fast and slow events respectively. The tn variable holds the scheduled time for the next event. From the definition of the lifetime function D of Min-Max DEVS a set of equations is deduced for tn, depending on the current state of the internal model build from the Min-Max DEVS model and the date of external fast and slow events.

The elapsed time e is deduced from the current simulation time and the last event time.

$$e = t - tl \,. \, (1)$$

```
Min_Max_DEVS_Simulator:
variables:
parent // parent coordinator
tl: real // time of last event
tf: real // time of last faster event (x,fast)
tl: real // time of last slower event (x,slow)
tn: real // time of next event
e: real // elapsed time
Min-max DEVS // associated model with total states
   when receive i-message (i,t) at time t
```

```
        tl=t
        tn=t+Min(D(initial state))
when receive *-message (*,t) at time t:
// the simulator computes the output function and
// carries out the internal transition for faster or
// slower systems according to the current state of
// the internal model
        If (t≠tn) then error: bad synchronization
        If (si=sj) then y=(λ(sj),fast)
              send y-message (y,t) at time t
              sj=δ_int(sj)
              the current state of the internal model:
              (si,sj,autonomous)
              tl=t
              tn=t+(Max(D(si))-Min(D(si)))+(ts-tf)
        If (si,sj,fast) then y=(λ(sj),fast)
              send y-message (y,t) at time t
              sj=δ_int(sj)
              the current state of the internal model:
              (si,sj,passive) if sj is passive state
              (si,sj,autonomous) if sj is active state
              tl=t
              tn=t+Min(sj)
        If ((si,sj,autonomous) / si ≠ sj) then
              y=(λ(si),slow)
              send y-message (y,t) at time t
              si=sj
              the current state of the internal model:
              (sj, sj, passive) if sj is a passive state
              (sj, sj, autonomous)if sj is an active state
              tl=t
              tn=t+Min(D(si))
when receive x-message ((x,fast),t) at time t:
// the simulator determines the current state of the
// internal model to carry out the external state
// transition for the faster system
        If not(tl≤t≤tn) then error: bad synchronisation
        e=t-tl
        If ((si, si, A) where si and δ_ext(si, e, x) are
        passive states) then
              current state of the internal model:
              (si,δ_ext(si, e, x), fast)
              tl=t
              tn=infinity
        If ((si, si, A) si is passive and δ_ext(si, e, x)
        is active) then
              current state of the internal model:
              (si,δ_ext(si, e, x), fast)
              tl=t
              tf=t
              tn=t+Min(δ_ext(si, e, x))
        If ((si, sj, {autonomous,fast})) then
              current state of the internal model:
              (Φ, Φ, passive)
              tl=t
              tn=?
when receive x-message ((x, slow), t) at time t:
// the simulator carries out the external
// transition for the slower system
        If not(tl ≤t ≤tn) then error: bad synchronization
        e=t-tl
```

```
    If ((si, sj, passive) where δ_ext(si, e, x) is a
    passive state) then
            current state of the internal model:
            (δ_ext(si, e, x), sj, passive)
            tl=t
            tn=infinity
    If ((si, sj, passive) where δ_ext(si, e, x) is an
    active state) then
            current state of the internal model:
            (δ_ext(si, e, x), sj, autonomous)
            tl=t
            tn=t+Max(δ_ext(si, e, x))
    If ((si, sj, fast) where si is passive and sj is
    active state) then
            current state of the internal model:
            (δ_ext(si, e, x), sj, autonomous)
            tl=t
            ts=t
            tn=tn
    If((si, sj, autonomous)) then
            current state of the internal model:
            (δ_ext(si, e, x), sj, autonomous)
            tl=t
            tn=t+Max(D(δ_ext(si, e, x)))
    If ((Φ, Φ, passive)) then
            current state of the internal model:
            (Φ, Φ, passive)
            tl=t
            tn=?
end Min_Max_DEVS_Simulator
```

As shown in simulator algorithm, first a initialization message *(i, t)* has to be received at the beginning of each simulation run. When a Min-Max DEVS simulator receives such an initialization message, it initializes its last event time tl by t. The time of the next event tn, is computed by adding the min value of the time advance $D(s)$ to the time of the last event tl. This time tn is sent to the parent coordinator to tell it when the fast internal event should be executed by this component simulator.

An internal state transition message *(*, t)* causes the execution of an internal event. When a *-message is received by a Min-Max DEVS simulator, it computes the output and carried out the internal transition function of the associated internal model, so that the new state of the Min-Max DEVS model can be deduced. The output is sent back to the parent coordinator in an output message *((y, fast), t)* or *((y, slow), t)* according to the current state of the internal model *(si = sj, si ≠ sj* respectively). Finally, the time of the last event is set to the current time and the time of the next event is set to the current time plus min value of the time advance function $D(sj)$ if all systems are in the same state, or plus the difference between max and min value of the time advance function $D(sj)$ plus the difference between the time occurrence of the slow and fast events if same systems are in the state *si* and other are in state *sj*.

An input message *((x, fast), t)* or *((x, slow), t)* informs the simulator of an arrival of an input event, *(x, fast)* or *(x, slow)*, at simulation time t. This causes the simulator to execute the external transition function of the internal model for the Min-Max DEVS model given *(x, fast)* or *(x, slow)* and the elapsed time *e*. The time of the last event is set to t and the time of the next event is set to a value determined by the algorithm.

5 Example and Simulation

Let us consider a simple example of a filling system, model with the Min-Max DEVS formalism. This system, a filling system, is composed of a tank with a valve, a conveyor and barrels (Fig. 1). The filling system has two input ports:

- control of the valve: Val = {open, close}
- control of the conveyor: In = {roll, stop}

and two sensor output ports:

- barrel level: Bl = {capacity}
- barrel position: Bp = {good}

The Min-Max DEVS formalism was chosen instead of the DEVS formalism because some lifetime of transitory states can not be known with accuracy. As an example, the time for adjusting the conveyor depends on:

- the speed of the conveyor,
- the length between barrels. So, the lifetime of the state Adcon0 (advance conveyor) can not be known with accuracy, it is expressed with a time interval limited by two real values $\sigma1$ and $\sigma2$. As for the state Adcon0, the filling of a barrel phase is expressed with a time interval [$\sigma3$, $\sigma4$]. This inaccuracy of estimating the lifetime of the state filling0 is due to liquid flow and the needed quantity to fill a barrel.

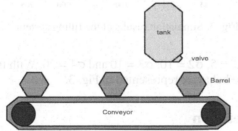

Fig. 1. A simple filling system

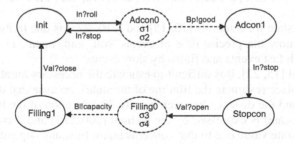

Fig. 2. Min-Max DEVS model of the filling system

The Min-Max DEVS model was built as shown in Fig. 2.

Now we suppose that the initial state of the internal model is (init, init, passive). Let us consider the following sequence of input events (In = (roll, fast), t=10), (In = (roll, slow), t=12), (In =(stop, ε), t=100), (Val =(open, fast), t=200), and (Val =(open,

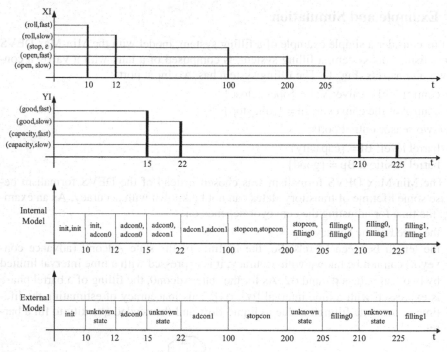

Fig. 3. Simulation results of the filling system

slow), t=205) where $\sigma1 = 5$, $\sigma2 = 10$, $\sigma3 = 10$ and $\sigma4 = 20$. With the Min-Max DEVS simulator we obtain the behavior represented in Fig. 3.

6 Result Interpretation

Fig. 3 summarizes the simulation of the model specified in 5. It is important to note that for intervals [10, 12], [15, 22], [200, 205] and [210, 225] the external model is in an unknown state.

The unknown state for intervals [10, 15] and [210, 225] is due to the fact that it is not possible to know the precise date of events 'roll' and 'open'; in addition these intervals start with fast events and finish by slow events.

For the interval [15, 22], it is difficult to estimate the necessary duration to put barrels in the good place (estimate the lifetime of the state), because that depends on the conveyor speed and the distance between barrels. The duration of the filling phase of a barrel (filling0 state) is expressed by the interval [200, 205] and it cannot be represented by an accurate value due to that concerns liquid flow and depends on the barrel volume to be filled. This example shows that it is impossible to obtain a precise value for the length of the transitory states of the real system and then, the unique realistic solution is to define these lengths with a minimal value and a maximal value.

Several simulations can be carried out to determine a set of values for every active state of the external model, until the obtaining an optimal configuration of the system according to the speed of the conveyor, distance between barrels, etc.

Remark 1. Some one can think that this ambiguous model is less accurate than a classical DEVS model because sometime the simulation results remain in the unknown value whereas the classical DEVS model will give an exact value. In fact, the information given by the Min-Max model is more accurate because it indicates with the given values of delays (which are more realistic than a mean value) it is impossible to know the state of the real system.

7 Conclusions

In this paper we have recalled the Min-Max DEVS formalism and how we build the internal model from the specification realized with ambiguous delays.

We have developed a Min-Max DEVS conceptual simulator, which allows the simulation of Min-Max DEVS model based on the internal Min-Max model semantics. The simulation of the model is aborted when the internal model reaches an unknown state (Φ, Φ, passive). The analysis of the simulation results in this case, concludes on the fact that the knowledge on the delay values is too imprecise to obtain exact information on the system behavior.

References

1. Breuer, M.A., Friedman, A. D.: Diagnosis and reliable design of digital systems. Computer science Press, 1976
2. Giambiasi, N., Miara, A.: SILOG: A practical tool for digital logic circuit simulation. 16th D.A.C, San Diego, 1976
3. Giambiasi, N., Miara, A. and Murach, D. (1979). SILOG: a Practical Tool for Logic Digital Network Simulation. 16th Design Automation Conference, San Diego
4. Ghosh S., Giambiasi, N.: On the need for consistency between the VHDL language constructs and the underlying hardware design. ESS 1996, Genoa, Italy
5. Walker, P. and S. Ghosh: On the nature and inadequacies of transport timing delay constructs in VHDL. IEEE conference on Computer Design, Austin, Texas, Oct. 1996
6. Smaili, M., Giambiasi, N. and Frydman, C.: Discrete Event Simulation with Fuzzy Dates. European Simulation Symposium 94 Istanbul, Turkey, and in: Smaili, M. (1994): Modélisation à retards flous de circuits logiques en vue de leur simulation. Ph.D Université de Montpellier II
7. Giambiasi, N. and Gosh, S.: Min-Max DEVS: A new formalism for the specification of discrete event models with Min-Max delays. Proc. of ESS'2001 European Simulation Symposium 18-20 Oct. 2001 at Marseille France 616-621
8. Seong, M.C. and Kim, T.G.: Real time DEVS simulation: Concurrent time-selective execution of combined RT-DEVS model and interactive environment. Proc. of SCSC-98, July 19-22, Reno, NV
9. Zeigler, B.: Multifaceted Modeling and Discrete Event Simulation. Academic Press, London, 1984
10. Zeigler, B., Praehofer, H. and Kim, T.G.: Theory of Modeling and Simulation. Academic Press, Second Edition, 2000-1976
11. IEEE standard VHDL language reference manual, IEEE, 1988
12. Ghosh, S., and Meng-Lin Yu.: A preemptive scheduling mechanism for accurate behavioral simulation of digital designs. IEEE trans. on Computers, Vol. 38, Nov. 1989
13. Magnhagen, B.: Probability based verification of time margins in digital design. Ph. D thesis, Linkoping, Sweden, 1977

Author Index